Modern Inorganic Chemistry

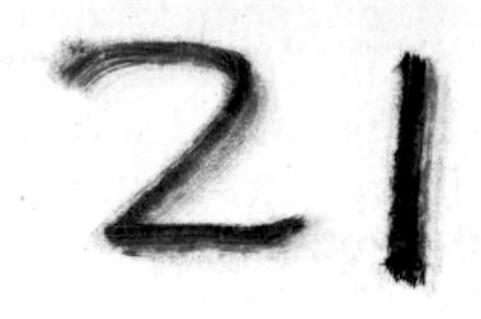

JAMES ALLEN'S GIRLS' SCHOOL

DATE	FORM	NAME OF BORROWER
30/11/86	LVI	Joanna Steele
SEPT '87	LVI	Naomi Hill-Archer.
Sept '89	LVI	VANESSA LAU
Sept 91	LVI	Katherine Woss
Sept 93	LVI	Emma Wedgeworth.
9/95	6H	Harriet Collinan
9/02	12Si	Amanuor Sagoodjor.
21/06/05	12K.	Sophie Kay.

Modern Chemistry Series

Under the supervisory editorship of D. J. Waddington, B Sc, ARCS, DIC, Ph D, Professor of Chemical Education, University of York, this series is specially designed to meet the demands of the new syllabuses for sixth form, introductory degree and technical college courses. It consists of self-contained major texts in the three principal divisions of the subject, supplemented by short readers and practical books.

Major Texts

Modern Inorganic Chemistry
G. F. Liptrot MA, Ph D

Modern Organic Chemistry
R. O. C. Norman MA, D Sc, C Chem, FRSC, FRS
D. J. Waddington, B Sc, ARCS, DIC, Ph D

Modern Physical Chemistry
G. F. Liptrot, MA, Ph D
J. J. Thompson, MA, Ph D
G. R. Walker, BA, B Sc

Readers

Organic Chemistry: a problem-solving approach
M. J. Tomlinson B Sc, C Chem, MRSC
M. C. V. Cane B Sc, Ph D, C Chem, MRSC

Investigation of Molecular Structure
B. C. Gilbert MA, D Phil, C Chem, FRSC

Mechanisms in Organic Chemistry: Case Studies
R. O. C. Norman, MA, D Sc, C Chem, FRSC, FRS
M. J. Tomlinson B Sc, C Chem, MRSC
D. J. Waddington B Sc, ARCS, DIC, Ph D

Practical Books

Inorganic Chemistry Through Experiment
G. F. Liptrot MA, Ph D

Organic Chemistry Through Experiment
D. J. Waddington B Sc, ARCS, DIC, Ph D
H. S. Finlay B Sc

G. F. Liptrot
MA, PhD
Eton College.

Modern Inorganic Chemistry

Foreword by
Professor R. N. Haszeldine
MA, PhD, DSc, ScD, MRSC, FRS
Formerly Principal, University of Manchester Institute of Science and Technology.

Bell & Hyman
London

Published by
BELL & HYMAN LIMITED
Denmark House
37–39 Queen Elizabeth Street
London SE1 2QB

First published in 1971 by Mills & Boon Limited

Reprinted 1972
Second edition 1974
Reprinted, 1975, 1977, 1978, 1980
Third edition 1981
Fourth edition, 1983

© G. F. Liptrot 1971, 1974, 1981 and 1983

All rights reserved. No part of this publication may be reproduced, stored in a retrieval system, or transmitted in any form or by any means, electronic, mechanical, photocopying, recording or otherwise, without the prior permission of Bell & Hyman Limited

British Library Cataloguing in Publication Data

Liptrot, G. F.
Modern inorganic chemistry.—4th ed.—(Modern chemistry series)
1. Chemistry, Inorganic
I. Title II. Series
546 QD151.2

ISBN 0 7135 1357 8
Typeset in Singapore by
Polyglot Pte Ltd. and
produced by Times Printers Sdn. Bhd., Singapore.

Contents

	Page
Foreword by Professor R. N. Haszeldine, MA, PhD, DSc, ScD, FRIC, FRS, Formerly Principal, University of Manchester Institute of Science and Technology	vi
Preface	vii
Chapter	
1 Early theoretical background of chemistry and the determination of relative atomic masses	1
2 The periodic classification of elements	16
3 The structure of the atom	23
4 The arrangement of electrons in atoms	34
5 Chemical bonds	51
6 Chemical geometry	70
7 Modern valency theory	84
8 The factors that determine whether a chemical reaction is possible	96
9 Chemical reactions—factors that influence their speed	120
10 Oxidation and reduction	128
11 Acids and bases	145
12 Occurrence and extraction of metals	152
13 Hydrogen	161
14 Group 0 the noble (inert) gases	169
15 Group 1A the alkali metals	174
16 Group 2A the alkaline earth metals	187
17 Group 3B boron, aluminium, gallium, indium and thallium	199
18 Group 4B carbon, silicon, germanium, tin and lead	213
19 Group 5B nitrogen, phosphorus, arsenic, antimony and bismuth	246
20 Group 6B oxygen, sulphur, selenium, tellurium and polonium	284
21 Group 7B The halogens (fluorine, chlorine, bromine, iodine and astatine)	325
22 The first transition series (scandium, titanium, vanadium, chromium, manganese, iron, cobalt and nickel)	349
23 Some more features of the first transition series	391
24 Group 1B copper, silver and gold	403
25 Group 2B zinc, cadmium and mercury	421
26 The first inner transition series (the lanthanides)	436
27 Nuclear chemistry	441
Answers to Questions	455
Index	457
Bibliography	465
Table of Relative Atomic Masses	466
Logarithm and Antilogarithm Tables	468
Acknowledgements	472

List of Plates

		Page
1.	Spectrum of atomic hydrogen—Balmer Series	35
2.	The structure of sodium chloride	52
3.	Ordered and disordered liquids	97
4.	Sequence of spectra or ClO after flash photolysis of a chlorine/oxygen mixture	120
5.	Xenon tetrafluoride crystals in a quartz container	171
6.	Production platform	225
7.	A microprocessor	227
8.	Amosite ore showing strand-like structure of asbestos	232
9.	Aerospace fasteners containing titanium	356
10.	Annealed carbon steel	375
11.	An electron linear accelerator	447
12.	Advanced gas-cooled reactor	448

Foreword

It is always a pleasure to welcome a new scientific text which effectively presents a clear, accurate and balanced view—as in the book that you are about to read. The pleasure is enhanced by the fact that Geoffrey Liptrot, now well-known through his other successful textbooks, is a friend and former student and colleague of the writer of this foreword.

Inorganic chemistry presents an undeniable challenge and attraction to those who have only just begun to realise the elegance and quality of its modern form; the challenge and attraction are also there, perhaps in a more subtle and intricate way, to those who have long been practitioners of an area of science that still delightfully requires for its thorough appraisal a combination of skill in the experimental art with deep insight over a wide range of our science. Chemistry is still largely a descriptive science, and inorganic chemistry is no exception—that is part of its attraction—but increasingly the ideas of physical and theoretical chemistry, physics, quantum mechanics, etc. help to provide detailed understanding of the phenomena observed and described. The inorganic chemist must thus take the approach of the partially-intuitive, often highly-inventive and -imaginative descriptive scientist, and blend it thoroughly with an ability to use effectively the ideas of physical and theoretical chemistry, without letting these ideas become over-predominant or merely ends in themselves. This book, which I have read with care, is a work of scholarship which succeeds admirably in these objectives. It has many novel, interesting and useful features of approach and presentation, but above all it captures the intrinsic intellectual appeal and excitement of the full range of modern inorganic chemistry. I commend it to you most warmly.

Manchester R. N. Haszeldine

Preface

During the last twenty or so years there has been a steady infiltration of physical chemistry into inorganic chemistry courses and this has meant that the subject can now be studied in a more rigorous and satisfactory manner than hitherto. For instance, it would now be unthinkable to write an inorganic text without first laying stress on structural and energy considerations—two of the kingpins upon which a satisfactory development of this subject rests. New A level syllabuses tend to reflect this trend and, in some at least, it is expected that sixth-formers will have some acquaintance with concepts such as entropy and free energy.

The first section of the book (Chapters 1 to 12) is concerned with principles which are applied to the descriptive chemistry of the elements (Chapters 13 to 27). In Chapter 1 some early theoretical chemistry is reviewed which culminated in the solution to the 'atomic weight' problem which bedevilled chemistry in the early nineteenth century; Chapter 2 is concerned with the Periodic Classification which forms the framework upon which a logical group by group discussion of the elements is based (Chapters 13 to 27). Chapter 3 reviews the efforts of the early twentieth century physicists in unravelling the structure of the atom, and Chapter 4 discusses the arrangement of the electrons in atoms, the spectrum of atomic hydrogen and ionisation energy values providing evidence for electronic levels; the *s*, *p*, *d*, etc. nomenclature is also discussed and electronic structure correlated with the Periodic Table. In Chapter 5 simple theories of chemical bonding are introduced but undue emphasis on the 'octet rule' is deliberately played down in favour of energetics. Chemical geometry is the subject matter of Chapter 6 and this chapter ends with the simple theory of 'electron pair repulsion' which is used here, and elsewhere in the book, to explain the shapes of covalent molecules. Chapter 7 is an optional one and discusses the wave nature of the electron and introduces the idea of molecular orbitals.

Chemical thermodynamics is introduced in Chapter 8 first from a purely descriptive point of view and then with a more mathematical slant, and Chapter 9 introduces the idea of an energy barrier which may oppose chemical reactions otherwise shown to be thermodynamically feasible. Chapter 10, on oxidation-reduction, introduces the concept of electrode potentials and discusses the determination of free energy changes from cell reactions; Chapter 11 is a theoretical account of acid-base behaviour and Chapter 12, on the extraction of metals, uses the electrode potential series and free energy changes to correlate chemical reactivity.

The descriptive chemistry of hydrogen and the noble gases is discussed in Chapters 13 and 14 respectively, and the next seven chapters deal with Groups 1A, 2A, 3B, 4B, 5B, 6B and 7B. In these chapters a distinction is made in the writing of electrovalent and covalent structures, compounds whose bonding is predominantly of the former type being written with plus and minus signs attached. Chapter 22 deals with the first transition series; first, the characteristic features of transition elements are discussed and then each metal is reviewed in turn. Chapter 23, which is optional, discusses the solution chemistry of the first transition series in terms of electrode potentials, and the origin of the colour of transition metal complexes. The remaining chapters deal with Groups 1B and 2B, the Lanthanides and Nuclear Chemistry.

The Stock nomenclature has been used throughout the book and thermochemical data has been quoted in terms of the joule (abbreviated J) and the kilojoule (abbreviated kJ), in keeping with modern practice.

The book deliberately goes beyond the immediate needs of existing A level syllabuses and because of this the sections are numbered so that the more advanced ones can be omitted on first readings. In addition to covering the A level syllabuses, this book contains all that should be required in inorganic chemistry for University Scholarship Examinations. It will also be useful for first year University students.

Eton College G. F. Liptrot

Preface to the Second Edition

The third major reprinting of this book in barely three years provides an opportunity for revising the nomenclature and I have followed the recommendations for Chemical Nomenclature in G.C.E. Chemistry Examination Papers, the joint statement recently made by the G.C.E. Boards. Some recent examination questions and up-to-date articles for Suggested Reading have been added.

Eton College G. F. Liptrot.

Preface to the Fourth Edition

A further opportunity has been taken to revise the book. While there are no major changes in layout, it was thought useful to attach state symbols to formulae used in equations. Improvements made on the industrial scene during the past ten years have been included and the section on the Contact process, for example, has been completely rewritten. The nomenclature has been further revised in keeping with modern practice. A number of recent examination questions and up-to-date articles for Suggested Reading have been added.

Eton College G. F. Liptrot

Chapter 1

Early theoretical background of chemistry and the determination of relative atomic masses

1.1 Introduction

The idea that matter is composed of tiny discrete particles was probably first put forward by the early Greek philosophers, while others suggested that matter was continuous, in much the same way that time is continuous. However, it was not until about 1807 that the atomic nature of matter finally became accepted as the most realistic theory. The evidence that led Dalton to put forward his atomic theory (pp. 1–2) contained a number of inaccurately verified chemical laws.

1.2 The fundamental laws of chemistry

The four basic laws, upon which the emergence of chemistry as a quantitative science rests, are adequately explained in more elementary texts, and we shall do no more than summarise them and show how they led Dalton to his atomic theory.

The Law of Conservation of Mass (Lavoisier 1774)

Matter is neither created nor destroyed in the course of a chemical reaction.

The Law of Constant Composition (Proust 1799)

All pure samples of the same chemical compound contain the same elements combined together in the same proportion by mass.

The Law of Multiple Proportions (Dalton 1803)

When two elements A and B combine together in more than one proportion, the different masses of A which separately combine with a fixed mass of B are in the ratio of small whole numbers.

The Law of Equivalents

Elements combine together or replace one another in the ratio of their relative equivalent masses.

This law evolved from the law of reciprocal proportions (Richter 1792) and the law of multiple proportions (Dalton 1803).

The relative equivalent mass of an element is defined to be 'the number of parts by mass of an element that combine with or replace 8 parts by mass of oxygen'.

On the O = 8 scale, the equivalent of hydrogen is 1·0080. The equivalent of chlorine is 35·457 since 35·457 parts by mass of chlorine will combine with 1·0080 parts by mass of hydrogen.

1.3 Dalton's atomic theory

Working on the assumption that matter is composed of numerous tiny particles and having four inaccurately verified chemical laws as evidence, Dalton proposed his atomic theory in 1807; in so doing he changed chemistry from a massive series of random observations into an orderly classification based on a small number of elements. Like most scientific generalisations the atomic theory is essentially simple, and can in modern terms be reduced to four main points:

(a) All matter is composed of atoms (the basic assumption).
(b) Atoms cannot be created, divided or destroyed (the law of conservation of mass).
(c) All the atoms of one element are alike, and different from those of any other element (the law of constant composition).
(d) Atoms combine together in the ratio of small whole numbers, and the compounds formed are held together by forces of chemical affinity (law of multiple proportions).

1.4 Modification of Dalton's atomic theory

None of the laws and theories on which chemistry was built in the nineteenth century has survived unchanged until now; indeed, as more experimental results are reported new theories are developed and older ones are either discarded or modified to embrace these findings. The discovery by Becquerel in 1896 that uranium atoms gave out a penetrating radiation implied that Dalton's hypothesis about the indestructibility of the atom was clearly at fault. This and other subsequent findings have necessitated some modification to his theory as outlined below.
(a) All matter is not composed of atoms. Although all substances that concern the chemist are made up of atoms there are sub-atomic particles as well, e.g. electrons, protons and neutrons.
(b) In all nuclear changes atoms are created, divided or destroyed, but never in purely chemical changes.
(c) All atoms of any one element are not the same (isotopes) but they are different from atoms of any other element.
(d) While it is still true to say that a single atom is never bound to more than a small whole number of other atoms by chemical bonds, giant molecules do exist containing thousands of atoms, owing to the ability of some atoms, e.g. carbon, to form rings and long chains.

1.5 The combining volumes of gases

Cavendish (1731–1810) was the first chemist to attempt quantitative experiments on volumes of reacting gases. He showed that hydrogen burnt in oxygen and that two volumes of hydrogen combined with one volume of oxygen to form water. Gay-Lussac confirmed Cavendish's result in 1805 and went on to examine other gaseous compounds. He was impressed by the simplicity of the figures he obtained for the ratio of the combining volumes of gases, compared with the cumbersome fractions involved in any determination of combining masses. In 1808 he expressed his findings in the law which bears his name:

The volumes of all gases that react and the volumes of the products if gaseous, when measured under the same conditions of temperature and pressure, are in the ratio of small whole numbers.

Typical experimental results are shown below:

$$\underset{2\text{ vol.}}{\text{Hydrogen}} + \underset{1\text{ vol.}}{\text{Oxygen}} \longrightarrow \underset{2\text{ vol.}}{\text{Steam}}$$

$$\underset{1\text{ vol.}}{\text{Hydrogen}} + \underset{1\text{ vol.}}{\text{Chlorine}} \longrightarrow \underset{2\text{ vol.}}{\text{Hydrogen chloride}}$$

$$\underset{1\text{ vol.}}{\text{Nitrogen}} + \underset{1\text{ vol.}}{\text{Oxygen}} \longrightarrow \underset{2\text{ vol.}}{\text{Nitrogen oxide}}$$

Gay-Lussac's law and its explanation by Avogadro (1811) enabled Cannizzaro (1858) to lay the foundations for the unambiguous determination of relative atomic masses (p. 8).

1.6 The problem of determining relative atomic masses

In 1808 Dalton produced a list of 'atomic masses'. In reality these were not atomic masses at all, but equivalent masses of the elements on a scale in which hydrogen was allocated the number 1 and the other elements were related to it. Dalton had no means of knowing the number of atoms in any compound, and could not therefore deduce any atomic masses from the known equivalents without making a further hypothesis. The hypothesis he adopted was that if only one compound was formed by the combination of two elements, this compound contained one atom of each element. Since he knew nothing of hydrogen peroxide, he assumed that water was HO and therefore the atomic mass of oxygen was 7, the same as his inaccurately measured equivalent. Similarly, ammonia was NH, and the atomic mass of nitrogen was 5, the same figure as his measured equivalent.

In the light of Gay-Lussac's law, the Swedish chemist Berzelius in 1813 suggested that equal volumes of gaseous elements contained equal numbers of atoms (Avogadro in 1811 had suggested that molecules of gaseous elements might contain more than one atom chemically combined but this hypothesis was not accepted until the work of Cannizzaro at a much later date). Using this hypothesis he determined the atomic masses of oxygen, nitrogen, chlorine and iodine. The values he obtained, when corrected to the $O = 16$ scale, are remarkably close to the accepted values; nevertheless the hypothesis upon which they were based is basically unsound (see p. 4). The modern equivalent of his arguments would probably have run as follows:

Since 2 volumes of hydrogen combine with 1 volume of oxygen, then 2 atoms of hydrogen combine with 1 atom of oxygen. The formula of water must therefore be H_2O and, since 1 g of hydrogen combines with 8 g of oxygen, the atomic masses of hydrogen and oxygen are 1 and 16 respectively. Knowing the combining volumes and combining masses of hydrogen with chlorine and iodine, and of nitrogen with oxygen, the atomic masses of chlorine, iodine and nitrogen were determined by similar reasoning.

Berzelius went on to determine the atomic masses of carbon, phosphorus and sulphur. In the case of carbon he assumed that the lower oxide could be represented as CO and since analysis showed that 12 g of carbon combined with 16 g of oxygen, the atomic mass of carbon was 12. The atomic mass of phosphorus was determined by analysis of the two chlorides (the atomic mass of chlorine being known from previous work). The analysis figures below indicate how this was done:

1·00 g of phosphorus combines with 3·44 g of chlorine to give the lower chloride of phosphorus.

1·00 g of phosphorus combines with 5·72 g of chlorine to give the higher chloride of phosphorus.

The ratio of chlorine that combines with a fixed amount of phosphorus in both compounds is:

$$\frac{5{\cdot}72}{3{\cdot}44} = 1{\cdot}666 \text{ or } 5/3$$

Therefore, assuming that a molecule of each compound contains only one atom of phosphorus, the respective formulae are PCl_3 and PCl_5. It is now an easy matter to calculate the atomic mass of phosphorus from the above analysis figures. The atomic mass of sulphur was determined by an analogous procedure using the two oxides, a careful analysis of which showed them to have the formulae SO_2 and SO_3. Again Berzelius was fortunate in that only one atom of the element whose atomic mass he was determining appeared in one molecule of the compound.

Berzelius's luck held still further when he went on to determine the atomic masses of some metals. He assumed that the formula MO represented their oxides and hence, in modern terms, determined the mass of metal that combined with 16 parts by mass of oxygen. Only when the valency of the metal happened to differ from that of oxygen, e.g. for sodium and silver, were his atomic masses appreciably different from our present day values.

The success of Berzelius's methods while based on one completely false and other shaky hypotheses can be attributed to a number of fortunate facts:

(a) The combination of hydrogen with oxygen, or with chlorine, or with iodine vapour, and the combination of oxygen with nitrogen all occur in the same proportions by atoms as by molecules (all these elements are now known to be diatomic).

(b) The compounds of carbon, phosphorus and sulphur which were analysed, in order to determine the atomic masses of these three elements, only contain one atom of these elements in one molecule of the compounds. Berzelius of course had no proof that this was so.

(c) The valency of many metals is the same as that of oxygen.

1.7 The connection between relative atomic mass and relative equivalent mass

While Dalton and Berzelius appreciated the fact that it was necessary to know the formula of a substance before they could deduce any atomic masses from the known equivalents, this information was lacking and they were forced back on guesswork. The significant connection between atomic masses and equivalents, namely valency was, however, not properly understood until the work of Frankland in 1853, and the name valency itself was first mentioned by Lothar Meyer in 1864.

Valency is defined as:

The number of atoms of hydrogen that will combine with or replace (either directly or indirectly) one atom of the element.

The connection between this term and the atomic and equivalent masses of an element is:

$$\textbf{Atomic mass} = \textbf{Equivalent mass} \times \textbf{Valency}$$

The above relationship may be derived as follows:

Suppose an element X combines with hydrogen to form a compound XH_n where n is the valency of the element X. Then the relative atomic mass of X combines with n times the relative atomic mass of hydrogen; but the relative equivalent mass of X combines with the relative equivalent mass of hydrogen. Therefore on the H = 1 scale:

$$\frac{\text{Atomic mass of X}}{\text{Equivalent mass of X}} = \frac{n}{1}$$

or $$\text{Atomic mass of X} = \text{Equivalent mass of X} \times \text{Valency.}$$

Once the valency of an element has been determined it is a comparatively simple matter to determine its atomic mass.

The construction of an accurate table of atomic masses demanded an accurate determination of equivalent masses in the first instance. Although the classical methods of determining accurate equivalent masses date from the late nineteenth and early twentieth centuries, i.e. after the problem of atomic mass determination was finally solved by Cannizzaro, some of them will be discussed now. The routine laboratory methods are described in elementary texts and will not be dealt with here. Indeed the topic nowadays is of purely historical interest, since the equivalents of all common elements are known, and really accurate atomic mass determinations can now be achieved by purely physical methods (p. 32).

1.8 Classical methods for determining the equivalents of hydrogen, chlorine and silver

The determination of the equivalent of hydrogen (Morley 1895)

Purified hydrogen and oxygen were weighed in glass globes, and burnt at platinum jets sealed into the glass vessel which had previously been evacuated. During the combustion, the vessel was immersed in cold water. The water formed during the experiment was now frozen and the residual mixture of gases pumped out and analysed. The water was also weighed so that Morley had a complete check on his experiment. From a series of twelve experiments in which 400 g of water were produced, Morley obtained the ratio:

$$\text{Oxygen} : \text{Hydrogen} :: 8 : 1{\cdot}0076$$

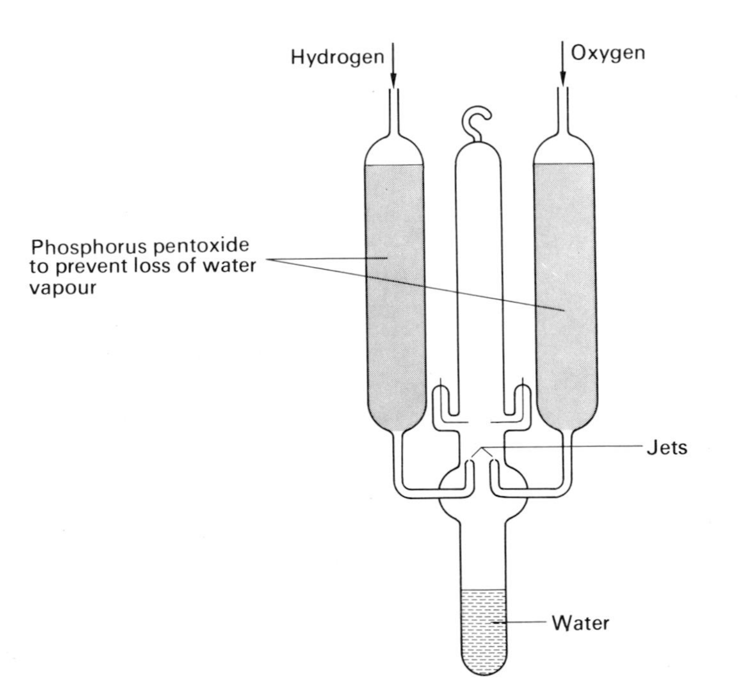

FIG. 1.1 *Morley's apparatus for determining the equivalent of hydrogen*

The determination of the equivalent of chlorine (Edgar 1908)

Purified chlorine (obtained by electrolysis of fused silver chloride) and purified hydrogen, absorbed on palladium, were both weighed. The hydrogen and chlorine were now sparked in an evacuated quartz vessel and the hydrogen chloride produced was liquefied (by means of liquid air) and then weighed.

The equivalent of chlorine is the number of parts by mass that combine with 1·008 parts by mass of hydrogen. The accepted value for this ratio is:

Hydrogen : Chlorine : : 1·008 : 35·457

The determination of the equivalent of silver (Stas 1860)

The method involves three stages as follows:

(a) A known mass of potassium chlorate was heated and the mass of potassium chloride residue was determined. The masses of potassium chlorate and potassium chloride equivalent to 48 g of oxygen (6 equivalents) are shown below.

$$\underset{122{\cdot}592\,g}{2KClO_3(s)} \longrightarrow \underset{74{\cdot}592\,g}{2KCl(s)} + \underset{48\,g}{3O_2(g)}$$

Potassium chlorate contains six equivalents of oxygen combined with one equivalent of potassium and chlorine; it therefore follows that the equivalent of potassium chloride is 74·592.

(b) An excess of silver nitrate solution was added to a known mass of potassium chloride dissolved in water. The precipitated silver chloride was weighed and the amount equivalent to 74·592 g of potassium chloride was calculated to be 143·397 g (the relative equivalent mass of silver chloride expressed in grammes)

$$\underset{74{\cdot}592\,g}{KCl(aq)} + AgNO_3(aq) \longrightarrow KNO_3(aq) + \underset{143{\cdot}397\,g}{AgCl(s)}$$

(c) A known mass of silver was heated in a stream of pure chlorine and the mass of silver required to produce 143·397 g of silver chloride was found to be 107·943 g. The equivalent of silver is therefore 107·943. The equivalent of chlorine is 143·397 − 107·943, i.e. 35·454, and the equivalent of potassium is 74·592 − 35·454, i.e. 39·138.

1.9 Avogadro's hypothesis—the existence of molecules of gaseous elements

In order to trace the steps which finally culminated in Cannizzaro's solution to the atomic mass problem it is necessary to go back to the hypothesis made by Berzelius—'equal volumes of gaseous elements (or compound atoms if the gas is a compound) contain equal numbers of atoms (or compound atoms)'. If Berzelius had been right it is at once impossible to explain the combining ratios of nitrogen and oxygen to give nitrogen oxide:

$$\underset{1\ \text{vol.}}{\text{Nitrogen}} + \underset{1\ \text{vol.}}{\text{Oxygen}} \longrightarrow \underset{2\ \text{vol.}}{\text{Nitrogen oxide}}$$

By Berzelius's hypothesis:

1 atom of nitrogen + 1 atom of oxygen ⟶
2 compound atoms of nitrogen oxide

But this is impossible since 1 compound atom of nitrogen oxide cannot contain half an atom of nitrogen and half an atom of oxygen (contrary to Dalton's postulate concerning the indivisibility of the atom).

Meanwhile several years earlier, in 1811, Avogadro had suggested the correct solution to the difficulty. Molecules (compound atoms) of a compound must by their very nature contain more than one atom, so it was not so ridiculous to suggest that molecules of an element might also exist, containing more than one atom of the same kind. By substituting the word 'molecule' for 'atom' in the assumption that the space occupied by one atom is the same for all gases, he gave the basis for the hypothesis which bears his name:

Equal volumes of all gases, if measured under similar conditions of temperature and pressure, contain equal numbers of molecules.

Though this statement is still commonly referred to as a hypothesis, it is backed by so much experimental evidence that it has all the weight of a chemical law.

Once it is clear that the molecule of a gas can contain two or more atoms it is at once possible to explain the combining volumes of nitrogen and oxygen:

$$\underset{\text{1 vol.}}{N_2(g)} + \underset{\text{1 vol.}}{O_2(g)} \longrightarrow \underset{\text{2 vol.}}{2NO(g)}$$

It is unfortunate that the full value of Avogadro's hypothesis was not realised until 1858 when it was revived by Cannizzaro, in whose hands it formed a powerful weapon for fixing atomic masses (p. 8).

1.10 Deductions from Avogadro's hypothesis applied to Gay-Lussac's law of combining volumes

The molecule of hydrogen is diatomic

It can be shown experimentally that:

1 vol. of hydrogen + 1 vol. of chlorine → 2 vol. of hydrogen chloride

Therefore by Avogadro's hypothesis

n molecules of hydrogen + *n* molecules of chlorine → 2*n* molecules of hydrogen chloride

or

1 molecule of hydrogen + 1 molecule of chlorine → 2 molecules of hydrogen chloride

Each molecule of hydrogen chloride must contain half a molecule of combined hydrogen, thus showing that a hydrogen molecule contains an even number of hydrogen atoms, e.g. H_2, H_4, H_6, etc. It is reasonable to assume that a molecule of hydrogen contains two atoms of hydrogen, since one molecule of hydrogen never does produce more than two molecules of any gaseous compound containing hydrogen.

With the later development of the kinetic theory of gases it has been shown theoretically that the ratio:

$$\frac{\text{Specific heat capacity of a gas at constant pressure}}{\text{Specific heat capacity of a gas at constant volume}}$$

is 5/3 for a monatomic gas and approximately 7/5 for diatomic gases. Hydrogen with a ratio of specific heat capacities of 7/5 obeys the rule for diatomic gases.

The vapour density of a gas is half its relative molecular mass
If the mass of a given volume of gas is known, its density in $g\,dm^{-3}$ can easily be found; but this density alters with atmospheric pressure and temperature, and it is often convenient to have a unit which is independent of these factors. Such a unit is the vapour density, defined as:

$$\textbf{Vapour density} = \frac{\textbf{Mass of any volume of gas}}{\textbf{Mass of the same volume hydrogen}}$$

at the same temperature and pressure.

The molecular mass on the hydrogen scale is defined as:

$$\frac{\text{Mass of one molecule of gas}}{\text{Mass of 1 atom of hydrogen}}$$

By Avogadro's hypothesis:

$$\frac{\text{Mass of any volume of gas}}{\text{Mass of same volume of hydrogen}} = \frac{\text{Mass of } n \text{ molecules of gas}}{\text{Mass of } n \text{ molecules of hydrogen}}$$

$$= \frac{\text{Mass of 1 molecule of gas}}{\text{Mass of 1 molecule of hydrogen}}$$

$$\frac{\text{Mass of 1 molecule of gas}}{\text{Mass of 1 molecule of hydrogen}} = \frac{\text{Mass of 1 molecule of gas}}{\text{Mass of 1 atom of hydrogen}} \times \frac{\text{Mass of 1 atom of hydrogen}}{\text{Mass of 1 molecule of hydrogen}}$$

$$\textbf{Vapour density} = \textbf{Relative molecular mass} \times \tfrac{1}{2}$$

If the O = 16 scale is taken as the standard to which atomic masses refer, then the relationship would have been

$$\textbf{Vapour density} = \textbf{Relative molecular mass} \times \frac{1}{2{\cdot}016}$$

The importance of the above relationship lies in the fact that it enables the molecular mass of a gas to be determined simply by comparing the masses of equal volumes of the gas and hydrogen at the same temperature and pressure.

One mole of any gas at s.t.p. occupies 22·4 dm³
By definition, the relative molecular mass of any gas (expressed in any units by mass) contains the same number of molecules, and therefore occupies the same volume at s.t.p. This volume, if the relative molecular mass is expressed in grams (one **mole**) is called the **standard molar volume** and is $22{\cdot}4\,dm^3$.

If the density in $g\,dm^{-3}$ at s.t.p. (normal density) of any gas has been determined experimentally, the relative molecular mass of the gas can be found immediately by multiplying by $22{\cdot}4\,dm^3$, i.e. by finding the mass of gas (in grams) that occupies $22{\cdot}4\,dm^3$.

1.11 Cannizzaro's method of fixing atomic masses

In the last section we have seen how Cannizzaro used Gay-Lussac's law of combining volumes and Avogadro's hypothesis to deduce that the hydrogen molecule was diatomic and further showed how it was possible to determine vapour densities (and hence molecular masses) of gases. He now proceeded to measure the vapour densities of 80 volatile compounds of carbon and to determine their gravimetric composition. Cannizzaro reasoned that, if enough volatile compounds of carbon were investigated, then it was overwhelmingly likely that at least some of the compounds would contain only one mole of carbon atoms in one mole of the compound. The table below shows the application of this method.

Table 1A Determination of the relative atomic mass of carbon by Cannizzaro's method

Compound	Vapour Density	Relative Molecular Mass	Percentage by mass of Carbon in Compound	Mass of Carbon in 1 mole of the compound
Methane	8	16	75·0	$\frac{75{\cdot}0 \times 16}{100} = 12$ g
Ethane	15	30	80·0	$\frac{80{\cdot}0 \times 30}{100} = 24$ g
Ethyne	13	26	92·3	$\frac{92{\cdot}3 \times 26}{100} = 24$ g
Ethene	14	28	85·7	$\frac{85{\cdot}7 \times 28}{100} = 24$ g
Propane	22	44	81·8	$\frac{81{\cdot}8 \times 44}{100} = 36$ g
Butane	29	58	82·7	$\frac{82{\cdot}7 \times 58}{100} = 48$ g
Cyanogen	26	52	46·2	$\frac{46{\cdot}2 \times 52}{100} = 24$ g
Carbon monoxide	14	28	42·9	$\frac{42{\cdot}9 \times 28}{100} = 12$ g
Carbon dioxide	22	44	27·3	$\frac{27{\cdot}3 \times 44}{100} = 12$ g

Since the smallest mass of carbon in one mole of a carbon compound is 12 g, it is likely that the atomic mass of carbon is 12.

Cannizzaro applied this same method to determine the atomic masses of other elements, e.g. bromine and iodine which form many volatile compounds. Atomic masses determined by this method are only approximate, since the expression

$$\text{Relative Molecular mass} = 2 \times \text{Vapour density}$$

is only strictly applicable to 'ideal gases'. However, an accurate determination of the equivalent of the element allows a value for the valency to be deduced, and hence an accurate value for the atomic mass can be calculated.

1.12 Dulong and Petit's method for determining relative atomic masses and Cannizzaro's contribution

About forty years before Cannizzaro made his great contribution, Dulong and Petit (1819) examined Berzelius's atomic mass table (which it will be remembered was built up from shaky hypotheses) and showed that in most cases:

$$\textbf{Relative atomic mass} \times \textbf{Specific heat capacity} \simeq \textbf{26·8*}$$

The product of atomic mass and specific heat capacity, with a numerical value in the region of 26·8, is called the atomic heat. In the few cases where this relationship broke down Dulong and Petit corrected the atomic masses of Berzelius, usually by dividing them by 2, e.g. with silver. As a means of determining atomic masses, Dulong and Petit's method showed little advance on the methods of Berzelius and so it remained until the problem was taken up by Cannizzaro.

Cannizzaro was interested in knowing whether the relationship

$$\text{Atomic mass} \times \text{Specific heat capacity} \simeq 26{\cdot}8^*$$

did indeed represent the truth, i.e. whether the 'atomic masses' obtained from an application of this expression were indeed atomic masses. He had already determined the atomic masses of bromine and iodine by analysing their volatile compounds (method described in section 1.11); now he proceeded to multiply the atomic masses of these two elements by their respective specific heat capacities (as solid elements) and obtained an atomic heat value remarkably close to 26·8 in both cases. Having thus shown that Dulong and Petit's law did indeed apply to the two elements bromine and iodine, he went on to determine the atomic mass of mercury in the following way.

Analysis of mercury vapour and the vapours of the two chlorides and bromides of mercury showed that 1 mole of each contained 200 g of mercury, which suggested that the atomic mass of mercury might be 200 (since only five substances were analysed, each of these substances might contain more than one atom of mercury per molecule, i.e. the atomic mass of mercury could be a sub-multiple of 200). Assuming mercury to have an atomic mass of 200, Cannizzaro then multiplied this value by the specific heat capacity of mercury and obtained a value for the atomic heat close to 26·8; thus he had no hesitation in stating that the atomic mass of mercury was indeed 200.

Once it had been demonstrated that Dulong and Petit's law was valid, it became a powerful means for determining the atomic masses of elements which could not be analysed by Cannizzaro's method (for those elements that do not form volatile compounds). The method cannot be used, however, to determine the atomic masses of solids of low atomic mass and high melting point, e.g. beryllium, boron, carbon and silicon, since their atomic heats are considerably less than 26·8; neither is it applicable to gaseous elements.

* An atomic heat value numerically equal to 26·8 is obtained if the specific heat capacity is measured in terms of the joule. Since 4·184 joules equal 1 calorie the value becomes 6·4 if the calorie is used.

EXAMPLE

A metal has an equivalent of 18·61 and its specific heat capacity is 0·46 J/g. Calculate its accurate atomic mass.

$$\text{Approximate atomic mass} = 26{\cdot}8/\text{specific heat capacity} = 26{\cdot}8/0{\cdot}46 = 58$$

$$\text{Valency} = \frac{\text{Approximate atomic mass}}{\text{Equivalent}} = \frac{58}{18{\cdot}61} = 3{\cdot}1$$

(or 3 to the nearest integer)

$$\text{Accurate atomic mass} = \text{Valency} \times \text{Equivalent} = 3 \times 18{\cdot}61 = 55{\cdot}83$$

$$\text{Accurate atomic mass} = 55{\cdot}83$$

It is important to realise that any method of determining atomic masses (including those which follow) must first be shown to 'work'. As with Dulong and Petit's method they must first be calibrated against Cannizzaro's method, either directly or indirectly. The different methods which may be employed to determine atomic masses depend, in the last analysis, upon the truth of Avogadro's hypothesis, so brilliantly applied by Cannizzaro.

1.13 Atomic masses by the application of Mitscherlich's law of isomorphism

At about the same time as Dulong and Petit were working, Mitscherlich, a pupil of Berzelius, noted that identical crystalline form (isomorphism) was exhibited by compounds which had similar chemical composition. For example, sodium phosphate is isomorphous with sodium arsenate, and one equivalent of both phosphorus and arsenic is combined with five equivalents of oxygen. The law of isomorphism was stated in the form:

The same number of elementary atoms combined in the same manner produce the same crystalline form.

There are, however, a large number of exceptions, even among soluble salts which form mixed crystals.

From the isomorphism of the chromates and sulphates, Berzelius was forced to the conclusion that the formula of the highest oxide of chromium was CrO_3, similar to the formula of sulphur trioxide which was known to be SO_3. Thus the lower oxide of chromium was Cr_2O_3, which fixed the formulae Fe_2O_3 and Al_2O_3, owing to the isomorphism of the alums which are double salts; for example, chrome (III) alum has the formula $KCr(SO_4)_2.12H_2O$ and iron (III) alum $KFe(SO_4)_2.12H_2O$ (p. 209). This in turn caused the adoption of FeO as the formula of the lower oxide of iron; oxides of metals were also allocated the simple formula XO, if the salts of these metals were isomorphous with the corresponding iron (II) salt, e.g. iron (II) sulphate and zinc sulphate are isomorphous, so the formula of zinc oxide is ZnO.

Although limited in its application, Mitscherlich's law of isomorphism led to the determination of the atomic mass of selenium following the discovery of the isomorphism of sodium sulphate and sodium selenate. Another interesting case of isomorphism led to the correction of the atomic mass of vanadium (Roscoe 1868). The minerals pyromorphite, mimetite and vanadinite are isomorphous; the formulae allotted to these minerals by Berzelius, together with the corrected one for vanadinite, are shown in Table 1B.

Table 1B The corrected formula for vanadinite

	Berzelius's Formula	Corrected Formula
Pyromorphite	$3Pb_3(PO_4)_2 \cdot PbCl_2$	$3Pb_3(PO_4)_2 \cdot PbCl_2$
Mimetite	$3Pb_3(AsO_4)_2 \cdot PbCl_2$	$3Pb_3(AsO_4)_2 \cdot PbCl_2$
Vanadinite	$3Pb_3V_2O_6 \cdot PbCl_2$	$3Pb_3(VO_4)_2 \cdot PbCl_2$

On the above evidence, the valency of vanadium in vanadinite is five, from which the atomic mass of vanadium can be calculated.

EXAMPLE

The oxide of an element Y reacts with potassium hydroxide solution to form a salt isomorphous with potassium manganate (VII). The oxide contains 61·2% of oxygen. What is (a) the formula of the oxide, (b) the atomic mass of the element Y? Assuming the formula of potassium manganate (VII) to be $KMnO_4$, the isomorphous compound will have a similar formula KYO_4. The oxide will be Y_2O_7 corresponding to Mn_2O_7 (the acidic oxide of manganic (VII) acid).

38·8 g of element Y combine with 61·2 g of oxygen
$2x$ g of element Y combine with 112 g of oxygen (from formula Y_2O_7) (where x is the atomic mass of Y)

$$x = \frac{112 \times 38{\cdot}8}{2 \times 61{\cdot}2} = 35{\cdot}5$$

Atomic mass of element Y is 35·5

1.14 Relative atomic masses by the method of limiting densities

This is a method of determining the accurate molecular mass of a gas. Accurate atomic masses can be derived from the measured molecular masses if the atomicity of the gas is known (for gaseous elements) or if the atomic masses of all the combined elements except one are known (for gaseous compounds).

The gas equation for 1 mole of an ideal gas is

$$pV = RT$$

where p is the pressure, V is the volume containing 1 mole, R is the gas constant and T is the temperature Kelvin. By careful experiments, in which allowances were made for the interaction between molecules, it has been found that 1 mole of an ideal gas should occupy 22·414 dm^3 ($2{\cdot}2414 \times 10^{-2}$ m^3) at 0°C (273·16 K) and 1 atmosphere pressure (101 325 N m^{-2}).

Then: $R = pV/T = (101\,325 \times 2{\cdot}2414 \times 10^{-2})/273{\cdot}16$ $= 8{\cdot}314$ J K^{-1} mol^{-1}

For n moles of gas the appropriate gas equation is:

$$pV = nRT$$

As $n = w/M_r$ where w is the mass of the gas in grams and M_r is its relative molecular mass:

$$pV = \frac{wRT}{M_r} \quad \text{or} \quad M_r = \frac{wRT}{pV} \qquad (1)$$

The molecular mass obtained by applying equation (1) is only approximate, since the quantity pV varies with pressure (at a fixed temperature) for all gases. However, a gas approaches ideal behaviour as the pressure is reduced, and the dilemma is resolved by determining a series of pV values for the gas at 0°C as the pressure is reduced; the pV values are plotted graphically and extrapolated to zero pressure. Suppose the extrapolated value of pV is denoted by p_0V_0 and its value at one atmosphere and 0°C is denoted by p_1V_1, then

$$\text{Molecular mass of gas at 1 atmosphere, } M_1 = \frac{wRT}{p_1V_1}$$

$$\text{Accurate molecular mass of gas at zero pressure, } M_0 = \frac{wRT}{p_0V_0}$$

$$M_0 = M_1 \times \frac{p_1V_1}{p_0V_0}$$

or, since $M_0/22{\cdot}414$ and $M_1/22{\cdot}414$ are respectively the limiting density and normal density of the gas:

$$\text{Limiting density} = \text{Normal density} \times \frac{p_1V_1}{p_0V_0}$$

The atomic mass of a gaseous element is easily calculated once its molecular mass has been determined, since it is given by the expression:

$$\textbf{Atomic mass} = \frac{\textbf{Molecular mass}}{\textbf{Atomicity}}$$

and the atomicity is easily determined from the ratio of its specific heat capacities at constant pressure and constant volume (p. 7). The determination of the atomic mass of an element present in a gaseous compound requires a knowledge of both the formula of the gas and the atomic masses of the other component elements, as the example below shows.

EXAMPLE
The normal density of hydrogen chloride is 1·639 15 g/dm^3. The values of p_1V_1 and p_0V_0 (found by extrapolation of a $pV - p$ graph) are respectively 54 803 and 55 213. Calculate the molecular mass of hydrogen chloride and the atomic mass of chlorine, assuming that of hydrogen to be 1·008

$$\text{Limiting density} = \frac{1{\cdot}639\,15 \times 54\,803}{55\,213} \text{ g dm}^{-3}$$

$$\text{Molecular mass of hydrogen chloride} = \text{limiting density} \times 22{\cdot}414$$
$$= \frac{1{\cdot}639\,15 \times 54\,803 \times 22{\cdot}414}{55\,213} = 36{\cdot}467$$

Since the formula of hydrogen chloride is HCl, the atomic mass of chlorine is 36·467—1·008 = 35·459

1.15 The changing atomic mass scale

Atomic masses were originally defined as:

$$\frac{\text{Mass of one atom of the element}}{\text{Mass of one atom of hydrogen}}$$

When the basis of the equivalent mass scale was altered from H = 1 to O = 8, it was logical to change the definition of atomic mass to

$$\frac{\textbf{Mass of one atom of the element}}{\frac{1}{16}\textbf{ the mass of one atom of oxygen}}$$

and recalculate the values with reference to oxygen. Whereas oxygen had been 15·88 on the H = 1 scale, hydrogen now became 1·008 on the O = 16 scale. Oxygen is now known to consist of three isotopes (p. 31) of mass 16, 17 and 18, and oxygen from different sources contains very slightly varying proportions of each isotope. Thus the atomic mass scale based on natural oxygen as 16 (the chemical atomic mass scale) has no truly fixed point, and chemical atomic masses quoted to an accuracy of greater than 1 part in 100 000 have no meaning. Since weighings to this degree of accuracy have only been attempted a few times in connection with atomic mass determinations, there is no practical disadvantage in still using this scale for ordinary chemical calculations.

However, it is essential for the understanding of nuclear reactions to be able to determine isotopic masses to an accuracy of at least one part in a million. Such is the refinement of modern mass spectrometric methods (p. 32) that this accuracy is readily achieved, and a physical atomic mass scale based on the ^{16}O isotope = 16 was introduced.

$$\text{Physical atomic mass} = \frac{\text{Mass of one atom of the element}}{\frac{1}{16}\text{ the mass of one atom of } ^{16}\text{O}}$$

Ordinary oxygen has an atomic mass of 16·004 3 on this scale and is the average atomic mass of the ^{16}O, ^{17}O and ^{18}O isotopes. The most accurate modern determination of atomic masses is made by measuring the molecular masses of volatile compounds of carbon containing a known number of atoms of the element under investigation in a mass spectrometer (p. 32). It is thus convenient to use the mass of the carbon 12 isotope as a standard and in 1961 the International Union of Pure and Applied Chemistry recommended that the ^{12}C = 12 physical atomic mass scale, defined as

$$\frac{\textbf{Mass of one atom of the element}}{\frac{1}{12}\textbf{ the mass of } ^{12}\textbf{C}}$$

should be adopted for both isotopic masses and chemical atomic masses. This scale will have the effect of reducing atomic masses determined on the ^{16}O scale by 43 parts per million. The new scale must be quoted as ^{12}C = 12 and not just as C = 12, since carbon has two other isotopes of masses 13 and 14.

1.16 Suggested reading

John Bradley, *From Organic to Inorganic Chemistry,* School Science Review, 1959, Vol. 15, No. 141.

1.17 Questions on chapter 1

Describe fully, using diagrams and equations, an experiment which you can carry out in the laboratory to verify the Law of Constant Composition. Explain what is meant by isotopes. To what extent does the existence of isotopes affect the exactness of this law as originally stated? (S)

2 'Although in the development of chemical theory Dalton's work stands pre-eminent, almost equal prominence must be given to the combined contribution of Avogadro and Cannizzaro since this provided the first correct methods for determining
(a) the relative molecular masses of gaseous substances,
(b) the relative atomic masses of certain non-gaseous elements,
(c) the empirical formulae of many compounds and the molecular formulae of many of these.'
Elaborate and justify this statement. (JMB)

3 State (a) Gay-Lussac's Law of Combining Volumes, (b) Avogadro's Law, and define (c) relative molecular mass, (d) vapour density.

Describe, with necessary practical detail, how you would demonstrate the volume composition of steam. Hence deduce the molecular formula of steam and state what assumptions have been made in this deduction.

The following data refer to four volatile compounds of an element **X**:

Compound	Vapour density of compound	% of **X** by mass in compound
A	33·9	15·93
B	26·6	81·21
C	37·4	86·64
D	13·8	78·27

Calculate the probable relative atomic mass of **X**. (JMB)

4 A crystalline solid which is isomorphous with potash alum contains water and the following, in percentages: a metal **X** 23·41%, aluminium 4·75% and sulphate radical 33·80%. The specific heat capacity of **X** is 0·202 J/g. Find the relative atomic mass of **X**.

5 Give an account of three important methods which have been used for finding the relative atomic masses of elements. 0·1 g of a metal **M** of specific heat capacity 0·88 J g^{-1} was dissolved in acid to give 173 cm^3 of hydrogen measured at 27°C and 600 mm; 0·1 g of the anhydrous chloride volatilised completely to 12·05 cm^3 of vapour measured under the same conditions. Calculate the formula of the chloride. (O & C)

6 Describe how measurements of the density of a gas may be used to determine accurately its relative molecular mass.

The density, ρ, of sulphur dioxide at 0°C is found to vary with pressure, p, in the following manner:

p (atm)	1	2/3	1/3
ρ(g/dm^3)	2·927 4	1·936 4	0·960 6

Calculate the relative molecular mass of sulphur dioxide to four figures.
(R = 8·314 J K^{-1} mol^{-1}; T K = T°C + 273·2) (Camb. Schol.)

7 Comment on the significance of the statement that relative equivalent mass may be determined with considerably greater precision than relative molecular mass.

A transition metal, **M**, forms a volatile chloride which has a vapour density of 94·9 and contains 74·75% of chlorine; 25 cm^3 of a solution of the chloride containing 4·00 g per dm^3 of the metal, after reduction with aluminium in an inert atmosphere, was treated with an excess of iron (III) sulphate solution. The iron (II) ion in this solution was titrated with 0·1 M cerium (IV) sulphate oxidant and required 20·9 cm^3 of this reagent. Deduce formulae for the original and the reduced chlorides.

$Fe^{2+} + Ce^{4+} \rightarrow Fe^{3+} + Ce^{3+}$ (Camb. Schol.)

8 Write an account of your reasons for believing in the existence of atoms, molecules and ions. (Oxford Schol.)

9 Just over a hundred years ago the relative atomic mass of carbon was thought to be 6 and that of oxygen 8, and the formula of water was written HO. What arguments would you use if you were asked (a) why the relative atomic mass of carbon is now considered to be 12 and not 6, (b) why the formula of water is now written H_2O? What significance is there in the fact that the relative atomic mass of carbon is nearer 12·1 than 12? What evidence is there that the molecular formula of water is not H_2O? (O & C)

Chapter 2

The periodic classification of elements

2.1 Some early attempts at developing a periodic system

Once atomic masses became available, attempts were made to discover if there was any pattern between these figures and the properties of the elements. Indeed, as early as 1815 Prout advanced the hypothesis that all elements were formed by the coalescence of hydrogen atoms, on the flimsy and inaccurate evidence that all atomic masses were whole numbers. It was left to Berzelius to show that the 'atomic mass' of chlorine was not 35 nor 36, but 35·5, i.e. as far from a whole number as it could possibly be; nevertheless, with the discovery of isotopes (Soddy 1913) it became apparent that chlorine was a mixture containing two different kinds of atoms of mass 35 and 37, so Prout's original hypothesis was not quite as far from the mark as had at first appeared.

Several years later, in 1829, Döbereiner noticed that particular groups of three elements with similar chemical properties (triads as he called them) had atomic masses, such that the atomic mass of the middle member of the triad was approximately the arithmetic mean of the other two. Some examples of these triads are shown in Table 2A:

Table 2A Some Döbereiner triads

Lithium 7	Sodium 23	Potassium 39	Sulphur 32	Selenium 79	Tellurium 127·5
Calcium 40	Strontium 87·5	Barium 137·5	Chlorine 35·5	Bromine 80	Iodine 127

No further constructive work was possible until the appearance of Cannizzaro's unambiguous atomic mass table in 1858, with the consequent allotment of valencies to atoms as a measure of combining power.

Newlands (1864) was the first to notice that if the known elements were written down in ascending order of atomic mass similar properties recurred in every eighth element, like notes in a musical scale. This system worked quite well for the lightest elements; for example it brought the lithium—sodium—potassium triad together but, by failing to allow for a bigger octave when dealing with the heavier elements, the theory broke down and was the object of a good deal of ridicule. The reason for this stemmed from the idea that atomic properties recurring at fixed intervals suggested some form of repeating structure inside the atom, a flat contradiction of the immensely successful atomic theory of Dalton. The most important step in developing a periodic classification of the elements was taken in 1869, when the Russian chemist, Mendeléeff, studied the relationship between the atomic masses of elements and their properties with special emphasis on their valencies. He was led to the conclusion that 'the properties of the elements are in periodic dependence on their atomic masses', a conclusion that had previously been hinted at by Newlands but never developed. Proof that one quantitative property of atoms really was periodic was provided by Lothar Meyer in 1870; his work will be described first, before dealing with that of Mendeléeff.

2.2 Lothar Meyer's atomic volume curve

Lothar Meyer calculated the atomic volumes of the known elements, i.e. the volume in cm^3 occupied by 1 mole of the elements in the solid state; thus:

$$\textbf{Atomic volume} = \frac{\textbf{Mass of one mole}}{\textbf{Density}}$$

When he plotted these atomic volumes against the corresponding atomic masses, a curve taking the form of sharp peaks and broad minima was obtained. Figure 2.1 shows a modern representation of this curve using atomic number (p. 19) rather than atomic mass, since this gives a more regular shape. Several generalisations can be made from the shape of the curve and are summarised on page 18.

FIG. 2.1. *Atomic volume plotted against atomic number*

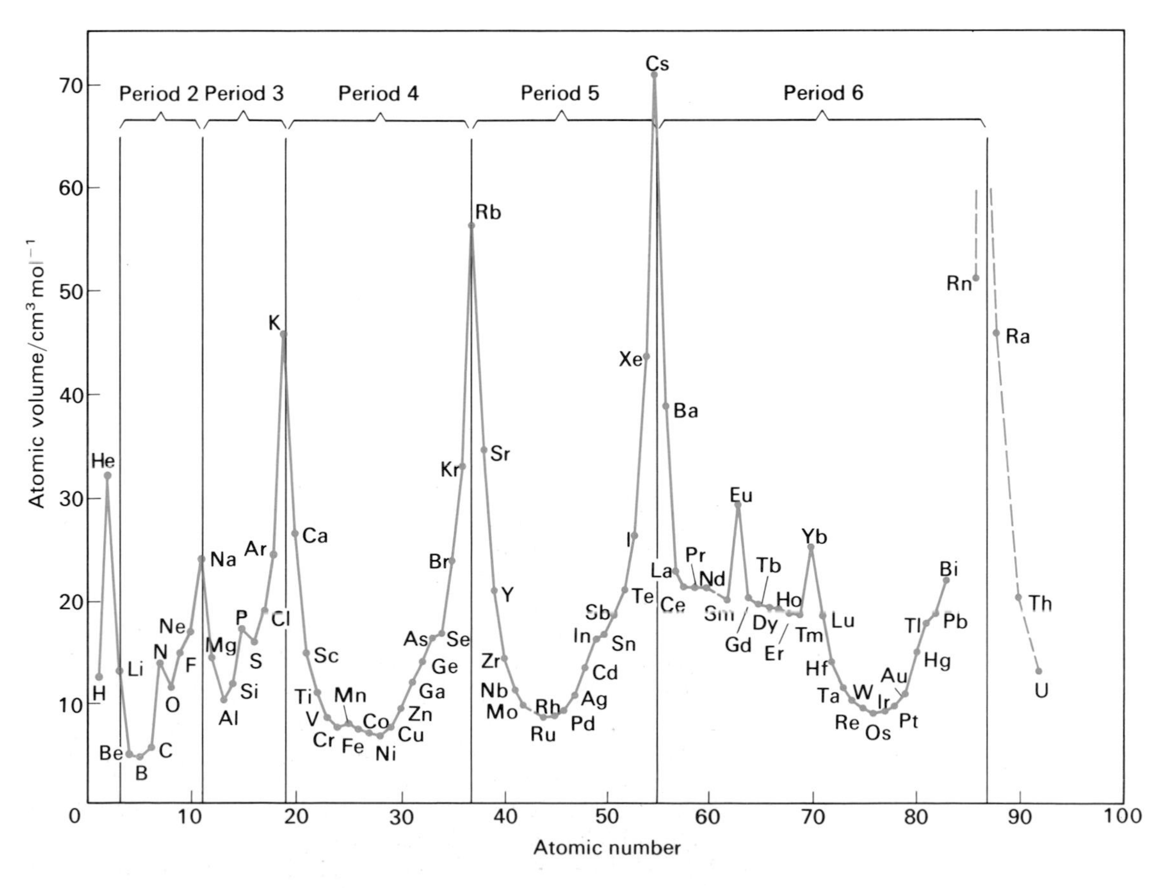

(a) Gaseous, volatile and readily fusible elements are located on the ascending portions of the curve or at the peaks.
(b) Elements with high melting points are found on the descending portions of the curve and at the broad minima.
(c) Chemically similar elements occupy similar positions on the curve; thus the alkali metals (sodium, potassium, rubidium and caesium) occur at the peaks, and the halogens (fluorine, chlorine, bromine and iodine) occur on the ascending portions, immediately followed by the noble gases (neon, argon, krypton and xenon).

Other properties of elements show a similar periodicity, e.g. first ionisation energies of atoms (p. 40).

2.3 Mendeléeff's classification of the elements

Mendeléeff wrote down the names of the elements in order of increasing atomic mass and noticed that the properties of the elements, with slight modifications, repeated themselves at intervals. Elements with similar properties were arranged vertically beneath each other and by this means a table was eventually constructed. There were, of necessity, many gaps in this table which Mendeléeff correctly predicted would eventually be filled by the discovery of more elements. No gaps now remain to be filled and indeed the properties of many of these elements are in close agreement with those predicted by Mendeléeff. A modern form of the Periodic Table (Table 2B) is shown on page 19, the ordering of the elements being based on atomic numbers rather than atomic masses (see section 2.4). It has survived the discovery of a whole new group (the noble gases), 14 rare earths (the first inner transition series called the lanthanides) and 11 man-made elements (part of the second inner transition series called the actinides).

The table is divided into a number of vertical groups and 7 horizontal periods. The Groups 1 to 7 are subdivided into A and B, and Group 8 contains three elements in a given period. It should not be thought that there is any special resemblance between the sub-groups, except perhaps one of valency; indeed the chemical differences between sub-groups are often so great that they are best considered separately. Furthermore, there is no special significance in, say, iron, cobalt and nickel being placed in Group 8; these three elements are best treated as members of the first transition series extending from scandium to copper. The division into sub-groups, etc., is merely the method used by Mendeléeff in his original table and no one has considered it necessary to alter it.

2.4 Some features of the Periodic Table

The ordering of elements based on their atomic numbers

Several pairs of elements have had to be interchanged in order to maintain chemical similarities in the same group; thus argon clearly belongs to Group 0 and potassium should immediately succeed it in Group 1, even though the atomic mass of argon is slightly greater than that of potassium. Similarly, tellurium (atomic mass 127·60) belongs to Group 6B, immediately followed by iodine (atomic mass 126·90). Mendeléeff himself was aware of examples of this kind and had carried out some corrections.

The existence of some more fundamental property of the atom, that characterised its behaviour much more satisfactorily than its atomic mass, was discovered by Moseley in 1913 following his work on the X-ray spectra of the elements (p. 30). He discovered that the X-ray spectra of elements could be explained by allocating a different number to each element and, furthermore, that the numbers increased by steps of unity from one element to the next. Hydrogen was given the number 1, helium 2, lithium 3, beryllium 4, etc. The anomalous pairs of elements now fitted in perfectly, since argon had the number 18 and potassium 19, likewise tellurium was characterised by the number 52 and iodine by 53. This number, which is characteristic of the particular element, is known as the atomic number and is numerically equal to the number of protons (or electrons) in the atom of the element.

The position of metals and non-metals in the Periodic Table

Metals are characterised by several distinctive properties, e.g. high electrical and thermal conductivities, 'metallic' lustre, malleability and ductility, whereas non-metals usually exhibit no such behaviour. Like all generalisations in chemistry, however, the borderline between the two classes of elements is far from rigid, and it is easy to discover exceptions.

Table 2B The Periodic Table

Groups / Periods	1A	2A	3A	4A	5A	6A	7A	8	8	8	1B	2B	3B	4B	5B	6B	7B	0
1	1 H																	2 He
2	3 Li	4 Be											5 B	6 C	7 N	8 O	9 F	10 Ne
3	11 Na	12 Mg	←—	TRANSITION ELEMENTS								—→	13 Al	14 Si	15 P	16 S	17 Cl	18 Ar
4	19 K	20 Ca	21 Sc	22 Ti	23 V	24 Cr	25 Mn	26 Fe	27 Co	28 Ni	29 Cu	30 Zn	31 Ga	32 Ge	33 As	34 Se	35 Br	36 Kr
5	37 Rb	38 Sr	39 Y	40 Zr	41 Nb	42 Mo	43 Tc	44 Ru	45 Rh	46 Pd	47 Ag	48 Cd	49 In	50 Sn	51 Sb	52 Te	53 I	54 Xe
6	55 Cs	56 Ba	57 La	72 Hf	73 Ta	74 W	75 Re	76 Os	77 Ir	78 Pt	79 Au	80 Hg	81 Tl	82 Pb	83 Bi	84 Po	85 At	86 Rn
7	87 Fr	88 Ra	89 Ac															

Lanthanides	58 Ce	59 Pr	60 Nd	61 Pm	62 Sm	63 Eu	64 Gd	65 Tb	66 Dy	67 Ho	68 Er	69 Tm	70 Yb	71 Lu
Actinides	90 Th	91 Pa	92 U	93 Np	94 Pu	95 Am	96 Cm	97 Bk	98 Cf	99 Es	100 Fm	101 Md	102 No	103 Lr

Perhaps the best distinguishing feature between metals and non-metals is the tendency of the former to form predominantly basic oxides, whereas the latter favour predominantly acidic oxides (although even here exceptions occur).

By far the greatest proportion of elements are classified as metals and include Groups 1A (except hydrogen), 2A, 3A, 1B, 2B, the transition elements, including the lanthanides and actinides. Metallic properties are more pronounced in the lower left-hand corner of the Periodic Table; non-metallic properties are more obvious in the upper right-hand corner of the table. The rather diffuse dividing line between metallic and non-metallic behaviour can be considered to run diagonally from upper left to lower right with the elements boron, silicon, arsenic, tellurium and polonium, and others to the right of this diagonal, showing predominantly non-metallic properties.

Trends across a period and down a group

There is a gradual change from metallic to non-metallic character in passing across a particular period. This is clearly seen by considering the oxides of period 3 elements.

$$\begin{array}{ccccccc} Na_2O & MgO & Al_2O_3 & (SiO_2)_x & P_4O_{10} & SO_3 & Cl_2O_7 \\ \longleftarrow \text{Basic} \longrightarrow & & \text{Amphoteric} & \longleftarrow & \text{Acidic} & & \longrightarrow \end{array}$$

The same kind of gradation in properties is also observed in the lower periods, if the interposing transition series is disregarded.

Metallic character increases down a particular group, as can be seen by comparing the 3-valent oxides of Group 5B elements.

$$\begin{array}{ccccc} N_2O_3 & P_4O_6 & As_4O_6 & Sb_4O_6 & Bi_2O_3 \\ \text{— Decreasingly acidic} \longrightarrow & & & \text{Amphoteric} & \text{Essentially basic} \end{array}$$

The increase in non-metallic properties, that occurs on traversing a particular period and ascending a particular group, is the reason why some of the period 2 elements have a similar chemistry to their diagonal neighbours in period 3, e.g. lithium and magnesium, beryllium and aluminium, boron and silicon.

The differences between the first and second members of a group

There are often significant differences between the first and second members of a periodic group. For instance, lithium differs more from sodium than does sodium from potassium (the third member of Group 1A). Similarly, the following pairs of elements differ appreciably in their chemistry: boron and aluminium, carbon and silicon, nitrogen and phosphorus, oxygen and sulphur, and fluorine and chlorine. These differences are discussed later in the book, when the chemistry of each group is described in detail.

Some valencies exhibited by elements

Some of the elements exhibit the valency corresponding to the group number, thus the alkali metals (Group 1A) and the alkaline earths (Group

2A) are exclusively 1-valent and 2-valent respectively. In some cases elements of a particular group can exhibit the group valency and eight minus the group valency; thus phosphorus is 5-valent in phosphorus pentachloride, PCl_5, and 3-valent in phosphorus trichloride, PCl_3. Similarly the valency of iodine in iodine heptafluoride, IF_7, is seven, and one in hydrogen iodide, HI. The chemistry of the transition elements, including the lanthanides and the actinides, is characterised by variable valency, and little can be deduced about their valencies from their positions in the Periodic Table. Other elements such as lead (valencies of 2 and 4, the former being the predominant valency state) have a valency state which is difficult to correlate with group number. The relationship between valency and group position is therefore only of significance for some groups in the Periodic Table.

The position of the noble gases

The noble gases occupy Group 0 in the Periodic Table, a position that implies they have zero valency. However, the heavier members have been shown to be less 'inert' than had hitherto been supposed; thus krypton and xenon have been shown to form well-established chemical compounds (principally fluorides and oxides), and there is evidence that radon forms a fluoride. Helium and argon, however, have so far resisted attempts to induce them to participate in chemical reactions and it is still true to say that, as a group, they form chemical compounds less readily than other elements. This fact has been utilised in developing a simple theory of valency (p. 51). The position they occupy in the Periodic Table is interesting in that they immediately follow a very reactive non-metal (halogen) and precede a very reactive metal (alkali metal).

The position of hydrogen

Hydrogen is often positioned in Group 1A; however, in no sense can it be said that hydrogen resembles a Group 1A metal, except perhaps in having a valency of one and in forming a hydrated positive ion, H_3O^+. Another possible position is in Group 7B, immediately followed by the noble gas helium. This too is very unsatisfactory, even allowing for the fact that hydrogen shows a covalency of one, like the halogens, and can under certain conditions form the negative hydride ion, H^-, cf. the fluoride ion, F^-. The properties of hydrogen are unique and no profitable comparisons can be made with other elements.

The transition elements

These are all dense metals with high melting points and the elements of a particular transition series have much in common with each other. Their chemistry is characterised by a display of variable valency which explains some features of their marked catalytic activity. Many of their salts are coloured, both as solids and in solution. For a given transition series, their atoms are similar in size, a factor which is of importance in understanding why these elements can be used in the manufacture of alloy steels. Their ions of a given valency state are also similar in size and this explains why they form so many isomorphous compounds, e.g. the alums of formula $KX(SO_4)_2.12H_2O$ where X may be Ti, V, Cr, Mn, Fe or Co in the 3-valent state.

2.5 Suggested reading

The Periodic Table, Nuffield chemistry background book, Longman/Penguin, 1966.

The Periodic Table and Chemical Bonding, Open University, S 100, Unit 8.

R. M. Cameron, *Mendeleev made easy*, Education in Chemistry, Vol. 18, 179, 1981.

2.6 Questions on chapter 2

1 Discuss the general arrangement of elements in the modern (extended) form of the Periodic Table, explaining the underlying principles of the classification, and describing the broad trends observed in the characteristics of the elements.

A metallic element forms two stable chlorides with formulae MCl_2 and MCl_4 respectively. Where would you expect the element to be placed in the classification? Give reasons. (S)

2 What is meant by the terms atomic number, relative atomic mass and isotopes?

Why do the chemical properties of elements correlate better with atomic numbers than with relative atomic masses? (O & C)

3 Explain the terms (a) atomic number, (b) relative atomic masses, (c) isotope, and show that a given element occupies a single fixed position in the Periodic Table.

Discuss the position of argon and potassium in the Periodic Table. (C)

4 'Metallic character of the elements in the Periodic Classification decreases in progressing from left to right in the table, but increases in progressing from top to bottom'. Summarise clearly and concisely in note form what you feel are the key points in favour of or against such a statement, illustrating your answer by carefully chosen examples where possible. (Camb. Schol.)

5 Discuss the main features of the Periodic Table with reference to (a) the position of metals and non-metals, (b) the valencies exhibited by elements, (c) the position of the noble gases, (d) the position of hydrogen and (e) transition elements.

6 Look up the m.p. of the first twenty elements and plot them graphically against atomic number. Is any trend discernible?

7 Mendeleev stated in 1869 that 'the elements, if arranged according to their relative atomic masses exhibit an evident periodicity of properties'. Discuss this statement with the aid of specific physical and chemical properties of the elements to illustrate your answer. How far can the periodicity of properties be explained in terms of structure and bonding?

A, **B** and **C** are three non-transition elements. State, giving your reasons, the group of the Periodic Table to which **each** element is likely to belong, given that

(*a*) element **A** forms a chloride which has a tetrahedral molecule.

(*b*) element **B** forms an oxide which has a sodium chloride type crystal structure,

(*c*) element **C** forms a dichloride which is a non-conducting liquid at room temperature. (C)

Chapter 3

The structure of the atom

3.1 Some fundamental particles of matter

As a result of a brilliant era in experimental physics which began towards the end of the nineteenth century and which extended into the 1930s, we now know that the atom is composed of three basic sub-atomic particles, namely the electron, the proton, and the neutron. The characteristics of these three particles are given in Table 3A.

Table 3A The three main sub-atomic particles

Particle	Mass	Charge	Symbol
Electron	1/1837 unit	−1 unit	*e*
Proton	1 unit	+1 unit	p
Neutron	1 unit	No charge	n

It is now known that many more sub-atomic particles exist, e.g. the positron, the neutrino, the hyperon, etc., but in chemistry only those listed in Table 3A generally need be considered. The discovery of these particles and the way in which the structure of the atom was worked out are discussed in subsequent pages of this chapter. The detailed arrangement of the extra-nuclear electrons is the subject matter of Chapter 4.

3.2 Evidence for the existence of electrons

Evidence that electrical charge was not continuous but existed in the form of discrete particles was obtained from Faraday's work on electrolysis (1834). He discovered that the amount of electricity needed to deposit L atoms during electrolysis (where L is the number of atoms in one mole and is called the **Avogadro Constant** or **Avogadro's number**) of any 1-valent element is exactly the same, irrespective of the size or mass of the atom involved. It follows, therefore, that a certain amount of electricity is associated with one atom of a 1-valent element. Despite the fact that the rest of their physical properties differ widely, hydrogen, potassium, and silver share a common valency of one and this basic quantity of electricity, and it is overwhelmingly likely that the two are related. Since only exact multiples of this quantity of electricity are found necessary for the deposition of one atom of a 2-valent or a 3-valent element, it is a good working hypothesis to assume that the quantity of electricity needed for the deposition of one atom of a 1-valent element is fundamental and cannot be divided. The term 'electron' was given to the smallest particle that could carry a negative charge equal in magnitude to the charge necessary to deposit one atom of a 1-valent element (Stoney 1891).

3.3 Electrons obtained by passing electrical discharges through gases

In 1879, Crookes discovered that when a high voltage is applied to a gas at low pressure streams of particles, which could communicate momentum, moved from the cathode to the anode. It did not seem to matter what gas was used and there was strong evidence to suppose that the particles were common to all elements. In a very high vacuum they could not be detected. The properties of these cathode rays (as they were then called) are given below:

(a) When a solid metal object is placed in a discharge tube in their path, a sharp shadow is cast on the end of the discharge tube, showing that they travel in straight lines.

(b) They can be deflected by magnetic and electric fields, the direction of deflection showing them to be negatively charged.

(c) A freely moving paddlewheel, placed in their path, is set in motion, showing that they possess momentum.

(d) They cause many substances to fluoresce, e.g. the familiar zinc sulphide coated television tube.

(e) They can penetrate thin sheets of metal.

J. J. Thomson (1897) extended these experiments and determined the velocity of these particles and their charge/mass ratio as follows. The particles from the cathode were made to pass through a slit in the anode and then through a second slit. They then passed between two aluminium plates spaced about 5 cm apart and eventually fell on to the end of the tube, producing a well-defined spot. The position of the spot was noted and the magnetic field was then switched on, causing the electron beam to move in a circular arc while under the influence of this field (fig. 3.1).

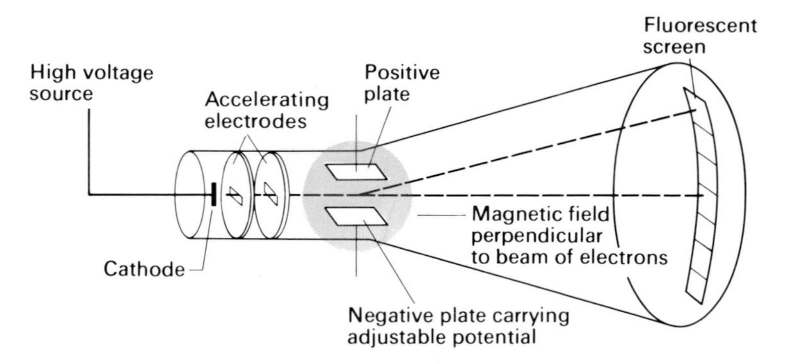

FIG. 3.1. *Thomson's apparatus for determining e/m for the electron*

An electric field was now applied in opposition to the magnetic field and gradually increased until the spot returned to its original position. If:

B = magnetic flux density
e = charge on the electron
v = velocity of the electron
m = mass of the electron
r = radius of the arc in which the electron moves

then the magnetic force, Bev, acting on each electron causes it to accelerate in the direction of the force, and thus to move along the arc of a circle.

Thus:

$$Bev = \frac{mv^2}{r} \quad \text{or} \quad e/m = \frac{v}{rB} \tag{1}$$

When an electric field is applied in opposition (E) so that the electric force on each electron balances the magnetic force:

$$Ee = Bev \quad \text{or} \quad v = \frac{E}{B} \tag{2}$$

The velocity of the electrons can be calculated from equation (2). It is found that they travel at about $3 \times 10^7 \text{ m s}^{-1}$, i.e. about $\frac{1}{10}$ the velocity of light. Substituting for v in equation (1) gives

$$\frac{e}{m} = \frac{E}{rB^2}$$

and since r can be determined from simple geometry knowing the dimensions of the apparatus, the value e/m can be evaluated. Its value $1{\cdot}76 \times 10^{11} \text{ C kg}^{-1}$ is quite independent of the nature of the residual gas in the apparatus, suggesting that electrons are constituents of all matter.

If the charge on the electron is assumed to be equal in magnitude to that on the hydrogen ion, the ratio of the mass of the electron to that of the hydrogen ion can be obtained:

$$\text{For an electron } e/m = 1{\cdot}76 \times 10^{11} \text{ C kg}^{-1}$$

$$\text{For an hydrogen ion } e/m = 96\,520/(1{\cdot}008 \times 10^{-3}) = 9{\cdot}57 \times 10^{7} \text{ C kg}^{-1}$$

$$\frac{\text{Mass of electron}}{\text{Mass of hydrogen ion}} = \frac{9{\cdot}57 \times 10^7}{1{\cdot}76 \times 10^{11}} = 1/1837$$

Thomson's experiments show electrons to be negatively charged and of minute mass (if the assumption is correct that they do exist as discrete particles). However, his experimental results could equally well be interpreted assuming the electrons to be a 'charged fluid'. Conclusive proof that electrons did have a particulate nature was finally obtained by Millikan.

3.4 Conclusive proof that electrons are particles

Evidence that electrons were discrete particles was obtained by Millikan during the years 1910–14, when he undertook a series of very careful experiments to determine the value of the electronic charge; his apparatus is shown diagramatically in fig. 3.2.

Small droplets of oil from an atomiser are blown into a still thermostatted air space between parallel plates, and the rate of fall of one of these droplets under gravity is observed, from which its mass can be calculated. The air space is now ionised with an X-ray beam, enabling the droplets to pick up charge by collision with the ionised air molecules. By applying a potential of several thousand volts across the parallel metal plates, the oil droplet can either be speeded up or made to rise, depending upon the direction of the electric field. Since the speed of the droplet can be related to its mass, the magnitude of the electric field, and the charge it picks up, the value of

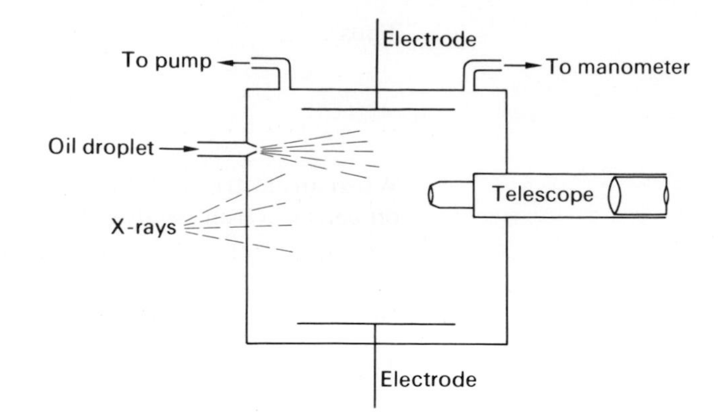

FIG. 3.2. *Millikan's apparatus for determining the value of the electronic charge*

the charge can be determined. Millikan observed that droplets of oil could pick up several different charges, but that the total charge was always an exact integral multiple of the smallest charge, i.e. the charge on the electron. The present-day accepted value for the charge on the electron is $1{\cdot}602 \times 10^{-19}$ C. When this value for e is compared with the most modern value of e/m, the mass of the electron becomes $9{\cdot}11 \times 10^{-31}$ kg.

3.5 Positive particles

If the conduction of electricity through gases is due to particles which are similar to those involved during electrolysis, it was to be expected that positive as well as negative ones would be involved, and that they would be drawn to the cathode. By using a discharge tube containing a perforated cathode, Goldstein (1886) had observed the formation of rays (shown to the right of the cathode in fig. 3.3). J. J. Thomson (1910) measured their

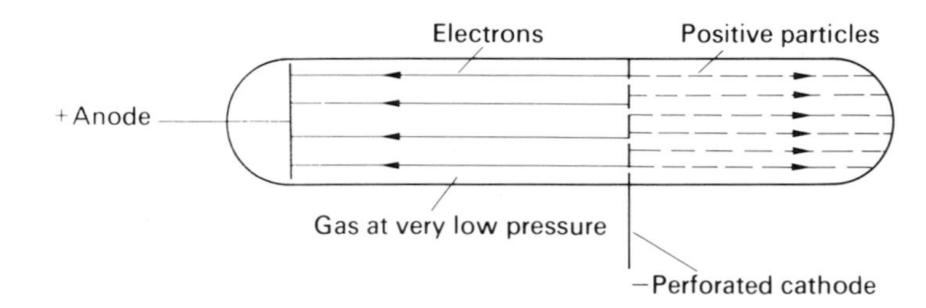

FIG. 3.3. *Apparatus for producing positive ions*

charge/mass ratio, from which he was able to deduce that the particles were positive ions, formed by the loss of electrons from the residual gas in the discharge tube. The proton is the smallest positively charged particle that can be obtained in the discharge tube. It carries a positive charge equal in magnitude to that on the electron and is formed from the hydrogen atom by the loss of an electron:

$$H \longrightarrow H^+ + e^{-1}$$

3.6 The discovery of X-rays and radioactivity

X-rays were discovered by Röntgen (1895) when he noticed that a penetrating radiation was emitted from discharge tubes and appeared to originate from the anode. The radiation had the following properties:

(a) It blackened wrapped photographic film.
(b) It ionised gases, so allowing them to conduct electricity.
(c) It made certain substances fluoresce, e.g. zinc sulphide.

Furthermore, the radiation was shown to carry no charge, since it could not be deflected by magnetic or electric fields. The true nature of X-radiation was not discovered until 1912, when it became apparent that its properties could be explained by assuming it to be wavelike in character, i.e. similar to light but of much smaller wavelength. It is now known that X-rays are produced whenever fast-moving electrons are stopped in their tracks by impinging on a target, the excess energy appearing mainly in the form of X-radiation.

The year after Röntgen discovered X-rays, Becquerel observed that uranium salts emitted a radiation with properties similar to those possessed by X-rays. The Curies followed up this work and discovered that the ore pitchblende was more radioactive than purified uranium oxide; this suggested that something more intensely radioactive than uranium was responsible for this increased activity, and eventually the Curies succeeded in isolating two new elements, called polonium and radium, which were responsible for this increased activity.

3.7 α-particles, β-particles and γ-rays

In 1899 Becquerel reported that the radiation from the element radium could be deflected by a magnetic field and in the same year Rutherford noticed that the radiation from uranium was composed of at least two distinct types. Subsequently it was shown that the radiation from both sources contained three distinct components, a diagrammatical representation of the apparatus used in this discovery being shown in fig. 3.4.

FIG. 3.4. *Effect of magnetic field on the radiation emitted by radium*

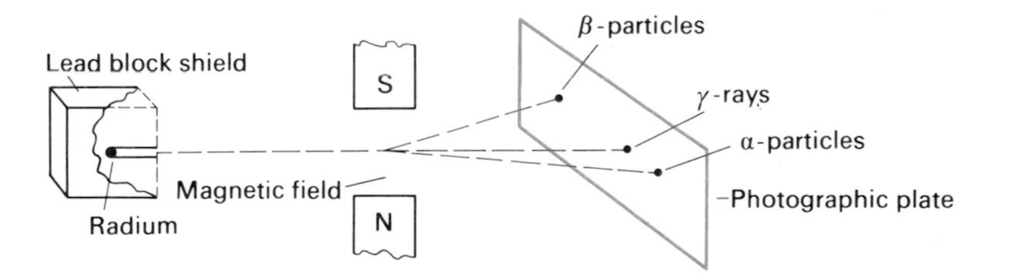

The radiation was confined to a narrow beam by allowing it to pass out of a small hole drilled in a lead block. On passing this radiation through a strong magnetic field it was resolved into three components (α-particles, β-particles, and γ-rays as they are now called) which blackened a photographic plate. From the positions of the blackened regions of the plate it was obvious that α-particles were positively charged and that β-particles were negatively charged and very much lighter than the former. Since the γ-rays were not deflected by the field, they must be uncharged; in fact they are known to be similar to X-rays but of even shorter wavelength.

Rutherford measured the charge/mass ratio of the α-particles, using a method similar to the one employed by J. J. Thomson but using much more intense magnetic and electric fields. His value for the charge/mass ratio was $4{\cdot}82 \times 10^7$ C kg^{-1}, almost exactly half that for an hydrogen ion. The actual charge carried by α-particles was now measured by Rutherford and Geiger and shown to be double that carried by the electron; simple calculations now showed that the mass of an α-particle was 4 units, the same as the relative atomic mass of helium, thus if

e = charge on the hydrogen ion (same as that on the electron)
m = mass of the hydrogen ion
q = charge on the α-particle
w = mass of the α-particle

then

$$\frac{e}{m} = \frac{2q}{w} \text{ and } q = 2e$$

or

$$\frac{e}{m} = \frac{4e}{w} \text{ or } \frac{w}{m} = 4$$

Mass of an α-particle = 4 × Mass of an hydrogen ion

Proof that α-particles were doubly ionised helium atoms, He^{2+}, came when Rutherford sealed pure radon gas (a source of α-particles) in a glass tube and surrounded it by another one which was evacuated. Several days later, helium gas was detected in the outer tube and, furthermore, Rutherford was able to establish that one atom of helium had originated from every α-particle that had passed through the wall between the two tubes.

β-particles were shown to be electrons by a measurement of their charge/mass ratio. Their speeds are variable, some approaching the velocity of light. They are capable of penetrating thin sheets of aluminium.

3.8 Bombardment of matter with α-particles and β-particles and the Rutherford theory of the atom

When a thin sheet of metal such as aluminium is placed in the path of β-particles, a divergent beam of particles emerges (fig. 3.5). Rutherford explained this effect as being due to the repulsive forces operating between the electrons in the metal and those in the beam. From the angle of

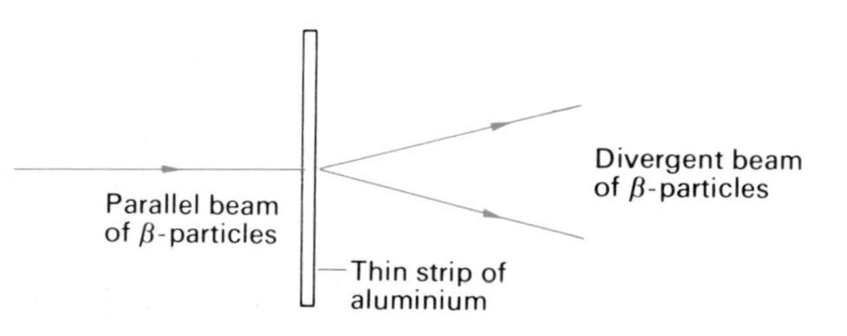

FIG. 3.5. *Bombardment of thin aluminium strip with β-particles*

divergence he was able to calculate the number of electrons in each atom of the metal and found that this number was generally about half the numerical value of the atomic mass of the particular metal, e.g. 13 for aluminium which has an atomic mass of 27. Since the mass of the electron was known to be very small, he reasoned that the main mass of the atom must reside in its positive nucleus. This view was supported by experiments performed by Geiger and Marsden, two of Rutherford's research students.

A thin parallel beam of α-particles was directed onto a thin strip of gold and the subsequent path of the particles was determined. Some particles were deflected off-course and diverged, but the majority passed straight through the gold strip with little or no disturbance. The exciting fact that emerged was that an incredibly few α-particles (about 1 in 20 000) were deflected backwards through angles greater than 90° (fig. 3.6). On increasing the thickness of the gold strip, more and more of the particles suffered the same fate.

Rutherford reasoned that the large angle scattering of the α-particles must be due to a collision or near collision with an incredibly small nucleus which carried a positive charge approximately numerically equal to half the atomic mass of the atom. Furthermore virtually all the mass resided

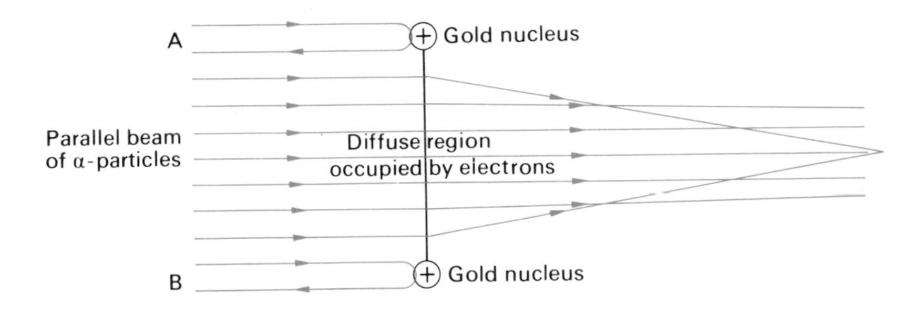

FIG. 3.6. *Representation of Geiger and Marsden's experiment. At* A *and* B *α-particles collide with gold nuclei*

in this nucleus. The atom was visualised as containing a small positive nucleus where practically all the mass resided, and itself compounded from a number of positively charged particles (later called protons) together with a number of electrons. Around this nucleus there was a rather diffuse region containing sufficient electrons to maintain electrical neutrality. It has been estimated that nuclei of atoms have radii in the order of 10^{-14} to 10^{-15} m and that the outermost electron clouds are about 10^{-10} m from the centre of the nucleus, i.e. an atom magnified to about 60 cm in diameter would contain a nucleus no larger than 0·003 cm in diameter, which is about the same size as a very fine grain of sand. On this basis, it has been calculated that less than 10^{-12} of the volume of the atom is occupied by material particles.

3.9 The discovery of the neutron

The neutron proved to be a very elusive particle to track down and its existence, predicted by Rutherford in 1920, was first noticed by Chadwick in 1932. Chadwick was bombarding the element beryllium with α-particles and noticed a particle of great penetrating power which was unaffected by magnetic and electric fields. It was found to have approximately the same mass as the proton (hydrogen ion). The reaction is represented as

$$^{9}_{4}\text{Be} + ^{4}_{2}\text{He} \longrightarrow ^{12}_{6}\text{C} + ^{1}_{0}\text{n}$$

where the superscript refers to the atomic mass and the subscript refers to the atomic number (the number of protons in the nucleus). Notice that a new element, carbon, emerges from this reaction.

With the discovery of the neutron it was no longer necessary to assume the presence of electrons in the nucleus of the atom; indeed it was simpler to consider the nucleus as containing protons with sufficient neutrons present to account for the observed atomic mass. To account for the emission of β-particles (electrons) from the nucleus during radioactive decay, it is assumed that a neutron can split to produce a proton and an electron thus:

$$\text{Neutron} \longrightarrow \text{Proton} + \beta\text{-particle (electron)}$$

The atomic mass of an atom is the combined masses of the protons and neutrons (1 unit of mass each), the number of extra-nuclear electrons being the same as the number of protons to ensure electrical neutrality of the atom.

3.10 Determination of nuclear charges—Moseley's experiments

When fast-moving electrons strike a solid target X-rays are produced with wavelengths characteristic of the particular elements present in the target. By using a series of solid elements in turn as targets, Moseley (1913) was able to resolve and photograph their X-ray spectra. The spectral lines occur in a series of groups now called the *K*, *L*, *M*, *N*, *O* series, and each particular series itself is capable of resolution into a number of closely spaced lines. The number of series and their complexity increase with increasing atomic mass, but for the lightest elements only the *K*-series is present.

By measuring the wavelengths of corresponding lines, belonging to a particular X-ray series, for a number of different elements, Moseley was able to show that the square roots of the frequency of these X-rays were directly proportional to integers which he could allocate to the different elements (these are now called atomic numbers) (fig. 3.7).

The atomic number, a fundamental property of the atom, was identified as the number of protons in the nucleus of the atom, i.e. it is numerically equal to the charge on the nucleus of the atom. Its discovery helped to resolve certain anomalies in the Periodic Table, e.g. the positions of the two pairs of elements argon and potassium, and tellurium and iodine (p. 19); furthermore the number of lanthanides (p. 436) was conclusively established as 14 on the basis of an examination of their *L*-series X-ray spectral lines. Figure 3.7 also shows that the atomic mass is not a fundamental property of the element.

FIG. 3.7. *Graph showing linear relationship between atomic numbers and the square root of the frequency for the K_α spectral lines*

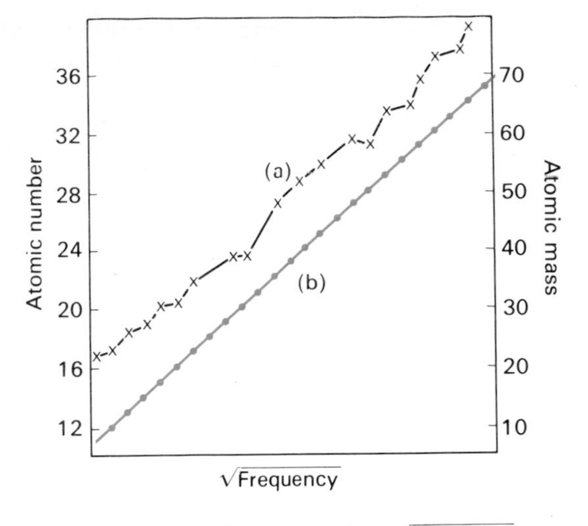

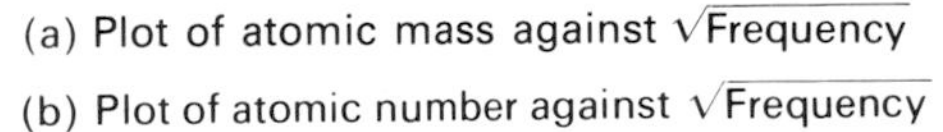

3.11 The existence of isotopes—the mass spectrometer

In 1913 Thomson produced positive ions from neon in a discharge tube, and passed the ions through simultaneously applied electric and magnetic fields. The effect was that ions of the same charge/mass ratio traced out a parabola on a photographic plate (the effect is due to ions of the same charge/mass ratio having different velocities). Thomson observed that in addition to a parabolic trace corresponding to singly ionised neon of atomic mass 20, there was also a much fainter trace corresponding to a singly ionised substance of atomic mass 22 or a doubly ionised substance of mass 44. Doubly ionised carbon dioxide, CO_2^{2+}, with a mass of 44 and a charge of 2 units, was at first thought to be the cause of this faint trace. However, in 1919, Aston re-examined the problem with a much improved apparatus (ions of the same charge/mass ratio produced a line on a photographic plate instead of a parabola). Using extremely pure neon, Aston observed that the faint trace was still present and concluded that neon exists in two chemically identical forms, one of mass 20 and the other of mass 22. These two species are called **isotopes** of neon and they differ in that one species has 10 neutrons in the nucleus while the other has a nucleus containing 12 neutrons:

Neon (atomic mass 20) 10 protons + 10 neutrons + 10 electrons
Neon (atomic mass 22) 10 protons + 12 neutrons + 10 electrons

Using this apparatus, called a mass spectrograph, Aston discovered isotopes of many other elements; thus chlorine of approximate atomic mass 35·5 was found to contain about 75 per cent of Cl atomic mass 35 (17 protons, 18 neutrons) and 25 per cent of Cl atomic mass 37 (17 protons, 20 neutrons). Isotopic species of the same atom contain the same number of protons and electrons but different numbers of neutrons.

Refinements to the mass spectrograph led to the replacement of the photographic plate by ion-current measuring instruments and this gave increased sensitivity. Modern instruments using this detecting device are called mass spectrometers.

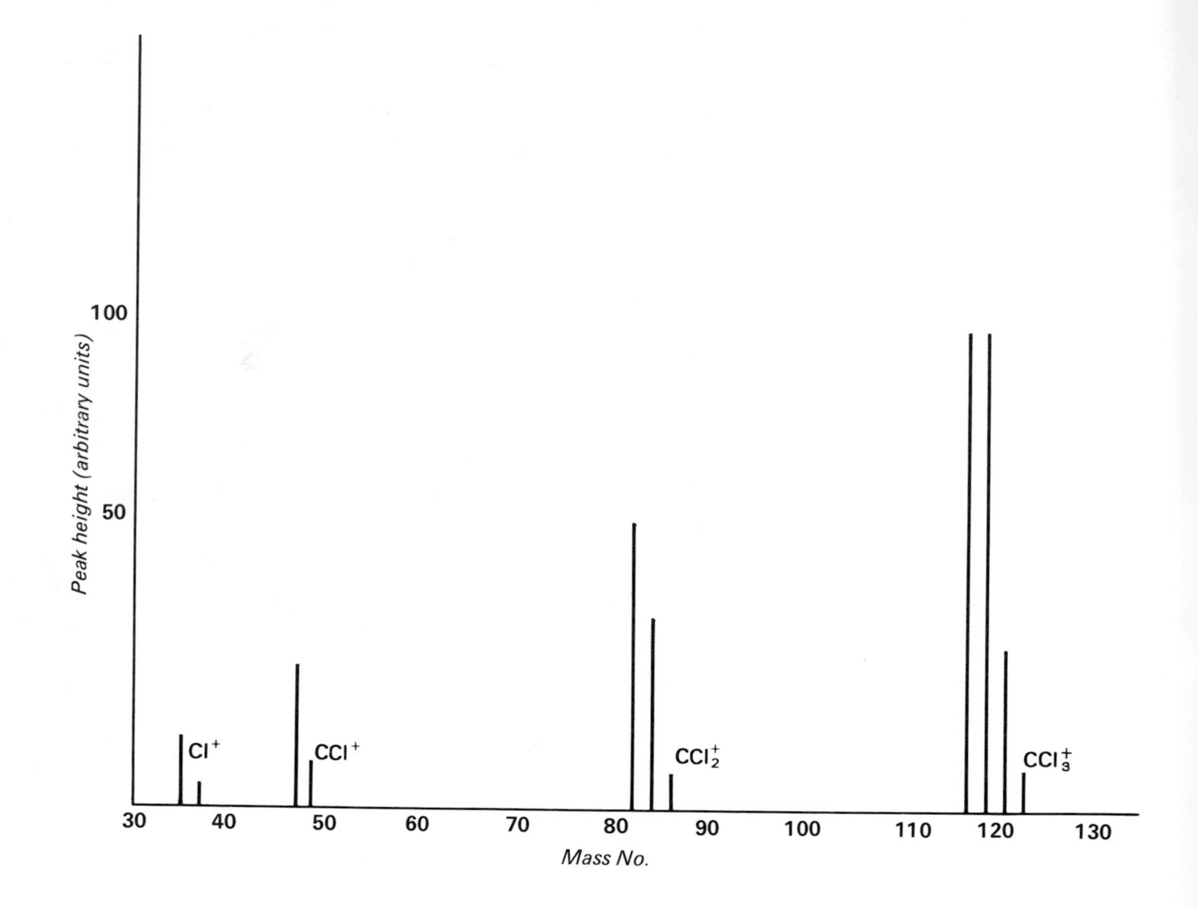

FIG. 3.8. *The mass spectrum of tetrachloromethane*

3.12 The determination of physical atomic masses

In addition to being a powerful instrument for detecting isotopes, the mass spectrometer is an invaluable instrument for determining atomic masses to a high degree of accuracy. For instance, the atomic mass of fluorine might be determined using the compound tetrafluoromethane, CF_4.

Tetrafluoromethane produces several ions in the mass spectrometer. No peak corresponding to CF_4^+ (the molecular ion) is observed and CF_3^+ is produced in greatest yield by the loss of a fluorine atom and an electron. The mass of CF_3^+ is found in this experiment to be 68·995 2, and the atomic mass of fluorine can be obtained by calculation:

Mass of CF_3^+ (using isotope ^{12}C) on $^{12}C = 12$ scale $= 68{\cdot}995\,2$
Mass of three fluorine atoms $= 68{\cdot}995\,2 - 12{\cdot}000\,0 = 56{\cdot}995\,2$
Mass of one fluorine atom $= 56{\cdot}995\,2/3 = 18{\cdot}998\,4$

The atomic mass of fluorine on the natural oxygen = 16 scale is 19·000, so the difference between the isotopic mass of fluorine and the chemically determined atomic mass is insignificantly small in any determinations involving weighings.

There is more difficulty if the element combining with carbon has more than one isotope (for example, chlorine). Figure 3.8 shows the mass spectrum of tetrachloromethane. The spectrum has been drawn as a **'stick diagram'**, the most abundant ion (the **'base peak'**) being given a height of 100 units. Again the molecular ion, this time CCl_4^+, is not observed. As chlorine has two isotopes, there is more than one peak corresponding to each ion; for example, there are three peaks for CCl_2^+, $C^{35}Cl^{35}Cl^+(82)$, $C^{35}Cl^{37}Cl^+(84)$ and $C^{37}Cl^{37}Cl^+(86)$. By measuring the relative heights of these lines in the stick diagram, the relative abundance of the isotopes and hence the atomic mass of chlorine is calculated.

3.13 Suggested reading

Atoms, Elements and Isotopes: Atomic structure, Open University, S 100, Unit 6.

J. F. J. Todd, *Modern Aspects of Mass Spectrometry*, Education in Chemistry, No. 3, Vol. 10, 1973.

3.14 Questions on chapter 3

1 What are the three fundamental units of all matter? Give their relative masses and the charges carried by each; give their location in the atom.

2 'All atoms contain electrons'. What evidence is there for this statement?

3 What contributions did the following scientists make in unravelling the structure of the atom:

(a) Crookes, (b) J. J. Thomson, (c) Millikan, (d) Becquerel, (e) Rutherford, (f) Chadwick, (g) Moseley, (h) Aston?

4 How do you account for the emission of β-particles (electrons) from the nucleus of an atom? An atom emits a β-particle, what change in atomic number will occur? If another atom emits an α-particle what resulting change in atomic number will occur?

5 List the three main fundamental particles which are constituents of atoms. Give their relative charges and masses.

Similarly name and differentiate between the radiations emitted by naturally-occurring radioactive elements.

Complete the following equations for nuclear reactions by using the Periodic Table to identify the elements **X, Y, Z, A** and **B** and add the atomic and mass numbers where they are missing.

$$^{207}_{82}Pb \longrightarrow {}_{83}\mathbf{X} + {}^{0}_{-1}e$$

$$^{27}_{13}Al + {}^{1}_{0}n \longrightarrow {}^{24}\mathbf{Y} + {}^{4}_{2}\mathbf{Z}$$

$$^{14}_{7}N + {}^{4}_{2}\mathbf{Z} \longrightarrow {}^{17}\mathbf{A} + {}_{1}\mathbf{B}$$

(JMB)

6 Neutron irradiation of aluminium gives an isotope **X**, each aluminium atom, $^{27}_{13}Al$, absorbing one neutron and then emitting an α-particle in the process. Isotope **X** then decays by β emission to another element **Y**. Deduce the mass numbers of **X** and **Y**. (C)

Chapter 4

The arrangement of electrons in atoms

4.1 Introduction

It is now known that the number and arrangement of electrons in a particular atom govern its chemical and physical properties. It is thus logical to examine the simplest atom first, namely the atom of hydrogen. Since much information is provided from a study of atomic spectra, that of hydrogen will now be discussed.

4.2 The spectrum of atomic hydrogen

When an electrical discharge is passed through hydrogen at low pressure, a pink-coloured glow is observed. If this is examined by a simple spectroscope, a few sharp, coloured lines are observed. Other lines occur in the ultraviolet and infrared regions of the electromagnetic spectrum and thus need special instruments for their detection (fig. 4.1). When the lines corresponding to a particular wavelength are examined it is seen that they all fit the equation:

$$\frac{1}{\lambda} = R\left[\frac{1}{n^2} - \frac{1}{m^2}\right]$$

FIG. 4.1. *The complete electromagnetic spectrum*

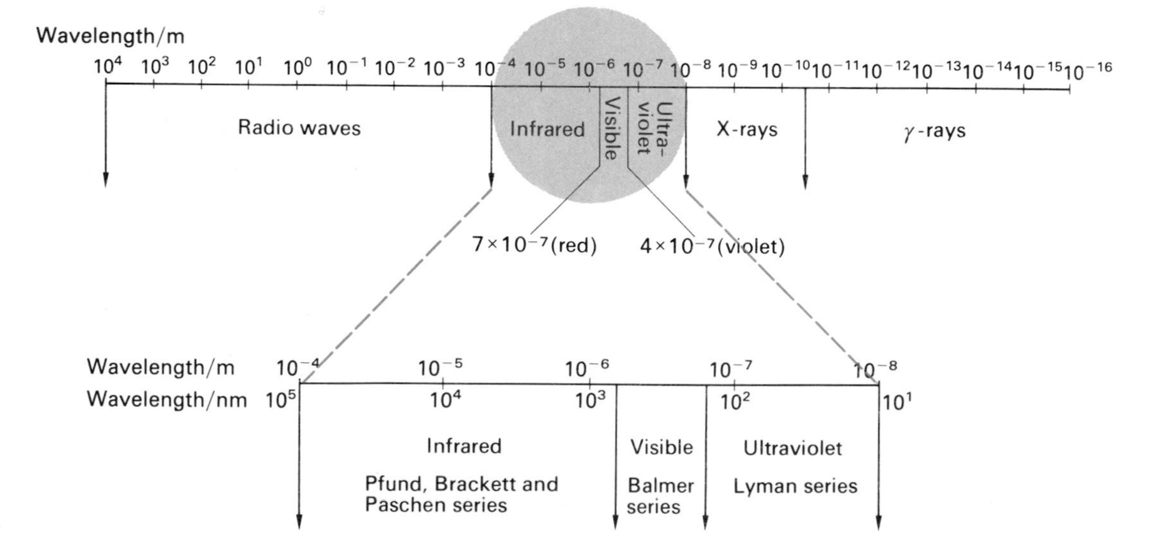

where λ is the wavelength of the particular line, R is Rydberg's constant, and n and m are integers (m always being greater than n). The complete atomic spectrum of hydrogen is resolvable into five definite series of lines, characterised by one particular value of n for each series with varying values for m. These five series, named after their respective discoverers, are shown in Table 4A:

Table 4A The spectrum of atomic hydrogen

Lyman series	$n = 1$	$m = 2, 3, 4$ etc.	in the ultraviolet region
Balmer series	$n = 2$	$m = 3, 4, 5$ etc.	in the visible region
Paschen series	$n = 3$	$m = 4, 5, 6$ etc.	in the infrared region
Brackett series	$n = 4$	$m = 5, 6, 7$ etc.	in the far infrared region
Pfund series	$n = 5$	$m = 6, 7, 8$ etc.	in the far infrared region

Plate 1. The spectrum of atomic hydrogen—the Balmer Series (By courtesy of Prof. M. J. S. Dewar and the Athlone Press)

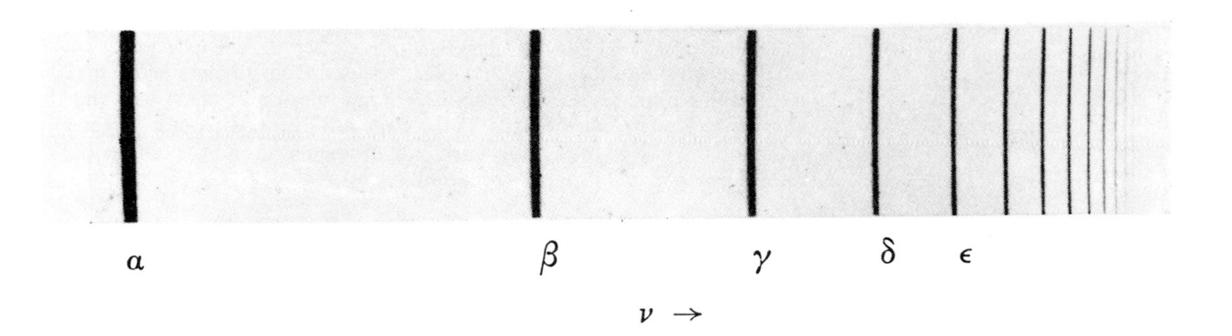

The origin of the atomic spectrum of hydrogen was explained by Bohr, applying the quantum theory first developed by Planck in 1900.

4.3 An outline of the quantum theory

As a result of an examination of the radiation emitted by hot bodies, Planck realised that the radiant energy was not emitted continuously but in 'packets', which he called quanta. These individual quanta have an energy given by the expression:

$$E = h\nu$$

where E is the energy in joules, ν is the frequency of the radiation (velocity of light/wavelength) and h is called Planck's constant, its value being $6{\cdot}625 \times 10^{-34}$ J s. A simple calculation will illustrate the use of this equation.

EXAMPLE

Calculate the energy of one quantum of light with a wavelength of 650 nm (1 nm = 10^{-9} m)

$$\text{The frequency of the light} = \frac{\text{velocity of light}}{\text{wavelength}} = \frac{3 \times 10^{8}}{650 \times 10^{-9}}$$

$$\text{Energy of quantum} = h\nu = \frac{6{\cdot}625 \times 10^{-34} \times 3 \times 10^{8}}{650 \times 10^{-9}} \text{J}$$

$$= 3{\cdot}06 \times 10^{-19}\,\text{J}$$

If the energy emitted is $9{\cdot}18 \times 10^{-19}$ J, 3 quanta have been involved.

A crude analogy that might help in grasping the fundamental ideas behind the quantum theory is given below.

EXAMPLE

n balls each of mass 2 g (2×10^{-3} kg) are released from a height of 20 cm (20×10^{-2} m) above ground level. Calculate the energy released by each ball and the total energy released by all the balls. If the balls are dropped from a height of 30 cm, calculate the energy released by each ball, all the balls, and the difference in energy released by one ball dropped from both levels.

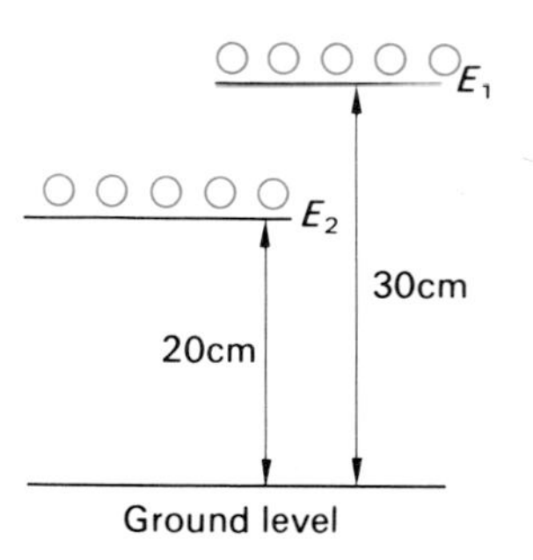

FIG. 4.2. *The potential energy E_1, of one ball at height 30 cm is greater than the energy, E_2, at 20 cm*

In fig. 4.2 E_1 and E_2 represent the potential energies of one ball at each level. This potential energy is released when each ball is dropped. The potential energy released by each ball is mgs where m is its mass, g is the acceleration due to gravity, and s is the distance above ground level. The energy released by one ball falling a distance of 20 cm is $(2 \times 10^{-3}) \times (9{\cdot}81) \times (20 \times 10^{-2}) = 3{\cdot}92 \times 10^{-3}$J. The energy released by n balls falling a distance of 20 cm is $3{\cdot}92\, n \times 10^{-3}$J. The smallest release of energy is $3{\cdot}92 \times 10^{-3}$J and the total energy is an integral multiple of this energy. The energy $3{\cdot}92 \times 10^{-3}$J is analogous to the very much smaller energy quantum of radiation, say $h\nu_2$. The energy released by one ball falling a distance of 30 cm is $(2 \times 10^{-3}) \times (9{\cdot}81) \times (30 \times 10^{-2}) = 5{\cdot}89 \times 10^{-3}$J, and the energy released by n balls falling this distance is $5{\cdot}89\, n \times 10^{-3}$J. The value $5{\cdot}89 \times 10^{-3}$J is analogous to the energy quantum of radiation, say $h\nu_1$ where $\nu_1 > \nu_2$. The difference in energy ΔE released by one ball falling from both levels is given by the equation:

$\Delta E = E_1 - E_2 = (5{\cdot}89 - 3{\cdot}92) \times 10^{-3} = 1{\cdot}97 \times 10^{-3}$J suggesting that in quantum theory $\Delta E = h\nu_1 - h\nu_2$, or that $\Delta E = h\nu$ where $\nu = \nu_1 - \nu_2$, an expression Bohr assumed in working out his theory of the hydrogen atom.

4.4 The Bohr atom

The following assumptions were made by Bohr in working out his theory of the hydrogen atom:

(a) The electron moves in an orbit around the central nucleus and only certain orbits are allowed.

(b) The electron does not radiate energy when in these orbits, and has associated with it a definite amount of energy in each orbit.

(c) Radiation is emitted when an electron undergoes a transition from one orbit to another one of lower energy. The frequency of the radiation is given by the expression $\Delta E = h\nu$, where ΔE is the difference in energies of the electron in the two levels, h is Planck's constant, and ν is the frequency of the radiation. In order to move an electron from one orbit to another one of higher energy, radiation of the frequency given by the above expression is of course absorbed.

By knowing the mass and charge of the electron, Bohr was able to calculate the radii of the possible orbits for the hydrogen atom. Furthermore, the energy of the electron in the possible orbits is given by the expression:

$$E_n = -\frac{Rch}{n^2}$$

where E_n is the energy of the electron in the nth orbit, R is Rydberg's constant (p. 34), c is the velocity of light, h is Planck's constant and n is an integer, called the principal quantum number, which can take the values 1, 2, 3, 4, 5, etc. It will be observed that the energy of the electron is a negative quantity; the reason for this is that the zero of energy is reckoned as being when the electron is removed to an infinitely large distance from the nucleus, i.e. when it is completely ionised.

It is now possible to construct an energy diagram for the hydrogen atom for various values of n using the Bohr equation, the zero energy value corresponding to the complete removal of the electron from the atom, i.e. complete ionisation. The energy of the electron in the various levels

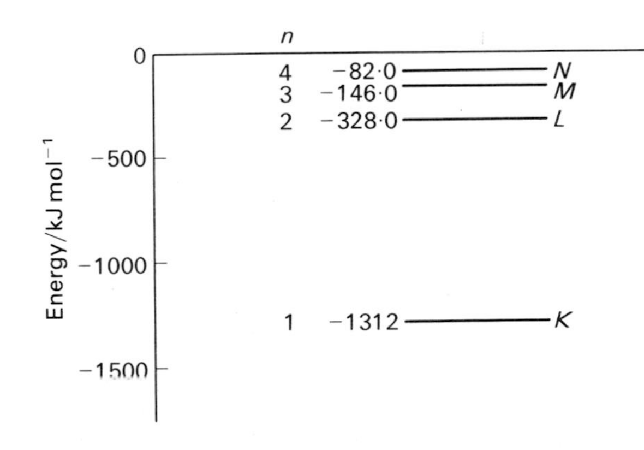

FIG. 4.3. *Energy levels in the hydrogen atom*

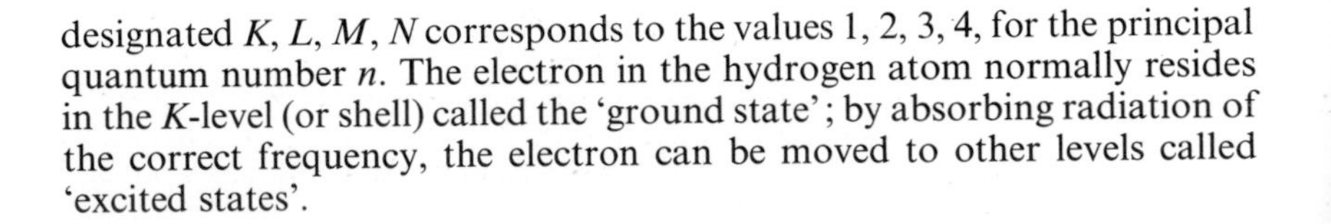
designated K, L, M, N corresponds to the values 1, 2, 3, 4, for the principal quantum number n. The electron in the hydrogen atom normally resides in the K-level (or shell) called the 'ground state'; by absorbing radiation of the correct frequency, the electron can be moved to other levels called 'excited states'.

A simple explanation of the observed lines in the atomic spectrum now becomes possible, using the energy level diagram (fig. 4.3). Thus suppose the electron in the hydrogen atom is raised from the K-level to the L-level, then on falling back to the lowest level it will emit $-328 - (-1312) = 984$ kJ mol^{-1} of energy. If the units of energy are converted into joules per electron, then the energy change is related to the frequency of the observed spectral line by the equation:

$$\Delta E = h\nu$$

ΔE being the change in energy and ν the frequency of the spectral line. Conversion to wavelength gives 121·6 nm as the calculated value, in excellent agreement with the experimentally determined value.

A general equation can be deduced from which all the spectral frequencies (or wavelengths) can be calculated. We begin with the Bohr equation:

$$E_n = -\frac{Rch}{n^2}$$

Suppose the electron is in a level designated by m and falls into a lower level n, then the energy emitted is:

$$\Delta E = E_m - E_n = -\frac{Rch}{m^2} + \frac{Rch}{n^2}$$

$$= Rch\left[\frac{1}{n^2} - \frac{1}{m^2}\right]$$

$$\text{but } \Delta E = h\nu \;\therefore\; h\nu = Rch\left[\frac{1}{n^2} - \frac{1}{m^2}\right]$$

$$\text{The wavelength } \lambda = c/\nu \;\therefore\; \frac{1}{\lambda} = R\left[\frac{1}{n^2} - \frac{1}{m^2}\right]$$

This equation is precisely the same as the one first arrived at empirically from a study of the spectral lines of atomic hydrogen (section 4.2 p. 34).

Some of the possible transitions between the various energy levels are shown in fig. 4.4.

The Bohr theory of the atom is able to account for the observed spectral lines of atomic hydrogen with exceptional precision. It needs a good deal of modification when applied to other atoms. The problem of how the electrons are arranged in other atoms is best approached by a consideration of ionisation energies.

FIG. 4.4. *Bohr orbits for the hydrogen atom (not to scale)*

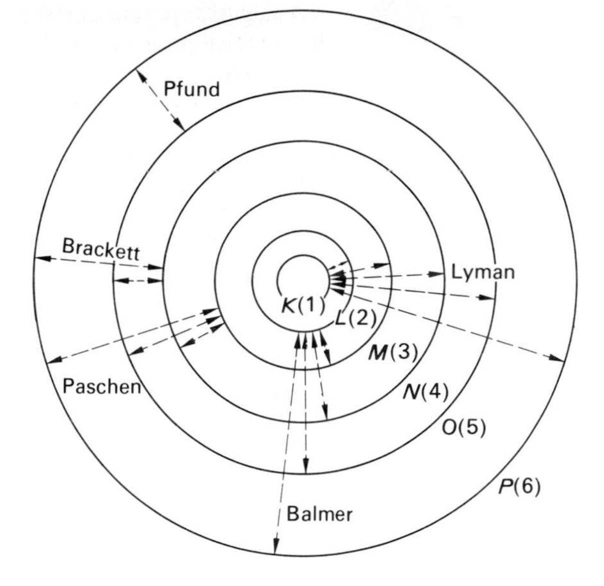

4.5 Ionisation energies

The energy required to remove an electron completely from an atom of an element is known as the first ionisation energy; the second ionisation energy is similarly the energy needed to remove completely the second electron from a singly charged ion, and so on. The values are quoted in kJ mol^{-1}, and for the potassium atom they are:

$$K(g) \longrightarrow K^{+}(g) + e^{-1} \quad \text{1st ionisation energy} = +418\ \text{kJ mol}^{-1}$$
$$K^{+}(g) \longrightarrow K^{2+}(g) + e^{-1} \quad \text{2nd ionisation energy} = +3070\ \text{kJ mol}^{-1}$$

Notice the large difference between the two values, due to the fact that it is obviously more difficult to ionise an electron if the atom already bears a positive charge. Ionisation energies are either determined spectroscopically or by means of a discharge tube, the simple device described below being suitable for measuring the first ionisation energy of xenon (fig. 4.5).

FIG. 4.5. *Determination of the first ionisation energy of xenon*

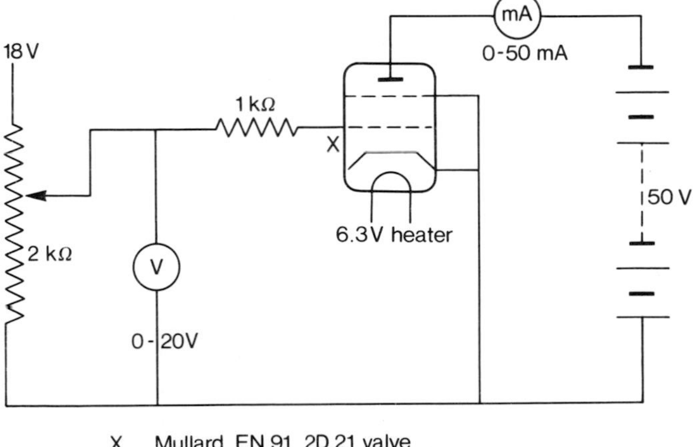

X Mullard EN 91 2D 21 valve
mA Milliammeter reading 0-50 mA
V Voltmeter reading 0-20 volts

A negative potential (50 volts) is applied to the anode of the valve, and the positive potential on the grid is gradually increased until a current is registered by the galvanometer. No current will flow until the potential on the grid is sufficient to produce singly ionised xenon Xe^{+}. Current should flow when the grid potential is 12·1 volts, from which the ionisation energy is calculated as 1170 kJ mol^{-1}.

The first ionisation energies of some elements plotted against their respective atomic numbers are shown in fig. 4.6. It is apparent that a definite pattern exists; thus the noble gases occupy peak positions and the Group 1A metals occur at minima.

From lithium to neon there is a pronounced increase in ionisation energy, with slight breaks occurring at positions occupied by boron and oxygen. An exactly similar trend occurs in the portion of the graph from sodium to argon. The sharp decrease in ionisation energy from helium to lithium is understandable, if it is assumed that the two electrons in the helium atom both occupy the *K*-shell; in lithium, however, it appears as though two electrons occupy the *K*-shell with the third in the *L*-shell further away from the nucleus and hence more readily removed. A logical extension of this reasoning would be to assume that from lithium to neon each additional electron goes into the *L*-shell, a new *M*-shell being started at sodium.

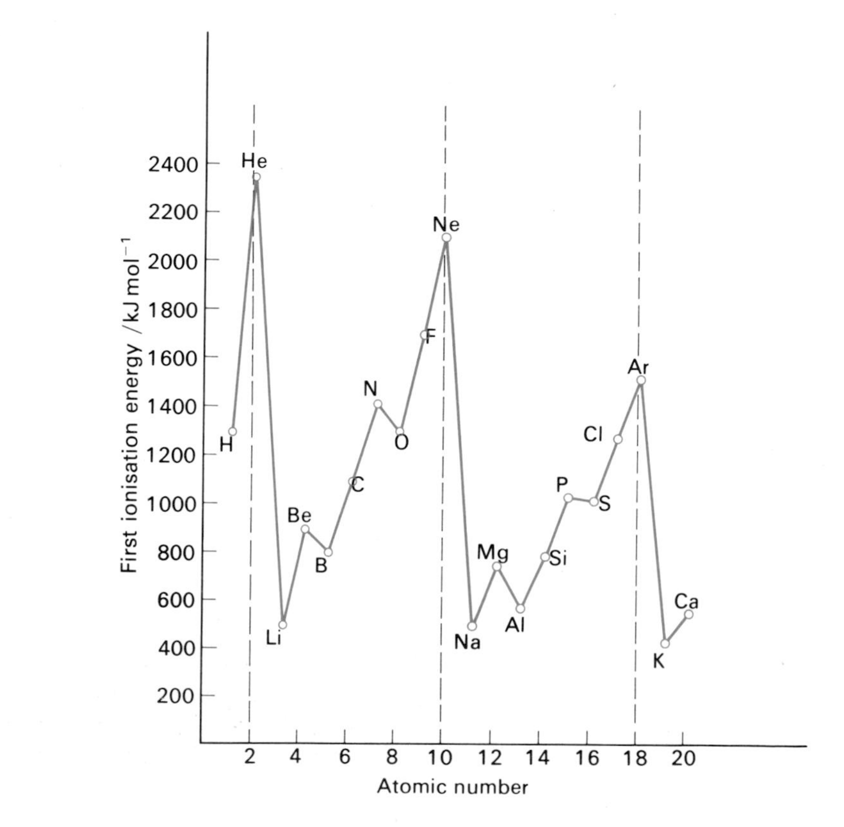

FIG. 4.6. *First ionisation energies of the first twenty elements*

Table 4B Some electronic configurations of atoms

	K	L	M	N
Helium	2			
Lithium	2	1		
Fluorine	2	7		
Neon	2	8		
Sodium	2	8	1	
Chlorine	2	8	7	
Argon	2	8	8	
Potassium	2	8	8	1

Furthermore, for a particular periodic group, the first ionisation energies become progressively smaller, e.g. the ionisation energies of helium, neon and argon are respectively 2372, 2080, and 1519 kJ mol^{-1}; for lithium, sodium and potassium the respective values are 520, 494, and 418 kJ mol^{-1}.

Some electronic configurations of atoms are given in Table 4B.

Very convincing evidence for the arrangement of electrons into definite shells is available from tabulated values of successive ionisation energies of atoms. Those for the potassium atom are shown in fig. 4.7; the ionisation energies cover a very wide range of values and for the convenience of graphical plotting their logarithmic values are used.

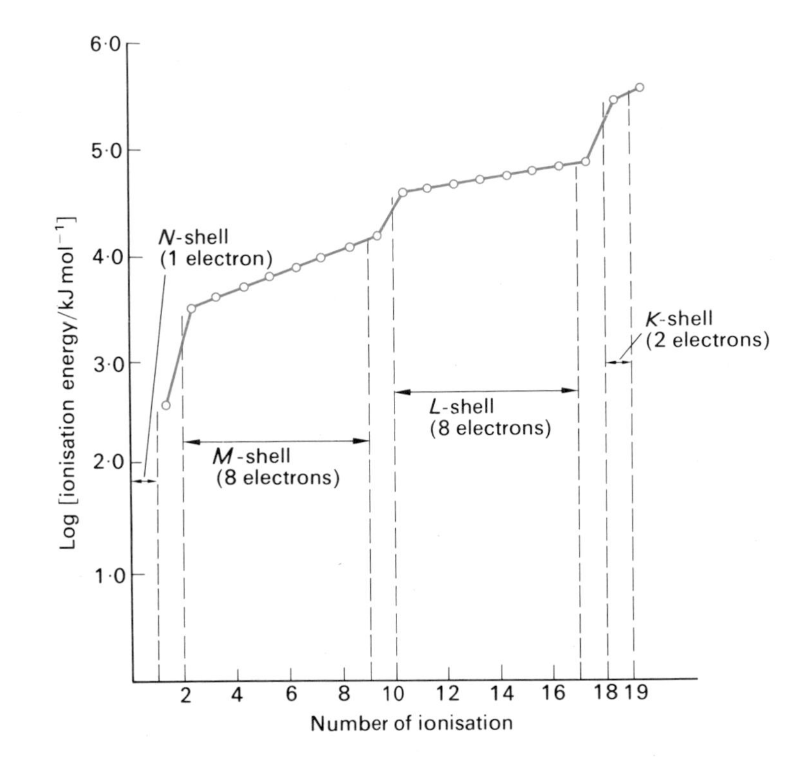

FIG. 4.7. *Successive ionisation energies (plotted logarithmetically) for the potassium atom*

Clear-cut increases in ionisation energies are observed when the 2nd, 10th and 18th electrons are involved, which would seem to indicate quite clearly that the nucleus of the potassium atom is surrounded by electrons grouped into a number of shells. Successive ionisation of the electrons becomes more difficult, since every time an electron is removed the atom carries one more unit of positive charge. However, the large increase in ionisation energy for the removal of the 2nd, 10th and 18th electrons immediately suggests that the 2nd electron is closer to the nucleus than the 1st electron; similarly the 10th and 18th electrons are nearer to the nucleus than the 9th and 17th electrons respectively. The grouping of electrons into shells is shown for the potassium atom below:

Shells	K	L	M	N
Nucleus 19p, 20n	2	8	8	1

4.6 Subdivision of the main energy levels

More detailed study (mainly spectroscopic) shows that the main energy levels of an atom designated by the principal quantum number $n = 1, 2, 3,$ 4, etc. and referred to earlier by the letters *K*, *L*, *M*, *N*, etc. are themselves capable of subdivision. For instance the 8 electrons in the *L*- and *M*-levels of potassium ($n = 2$ and 3 respectively) are distributed between two sub-levels containing 2 and 6 electrons respectively. It is possible to characterise the energy of an electron in an atom by four quantum numbers as follows:

(a) The principal quantum number n has integral values 1, 2, 3, 4, etc. As an electron with the largest value of n has the most energy, it is the one that requires the least input of energy to ionise it (i.e. it is the one most readily ionised).

(b) The subsidiary quantum number l has integral values ranging from $0, 1, 2, \ldots (n - 1)$. For a given principal quantum level, an electron with the largest value of l is the one most readily ionised.

(c) The third quantum number m has integral values ranging from $-l$, $-(l - 1)$, $-(l - 2) \ldots 0 \ldots (l - 2)$, $(l - 1)$, l. This number arises because some levels, which are normally of the same energy (degenerate), have slightly different energies when the atom is exposed to a strong inhomogeneous magnetic field.

(d) The spin quantum number has values of $-\frac{1}{2}$ and $+\frac{1}{2}$. The electron can be regarded as spinning on its axis, like a top, in a clockwise and anticlockwise direction.

It is convenient to refer to the electrons with different subsidiary quantum numbers by the letters *s*, *p*, *d*, and *f* (the letters originally used to describe particular spectral series). Thus when the subsidiary quantum number $l = 0, 1, 2, 3$ the electrons are referred to as *s*, *p*, *d*, and *f* electrons respectively.

Before it is possible to apply the quantum numbers to express the electronic configurations of atoms, it is necessary to state the Pauli exclusion principle:

No two electrons in the same atom can have the same values for all four quantum numbers.

This amounts to saying that no two electrons in any one atom behave in an identical manner. Thus consider the helium atom in its lowest energy state (ground state); the two electrons are assigned the quantum numbers $n = 1$, $l = 0$, and $m = 0$, and, since their spins cannot be the same (Pauli exclusion principle), one electron has a spin quantum number of $-\frac{1}{2}$ and the other one has a spin quantum number of $+\frac{1}{2}$. The electronic configuration of helium in the ground state is written as $1s^2$, the first numeral being the value of the principal quantum number n, the letter *s* denoting that the subsidiary quantum number l is 0, and the superscript indicating that there are two electrons in the same level with opposed spins. To indicate a pairing of two electrons, as the above condition is called, the notation $\boxed{\uparrow\downarrow}$ is convenient.

4.7 The electronic configurations of the first ten elements

The electronic configurations of the elements are constructed by assuming that electrons occupy the lowest possible energy levels available, the number of electrons in any one level being determined by the four quantum numbers and the Pauli exclusion principle. The electronic configurations of the first ten elements are given in Table 4C.

Table 4C The electronic configurations of the atoms of the first ten elements

The Four Quantum Numbers				Maximum Number of Electrons in each sublevel	Maximum Number of Electrons in each principal level	Electronic Configuration of the atoms in their ground state
Principal	Subsidiary	Third	Spin			
$n = 1$	$l = 0$	$m = 0$	$\pm\frac{1}{2}$	2 *s* electrons	2	Hydrogen $1s^1$ Helium $1s^2$
$n = 2$	$l = 0$	$m = 0$	$\pm\frac{1}{2}$	2 *s* electrons	8	Lithium $1s^2 2s^1$ Beryllium $1s^2 2s^2$
	$l = 1$	$m = 1$	$\pm\frac{1}{2}$	6 *p* electrons		Boron $1s^2 2s^2 2p^1$ Carbon $1s^2 2s^2 2p^2$ Nitrogen $1s^2 2s^2 2p^3$ Oxygen $1s^2 2s^2 2p^4$ Fluorine $1s^2 2s^2 2p^5$ Neon $1s^2 2s^2 2p^6$
		$m = 0$	$\pm\frac{1}{2}$			
		$m = -1$	$\pm\frac{1}{2}$			

It can be seen that for $l = 1$ there are three levels corresponding to $m = 1, 0, -1$ which are degenerate (have the same energy), unless the atom is placed in a strong magnetic field (p. 42). These three levels can accommodate two electrons each, with opposed spins (a maximum of 6 electrons). The question now arises: how will the electrons in these three levels of equal energy be arranged, i.e. if there are two, three or four electrons to be accommodated, will one level fill completely, holding two electrons with opposite spins, or will the electrons occupy each level singly before electron pairing takes place? The answer to this question is that electrons occupy each level singly before electron pairing takes place (because of their mutual repulsion), and only then does electron pairing take place. This principle is known as Hund's rule. Thus the electronic configuration of the nitrogen atom in the ground state can be represented as:

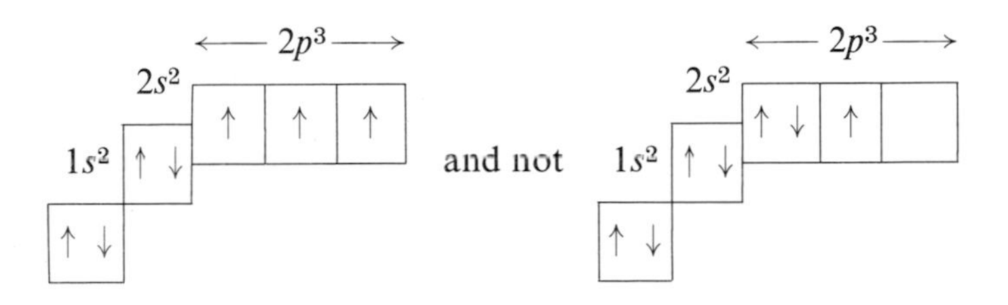

With this principle in mind, we can now display in a more detailed manner the electronic configurations of the atoms of the first ten elements in their ground states:

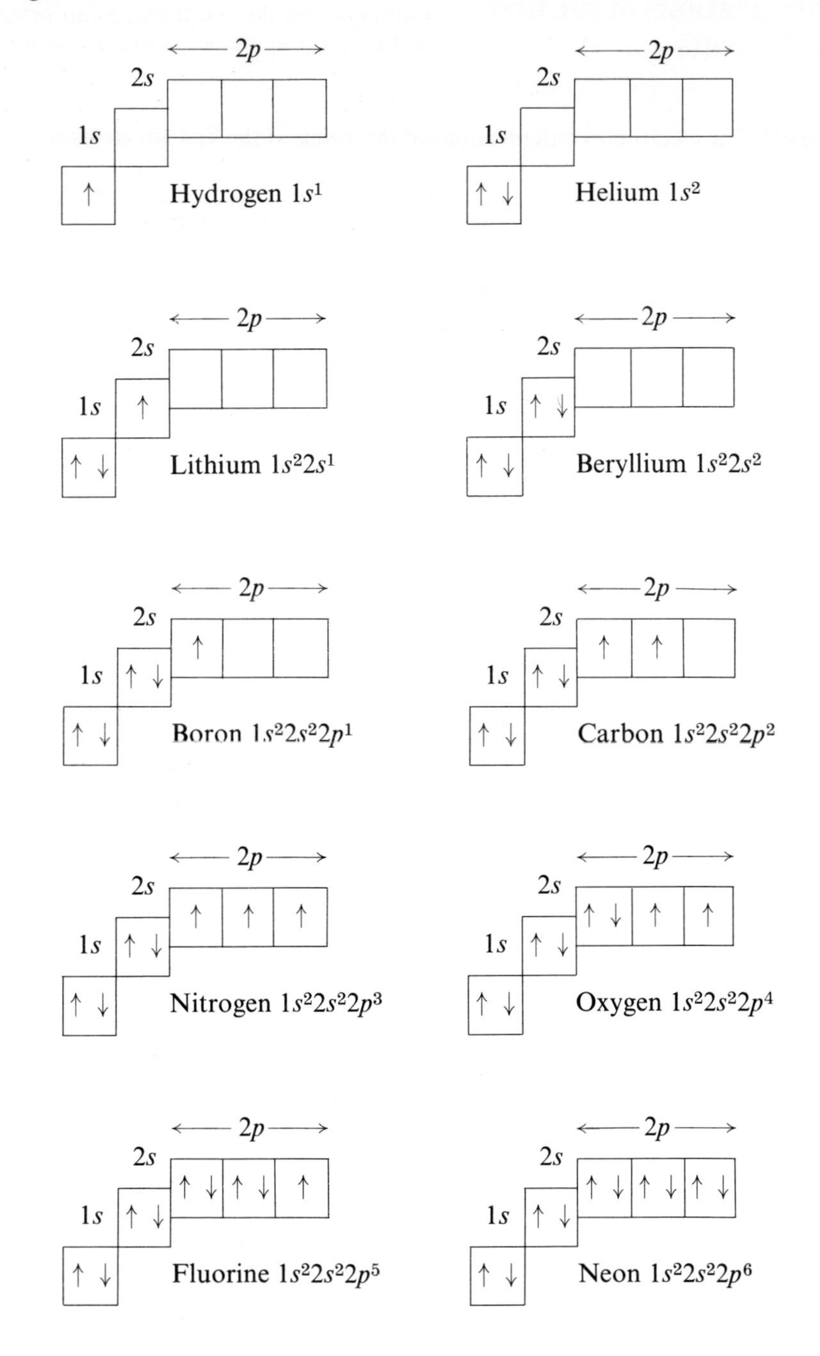

4.8 The electronic configurations of some heavier atoms

The atom of sodium has one more electron than the atom of neon, and this electron is accommodated in the level characterised by a principal quantum number $n = 3$. It is found that the maximum number of electrons that can be accommodated in any principal quantum level is $2n^2$ where n is the value of the principal quantum number, so that when $n = 3$, the quantum level can hold a maximum of 18 electrons. In practice, it is found that after 8 are present, the next 2 enter the principal quantum level $n = 4$ before the preceding one expands from 8 to 18 electrons. The electronic configurations of the elements from sodium to zinc are shown in Table 4D. Reference to p. 42 will show how to determine the number of subsidiary levels etc., corresponding to a particular quantum number.

Table 4D The electronic configurations of atoms from sodium to zinc

The Four Quantum Numbers				Maximum number of electrons in each sublevel	Maximum number of electrons in each principal level	Atomic number	Electronic configurations of the atoms from sodium to zinc in their ground states
Principal	Subsidiary	Third	Spin				
	$l = 0$	$m = 0$	$\pm\frac{1}{2}$	2 s electrons		11	Sodium $1s^2 2s^2 2p^6 3s^1$
						12	Magnesium $1s^2 2s^2 2p^6 3s^2$
						13	Aluminium $1s^2 2s^2 2p^6 3s^2 3p^1$
		$m = 1$	$\pm\frac{1}{2}$			14	Silicon $1s^2 2s^2 2p^6 3s^2 3p^2$
						15	Phosphorus $1s^2 2s^2 2p^6 3s^2 3p^3$
	$l = 1$	$m = 0$	$\pm\frac{1}{2}$	6 p electrons		16	Sulphur $1s^2 2s^2 2p^6 3s^2 3p^4$
		$m = -1$	$\pm\frac{1}{2}$			17	Chlorine $1s^2 2s^2 2p^6 3s^2 3p^5$
						18	Argon $1s^2 2s^2 2p^6 3s^2 3p^6$
$n = 3$					18	21	Scandium $1s^2 2s^2 2p^6 3s^2 3p^6 3d^1 4s^2$
						22	Titanium $1s^2 2s^2 2p^6 3s^2 3p^6 3d^2 4s^2$
						23	Vanadium $1s^2 2s^2 2p^6 3s^2 3p^6 3d^3 4s^2$
		$m = 2$	$\pm\frac{1}{2}$			24	Chromium $1s^2 2s^2 2p^6 3s^2 3p^6 3d^5 4s^1$
		$m = 1$	$\pm\frac{1}{2}$			25	Manganese $1s^2 2s^2 2p^6 3s^2 3p^6 3d^5 4s^2$
	$l = 2$	$m = 0$	$\pm\frac{1}{2}$	10 d electrons		26	Iron $1s^2 2s^2 2p^6 3s^2 3p^6 3d^6 4s^2$
		$m = -1$	$\pm\frac{1}{2}$			27	Cobalt $1s^2 2s^2 2p^6 3s^2 3p^6 3d^7 4s^2$
		$m = -2$	$\pm\frac{1}{2}$			28	Nickel $1s^2 2s^2 2p^6 3s^2 3p^6 3d^8 4s^2$
						29	Copper $1s^2 2s^2 2p^6 3s^2 3p^6 3d^{10} 4s^1$
						30	Zinc $1s^2 2s^2 2p^6 3s^2 3p^6 3d^{10} 4s^2$
$n = 4$	$l = 0$ etc.	$m = 0$	$\pm\frac{1}{2}$	2 s electrons	2	19	Potassium $1s^2 2s^2 2p^6 3s^2 3p^6 4s^1$
						20	Calcium $1s^2 2s^2 2p^6 3s^2 3p^6 4s^2$

This table shows the subdivision of the principal quantum level $n = 3$ and part of the level $n = 4$. The $4s$ level is of lower energy than the $3d$ level for the potassium and calcium atoms. The chromium and copper atoms have the configurations $\ldots 3d^5 4s^1$ and $\ldots 3d^{10} 4s^1$ respectively, instead of the expected $\ldots 3d^4 4s^2$ and $\ldots 3d^9 4s^2$ configurations.

4.9 Electronic structures and the Periodic Table

The modern Periodic Table consists of seven horizontal periods whose complexity increases with increasing atomic number (p. 19). There is a very close correlation between the number of periods and the electronic structures of the elements; thus

Period 1 contains only hydrogen and helium, and the atoms of these two elements have the respective electronic structures $1s^1$ and $1s^2$, i.e. the principal quantum level $n = 1$ is full at helium.

Period 2 starts with lithium ($1s^22s^1$) and contains eight elements, ending with neon ($1s^22s^22p^6$), i.e. the principal quantum level $n = 2$, which can hold 8 electrons, is full at neon.

Period 3 is similar to period 2, the first element being sodium ($1s^22s^22p^63s^1$) and the last element being argon ($1s^22s^22p^63s^23p^6$). However the principal quantum level $n = 3$ can accommodate a maximum of 18 electrons but it does not immediately expand from 8 to 18 until scandium is reached.

Period 4 begins with potassium ($1s^22s^22p^63s^23p^64s^1$) and is followed by calcium ($1s^22s^22p^63s^23p^64s^2$); with scandium the next electron enters principal quantum level $n = 3$, i.e. it now begins to fill from 8 to 18. There are ten elements from scandium ($1s^22s^22p^63s^23p^63d^14s^2$) to zinc ($1s^22s^22p^63s^23p^63d^{10}4s^2$). Period 4 now continues, with the extra electrons entering the principal quantum level $n = 4$ which already contains 2 electrons; gallium is thus $1s^22s^22p^63s^23p^63d^{10}4s^24p^1$ and krypton which ends this period is $1s^22s^22p^63s^33p^63d^{10}4s^24p^6$.

Period 5 begins like period 4; thus rubidium is . . . $4s^24p^65s^1$ and strontium is . . . $4s^24p^65s^2$. The next element yttrium has the extra electron in the level $n = 4$ and the filling up of this level continues until cadmium is reached, i.e. yttrium is . . . $4s^24p^64d^15s^2$ and cadmium is . . . $4s^24p^64d^{10}5s^2$. The next element indium has the electronic structure . . . $4s^24p^64d^{10}5s^25p^1$ and the period ends with xenon (. . . $4s^24p^64d^{10}5s^25p^6$).

Period 6 begins with caesium (. . . $5s^25p^66s^1$) and barium (. . . $5s^25p^66s^2$), and the next element lanthanum has the extra electron in the level $n = 5$, as its electronic structure is . . . $5s^25p^65d^16s^2$. A group of fourteen elements now appears in which the level $n = 4$ fills from 18 to 32, as the principal quantum level $n = 4$ can hold a maximum of $2n^2$ or 32 electrons. The first of these elements is cerium (. . . $4f^25s^25p^65d^06s^2$) and the last is lutetium (. . . $4f^{14}5s^25p^65d^16s^2$). The principal quantum level $n = 5$ now begins to expand from 9 to 18, beginning at hafnium (. . . $4f^{14}5s^25p^65d^26s^2$) and ending with mercury (. . . $4f^{14}5s^25p^65d^{10}6s^2$). Period 6 ends with six elements beginning at thallium (. . . $4f^{14}5s^25p^65d^{10}6s^26p^1$) and ending with radon (. . . $4f^{14}5s^25p^65d^{10}6s^26p^6$) in which principal quantum level $n = 6$ fills from 2 to 8.

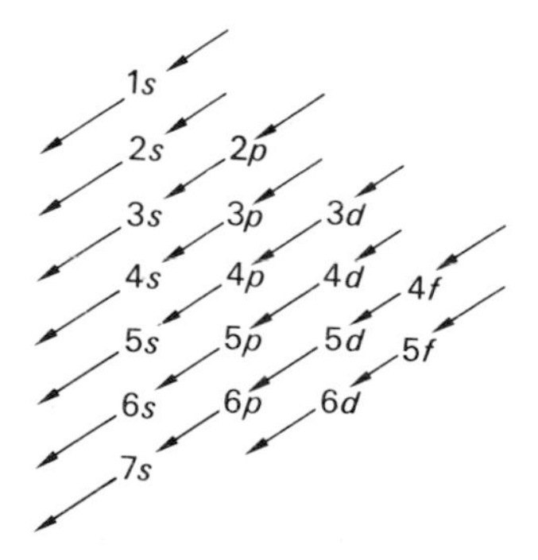

FIG. 4.8. *Method by which the electronic structure of elements can be worked out*

Period 7 begins with francium (. . . $6s^26p^67s^1$) and radium (. . . $6s^26p^67s^2$); the next elements are actinium (. . . $6s^26p^66d^17s^2$) and thorium (. . . $6s^26p^66d^27s^2$). The principal quantum level $n = 5$ now proceeds to expand from 18 to 32. Thus protactinium (atomic number 91) has the structure . . . $5f^26s^26p^66d^17s^2$ and nobelium (atomic number 102) the structure . . . $5f^{14}6s^26p^66d^07s^2$.

With few exceptions, which are not significant, the electronic structures of the elements can be worked out from the sequence given in fig. 4.8. The electronic configurations of the elements are given in Table 4E.

Table 4E
The electronic configurations of the atoms of the elements

Element	Atomic Number	1s	2s	2p	3s	3p	3d	4s	4p	4d	4f	5s	5p	5d	5f	6s	6p	6d	6f	7s
H	1	1																		
He	2	2																		
Li	3	2	1																	
Be	4	2	2																	
B	5	2	2	1																
C	6	2	2	2																
N	7	2	2	3																
O	8	2	2	4																
F	9	2	2	5																
Ne	10	2	2	6																
Na	11	2	2	6	1															
Mg	12				2															
Al	13				2	1														
Si	14		10		2	2														
P	15		electrons		2	3														
S	16				2	4														
Cl	17				2	5														
Ar	18	2	2	6	2	6														
K	19	2	2	6	2	6		1												
Ca	20							2												
Sc	21						1	2												
Ti	22						2	2												
V	23						3	2												
Cr	24						5	1												
Mn	25						5	2												
Fe	26						6	2												
Co	27		18				7	2												
Ni	28		electrons				8	2												
Cu	29						10	1												
Zn	30						10	2												
Ga	31						10	2	1											
Ge	32						10	2	2											
As	33						10	2	3											
Se	34						10	2	4											
Br	35						10	2	5											
Kr	36	2	2	6	2	6	10	2	6											
Rb	37	2	2	6	2	6	10	2	6			1								
Sr	38											2								
Y	39									1		2								
Zr	40									2		2								
Nb	41									4		1								
Mo	42									5		1								
Tc	43									5		2								
Ru	44									7		1								
Rh	45									8		1								
Pd	46				36					10		0								
Ag	47				electrons					10		1								
Cd	48									10		2								
In	49									10		2	1							
Sn	50									10		2	2							
Sb	51									10		2	3							
Te	52									10		2	4							
I	53									10		2	5							
Xe	54	2	2	6	2	6	10	2	6	10		2	6							

Table 4E (contd.)

Element	Atomic Number	1*s*	2*s*	2*p*	3*s*	3*p*	3*d*	4*s*	4*p*	4*d*	4*f*	5*s*	5*p*	5*d*	5*f*	6*s*	6*p*	6*d*	6*f*	7*s*
Cs	55	2	2	6	2	6	10	2	6	10		2	6			1				
Ba	56											2	6			2				
La	57											2	6	1		2				
Ce	58										2	2	6			2				
Pr	59										3	2	6			2				
Nd	60										4	2	6			2				
Pm	61										5	2	6			2				
Sm	62										6	2	6			2				
Eu	63										7	2	6			2				
Gd	64										7	2	6	1		2				
Tb	65										8	2	6	1		2				
Dy	66										10	2	6			2				
Ho	67										11	2	6			2				
Er	68										12	2	6			2				
Tm	69										13	2	6			2				
Yb	70										14	2	6			2				
Lu	71				46						14	2	6	1		2				
Hf	72				electrons						14	2	6	2		2				
Ta	73										14	2	6	3		2				
W	74										14	2	6	4		2				
Re	75										14	2	6	5		2				
Os	76										14	2	6	6		2				
Ir	77										14	2	6	7		2				
Pt	78										14	2	6	9		1				
Au	79										14	2	6	10		1				
Hg	80										14	2	6	10		2				
Tl	81										14	2	6	10		2	1			
Pb	82										14	2	6	10		2	2			
Bi	83										14	2	6	10		2	3			
Po	84										14	2	6	10		2	4			
At	85										14	2	6	10		2	5			
Rn	86	2	2	6	2	6	10	2	6	10	14	2	6	10		2	6			
Fr	87	2	2	6	2	6	10	2	6	10	14	2	6	10		2	6			1
Ra	88															2	6			2
Ac	89															2	6	1		2
Th	90															2	6	2		2
Pa	91														2	2	6	1		2
U	92														3	2	6	1		2
Np	93														5	2	6			2
Pu	94														6	2	6			2
Am	95					78									7	2	6			2
Cm	96					electrons									7	2	6	1		2
Bk	97														7	2	6	2		2
Cf	98														9	2	6	1		2
Es	99																			
Fm	100																			
Md	101																			
No	102																			
Lr	103																			

4.10 The electron as a cloud of negative charge

Modern theory abandons the hopeless task of attempting to determine the velocity and the position of a particular electron in an atom at any given moment in time. Instead it uses the language of probability and asks: where is a particular electron most likely to be found in an atom? A visual, though oversimplified, picture can be gained in the following manner. Suppose we imagine that it is possible to locate the position of an electron at any given moment, and that we can, furthermore, represent its position by a dot on a sheet of paper. Now suppose we can follow the track of the electron and record its positions say several million times, representing these positions by similar dots on paper. Then for the 1*s* electron the effect will be something akin to that shown in fig. 4.9 (for other electrons the effects are rather more complicated). The figure shows that the electron spends some time at large distances from the nucleus but for most of the time it is close to the nucleus; and the general effect is like that of a smeared-out negative cloud with diffuse edges (for the 1*s* electron the negative cloud is spherical. Figure 4.9 simply shows a section through the sphere).

FIG. 4.9. *The negative cloud representation of the hydrogen* 1*s* *electron*

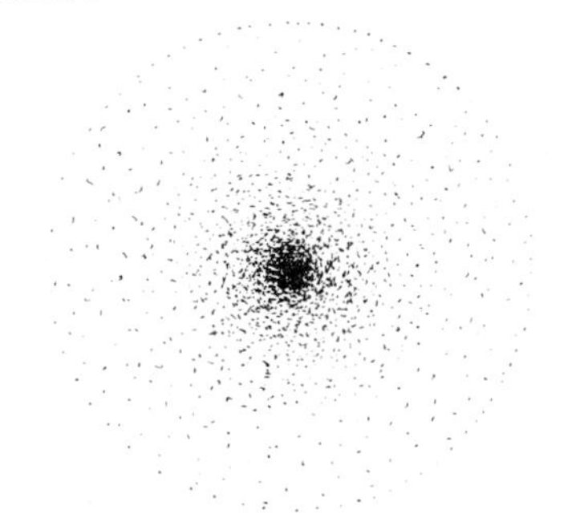

A more useful idea is to enquire how much of this negative charge is likely to be found at certain distances from the nucleus. If $\rho(r)$ represents the density of the electron cloud at varying distances r from the nucleus (density is a direct measure of probability), then the radial density is defined to be the density $\rho(r)$ multiplied by the area over which this charge is spread, $4\pi r^2$. Thus:

$$\text{Radial density} = 4\pi r^2 \rho(r)$$

FIG. 4.10. *Radial density/radius plot for the hydrogen* 1*s* *electron*

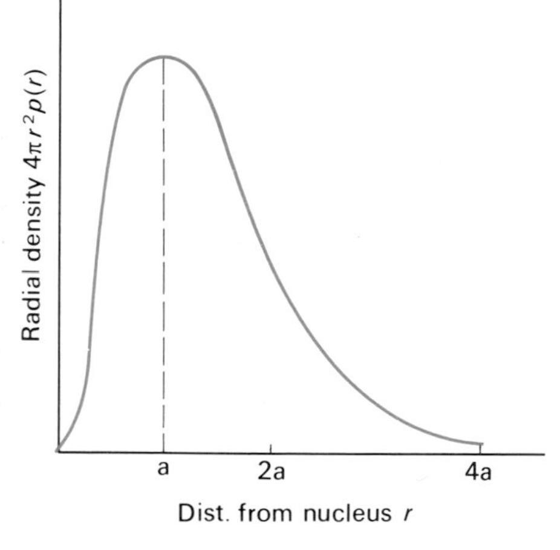

Values for the radial density of the 1*s* electron in the hydrogen atom can be calculated by the application of specialised mathematics (wave mechanics), and the general shape of the radial density/radius plot is shown in fig. 4.10 for the hydrogen 1*s* electron. The maximum radial density occurs at a value $r = a$, which is precisely the same value as that calculated by the Bohr theory (p. 37). However, as the plot shows, the radial density is appreciable over a wide range of values, i.e. we are forced to replace the precise orbits of the Bohr theory by more diffuse ones which are called orbitals. Although in much of the book it will be convenient to represent the electron as though it were a point charge, the word orbital will be used instead of orbit.

4.11 Suggested reading

M. J. Smith, *Ionisation potentials, periodicity and electronic configuration,* School Science Review, No. 170, Vol. 49, 1968.

K. J. Tetlow, *Atomic Energy Levels—Hydrogen and Sodium*, School Science Review, No. 183, Vol. 53, 1971.

J. J. Morwick, *What is the electron, Really*? Journal of Chemical Education, Vol. 55, 662, 1978.

4.12 Questions on chapter 4

1 The Balmer series of spectral lines in the spectrum of atomic hydrogen arises from electronic transitions from the level $n = 2$ to $m = 3, 4, 5$, etc. These lines fit the general equation:

$$\frac{1}{\lambda} = R\left[\frac{1}{n^2} - \frac{1}{m^2}\right]$$

where R is Rydberg's constant and has the value 109 677 cm^{-1} and λ is the wavelength in cm. Calculate the wavelengths of the first three spectral lines in this series in nm).

The Lyman series of spectral lines arise from electronic transitions from $n = 1$ (the ground level) to $m = 2, 3, 4$, etc. Calculate the series limit (which corresponds to complete removal of the electron). Hence calculate the energy in joules needed to remove the electron completely. (Planck's constant is $6{\cdot}625 \times 10^{-34}$ J s and the velocity of light is 3×10^8 m s^{-1}).

Calculate the ionisation energy of the hydrogen atom in kJ mol^{-1}.

2 Electromagnetic radiation of wavelength 242 nm is just sufficient to ionise the outermost electron of the sodium atom. Calculate the ionisation energy of the sodium atom in kJ mol^{-1} (Planck's constant is $6{\cdot}625 \times 10^{-34}$ J s the velocity of electromagnetic radiation is 3×10^8 m s^{-1}).

3 What is meant by the ionisation energy of an atom of an element? The first three ionisation energies of the sodium atom are 494, 4564 and 6924 kJ/mol^{-1}; those for the magnesium atom are 744, 1464 and 7750 kJ/mol^{-1}; similarly the first four ionisation energies of the aluminium atom are 578, 1811, 2745 and 11 540 kJ/mol^{-1}. Plot the logarithm of these successive ionisation energies against the number of ionisation, i.e. one, two, three, etc, and comment on your results.

4 Explain the formation of the line spectra of hydrogen and show how the line spectra are related to the electronic structure and ionisation energy of the atom. What evidence concerning atomic structure is provided by X-ray spectra? (C)

5 Explain briefly, in terms of electron structures, the occurrence of Periods, Groups and Transition Series in the Periodic Table. (O & C)

6 Explain, using the electronic structure of the oxygen atom as an example, what is meant by (a) the four quantum numbers, (b) the Pauli exclusion principle, (c) Hund's rule.

7 The electronic structure of the sodium atom is 2.8.1 or more precisely $1s^2 2s^2 2p^6 3s^1$. Write down, in a similar way, the electronic structures of the atoms of boron, nitrogen, fluorine, aluminium, sulphur and argon.

8 Without referring to the Periodic Table name the elements, give their electronic structure and name the group in the Periodic Table to which elements with the following atomic numbers belong: 4, 9, 15, 17, 20, 28, 37, 50 and 54.

9 Discuss the atomic spectrum of hydrogen and its relation to our understanding of the electronic structure of atoms.

Suggest explanations for the following observations:

(*a*) The atomic spectrum of hydrogen contains lines in the radio-frequency region of the electromagnetic spectrum.

(*b*) A line in the spectrum of atomic hydrogen on a distant object in the universe occurs at a wavelength of 300 nm though it is known to occur in the laboratory at 121·6 nm. (Oxford Schol.)

10 On the basis of current views on atomic structure explain the meaning and significance of the following statements with, where possible, illustrative examples:

(a) there is no precise arithmetical relationship between atomic number and relative atomic mass;

(b) the reactivity of an atom is determined by its electronic configuration;

(c) a monatomic ion has the same atomic number and relative atomic mass as a neutral atom;

(d) isotopes of an element have the same atomic number. (JMB)

Chapter 5

Chemical bonds

5.1 Introduction

Once the nuclear theory of the atom (Rutherford) and the structure of the hydrogen atom (Bohr) had been worked out, the time was ripe for an explanation of chemical bonding. Indeed, even before the general principles of atomic structure had been fully worked out, Kossel and Lewis had independently put forward the view that elements tended to react together to attain the stable electronic configurations of the noble gases. While Kossel's work was mainly directed towards explaining the nature of the bonds in electrolytes (electrovalent or ionic bonds), Lewis was concerned with developing a theory to account for the bonding in non-electrolytes (covalent bonds). All modern refinements can be traced back to two important research papers published independently in 1916.

5.2 The electrovalent or ionic bond

Although it has now been established that the noble gases krypton and xenon can be induced to form chemical compounds, it is still true to say that, as a group, the noble gases enter into chemical combination far less readily than other elements. It is therefore reasonable to assume that the unreactive nature of this group of elements is due to their particular electronic configurations (except for helium they all have an outer shell containing 8 electrons). Kossel pointed out that an alkali metal atom could attain this supposedly stable electronic configuration by the loss of one electron and an halogen atom by the gain of one electron. He visualised the hypothetical reaction between an atom of sodium and one of chlorine as a complete transfer of one electron from the sodium atom to the atom of chlorine thus:

$$\text{Na (2.8.1)} + \text{Cl (2.8.7)} \longrightarrow \underset{\text{electronic configuration of neon}}{\text{Na}^+ \text{(2.8)}} + \underset{\text{electronic configuration of argon}}{\text{Cl}^- \text{(2.8.8)}}$$

The resulting charged species are ions, which are held together by electrostatic attraction; the bonding is said to be electrovalent or ionic.

Electrovalent compounds are formed between the most reactive metals, e.g. Groups 1A and 2A of the Periodic Table, and the most reactive non-metals, e.g. Groups 6B and 7B. The following are a few more examples:

$$\text{Ca (2.8.8.2)} + \text{O (2.6)} \longrightarrow \underset{\text{electronic configuration of argon}}{\text{Ca}^{2+} \text{(2.8.8)}} + \underset{\text{electronic configuration of neon}}{\text{O}^{2-} \text{(2.8)}}$$

$$2K\ (2.8.8.1) + S\ (2.8.6) \longrightarrow 2K^+\ (2.8.8) + S^{2-}\ (2.8.8)$$

electronic configuration of argon | electronic configuration of argon

$$Li\ (2.1) + F\ (2.7) \longrightarrow Li^+\ (2) + F^-\ (2.8)$$

electronic configuration of helium | electronic configuration of neon

In modern notation the latter example can be represented thus:

$$Li\ (1s^2 2s^1) + F\ (1s^2 2s^2 2p^5) \longrightarrow Li^+\ (1s^2) + F^-\ (1s^2 2s^2 2p^6)$$

Electrovalent compounds exist as hard crystals, of high melting point, with ions of opposite charge arranged in a symmetrical array (p. 73); thus X-ray examination of sodium chloride shows that each sodium ion is surrounded by six equidistant chloride ions and that each chloride ion is surrounded by six equidistant sodium ions. The actual positions taken up by the ions in a crystal are determined by their charges and relative sizes (p. 73). In no sense can a single molecule of an electrovalent compound be said to exist, e.g. Na^+Cl^- simply represents the empirical formula of sodium chloride, the whole crystal being one giant molecule or macromolecule.

Plate 2. A model showing the structure of sodium chloride. The small spheres represent the Na^+ ion and the larger spheres the Cl^- ion. Note that each ion has a co-ordination number of 6
(By courtesy of Catalin Limited, Waltham Abbey)

5.3 The covalent bond

An electrovalent bond is formed by the transfer of one or more electrons from one atom to another so that the ions formed have the electronic configurations of one of the noble gases. Lewis recognised that in simple molecules another form of bonding had to operate and he suggested that electrons might be shared in pairs so that each atom could be said to have attained the stable electronic configuration of a noble gas. Thus, consider the chlorine atom which has seven electrons in its outer shell; if one electron is provided by each atom and shared equally, then each chlorine atom can acquire a share in eight electrons—a completed octet as it is sometimes called.

$$:\underset{\cdot\cdot}{\ddot{\text{Cl}}}\cdot + \cdot\underset{\cdot\cdot}{\ddot{\text{Cl}}}: \longrightarrow :\underset{\cdot\cdot}{\ddot{\text{Cl}}}:\underset{\cdot\cdot}{\ddot{\text{Cl}}}:$$

The sharing of two electrons, one electron being provided by each atom, constitutes a single covalent bond; it is usually represented by a single line joining the two atoms together. For example, the chlorine molecule can be represented thus:

$$\text{Cl—Cl}$$

The oxygen atom has six electrons in its outer shell, and in the oxygen molecule, O_2, a stable electronic configuration can be assumed to be attained by the sharing of four electrons, two being provided by each atom. The oxygen atoms are bound together by a double covalent bond and this is represented by a double line:

$$:\underset{\cdot\cdot}{\ddot{\text{O}}} + \underset{\cdot\cdot}{\ddot{\text{O}}}: \longrightarrow \dot{.}\text{O}\vdots\text{O}\dot{.} \qquad \text{or} \quad \text{O}{=}\text{O}$$

Similarly, the nitrogen molecule contains a triple bond, which involves the sharing of six electrons, three being provided by each nitrogen atom:

$$\cdot\underset{\cdot\cdot}{\dot{\text{N}}} + \underset{\cdot\cdot}{\dot{\text{N}}}\cdot \longrightarrow :\text{N}\,\vdots\!\vdots\,\text{N}: \qquad \text{or} \quad \text{N}\equiv\text{N}$$

When the two atoms that are bound together by covalent bonds are different, the electrons are not equally shared. For instance, the chlorine atom has a greater affinity for electrons than does the hydrogen atom in hydrogen chloride, and the electron pair constituting the single covalent bond is displaced towards the chlorine atom. The molecule is said to possess an electrical dipole and this is responsible for the substance having a boiling point higher than it would otherwise have had.

$$\text{H}\cdot + {}_{\times}\underset{\times\times}{\overset{\times\times}{\text{Cl}}}{}^{\times}_{\times} \longrightarrow \text{H}\;{}^{\cdot}_{\times}\;\underset{\times\times}{\overset{\times\times}{\text{Cl}}}{}^{\times}_{\times} \qquad \text{or} \quad \text{H—Cl}$$

To indicate the presence of an electrical dipole in the molecule, the notation $\overset{\delta+}{\text{A}}\text{—}\overset{\delta-}{\text{B}}$ is often used, e.g. $\overset{\delta+}{\text{H}}\text{—}\overset{\delta-}{\text{Cl}}$.

Three more examples of covalent compounds are given below:

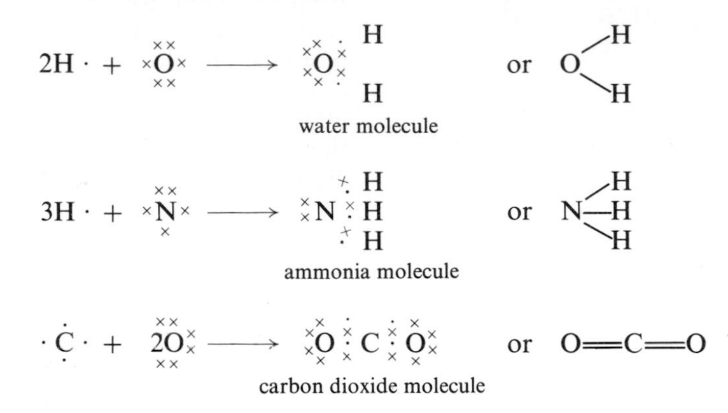

The covalent compounds discussed above exist in the form of discrete molecules with little force of attraction between the individual molecules. This is true of many covalent compounds and accounts for many of them being gases, volatile liquids or easily fusible solids. There are some giant molecules or macromolecules, however, e.g. diamond and silicon dioxide, in which directional covalent bonds extend throughout the whole structures. Compounds of this type are, of course, solids with high melting and boiling points.

5.4 The co-ordinate or dative covalent bond

This is similar to the covalent bond, except that only one atom provides the two electrons that are shared. It is present in the stable complex $BCl_3.NH_3$ formed between boron trichloride and ammonia. The nitrogen atom in the ammonia molecule contains two electrons not involved in bonding (a lone pair of electrons) while the boron atom in the boron trichloride molecule is two electrons short of the stable octet (an example of a compound disobeying the octet rule) (p. 53). The octet can be completed thus:

H Cl H Cl
H : N + B : Cl ⟶ H : N : B : Cl
H Cl H Cl

or

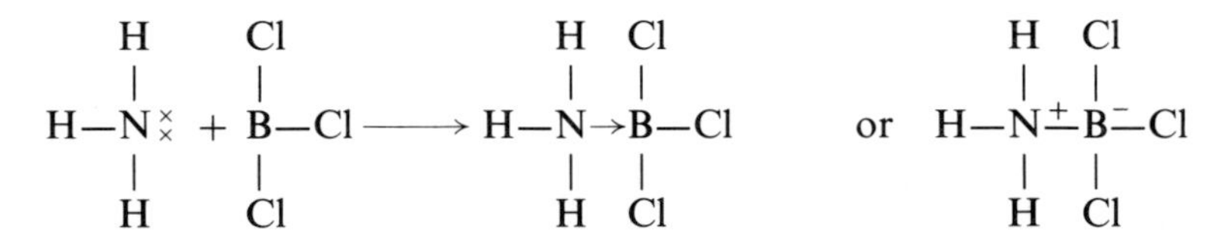

A co-ordinate bond is represented by an arrow pointing from the donor to the acceptor atom, or by a single line with a negative charge on the acceptor atom and an equal but opposite charge on the donor atom.

Other examples of compounds containing co-ordinate bonds are carbon monoxide and nitric acid:

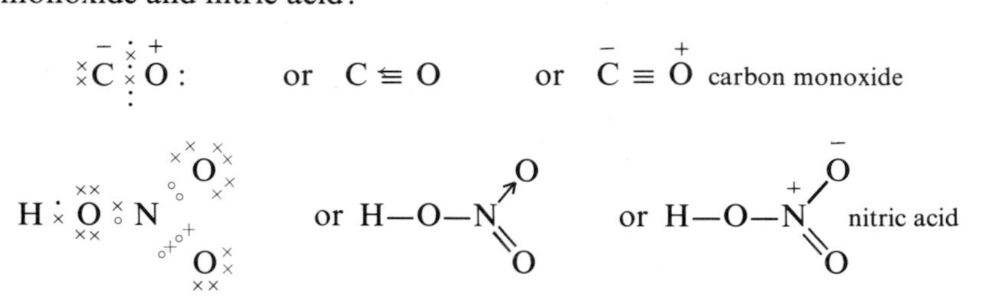

Ammonium chloride is an electrovalent compound $NH_4^+Cl^-$ whose formation from ammonia and hydrogen chloride can be considered to involve the formation of a co-ordinate bond thus:

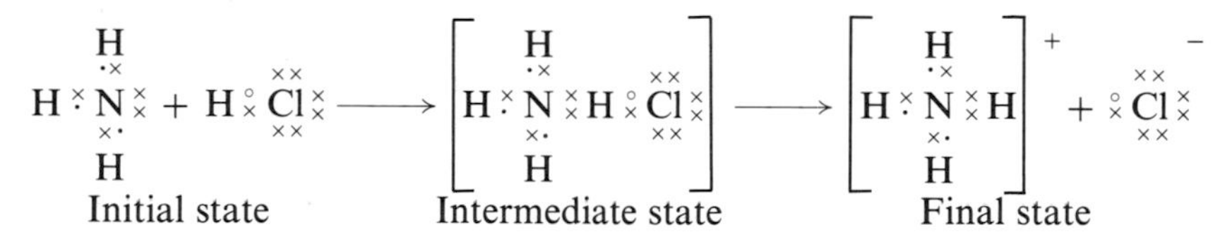

5.5 The metallic bond

The complete loss or gain of electrons (electrovalency) and the sharing of electrons (covalency and dative covalency), in order to attain a stable noble gas configuration, are useful theories that are helpful in explaining the formation of bonds between metal and non-metallic atoms (e.g. as in sodium chloride) and between many non-metallic atoms (e.g. as in nitrogen, water, etc.). These theories cannot, however, account for the strong bonds that are formed between metal atoms.

X-ray analysis shows that the atoms of many metals tend to pack together as closely as possible in three dimensions, in much the same kind of way that marbles would arrange themselves if put into a box and shaken. The close packing of spheres in two dimensions is shown in fig. 5.1.

FIG. 5.1. *Close packing of spheres in two dimensions*

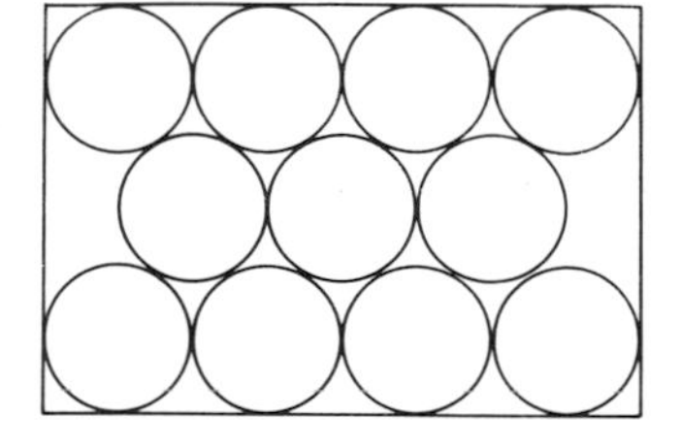

This physical picture of metallic structure accounts quite adequately for the two properties, malleability and ductility, since it is easy to visualise the atoms altering their positions, as though they had slipped over each other, when subjected to stress or strain. It is assumed that each metal atom in the crystal pools its valency electrons and that these 'free' electrons 'cement' the positive ions together (in this respect there is some resemblance between metallic bonding and electrovalency, as is shown diagrammatically in two dimensions in fig. 5.2).

FIG. 5.2. *The structures of sodium chloride and a metal (in two dimensions)*

Na^+ Cl^- Na^+ Cl^-
Cl^- Na^+ Cl^- Na^+
Na^+ Cl^- Na^+ Cl^-
Cl^- Na^+ Cl^- Na^+

e^- e^- e^-
M^{2+} e^- M^{2+} e^- M^{2+} e^-
e^- e^- e^-
M^{2+} e^- M^{2+} e^- M^{2+} e^-
e^- e^- e^-
M^{2+} e^- M^{2+} e^- M^{2+} e^-

M^{2+} = 2-valent metal ion
e^- = electron

The high thermal and electrical conductivities of metals are due to the presence of 'free' electrons which can move through the metal structure when it is connected to a source of electricity, since the electron is very small by comparison with the gaps between positive ions.

5.6 Energetics of electrovalent compound formation

Consider the reaction between sodium and chlorine; experiment shows that 411 kJ mol⁻¹ of heat is evolved when sodium chloride is formed from solid sodium and gaseous chlorine at 298 K and 1 atmosphere pressure according to the equation:

$$Na(s) + \tfrac{1}{2}Cl_2(g) \longrightarrow Na^+Cl^-(s) \qquad \Delta H^\ominus(298\,K) = -411\,kJ\,mol^{-1}$$

(energy evolved is shown with a negative sign, i.e. $\Delta H^\ominus$ is negative, energy absorbed is shown with a positive sign, i.e. $\Delta H^\ominus$ is positive).

The reaction can be broken down into a number of hypothetical stages, each one involving an energy change as outlined below. The algebraic sum of the individual energy changes must be the same as the experimentally determined value (principle of the conservation of energy).

Conversion of solid sodium into gaseous sodium atoms
Energy is needed to break down the metallic lattice of sodium. This is called the **molar enthalpy of sublimation,** *S*, and is the energy required to convert 1 mole of solid sodium into gaseous sodium atoms.

$$Na(s) \longrightarrow Na(g) \qquad S = +108{\cdot}4\,kJ\,mol^{-1}$$

Removal of the valency electron of the sodium atom to give a sodium ion
The energy required for this is the **first ionisation energy,** *I*, (p. 39).

$$Na(g) \longrightarrow Na^+(g) + e^{-1} \qquad I = +496\,kJ\,mol^{-1}$$

Dissociation of the chlorine molecule into chlorine atoms
Energy is needed to break the covalent bond. The **dissociation energy,** *D*, is the energy needed to convert 1 mole of chlorine molecules into 2 moles of chlorine atoms in their state of lowest energy. To give 1 mole of chlorine atoms, half the dissociation energy, $D/2$, is required.

$$\tfrac{1}{2}Cl_2(g) \longrightarrow Cl(g) \qquad D/2 = +121{\cdot}1\,kJ\,mol^{-1}$$

Addition of an electron to the chlorine atom to give a chloride ion
This process for chlorine is actually exothermic. The energy involved in adding an electron to an atom of an element is called the **electron affinity,** *E*.

$$Cl(g) + e^- \longrightarrow Cl^-(g) \qquad E = -348\,kJ\,mol^{-1}$$

The bringing together of the sodium and chloride ions to form solid sodium chloride
This process results in the liberation of energy, known as the **lattice energy,** *U*.

$$Na^+(g) + Cl^-(g) \longrightarrow Na^+Cl^-(s) \qquad U = -788{\cdot}5\,kJ\,mol^{-1}$$

The overall energy change in the complete reaction, $\Delta H_t^\ominus$, is thus:

$$\Delta H_t^\ominus = (+108{\cdot}4 + 496 + 121{\cdot}1 - 348 - 788{\cdot}5)\ \mathrm{kJ\ mol^{-1}}$$
$$= -411\ \mathrm{kJ\ mol^{-1}}$$

($\Delta H_t^\ominus$ must equal $\Delta H^\ominus$ by the law of conservation of energy)

The individual energy changes involved in this reaction are set out in fig. 5.3.

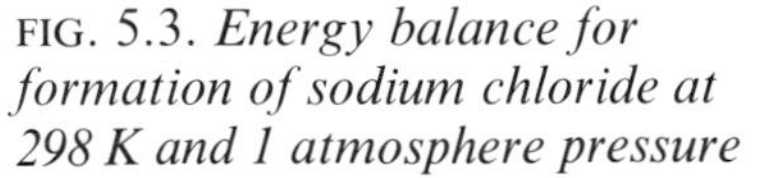

FIG. 5.3. *Energy balance for formation of sodium chloride at 298 K and 1 atmosphere pressure*

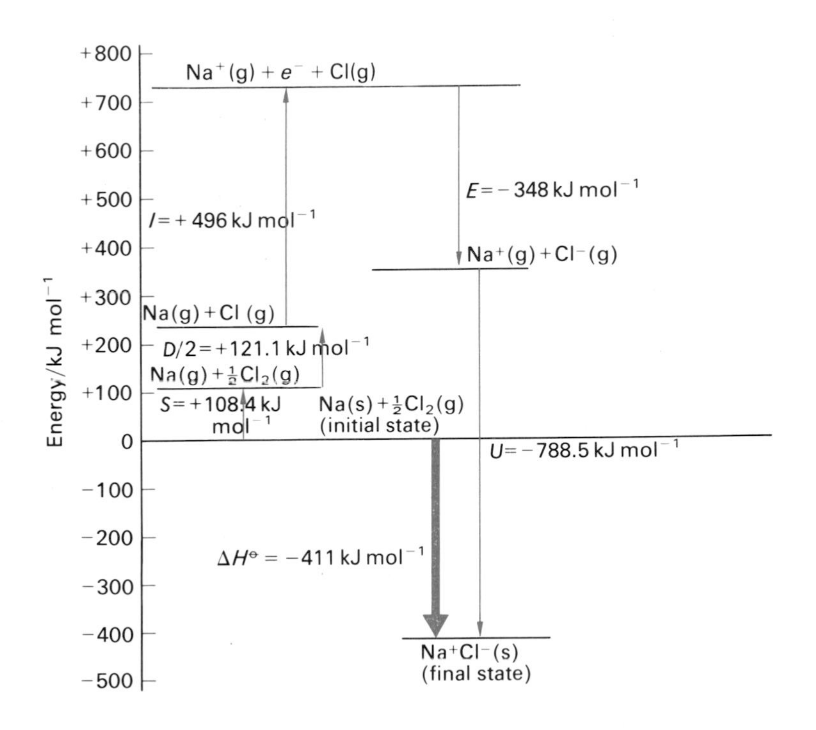

A compound is likely to be electrovalent if the calculated heat of formation—assuming an electrovalent structure—is highly exothermic. Otherwise a structure involving covalent bonding is energetically more favourable. As is clear from fig. 5.3, the three most important terms that decide whether a compound is likely to be electrovalent are: ionisation energy, electron affinity and lattice energy.

An electrovalent compound is likely to be formed if:

(a) the ionisation energy of one atom is relatively low,

(b) the electron affinity of the other atom is 'high' (only atoms that give 1-valent anions do so with the liberation of energy; 'high' is meant in the sense 'as least endothermic as possible').

(c) the lattice energy is high.

The lattice energy is very often the controlling factor. Thus consider the highly exothermic reaction between magnesium and oxygen at 298 K and 1 atmosphere pressure:

$$Mg(s) + \tfrac{1}{2}O_2(g) \longrightarrow Mg^{2+}O^{2-}(s) \qquad \Delta H^\ominus(298\ \mathrm{K}) = -602\ \mathrm{kJ\ mol^{-1}}$$

The energy terms for this reaction are given in fig. 5.4.

FIG. 5.4. *Energy balance for formation of magnesium oxide at 298 K and 1 atmosphere pressure*

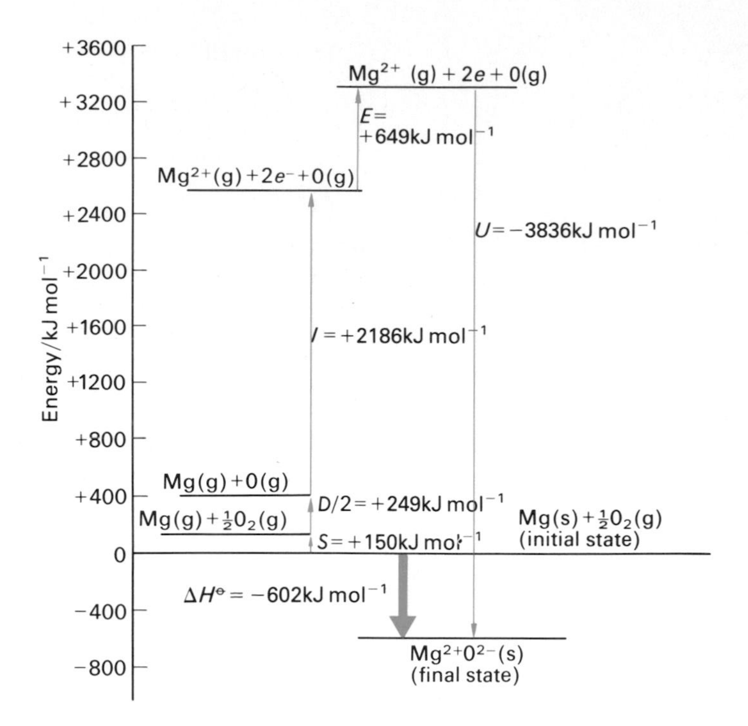

As can be seen from fig. 5.4, all the energy terms except that for the crystal lattice energy are endothermic, and the high lattice energy of magnesium oxide is the prime reason why this compound is electrovalent.

It is relevant at this stage to ask the question: 'Why is the formula of magnesium oxide $Mg^{2+}O^{2-}$ and not $Mg^{+}O^{-}$?' Clearly less energy is required to ionise one electron from the magnesium atom than two; similarly, the energy needed to add one electron to the oxygen atom is less than that needed to add two, since negative charges repel one another. In fact, the addition of the first electron is an exothermic process:

$$O(g) + e^- \longrightarrow O^-(g) \qquad E' = -142\ \text{kJ mol}^{-1}$$
$$O^-(g) + e^- \longrightarrow O^{2-}(g) \qquad E'' = +791\ \text{kJ mol}^{-1}$$

The deciding factor must be the high lattice energy term, i.e. the lattice energy of the structure $Mg^{2+}O^{2-}$ must be very much higher than that of the hypothetical $Mg^{+}O^{-}$ structure, a consequence of the fact that the force of attraction between ions of opposite charge increases with the magnitude of the charge they carry.

5.7 Distinguishing features between electrovalent and covalent compounds

Electrovalency and covalency are two extremes of chemical bonding; consequently many compounds have bonding of intermediate character. Thus whenever an electron pair is not equally shared between two bonding atoms in a covalent compound, a certain amount of 'ionic character' is introduced, e.g. hydrogen chloride is polarised in the sense $\overset{\delta+}{H}—\overset{\delta-}{Cl}$. Whenever a compound is described as being either electrovalent or covalent, the wording is intended to convey the idea that the bonding is predominantly of one particular type.

The most important distinguishing features between electrovalent and covalent compounds are discussed below under several headings. Like all classifications in chemistry many exceptions do exist.

Structure

Covalent compounds contain bonds which have directional properties and the overall molecules have a definite shape (p. 75). The molecules themselves are often small and intermolecular attractions very small (van der Waals' forces, p. 67); thus the individual molecules are easily separated, so that covalent compounds are often gases, liquids, or soft solids. There are, nevertheless, a large number of exceptions, e.g. diamond (p. 79) and silicon dioxide (p. 79), which are giant molecules with covalent bonding running completely through the crystal structures.

Electrovalent compounds exist as high melting-point brittle solids in which a symmetrical arrangement of ions within the crystal is apparent. The arrangement of the ions within the crystal is determined by the magnitude of their charges and their relative sizes (p. 73). The brittle nature of ionic crystals is easily visualised using the models shown in fig. 5.5.

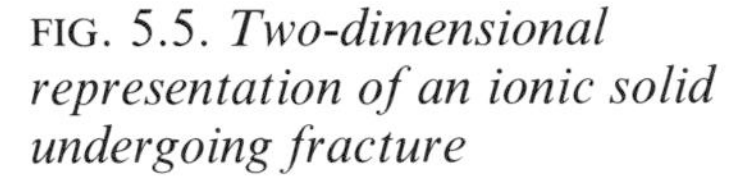

FIG. 5.5. *Two-dimensional representation of an ionic solid undergoing fracture*

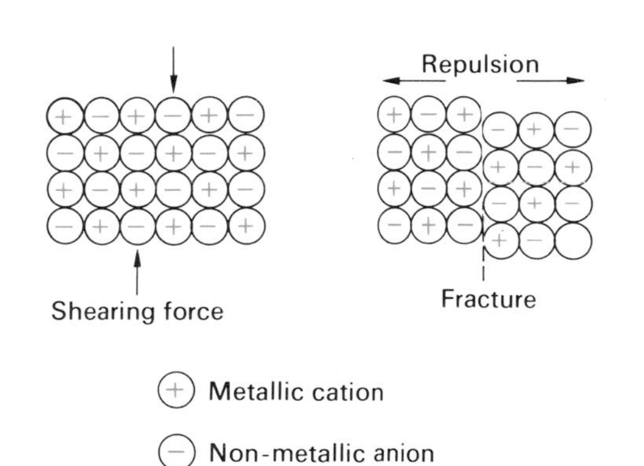

Since no one ion in an electrovalent compound is associated with any single oppositely charged ion, the bonding is non-directional. There is no such entity as a single molecule unless the whole crystal itself is regarded as a giant molecule.

Conduction of electricity

Covalent compounds are generally non-electrolytes. There are some, however, that conduct if they dissolve in polar solvents, i.e. solvents of high dielectric constant such as water. Hydrogen chloride is a typical covalent gas which dissolves in water, a typical covalent liquid, to produce a conducting solution. The reason for this is that the covalent bond in the hydrogen chloride molecule is broken by chemical interaction with the water, thus:

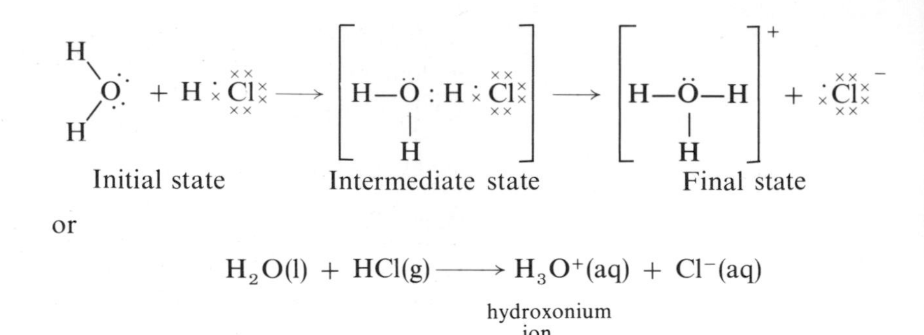

or

$$H_2O(l) + HCl(g) \longrightarrow H_3O^+(aq) + Cl^-(aq)$$

hydroxonium ion

Electrovalent compounds are electrolytes when fused or in solution, since the ions, held rigid in the solid, become mobile.

Solubility

Many electrovalent compounds are soluble in water, the separated ions being surrounded by water molecules (a process called hydration). Thus, consider the dissolution of an electrolyte A^+B^- in water which is itself polarised in the sense:

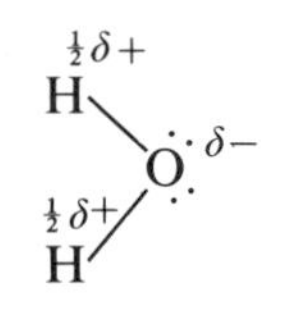

The negative end of the water dipole points towards the cation A^+, and the positive end points towards the anion B^-. The hydrated cation can be represented thus:

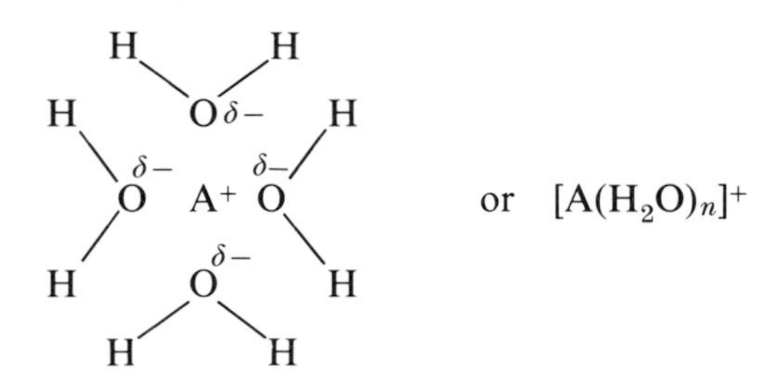

or $[A(H_2O)_n]^+$

Hydration is an exothermic process (weak chemical bonds are formed), the energy released being known as **hydration energy**. This energy release helps to offset the large lattice energy that must be absorbed in the process of separating the ions in the crystal. Electrovalent compounds, such as barium sulphate, which are virtually insoluble in water, have very high lattice energies, i.e. the very large amount of energy that must be absorbed by the crystal to separate the ions is not offset to any great extent by the energy released when the ions hydrate.

Electrovalent compounds are insoluble in non-polar solvents such as benzene and tetrachloromethane, since there can be little or no interaction between solvent molecules and the ions.

Covalent compounds are generally soluble in non-polar solvents. Those that dissolve in polar solvents such as water either do so because chemical reaction changes their structure, e.g. the ionisation of hydrogen chloride in water, or because both solute and solvent contain similar chemical groupings, e.g. ethanol is completely soluble in water and this is principally because both compounds contain hydroxyl, OH, groups that can interact.

The nature of their reactions

The reactions of electrovalent compounds in aqueous solution are those of their hydrated ions and are virtually instantaneous, because chemical bonds do not have to be broken initially, e.g. all ionic chlorides react with silver salts in aqueous solution to precipitate silver chloride:

$$Ag^+(aq) + Cl^-(aq) \longrightarrow AgCl(s)$$

Most covalent compounds, particularly organic ones, react at a much slower rate, since chemical bonds must be broken in addition to new ones being formed.

5.8 Electrovalent or covalent—the concept of electronegativity

Electronegativity is a measure of the power of an atom to attract electrons; it involves the two terms, ionisation energy and electron affinity. Mulliken has defined electronegativity as the arithmetical mean of the first ionisation energy and the electron affinity of an atom. Pauling's definition is somewhat different and it is his electronegativity values that are given in Table 5A. Fluorine, the most electronegative element, is given an arbitrary value of 4·0 and the electronegativities of the atoms of other elements are related to it.

Table 5A The Pauling electronegativity values of some elements

			H 2·1			
Li 1·0	Be 1·5	B 2·0	C 2·5	N 3·0	O 3·5	F 4·0
Na 0·9	Mg 1·2	Al 1·5	Si 1·8	P 2·1	S 2·5	Cl 3·0
K 0·8	Ca 1·0	—	Ge 1·8	As 2·0	Se 2·4	Br 2·8
Rb 0·8	Sr 1·0	—	Sn 1·8	Sb 1·9	Te 2·1	I 2·5
Cs 0·7	Ba 0·9					

As Table 5A shows, electronegativity decreases down a particular group. This is because the atom becomes progressively larger and the inner electron shells decrease the attraction between the positive nucleus and the peripheral electrons. Electronegativity increases across a particular period from alkali metal to halogen, since the progressive increase in nuclear charge exerts a contracting effect on the electron shells (in the same period, additional electrons reside in the same shell).

A bond formed between two atoms of similar electronegativity will be essentially covalent. Increase in electronegativity of one atom will result in that atom having greater control over the bonding pair of electrons, i.e. the covalent bond will be polarised. Further increase in electronegativity will result in increased polarity of the bond until eventually the electron pair can be considered to reside almost entirely on one atom, i.e. an electrovalent bond will be established, thus:

A : X	$\overset{\delta+\ \delta-}{A : Y}$	$A^+ : Z^-$
covalent bond	polarised covalent bond	electrovalent bond

electronegativities of the atoms increase in the order $X < Y < Z$.

It now becomes clear why Group 1A and 2A elements react with Group 6B and 7B elements to give essentially electrovalent compounds, caesium fluoride, Cs^+F^-, being the most 'ionic' compound. On the other hand non-metals react by forming essentially covalent bonds.

The reason why aluminium fluoride is an electrovalent compound, $Al^{3+}(F^-)_3$, whereas aluminium chloride, Al_2Cl_6, is covalent is probably because the fluorine atom is more electronegative than the chlorine atom. An explanation in terms of lattice energy would run as follows: since the fluoride ion is smaller than the chloride ion, the lattice energy of the $Al^{3+}(F^-)_3$ crystal structure would be greater than that of a similar $Al^{3+}(Cl^-)_3$ structure by virtue of the closer approach of the oppositely charged ions (other things being equal). In this instance the two explanations amount to practically the same thing, since the fluorine atom is more electronegative than the chlorine atom because it is smaller and consequently gives rise to a smaller anion.

5.9 Electrovalent compounds which violate the octet rule

Non-metals that form negative ions (anions) do so by gaining sufficient electrons from metal atoms to attain the electronic configuration of a noble gas. The following are a few such structures:

N^{3-}, O^{2-}, F^-	(neon structure 2.8)
S^{2-}, Cl^-	(argon structure 2.8.8)
Se^{2-}, Br^-	(krypton structure 2.8.18.8)
Te^{2-}, I^-	(xenon structure 2.8.18.18.8)

Many simple cations also have noble gas structures, e.g. those formed by the metals in Groups 1A and 2A. There are, however, many cations whose electronic structures are in no way related to those of the noble gases. The following three classes are apparent.

Cations having 18 electrons in their outer shell

Examples of cations assuming this configuration are the 1-valent ions of copper and silver, and the 2-valent ions of zinc, cadmium and mercury. The electronic configurations of their atoms and ions are shown in Table 5B:

Table 5B Examples of cations having 18 electrons in their outer shell

Atoms	Ions
Cu (2.8.18.1)	Cu^{+} (2.8.18)
Zn (2.8.18.2)	Zn^{2+} (2.8.18)
Ag (2.8.18.18.1)	Ag^{+} (2.8.18.18)
Cd (2.8.18.18.2)	Cd^{2+} (2.8.18.18)
Hg (2.8.18.32.18.2)	Hg^{2+} (2.8.18.32.18)

Cations having 2 electrons in the outer shell preceded by one containing 18 electrons

Examples of cations having this configuration are those of thallium, lead and bismuth, Tl^{+}, Pb^{2+} and Bi^{3+} respectively (Table 5C). The outer sub-shell containing 2 electrons (*s* electrons) not used for valency purposes is known as the 'inert pair'. The 'inert pair' effect is a feature of the chemistry of the heavier metals in Groups 3B, 4B, and 5B and is discussed later on in the book.

Table 5C Examples of cations having 2 electrons in their outer shell

Atoms	Ions
Tl (2.8.18.32.18.3)	Tl^{+} (2.8.18.32.18.2)
Pb (2.8.18.32.18.4)	Pb^{2+} (2.8.18.32.18.2)
Bi (2.8.18.32.18.5)	Bi^{3+} (2.8.18.32.18.2)

Cations of transition and inner transition metals

For transition metals, cations having from 9 to 17 electrons in the outer shell are possible. Examples include the following:

V^{2+} (2.8.11) Cr^{2+} (2.8.12) Fe^{2+} (2.8.14)
V^{3+} (2.8.10) Cr^{3+} (2.8.11) Fe^{3+} (2.8.13)

None of the cations falling into the above three categories is as stable as one with a noble gas configuration. This can be seen by comparing the successive ionisation energies ($kJ\,mol^{-1}$) of an atom that forms a noble gas-like ion, e.g. aluminium, with those of, say, a transition metal, e.g. vanadium:

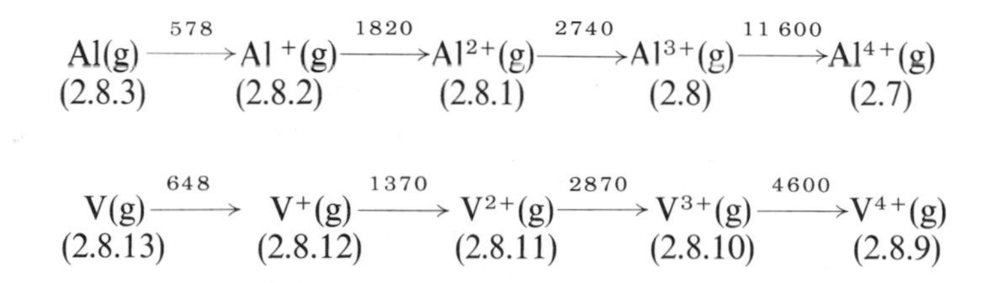

$$\underset{(2.8.3)}{Al(g)} \xrightarrow{578} \underset{(2.8.2)}{Al^{+}(g)} \xrightarrow{1820} \underset{(2.8.1)}{Al^{2+}(g)} \xrightarrow{2740} \underset{(2.8)}{Al^{3+}(g)} \xrightarrow{11\,600} \underset{(2.7)}{Al^{4+}(g)}$$

$$\underset{(2.8.13)}{V(g)} \xrightarrow{648} \underset{(2.8.12)}{V^{+}(g)} \xrightarrow{1370} \underset{(2.8.11)}{V^{2+}(g)} \xrightarrow{2870} \underset{(2.8.10)}{V^{3+}(g)} \xrightarrow{4600} \underset{(2.8.9)}{V^{4+}(g)}$$

For aluminium, the formation of Al^{3+} is possible but the removal of a fourth electron from the stable noble gas-like ion Al^{3+} is energetically impossible in chemical reactions. The ionisation energies of the vanadium atom show a more gradual increase with no pronounced break.

Since metals and non-metals that mutually combine to form electrovalent compounds with noble gas configurations are just those which are not too far removed from the noble gases in the Periodic Table, it is not really surprising that these configurations are attained. However, since many cations do exist which have configurations in no way related to those of the noble gas-like ions, over-emphasis of the octet rule should be avoided; energy changes are of more fundamental importance.

5.10 Electrovalent compounds containing more complex anions

Anions containing more than two combined atoms are common and include the carbonate, nitrite, nitrate, sulphite and sulphate anions. Consider the structure of the carbonate ion, CO_3^{2-}. Two electrons from a metal atom (or atoms) can be transferred to two oxygen atoms to give the configuration $O^-(2.7)$, the carbon octet can now be completed by the formation of two C—O⁻ covalent bonds and one C=O double covalent bond thus:

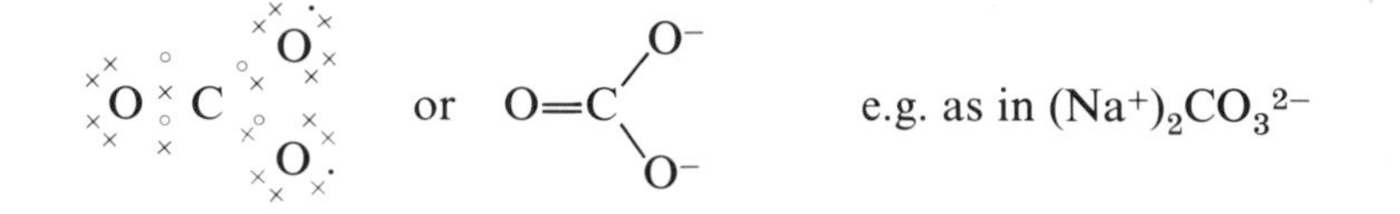

The structures of other anions are constructed similarly, and can be represented as, for example,

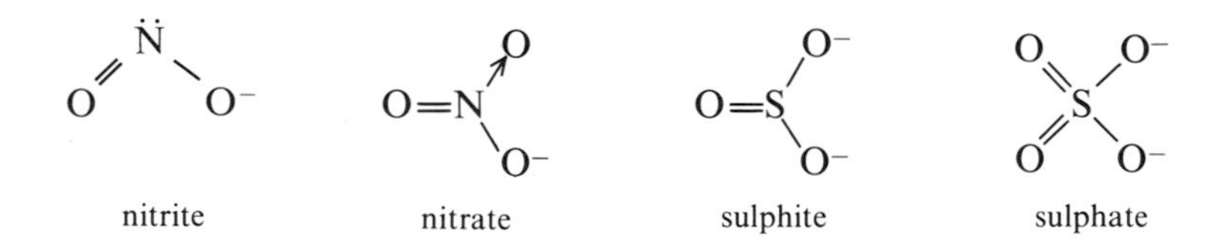

Note that the sulphur atom in the sulphite ion is surrounded by 10 electrons, and that in the sulphate ion it is surrounded by 12 electrons. The octets could be preserved by writing the S=O bonds as dative bonds, the sulphur being the donor atom. However, there is evidence to show that the former is the best representation of these structures, i.e. sulphur can expand its octet (see section 5.11).

5.11 Covalent compounds which violate the octet rule

Although the atoms of many elements which form covalent bonds can be considered to attain the electronic configuration of a noble gas, e.g. the structures of hydrogen, oxygen, hydrogen chloride, water, etc., many exceptions exist and include $BeCl_2$, BF_3, PF_5 and SF_6 together with SO_3^{2-} and SO_4^{2-} (mentioned in the previous section). Modern valency theory focuses attention on electron pairing rather than on completion of the octet. Consider beryllium chloride: the beryllium atom has the configuration $1s^22s^2$ (no unpaired electrons) and the chlorine atom has the configuration $1s^22s^22p^63s^23p^5$ (1 unpaired electron). When the beryllium atom enters into chemical combination an electron from the 2*s* level is considered to be promoted to the higher 2*p* level with the absorption of energy (promotion energy absorbed) and each single electron is paired off with the unpaired electrons of two chlorine atoms.

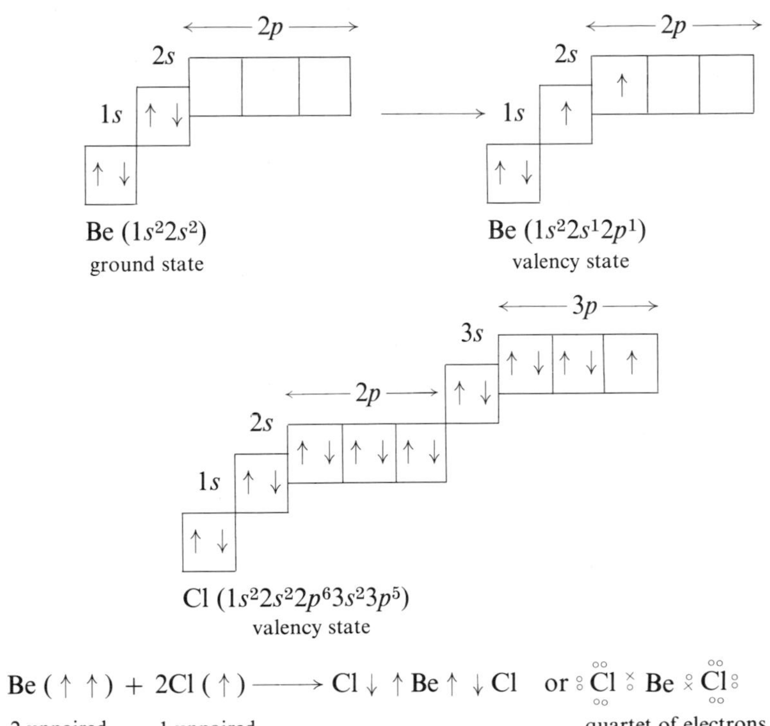

Be (↑ ↑) + 2Cl (↑) ⟶ Cl ↓ ↑ Be ↑ ↓ Cl or Cl Be Cl

2 unpaired electrons　　1 unpaired electron　　quartet of electrons only

The promotion energy is less than the energy released in forming bonds with the unpaired electrons of the chlorine atom; indeed if this were not so, then beryllium would be devoid of chemical properties like the noble gas helium.

The formation of boron trifluoride can be explained in a similar way (the three equivalent *p* levels are designated p_x, p_y and p_z):

B ($1s^22s^22p^1$) ⟶ B ($1s^22s^12p_x{}^12p_y{}^1$)

ground state, 1 unpaired electron　　valency state, 3 unpaired electrons

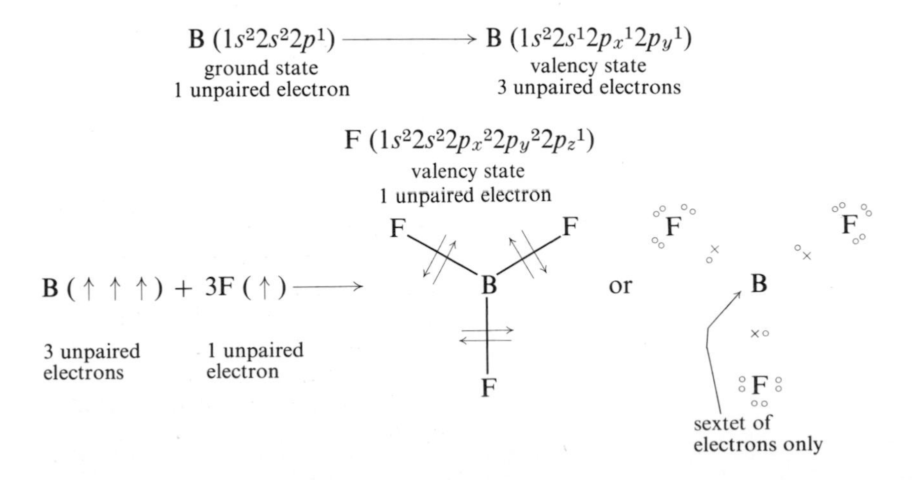

Phosphorus pentafluoride, PF_5, and sulphur hexafluoride, SF_6, are also compounds in which the octet of electrons around the central atom is exceeded. The same reasoning as before applies; thus 3*d* levels are available to phosphorus and sulphur which can accommodate electrons promoted from the ground state:

$$P\ (1s^22s^22p^63s^23p^3) \longrightarrow P\ (1s^22s^22p^63s^13p^33d^1)$$

3 unpaired electrons → 5 unpaired electrons, highest valency state

$$S\ (1s^22s^22p^63s^23p^4) \longrightarrow S\ (1s^22s^22p^63s^23p^33d^1) \longrightarrow S\ (1s^22s^22p^63s^13p^33d^2)$$

2 unpaired electrons → 4 unpaired electrons → 6 unpaired electrons, highest valency state

Phosphorus pentafluoride and sulphur hexafluoride can be prepared because the energy released in bond formation with fluorine is more than sufficient to produce a phosphorus atom with 5, and a sulphur atom with 6 unpaired electrons.

It is interesting to note that, whereas phosphorus has valencies of three and five, e.g. in PF_3 and PF_5, and sulphur has valencies of two, four and six, e.g. in H_2S, SF_4 and SF_6, nitrogen is limited to a valency of three and oxygen to a valency of two. This is because *d* levels are not easily accessible for the elements from hydrogen to neon in the Periodic Table, i.e. *d* levels first become available for atoms with a principal quantum number $n = 3$ (see fig. 4.9. p. 45).

It must be pointed out that generally only electron promotion is possible between sublevels characterised by the same principal quantum number, e.g. $2s \rightarrow 2p$, $3s \rightarrow 3p \rightarrow 3d$, etc., are possible, since the energy required can be provided by chemical reaction. But $2s \rightarrow 3s$ or $2s \rightarrow 3p$, etc., are impossible since the energy jumps are too high. One exception occurs in the chemistry of transition metals where the $(n-1)d$ and *ns* levels are very nearly of the same energies.

5.12 Hydrogen bonds and molecular bonds

By comparison with electrovalent and covalent bonds these are very much weaker; nevertheless hydrogen bonding is of enormous biochemical importance.

Hydrogen bonds

The melting and boiling points of the hydrides of Group 4B elements, i.e. those of carbon, silicon, germanium and tin, increase regularly with increasing molecular weight. However, the melting and boiling points of ammonia, water and hydrogen fluoride are abnormally high by comparison with those of the hydrides of corresponding elements in the same periodic group (fig. 5.6). The phenomenon is explained by postulating the formation of hydrogen bonds.

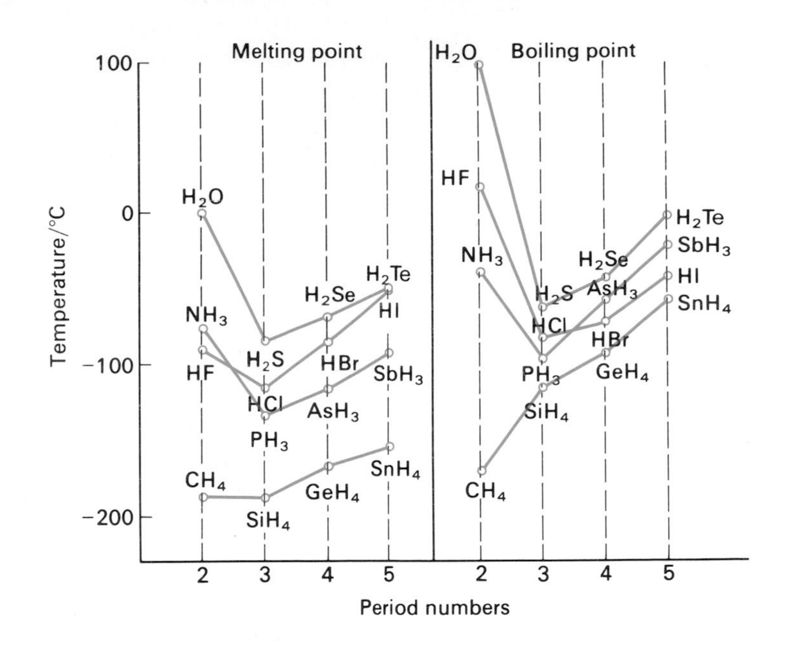

FIG. 5.6. *The abnormal melting and boiling points of ammonia, water and hydrogen fluoride*

The electronegative atoms fluorine, oxygen and nitrogen strongly attract the electron pair constituting the bond with the hydrogen atom; consequently the hydrogen atom carries an appreciable net positive charge. Since the hydrogen atom is smaller than other atoms and since its nucleus is only slightly screened, it is able to attract the negative end of another molecule reasonably strongly. The lone pair of electrons on the electronegative element determines the strength of the bond and the orientation of the lone pairs determines the direction of the bonding, e.g. hydrogen fluoride vapour contains some $(HF)_5$ molecules of the shape depicted below:

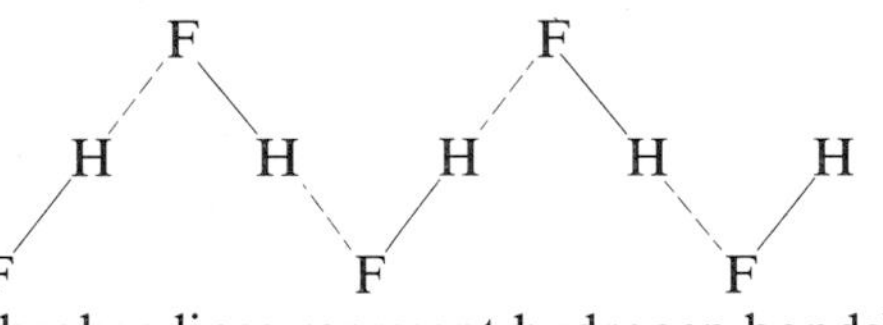

(broken lines represent hydrogen bonds)

Hydrogen bonding generally only occurs between hydrogen and the electronegative elements fluorine, oxygen and nitrogen. Although only a weak bond by comparison with conventional chemical bonds, hundreds of hydrogen bonds may form along the length of two adjacent protein molecules, thereby producing a very large cumulative effect. Hydrogen bonding is also responsible for the holding together of cellulose molecules in the form of fibres.

Molecular or van der Waals' bonds

The liquefaction and eventual solidification of the noble gases is explained by postulating so-called van der Waals' forces (the same forces that account for the deviation of actual gases from ideal behaviour). They are essentially forces operating between electrical dipoles. In the case of the noble gases, which clearly have no permanent dipoles, it is suggested that electron motion around a central nucleus can establish fluctuating dipoles; such forces that are set up by this means must, of necessity, be weak ones. van der Waals' forces operate between individual molecules in so-called molecular crystals, e.g. iodine (p. 325) and naphthalene, which are characterised by low melting points.

5.13 Suggested reading

N. F. Hall, *The Concept of Electronegativity,* School Science Review, No. 161, Vol. 47, 1965.
J. H. J. Peet, *The Hydrogen Bond*, Education in Chemistry, Vol. 50, 833, 1969.
G. J. Moody and J. D. R. Thomas, *Energy Relationships in Chemistry, The Born-Haber Cycle*, School Science Review, No. 173, Vol. 50, 1969.
R. S. Nyholm, *The Stability of Inorganic Compounds*, Chap. 3, Modern Chemistry and the Sixth Form, edited by D. J. Millen, Collins, 1965.

5.14 Questions on chapter 5

1 Describe and explain in terms of the electronic theory of valency, the structure of each of the following compounds: ammonia, carbon dioxide, ammonium chloride, copper tetra-ammine sulphate monohydrate.

Account for the differences in properties between hydrogen chloride, sodium chloride and phosphorus trichloride. (S)

2 Outline modern views on valency using as examples NaCl, CCl_4, CO and K_2SO_4.

Discuss the differences between the structure of a single crystal of potassium chloride and that of diamond. How do these structures account for differences in the physical properties of these substances? (O & C)

3 Explain what is meant by the terms covalency and electrovalency? Using as examples hydrogen chloride, potassium hexacyanoferrate (II), $K_4Fe(CN)_6$, and diamond:

(a) explain briefly how the electronic structure of the constituent atoms of a compound determine the type of bond formed between them;

(b) discuss to what extent chemical and physical properties are related to bond type. (O & C)

4 (a) Define the terms atom, atomic number, mass number.

(b) Name and give the relative masses and charges of the three fundamental particles which are significant in chemistry.

(c) Deduce the numbers of these particles in atoms of the principal natural forms of the elements hydrogen, chlorine, argon and potassium and call attention to any specially interesting features.

(d) Contrast the electronic structures of chlorine and potassium chloride and correlate these with the facts that one is a gas and the other a high melting point solid. (JMB)

5 Discuss the nature of the metallic bond and show how it accounts for the physical properties of metals.

6 'The valency of elements in the first row can usually be described by the octet rule'. Give examples of the success and failure of this rule. Does the rule apply to compounds of the second row elements? (Oxford Schol.)

7 Give a short account of the stable electronic configurations of the elements as ions and as atoms in simple molecules, and discuss the view that electron-pairing is a more important concept in chemistry than the rule of eight (noble gas structure concept). (Camb. Schol.)

8 According to Fajans' theory, the tendency towards covalency is dependent upon ionic size and charge. Discuss the usefulness of this theory, giving examples. (Oxford Schol.)

9 What explanations can you offer for the following?

(a) Aluminium chloride is essentially covalent, whereas aluminium fluoride is predominantly ionic.

(b) Aluminium is exclusively 3-valent, whereas transition metals display several valency states.

(c) The boiling points of the hydrogen halides are:

HF	HCl	HBr	III
19°C	−85°C	−67°C	−35°C

(d) Calcium oxide has the formula $Ca^{2+}O^{2-}$ and not $Ca^{+}O^{-}$.

10 By using the data given below, show that the energy liberated when 1 mole of solid calcium reacts with chlorine molecules to give $Ca^{2+}(Cl^{-})_2$ is greater than that for the hypothetical formula $Ca^{+}Cl^{-}$. What is the most important energy term which favours $Ca^{2+}(Cl^{-})_2$ rather than $Ca^{+}Cl^{-}$?

$Ca(s) \longrightarrow Ca(g)$ $\quad S = +176{\cdot}6\ kJ\ mol^{-1}$

$Ca(g) \longrightarrow Ca^{+}(g) + e^{-}$ $\quad I_1 = +585{\cdot}7\ kJ\ mol^{-1}$

$Ca^{+}(g) \longrightarrow Ca^{2+}(g) + e^{-}$ $\quad I_2 = +1147\ kJ\ mol^{-1}$

$Cl_2(g) \longrightarrow 2Cl(g)$ $\quad D = +241{\cdot}8\ kJ\ mol^{-1}$

$Cl(g) + e^{-} \longrightarrow Cl^{-}(g)$ $\quad E = -379{\cdot}5\ kJ\ mol^{-1}$

$Ca^{2+}(g) + 2Cl^{-}(g) \longrightarrow Ca^{2+}(Cl^{-})_2(s)$ $\quad U_1 = -2233\ kJ\ mol^{-1}$

$Ca^{+}(g) + Cl^{-}(g) \longrightarrow Ca^{+}Cl^{-}(s)$ $\quad U_2 = -711{\cdot}2\ kJ\ mol^{-1}$

11 'The sodium atom has one more electron than that of a noble gas and the chlorine atom has one electron short of a noble gas structure. Therefore sodium reacts with chlorine, an electron being transferred from the sodium atom to the chlorine atom because by this means both ions produced have noble gas structures'. By considering the data given below show that this statement is an inadequate approximation to the truth.

$Na(s) \longrightarrow Na(g)$ $\quad S = +108{\cdot}7\ kJ\ mol^{-1}$

$Na(g) \longrightarrow Na^{+}(g) + e^{-}$ $\quad I = +493{\cdot}8\ kJ\ mol^{-1}$

$Cl_2(g) \longrightarrow 2Cl(g)$ $\quad D = +241{\cdot}8\ kJ\ mol^{-1}$

$Cl(g) + e^{-} \longrightarrow Cl^{-}(g)$ $\quad E = -379{\cdot}5\ kJ\ mol^{-1}$

Give what you consider to be a more realistic reason for sodium reacting with chlorine to give an electrovalent compound.

12 Sulphur can show valencies of 2, 4 and 6 whereas oxygen is restricted to a valency of 2; similarly, phosphorus can show valencies of 3 and 5 whereas nitrogen is restricted to a valency of 3. Give what explanation you can for these two statements. Why do you think the highest chloride of sulphur is SCl_4 whereas the highest fluoride is SF_6?

13 The electron distribution around the nuclei of the noble gases is symmetrical, i.e. they have no permanent dipoles, yet at sufficiently low temperatures these gases liquefy and eventually solidify, indicating that some force of attraction exists between their atoms. What explanation has been given to account for these weak forces of attraction?

14 Explain concisely, giving one example in each case, the meaning of: (a) atomic nucleus; (b) atomic number; (c) isotope.

The element of atomic number 17 forms one compound with the element of atomic number 12 and another compound with the element of atomic number 6.

Represent by conventional diagrams the electronic structures of the compounds formed and name the type of valency displayed. For each compound state whether you would expect it to be gaseous, liquid or solid and give three other properties which you would expect it to possess. (JMB)

15 Explain concisely what is meant by the terms: electrovalent bond, covalent bond, co-ordinate bond. Compare and contrast the physical properties usually displayed by compounds containing electrovalent and covalent bonds.

Write electronic formulae for the following compounds:

(i) magnesium chloride; (ii) ethene; (iii) ammonium chloride; and comment on the types of bonds involved. (L)

Chapter 6

Chemical geometry

6.1 Introduction

Once the essential differences between electrovalent and covalent compounds are appreciated, it is natural to probe a little deeper and ask, for instance, why the structure of sodium chloride, Na^+Cl^-, is different from that of caesium chloride, Cs^+Cl^-, or what are the factors that determine the shapes of covalent molecules, e.g. why is the shape of the ammonia molecule, NH_3, so different from that of boron trifluoride, BF_3. A complete answer to these and related problems requires sophisticated treatment but with the help of a few simple ideas it is possible to go some way towards answering these questions.

6.2 Atomic or covalent radii

Employing spectroscopic or X-ray and electron diffraction techniques it is possible to determine the distance between the nuclei of bonded atoms. For the chlorine molecule, the Cl—Cl bond length is 0·198 nm, giving a covalent or atomic radius of 0·099 nm for the chlorine atom. The C—C bond distance in diamond is 0·154 nm, almost the same value as that observed in a variety of saturated hydrocarbons, giving an atomic radius of 0·077 nm for carbon. The interesting point is that atomic radii are additive, e.g. the C—Cl bond distance observed in a variety of compounds is almost the same as that obtained by adding together the atomic radius of carbon (0·077 nm) and the atomic radius of chlorine (0·099 nm), i.e. 0·176 nm.

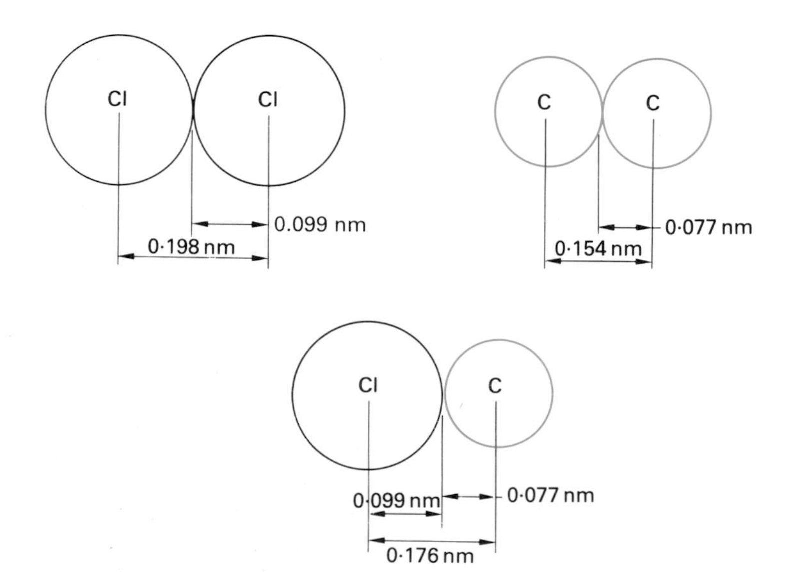

FIG. 6.1. *Additivity of atomic radii*

Values of atomic (or covalent) radii for most elements are now available (Table 6A, p. 72); they have been evaluated by determining bond lengths in covalent molecules and applying the principle of additivity. For metals, the inter-nuclear distance in the crystal has been shown to be approximately the same as that for a single metal-metal covalent bond, where compounds containing the latter type bond are known; the values of these bond distances in solids, however, do depend to some extent on the way the atoms are arranged in the crystal. It is essential to realise that bond lengths in covalent compounds are dependent upon the number of covalent bonds linking two atoms together, e.g. there is a decrease in bond length along the series C—C, C=C, C≡C, and compounds containing single covalent bonds must be used to compute atomic radii. Indeed, whenever the principle of bond additivity breaks down it is usually an indication that the particular bond contains some multiple bond character.

6.3 Ionic radii

A crystal of sodium chloride consists of a symmetrical arrangement of sodium and chloride ions, each sodium ion being surrounded by six chloride ions and vice versa (Bragg 1912). The interionic distance (the distance from the centre of one ion to the centre of its nearest neighbour of opposite charge) was found to be 0·281 nm for sodium chloride. Other alkali metal halides adopt the sodium chloride structure but with differing interionic distances, e.g.

K^+Cl^-	0·314 nm	Na^+Cl^-	0·281 nm	Difference = 0·033 nm
K^+F^-	0·266 nm	Na^+F^-	0·231 nm	Difference = 0·035 nm
Difference	0·048 nm		0·050 nm	

The almost constant differences are explained if we assume that each ion acts as a sphere of constant radius, the measured interionic distance being represented by two spheres that are just in contact.

In order to determine the actual radius of an ion, we must obviously have more information; several methods of approach have been used, one of the earliest and most readily understood being that introduced by Landé (1920).

Landé's method may be illustrated by the following example. The interionic distances in the oxides and sulphides of magnesium and manganese are given below (all four compounds adopt the sodium chloride structure):

$Mg^{2+}O^{2-}$	0·210 nm	$Mn^{2+}O^{2-}$	0·224 nm
$Mg^{2+}S^{2-}$	0·260 nm	$Mn^{2+}S^{2-}$	0·259 nm

FIG. 6.2. *Calculation of ionic radii (Landé's method)*

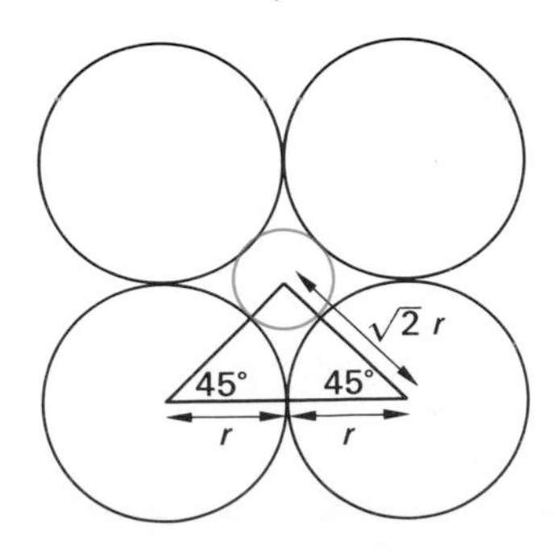

Since the interionic distance for the two sulphides is almost the same, the assumption is made that the sulphide ions are just in contact with each other. An ionic radius of 0·184 nm for the S^{2-} ion ($\sqrt{2}r$ = 0·259 nm, or r = 0·184 nm) is now easily deduced by simple geometry.

Once the ionic radius of one ion has been fixed, it becomes possible to determine the absolute values for other ions. The values used to construct Table 6A, together with values of atomic radii, are those due to Pauling, who used more refined but more complex methods. Basically his method involved the determination of the interionic distance in potassium chloride, in which both ions are isoelectronic (have the same number of electrons). After correcting for the screening of the nucleus by the electrons.

he was able to compute values for the effective nuclear charges of the K^+ and Cl^- ions; he now assumed that the interionic distance could be divided in the inverse ratio of these computed nuclear charges, and so arrived at values for the ionic radii of both K^+ and Cl^- from which other values of ionic radii could be derived.

Ionic radii, like atomic radii, are subject to slight variation and their values depend to some extent on the manner in which the ions are packed together in the crystal.

Table 6A Atomic and ionic radii

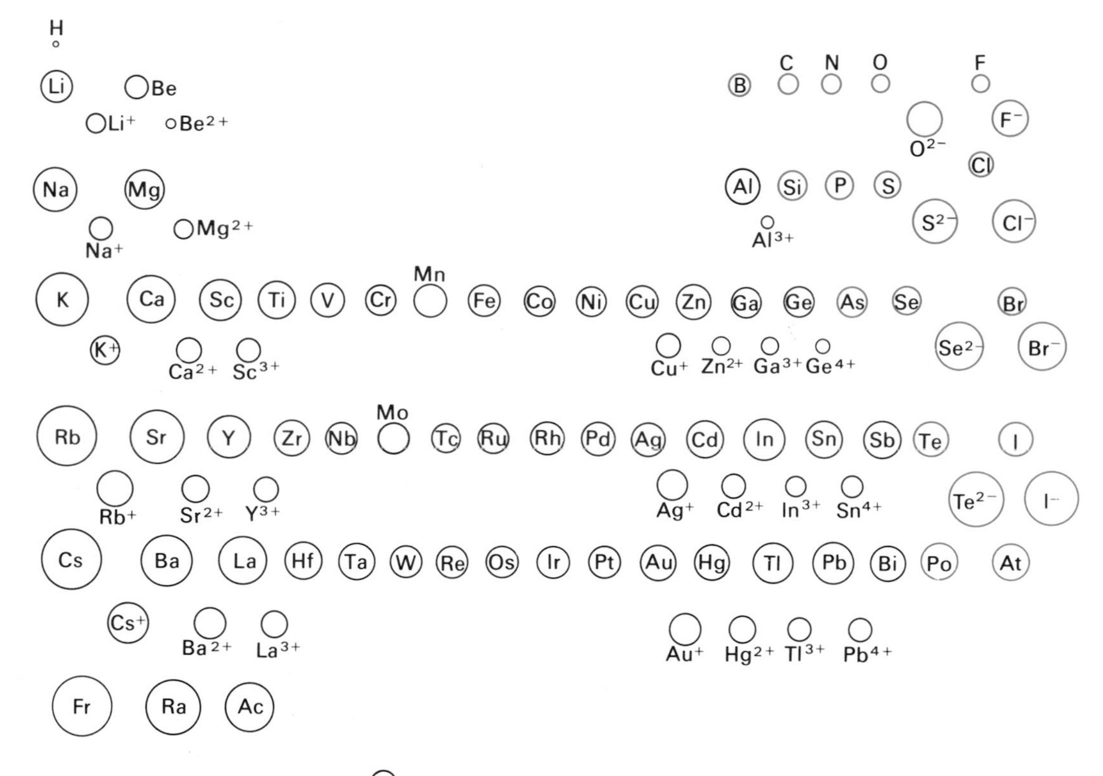

6.4 Variation of atomic and ionic radii

Several distinctive features are apparent from Table 6A.

(a) Atomic radii are very much smaller than anionic radii but very much larger than cationic radii, e.g.

Cl	0·099 nm	Cl^-	0·181 nm
Na	0·157 nm	Na^+	0·095 nm

(b) Atomic and ionic radii increase down a particular periodic group since the number of electronic shells increases, and the increase in nuclear charge is insufficient to counteract this expansion, e.g.

F	0·072 nm	Cl	0·099 nm	Br	0·114 nm	I	0·133 nm
F^-	0·136 nm	Cl^-	0·181 nm	Br^-	0·195 nm	I^-	0·216 nm

For transition metals this increase is very much less.

(c) Atomic radii decrease across a particular period, since additional electrons enter the same shell and the increase in nuclear charge exerts a contracting effect, e.g.

Li 0·133 nm	Be 0·089 nm	B 0·080 nm	C 0.077 nm
N 0·074 nm	O 0·074 nm	F 0·072 nm	

(d) The radii of isoelectronic positive ions, i.e. ions with the same number of electrons, decrease with increasing charge. There is only a small decrease in radii for isoelectronic negative ions with decreasing negative charge.

Li^+	0·060 nm	S^{2-}	0·184 nm
Be^{2+}	0·031 nm	Cl^-	0·181 nm

(e) Transition metals show little variation in atomic size. Along a given transition series there is a small decrease in size of the atom (at least for the first five or six members) as additional electrons enter the penultimate shell and the *d* level fills.

6.5 The structures of some ionic compounds

Since an electrovalent bond is non-directional, an ionic solid will consist of a symmetrical array of positive and negative ions. As many positive ions surround the negative ion, and vice versa, as is consistent with

(a) maintaining electrical neutrality,

(b) ensuring a maximum lattice energy (p. 56).

Spheres of equal radius can be packed so that each sphere in a particular layer is surrounded by six more in the same layer; thus in fig. 6.3 the sphere labelled A is surrounded by six spheres labelled B in the same layer. It is possible to place three spheres above this first layer (labelled B in fig. 6.3) so that they are in contact with sphere A in the first layer. Similarly, three spheres can be placed below this first layer so that they are each in contact with A. There are therefore twelve B spheres in contact with sphere A and the co-ordination number is said to be twelve, a feature of many metallic structures (p. 187–188).

FIG. 6.3. *The close packing of spheres of equal size*

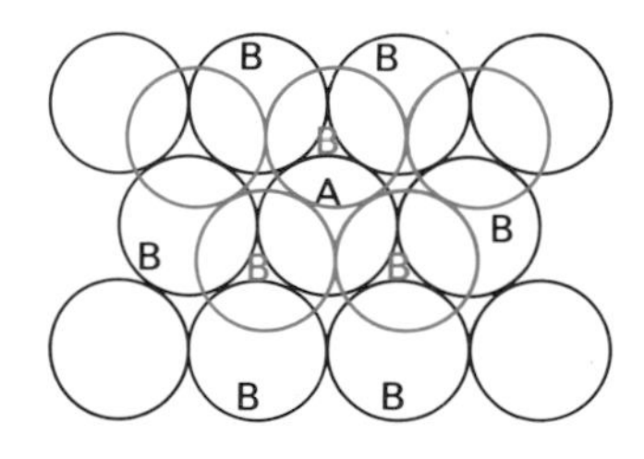

Consider now the hypothetical case of an ionic solid A^+B^- in which both ions are of identical size. It is geometrically impossible to arrange for twelve B^- ions to surround each A^+ ion and vice versa, in order to maintain electrical neutrality, so ionic structures must, of necessity, have co-ordination numbers less than twelve. As a general rule, however, it can be stated that ions of opposite charge, which are nearest neighbours, tend to touch each other and, in so doing, a maximum force of attraction between them is achieved.

In practice, ions of opposite charge are not of equal size (cations are generally smaller than anions) and this is an additional reason why ionic compounds of the type A^+B^- have co-ordination numbers less than twelve.

Consider the sodium chloride structure where each ion has a co-ordination number of six, i.e. each ion is surrounded by four coplanar ones of opposite charge with one above this plane and one below it; by considering the situation depicted in fig. 6.4, where the size of the anion is gradually increased, it is possible to determine the critical ratio (radius of cation/radius of anion), or the radius ratio as it is called, when a lower co-ordination number must be adopted. The critical situation is shown in fig. 6.4(b) where an anion is just in contact with other anions and with a cation. By simple geometry it can be seen that:

$$ab = cb\cos 45°$$

or

$$\text{radius of anion} = \frac{\text{sum of the radii of cation and anion}}{\sqrt{2}}$$

or

$$\frac{\text{radius of cation}}{\text{radius of anion}} = \sqrt{2} - 1 = 0{\cdot}414$$

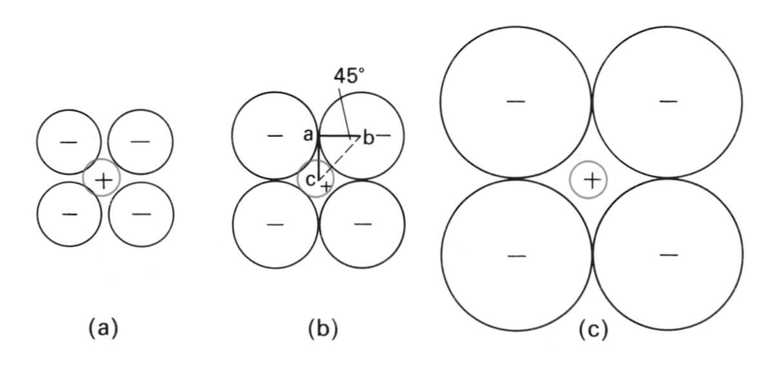

FIG. 6.4. *The limiting condition for 6:6 co-ordination*

If the radius ratio is less than 0·414 a binary ionic compound must adopt a structure with a co-ordination number less than six; on the other hand a co-ordination number greater than six can be achieved if the radius ratio exceeds 0·732, e.g. the caesium chloride structure which has a co-ordination number of eight. Table 6B lists some compounds having the sodium chloride and caesium chloride structures, the radius ratio criteria being deduced by simple geometrical arguments.

Table 6B Some ionic structures

Type of structure	Examples	Radius ratio	Co-ordination number of the positive ion	Co-ordination number of the negative ion
Caesium chloride	Cs^+Cl^-, Cs^+Br^- Cs^+I^-	$>0{\cdot}732$	8	8
Sodium chloride	Na^+Cl^-, Na^+Br^- K^+Cl^-, K^+Br^-	$<0{\cdot}732$ $>0{\cdot}414$	6	6

For a ternary ionic compound, e.g. calcium fluoride, $Ca^{2+}(F^-)_2$, the co-ordination number of the cation must be twice that of the anion in order to maintain electrical neutrality, in this example eight and four respectively; in calcium fluoride each Ca^{2+} ion is surrounded octahedrally by eight F^- ions, and each F^- ion is surrounded tetrahedrally by four Ca^{2+} ions.

Figure 6.5 shows the sodium chloride and caesium chloride structures (6 and 8 co-ordination respectively).

FIG. 6.5. *Structures of (a) sodium chloride and (b) caesium chloride*

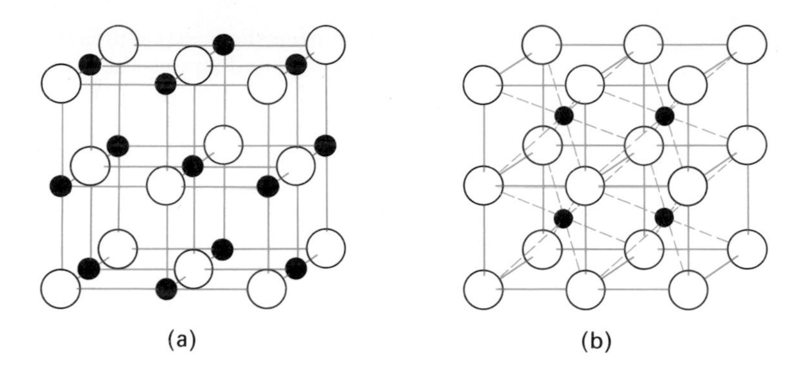

6.6 The shape of covalent molecules—a simple theory

In contrast to the bonding in ionic solids, that operating between atoms in a covalent molecule is directional in character; thus a covalent molecule has an overall shape. A simple explanation was advanced by Sidgwick and Powell (1940) who pointed out that the arrangements in space of the covalent bonds in multi-covalent atoms are very simply related to the number of electrons in their valency shells. It was assumed that electron pairs would arrange themselves so as to be as far apart as possible, because of their mutual repulsion; thus two electron pairs would arrange themselves linearly, three pairs in the form of a plane triangle, and four pairs would arrange themselves tetrahedrally. This arrangement applies irrespective of whether the electron pair constitutes a covalent bond (bonding pair) or is simply non-bonding (lone pair). For five and seven electron pairs the geometrical arrangement is less symmetrical, as is shown in fig. 6.6.

FIG. 6.6. *Arrangement of electron pairs*

linear
2 electron pairs,
bond angle 180°

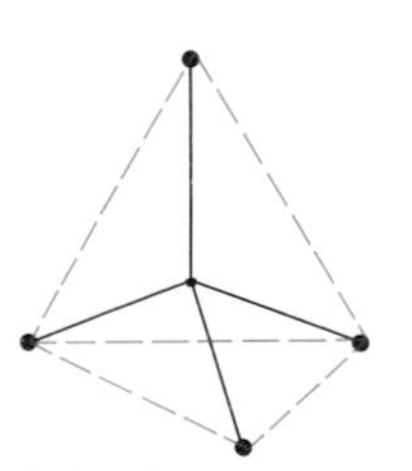

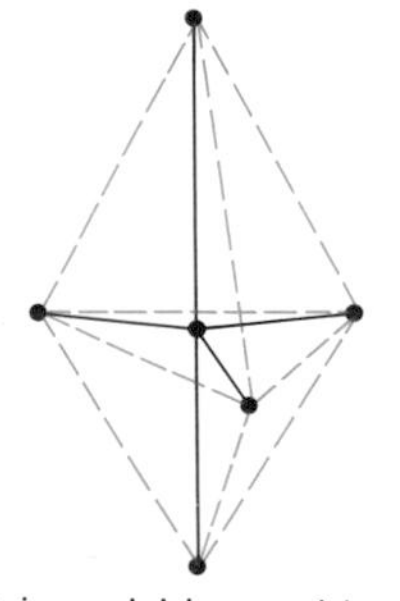

triangular plane
3 electron pairs, bond
angles 120°

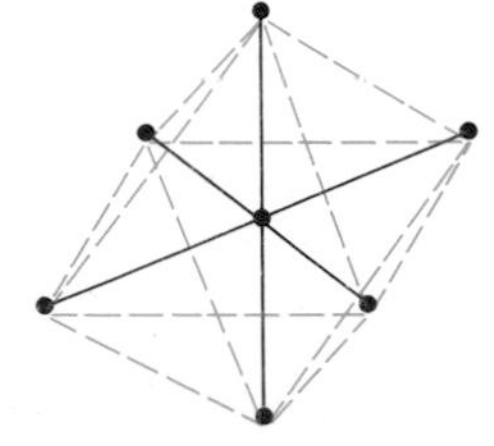

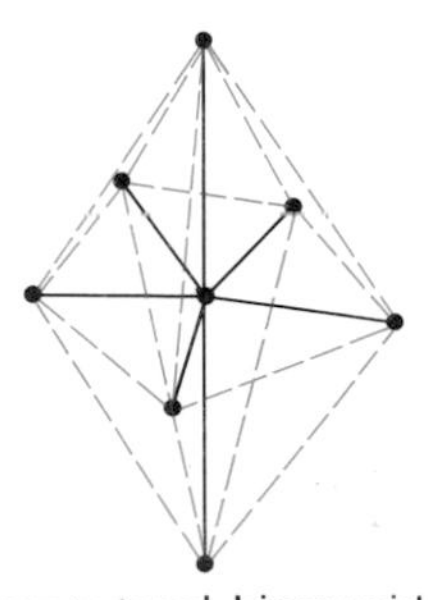

Apart from slight departures from these arrangements few exceptions have yet been found among the non-transition elements of molecules adopting alternative shapes (fig. 6.6).

The regular arrangements shown in fig. 6.6 are only realised in practice if the electron pairs are engaged in bonding to identical atoms or groups; for instance, all the valency bonds in methane, CH_4, are tetrahedrally arranged, whereas in chloromethane, CH_3Cl, the H—C—H angles are slightly greater than the tetrahedral angle of 109° 28′. Deviations from regular configurations also occur if lone pair electrons are present; thus methane, CH_4, ammonia, NH_3, and water, H_2O, which have respectively none, one, and two lone pairs of electrons have bond angles which decrease in the order 109° 28′, 106° 7′, and 104° 27′. To account for this latter type of deviation, Gillespie and Nyholm (1957) have extended the simple theory and have suggested that repulsion between electron pairs decreases in the order:

lone pair/lone pair $>$ lone pair/bond pair $>$ bond pair/bond pair

Since lone pairs will be closer to the nucleus than bonding pairs, they will repel each other more strongly and hence slightly force together the bonding pairs. In this way it is possible to explain the decrease in bond angles in passing from methane (no lone pairs) to ammonia (one lone pair) to water (two lone pairs). It is possible to explain the shapes of simple molecules and ions in terms of the number of valency electrons, with this simple theory in mind.

6.7 The shapes of simple molecules and ions

The structures will be considered under several headings according to the number of valency electrons surrounding the central atom in the compound.

Four valency electrons—linear shape

Not very many compounds have this configuration; two examples are beryllium chloride, $BeCl_2$, and mercury (II) chloride, $HgCl_2$. Both compounds are covalent and have linear molecules, e.g.

Cl—Be—Cl

It is, however, a common configuration of molecules containing double and triple bonds where it can be assumed that the electron pairs in a double and triple bond together occupy the position of one electron pair in a single bond, e.g.

H—C≡C—H H—C≡N O=C=O

Six valency electrons—triangular plane shape

An example of a compound assuming this configuration is boron trifluoride, BF_3, a planar molecule with bond angles of 120°.

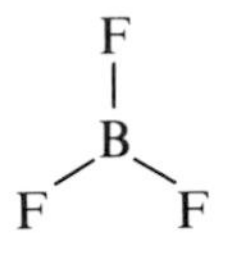

This structure is also adopted by several molecules and ions that contain double bonds, e.g.

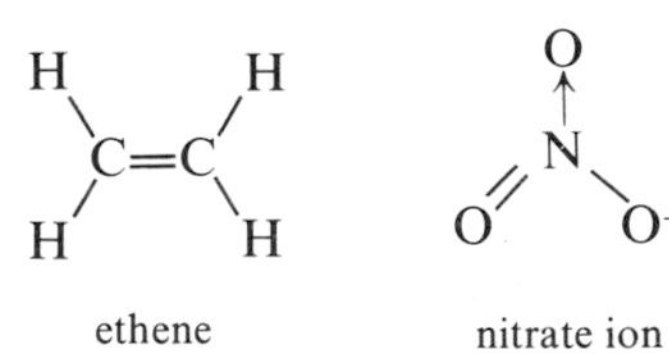

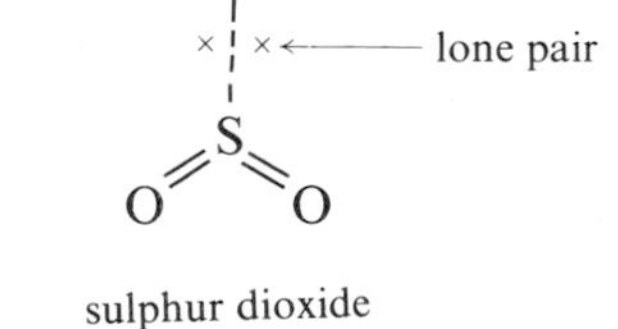

Eight valency electrons—tetrahedron

This is a very common configuration for molecules and ions; examples include the following:

FIG. 6.7. *The shapes of some molecules and ions based on the tetrahedron*

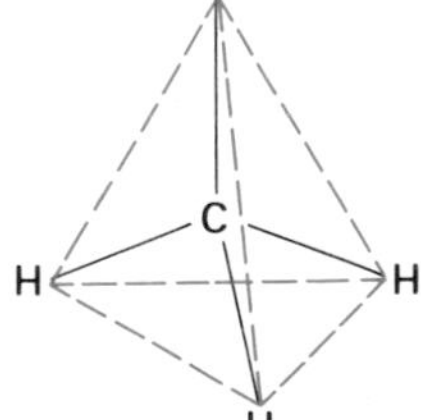

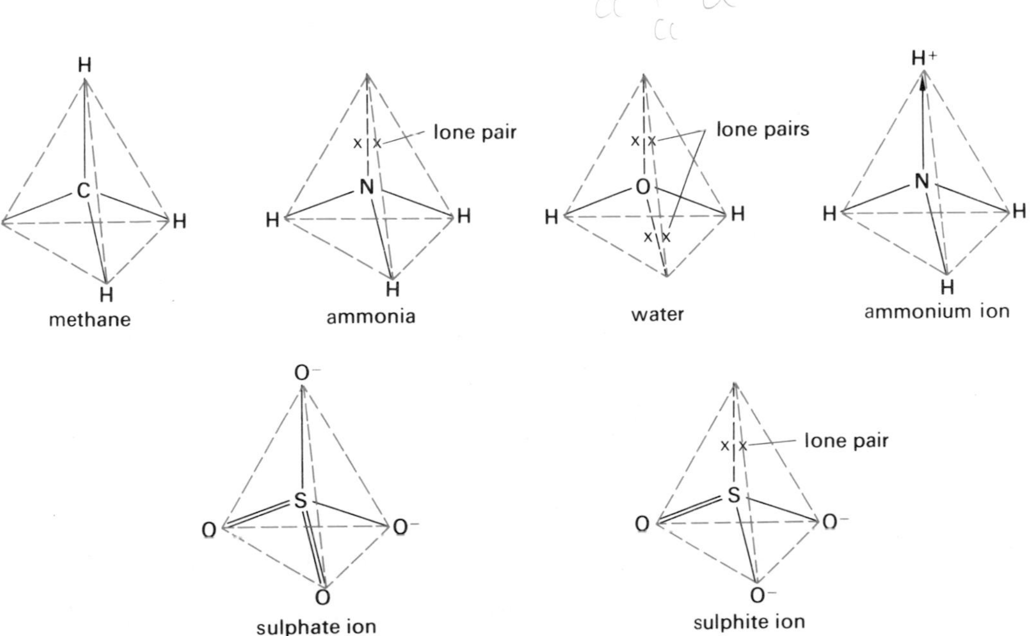

Note that in the sulphate and sulphite ions there are respectively twelve and ten electrons in the valency shells of the sulphur atoms, i.e. sulphur expands its octet (p. 64).

Ten valency electrons—trigonal bipyramid

For this configuration, and configurations involving more than ten electrons, expansion of the octet is necessary. Elements in the first row of the Periodic Table cannot be the central atom in such structures. Examples include phosphorus pentafluoride, PF_5, sulphur tetrafluoride, SF_4, and chlorine trifluoride, ClF_3. The structures of the latter two molecules cannot be predicted uniquely (alternatives are available using this simple theory of electron pair repulsion) and the experimentally determined structures are shown below:

FIG. 6.8. *The shapes of some molecules based on the trigonal bipyramid*

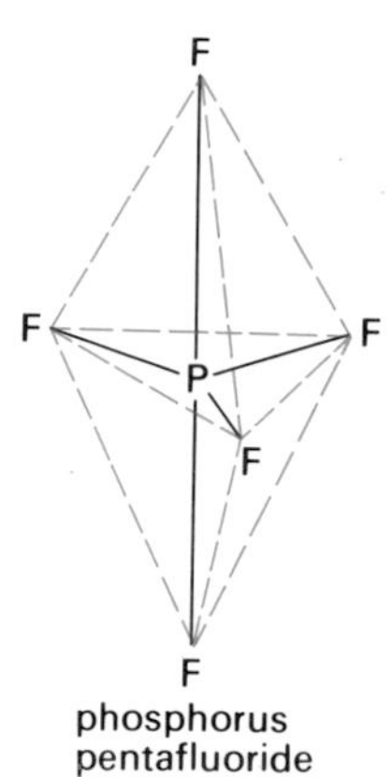

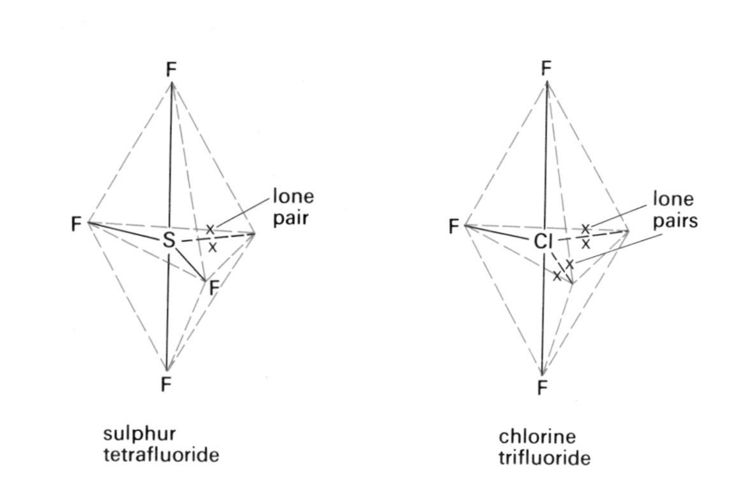

Twelve valency electrons—octahedron

Two compounds adopting this structure are sulphur hexafluoride, SF_6, and iodine pentafluoride, IF_5, as shown below:

FIG. 6.9. *The shapes of two molecules based on the octahedron*

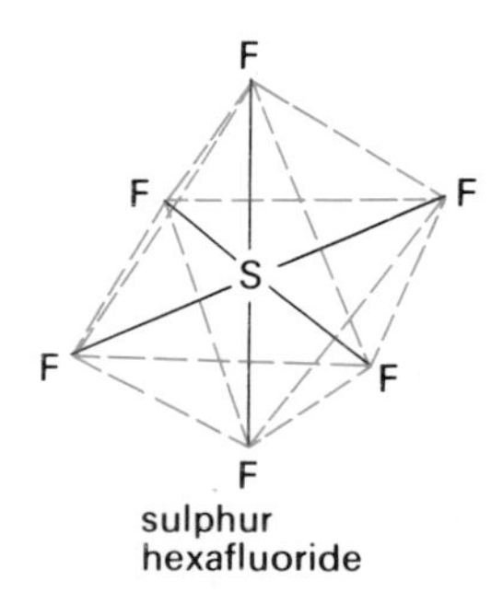

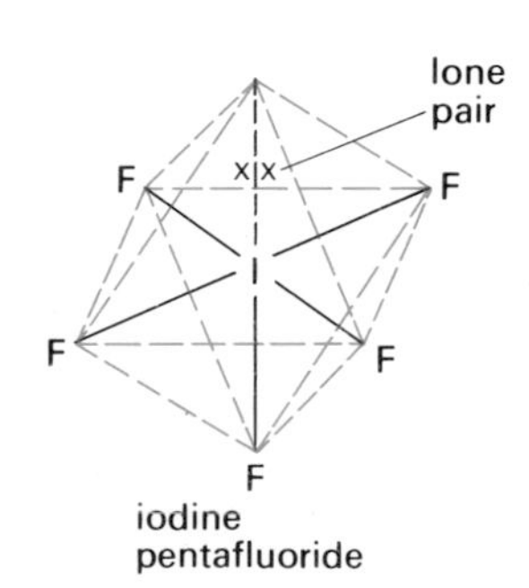

Fourteen valency electrons—pentagonal bipyramid

Iodine heptafluoride, IF_7, is an example of a compound having this structure.

FIG. 6.10. *The pentagonal bipyramidal shape of the iodine heptafluoride molecule*

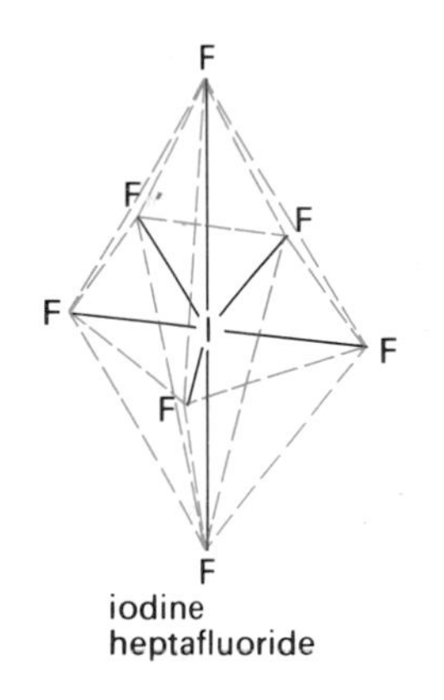

6.8 The shapes of some giant covalent solids

Three-dimensional structures

Typical three-dimensional structures include those of diamond, silicon carbide, boron nitride (one polymorphic form), zinc blende, and silicon dioxide.

The diamond allotrope of carbon contains a three-dimensional array of carbon atoms with the valency bonds directed towards the apices of a regular tetrahedron (electron pair repulsion). The whole structure is one giant molecule and, because the bonding is strong and extended in three dimensions, diamond is exceptionally hard (see fig. 6.11 for the structure of diamond).

Silicon carbide, $(SiC)_n$, is nearly as hard as diamond itself and has the same type of structure with silicon and carbon atoms occupying alternate positions in the crystal lattice. One form of boron nitride, $(BN)_n$, adopts the diamond structure and this substance also is extremely hard; since this compound contains the boron atom (one electron less than the carbon atom) and the nitrogen atom (one electron more than the carbon atom) the number of electrons on average per atom is the same as that in diamond. The covalent solid zinc blende, ZnS, is yet another compound with the diamond structure, alternate positions in the lattice being occupied by zinc and sulphur atoms.

Silicon dioxide is also a three-dimensional giant molecule and one form of this compound (cristobalite) has a structure in some ways similar to that of diamond. Each silicon atom is surrounded tetrahedrally by four oxygen atoms, each of these oxygen atoms being linked to other silicon atoms, i.e. each oxygen atom is shared equally between two silicon atoms giving the empirical formula SiO_2. A two-dimensional representation of this structure is shown opposite.

FIG. 6.11. *Part of the structure of diamond showing the tetrahedral valency bonds*

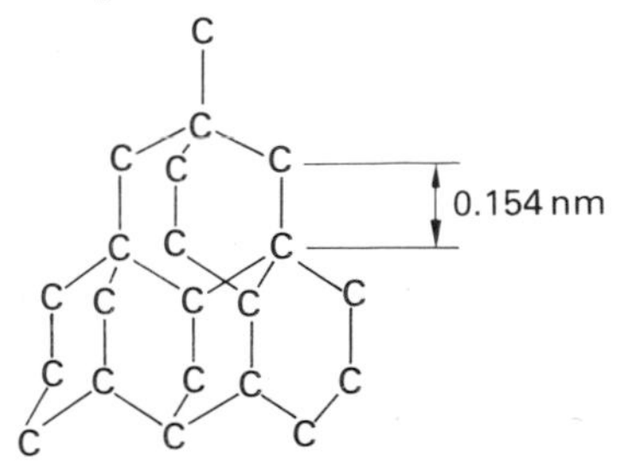

FIG. 6.12. *The structure of silicon dioxide represented in two dimensions*

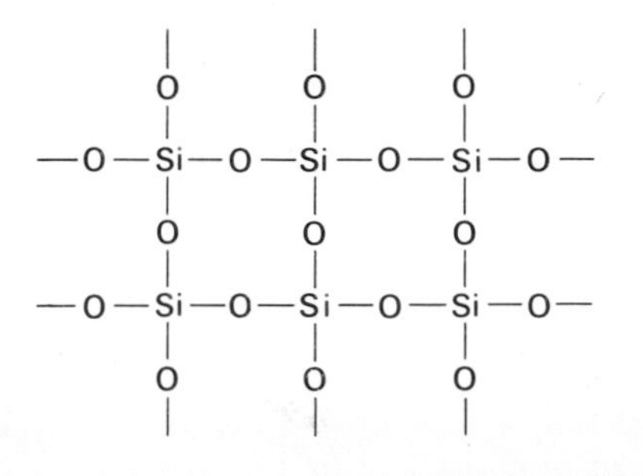

Two-dimensional structures and one-dimensional structures

Graphite, the other allotropic form of carbon, has a completely different structure from that of diamond. It contains layers of carbon atoms, each carbon atom in a particular layer being covalently bonded to three others, giving C—C—C bond angles of 120°. In order to represent the structure by conventional valencies, alternate double and single C—C bonds must be drawn, but in practice all C—C bonds within each layer are equal (see the theory of resonance, p. 81). The individual layers are held together by van der Waals' forces (p. 67); thus the slippery nature of graphite is adequately explained.

It is interesting to note that boron nitride also exists in a form which resembles the graphite structure; however, it differs from graphite in having no 'metallic lustre' and in being only a semiconductor of electricity. Its structure involves co-ordinate bonds from nitrogen to boron. Parts of the structures of graphite and boron nitride are shown in fig. 6.13.

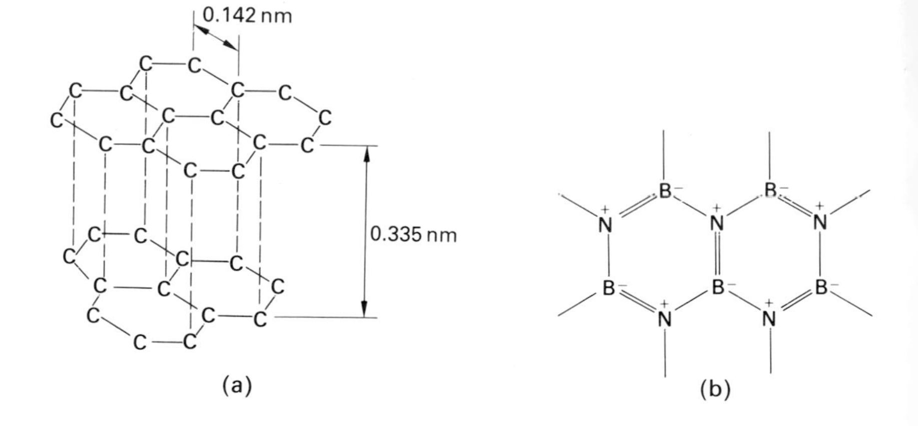

FIG. 6.13. *The structures of (a) graphite and (b) part of the 'graphite form' of boron nitride*

Some examples of one-dimensional structures include those of silicon disulphide, $(SiS_2)_n$, and sulphur trioxide (one form, see p. 309). The chain structures of these compounds (see below) explain why they crystallise in the form of long silky needles.

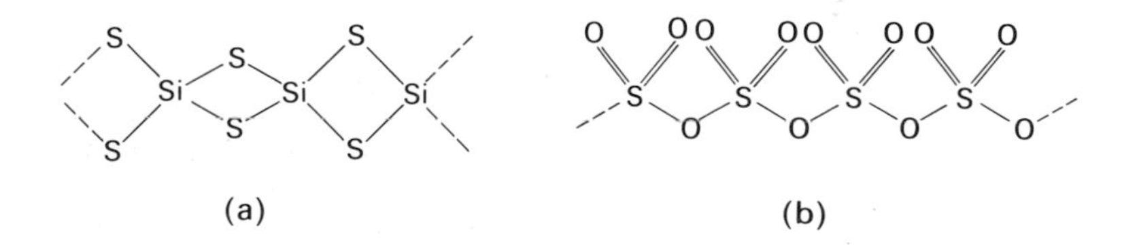

FIG. 6.14. *The structures of (a) silicon disulphide and (b) sulphur trioxide*

6.9 The concept of resonance

Two alternative structural formulae can be drawn to represent the benzene molecule, the only difference being in the positions of the single and double bonds, e.g. carbon atoms labelled 1 and 2 can be drawn linked together by either a single bond or a double bond:

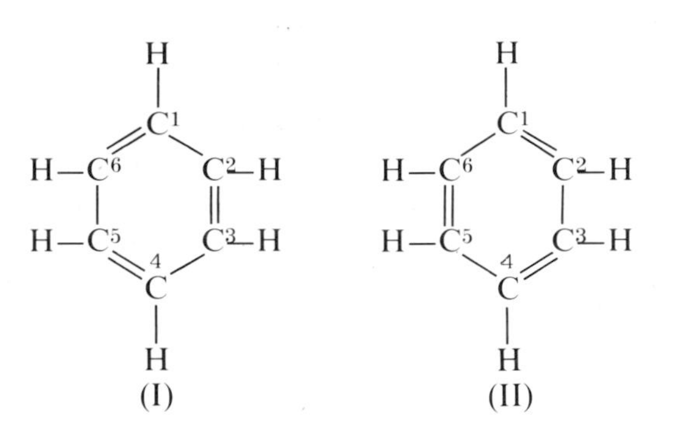

In each formula every carbon atom completes its octet, but neither structure is correct since experimentally it has been found that all the carbon—carbon bonds in the molecule are of equal length (0·139 nm). If the benzene molecule had alternating double and single bonds, there would be two different bond lengths (C—C bond length = 0·154 nm, C = C bond length = 0·134 nm). Since the carbon-carbon bond length is intermediate between these two extremes, the structure of the benzene molecule could be represented as below the dotted line indicating a fractional order bond:

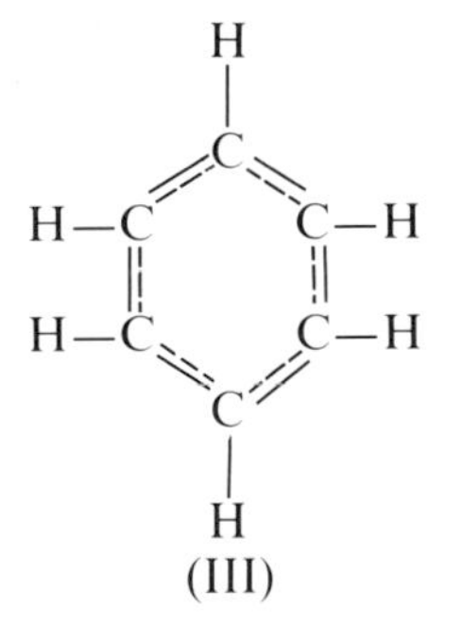

Although structure (III) is a better representation of the benzene molecule than either of structures (I) or (II), it is easier to visualise the molecule in terms of the two limiting structures (I) and (II), where conventional valency bonds can be drawn, rather than in terms of a diagram involving fractional order bonds. Benzene is an example of a compound which exhibits resonance, i.e. its actual structure has the characteristics of both limiting structures (I) and (II) but neither alone accounts for the properties of the benzene molecule. The two limiting structures contribute equally in the case of the benzene molecule, and the actual structure is said to be a resonance hybrid of these two limiting forms.

Three more examples of resonance hybrids are the planar carbonate and nitrate anions and the linear carbon dioxide molecule. Experimental evidence which suggests resonance is the equivalence of all three bonds in the anions, and a carbon-oxygen bond shorter than a C=O in carbon dioxide:

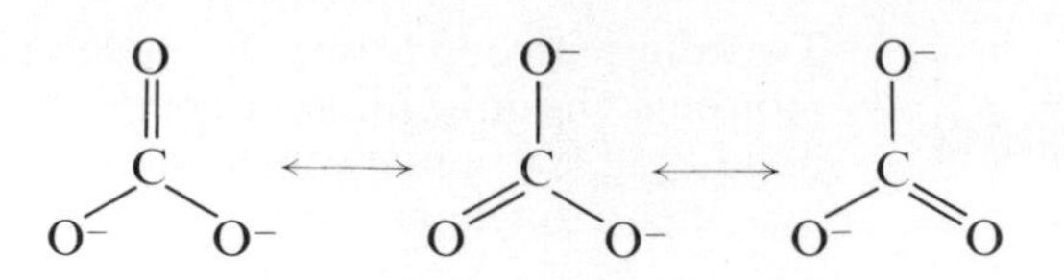

each structure contributes equally

or

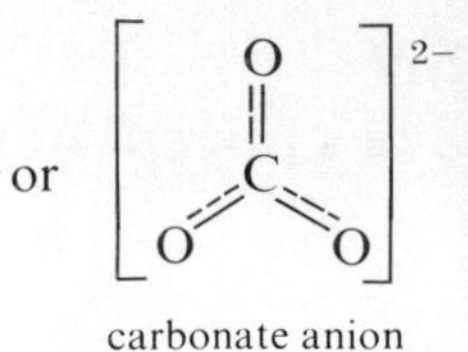

carbonate anion

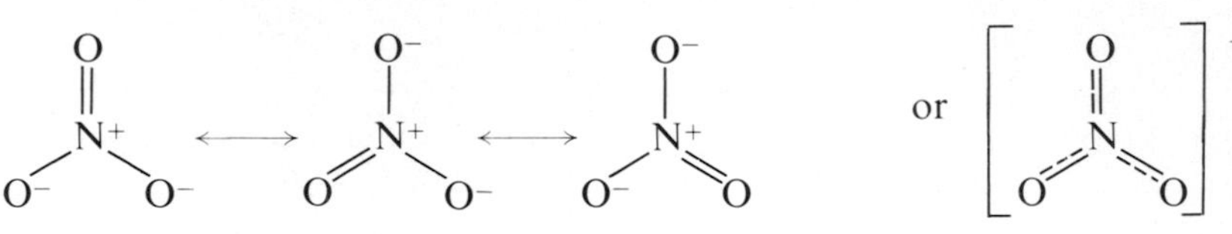

each structure contributes equally

or

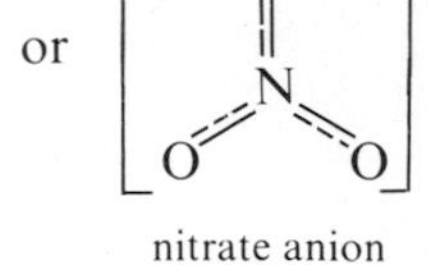

nitrate anion

$$\overset{-}{\mathrm{O}}-\mathrm{C}\equiv\overset{+}{\mathrm{O}} \longleftrightarrow \mathrm{O}=\mathrm{C}=\mathrm{O} \longleftrightarrow \overset{+}{\mathrm{O}}\equiv\mathrm{C}-\overset{-}{\mathrm{O}}$$

carbon dioxide molecule

By convention, the limiting structures which contribute to a resonance hybrid are drawn with a double-headed arrow between them, as above. It is important to realise that the actual resonance hybrid does not change from one limiting structure to another; the limiting structures do not exist but help one to visualise the real structure of a molecule which cannot be written down on paper using conventional bonding.

The effect of resonance is to increase the stability of the compound in which it occurs, the difference in energy between the actual molecule and that calculated for the most stable of the limiting structures being known as the resonance energy or stabilisation energy. For benzene the stabilisation energy is 172 kJ mol^{-1}.

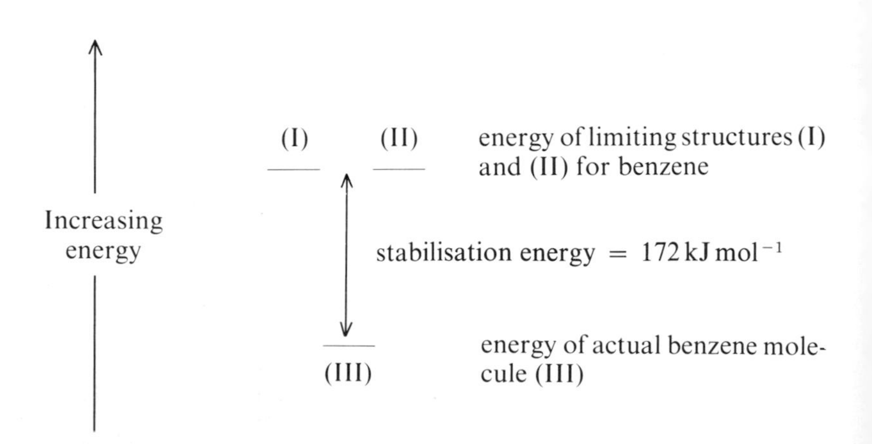

6.10 Suggested reading

R. J. Gillespie, *Electron-pair Repulsion Model for Molecular Geometry*, J. Chem. Education, Vol. 47, 18, 1970.
C. Glidewell, Some exceptions to the VSEPR rule, Education in Chemistry, No. 5, Vol. 16, 1979.

6.11 Questions on chapter 6

1 What do you understand by the 'octet rule' of valency? Summarise the limitations of this rule and indicate the principles on which modern valency theory is based. Illustrate your answer by reference to at least two compounds from each of the following three groups.

(a) $ZnCl_2$ $PbCl_2$ $FeCl_2$
(b) $BeCl_2$ BF_3 PF_3 SF_6
(c) Na_2SO_3 Na_2SO_4 H_3PO_4 (L)

2 The oxygen—oxygen bond distance in the oxygen molecule and in hydrogen peroxide are respectively 0·121 nm and 0·146 nm. How do you account for this difference?

3 The atomic radii of sodium and chlorine are 0·157 nm and 0·099 nm respectively, whereas the ionic radii of Na^+ and Cl^- are 0·095 nm and 0·181 nm respectively. Comment on these values.

4 Discuss the bonding in each of the following:

(a) $Ag(CN)_2^-$ (b) CsCl (c) SiO_2
(d) Cu (e) HF (f) I_2

In the case of (e) and (f), make some comment on the nature of the intermolecular forces. In the case of (b), (c) and (d), make some comment on the nature of the overall structure. (S)

5 (a) What are the shapes of the following species: CO_3^{2-}; PCl_5; PCl_6^-; NO_2^-; NO_2^+; P_4? For each species, your answer should consist of a clear sketch, appropriately labelled, and a brief description. Details of bonding are **not** required.

(b) Discuss the nature of the bonding in (i) sodium hydride, (ii) tetrachloromethane, (iii) conducting metals. (O)

6 Determine the limiting value for the radius ratio when both anion and cation have a co-ordination number of 3.

7 Using the radius ratio criterion, determine the co-ordination numbers of both anions and cations in the following compounds: Cs^+Br^-, $Sr^{2+}(F^-)_2$ and $Ca^{2+}O^{2-}$ The radii of the ions (nm) are as follows:
Cs^+ = 0·169, Br^- = 0·195, Sr^{2+} = 0·113, F^- = 0·136, Ca^{2+} = 0·097, O^{2-} = 0·140.

8 Discuss the theory of 'electron pair repulsion' and use it to predict the shapes of the following molecules:
BCl_3, PH_3, $SnCl_4$, SO_2, CO_2 and PF_5.

9 Predict the shape of the molecule of N_2F_2 which contains an N=N bond. If free rotation about this N=N bond cannot occur, show that two possible structures are possible. Which of the two possibilities would you expect to be the more stable and why?

10 (a) State the principles of the Sidgwick–Powell theory of electron pair repulsion and molecular shape.

(b) Illustrate the application of this theory by considering how it accounts for the shapes of the following species: $BeCl_2$; $SnCl_2$; BCl_3; PCl_3. [For each species your answer should include a clear sketch, appropriately labelled, showing the spatial arrangements of the atoms.]

(c) Which of the following molecules have dipole moments:

SO_2; CO_2; CH_3Cl; CH_4? (O)

11 Boron nitride, $(BN)_n$, exists in two polymorphic forms, one form being extremely hard like diamond and the other form being slippery like graphite. Show how the structures of these two modifications of boron nitride account for these properties.

12 Outline the lattice structures of (a) graphite, (b) copper, and show how they account for the characteristic physical properties of these substances.

Explain qualitatively what information is required about crystalline sodium chloride to allow the determination of Avogadro's Number (the Avogadro Constant). (C)

13 How can the shapes of simple molecules be explained in terms of electron pair repulsions? Your answer should include at least one example from each of four different shapes.

What effect does the presence of a lone pair of electrons on the nitrogen atom have on:

(a) the H-N-H angle in ammonia,
(b) the properties of the ammonia molecule? (JMB)

Chapter 7

Modern valency theory

7.1 The electron as a particle or wave

The Bohr theory of the hydrogen atom presupposed that the electron could be regarded as a discrete particle whose position and velocity could be measured simultaneously and with high accuracy. However, for something as small as an electron this is impossible; suppose we could devise a method of determining the position of an electron by directing short wavelength radiation onto it, then the effect on the radiation would allow us to determine the electron's position, but the energy associated with a photon of radiation is comparable with the energy of an electron and the velocity of the electron would change, i.e. the velocity or momentum (product of the mass and velocity) of the electron would now be 'uncertain'. Heisenberg (1927) was the first to realise this difficulty, which he expressed in his famous 'Uncertainty Principle'.

If the position of an electron is known with great precision its velocity (and hence its momentum) is known with far less accuracy, and likewise if the velocity (and hence the momentum) of an electron is known with great precision its position becomes uncertain.

This principle is often expressed as:

$$(\Delta x)(\Delta mv) \simeq h$$

where Δx is the uncertainty in position, Δmv is the uncertainty in momentum and h is Planck's constant ($6{\cdot}625 \times 10^{-34}$ J s). The product of these two uncertainties is incredibly small and need not concern us in our dealings with macroscopic systems, but for minute objects like electrons this relationship is of great importance. A direct consequence of the Uncertainty Principle is that precise statements about the position and velocity of an electron must be replaced by less precise ones. The best we can do is to state the probable position and velocity of a particular electron.

Three years before Heisenberg put forward his Uncertainty Principle, the French physicist de Broglie had predicted that waves should be associated with moving electrons, and, furthermore, that the wavelength associated with electrons could be calculated from the equation:

$$\lambda = h/mv$$

where h is Planck's constant, m is the mass of the electron, and v is the velocity of the electron. This prediction was verified by Davisson and Germer (1927) when they demonstrated that an electron beam was diffracted by a nickel crystal. Since diffraction phenomena can only be explained on a wave theory, we must conclude that electrons have wavelike characteristics associated with them under certain conditions. Are electrons particulate in nature or is their behaviour best interpreted in terms of waves? Certainly the experiments carried out to determine the mass and the charge carried by an electron are explicable in terms of a particulate nature but diffraction experiments need a wave theory. Little is gained by arguing for or against either of these theories and it must be accepted that electrons, like light, show a dual behaviour. Some experiments are, therefore, best interpreted by assuming electrons to be discrete particles, while others are more in accordance with a wave theory.

7.2 The wave equation for electrons

In 1927 Schrödinger expressed the wavelike characteristics of the electron in terms of a differential equation. This equation, given below, cannot be derived and Schrödinger himself arrived at it mainly by mathematical intuition. It can be solved exactly for one-electron systems, e.g. for the hydrogen atom, and for such systems the predicted energy levels of the electron are in complete agreement with experimental results. The equation is:

$$\frac{\partial^2\psi}{\partial x^2} + \frac{\partial^2\psi}{\partial y^2} + \frac{\partial^2\psi}{\partial z^2} + \frac{8\pi^2 m}{h^2}(E - V)\psi = 0$$

where ψ is the wave function associated with the electron, m, E and V are respectively the mass, the total energy and the potential energy of the electron, h is Planck's constant, and x, y and z are the normal Cartesian co-ordinates of the electron, the nucleus of the atom representing the origin of the system. The wave functions, ψ, have no physical reality themselves but the quantity $\psi^2 dxdydz$ is a direct measure of the probability of finding the electron in the volume $dxdydz$ (i.e. in volume dv) with an energy E.

7.3 Solution of the wave equation

The solution of the Schrödinger equation which gives the lowest value of E for the electron, i.e. the ground state, is of the form:

$$\psi = \exp^{-kr}$$

where $r^2 = x^2 + y^2 + z^2$ (r being the distance of the electron from the nucleus) and k is a constant. Since the solution is of exponential form, the value of ψ falls off with increasing values of r and has its largest value when $r = 0$. Since this function is spherically symmetrical about the origin, concentric spheres drawn with the origin as centre represent points where ψ is constant, i.e. points on any one spherical surface are positions of constant ψ. A most important point here is that this wave function is always positive. As we have mentioned above, ψ has no physical reality but ψ^2 is related to the probability of finding the electron at a particular point; the different ways of representing ψ and ψ^2 are shown in the next section.

7.4 Ways of representing the wave function for the ground state of the hydrogen atom

The wave function can be said to 'describe' the electron in the atom, the wave function corresponding to the lowest value of E 'describing' the 1s atomic orbital. Similarly the wave function corresponding to the next energy level is called the 2s orbital, etc. Three different ways of representing the 1s orbital of the hydrogen atom are used.

The wave function $\psi = \exp^{-kr}$ can be represented graphically in the normal manner, or concentric circles can be drawn, points on any one circle being positions of constant ψ (in three dimensions the concentric circles would be replaced by concentric spheres). The third, though less precise, way of representing this wave function is to draw a boundary surface (in three dimensions a sphere) so that the surface encloses values of ψ greater than a chosen value (fig. 7.1).

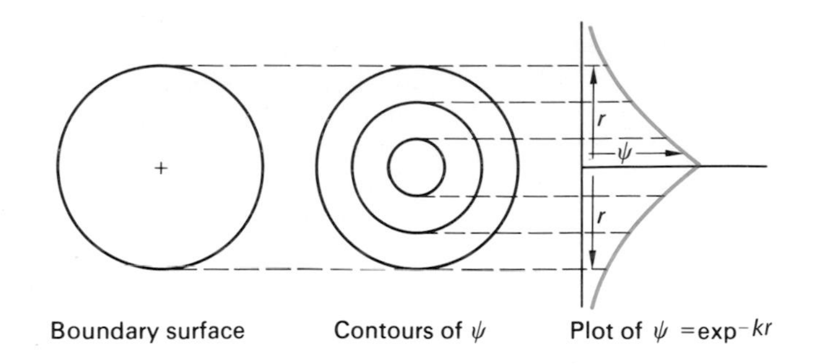

FIG. 7.1. *The three ways of representing the* 1*s orbital of the hydrogen atom*

The sign attached to the wave function is of the greatest importance when considering the formation of bonds between atoms (see later).

Since ψ^2 rather than ψ itself is of more physical significance the ways in which ψ^2 can be represented will now be discussed. Remember that ψ^2 is related to the probability of finding the electron at a particular point. The values of ψ^2 can be plotted for various values of r or alternatively a boundary surface can be drawn which encloses, say, more than 95 per cent of the electronic charge, i.e. the region enclosed by this surface corresponds to a 95 per cent chance of finding the electron in this volume. The charge cloud representation of the electron is a useful pictorial representation (see section 4.10, p. 49). Here it is imagined that the electron can be photographed, say, several million times and each position represented by a dot on a sheet of paper. The general effect is something akin to that of a smeared-out negative charge with rather diffuse edges (fig. 7.2).

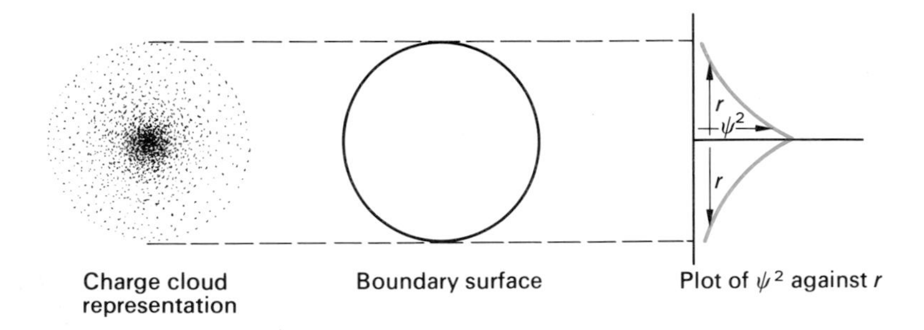

FIG. 7.2. *The three ways of representing electron probability for the* 1*s orbital of the hydrogen atom*

As the graph of ψ^2 against r shows, there is the greatest chance of finding the electron at the nucleus of the hydrogen atom (at $r = 0$). The probability of finding the electron at further and further distances from the nucleus falls; nevertheless there is a small but finite chance of finding the electron at very great distances from the nucleus. Since the 1s orbital is spherically symmetrical about the nucleus of the atom it is more useful to enquire as to the probability of finding the electron at a distance r from the nucleus and in a volume bounded by two spheres of radii r and $r + dr$. The volume of this spherical shell is the area of the sphere radius r, that is

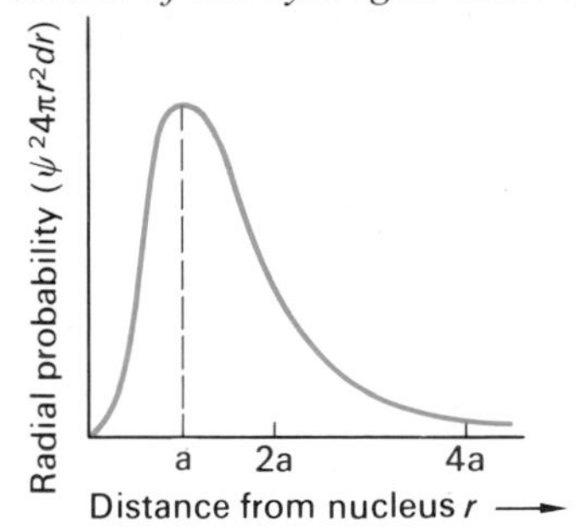

FIG. 7.3. *Radial probability of the 1s orbital of the hydrogen atom*

$4\pi r^2$, multiplied by the distance between the two spheres dr, i.e. the volume is $4\pi r^2 dr$. The product of ψ^2 and $4\pi r^2 dr$ is defined to be the radial probability. Since ψ^2 decreases with increasing r but the volume of the spherical shell increases with r (the volume is proportional to r^2), the position of maximum radial probability will not be at the nucleus but at some distance away from the nucleus. Figure 7.3 shows a graph of radial probability against r for the 1*s* orbital of the hydrogen atom. The maximum radial probability occurs at a value $r = a$ which is precisely the same value as that calculated for the radius of the lowest energy electron orbit (the Bohr theory, p. 37). However, the radial probability is appreciable over a wide range of values whereas the Bohr theory assumed that the electron could be located with extreme precision.

Another way of looking at radial probability is to consider the charge cloud representation of the 1*s* orbital. We now calculate the number of dots lying between radial shells of thickness dr and area $4\pi r^2$. This is the number of dots in the volume $4\pi r^2 dr$, the dots representing imaginary photographs of the electron's position over a period of time. Since the dot representation of the electron's position represents radial density, this term and radial probability are one and the same thing.

7.5 Higher energy level orbitals of the hydrogen atom

Other solutions of the wave equation give values for the wave function, ψ, which correspond to higher energy levels, i.e. higher E values. The 2*s* orbital, like the 1*s* orbital, is spherically symmetrical about the origin, but, unlike the 1*s* orbital, the value of ψ corresponding to this higher energy orbital can be negative. From the graphical plot of ψ_{2s} and the corresponding radial probability of the 2*s* orbital, it can be seen that maximum radial probability occurs at a value $r = 4a$ (fig. 7.4).

There are three more solutions to the wave equation which have the same energy as the 2*s* orbital but they are more difficult to interpret than the latter atomic orbital. Unlike the 2*s* orbital these three wave functions have values which depend upon both radial distance from the nucleus and angular distance from the nucleus, i.e. they are not spherically symmetrical about the origin. The actual wave functions can be regarded as the product of a radial wave function ψ_{radial} and an angular wave function ψ_{angular}. These three wave functions only differ in that one is directed along the *x*-axis, another is directed along the *y*-axis, and the third is directed along the *z*-axis; they are the 2*p* atomic orbitals and are designated $2p_x$, $2p_y$, and

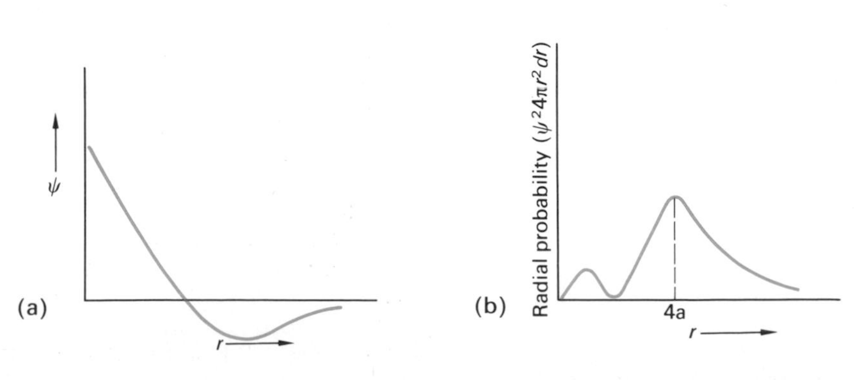

FIG. 7.4. *Plots of (a) ψ against r and (b) radial probability against r, for the 2s orbital of the hydrogen atom*

$2p_z$ respectively. The radial wave function ψ_{radial} and the angular wave function ψ_{angular} are shown for the $2p$ orbital (fig. 7.5). The angular wave function is shown plotted as a polar diagram along the x-axis where the angle θ is the angle from the origin for the $2p_x$ orbital.

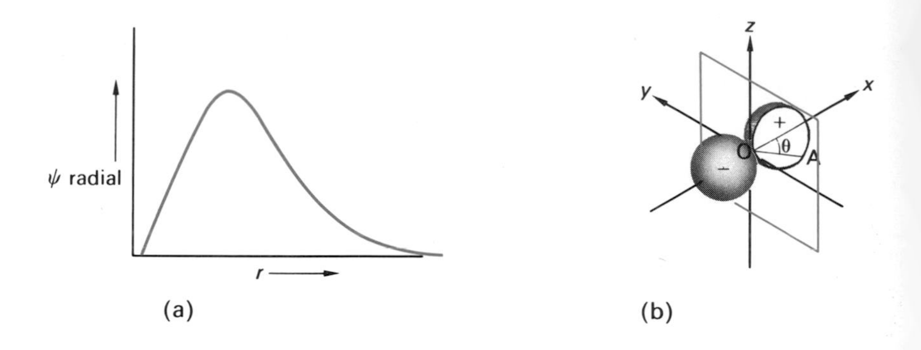

FIG. 7.5. *(a) The radial wave function and (b) the angular wave function for the 2p orbital of the hydrogen atom*

The ψ_{2p} radial wave function is positive for all values of r, whereas the angular wave function for the $2p_x$ orbital can be both positive and negative. The maximum value of the angular wave function occurs when θ is 0 or 180°, having positive and negative values respectively. For $\theta = 90°$ or 270° the angular wave function is zero and since the complete wave function is the product of ψ_{radial} and ψ_{angular} and electron probability is measured by the square of the complete wave function, there is no chance of finding the $2p_x$ electron along the z-axis. Since the complete angular wave function consists of two tangential spheres, i.e. the solid shape generated by revolution of fig. 7.5(b) about the x-axis, it follows that there is also no chance of finding the electron along the y-axis. At an angle θ the magnitude of the angular wave function is proportional to the distance OA (see fig. 7.5(b)). The angular wave functions for the $2p_x$, $2p_y$ and $2p_z$ orbitals are shown in fig. 7.6, the positive and negative regions being marked.

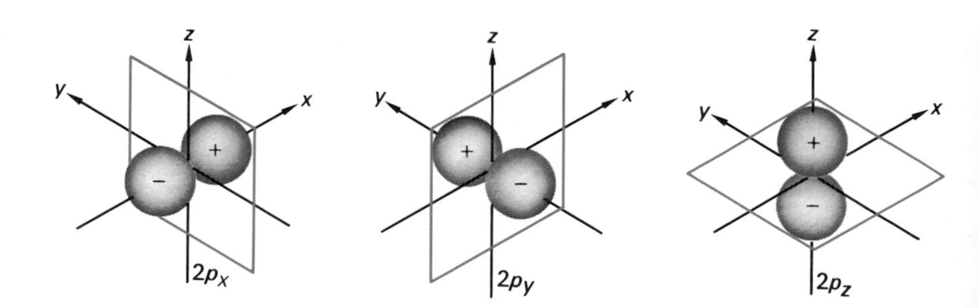

FIG. 7.6. *The boundary surface of the $2p_x$, $2p_y$ and $2p_z$* atomic *orbitals*

The directional properties of $2p$ orbitals are of importance when discussing the formation of bonds between atoms and so too is the sign of the angular wave function. Therefore it is not usually necessary to consider the radial wave function, although it should be appreciated that an approximation is being made.

Further solutions of the wave equation give nine separate wave functions all corresponding to the same energy; these are the $3s$, the $3p$ ($3p_x$, $3p_y$ and $3p_z$), and the $3d$ orbitals. The latter are even more difficult to visualise than the p orbitals and are designated $3d_{xy}$, $3d_{xz}$, $3d_{yz}$, $3d_{x^2-y^2}$ and $3d_{z^2}$.

Like the p orbitals they can be described as the product of a radial wave function and an angular wave function. Figure 7.7 shows the directional nature of the $3d$ angular wave functions together with the positive and negative regions. Notice that the $3d_{z^2}$ orbital is a different shape from the other four.

FIG. 7.7. *The boundary surfaces of the five 3d atomic orbitals*

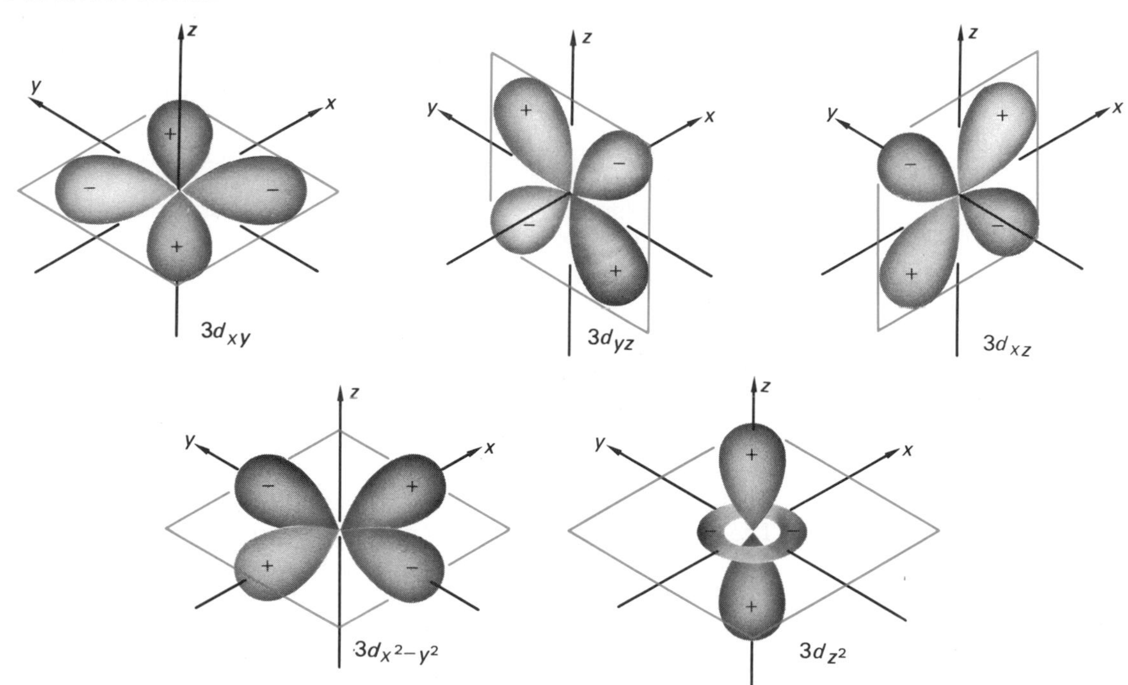

The general shape of the $3d_{xy}$ angular wave function means that the electron is most likely to be found along two mutually perpendicular straight lines which bisect the xy-plane; similarly for the $3d_{x^2-y^2}$ angular wave function the electron is most likely to be found along the x- and y-axes. As in the case of p orbitals the angular wave functions corresponding to the $3d$ orbitals are of importance when discussing bond formation, but the complete wave function is the product of both the radial wave function and the angular wave function.

7.6 Solution of the wave equation for a many-electron atom

Exact solutions of the wave equation for the hydrogen atom are possible and these solutions quite naturally give rise to the quantum numbers that were first mentioned in section 4.6, p. 42. For example, the wave functions for an excited hydrogen atom (the electron being characterised by principal quantum number $n = 2$) correspond to the $2s$, $2p_x$, $2p_y$ and $2p_z$ orbitals. The shape of the $2s$ orbital is different from the $2p$ orbitals and is characterised by the subsidiary quantum number $l = 0$, whereas the three equivalent $2p$ orbitals are characterised by the subsidiary quantum number $l = 1$. In order to distinguish between the three mutually perpendicular $2p$ orbitals a third quantum number denoted by m is required (see

Table 4C, p. 43). The wave equation cannot be solved exactly for a system which contains more than one electron because, in addition to electron-nucleus attractive forces, we must consider electron-electron repulsion and the resulting mathematics is quite complex and not amenable to exact solution. Nevertheless, approximate solutions can be obtained which indicate that the electrons in many-electron atoms can be considered to be classified in terms of the *s*, *p*, *d*, etc., atomic orbitals. However, whereas the 2*s* and the equivalent three 2*p* orbitals of the excited hydrogen atom are all of the same energy, for other atoms the 2*p* orbitals (which are still of the same energy) are of higher energy than the 2*s* orbital. Similarly for atoms other than hydrogen, the five equivalent 3*d* orbitals are of higher energy than the three equivalent 3*p* orbitals which, in turn, are of higher energy than the 3*s* orbital. In the first transition series there is little difference in energy between the 4*s* orbital and the 3*d* orbitals and this is of importance when considering the variable valency displayed by transition metals.

The directional properties of the 3*d* orbitals are of importance in discussing a simple theory to account for the colour of transition metal ion complexes (p. 395).

7.7 The covalent bond—the molecular orbital theory

The molecular orbital theory of covalent bonding considers the ways in which electrons are influenced by the presence of two or more nuclei, the electrons in bonded atoms being said to occupy molecular orbitals. In the case of the hydrogen molecule there will be repulsive forces operating between the two nuclei and between the two electrons; in addition there will be attractive forces operating between the nuclei and the two electrons. These complications mean that precise solutions of the energies and wave functions corresponding to molecular orbitals are out of the question, but approximate solutions are possible. We shall do no more than consider the molecular orbital theory in a purely qualitative sense.

Let us first consider the hydrogen molecule. If two hydrogen atoms are first considered to be widely separated, then each electron is only influenced by its own nucleus, each electron occupying a 1*s* atomic orbital. As the hydrogen atoms approach one another both electrons come under the influence of the two hydrogen nuclei and eventually a chemical bond is established; under these conditions the separate atomic orbitals can be said to merge and constitute molecular orbitals. An important point here is that the number of possible molecular orbitals is equal to the number of separate atomic orbitals in the separated atoms. For the hydrogen molecule there are two possible molecular orbitals. The two molecular orbitals are obtained by adding and subtracting the two atomic orbitals (the linear combination of atomic orbitals, abbreviated as LCAO method). The two possible molecular orbitals are represented thus:

$$\psi_{\text{bonding}} = \psi_{\text{A.1}s} + \psi_{\text{B. 1}s}$$
$$\psi_{\text{antibonding}} = \psi_{\text{A.1}s} - \psi_{\text{B. 1}s}$$

where the letters A and B are used to distinguish the two hydrogen atoms. A bonding molecular orbital is obtained by adding together the separate atomic orbitals and this results in the build-up of negative charge between the two nuclei; an antibonding molecular orbital is obtained by subtracting the two atomic orbitals and this leads to a draining away of negative charge between the two nuclei. A visual picture of what is involved can be gained by considering the overlap of the two atomic orbitals (fig. 7.8).

FIG. 7.8. *The combination of hydrogen* 1*s atomic orbitals to give a bonding and an anti-bonding molecular orbital*

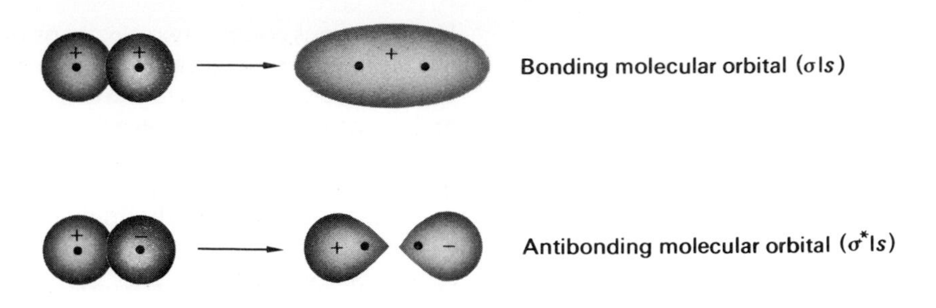

In the case of hydrogen the wave function corresponding to the two 1*s* orbitals is positive and the wave function corresponding to the bonding molecular orbital is also positive (abbreviated to $\sigma 1s$). Subtraction of the two atomic orbitals gives a molecular orbital which has positive and negative regions (abbreviated $\sigma^* 1s$).

In general, a bonding molecular orbital is of lower energy than the two separate atomic orbitals, while the reverse is true for an antibonding orbital. Furthermore, two electrons with opposite spins can occupy each molecular orbital (two electrons with opposite spins can occupy each atomic orbital by the Pauli exclusion principle, p. 42). Thus, in the hydrogen molecule, both electrons occupy the bonding molecular orbital, a chemical bond being established because this bonding molecular orbital is of lower energy than the two singly occupied atomic orbitals of the hydrogen atoms (fig. 7.9).

FIG. 7.9. *The energy difference between bonding and antibonding molecular orbitals*

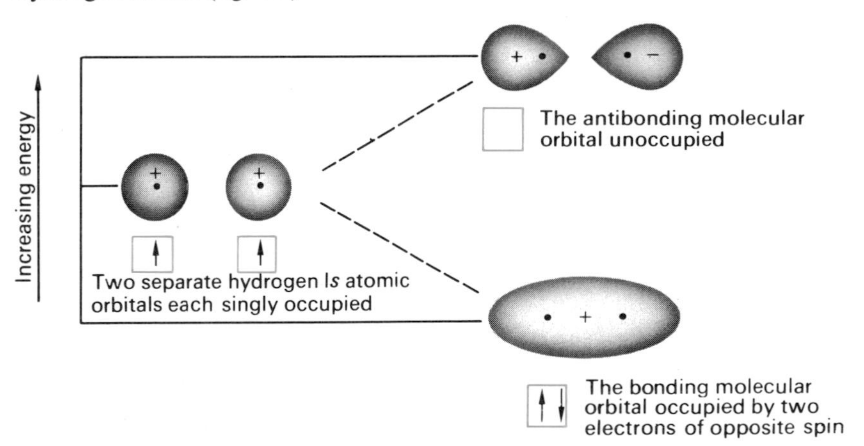

Before proceeding to discuss the application of molecular orbital theory to more complex molecules it is necessary to state that atomic orbitals will only combine to give molecular orbitals under certain conditions:

(a) The atomic orbitals must overlap to an appreciable extent, otherwise there will be no significant build-up of charge between the nuclei and therefore no chemical bond will be established.
(b) In the region of overlap the wave functions of the separate atomic orbitals must be of the same sign, otherwise there will be no build-up of charge between the two nuclei.
(c) The energies of the separate atomic orbitals must be of comparable magnitude.

It will be seen that these conditions are fulfilled for the simple example we have been discussing.

7.8 The helium molecule and helium molecule-ion

Helium atoms contain two electrons occupying the 1s atomic orbitals, and the electronic configuration of the helium atom is 1s^2. If these atomic orbitals are combined they will give one bonding and one antibonding molecular orbital and, since four electrons are involved, both of these orbitals will be doubly occupied. However, the lower energy of the bonding molecular orbital (as compared with the separate atomic orbitals) is effectively cancelled out by the higher energy of the antibonding molecular orbital and no chemical bond is established; helium molecules, He_2, cannot be formed (fig. 7.10).

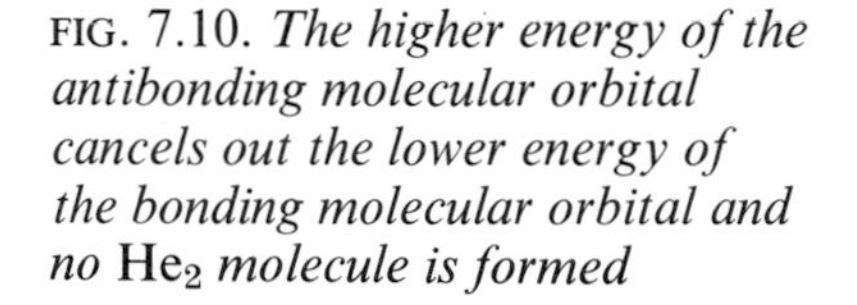

FIG. 7.10. *The higher energy of the antibonding molecular orbital cancels out the lower energy of the bonding molecular orbital and no* He_2 *molecule is formed*

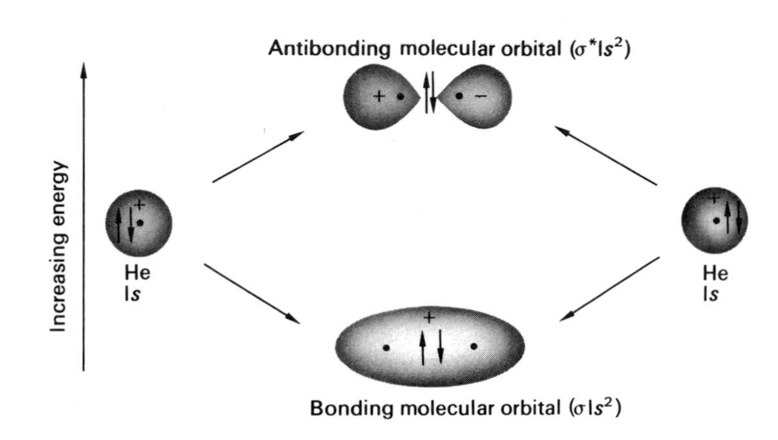

The entity He_2^+ molecule ion can be obtained by subjecting helium gas at a low pressure to a high energy discharge. One of the successes of molecular orbital theory is that it explains the existence of this species. The He_2^+ species contains three electrons and two can be accommodated in a bonding molecular orbital, while the third can occupy the antibonding molecular orbital; there is some overall bonding and this species can exist.

7.9 Molecular orbitals formed by the overlap of *p* atomic orbitals

The criterion that molecular orbitals can be formed by the overlapping of atomic orbitals, provided their wave functions are of the same sign in the region of overlap, implies that p-type atomic orbitals can overlap in two distinct ways. Thus two $2p_x$ orbitals can overlap along the x-axis and addition and subtraction of their wave functions gives rise to a bonding and an antibonding molecular orbital. The two $2p_x$ atomic orbitals can also overlap laterally and again two molecular orbitals (one bonding and the other one antibonding) can be obtained. These methods of overlap are shown in fig. 7.11 and in fig. 7.12.

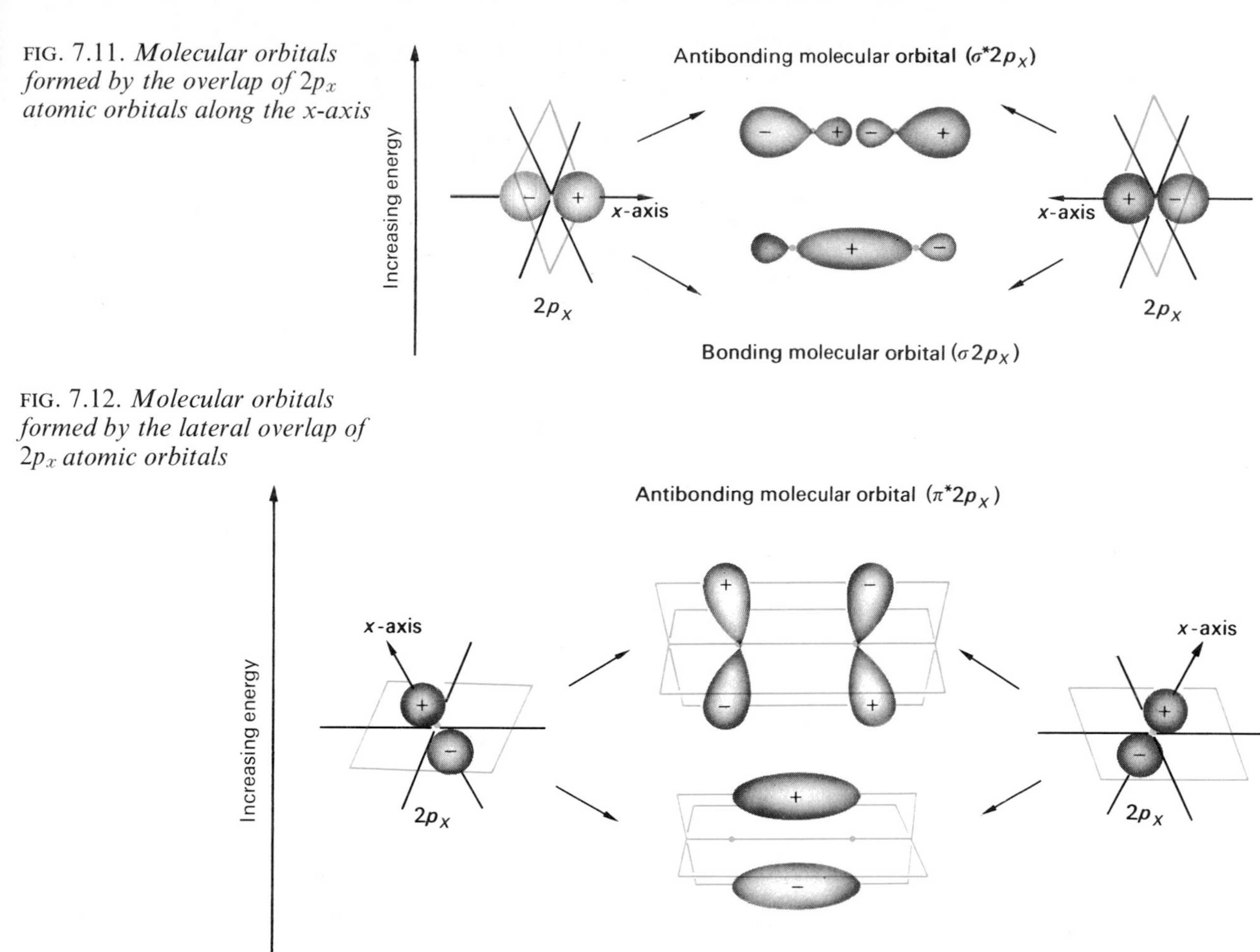

FIG. 7.11. *Molecular orbitals formed by the overlap of* $2p_x$ *atomic orbitals along the x-axis*

FIG. 7.12. *Molecular orbitals formed by the lateral overlap of* $2p_x$ *atomic orbitals*

In general the energy of the $\sigma 2p_x$ molecular orbital is of lower energy than the $\pi 2p_x$ molecular orbital. The energy of the $\pi^* 2p_x$ molecular orbital is of lower energy than the $\sigma^* 2p_x$ molecular orbital. Thus the energy increases in the order:

$$\sigma 2p_x < \pi 2p_x < \pi^* 2p_x < \sigma^* 2p_x$$

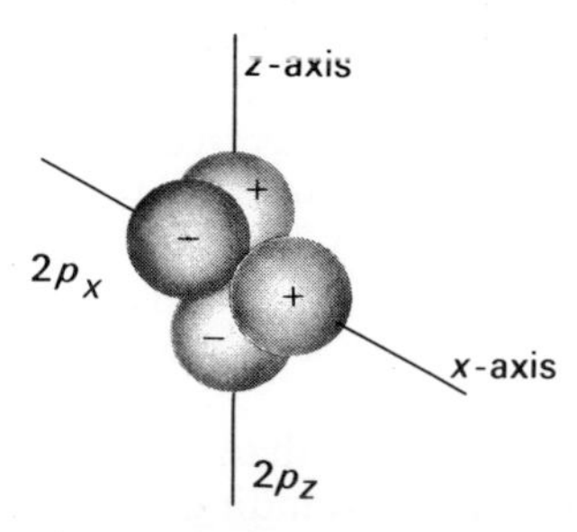

FIG. 7.13. *A* $2p_x$ *and a* $2p_z$ *atomic orbital cannot overlap to form a molecular orbital*

It is important to realise that a $2p_x$ atomic orbital cannot be combined with, say, a $2p_z$ atomic orbital of another atom since the wave functions are not of the same sign in the region of overlap; the negative portion cancels out the positive portion.

We have considered the formation of molecular orbitals by the overlapping of $2p$ atomic orbitals and it only remains to see how $2s$ atomic orbitals overlap. The type of overlapping here is similar to that for the overlapping of two $1s$ atomic orbitals, thus a bonding molecular orbital ($\sigma 2s$) and an antibonding molecular orbital ($\sigma^* 2s$) are formed. The energy sequence of these molecular orbitals in a homonuclear diatomic molecule, i.e. a molecule containing two like atoms, is given overleaf.

FIG. 7.14. *The energy sequence for molecular orbitals (the bond direction is taken to be along the x-axis). In some instances the levels $\sigma 2p_x$ and $\pi 2p_y/\pi 2p_z$ are interchanged.*

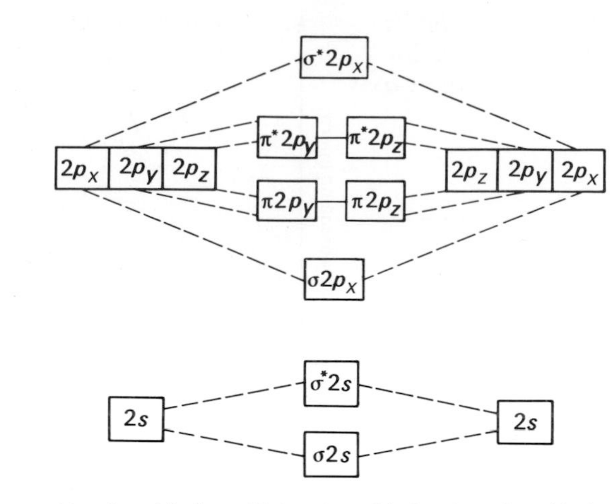

7.10 The nitrogen, oxygen and fluorine molecules

The electronic configurations of the atoms of nitrogen, oxygen and fluorine are respectively $1s^22s^22p^3$, $1s^22s^22p^4$ and $1s^22s^22p^5$. Consider two nitrogen atoms approaching each other; eventually a chemical bond is established and the fourteen electrons come under the influence of both nuclei. The four $1s$ electrons occupy a bonding and an antibonding molecular orbital, i.e. no net bond results, and this is the justification for neglecting inner shell electrons when discussing bond formation. There are now ten electrons to accommodate in the various molecular orbitals, and the four $2s$ electrons, like the four $1s$ electrons, occupy a bonding ($\sigma 2s$) and an antibonding (σ^*2s) molecular orbital; again no net bonding results. Six electrons are now left, namely two $2p_x$, two $2p_y$ and two $2p_z$ atomic orbitals; since the bond is taken to be along the x-axis the two $2p_x$ atomic orbitals overlap along this axis and a bonding $\sigma 2p_x$ molecular orbital is formed and is occupied by both electrons. The separate $2p_y$ and $2p_z$ atomic orbitals of each nitrogen atom cannot overlap along their axes but they can overlap laterally to give respectively bonding $\pi 2p_y$ and $\pi 2p_z$ molecular orbitals, each of which is doubly filled. The nitrogen molecule thus contains one σ bond and two π bonds; the antibonding orbitals formed from the $2p$ atomic orbitals are left unoccupied and the net result is a triple bond sometimes represented as $N \equiv N$.

The oxygen molecule contains two more electrons than the nitrogen molecule and these occupy the next higher energy molecular orbital. In fact there are two such molecular orbitals having the same energy and they are antibonding (π^*2p_y and π^*2p_z, see fig. 7.14). Each of these two molecular orbitals is singly occupied (cf. atomic orbitals of the same energy are singly occupied before electron pairing occurs, see Hund's rule p. 43). Since the higher energy of these antibonding orbitals effectively cancels out the lower energy of one of the $\pi 2p$ bonding molecular orbitals (which is occupied by two electrons) the bond in the oxygen molecule can be considered to be a double bond. Molecular orbital theory explains why the oxygen molecule is paramagnetic (see p. 355) since this phenomenon is associated with the presence of unpaired electron spins.

The fluorine molecule contains two more electrons than the oxygen molecule and it will be seen from fig. 7.14 that the antibonding molecular orbitals π^*2p_y and π^*2p_z are both doubly occupied. There are, therefore, three bonding and two antibonding molecular orbitals occupied and the fluorine molecule can be considered to contain a single bond.

Caution is needed when the energy level diagram (fig. 7.14) is applied to diatomic molecules which contain different atoms, since slight changes in the energy level sequence sometimes occur. This is not really surprising since the energy levels of the separate atoms will now no longer be identical. However, the nature of the bonding in the nitrogen oxide molecule is worth considering from the sequence of energies given in fig. 7.14. The nitrogen oxide molecule contains one more electron than the nitrogen molecule and one fewer electron than the oxygen molecule; thus one electron occupies a π^*2p_y molecular orbital which is antibonding. The nitrogen oxide molecule thus contains three bonding and a singly occupied antibonding molecular orbital, and the bond order can be considered to be 2·5. There is thus no difficulty in accounting for the existence of molecules which contain an odd number of electrons when molecular orbital theory is used.

7.11 Suggested reading

N. N. Greenwood, *Chemical bonds* Education in Chemistry No. 4, Vol. 4, 1967.

P. G. Perkins, *A Simple Approach to Molecular Bonding Theory: I—Hybridisation*, School Science Review, No. 165, Vol. 48, 1967.

—*A Simple Approach to Molecular Bonding Theory: II—the Relation of Orbital Overlap and Bonding,* School Science Review, No. 166, Vol. 48, 1967.

— *Orbitals without Tears,* School Science Review, No. 164, Vol. 48, 1966.

H.S. Pickering, *Multiple Covalent bonds*, School Science Review, No. 212, Vol. 60, 1979.

7.12 Questions on chapter 7

1. Why is it impossible to determine both the position and velocity of an electron at any instance with high precision? How has this led to a modification of the Bohr theory of the hydrogen atom?
2. Using de Broglie's equation determine the wavelength associated with an electron whose mass is 9×10^{-28} g and whose velocity is 3×10^7 m s^{-1}. (Planck's constant is $6{\cdot}625 \times 10^{-34}$ J s).
3. How are the two separate atomic orbitals of hydrogen atoms combined to give (a) a bonding molecular orbital and (b) an antibonding molecular orbital? How are the energies of the bonding and antibonding molecular orbitals related to the energies of the separate atomic orbitals?
4. Helium atoms do not combine to give the molecule He_2 but a helium atom can bond to a helium ion to give the entity He_2^+. How can these facts be explained using molecular orbital theory? Would you expect the H_2^+ species to exist under certain conditions? Give your reasons.
5. What justification is there for the statement: 'inner shell electrons do not contribute to chemical bonding'. Using molecular orbital theory, show that the fluorine molecule can be considered to contain a single covalent bond.
6. What is meant by a σ bond and a π bond? Show how molecular orbital theory is successful in explaining the presence of two unpaired electrons in the oxygen molecule, i.e. it is paramagnetic. Sulphur vapour contains some S_2 molecules; would you expect these molecules to be paramagnetic?
7. By citing one example, show how molecular orbital theory can successfully explain the existence of molecules containing an odd number of electrons.
8. Would you expect the O_2^+ molecule ion to contain stronger or weaker bonding than the O_2 molecule? Similarly, compare the strengths of the bonding in N_2^+ and N_2. (Hint: Count the number of electrons in each species and then fill the molecular orbitals in the order given in fig. 7.14 p. 94. Remember that the 1*s* electrons (four in all) will not contribute to the bonding.)

Chapter 8

The factors that determine whether a chemical reaction is possible

8.1 Introduction

From a study of a wide range of chemical reactions it has been found that an energy change is invariably involved. Since reactions which proceed with the liberation of heat (exothermic reactions) are generally those which take place most readily, it is tempting to suggest that the tendency for chemicals to react in order to evolve energy is the factor which decides whether a particular reaction is possible. However, many reactions are endothermic and must therefore extract energy from the surroundings, e.g. thermal decomposition reactions which generally take place more readily at higher temperatures, and the dissolution of many salts in water. Furthermore, many reactions are easily reversed and thus, if exothermic in one direction, must be endothermic to the same extent in the other direction (law of conservation of energy). Consider the Haber process:

$$N_2(g) + 3H_2(g) \rightleftharpoons 2NH_3(g)$$

By the application of high pressure and a moderately high temperature in the presence of a suitable catalyst, the reaction can be made to proceed from left to right at a convenient rate, particularly if the ammonia is liquefied and removed. For every mole of ammonia that is formed 46·0 kJ of heat are evolved:

$$\tfrac{1}{2}N_2(g) + 1\tfrac{1}{2}H_2(g) \longrightarrow NH_3(g) \qquad \Delta H = -46{\cdot}0\,\text{kJ mol}^{-1}$$

Ammonia can be made to decompose into nitrogen and hydrogen and this reaction is endothermic to the extent of 46·0 kJ for every mole of ammonia decomposed:

$$NH_3(g) \longrightarrow \tfrac{1}{2}N_2(g) + 1\tfrac{1}{2}H_2(g) \qquad \Delta H = +46{\cdot}0\,\text{kJ mol}^{-1}$$

It is clear from this one example that at least one more factor, other than a tendency for a chemical system to assume a lower energy state by transferring some energy to the surroundings, is involved, otherwise no endothermic reactions would be possible. It is interesting to note that in the breakdown of the ammonia molecule into nitrogen and hydrogen there is an increase in the number of molecules, i.e. the more ordered ammonia molecule gives simpler molecules of nitrogen and hydrogen. In fact every endothermic reaction is accompanied by some decrease in order, whether this be in the actual chemical system itself or in the surroundings.

Order is associated with molecular structure and solids are more ordered than the liquids into which they pass on melting, since the atoms or ions which vibrate about fixed positions in the crystal lattice become more mobile in the liquid phase. Similarly, liquids are more ordered than gases into which they pass on boiling, since gaseous molecules can move about at random. Polyatomic gases are likewise more ordered than the simpler ones into which they may decompose under suitable conditions.

Plate 3. Snowflakes show almost perfect order. Still waters are more ordered than a rough sea (By courtesy of Radio Times Hulton Picture Library)

It is now known that the possibility of a chemical reaction taking place under conditions which allow a transfer of energy to and from the surroundings, but do not allow a transfer of matter, (closed system conditions, see later) is governed by two factors:

(a) A tendency for the chemical system to move to minimum energy, e.g. evolve heat to the surroundings. The energy lost by the chemical system is, of course, equal to the energy gained by the surroundings (law of conservation of energy).

(b) A tendency for the chemical system to move in such a direction which results in a decrease in order, i.e. an increase in disorder. Entropy is a measure of order/disorder and there is reason to believe that at absolute zero the entropy of a crystalline solid is zero. At this temperature a crystalline solid possesses perfect order. An increase in entropy is equated with an increase in disorder.

Whether a reaction is possible is determined by an interplay of these two effects. However, it is important to realise that it is not possible to predict how fast a particular reaction will take place, even though both factors (a) and (b) may be favourable. For instance, carbon shows no signs of reacting with oxygen at room temperature to give carbon monoxide, even though it can be shown that the reaction should be very exothermic at this temperature and should result in an increase in disorder, i.e. one extra mole of gas is produced for each mole of oxygen:

$$2C(s) + \underset{1\ \text{mole}}{O_2(g)} \rightarrow \underset{2\ \text{moles}}{2CO(g)}$$

Carbon and oxygen are said to be thermodynamically unstable with respect to carbon monoxide, the reaction being opposed by another effect (an activation energy barrier, see chapter 9).

The two effects, i.e. energy considerations and order/disorder effects, will first be considered separately, and then the concept of free energy, which takes into account both effects, will be introduced.

8.2 Internal energies, enthalpies and heats of reaction

The first law of thermodynamics (or the law of conservation of energy) simply states that energy cannot be destroyed. If one form of energy is consumed then another form of energy in equivalent quantities takes its place. For instance, if a gas is heated, it can absorb this heat energy by increasing the energies of vibration, rotation and translation of its molecules.

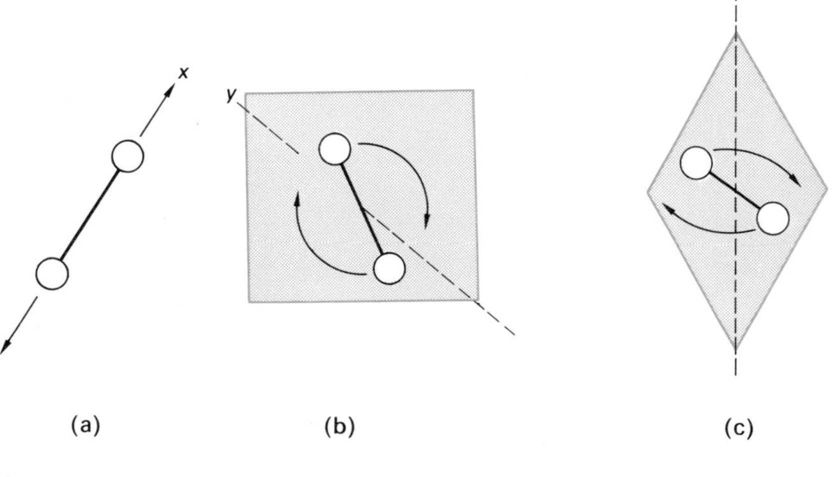

FIG. 8.1. *There is one way in which a diatomic molecule can vibrate (a) but there are two axes of rotation which are mutually perpendicular to one another (b) and (c)*

In addition to this, it can expand and thus do work against the atmosphere. By the first law of thermodynamics:

$$q = \Delta U + w$$

where q is the heat absorbed, ΔU is the change in internal energy (made up of changes in the vibrational, rotational and translational energies etc., of the gas), and w is the work done by the gas. Absolute values of internal energies of substances cannot be evaluated but changes in internal energies, ΔU, are readily obtained.

Suppose a chemical reaction takes place at constant volume. No work is done, and the heat change, q_v, can be equated to the change in internal energy, ΔU.

$$q_v = \Delta U$$

However, a chemical reaction carried out at constant pressure may result in work done against the atmosphere and in this case the heat change, q_p, results in changes in the internal energy, ΔU, and work done against the atmosphere, $p\Delta V$.

$$q_p = \Delta U + p\Delta V$$

The convention used in this book is that heat absorbed by a system is positive, increases in internal energies are positive and work done by the system is positive. For heat evolution, decreases in internal energies and work done on the system, a negative sign is used.

The heat change of a chemical reaction carried out at constant pressure is called an enthalpy change and is represented by ΔH.

$$q_p = \Delta H \qquad \text{or} \qquad \Delta H = \Delta U + p\Delta V$$

If heat is evolved by a system at constant pressure then ΔH is negative; similarly if heat is absorbed, then ΔH is positive.

Consider the heat change that occurs when 1 mole of molar sodium hydroxide solution is reacted with 1 mole of molar hydrochloric acid; in this reaction a temperature rise is observed. By the law of conservation of energy, the total energy of the unmixed alkali and acid at room temperature is the same as the energy of the sodium chloride solution produced at the higher temperature (negligible volume change occurs and hence no work is done on or by the atmosphere). However, when the sodium chloride solution cools down to room temperature, energy in the form of heat is lost to the surroundings and this can easily be determined. The energy content of the sodium chloride solution is less than that of the unmixed alkali and acid. The heat evolved is known as the heat of reaction.

Tabulated values of many heat changes are available and it is possible to compute heat changes for reactions that are difficult or impossible to carry out experimentally (see later in this section). Heats of reaction can either be measured at constant volume or, more generally, at constant pressure but, except in the case of gaseous reactions where there may be an increase or decrease in the number of moles and hence work done on or by the atmosphere, the difference between the two is negligible. Standard heats of formation are generally quoted at 298 K and 1 atmosphere pressure (it does not matter what temperature and pressure are attained during a reaction, provided that the initial and final temperature and pressure are respectively 298 K and 1 atmosphere).

When oxygen is one of the reactants the heat change is known as the enthalpy of combustion, which is defined to be:

The amount of heat involved when 1 mole of the substance is burnt completely in oxygen, reactants and products being at 25°C (298K) and 1 atmosphere pressure, e.g.

$$\underset{\text{graphite}}{C(s)} + O_2(g) \rightarrow CO_2(g) \qquad \Delta H^\ominus(298\,K) = -393{\cdot}4\,kJ\,mol^{-1}$$

Let us now analyse the above reaction in a little more detail so that we fully understand what information is conveyed by the chemical equation. The energy content (enthalpy) of an element in its standard state, i.e. in its most stable physical form at a specified temperature (generally 298 K) and 1 atmosphere pressure, is arbitrarily fixed at zero, since no means of determining the absolute value of the enthalpy of an element is available. For instance, the standard enthalpy of graphite is represented as $H^\ominus(298\,K) = 0$, but the standard enthalpy of diamond is greater than zero since diamond is more chemically reactive than graphite. Similarly, the standard enthalpy of oxygen is also zero. The standard enthalpy of carbon dioxide is less than the combined standard enthalpies of carbon (graphite) and oxygen, since heat is evolved when it is formed from its elements, i.e. the standard enthalpy of carbon dioxide is a negative quantity and is found experimentally to be $-393{\cdot}4\,kJ\,mol^{-1}$. The standard enthalpy change in this reaction is denoted as $\Delta H^\ominus(298\,K) = -393{\cdot}4\,kJ\,mol^{-1}$ and represented diagrammatically in fig. 8.2.

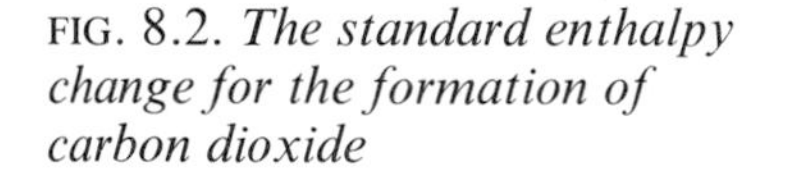

FIG. 8.2. *The standard enthalpy change for the formation of carbon dioxide*

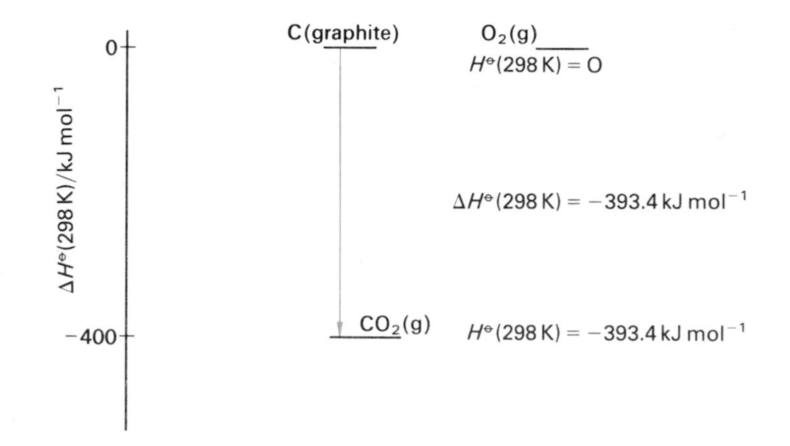

The heat of formation of a substance is defined to be:

The amount of heat involved when 1 mole of a compound is formed from its elements under standard conditions, i.e. at a specified temperature (generally 298 K) and 1 atmosphere pressure, e.g.

$$\tfrac{1}{2}H_2(g) + \tfrac{1}{2}Cl_2(g) \rightarrow HCl(g) \qquad \Delta H^\ominus(298\,K) = -92\,kJ\,mol^{-1}$$

It is possible to determine enthalpies of reaction that are difficult or impossible to carry out experimentally by applying Hess's law of constant heat summation (one particular form of the law of conservation of energy):

The change in enthalpy in a chemical reaction is the same irrespective of the number of stages involved in the reaction.

The enthalpy of formation of methane can thus be computed knowing the enthalpies of combustion of carbon, hydrogen and methane:

$$\underset{\text{graphite}}{C(s)} + 2H_2(g) \rightarrow CH_4(g) \qquad \Delta H^{\ominus}(298\,K) = x\,kJ\,mol^{-1}$$

$$\underset{\text{graphite}}{C(s)} + O_2(g) \rightarrow CO_2(g) \qquad \Delta H^{\ominus}(298\,K) = -393{\cdot}4\,kJ\,mol^{-1}$$

$$H_2(g) + \tfrac{1}{2}O_2(g) \rightarrow H_2O(l) \qquad \Delta H^{\ominus}(298\,K) = -285{\cdot}8\,kJ\,mol^{-1}$$

$$CH_4(g) + 2O_2(g) \rightarrow CO_2(g) + 2H_2O(l) \qquad \Delta H^{\ominus}(298\,K) = -890{\cdot}2\,kJ\,mol^{-1}$$

By algebraic manipulation (multiplying the third equation by 2, adding to the second equation and subtracting the fourth equation) the following result is obtained:

$$C(s) + 2H_2(g) \rightarrow CH_4(g) \quad \Delta H^{\ominus}(298\,K) = -393{\cdot}4 + (2 \times -285{\cdot}8) - (-890{\cdot}2)\,kJ\,mol^{-1}$$

Therefore the enthalpy of formation of methane is $-74{\cdot}8\,kJ\,mol^{-1}$.

8.3 Bond energies

Chemical reactions involve the breaking of bonds in the reactants and the formation of new bonds in the products; it is, therefore, necessary to have some quantitative information about the strengths of chemical bonds in order to discuss chemical reactivity. The strength of a bond joining two atoms together depends upon its environment; for instance, the energy needed to break the O—H bond in the water molecule to give O—H and H is $494\,kJ\,mol^{-1}$, while the energy input necessary to break the O—H bond in the hydroxyl radical to give an oxygen and hydrogen atom is $430\,kJ\,mol^{-1}$.

$$H_2O(g) \rightarrow OH(g) + H(g) \qquad \Delta H^{\ominus}(298\,K) = +494\,kJ\,mol^{-1}$$
$$OH(g) \rightarrow O(g) + H(g) \qquad \Delta H^{\ominus}(298\,K) = +430\,kJ\,mol^{-1}$$

Despite the fact that there is no such quantity as a constant A—B bond energy, it is possible to arrive at some average bond energies which do not deviate too widely. In the above example, the average O—H bond energy is the mean of the two separate bond dissociation energies, i.e. (494 + 430/2 or $462\,kJ\,mol^{-1}$. Average bond energies are generally quoted at 298 K and 1 atmosphere pressure; they refer to the breaking of bonds in the gaseous molecule, and they do not include the energy necessary to vaporise liquids enthalpy of vaporisation) or sublime solids (enthalpy of sublimation).

Bond energies can be determined by a variety of techniques, including calorimetry and spectroscopy. The H—H bond energy has been determined by heating hydrogen to several thousand degrees, when atomic hydrogen is formed; combination of hydrogen atoms is now allowed to occur in a calorimeter and the energy released is measured ($435{\cdot}9\,kJ\,mol^{-1}$).

The H—H bond energy is the same as the bond dissociation energy, since the breaking of one bond disrupts the whole molecule.

$$H_2(g) \rightarrow 2H(g) \qquad \Delta H^{\ominus}(298\,K) = +435{\cdot}9\,kJ\,mol^{-1}$$

The bond dissociation energy of methane, i.e. the energy needed to convert 1 mole of methane into methyl radicals and hydrogen atoms, can be determined by bombarding methane with electrons in a mass spectrometer. The energy of the electrons is gradually increased until the C—H bond is broken. Since methane contains four C—H bonds, there are four dissociation energies which are given below (the values are known only approximately):

$$CH_4(g) \rightarrow CH_3(g) + H(g) \qquad \Delta H^{\ominus}(298\,K) = +425\,kJ\,mol^{-1}$$
$$CH_3(g) \rightarrow CH_2(g) + H(g) \qquad \Delta H^{\ominus}(298\,K) = +470\,kJ\,mol^{-1}$$
$$CH_2(g) \rightarrow CH(g) + H(g) \qquad \Delta H^{\ominus}(298\,K) = +416\,kJ\,mol^{-1}$$
$$CH(g) \rightarrow C(g) + H(g) \qquad \Delta H^{\ominus}(298\,K) = +335\,kJ\,mol^{-1}$$

The average C—H bond energy is the average of the four bond dissociation energies, i.e. $(425 + 470 + 416 + 335)/4$ or $412\,kJ\,mol^{-1}$, which is in close agreement with the value obtained by a different method (see next section).

8.4 The determination of some bond energies in compounds

The C—H bond energy

In order to determine the average C—H bond energy it is necessary to know:

(a) *The enthalpy of atomisation of graphite.* This is the heat necessary to convert 1 mole of graphite into free atoms, i.e.

$$\underset{\text{graphite}}{C(s)} \rightarrow C(g) \qquad \Delta H^{\ominus}(298\,K) = +715\,kJ\,mol^{-1}$$

(b) *The enthalpy of formation of methane* (given in section 8.2)

$$\underset{\text{graphite}}{C(s)} + 2H_2(g) \rightarrow CH_4(g) \qquad \Delta H^{\ominus}(298\,K) = -75\,kJ\,mol^{-1}$$

(c) *The* H—H *bond energy* (given in section 8.3)

$$H_2(g) \rightarrow 2H(g) \qquad \Delta H^{\ominus}(298\,K) = +436\,kJ\,mol^{-1}$$

By algebraic manipulation—multiplying equation (c) by 2, adding to (a) and subtracting (b)—the following result is obtained:

$$CH_4(g) \rightarrow C(g) + 4H(g) \qquad \Delta H^{\ominus}(298\,K) = +1662\,kJ\,mol^{-1}$$

Since there are four C—H bonds in methane, the average C—H bond energy is 1662/4 or $415\,kJ\,mol^{-1}$

The C—C bond energy

This bond energy can be determined from the following data:

(a) *The enthalpy of atomisation of graphite*

$$\underset{\text{graphite}}{C(s)} \rightarrow C(g) \qquad \Delta H^{\ominus}(298\,K) = +715\,kJ\,mol^{-1}$$

(b) *The* H—H *bond energy*

$$H_2(g) \rightarrow 2H(g) \qquad \Delta H^{\ominus}(298\,K) = +436\,kJ\,mol^{-1}$$

(c) *The enthalpy of formation of ethane*

This can be determined by arguments similar to those used to evaluate the enthalpy of formation of methane (p. 101), knowing the enthalpies of combustion of carbon, hydrogen and ethane:

$$\underset{\text{graphite}}{2C(s)} + 3H_2(g) \rightarrow C_2H_6(g) \qquad \Delta H^{\ominus}(298\,K) = -85\,kJ\,mol^{-1}$$

Adding 2 × (a) and 3 × (b) and subtracting (c) gives:

$$C_2H_6(g) \rightarrow 2C(g) + 6H(g) \qquad \Delta H^{\ominus}(298\,K) = +2823\,kJ\,mol^{-1}$$

When the ethane molecule is broken down, six C—H and one C—C bonds are broken; thus by subtracting the energy necessary to break six C—H bonds in one mole of ethane the C—C bond energy is

$$2823 - (6 \times 415) = 333\,kJ\,mol^{-1}$$

By the application of similar methods it is possible to evaluate the C=C and C≡C bond energies, knowing the enthalpies of combustion of ethene and ethyne respectively.

8.5 The move towards disorder—the concept of entropy

The interaction of chemical substances involves the breaking of bonds (an endothermic process) and the forming of new ones (exothermic process). The enthalpy of the chemical system will be lowered if the process of making bonds to give product molecules is more exothermic than the process of breaking bonds in the reactants is endothermic, provided of course that the energy so released is allowed to flow out of the system into the surroundings (by the law of conservation of energy there is no overall change in energy, since the energy lost by the chemical system exactly matches that gained by the surroundings). Thus, consider the synthesis of hydrogen chloride from hydrogen and chlorine, which involves the breaking of H—H and Cl—Cl bonds and the making of H—Cl bonds, irrespective of the actual detailed mechanism involved in this reaction:

$$H_2(g) \rightarrow 2H(g) \qquad \Delta H^{\ominus}(298\,K) = +435{\cdot}9\,kJ\,mol^{-1}$$
$$Cl_2(g) \rightarrow 2Cl(g) \qquad \Delta H^{\ominus}(298\,K) = +241{\cdot}8\,kJ\,mol^{-1}$$
$$2H(g) + 2Cl(g) \rightarrow 2HCl(g) \qquad \Delta H^{\ominus}(298\,K) = -(2 \times 431)\,kJ\,mol^{-1}$$

Overall process:

$$H_2(g) + Cl_2(g) \rightarrow 2HCl(g) \qquad \Delta H^{\ominus}(298\,K) = -184{\cdot}3\,kJ\,mol^{-1}$$

The reaction is exothermic to the extent of 184·3 kJ mol^{-1} for every 2 moles of hydrogen chloride produced.

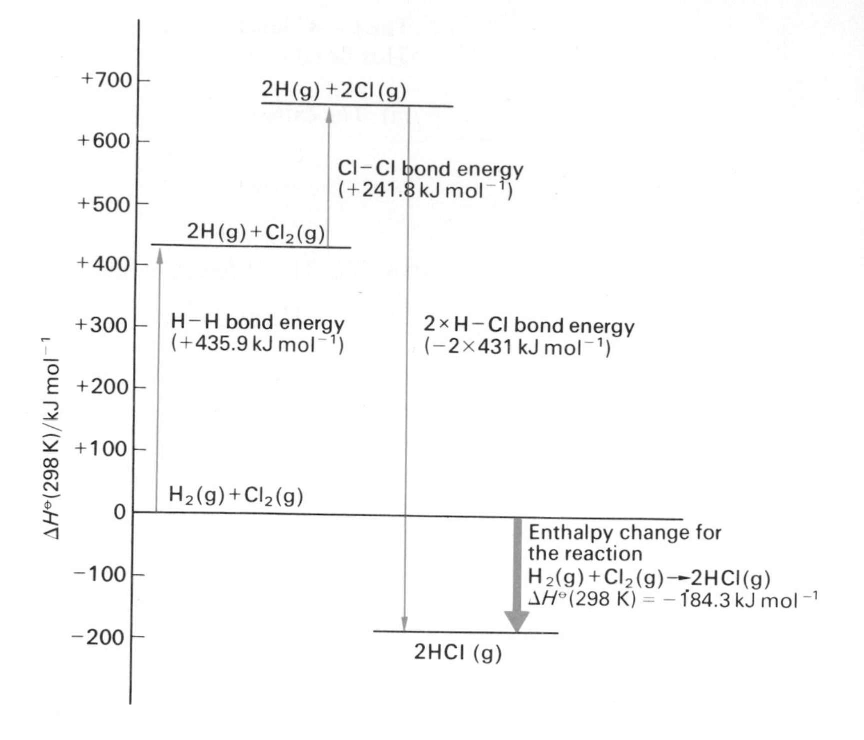

FIG. 8.3. *The standard enthalpy change for the formation of hydrogen chloride*

The tendency for change to occur in such a manner that order is lowered (or disorder is increased) is often at conflict with the move towards minimum energy (with the consequent flow of energy into the surroundings), but in one system at least, i.e. an isolated system, a move towards greater disorder becomes the all-important criterion for change to occur.

Changes occurring in an isolated system

An isolated system is one in which neither energy nor matter can enter or leave. The universe is considered to be one, and the nearest approach to an isolated system in the laboratory is a closed Dewar container.

Consider an isolated system which is divided into two equal volumes by a partition and which contains helium in one compartment and neon in the other, at the same temperature and pressure. When the partition is removed the gases diffuse throughout all the available space and become completely mixed. Since noble gases have been specified there can be little or no interaction between their atoms; furthermore, if interaction did take place to a slight extent there could not be any overall energy change since the system is an isolated one. However, the molecules of each gas can now move throughout twice the original volume or, in other words, the gases are now in a more disordered or random state. Disorder can be measured quantitatively and is given the name 'entropy'. An increase in disorder is an increase in entropy and this is denoted as ΔS, which is positive; conversely, when a decrease in entropy occurs ΔS is negative. The increase in entropy which occurs when gases mix without interaction is given by the mathematical equation:

$$\Delta S = \text{Constant} \times \ln \frac{\text{product of final volumes of each gas}}{\text{product of initial volumes of each gas}}$$

The constant is the gas constant R when 1 mole of each gas is involved; thus 1 mole of helium and 1 mole of neon, initially occupying a volume V_1 each will eventually occupy a volume $2V_1$ each after mixing and the increase in entropy will be:

$$\Delta S = R\ln\frac{2V_1 \times 2V_1}{V_1 \times V_1} = R\ln 4\,\mathrm{J\,K^{-1}\,mol^{-1}}$$

Note that the units of entropy are joules per degree, i.e. thermal units are involved even though in this example no heat transfer has taken place.

What are the chances that helium and neon will unmix? Certainly on energetic grounds alone this is not impossible, but statistically the process becomes more and more unlikely with increasing numbers of molecules.

Consider two molecules of helium labelled He_1 and He_2, and two molecules of neon labelled Ne_1 and Ne_2 'mixed' in the system of total volume $2V_1$; for 'unmixing' to occur Ne_1 must exchange with He_2 (fig. 8.4).

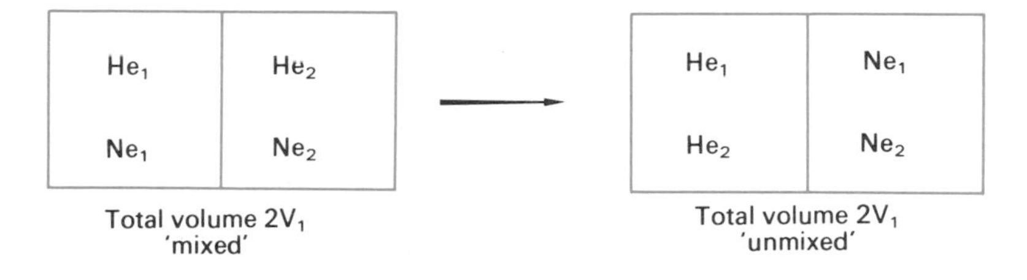

FIG. 8.4. *The probability of a mixture of helium and neon unmixing is an unlikely event statistically*

There are six different ways in which two molecules of helium and two molecules of neon can be distributed evenly in two compartments A and B and these are shown below.

Compartment A	Compartment B
He_1 He_2	Ne_1 Ne_2
He_1 Ne_1	Ne_2 He_2
He_1 Ne_2	Ne_1 He_2
He_2 Ne_1	Ne_2 He_1
He_2 Ne_2	Ne_1 He_1
Ne_1 Ne_2	He_1 He_2

In only one of these arrangements do both helium molecules appear in compartment A and the probability of the system returning to its original state is 1/6. In four of the arrangements a helium molecule is paired with one of neon and the probability of the system remaining completely mixed is 4/6 or 2/3.

Consider now four molecules of helium and four molecules of neon 'mixed' in a similar system; the probability of finding all four molecules of helium in compartment A and all four molecules of neon in compartment B is now 1/70. For eight molecules of each kind the probability reduces to 1/12 870.

The number of ways in which n molecules can be distributed so that m appear in one compartment and $(n—m)$ appear in the other is given by the formula:

$$\frac{n!}{m!(n-m)!}$$

where $n!$ means factorial n and is $n(n-1)(n-2)(n-3) \ldots\ldots 1$. In the examples we have been considering $m = n/2$ and $(n-m)$ has also been $n/2$.

If we now go back to the original problem which involved 1 mole of helium and 1 mole of neon, i.e. Avogadro's constant or approximately 6×10^{23} molecules of each kind, then the probability of 'unmixing' occurring is infinitesimally small. The most probable state is one of complete mixing (greater disorder): thus the most probable state of a system is one of high entropy.

Consider now the dissolution of ammonium nitrate in water, carried out in an isolated system. On dissolving, the highly ordered structure of the ammonium nitrate is broken down and dispersed throughout the liquid, and there is an overall increase in entropy, but the energy is unchanged since the system is an isolated one. It is instructive to examine this system in more detail as follows: the breakdown of the solid ammonium nitrate to form a solution results in an increase in entropy; but the process is endothermic and this results in the air above the solution being cooled down. The air molecules move more slowly and hence there is a decrease in the entropy of the air. A decrease in entropy also occurs owing to the water molecules being more ordered when hydration of the ammonium nitrate occurs, the ammonium and nitrate ions binding water molecules. However, there is an overall increase in entropy and dissolution of ammonium nitrate ceases when no further increase in entropy can occur.

In the above example there has been no overall change of energy since the system is an isolated one but, since the process is endothermic, thermal energy of the air has been transformed into chemical energy of the hydrated ammonium and nitrate ions. Here, and in chemical reactions in general, high entropy is associated with the maximum number of ways in which the energy of the system can be distributed.

By the quantum theory (p. 35) energy is held in a series of definite energy levels; for a given energy range there are more translational energy levels than rotational energy levels; similarly there are more rotational energy levels than vibrational energy levels. This means that there are far more ways of distributing a given energy among the various translational energy levels than among either rotational or vibrational energy levels. Thus, in a purely qualitative sense, we can visualise the increase in entropy (the spread of energy) that occurs when ammonium nitrate dissolves in water in the following manner:

(a) The solid structure of ammonium nitrate is broken down and the hydrated ions are dispersed in solution. Energy of vibration is replaced by energy of translation and there is a very large increase in the number of energy levels now available. The entropy therefore increases.

(b) The air is cooled down and this means that there is less energy to be distributed among the various energy levels (this applies to translational, rotational and vibrational energy levels but the first is the more important). The entropy therefore decreases.

(c) Water molecules are bound to the ammonium and nitrate ions, i.e. their translational and rotational motion is restricted, and there is a decrease in entropy.

However, the conversion of vibrational energy of the ammonium nitrate into translational energy of the hydrated ions results in a large increase in entropy and this completely overshadows the decrease in entropy that occurs in stages (b) and (c). The overall entropy therefore increases.

In order to visualise the increase in the number of ways in which the energy of a system can be distributed when a change occurs, let us consider a simple but purely hypothetical situation. Suppose we have 3 molecules of a perfect gas with a total energy of 2 units which can be distributed between energy levels 1 unit apart, i.e. the available energy levels can be labelled 0, 1, 2 etc. There are six ways in which these two units of energy can be distributed between the three molecules (fig. 8.5).

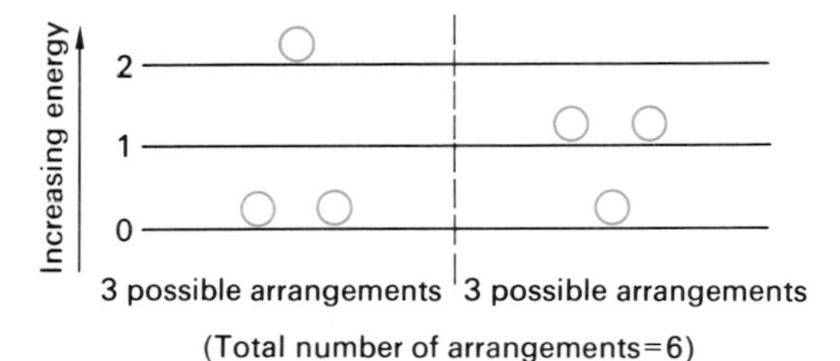

FIG. 8.5. *The possible ways in which two units of energy can be distributed between three molecules*

One molecule can be given two units of energy and the other two molecules will now have zero energy; there are three ways of arranging this. Alternatively two molecules can be given one unit of energy each, the remaining molecule having zero energy; there are also three ways of arranging this, making a total of six possible ways of distributing the available two units of energy.

Suppose we also have three molecules of a perfect gas with a total of four units of energy. There are fifteen ways in which this energy can be distributed between the three molecules (fig. 8.6).

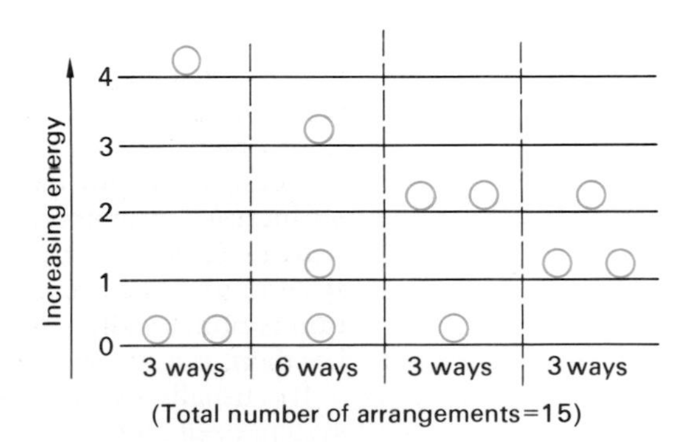

FIG. 8.6. *The possible ways in which four units of energy can be distributed between three molecules*

If the three molecules having a total energy of two units and the three molecules having a total of four units of energy are placed in separate halves of an isolated system and separated from each other by a perfectly conducting partition, experience tells us that there will be a transfer of thermal energy from one to the other. The three molecules having a total energy of two units will now gain one unit of energy at the expense of the three molecules initially having a total of four units of energy. It can easily be shown that the number of ways in which three units of energy can be distributed between three molecules is ten, and since we have two lots of such molecules the total number of possible distributions of the total energy is 10×10 or 100. Before the gases were brought into thermal contact with each other there were 6×15 or 90 ways of distributing this same amount of energy. Energy transfer takes place from one to the other because this results in an increase in the number of ways of distributing the energy.

If the thermally conducting partition is now removed there is a further increase in the number of ways of distributing the total energy. The problem now is to determine how many arrangements there are for distributing a total of six units of energy between six molecules. There are, in fact, 462 possible ways and this means that mixing of the gases will take place when the partition is removed.

We can summarise the situation as follows:

Changes take place in an isolated system in a direction such that when equilibrium is achieved the number of ways in which the total energy of the system can be distributed is a maximum (the entropy of the system is a maximum).

Changes occurring in a closed system

A closed system may be defined as one that allows a free exchange of energy between itself and the surroundings but which does not allow matter to enter or leave. A thermally conducting closed container is an example of such a system.

Consider the case of the spontaneous crystallisation of a supersaturated solution of sodium thiosulphate (obtained by heating some crystals in a test-tube until they liquefy and then plugging the mouth of the tube and allowing it to cool to room temperature undisturbed). If a small seed crystal of sodium thiosulphate is added to the supersaturated solution, a large amount of heat is released (ΔH is negative) and eventually the whole contents solidify; there is thus an increase in order in the closed system and ΔS is negative, i.e. energy is lost from the system and there are fewer ways of distributing the remaining energy among the vibrational energy levels of solid sodium thiosulphate than there were of distributing a higher energy among the translational energy levels of the hydrated sodium and thiosulphate ions in the liquid state. In this closed system there is thus a decrease in entropy but the energy released to the surroundings results in an increase in temperature of the surroundings and thus an increase in entropy here. If the closed system is considered as part of an isolated system, then the decrease in entropy which occurs during the crystallisation is numerically less than the increase in entropy of the surroundings, i.e. there is a net increase in entropy.

It is usually simpler to consider only the changes occurring in the closed system itself and, for such a system at constant temperature and pressure, a decrease in free energy must occur if a change is to take place spontaneously. A change in free energy at constant temperature and pressure is defined by the equation:

$$\Delta G = \Delta H - T\Delta S$$

where ΔH is the enthalpy change, T is the temperature Kelvin, and ΔS is the change in entropy. For spontaneous changes in a closed system, ΔG must be negative. Like enthalpy changes, free energy changes and entropy changes are additive; the notations $\Delta G^{\ominus}(298\,K)$ and $\Delta S^{\ominus}(298\,K)$ refer to changes under standard conditions, i.e. those carried out at 1 atmosphere pressure and 25°C (298 K).

Since ΔG must be negative for the spontaneous crystallisation of sodium thiosulphate to occur, ΔH must be more negative than $T\Delta S$. However, the process becomes less and less likely to occur at higher temperatures since ΔH and ΔS only vary slowly with temperature (unless there is a change of state, e.g. liquid to gas and vice versa) whereas ΔG varies rapidly because of the T factor in the expression:

$$\Delta G = \Delta H - T\Delta S$$

In fact the reverse change, i.e. the liquefaction of sodium thiosulphate, takes place at higher and higher temperatures, for this is the direction of change which results in a free energy decrease.

A change in free energy provides a measure of the maximum work that a system is capable of doing other than work due to any volume changes, e.g. expansion against an opposing atmosphere (see p. 99).

8.6 Quantitative aspects of entropy and free energy

A definition of entropy change

So far entropy has been rather loosely equated with order/disorder, an increase in disorder meaning an increase in the number of ways in which the energy of a system can be distributed amongst its energy levels and corresponding to an increase in entropy. We shall now consider the entropy change which results from a reversible and isothermal expansion of a gas, and define the entropy change in precise terms.

Consider an ideal gas initially at pressure p which can be supplied with heat and allowed to expand. If the gas is opposed by a pressure only infinitesimally less than the gas pressure at all stages of the expansion, then the maximum amount of work is extracted from the expanding gas; this is denoted by the symbol w_{rev} and the conditions are said to be reversible ones. By increasing the opposing pressure very slightly the gas can be made to contract. Since the gas is considered to be ideal, in that intermolecular attractions do not exist, and since the expansion is considered to be carried out under isothermal conditions, i.e. the temperature of the gas remaining constant, all the heat put into the gas is converted into work of expansion. For a small volume change dV the work done is pdV, and for a volume change from V_1 to V_2 the work done is w_{rev} where:

$$w_{rev} = \int_{V_1}^{V_2} p\,dV$$

For 1 mole of an ideal gas $pV = RT$, therefore

$$w_{\text{rev}} = \int_{V_1}^{V_2} \frac{RT}{V}\,dV = RT\ln\frac{V_2}{V_1}$$

The amount of heat put into the gas, q_{rev}, equals the work done, w_{rev}, therefore

$$q_{\text{rev}} = w_{\text{rev}} = RT\ln\frac{V_2}{V_1}$$

or

$$\frac{q_{\text{rev}}}{T} = R\ln\frac{V_2}{V_1}$$

The quantity q_{rev}/T is defined to be the entropy change (an increase since heat is absorbed). Thus

$$\Delta S = \frac{q_{\text{rev}}}{T} = R\ln\frac{V_2}{V_1} \qquad \text{(for 1 mole of ideal gas)}$$

If the 1 mole of ideal gas had expanded from an initial volume V_1 into a vacuum so that the final volume was V_2, then of course no heat would have been absorbed; nevertheless, the entropy change is still given by the expression q_{rev}/T, since the final state of the gas is the same irrespective of whether it expands reversibly and isothermally or into a vacuum where it does no work.

A definition of free energy change

Consider the reaction between zinc and a molar solution of hydrochloric acid:

$$\text{Zn(s)} + 2\text{H}^+\text{(aq)} \rightarrow \text{Zn}^{2+}\text{(aq)} + \text{H}_2\text{(g)}$$

This reaction can take place in a calorimeter at constant pressure in which there is a heat change and work done against the atmosphere by the expanding hydrogen gas. Alternatively the reaction can be carried out reversibly in an electrochemical cell (p. 138), when the maximum amount of work is done, w_{rev}. In this example the maximum work, w_{rev}, is the sum of the electrical work denoted as w_{useful} and the work done by the hydrogen against the atmosphere; the heat involved under these conditions is q_{rev} (see above).

Since the chemical system attains the same final state irrespective of whether the reaction takes place in a constant pressure calorimeter or in an electrochemical cell, the change in internal energy, ΔU, is the same in both cases (internal energy changes are additive like enthalpy changes and represent changes in the total amount of energy stored in chemical systems). Only differences in internal energies can be measured since there is no way of determining the absolute internal energy associated with an element. If a chemical system absorbs heat, q, and does work, w, the change in internal energy, ΔU, is given by the equation:

$$q = \Delta U + w \qquad \text{or} \qquad \Delta U = q - w$$

The above expression is a statement of the law of conservation of energy.

When zinc is reacted with molar hydrochloric acid under constant pressure, the heat involved is the enthalpy change, ΔH, and the work done is that due to opposing the external pressure, $p\Delta V$. Therefore under these conditions:

$$\Delta U = \Delta H - p\Delta V \tag{1}$$

Similarly, when zinc reacts reversibly with molar hydrochloric acid in an electrochemical cell, the heat involved is q_{rev} and the work done is w_{rev}; this latter work comprises electrical work, w_{useful}, and work done against the atmosphere, $p\Delta V$, therefore under those conditions:

$$\begin{aligned}\Delta U &= q_{rev} - w_{rev} \\ &= q_{rev} - (w_{useful} + p\Delta V)\end{aligned}$$

The entropy change ΔS is equal to q_{rev}/T (p. 110) therefore:

$$\Delta U = T\Delta S - (w_{useful} + p\Delta V) \tag{2}$$

Equating equations (1) and (2) we have:

$$\Delta H - p\Delta V = T\Delta S - (w_{useful} + p\Delta V)$$

or

$$\Delta H = T\Delta S - w_{useful}$$

Let us now define a free energy change ΔG so that a decrease in free energy, $-\Delta G$, occurs when the maximum useful work is done, then:

$$-\Delta G = w_{useful}$$

or

$$\Delta G = \Delta H - T\Delta S \tag{3}$$

If ΔG is negative for a chemical reaction, then this reaction can be made to perform useful work, i.e. the reaction is feasible; however, if ΔG is positive, no useful work can be obtained from the system, i.e. the reaction is not feasible. It is important to realise that equation (3) applies to reactions carried out at constant temperature and pressure. When reactions are done under standard conditions, i.e. at 1 atmosphere pressure and a specified temperature, the free energy change, $\Delta G^{\ominus}$, is given by the expression

$$\Delta G^{\ominus} = \Delta H^{\ominus} - T\Delta S^{\ominus}$$

When zinc reacts with molar hydrochloric acid 152·4 kJ mol^{-1} of heat are evolved for every mole of hydrogen produced. The maximum useful work done if the reaction is carried out reversibly is 147·2 kJ mol^{-1} for every mole of hydrogen produced. Since both values refer to standard conditions at 25°C (298 K),

$$\Delta H^{\ominus}(298\,\text{K}) = -152{\cdot}4\,\text{kJ mol}^{-1},\ \Delta G^{\ominus}(298\,\text{K}) = -147{\cdot}2\,\text{kJ mol}^{-1}$$

therefore,

$$-147{\cdot}2 = -152{\cdot}4 - T\Delta S^{\ominus}(298\,\text{K})$$

$$\Delta S^{\ominus}(298\,\text{K}) = \frac{+147{\cdot}2 - 152{\cdot}4}{298} = -0{\cdot}017\,45\,\text{kJ K}^{-1}\,\text{mol}^{-1}$$

Since the units of entropy are $\text{J K}^{-1}\,\text{mol}^{-1}$, the entropy change in this reaction is $-17{\cdot}45\,\text{J K}^{-1}\,\text{mol}^{-1}$. Since the value carries a negative sign, this amounts to a decrease of entropy in the chemical system.

8.7 Further considerations involving the free energy equation

A qualitative treatment of free energy

Even without the help of tabulated data, the equation

$$\Delta G = \Delta H - T\Delta S$$

can be used to correlate a diverse range of physical and chemical behaviour.

(a) Changes occurring at low temperatures

At lower and lower temperatures the factor $T\Delta S$ becomes less and less important and at sufficiently low temperatures the feasibility of change occurring is controlled by the sign of ΔH, i.e. for ΔG to be negative, ΔH must also be negative, thus exothermic reactions are the rule. Furthermore, for purely physical processes such as changes of state, lowering of temperature results in the condensation of gases and the solidification of liquids, i.e. the creation of order in the closed system. For this to occur ΔH must be more negative than $T\Delta S$, and gases and liquids evolve a large amount of heat when a change of state occurs.

(b) Changes occurring at higher temperatures

Exothermic reactions which produce disorder, for example explosive reactions, are highly favoured since ΔH is negative and ΔS is positive, resulting in ΔG being very negative. As the temperature is raised, more and more endothermic reactions become possible and these are all reactions which result in an increase in disorder. An interesting endothermic reaction which readily takes place at room temperature is that between hydrated cobalt (II) chloride and thionyl chloride:

$$CoCl_2{\cdot}6H_2O(s) + 6SOCl_2(l) \longrightarrow CoCl_2(s) + 6SO_2(g) + 12HCl(g)$$

In the above reaction there is a large increase in entropy as the gases sulphur dioxide and hydrogen chloride are released.

At extremely high temperatures all chemical bonds can be broken and endothermic reactions are the rule.

(c) Changes occurring with little or no heat change

The diffusion of gases is an example of this change. This process is possible since ΔS is positive, i.e. a greater freedom of movement. When $\Delta H = 0$, i.e. for the diffusion of the noble gases, the free energy equation becomes

$$\Delta G = -T\Delta S$$

A quantitative treatment of free energy

Tabulated thermodynamical data for many chemicals is now available, i.e. standard enthalpies and entropies, from which standard free energy changes can be calculated. The feasibility of a reaction occurring under standard conditions can thus be readily determined; of course for reactions at pressures other than 1 atmosphere different data must be used. We shall examine one reaction taking place under standard conditions and at 298 K, e.g. that between a molar solution of copper (II) ions and zinc:

	$Zn(s)$ +	$Cu^{2+}(aq)$	$Zn^{2+}(aq)$ +	$Cu(s)$	$T = 298\,K$
$H^{\ominus}(298\,K)$	0	64·39	−152·4	0	$kJ\,mol^{-1}$
$S^{\ominus}(298\,K)$	41·63	−98·74	−106·5	33·31	$J\,K^{-1}\,mol^{-1}$

$$\Delta G^{\ominus} = \Delta H^{\ominus} - T\Delta S^{\ominus}$$

$$\Delta G^{\ominus}(298\,K) = (-152{\cdot}4 - 64{\cdot}39) - 298/1000$$

$$\{(-106{\cdot}5 + 33{\cdot}31) - (41{\cdot}63 - 98{\cdot}74)\}$$

$$= -216{\cdot}8 + 4{\cdot}77 = -212{\cdot}0$$

$$\Delta G^{\ominus}(298\,K) = -212{\cdot}0\,kJ\,mol^{-1}$$

Since $\Delta G^{\ominus}(298\,K)$ is negative, the reaction is feasible; for the reverse reaction $\Delta G^{\ominus}(298\,K)$ would be $+212{\cdot}0\,kJ\,mol^{-1}$, i.e. the reverse reaction is not possible under these conditions.

The changes in $\Delta H^{\ominus}$, $T\Delta S^{\ominus}$, and $\Delta G^{\ominus}$ at 298 K for this reaction are shown in fig. 8.7.

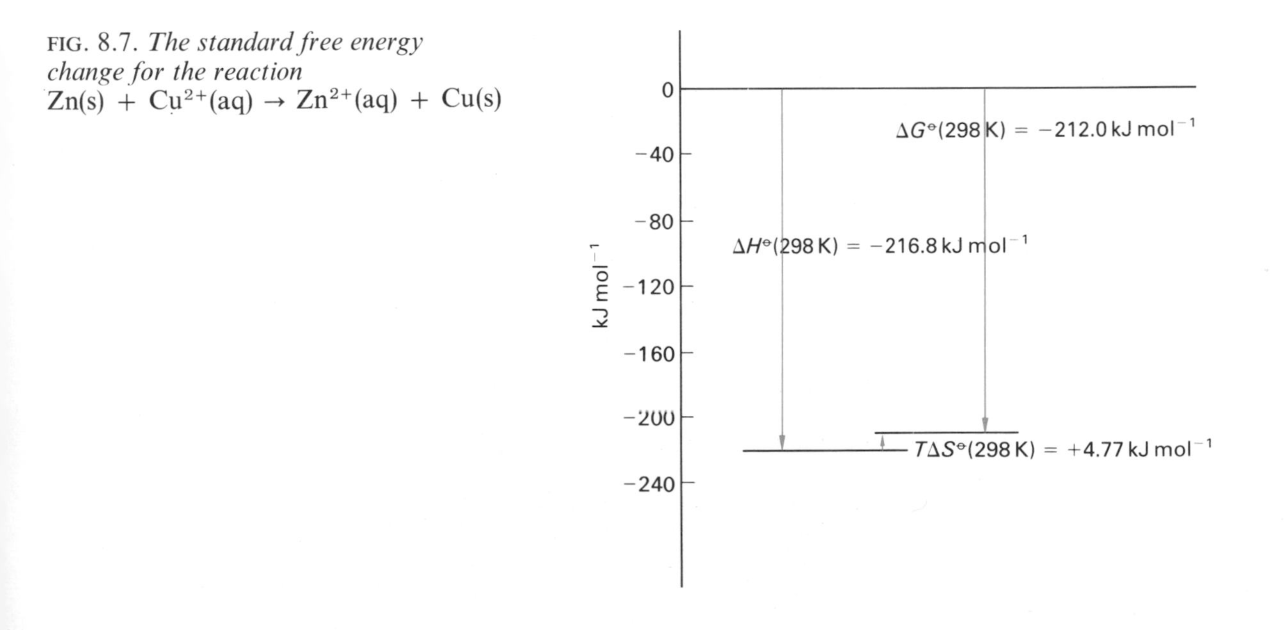

FIG. 8.7. *The standard free energy change for the reaction* $Zn(s) + Cu^{2+}(aq) \rightarrow Zn^{2+}(aq) + Cu(s)$

8.8 The effect of pressure changes on free energy changes—reversible reactions

When reactions between gases take place there is often a change of pressure and it is necessary to examine how free energy changes are influenced by changes in pressure. We begin with the equation:

$$\Delta G = \Delta H - T\Delta S$$

where ΔG represents the change G_1 to G_2, ΔH represents the change H_1 to H_2, etc. Thus:

$$(G_2 - G_1) = (H_2 - H_1) - T(S_2 - S_1)$$

or

$$G_2 = H_2 - TS_2 \qquad \text{or} \quad G = H - TS \text{ (subscripts omitted)}$$

Similarly, the equation $\Delta H = \Delta U + p\Delta V$ can be expressed in the form:

$$(H_2 - H_1) = (U_2 - U_1) + p(V_2 - V_1)$$

or

$$H_2 = U_2 + pV_2 \qquad \text{or} \quad H = U + pV \text{ (subscripts omitted)}$$

Combining the equations $G = H - TS$ and $H = U + pV$ we have:

$$G = U + pV - TS$$

Suppose now there is an infinitesimal change in free energy at constant temperature, due to similar infinitesimal changes in U, p, V, and S. Using the usual calculus notation this can be represented as:

$$dG = dU + pdV + Vdp - TdS \qquad (1)$$

For a reversible change in which the only work done is that against an opposing pressure we have:

$$dU = q_{rev} - pdV$$

or

$$dU = TdS - pdV \text{ (since } q_{rev} = TdS)$$

Substituting for dU in equation (1) we have:

$$dG = Vdp \text{ (this equation applies at constant temperature)}$$

Consider 1 mole of an ideal gas initially at pressure p_1 changing to pressure p_2; the change in free energy can be represented as a change from G_1 to G_2 thus:

$$G_2 - G_1 = \int_{p_1}^{p_2} Vdp$$

$$= \int_{p_1}^{p_2} \frac{RT}{p}\,dp \quad \text{(since } pV = RT)$$

$$= RT \ln \frac{p_2}{p_1}$$

For a pressure change from $p_1 = 1$ to p the change in free energy becomes:

$$G - G^{\ominus} = RT\ln p \qquad \text{(where } G^{\ominus} \text{ is the standard free energy)}$$

Similarly for n moles of ideal gas the free energy change for a pressure change from 1 atmosphere to a pressure p is:

$$n(G - G^{\ominus}) = nRT\ln p = RT\ln p^n \tag{2}$$

Let us now consider a specific gaseous reversible reaction, e.g. the equilibrium set up between dinitrogen tetroxide and nitrogen dioxide:

$$N_2O_4(g) \rightleftharpoons 2NO_2(g)$$

From equation (2) it can be seen that the free energy of n moles of ideal gas is:

$$nG = nG^{\ominus} + RT\ln p^n$$

When dinitrogen tetroxide is converted into nitrogen dioxide, 1 mole of the former gas gives 2 moles of the latter gas, thus:

$$2G_{NO_2} = 2G^{\ominus}{}_{NO_2} + RT\ln p^2{}_{NO_2}$$

$$G_{N_2O_4} = G^{\ominus}{}_{N_2O_4} + RT\ln p_{N_2O_4}$$

where p_{NO_2} and $p_{N_2O_4}$ are the partial pressures of nitrogen dioxide and dinitrogen tetroxide respectively in the gas mixture. The free energy change in this reaction ΔG is

$$2G_{NO_2} - G_{N_2O_4} = 2G^{\ominus}{}_{NO_2} - G^{\ominus}{}_{N_2O_4} + RT\ln p^2{}_{NO_2} - RT\ln p_{N_2O_4}$$

or

$$\Delta G = \Delta G^{\ominus} + RT\ln p^2{}_{NO_2}/p_{N_2O_4} \tag{3}$$

If ΔG is negative the reaction will proceed; the partial pressure of nitrogen dioxide will increase and the partial pressure of dinitrogen tetroxide will decrease. A stage is eventually reached when $\Delta G = 0$, i.e. the system is in equilibrium; under these conditions the partial pressures of nitrogen dioxide and dinitrogen tetroxide are their equilibrium pressures, represented as p'_{NO_2} and $p'_{N_2O_4}$ respectively:

At equilibrium $\Delta G = 0 \quad \therefore \Delta G^{\ominus} = -RT\ln p'^2{}_{NO_2}/p'_{N_2O_4}$

or $\qquad \Delta G^{\ominus} = -RT\ln K_p$ where K_p is the equilibrium constant in terms of partial pressures.

Equation (3) is applicable to all gaseous reactions provided that the gases behave ideally; thus for the reaction

$$aA + bB + \ldots \rightleftharpoons yY + zZ + \ldots.$$

$$\Delta G = \Delta G^{\ominus} + RT\ln(p^y_Y p^z_Z)/(p^a_A p^b_B) \tag{4}$$

8.9 A more detailed examination of a reversible reaction

It turns out that equation (4) is applicable to reactions that take place in solution; under these conditions the concentrations of the substances present are expressed in moles per unit volume.

Consider the reaction between ethanoic acid and ethanol to form ethyl ethanoate and water. The equilibrium constant K for this reaction at 25°C is approximately 4.

$$CH_3COOH(l) + C_2H_5OH(l) \rightleftharpoons CH_3COOC_2H_5(l) + H_2O(l) \quad (K = 4)$$

The free energy change in this reaction is given by the expression:

$$\Delta G = \Delta G^{\ominus} + 2{\cdot}303RT\log_{10}\frac{[CH_3COOC_2H_5][H_2O]}{[CH_3COOH][C_2H_5OH]} \quad (5)$$

At equilibrium $\Delta G = 0$ and the concentrations of the substances present are their equilibrium concentrations, thus:

$$\Delta G^{\ominus} = -2{\cdot}303RT\log_{10}4$$

i.e. the logarithmic term reduces to the logarithm of the equilibrium constant which is $K = 4$. Therefore,

$$\begin{aligned}\Delta G^{\ominus} &= -2{\cdot}303 \times 8{\cdot}31 \times 298 \times \log_{10}4\\ &= -3473\,\text{J mol}^{-1} \quad \text{or} \quad -3{\cdot}473\,\text{kJ mol}^{-1}\end{aligned}$$

Equation (5) therefore becomes:

$$\Delta G = -3{\cdot}473 + 5{\cdot}73\log_{10}\frac{[CH_3COOC_2H_5][H_2O]}{[CH_3COOH][C_2H_5OH]} \quad (6)$$

Consider a number of reaction systems made up to contain the same concentrations of ethanoic acid and ethanol and the same concentrations of ethyl ethanoate and water (the total concentrations of all four substances being 4 mol dm^{-3}) as follows:

Conc./mol dm^{-3}							
$[CH_3COOH] = [C_2H_5OH]$	0·25	0·50	0·75	1·00	1·25	1·50	1·75
$[CH_3COOC_2H_5] = [H_2O]$	1·75	1·50	1·25	1·00	0·75	0·50	0·25

The values of ΔG can easily be evaluated for each mixture, using equation (6), and are given below for the reaction:

$$CH_3COOH(l) + C_2H_5OH(l) \rightarrow CH_3COOC_2H_5(l) + H_2O(l)$$

Conc./mol dm^{-3}							
$[CH_3COOH]$	0·25	0·50	0·75	1·00	1·25	1·50	1·75
$[C_2H_5OH]$	0·25	0·50	0·75	1·00	1·25	1·50	1·75
$[CH_3COOC_2H_5]$	1·75	1·50	1·25	1·00	0·75	0·50	0·25
$[H_2O]$	1·75	1·50	1·25	1·00	0·75	0·50	0·25
ΔG/kJ mol^{-1}	+6·23	+2·01	−0·96	−3·47	−5·98	−8·95	−13·18

A positive value for ΔG means that the reaction will not proceed, i.e. a mixture containing 0·50 mol dm^{-3} each of ethanoic acid and ethanol, and 1·50 mol dm^{-3} each of ethyl ethanoate and water, will not react to form more ethyl ethanoate and water. A zero value for ΔG means that the reaction has reached equilibrium (inspection of the graph shows that the concentrations of ethanoic acid and ethanol are each 0·67 mol dm^{-3}, and the concentrations of ethyl ethanoate and water are each 1·33 mol dm^{-3} (fig. 8.8). A negative value of ΔG means that the reaction will proceed to form more ethyl ethanoate and water.

If the reaction of ethyl ethanoate and water to form ethanoic acid and ethanol is considered, the value of ΔG will be numerically the same with the sign reversed; this is the case since the equilibrium constant (considered from this point of view) is the reciprocal of $K = 4$, used above, i.e. it is $\frac{1}{4}$.

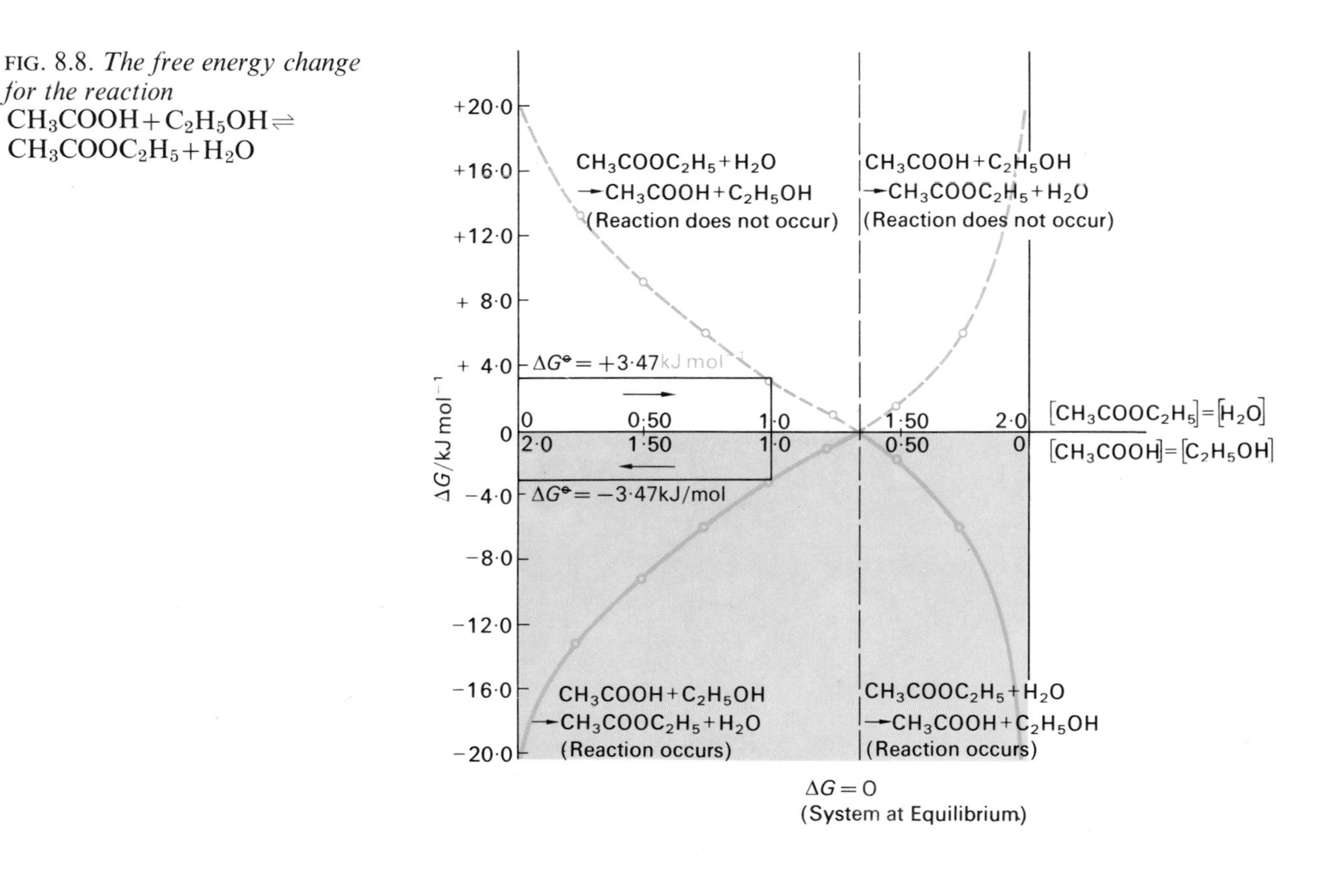

FIG. 8.8. *The free energy change for the reaction* $CH_3COOH + C_2H_5OH \rightleftharpoons CH_3COOC_2H_5 + H_2O$

8.10 Suggested reading

J. A. Campbell, *Why do Chemical Reactions Occur?*, Prentice-Hall, 1965.

J. E. Spice, *Teaching Thermodynamics to Sixth-formers*, Education in Chemistry, No. 1, Vol. 3, 1966.

L. E. Strong & W. J. Stratton, *Chemical Energy*, Van Nostrand Reinhold, 1966.

W. E. Dasent, *Inorganic Energetics*, Penguin, 1970. Second edition, Cambridge University Press 1982.

G. F. Liptrot, J. J. Thompson and G. R. Walker, *Modern Physical Chemistry*, Bell & Hyman, 1982.

8.11 Questions on chapter 8

1 Define heat of combustion and heat of formation. State Hess's law

(a) For the reaction

$$3C_2H_2(g) \longrightarrow C_6H_6(g)$$

calculate the heat of reaction, stating whether heat is absorbed or evolved.

(b) Calculate the heat of combustion of ethyne gas.

You are given the following heats of formation in kJ/mol: $C_6H_6(g)$, 82·9 absorbed; $CO_2(g)$, 393·4 evolved; $C_2H_2(g)$, 226·8 absorbed; $H_2O(g)$, 241·8 evolved. (O & C)

2 The heat of formation of the C—C link (as in ethane) is —341·4 kJ, and that of the C = C link (as in ethene) is −610·9 kJ, and that of the C ≡ C link (as in ethyne) is −803·7 kJ. How would you expect the reactivity of these three hydrocarbons to differ in the light of these data? (O & C)

3 Use the following table of bond-energies (in kJ/mol) to predict whether C_2H_2, C_2H_4, N_2 and P_2 should form polymers (including dimers, trimers, etc.). Do your predictions agree with experiment?

	C—H	C—C	N—N	P—P
Single bond	414	347	159	213
Double bond	—	611	418	335
Triple bond	—	837	946	485

(Oxford Schol.)

4 If chemical change always proceeds in the direction of greater stability, why is it that endothermic reactions occur at all? Illustrate your answer with some specific examples. (Oxford Schol.)

5 Make what qualitative deductions you can from the following list of absolute entropies at 298 K. ($S^{\ominus}$ in $J\,K^{-1}\,mol^{-1}$)

(*a*)	$F_2(g)$	$Cl_2(g)$	$Br_2(g)$	$I_2(g)$
	203·3	223·0	245·6	260·7
			$Br_2(l)$	$I_2(s)$
			152·3	116·8
(*b*)	$H_2O(g)$	$H_2S(g)$	$H_2Se(g)$	$H_2Te(g)$
	188·7	205·9	221·4	234·3
	$H_2O(l)$			
	69·87			
(*c*)	$N_2(g)$	$H_2(g)$	$NH_3(g)$	
	191·6	130·5	192·5	

In the case of (*c*) also calculate the standard entropy change when 1 mole of ammonia is formed from its elements at 1 atmosphere pressure and 298 K. Explain your result.

6 Calculate the standard free energy change for the following reaction at 1 atmosphere and 298 K from the following data.

	$H_2(g)$ +	$Cl_2(g)$ ⟶	$2HCl(g)$	
$H^{\ominus}(298\,K)$	0	0	−92·5	$kJ\,mol^{-1}$
$S^{\ominus}(298\,K)$	131	223	187	$J\,K^{-1}\,mol^{-1}$

7 The neutralisation of aqueous solutions of strong acids by strong alkalis can be represented by the equation:

$H^+(aq) + OH^-(aq) \longrightarrow H_2O(l) \qquad \Delta H^{\ominus}(298\,K) = -56{\cdot}1\,kJ\,mol^{-1}$

The standard entropies of $H^+(aq)$, $OH^-(aq)$ and $H_2O(l)$ are respectively 0, −10·5 and 69·9 $J\,K^{-1}\,mol^{-1}$.

Calculate the standard entropy change $\Delta S^{\ominus}(298\,K)$, for this reaction and comment on, and attempt to explain, the sign attached to this entropy change. Hence calculate the standard free energy change $\Delta G^{\ominus}(298\,K)$, for the reaction.

Note: The standard entropy of $H^+(aq)$ is arbitrarily assigned a value of zero and the standard entropies of other hydrated ions are related to this.

8 Calculate the standard free energy change, $\Delta G^{\ominus}(298\,K)$ and the equilibrium constant for the following reaction:

	$2SO_2(g)$	$+\ O_2(g)$	$\rightleftharpoons\ 2SO_3(g)$
$H^{\ominus}(298\,K)$	$-296{\cdot}9$	0	$-395{\cdot}2\,kJ\,mol^{-1}$
$S^{\ominus}(298\,K)$	$248{\cdot}5$	$205{\cdot}0$	$256{\cdot}3\,J\,K^{-1}\,mol^{-1}$

9 The free energy change for the reaction

$$C(s) + H_2O(g) \rightleftharpoons CO(g) + H_2(g)$$

at 1 atmosphere pressure is given by the equation:

$$\Delta G = \Delta H - T\Delta S$$

The enthalpy change, ΔH, for the reaction is very nearly constant at a value of $+126$ $kJ\,mol^{-1}$, while $T\Delta S$ increases nearly linearly from a value of 0 at 0 K to $+502\,kJ\,mol^{-1}$ at 3500 K. By graphical plotting determine the approximate temperature at which the reaction can occur.

10 The equilibrium constant for the reaction

$$CO_2(g) + H_2(g) \rightleftharpoons CO(g) + H_2O(g)$$

is 0·64 at 700°C (973K). Determine the standard free energy change for the forward reaction at this temperature.

A number of reaction systems are made up to contain the same partial pressures of carbon dioxide and hydrogen and the same partial pressures of carbon monoxide and steam (total pressure being 4 atmospheres) as follows:

$p_{CO} = p_{H_2O}$	0·25	0·50	0·75	1·00	1·25	1·50	1·75
$p_{CO_2} = p_{H_2}$	1·75	1·50	1·25	1·00	0·75	0·50	0·25

Determine the free energy change at each series of partial pressures, and then plot these values against the composition of the reaction mixture. Mark in the region where the forward reaction can occur and the region where the backward reaction is feasible.

(Hint: The problem is similar to the one worked out in section 8.9. p. 116).

Chapter 9

Chemical reactions—factors that influence their speed

9.1 Introduction

At the end of the last chapter it was shown that reactions in which there is a large evolution of heat and an increase in disorder are highly likely to occur; yet glyceryl trinitrate shows no tendency to decompose if it is treated with due respect, and coal shows no signs of reacting with the oxygen in the air at room temperature. However, glyceryl trinitrate decomposes with spectacular success if it is dropped and once the combustion of coal is under way the reaction is self-sustaining. In these two cases some energy is needed to initiate reaction even though the reactants are thermodynamically unstable with respect to their products. At the other extreme there are reactions that take place so rapidly that to all intents and purposes they can be considered to be instantaneous; thus silver chloride precipitates as fast as silver and chloride ions are mixed. In between these two extremes there is a complete spectrum of reaction velocities, and the branch of chemistry that attempts to explain these observations is known as reaction kinetics.

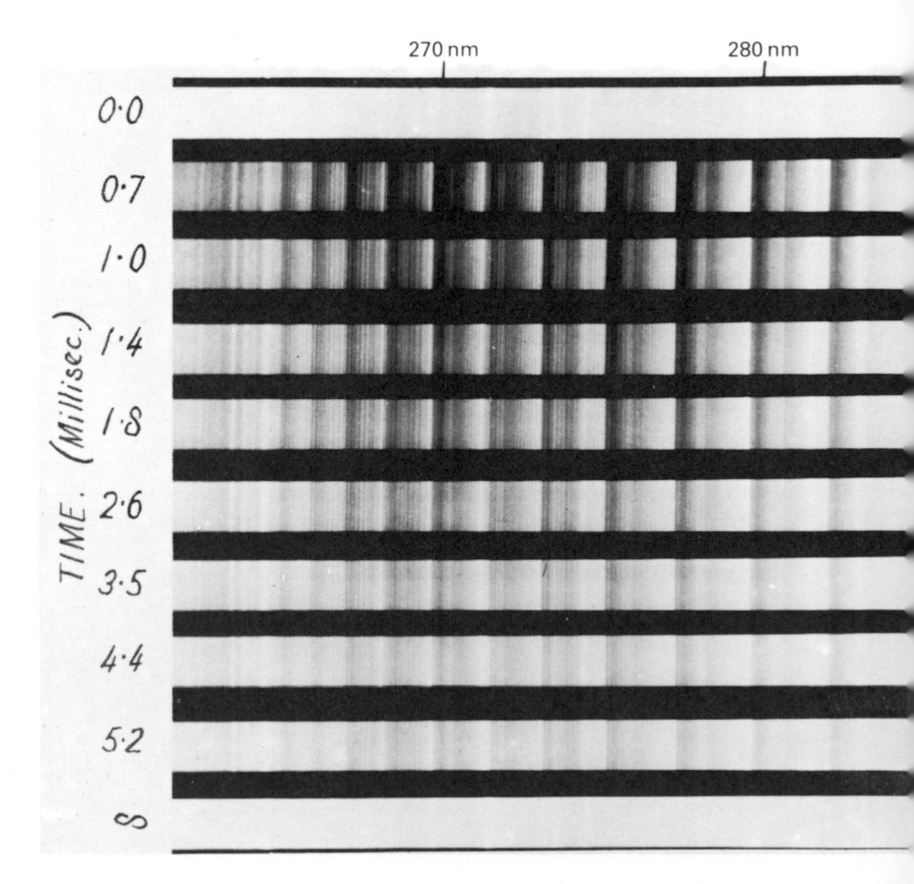

Plate 4. The technique known as flash photolysis is used for studying fast reactions. The photograph shows a sequence of spectra of the radical ClO, after flash photolysis of a chlorine/oxygen mixture, which shows bimolecular decay (By courtesy of Professor G. Porter, Royal Institution)

9.2 The rates of chemical reactions

Chemical reactions seldom take place in accordance with the stoichiometric equation and generally reaction proceeds through a number of stages in which one step—the slowest—controls the rate at which reactants are consumed and products formed. However, for the purpose of this discussion we shall limit ourselves to one reaction, that between hydrogen and iodine in the gas phase. The stoichiometric equation is:

$$H_2(g) + I_2(g) \longrightarrow 2HI(g)$$

Experiments show that, when due regard is taken of the reverse reaction, owing to the decomposition of hydrogen iodide, this reaction proceeds at a rate which is proportional to the concentration of hydrogen and to the concentration of iodine:

$$\text{Rate of reaction} \propto \text{Conc. of } H_2 \times \text{Conc. of } I_2$$
$$\text{or Rate of reaction} \propto [H_2][I_2]$$

where $[H_2]$ and $[I_2]$ are the concentrations expressed either in terms of moles per dm^3 or in terms of partial pressures. It follows therefore that:

$$\text{Rate of reaction} = k[H_2][I_2] \qquad (1)$$

where k is a constant at a particular temperature and is known as the rate constant for the reaction. The rate of the reaction can either be measured by observing the decrease in concentration of hydrogen or iodine (which are of course the same) or by observing the increase in concentration of hydrogen iodide:

$$-d[H_2]/dt = -d[I_2]/dt = \tfrac{1}{2}d[HI]/dt = k[H_2][I_2]$$

the negative sign implying a decrease in concentration and d/dt being the usual calculus notation for rate of change. It is important to realise that the rate equation by itself tells us little about the mechanism of the reaction.

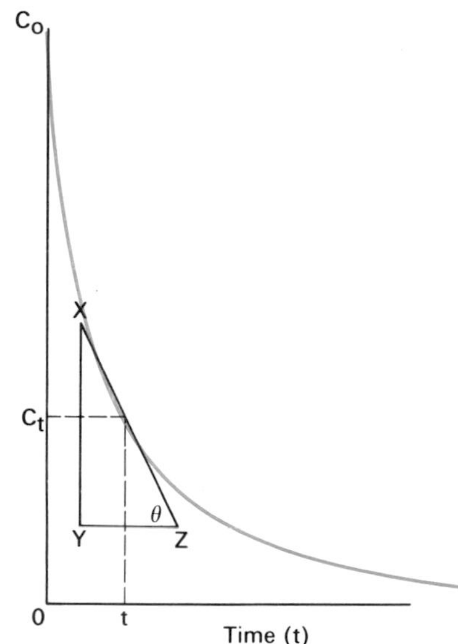

FIG. 9.1. *A typical rate of reaction curve*

C_o = *initial conc. of* H_2
C_t = *Conc. of* H_2 *at time t*
Rate of reaction at time t
= *slope of curve at time t*
$= \frac{XY}{YZ} = \tan\theta$
The slope of the curve is negative as the concentration of hydrogen is decreasing.

Rates of reaction naturally fall off with time and results can be expressed graphically (fig. 9.1). The rates of reaction at any particular time can be obtained by drawing tangents to the curve.

If the rate of reaction at varying times is determined by drawing a series of tangents to the curve, and the results are substituted in equation (1) the rate constant of the reaction at a particular temperature can be evaluated. Since the rate of reaction between hydrogen and iodine to give hydrogen iodide (after applying corrections for the back reaction) depends directly on the concentration of hydrogen and of iodine, the reaction is said to be first order with respect to hydrogen and first order with respect to iodine; the reaction is said to be second order overall. The order of a reaction is an experimentally determined quantity and seldom has any relation to the stoichiometric equation summarising the overall change, except for a few simple reactions such as this one. The order of a reaction need not be a whole number. Reactions having a fractional order are known.

9.3 The collision theory of reaction rates

This theory is based on the reasonable assumption that chemical reactions take place as a result of reacting molecules colliding. The kinetic theory of gases, however, shows that the frequency of collision is so high that reactions would be virtually instantaneous if every collision led to reaction. The suggestion was therefore made by Arrhenius that only molecules which possess more than a certain critical energy, called the activation energy, are able to react.

The kinetic theory of gases shows that the kinetic energies of gas molecules cover a very wide range. Some have very small, others have intermediate, while a very few have very high energies. It is not difficult to see in a qualitative manner why this should be so for, suppose that, at a particular instant, all the molecules in a gas had the same kinetic energy, random collisions would tend to speed up some molecules and slow others down and in no time at all a complete range of speeds, and hence kinetic energies, would result. The spread of energies follows the Maxwell-Boltzmann distribution shown for two different temperatures in fig. 9.2.

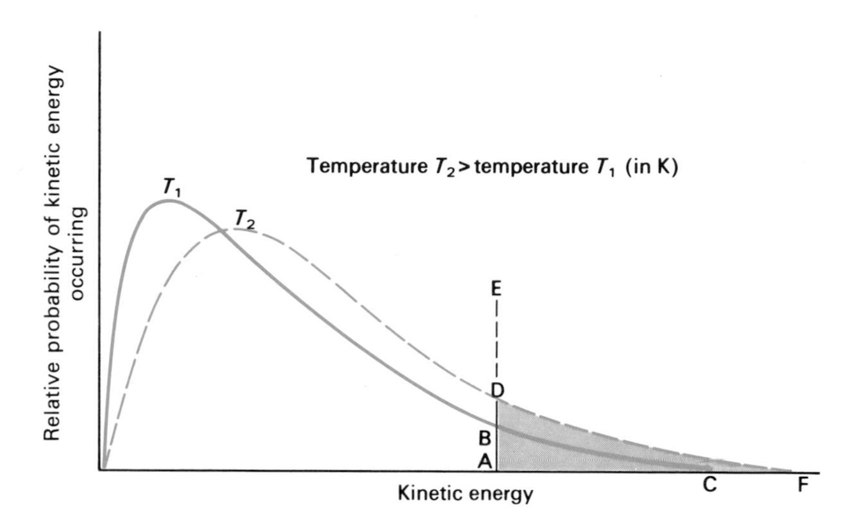

FIG. 9.2. *The Maxwell-Boltzmann distribution of kinetic energies at two different temperatures*

The number of molecules having energies greater than E will be represented by the area (ABC) at temperature T_1 and by the area ADF at temperature T_2, i.e. the number of molecules having high energies increases markedly as the temperature rises. This result can be expressed mathematically by the equation:

$$n = n_0 e^{-E/RT} \qquad (2)$$

where n_0 is the total number of molecules and n is the number of molecules with energies greater than E.

By combining this equation (2) with the rate equation (1) (p. 121) it is possible to arrive at the Arrhenius equation; thus suppose the concentrations of hydrogen and iodine are kept constant at one mole per dm^3; then the rate of reaction is numerically equal to the value of the rate constant, but the rate of reaction is also proportional to the number of molecules with energies at least equal to the activation energy E. Thus:

$$k \propto n_0 e^{-E/RT}$$

or

$$k = Ae^{-E/RT}$$

where A is a constant.

When expressed logarithmically the Arrhenius equation is:

$$\ln k = \text{const.} - E/RT$$
$$\text{or} \quad 2{\cdot}303\log_{10}k = \text{const.} - E/RT$$

A plot of $\log_{10}k$ against the reciprocal of the temperature K ($1/T$) should be a straight line whose slope is $E/2{\cdot}303R$, from which the energy of activation E can be evaluated (fig. 9.3).

FIG. 9.3. *The Arrhenius plots for the hydrogen/iodine reaction*

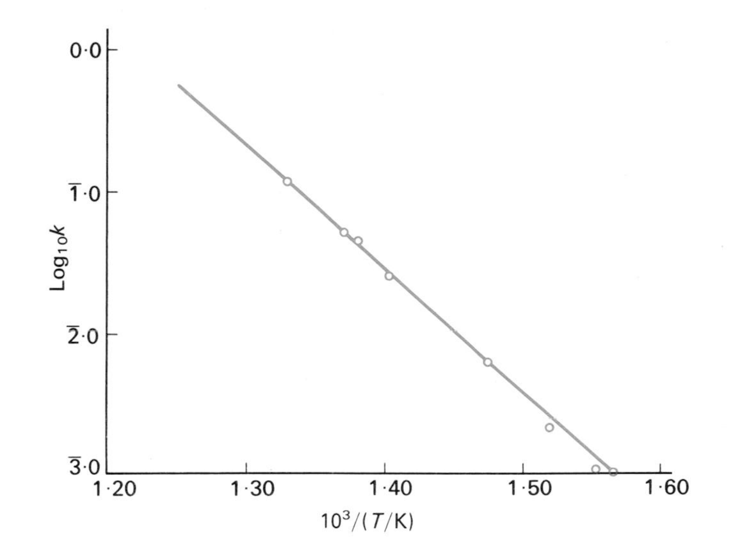

The Arrhenius plots for the hydrogen/iodine reaction (fig. 9.3) give a very good straight-line fit, the activation energy for this reaction being $165{\cdot}3\ \text{kJ mol}^{-1}$ which can be visualised as an energy barrier to be surmounted before reaction can proceed (fig. 9.4).

FIG. 9.4. *An energy barrier must be overcome before reactions can take place*

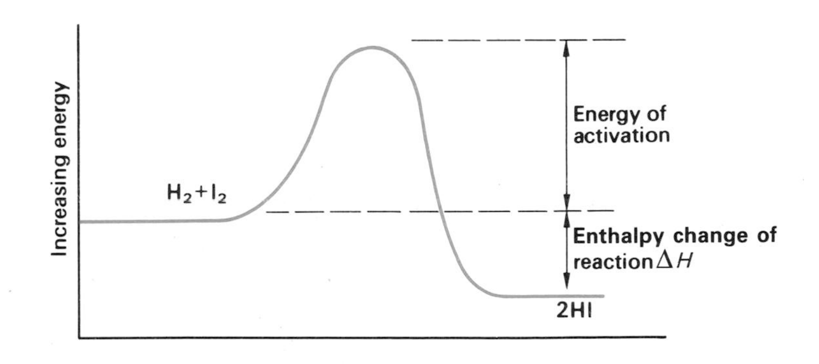

Note that for the reverse reaction, which is endothermic, the activation energy is larger by the amount ΔH:

$$E_{\text{reverse}} = E_{\text{forward}} + \Delta H$$

A crude analogy that might help in appreciating the concept of activation energy is the problem of how to transfer water over an intervening barrier from a higher to a lower level. The obvious solution is to use a siphoning tube; once some of the water has been raised over the barrier and energy thus expended, the rest flows spontaneously from the higher to the lower potential energy level.

The collision theory when developed fully is able to justify the value of A—the constant in the Arrhenius equation—on the assumption that bimolecular collisions are the prerequisite for chemical reaction. While theory and experiment show a reasonable measure of agreement for some simple gaseous reactions, and even for some that take place in solution, there are many reactions where rates differ by many orders of magnitude from those predicted by theory. Notwithstanding this discrepancy between theory and experiment as regards the A factor in the Arrhenius equation, the equation itself is obeyed by all simple one-stage reactions, and any marked deviation is interpreted as evidence of complex reactions.

9.4 The transition state theory of reaction rates

This theory is more sophisticated than the one just described and in particular focuses attention on the processes that are likely to occur just prior to reaction. Chemical reactions are assumed to take place via a transition state in which reactants come together as an activated complex in a particular orientation. The essential idea in this theory is that bond breaking and making are not instantaneous processes but occur continuously and simultaneously. For instance, the reaction between a hydrogen molecule and a chlorine atom is considered to occur by a gradual stretching of the H—H bond when two favourably orientated species come together to form an activated complex. The H------Cl bond in the complex gradually shortens and the complex decomposes to give an H—Cl molecule.

$$\mathrm{H-H+Cl} \rightleftharpoons \underset{\substack{\text{activated complex}\\ \text{in equilibrium with}\\ \text{the reactants}}}{\mathrm{H{-}{-}{-}{-}H{-}{-}{-}{-}Cl}} \rightarrow \mathrm{H+H-Cl}$$

This theory, when fully developed, leads to an equation of the form:

$$k = \text{const. } e^{-\Delta G_c/RT}$$

where the term ΔG_c is the free energy change in forming the activated complex from the reactants; but, since $\Delta G_c = \Delta H_c - T\Delta S_c$ (see previous chapter) where ΔH_c and ΔS_c are respectively the enthalpy change and entropy change in forming the activated complex from the reactants, the equation can be written:

$$k = \text{const. } e^{\Delta S_c/R}.\ e^{-\Delta H_c/RT}$$
$$\simeq \text{const. } e^{\Delta S_c/R}.\ e^{-E/RT}$$

Thus the constant A in the Arrhenius equation is replaced by the term:

$$\text{const. } e^{\Delta S_c/R}$$

For gas reactions involving two reactants, the change in entropy ΔS_c in forming the activated complex is usually negative and often large, since the formation of the complex involves the association of molecules and hence an increase in order.

Its advantage over the collision theory lies in the fact that the transition state theory gives reasonable quantitative agreement with experiment for simple reacting systems, and even for more complicated systems the exponential entropy factor is able to account for variations in rate—even if only in a qualitative manner.

The effect of temperature on reaction rates

9.5 The effect of temperature and catalysts on reaction rates

Many common reactions have an energy of activation in the region of $85\,kJ\,mol^{-1}$. Simple substitution in the Arrhenius equation predicts that in such cases a rise in temperature of 10° in the range 20°—60°C should approximately double the rate constant and hence the rate of reaction; this is in fact true for many reactions. A reaction carried out at 60°C should therefore go 2^4 or 16 times faster than at 20°C. It is little wonder that exothermic reactions carried out in a confined space very quickly get out of hand: since the heat cannot pass out of the system fast enough, the temperature rises and the reaction accelerates alarmingly.

The same principle is used in reverse by the frozen food industries. In extreme cold not only are bacteria dormant, but the aging processes that cause food to deteriorate are slowed down approximately 2^5 or 32 times on reducing the temperature from 20°C to −30°C. Nevertheless the aging processes are not arrested altogether and food will still go bad in a deep-freeze if it is kept there long enough.

The effect of catalysts on reaction rates

A catalyst is a substance which speeds up a chemical reaction and is reformed at the end; for reversible reactions it accelerates both forward and backward reactions to the same extent, and it cannot initiate a reaction that is shown to be thermodynamically impossible.

No one theory of catalytic activity can hope to explain the wide variety of such reactions but, essentially, catalysts operate by lowering the energy of activation. Reactions at a catalytic surface are thought to involve the intermediate formation of weak catalyst-reactant bonds which have the effect of loosening up the bonds in the reactants prior to reaction. Many catalytic reactions taking place in homogeneous solution (both catalyst and reactants in one phase) have been interpreted in terms of the formation of an intermediate compound as depicted in fig. 9.5.

9.6 What makes chemical reactions occur

Any research worker confronted with the problem of attempting to synthesise a new compound is immediately confronted with two questions: Is the reaction theoretically possible? If the answer is yes, he is encouraged to ask, 'Will it actually take place?' The answer to the first question was provided in chapter 8, where it was shown that in a closed system—the conditions under which a chemist normally operates—reactions are possible if the change in free energy ΔG is negative, but impossible if the free energy change is calculated to be positive. Reactions which are exothermic and create more disorder are thus more favoured. The answer to the second question can only be answered in the laboratory since, as we have seen in the previous pages, an energy barrier has to be surmounted before chemical reaction can occur. In attempting to induce a chemical reaction to take place, once it has been shown to be thermodynamically possible, a chemist would consider the possibilities of increasing the concentration of reactants and stepping up the temperature, for these are

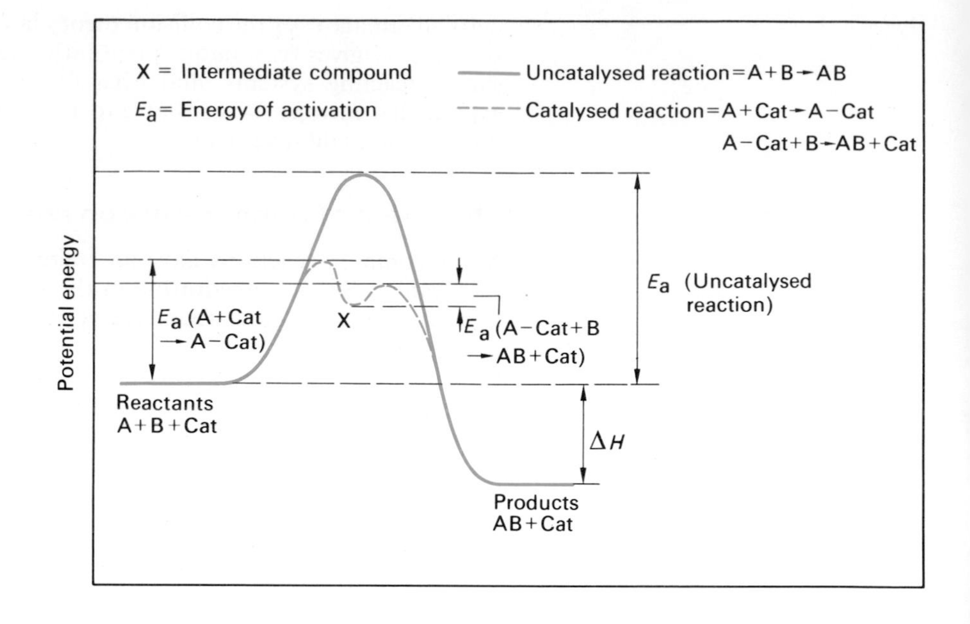

FIG. 9.5. *A catalyst lowers the energy of activation*

variables that increase the number of encounters between reactant molecules. As a last resort he might begin to search for a suitable catalyst to enable an easier reaction path to be followed and thus allow a lower temperature to be employed.

In order to illustrate the points mentioned in the above paragraph, consider the industrial synthesis of methanol from a mixture of carbon monoxide and hydrogen:

$$CO(g) + 2H_2(g) \longrightarrow CH_3OH(g) \qquad \Delta H^\ominus(298\,K) = -90{\cdot}8\,kJ\,mol^{-1}$$

Since an increase in order occurs in this reaction (3 moles of gas are converted into 1 mole of methanol) the entropy change at 298 K and 1 atmosphere pressure, $\Delta S^\ominus$, is a negative quantity. The free energy change for this reaction at 298 K and 1 atmosphere pressure, $\Delta G^\ominus$, is obtained by substitution in the equation:

$$\Delta G^\ominus = \Delta H^\ominus - T\Delta S^\ominus$$

When the known values of $\Delta H^\ominus$ and $\Delta S^\ominus$ are substituted, the free energy change under these conditions is calculated to be $-24{\cdot}7\,kJ\,mol^{-1}$, thus the reaction is thermodynamically possible at this temperature. The reaction does not in practice take place at 298 K because an energy barrier opposes it. In order to surmount this barrier an increase in temperature and pressure is necessary, since these are variables that increase the number of encounters between reactant molecules; but the temperature must not be made too high since the $T\Delta S$ term would eventually make ΔG zero and then positive. In practice a catalyst is also required to enable the reaction to proceed at a convenient rate.

The conditions actually used for the synthesis of methanol are a temperature of 400°C (673 K), a pressure of about 300 atmospheres and a catalyst containing zinc oxide and chromium (III) oxide.

9.7 Suggested reading

M. A. Atherton and J. K. Lawrence, *An Experimental Introduction to Reaction Kinetics*, Longman, 1970.
G. F. Liptrot, J. J. Thompson and G. R. Walker, *Modern Physical Chemistry*, Bell & Hyman, 1982.

9.8 Questions on chapter 9

1 The stoichiometric equations for the hydrolysis of two alkyl bromides are shown below:

$$(CH_3)_3CBr + OH^- \longrightarrow (CH_3)_3COH + Br^- \quad (1)$$

$$CH_3Br + OH^- \longrightarrow CH_3OH + Br^- \quad (2)$$

In the first equation, doubling the concentrations of both reactants increases the rate of reaction by a factor of two, while doubling the concentration of the alkyl bromide alone increases the rate by a factor of two.

In the second equation, doubling the concentrations of both reactants increases the rate of reaction by a factor of four, while doubling the concentration of the alkyl bromide alone increases the rate by a factor of two.

What do you deduce from this information?

2 The stoichiometric equation for the reaction between nitrogen oxide and oxygen is given below:

$$2NO + O_2 \longrightarrow 2NO_2$$

It is found that the reaction rate increases by a factor of eight if the concentration of both reactants is doubled but only by a factor of two if the oxygen concentration only is doubled.

What do you deduce from this information? What would be the effect of doubling the nitrogen oxide concentration only?

3 The hypothetical gaseous reaction $A + B \rightarrow$ products is exothermic and proceeds 100 times faster at 400 K than at 300 K.

(a) With the aid of suitable sketch graphs, discuss the effect of temperature on the reaction.

(b) The reaction is catalysed by a metal M. Sketch and label the energy profiles for the reaction of A with B with and without a catalyst. Attempt to explain the role of the catalyst.

(c) Using the following logarithmic expression of the Arrhenius equation

$$2{,}303 \log_{10} k = \text{constant} - E_a/RT$$

(where k = rate constant, R is the gas constant = $8{\cdot}31\,J\,K^{-1}\,mol^{-1}$ and T = temperature/K), determine graphically or otherwise the activation energy E_a of the reaction, assuming that A and B both have a concentration of $1{\cdot}0\,mol\,dm^{-3}$. (L)

4 The rate constants, at a series of temperatures, for the decomposition of nitrogen pentoxide into nitrogen dioxide and oxygen are given below:

$$2N_2O_5 \longrightarrow 4NO_2 + O_2$$

Temp./K	k/s^{-1}	Temp./K	k/s^{-1}
338	$4{\cdot}87 \times 10^{-3}$	308	$1{\cdot}35 \times 10^{-4}$
328	1.50×10^{-3}	298	$3{\cdot}46 \times 10^{-5}$
318	$4{\cdot}98 \times 10^{-4}$	273	$7{\cdot}87 \times 10^{-7}$

By drawing a suitable graph, determine the energy of activation for this reaction.

5 In what respects is the transition state theory of reaction rates superior to the Arrhenius theory?

Chapter 10

Oxidation and reduction

10.1 Early definition of oxidation and reduction

The simple definition of oxidation as addition of oxygen or removal of hydrogen, and reduction as addition of hydrogen or removal of oxygen has frequently been used to interpret chemical reactions. The two processes are complementary. No oxidation process can take place without a corresponding reduction. Consider the following reactions:

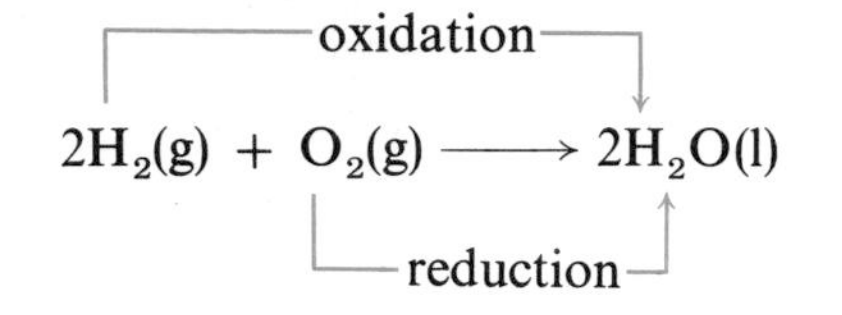

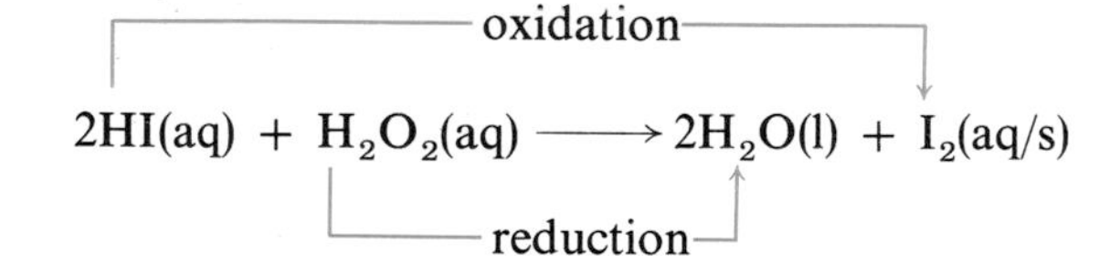

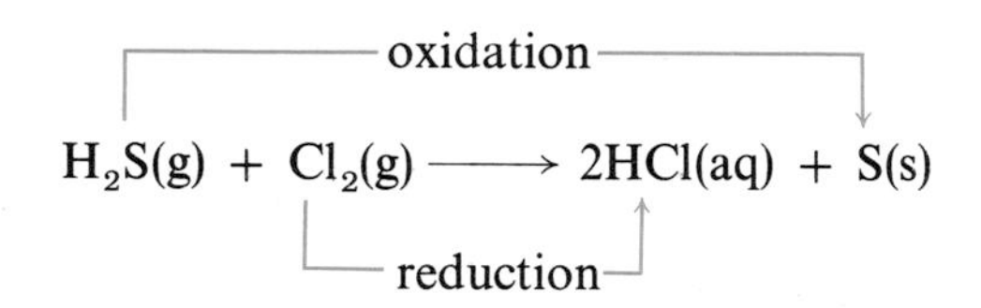

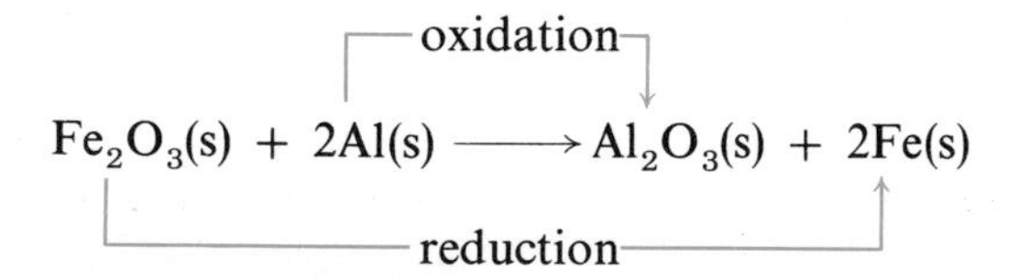

The substance that provides the oxygen or removes the hydrogen, e.g. oxygen, hydrogen peroxide and chlorine, and so becomes reduced, is the oxidising agent; similarly, the substance that provides the hydrogen or removes the oxygen, e.g. hydrogen, hydrogen iodide and aluminium, and so becomes oxidised, is the reducing agent.

10.2 Extension of early ideas

Consider the reaction between hydrogen sulphide and an aqueous solution of iron (III) chloride. Sulphur is precipitated, so there is no doubt that the hydrogen sulphide is oxidised; but there is only one substance that can play the role of oxidising agent—the iron (III) chloride—so this must be reduced. Accordingly the reaction can be written

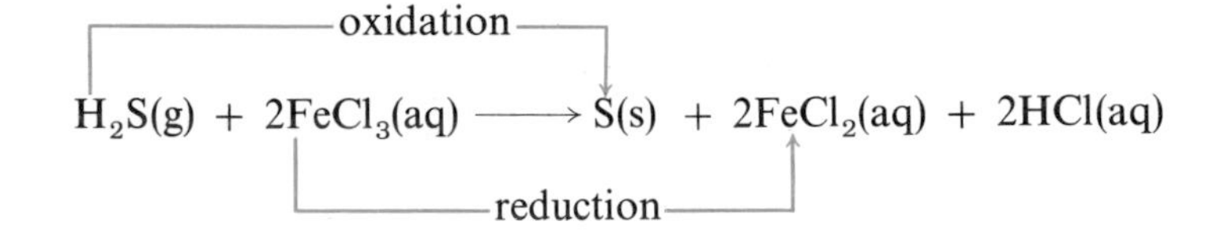

and the iron (III) chloride is reduced to iron (II) chloride. Since this reaction takes place in aqueous solution where it is known that iron (III) chloride is in the form of ions, it is possible to write a simplified ionic equation to summarise the overall result:

$$H_2S(g) + 2Fe^{3+}(aq) \longrightarrow S(s) + 2Fe^{2+}(aq) + 2H^+(aq)$$

The iron (III) ions are reduced to iron (II) ions by gain of electrons; likewise the hydrogen sulphide is oxidised to sulphur and hydrogen ions by the loss of electrons. This can be seen clearly by writing the partial equations:

$$2Fe^{3+}(aq) + 2e^- \longrightarrow 2Fe^{2+}(aq) \quad \text{reduction}$$
$$H_2S(g) \longrightarrow S(s) + 2H^+(aq) + 2e^- \text{oxidation}$$

All reactions involving ions in which electron transfer takes place are classified as oxidation-reduction—redox—reactions. Thus the displacement of copper from a solution of copper (II) ions by zinc can be expressed by the ionic equation:

$$Zn(s) + Cu^{2+}(aq) \longrightarrow Zn^{2+}(aq) + Cu(s)$$

or in terms of partial equations:

$$Zn(s) \longrightarrow Zn^{2+}(aq) + 2e^- \quad \text{oxidation}$$
$$Cu^{2+}(aq) + 2e^- \longrightarrow Cu(s) \quad \text{reduction}$$

Copper (II) ions (oxidising agent) are reduced to copper by zinc (reducing agent) which is itself oxidised to zinc ions.

For reactions in which ionic compounds participate, the scope of oxidation-reduction is thus extended and oxidation is defined as loss of electrons and reduction as gain of electrons.

10.3 Experimental evidence for electron transfer

All ionic reactions which involve loss and gain of electrons can be performed under conditions in which an electric current is generated. Such an arrangement is called an electrochemical cell. Two beakers, one containing potassium iodide solution and the other a solution of iron (III) chloride, are connected by a salt bridge containing potassium chloride (a device that prevents excessive mixing of the two solutions). A platinum electrode is dipped into each solution and joined together through a

FIG. 10.1. *Reaction between ionic substances in solution can be made to generate an electric current*

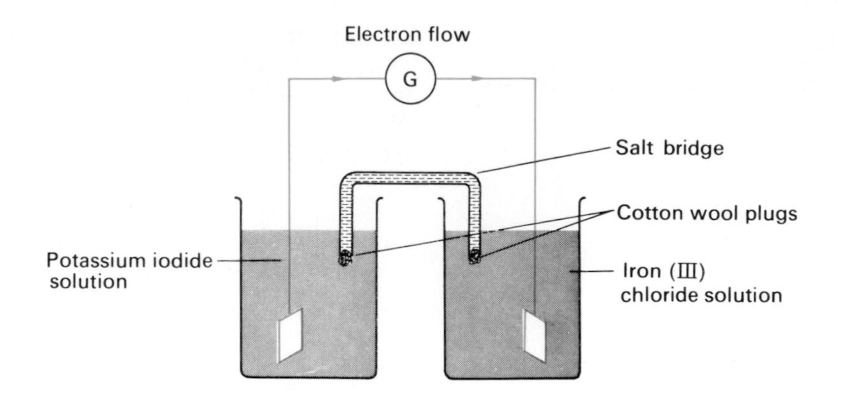

sensitive galvanometer (fig. 10.1). The galvanometer will show a deflection indicating the flow of a current; at the same time the yellow colour of liberated iodine will appear around the electrode dipping into the potassium iodide solution and iron (II) ions (detected by adding potassium hexacyanoferrate (III) solution) will appear at the other electrode. The reaction is thus:

$$2I^-(aq) \longrightarrow I_2(aq) + 2e^- \quad \text{oxidation}$$
$$2Fe^{3+}(aq) + 2e^- \longrightarrow 2Fe^{2+}(aq) \quad \text{reduction}$$

Electrons flow from the electrode dipping into the potassium iodide solution to the other through the external circuit (opposite to conventional current flow), current being conducted through the solution itself by the slow migration of ions.

Although we cannot be sure that the reaction carried out under these conditions involves the same mechanism as that occurring when the two solutions are mixed in a beaker, the overall chemical change is the same, namely:

$$2Fe^{3+}(aq) + 2I^-(aq) \longrightarrow 2Fe^{2+}(aq) + I_2(aq)$$

10.4 Redox reactions involving electron transfer and bond breaking

In the presence of hydrogen ions, the volumetric reagent potassium manganate (VII) quantitatively oxidises iron (II) ions to iron (III) ions and is itself reduced to manganese (II) ions. The two partial equations which summarise this change are:

$$5Fe^{2+}(aq) \longrightarrow 5Fe^{3+}(aq) + 5e^- \quad \text{oxidation}$$
$$MnO_4^-(aq) + 8H^+(aq) + 5e^- \longrightarrow Mn^{2+}(aq) + 4H_2O(l) \quad \text{reduction}$$
$$MnO_4^-(aq) + 8H^+(aq) + 5Fe^{2+}(aq) \longrightarrow 5Fe^{3+}(aq) + Mn^{2+}(aq) + 4H_2O(l) \quad \text{overall}$$

This reaction involves the breaking of manganese-oxygen bonds and the hydrogen ions are required to combine with the oxygen to give water. This reaction can be shown to generate an electric current in the apparatus of fig. 10.1 by placing an acidified potassium manganate (VII) solution in one beaker and an iron (II) solution in the other.

The reaction between dichromate (VI) ions and iron (II) ions is similar:

$$
\begin{array}{lll}
 & 6Fe^{2+}(aq) \longrightarrow 6Fe^{3+}(aq) + 6e^{-} & \text{oxidation} \\
Cr_2O_7^{2-}(aq) + 14^{+}(aq) + & 6e^{-} \longrightarrow 2Cr^{3+}(aq) + 7H_2O(l) & \text{reduction} \\
\hline
Cr_2O_7^{2-}(aq) + 14H^{+}(aq) + & 6Fe^{2+}(aq) \longrightarrow 6Fe^{3+}(aq) + 2Cr^{3+}(aq) + 7H_2O(l) & \text{overall}
\end{array}
$$

10.5 Electron transfer involving essentially covalent molecules

The reaction between essentially covalent substances to produce a predominantly covalent product can never involve the transfer of electrons; but because the electron pair linking two atoms together is drawn closer to the more electronegative atom, a polarised molecule results, with the partial shift of electrons. Thus consider the reaction between hydrogen and chlorine; the two electrons bonding the hydrogen and chlorine atoms together are drawn closer to the chlorine atom, and the chlorine can be considered to be reduced by partial gain of negative charge, and the hydrogen oxidised by partial loss of an equal negative charge:

$$H : H + Cl \overset{\times}{\underset{\times}{}} Cl \longrightarrow 2\overset{\delta+}{H} \overset{\times}{\cdot} \overset{\delta-}{Cl}$$

The arguments developed for ionic redox reactions should not, however, be carried over too literally when discussing reactions involving covalent compounds. Thus the formation of carbon monoxide from carbon and oxygen is certainly an oxidation of carbon; yet experiment shows that a very small partial negative charge resides on the carbon atom, despite the fact that the oxygen atom is more electronegative than carbon. Carbon monoxide is considered to have a structure involving the three resonance forms:

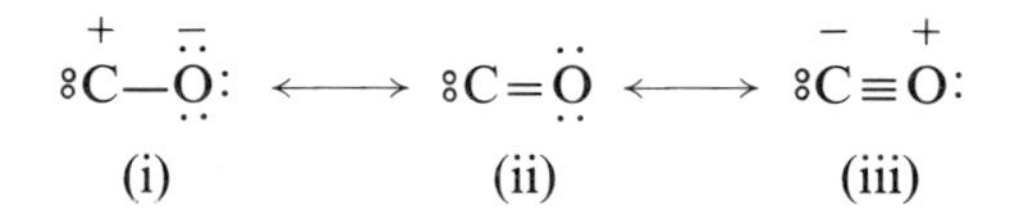

The anomaly is explained as being due to an appreciable contribution from structure (iii) in the resonance hybrid.

10.6 Quantitative treatment of redox systems in aqueous solution

When a metal is placed in a solution of its ions a potential difference is set up between the metal and the solution. The simple model of a metal as an assembly of positive ions held together by a kind of 'electron glue' (p. 55) is useful in visualising what has happened. There is a tendency for metal ions to leave the metal lattice and go into solution, thus leaving an excess of electrons and hence a negative charge on the metal; there is also a reverse tendency for metal ions from the solution to deposit on the metal leading to a positive charge on the metal. In practice one effect is greater than the other, so a potential difference is set up between the metal and the solution. The value of this potential difference for a particular metal depends upon the concentration of the metal ions and the temperature, and in practice a 1 M solution of metal ions and a temperature of 298 K are specified. The potential difference set up under these conditions is called a **standard electrode potential**.

Zinc in contact with a 1 M solution of zinc ions at 298 K has a negative electrode potential, while copper in contact with a 1 M solution of copper (II) ions at 298 K has a positive electrode potential (fig. 10.2).

FIG. 10.2. *A potential difference is set up when metals are placed in contact with a solution of their ions*

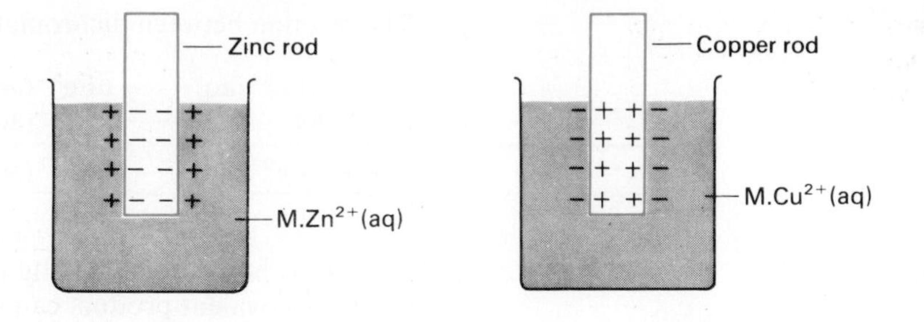

It is not possible to measure standard electrode potentials absolutely, since the very act of carrying out a measurement would introduce another metal into the solution which would set up its own electrode potential. Standard electrode potentials therefore have to be measured against some reference standard, and the one adopted is the hydrogen electrode. This consists of hydrogen gas at one atmosphere pressure in contact with a 1 M solution of its ions at 298 K; a platinum electrode coated with platinum black is incorporated into the set-up and catalyses the attainment of equilibrium between the hydrogen gas and hydrogen ions in the solution (fig. 10.3). The standard electrode potential of this system is arbitrarily assigned a value of zero.

FIG. 10.3. *The standard hydrogen electrode*

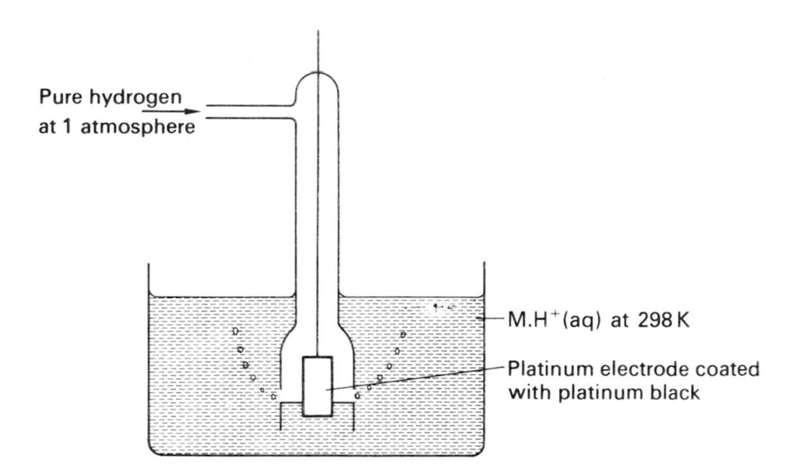

The standard electrode potential for the system M^{n+}(aq)/M(s) is found by connecting it to a hydrogen electrode, via a potassium chloride salt bridge, and reading the potential difference developed, either on a previously calibrated potentiometer or on a high resistance voltmeter, i.e. a valve voltmeter (so that no current flows) (fig. 10.4). The negative pole of the cell is allotted a negative electrode potential (this is the convention recommended by the International Union of Pure and Applied Chemistry—IUPAC—and is the one used in this book). The standard electrode potential for

FIG. 10.4. *Apparatus for measuring standard electrode potentials*

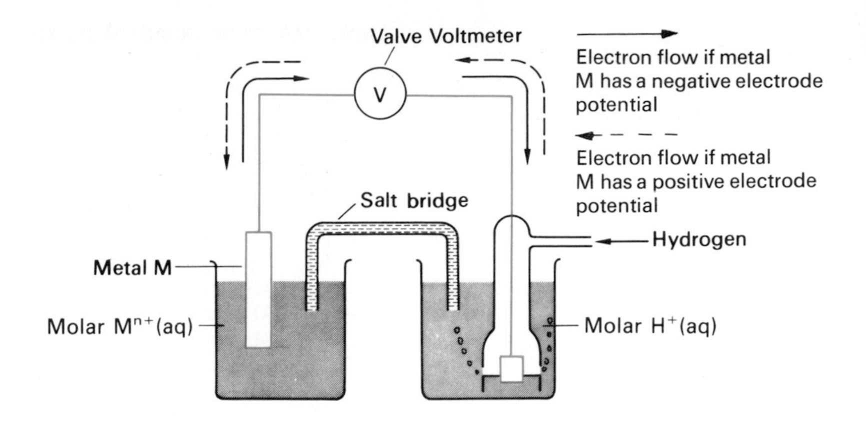

Zn^{2+}(aq)/Zn(s) is $-0{\cdot}76$ V ($E^{\ominus} = -0{\cdot}76$ V) and for Cu^{2+}(aq)/Cu(s) is $+0{\cdot}34$ V ($E^{\ominus} = +0{\cdot}34$ V); thus the zinc electrode is the negative and the copper electrode the positive pole when they are both connected separately to a hydrogen electrode (fig. 10.5)

FIG. 10.5. *Zinc forms the negative pole and copper forms the positive pole when both are connected separately to a standard hydrogen electrode*

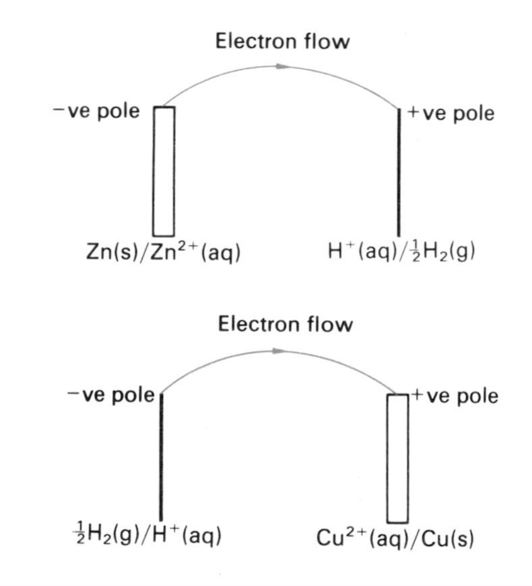

The values of some standard electrode potentials at 298 K are given in Table 10A; they are often called standard redox potentials.

Table 10A Some standard electrode (redox) potentials

Reaction	$E^\ominus$/V
$Li^+(aq) + e^- \longrightarrow Li(s)$	$-3{\cdot}04$
$K^+(aq) + e^- \longrightarrow K(s)$	$-2{\cdot}92$
$Ba^{2+}(aq) + 2e^- \longrightarrow Ba(s)$	$-2{\cdot}90$
$Ca^{2+}(aq) + 2e^- \longrightarrow Ca(s)$	$-2{\cdot}87$
$Na^+(aq) + e^- \longrightarrow Na(s)$	$-2{\cdot}71$
$Mg^{2+}(aq) + 2e^- \longrightarrow Mg(s)$	$-2{\cdot}37$
$Al^{3+}(aq) + 3e^- \longrightarrow Al(s)$	$-1{\cdot}66$
$Mn^{2+}(aq) + 2e^- \longrightarrow Mn(s)$	$-1{\cdot}18$
$Zn^{2+}(aq) + 2e^- \longrightarrow Zn(s)$	$-0{\cdot}76$
$Cr^{3+}(aq) + 3e^- \longrightarrow Cr(s)$	$-0{\cdot}74$
$Fe^{2+}(aq) + 2e^- \longrightarrow Fe(s)$	$-0{\cdot}44$
$Co^{2+}(aq) + 2e^- \longrightarrow Co(s)$	$-0{\cdot}28$
$Ni^{2+}(aq) + 2e^- \longrightarrow Ni(s)$	$-0{\cdot}25$
$Sn^{2+}(aq) + 2e^- \longrightarrow Sn(s)$	$-0{\cdot}14$
$Pb^{2+}(aq) + 2e^- \longrightarrow Pb(s)$	$-0{\cdot}13$
$H^+(aq) + e^- \longrightarrow \frac{1}{2}H_2(g)$	$0{\cdot}00$
$Cu^{2+}(aq) + 2e^- \longrightarrow Cu(s)$	$+0{\cdot}34$
$Ag^+(aq) + e^- \longrightarrow Ag(s)$	$+0{\cdot}80$
$Au^{3+}(aq) + 3e^- \longrightarrow Au(s)$	$+1{\cdot}50$

INCREASINGLY POWERFUL OXIDISING AGENTS (↓, left-hand side)

INCREASINGLY POWERFUL REDUCING AGENTS (↑, right-hand side)

A metal higher up in this series is a better reducing agent in aqueous solution than one that lies below it; thus zinc will displace both tin and copper from aqueous solutions of their salts and tin will displace copper but not zinc under the same conditions.

A metal in contact with its ions is known as a half-cell; when two such half cells are joined together through a salt bridge the e.m.f. of the cell is the algebraic difference between the two electrode potentials. For a system such as:

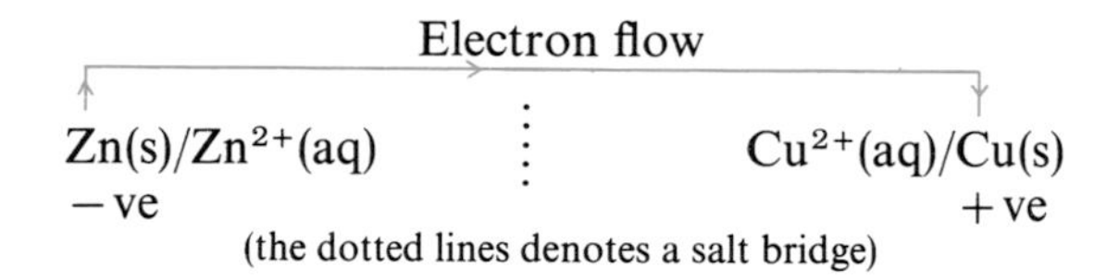

(the dotted lines denotes a salt bridge)

the e.m.f. of the cell ($E^\ominus_{total}$) is defined to be the standard electrode potential of the right-hand electrode minus the standard electrode potential of the left-hand electrode.

$$E^\ominus_{total} = E^\ominus_{R.H.S.} - E^\ominus_{L.H.S.} = +0{\cdot}34 - (-0{\cdot}76) = +1{\cdot}10\,V$$

Using this convention, a positive e.m.f. means that the left-hand electrode (zinc) is potentially capable of reducing copper (II) ions to copper, as is easily demonstrated in a test-tube:

$$Zn(s) + Cu^{2+}(aq) \longrightarrow Zn^{2+}(aq) + Cu(s)$$

If the cell had been written down as:

Electron flow

$$\underset{+ve}{Cu(s)/Cu^{2+}(aq)} \quad \vdots \quad \underset{-ve}{Zn^{2+}(aq)/Zn(s)}$$

the e.m.f. would be $-0{\cdot}76 - (+0{\cdot}34) = -1{\cdot}10$ V, indicating that the left-hand electrode (copper) cannot reduce zinc ions to zinc. In general, using this convention, a positive e.m.f. indicates that a reaction is thermodynamically possible as written down from left to right, but a negative e.m.f. implies that a reaction is thermodynamically impossible.

10.7 Factors affecting the values of standard electrode potentials

The electrode potentials of the Group 1A metals decrease in the order lithium > potassium > sodium; thus lithium appears to occupy an anomalous position since an increase in metallic character—taken as an increasing tendency for the metal atom to lose electrons and measured quantitatively as ionisation energy (p. 39)—increases in the order lithium > sodium > potassium. In order to see why lithium has an unexpectedly high electrode potential it is necessary to consider the individual energy changes that take place in the process:

$$Li(s) \longrightarrow Li^+(aq) + e^-$$

The energy change involved when one mole of lithium metal passes into solution as hydrated lithium ions can be determined by considering the process to take place in a series of hypothetical stages each one of which involves an energy change (cf. energetics of electrovalent compound formation p. 56).

(a) Conversion of solid lithium into gaseous lithium atoms. This is an endothermic process and the molar enthalpy of sublimation—the energy required to convert one mole of solid lithium into gaseous atoms—is absorbed:

$$Li(s) \longrightarrow Li(g) \qquad S = +159{\cdot}0\ \text{kJ mol}^{-1}$$

(b) Removal of the outer electron of the lithium atom to give a lithium ion. The energy required for this is the ionisation energy, I, (p. 39).

$$Li(g) \longrightarrow Li^+(g) + e^- \qquad I = +520{\cdot}0\ \text{kJ mol}^{-1}$$

(c) Hydration of the lithium ion.
This is an exothermic process and the heat liberated is known as the enthalpy of hydration, $\Delta H_h^{\ominus}$.

$$Li^+(g) + \text{water} \longrightarrow Li^+(aq) \qquad \Delta H_h^{\ominus} = -507{\cdot}1\ \text{kJ mol}^{-1}$$

The overall change in the complete reaction is the sum of the separate energy changes (law of conservation of energy) thus:

$$\begin{aligned}\Delta H^{\ominus} &= S + I + \Delta H_h^{\ominus} \\ &= (+159{\cdot}0 + 520{\cdot}0 - 507{\cdot}1)\ \text{kJ mol}^{-1} \\ &= +171{\cdot}9\ \text{kJ mol}^{-1}\end{aligned}$$

The change $Li(s) \longrightarrow Li^+(aq) + e^-$ is endothermic to the extent of 171·9 kJ mol^{-1}. A similar series of calculations could be done for sodium and potassium; the appropriate energy changes, together with those for lithium, are set out in Table 10B.

Table 10B Energy change for the process $M(s) \rightarrow M^{n+}(aq) + ne^-$

Enthalpy changes (kJ mol^{-1})	Lithium	Sodium	Potassium
Molar enthalpy of sublimation (S)	+159·0	+108·7	+90·00
Ionisation energy (I)	+520·0	+493·8	+418·4
Enthalpy of hydration ($\Delta H_h^\ominus$)	−507·1	−395·9	−317·2
Overall enthalpy change	+171·9	+206·6	+191·2

The process of an alkali metal passing into solution as its hydrated ions thus becomes less endothermic and, neglecting entropy effects, more likely along the series sodium, potassium, lithium in accordance with their standard electrode potentials. The anomalous behaviour of lithium is therefore due to the very high enthalpy of hydration of its ions, a consequence of the fact that smaller ions can bind water molecules more tightly than larger ones.

10.8 Standard redox potentials for other systems

Standard redox potentials for non-metals that produce negative ions in aqueous solution can be determined in a similar manner to that used for metals. Thus the standard electrode potential of chlorine can be determined using an electrode consisting of chlorine gas at one atmosphere pressure in equilibrium with a 1 M solution of chloride ions. Similarly for metals of variable valency, redox potentials for one ion in equilibrium with another of different charge can be determined. The redox potential of the system $Fe^{3+}(aq)/Fe^{2+}(aq)$ is obtained by coupling a half-cell containing 1 M iron (III) and 1 M iron (II) ions, into which a platinum wire is dipping, to a standard hydrogen electrode. The function of the platinum wire is that of a conductor and a catalyst. Some other values of standard redox potentials are tabulated in Table 10C.

Table 10C Some more standard redox potentials

Reaction (INCREASINGLY POWERFUL OXIDISING AGENTS ↓; INCREASINGLY POWERFUL REDUCING AGENTS ↑)		$E^\ominus$/V
$S(s) + 2H^+(aq) + 2e^-$	$\longrightarrow H_2S(g)$	+0·14
$Sn^{4+}(aq) + 2e^-$	$\longrightarrow Sn^{2+}(aq)$	+0·15
$Cu^{2+}(aq) + e^-$	$\longrightarrow Cu^+(aq)$	+0·15
$\frac{1}{2}I_2(s) + e^-$	$\longrightarrow I^-(aq)$	+0·54
$Fe^{3+}(aq) + e^-$	$\longrightarrow Fe^{2+}(aq)$	+0·76
$\frac{1}{2}Br_2(l) + e^-$	$\longrightarrow Br^-(aq)$	+1·07
$Cr_2O_7^{2-}(aq) + 14H^+(aq) + 6e^-$	$\longrightarrow 2Cr^{3+}(aq) + 7H_2O(l)$	+1·33
$\frac{1}{2}Cl_2(g) + e^-$	$\longrightarrow Cl^-(aq)$	+1·36
$MnO_4^-(aq) + 8H^+(aq) + 5e^-$	$\longrightarrow Mn^{2+}(aq) + 4H_2O(l)$	+1·52
$\frac{1}{2}S_2O_8^{2-}(aq) + e^-$	$\longrightarrow SO_4^{2-}(aq)$	+2·01
$\frac{1}{2}F_2(g) + e^-$	$\longrightarrow F^-(aq)$	+2·80

The acidity of the solution is seen to influence the redox potentials of three systems in Table 10C. A substance on the right of the table is potentially capable of reducing any substance on the left of the table that lies below it; thus iodide ions will reduce iron (III) ions to the iron (II) state:

$$2Fe^{3+}(aq) + 2I^-(aq) \longrightarrow 2Fe^{2+}(aq) + I_2(aq)$$

Similarly, a substance on the left of the table is potentially capable of oxidising any substance on the right of the table that lies above it; thus chlorine will oxidise bromide and iodide ions to bromine and iodine respectively, and bromine will oxidise iodide ions to iodine:

$$Cl_2(g) + 2Br^-(aq) \longrightarrow 2Cl^-(aq) + Br_2(l)$$
$$Cl_2(g) + 2I^-(aq) \longrightarrow 2Cl^-(aq) + I_2(s)$$
$$Br_2(l) + 2I^-(aq) \longrightarrow 2Br^-(aq) + I_2(s)$$

It thus becomes clear why substances can behave as oxidising agents under one set of conditions and as reducing agents under others.

The table also shows that solutions containing chloride ions react with manganate (VII) ions but not with dichromate (VI) ions, and this explains why hydrochloric acid may not be used to acidify potassium manganate (VII) solution when carrying out gas tests.

10.9 Tests for oxidising agents

All oxidising agents will react in one or more of the following ways:

Insoluble oxidising agents

(a) Heat strongly, oxygen is evolved which relights a glowing splint.

(b) Warm with concentrated hydrochloric acid; chlorine is evolved, which bleaches moist litmus paper.

Aqueous solutions of oxidising agents

(c) Hydrogen sulphide gives a milky yellow precipitate of sulphur with a solution of an oxidising agent:

$$H_2S(g) \longrightarrow S(s) + 2H^+(aq) + 2e^-$$

(d) Acidified potassium iodide solution reacts with a solution of an oxidising agent to give iodine. If starch solution is added an intense blue colour is obtained:

$$2I^-(aq) \longrightarrow I_2(aq) + 2e^-$$

(e) A few drops of a freshly prepared solution of iron (II) ammonium sulphate, acidified with dilute sulphuric acid, are converted to iron (III) ammonium sulphate by the addition of a solution of an oxidising agent:

$$Fe^{2+}(aq) \longrightarrow Fe^{3+}(aq) + e^-$$

The iron (III) ions can be detected by adding potassium thiocyanate solution, when a deep red coloration is produced.

10.10 Tests for reducing agents

All reducing agents will act in one or more of the following ways:

Insoluble reducing agents

(a) Heat with a few drops of concentrated nitric acid; brown fumes of nitrogen dioxide are evolved.

(b) Heat with powdered copper (II) oxide; a red deposit of copper is formed which will not dissolve in warm dilute sulphuric acid.

Aqueous solutions of reducing agents

(c) A few drops of a solution of potassium manganate (VII), acidified with dilute sulphuric acid, are decolorised when added to an excess of the reducing solution

$$MnO_4^-(aq) + 8H^+(aq) + 5e^- \longrightarrow Mn^{2+}(aq) + 4H_2O(l)$$

(d) A few drops of a solution of potassium dichromate (VI) solution, acidified with dilute sulphuric acid, will go green when added to an excess of the reducing solution.

$$Cr_2O_7^{2-}(aq) + 14H^+(aq) + 6e^- \longrightarrow 2Cr^{3+}(aq) + 7H_2O(l)$$

(e) A few drops of iron (III) chloride solution are reduced to the iron (II) state in the presence of an excess of a reducing solution.

$$Fe^{3+}(aq) + e^- \longrightarrow Fe^{2+}(aq)$$

On adding potassium hexacyanoferrate (III) solution a deep blue coloration or precipitate appears.

10.11 Determination of free energy changes using cell reactions

Consider the reaction between metallic zinc and a molar solution of copper (II) ions:

$$Zn(s) + Cu^{2+}(aq) \longrightarrow Zn^{2+}(aq) + Cu(s)$$

This reaction can be made to take place in an electrochemical cell using $Zn(s)/Zn^{2+}(aq)$ (1 M) as one half-cell and $Cu(s)/Cu^{2+}(aq)$ (1 M) as the other, the two half-cells being joined in the usual way by a salt bridge. The e.m.f. of this cell can be measured using the apparatus shown in fig. 10.6 below.

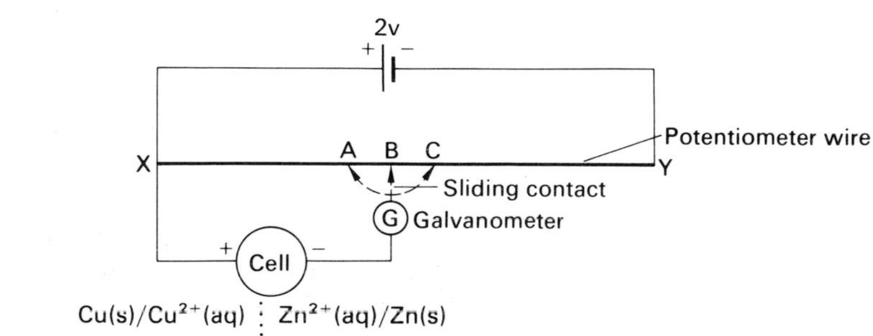

FIG. 10.6. *A method of determining the e.m.f. of an electrochemical cell*

If the galvanometer shows zero deflection when the sliding contact is at position B, i.e. no current is drawn from the cell, the e.m.f. of the cell is given by the expression:

$$E^{\ominus} = 2XB/XY = 1{\cdot}10\,\text{V (the cell being at 298 K)}$$

If now the contact is made at position A, then current is drawn from the cell; similarly, with the contact at position C, the accumulator drives current through the cell in the opposite direction, i.e. the normal cell reaction is reversed. Slight variations in position of the sliding contact thus allow the cell to operate or to be driven backwards, i.e. a reversible change can be made to occur. With the sliding contact at position B the cell is theoretically capable of delivering the maximum amount of electrical work. In practice, however, no current is drawn from the cell under these conditions and hence no chemical reaction can occur; but if the sliding contact is only fractionally off balance, an extremely small current can be drawn from the cell and the amount of electrical work obtained can be made to approach the theoretical maximum.

If we consider the cell to operate with the sliding contact fractionally off balance then, as chemical reaction occurs, the zinc ion concentration will rise and the copper (II) ion concentration will fall, and the e.m.f. of the cell will decrease. In order to maintain a constant e.m.f., zinc ions must be removed from and copper (II) ions added to the respective half-cells to maintain these ions at 1 M concentration. The maximum amount of electrical work obtained under these conditions for the simultaneous conversion of 1 mole of zinc into zinc ions and 1 mole of copper (II) ions into copper can easily be calculated. Thus electrical work (in joules) is the product of potential difference (in volts) and charge flowing (in coulombs):

$$w_{\text{elec}} = nFE$$

where w_{elec} is the work obtained, n is the number of moles of electrons that flow during the chemical reaction, and F is the Faraday (96 500 coulombs/mol of electrons).

Of course, in practice it is not necessary to remove zinc ions from one half-cell and add copper (II) ions to the other half-cell, since the e.m.f. is determined under conditions when no current flows. For the cell reaction under consideration $n = 2$ and the maximum amount of electrical work that can be obtained, w_{useful} (see p. 110), is:

$$\begin{aligned} w_{\text{useful}} = 2 \times 96\,500 \times 1{\cdot}10 &= 212\,300\,\text{J} \\ &= 212{\cdot}3\,\text{kJ mol}^{-1} \end{aligned}$$

A decrease in free energy, $-\Delta G$, was defined to be equal to the maximum useful work that a system could perform (see chapter 8, p. 111). Thus:

$$-\Delta G = w_{\text{useful}} = 212{\cdot}3\,\text{kJ mol}^{-1}$$

or

$$\Delta G = -212{\cdot}3\,\text{kJ mol}^{-1}$$

Since the electrochemical cell contains the substances in their standard states, i.e. the zinc and copper (II) ions are at unit concentration and the temperature is 298 K, the free energy change is the standard free energy change $\Delta G^{\ominus}$, therefore,

$$\Delta G^{\ominus} = -212{\cdot}3\,\text{kJ mol}^{-1}$$

10.12 Standard free energies of half-cell reactions

If we define the standard free energy of a half-cell reaction to be:

$$G^{\ominus} = -nFE^{\ominus}$$

where n is the number of moles of electrons involved, F is the Faraday, and $E^{\ominus}$ is the standard electrode potential, then the free energy change for any cell reaction is easily calculated. Since the standard electrode potential of hydrogen is arbitrarily fixed as zero (p. 132), the standard free energy of the hydrogen electrode is also zero. Consider the following two half-cell reactions:

$$2Ag^{+}(aq) + 2e^{-} \longrightarrow 2Ag(s) \quad G^{\ominus}_{Ag} = -2F(+0{\cdot}80) = -1{\cdot}60 \text{ Faraday-volts}$$
$$Cu^{2+}(aq) + 2e^{-} \longrightarrow Cu(s) \quad G^{\ominus}_{Cu} = -2F(+0{\cdot}34) = -0{\cdot}68 \text{ Faraday-volts}$$

By subtraction and rearrangement we have:

$$Cu(s) + 2Ag^{+}(aq) \longrightarrow Cu^{2+}(aq) + 2Ag(s) \quad \Delta G^{\ominus} = G^{\ominus}_{Ag} - G^{\ominus}_{Cu} = -1{\cdot}60 - (-0{\cdot}68) = -0{\cdot}92 \text{ Faraday-volts}$$

The negative value for the change in free energy means that the reaction as written will take place from left to right. By using the appropriate conversion factor, the free energy change can be converted into kJ mol^{-1}.

Consider now the two separate half-cell reactions:

$$Fe^{3+}(aq) + 3e^{-} \longrightarrow Fe(s) \quad G^{\ominus}_{Fe^{3+}/Fe} = -3F(-0{\cdot}04) = +0{\cdot}12 \text{ Faraday-volts}$$
$$Fe^{2+}(aq) + 2e^{-} \longrightarrow Fe(s) \quad G^{\ominus}_{Fe^{2+}/Fe} = -2F(-0{\cdot}44) = +0{\cdot}88 \text{ Faraday-volts}$$

By subtraction we have:

$$Fe^{3+}(aq) + e^{-} \longrightarrow Fe^{2+}(aq) \quad \Delta G^{\ominus} = G^{\ominus}_{Fe^{3+}/Fe} - G^{\ominus}_{Fe^{2+}/Fe} = +0{\cdot}12 - (+0{\cdot}88) = -0{\cdot}76 \text{ Faraday-volts}$$

The value $-0{\cdot}76$ Faraday-volts is the standard free energy for the $Fe^{3+}(aq)/Fe^{2+}(aq)$ electrode:

$$Fe^{3+}(aq) + e^{-} \longrightarrow Fe^{2+}(aq)$$

Since only one electron is involved in this change, the standard redox potential of this system is $+0{\cdot}76$ V. The use of standard free energies of half-cell reactions explains why electrode potentials cannot always be combined by simple addition and subtraction. Thus:

$$Fe^{3+}(aq) + 3e^{-} \longrightarrow Fe(s) \quad E^{\ominus} = -0{\cdot}04 \text{ V}$$
$$Fe^{2+}(aq) + 2e^{-} \longrightarrow Fe(s) \quad E^{\ominus} = -0{\cdot}44 \text{ V}$$

By subtraction we have:

$$Fe^{3+}(aq) + e^{-} \longrightarrow Fe^{2+}(aq)$$

but the standard redox potential of this system cannot be obtained by subtraction of the standard redox potentials, i.e. it is not $+0{\cdot}40$ V but $+0{\cdot}76$ V as we have seen above. In order to obtain the correct value, the numbers of electrons involved in each half-cell reaction must also be considered.

10.13 Dependence of redox potentials on ionic concentration and on temperature

When a metal is placed in a solution of its ions there is a tendency for metal ions to leave the metal lattice and pass into solution, thus leaving an excess of electrons and hence a negative charge on the metal. There is also a reverse tendency for metal ions from solution to deposit on the metal, leading to a positive charge on the metal. In practice one effect is greater than the other, so a potential difference is set up between the metal and a solution of its ions. If the concentration of metal ions in contact with the metal is decreased there will be a smaller tendency for metal ions to deposit on the metal, i.e. the electrode potential of the metal will become less positive. Similarly, if the metal ion concentration is increased, the electrode potential will become more positive.

The way in which the electrode potential of a metal is related to the metal ion concentration and to the temperature is given by the Nernst equation:

$$E = E^{\ominus} + \frac{RT}{nF} \ln[M^{n+}(aq)]$$

where E is the electrode potential, $E^{\ominus}$ is the standard electrode potential (for a molar solution of metal ions at 298 K), R is the gas constant and is approximately 8·31 J K^{-1} mol^{-1}. T is the temperature in K, n is the valency of the metal, and F is the Faraday. The above equation can be verified by measuring the electrode potential of a metal, at varying metal ion concentrations and at a given temperature, by connecting the metal/metal ion half-cell via a salt bridge to a standard hydrogen electrode. The e.m.fs. of the various cells can be measured on an electronic voltmeter, and a graphical plot of the various E values against $\ln[M^{n+}(aq)]$ will be a straight line.

10.14 Equilibrium constants for reactions that can be carried out in electrochemical cells

Consider the reaction between copper and a molar solution of silver ions:

$$Cu(s) + 2Ag^{+}(aq) \longrightarrow Cu^{2+}(aq) + 2Ag(s)$$

This reaction can be carried out in an electrochemical cell by using $Cu(s)/Cu^{2+}(aq)$ (1 M) as one half-cell and $Ag(s)/Ag^{+}(aq)$ (1 M) as the other half-cell. The standard free energy change for this reaction at 298 K is −0·92 Faraday-volts (see section 10.12).

$$\Delta G^{\ominus} = -0{\cdot}92 \text{ Faraday-volts} = \frac{-0{\cdot}92 \times 96\,500}{1000} \text{ kJ mol}^{-1}$$

$$= -88{\cdot}74 \text{ kJ mol}^{-1}$$

The electrode potentials of the silver and copper half-cells are given by the Nernst equation, i.e.

$$E_{Ag} = E^{\ominus}_{Ag} + \frac{RT}{F} \ln[Ag^{+}(aq)] \qquad (n = 1)$$

or

$$E_{Ag} = E^{\ominus}_{Ag} + \frac{RT}{2F} \ln[Ag^{+}(aq)]^2$$

$$E_{Cu} = E^{\ominus}_{Cu} + \frac{RT}{2F} \ln[Cu^{2+}(aq)] \qquad (n = 2)$$

If this cell is allowed to deliver current, silver is deposited and copper (II) ions are formed; the e.m.f. of the cell under these conditions is given by,

$$(E_{Ag} - E_{Cu}) = (E^{\ominus}_{Ag} - E^{\ominus}_{Cu}) + \frac{RT}{2F} \ln \frac{[Ag^+(aq)]^2}{[Cu^{2+}(aq)]}$$

Multiplying each side of the equation by $-2F$ we have.

$$-2F(E_{Ag} - E_{Cu}) = -2F(E^{\ominus}_{Ag} - E^{\ominus}_{Cu}) - RT\ln \frac{[Ag^+(aq)]^2}{[Cu^{2+}(aq)]}$$

but $-2F(E_{Ag} - E_{Cu})$ is the free energy change ΔG, and $-2F(E^{\ominus}_{Ag} - E^{\ominus}_{Cu})$ is the standard free energy change $\Delta G^{\ominus}$, thus,

$$\Delta G = \Delta G^{\ominus} - RT\ln \frac{[Ag^+(aq)]^2}{[Cu^{2+}(aq)]}$$

At equilibrium $\Delta G = 0$, i.e. the cell is fully discharged and its e.m.f. is zero, therefore,

$$\Delta G^{\ominus} = RT\ln \frac{[Ag^+(aq)_{equil.}]^2}{[Cu^{2+}(aq)_{equil.}]} = -RT\ln \frac{[Cu^{2+}(aq)_{equil.}]}{[Ag^+(aq)_{equil.}]^2}$$

or

$$\Delta G^{\ominus} = -2{\cdot}303RT\log_{10} \frac{[Cu^{2+}(aq)_{equil.}]}{[Ag^+(aq)_{equil.}]^2} = -2{\cdot}303RT\log_{10}K$$

Since R is approximately $8{\cdot}31\ J\ K^{-1}\ mol^{-1}$ (or $0{\cdot}008\ 31\ kJ\ K^{-1}\ mol^{-1}$) and the temperature is taken to be 25°C (298 K) we have:

$$-88{\cdot}74 = -2{\cdot}303 \times 0{\cdot}00831 \times 298\log_{10}K$$

or

$$\log_{10}K = \frac{88{\cdot}74}{2{\cdot}303 \times 0{\cdot}00831 \times 298} = 15{\cdot}45$$

$$K = 10^{15{\cdot}45}\ mol^{-1}\ dm^3$$

This exceedingly large value of K means that the reaction between copper and silver ions can be considered to go to completion.

10.15 Suggested reading

G. F. Liptrot, J. J. Thompson and G. R. Walker, *Modern Physical Chemistry*, Bell & Hyman, 1982.

G. J. Moody and J. D. R. Thomas, *Energy Relationships in Chemistry, Oxidation and Reduction*, School Science Review, No. 174, Vol. 51, 1969.

A. G. Sharpe, *Principles of Oxidation and Reduction,* Royal Institute of Chemistry, Monographs for Teachers, No. 2, 1960.

10.16 Questions on chapter 10

1 Define the terms 'oxidation' and 'reduction' in as many ways as you can. Identify the oxidising and reducing agents in each of the following processes, and discuss the application of your definitions to them:
(a) the reaction of hydrogen peroxide with acidified potassium iodide,
(b) the reaction of hydrogen peroxide with acidified potassium permanganate,
(c) the passage of hydrogen over heated sodium,
(d) the addition of calcium to water,
(e) the reaction of hydrogen sulphide with moist sulphur dioxide. (L)

2 Explain fully the meaning of the term reduction.
Describe the reactions which take place between the following pairs of substances and clearly distinguish between the oxidation and reduction processes in each case:
(a) zinc and copper (II) sulphate,
(b) copper (II) sulphate and potassium iodide,
(c) iodine and hydrogen sulphide,
(d) tin (II) chloride and mercury (II) chloride,
(e) potassium hexacyanoferrate (II) and chlorine. (AEB)

3 (a) Explain the meaning of oxidation and reduction in terms of the transfer of electrons.
(b) The following stoichiometric equations represent redox reactions which are commonly employed in volumetric analysis. In each case, elucidate the fundamental oxidation and reduction processes in the light of what you have written in answer to (a).
(i) $2CuSO_4 + 4KI \longrightarrow 2CuI + I_2 + 2K_2SO_4$
(ii) $6FeCl_2 + K_2Cr_2O_7 + 14HCl \longrightarrow 6FeCl_3 + 2CrCl_3 + 2KCl + 7H_2O$
(iii) $KBrO_3 + 5KBr + 3H_2SO_4 \longrightarrow 3K_2SO_4 + 3Br_2 + 3H_2O$
(O & C)

4 What is meant by a standard electrode (or redox) potential? Explain how you would measure the standard electrode potential of silver. Zinc has a standard electrode potential of $-0{\cdot}76$ V while copper has a standard electrode potential of $+0{\cdot}34$ V. Explain carefully what these values mean and in particular the nature of the sign attached to each.

5 Read section 10·6 (p. 131) again and then try and decide how the electrode potential of zinc would alter if the solution of its ions was gradually diluted, starting with a molar solution. How would the electrode potential of copper alter under similar conditions?

6 The standard electrode potential of sodium is $-2{\cdot}71$ V yet this potential cannot be measured directly by dipping a piece of sodium into a molar solution of sodium ions, since evolution of hydrogen would occur by reaction with the water. Can you suggest a way of measuring the standard electrode potential of sodium?

7 What would be the e.m.f. of the following cell combinations?
(a) $Zn(s)/Zn^{2+}(aq)$ (molar) ⋮ $Cu^{2+}(aq)$ (molar)$/Cu(s)$
(b) $Cr(s)/Cr^{3+}(aq)$ (molar) ⋮ $Co^{2+}(aq)$ (molar)$/Co(s)$
(c) $Ni(s)/Ni^{2+}(aq)$ (molar) ⋮ $Au^{3+}(aq)$ (molar)$/Au(s)$
Draw diagrams of each cell and show (i) the negative and positive poles, (ii) the direction of electron flow through the external circuit. (Use the standard electrode potentials given on p. 134).

8 What explanation can you give for the anomalously large negative standard electrode potential of lithium?

9 From the following standard redox potentials,

$$Fe^{3+}(aq) + e^- \longrightarrow Fe^{2+}(aq) \qquad E^\ominus = +0{\cdot}76\text{ V}$$
$$I_2(aq) + 2e^- \longrightarrow 2I^-(aq) \qquad E^\ominus = +0{\cdot}54\text{ V}$$

determine:
(a) the cell e.m.f.
(b) the reaction that proceeds as the cell is operated.

10 Sketch the electrochemical cell you could set up and whose reaction is:

$$Ag^+(aq)(molar) + Fe^{2+}(aq)(molar) \longrightarrow Ag(s) + Fe^{3+}(aq)(molar)$$

From the following standard redox potentials at 25°C

$$Ag^+(aq) + e^- \longrightarrow Ag(s) \qquad E^\ominus = +0{\cdot}80\ V$$
$$Fe^{3+}(aq) + e^- \longrightarrow Fe^{2+}(aq) \qquad E^\ominus = +0{\cdot}76\ V$$

calculate:

(a) the e.m.f. of the cell,

(b) the standard free energy change (in $kJ\ mol^{-1}$),

(c) the value of the equilibrium constant.

(96 500 coulombs = 1 Faraday).

11 By considering the appropriate standard electrode potentials given in Table 10A (p. 134), show that the reaction at 25°C

$$Zn(s) + Pb^{2+}(aq)(molar) \longrightarrow Pb(s) + Zn^{2+}(aq)(molar)$$

is to all intents and purposes irreversible, i.e. has an extremely large equilibrium constant for the reaction as written from left to right.

12 What do you understand by the terms electrode potential and standard electrode potential? How are they related?

Indicate how the electrode potential of hydrogen could be measured. What significance does it have?

A fuel cell, with a molar solution of potassium hydroxide in water at 25°C as electrolyte, has the following electrode reactions and potentials:

Hydrogen electrode (anode) $2H_2(g) + 4OH^-(aq) \longrightarrow 4H_2O(l) + 4e^-$ $\quad -0{\cdot}828\ V$

Oxygen electrode (cathode) $O_2(g) + 2H_2O(l) + 4e^- \longrightarrow 4OH^-(aq)$ $\quad +0{\cdot}401\ V$

Calculate the e.m.f. of such a fuel cell.

Calculate the maximum energy in joules produced when one mole of hydrogen is consumed at 25°C.

If the standard enthalpy change for the formation of water at 25°C is $-285{\cdot}8\ kJ\ mol^{-1}$, calculate the standard entropy change for the reaction at that temperature.

(1 Faraday = 96 500 coulombs) (O & C)

13 (a) Describe, with the aid of a labelled diagram, an experiment to measure the standard electrode potential of silver and write an equation representing the cell reaction.

(b) Construct a cycle of the Born-Haber type for the formation of silver ions in aqueous solution from solid silver. Name the enthalpy change in each step and indicate its sign.

(c) By reference to the following data, discuss possible methods of preparation of fluorine and chlorine.

$$\tfrac{1}{2}F_2(g) + e^- \longrightarrow F^-(aq) \qquad E^\ominus = +2{\cdot}87\ V$$
$$\tfrac{1}{2}Cl_2(g) + e^- \longrightarrow Cl^-(aq) \qquad E^\ominus = +1{\cdot}36\ V$$
$$MnO_4^-(aq) + 8H^+(aq) + 5e^- \longrightarrow Mn^{2+}(aq) + 4H_2O \qquad E^\ominus = +1{\cdot}51\ V$$
$$MnO_2(s) + 4H^+(aq) + 2e^- \longrightarrow Mn^{2+}(aq) + 2H_2O \qquad E^\ominus = +1{\cdot}23\ V$$

(JMB)

Chapter 11

Acids and bases

11.1 Early ideas

As early as 1663 Robert Boyle noted that a class of chemical compounds existed that had a similar set of properties, different from those of other compounds. These substances are called acids, and among their properties can be listed the following:

(a) Sour taste
(b) Ability to change the colour of vegetable dyes (indicators)
(c) High solvent power (dissolving ability)
(d) Tendency (for some) to react with certain metals to evolve hydrogen.

Other chemical compounds, called alkalis, were noted for their ability to produce a 'soapy feel' in water, affect vegetable dyes, but above all to neutralise acids to produce a salt and water. It soon became apparent that other substances shared this last property with alkalis—the oxides and hydroxides of metals insoluble in water—and the term base, introduced by Rouelle in 1754, was used to cover all compounds that on reaction with an acid formed a salt and water only.

11.2 The Arrhenius theory of acids and bases

In 1884 Arrhenius suggested that substances which contained hydrogen and gave hydrogen ions in aqueous solution were acids; similarly, bases were substances which produced hydroxyl ions when dissolved in water. It is now known that the hydrogen ion is hydrated in solution as the hydroxonium ion, H_3O^+, and indeed some recent work has suggested that hydration may proceed further to give $H^+(H_2O)_4$ or $H_9O_4^+$ ions. In this book hydrogen ions in aqueous solution will be represented as $H_3O^+(aq)$ ions or sometimes simply as $H^+(aq)$ ions.

The ionisation of hydrogen chloride and nitric acid, both covalent compounds, in water can be represented thus:

$$HCl(aq) + H_2O(l) \longrightarrow H_3O^+(aq) + Cl^-(aq)$$
$$HNO_3(aq) + H_2O(l) \longrightarrow H_3O^+(aq) + NO_3^-(aq)$$

The dissociation of a base in aqueous solution follows a similar course. Thus the dissociation of potassium and sodium hydroxides, both existing in the form of ions in the solid state (electrovalent compounds), can be represented as:

$$K^+OH^-(s) + H_2O(l) \longrightarrow K^+(aq) + OH^-(aq)$$
$$Na^+OH^-(s) + H_2O(l) \longrightarrow Na^+(aq) + OH^-(aq)$$

The neutralisation of an acid by a base, for example the reaction between hydrochloric acid and sodium hydroxide in aqueous solution, can be represented thus:

$$H_3O^+(aq) + Cl^-(aq) + Na^+(aq) + OH^-(aq) \longrightarrow 2H_2O(l) + Cl^-(aq) + Na^+(aq)$$

or, since chloride and sodium ions occur on both sides of the equation, more simply as:

$$H_3O^+(aq) + OH^-(aq) \longrightarrow 2H_2O(l)$$

The Arrhenius concept of acids and bases thus allows the process of neutralisation to be expressed by one simple equation. Furthermore, it predicts that interaction between dilute strong acids and bases, i.e. those that are completely dissociated in solution, should evolve a constant amount of heat. This prediction is borne out in practice when it is found that any strong acid neutralises a strong base in dilute solution to form one mole of water with the evolution of 57·3 kJ mol^{-1}.

11.3 The Brønsted-Lowry theory of acids and bases

The Arrhenius theory of acid-base reactions only applies to reactions taking place in water. Many reactions that occur in other solvents are of a similar nature and it is logical to extend this simple theory to embrace as large a field of similar reactions as possible. The Brønsted-Lowry theory does this.

Before discussing the theory, it is worth while considering what part the solvent plays in the formation of solvated hydrogen ions. Experiment shows that a solution of hydrogen chloride in toluene will not conduct an electric current—so presumably no ions are present—nor will the solution affect indicator paper; yet the introduction of a little water immediately allows a current to flow and litmus paper turns red. An explanation of these results lies in the different natures of toluene and water; thus, whereas toluene is non-polar, water is extensively polarised. Water molecules are considered to interact with hydrogen chloride to abstract a proton by the following scheme:

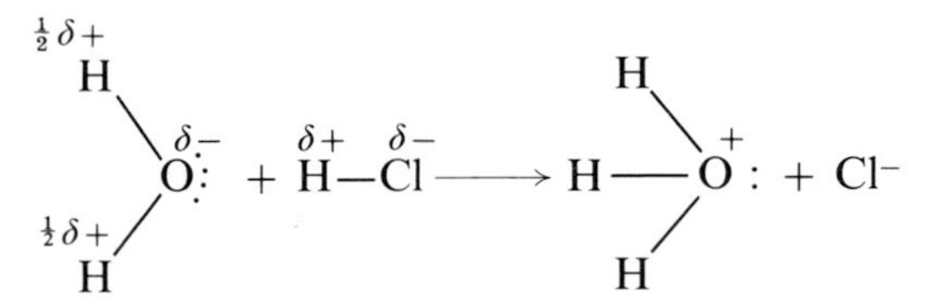

Other polar solvents are able to produce solvated hydrogen ions by an exactly similar type of reaction.

The Brønsted-Lowry theory starts from the definition of an acid as 'a substance existing as molecules or ions that can donate protons, i.e. a proton donor'; a base is consequently 'any substance existing as molecules or ions that can accept protons, i.e. a proton acceptor'. Consider the interaction of ethanoic acid and water where it is known that an equilibrium is set up:

$$\underset{\text{Acid (1)}}{CH_3COOH(aq)} + \underset{\text{Base (2)}}{H_2O(l)} \rightleftharpoons \underset{\text{Base (1)}}{CH_3COO^-(aq)} + \underset{\text{Acid (2)}}{H_3O^+(aq)}$$

Ethanoic acid is a proton donor (acid (1)) and water the proton acceptor is therefore a base (base (2)); but hydroxonium ions are also proton donors (acid (2)), the proton acceptor in this case being ethanoate ions (base (1)). In examples of this kind Acid (1)/Base (1) and Acid (2)/Base (2) constitute what are called conjugate acid-base pairs; ethanoic acid and ethanoate ions constitute one conjugate acid-base pair and hydroxonium ions and water another. Since water and ethanoate ions are both bases on this theory, the extent to which they compete for protons will determine the extent of the equilibrium; the equilibrium is known to lie far over to the left, so that ethanoate ions are a stronger base than water, i.e. ethanoate ions are better proton acceptors.

For the hydrogen chloride-water system, the equilibrium is far over to the right, so by the same argument water is a better base than are chloride ions:

$$\underset{\text{Acid (1)}}{HCl(aq)} + \underset{\text{Base (2)}}{H_2O(l)} \rightleftharpoons \underset{\text{Base (1)}}{Cl^-(aq)} + \underset{\text{Acid (2)}}{H_3O^+(aq)}$$

In this example hydrogen chloride and chloride ions constitute one conjugate acid-base pair, the other being hydroxonium ions and water.

So far two examples in which water is involved have been discussed; but the power of this theory lies in the logical extension of acid-base reactions in non-aqueous solvents. For hydrogen chloride dissolved in pure ethanoic acid an equilibrium is set up containing appreciable quantities of unionised hydrogen chloride:

$$\underset{\text{Acid (1)}}{HCl} + \underset{\text{Base (2)}}{CH_3COOH} \rightleftharpoons \underset{\text{Base (1)}}{Cl^-} + \underset{\text{Acid (2)}}{CH_3COOH_2^+}$$

Hydrogen chloride is therefore not such a strong acid in pure ethanoic acid because the latter is not such a good proton acceptor as water, i.e. ethanoic acid competes with chloride ions for the protons less effectively than does water. In fact, acids which appear equally strong in water can usually be differentiated in weaker bases like ethanoic acid. For instance, conductivity measurements on their solutions in ethanoic acid show that the strengths of the following mineral acids decrease in the order:

$$HClO_4 > HBr > H_2SO_4 > HCl > HNO_3$$

The reversible thermal dissociation of ammonium chloride, in which no solvent is involved, and similar reactions, come under the scope of the Brønsted-Lowry treatment as seen below:

$$\underset{\text{Base (1)}}{NH_3} + \underset{\text{Acid (2)}}{HCl} \rightleftharpoons \underset{\text{Acid (1)}}{NH_4^+} + \underset{\text{Base (2)}}{Cl^-}$$

(ions shown separately for convenience)

11.4 The Brønsted-Lowry theory applied to salt hydrolysis

Water is a very poor conductor of electricity; nevertheless the fact that it allows small currents to pass means that it must contain some ions. Experiment shows that at 298 K one dm^3 of pure water contains 10^{-7} moles of $H_3O^+(aq)$ and an equal quantity of $OH^-(aq)$. It is therefore clear that in the self-ionisation of water some molecules are acting as an acid and an equal number as a base:

$$\underset{\text{Acid (1)}}{H_2O(l)} + \underset{\text{Base (2)}}{H_2O(l)} \rightleftharpoons \underset{\text{Base (1)}}{OH^-(aq)} + \underset{\text{Acid (2)}}{H_3O^+(aq)}$$

When water is added to a substance that is capable of acting as a proton acceptor it will act as an acid; similarly, in the presence of a proton donor, water will act as a base.

An aqueous solution of sodium carbonate shows an alkaline reaction to indicators (owing to the presence of $OH^-(aq)$ ions). In solution sodium and carbonate ions are present, the latter being a strong proton acceptor; the alkalinity of the solution is thus due to the reaction:

$$\underset{\text{Acid (1)}}{H_2O(l)} + \underset{\text{Base (2)}}{CO_3^{2-}(aq)} \rightleftharpoons \underset{\text{Base (1)}}{OH^-(aq)} + \underset{\text{Acid (2)}}{HCO_3^-(aq)}$$

The highly charged small cations of metals, e.g. those of transition metals, are strongly hydrated in aqueous solution. Furthermore, the small, highly charged cations exert a considerable attraction on the oxygen atoms of the water molecules, thereby weakening the links between the hydrogen and oxygen atoms. Under these conditions solvent water molecules are able to act as a base and thus give rise to an acid solution. Aqueous solutions of iron (III) chloride, nickel (II) sulphate and copper (II) nitrate, for instance, show acid reactions:

$$\underset{\text{Acid (1)}}{Fe(H_2O)_6^{3+}(aq)} + \underset{\text{Base (2)}}{H_2O(l)} \rightleftharpoons \underset{\text{Base (1)}}{Fe(H_2O)_5(OH)^{2+}(aq)} + \underset{\text{Acid (2)}}{H_3O^+(aq)}$$

There is little tendency for the chloride, sulphate and nitrate anions to detach hydrogen ions from water molecules, and thereby counteract this acidity, because they are poor proton acceptors.

There are some salts that show a neutral reaction in aqueous solution, principally those of Group 1A and 2A metals in combination with the chloride, sulphate and nitrate radicals. The reason for this is that the cations are poor proton donors when hydrated and the anions are poor proton acceptors.

11.5 Alkalis, acids and amphoteric hydroxides

Alkalis, oxyacids and amphoteric hydroxides contain one or more OH groups in association with an element E which may be united with other atoms. In the presence of a polar solvent such as water, dissociation of E—O—H can take place in two ways. If the element E is one of the Group 1A or 2A metals (very electropositive) dissociation into E^+ and OH^- ions occurs; indeed these hydroxides are predominantly ionic in the solid state.

$$\text{(Alkali) } E{-}O{-}H(aq) + H_2O(l) \longrightarrow E^+(aq) + OH^-(aq)$$

If, however, the group OH is bound to a non-metal (more electronegative than metals) electrons are withdrawn from the oxygen atom towards the element E and in the presence of water ionisation into EO^- anions and H_3O^+ cations occurs:

$$\text{(Acid) } E—O—H(aq) + H_2O(l) \longrightarrow EO^-(aq) + H_3O^+(aq)$$

The nature of the element E thus has a pronounced effect on the acid-base properties of hydroxides.

For an element E of intermediate electronegativity there may be occasions when the hydroxide behaves as an alkali and others when it more resembles an oxyacid, and this is so for some hydroxides, e.g. zinc hydroxide and lead (II) hydroxide. Both these hydroxides are only sparingly soluble in water; even so the resulting solutions slowly turn red litmus blue, leaving little doubt that they act as alkalis under these conditions. However, in the presence of a strong alkali, e.g. sodium hydroxide, sodium salts are formed which can only be interpreted in terms of acidic tendencies. Hydroxides such as these are said to be amphoteric and their behaviour is easily explained by assuming an equilibrium to be set up in solution; thus for zinc hydroxide:

$$\begin{array}{l} 2OH^-(aq) + Zn^{2+}(aq) \rightleftharpoons Zn(OH)_2(s) + H_2O(l) \rightleftharpoons Zn(OH)_4{}^{2-}(aq) + 2H_3O^+(aq) \\ \updownarrow 2H_3O^+(aq) \qquad\qquad\qquad\qquad\qquad\qquad\qquad\qquad \updownarrow 2OH^-(aq) \\ 4H_2O(l) \qquad\qquad\qquad\qquad\qquad\qquad\qquad\qquad\qquad 4H_2O(l) \end{array}$$

In water the equilibrium lies over to the left, i.e. there is an excess of hydroxyl ions compared with hydroxonium ions and hence an alkaline reaction. The equilibrium can be driven further over to the left by adding hydroxonium ions (a strong acid), which combine with the hydroxyl ions to give water:

$$2OH^-(aq) + 2H_3O^+(aq) \rightleftharpoons 4H_2O(l)$$

The zinc hydroxide consequently dissolves completely and the solution contains zinc ions. If hydroxyl ions are added (a strong alkali), the equilibrium is shifted over to the right as hydroxonium ions in the equilibrium mixture are removed to produce water:

$$2H_3O^+(aq) + 2OH^-(aq) \rightleftharpoons 4H_2O(l)$$

The zinc hydroxide again dissolves completely, but in this instance the solution contains zincate, $Zn(OH)_4{}^{2-}(aq)$, ions.

11.6 The strengths of oxyacids in aqueous solution

As the last section has shown, the tendency for an hydroxide to ionise into anions and hydrogen ions (hydroxonium ions in water) is enhanced if the atom to which the OH group is bonded is very electronegative; thus the inclusion of other highly electronegative atoms in the molecule should lead to an increase in acid strength. The strength of an acid in water is measured by its tendency to produce hydroxonium ions; the following equation describes this tendency:

$$HOE(aq) + H_2O(l) \rightleftharpoons H_3O^+(aq) + OE^-(aq)$$

The dissociation constant for this reaction—the first dissociation constant if the acid contains more than one acidic hydrogen atom—is given by K in the usual expression:

$$K = \frac{[H_3O^+(aq)][OE^-(aq)]}{[HOE(aq)]}$$

The concentration of water remains virtually constant for dilute solution. A high value of K means that the particular acid is strong and a low value that the acid is weak. For a series of acids containing the same number of hydroxide groups, K is found to increase in the order:

$$H{-}O{-}Cl < H{-}O{-}Cl \rightarrow O < H{-}O{-}\overset{\overset{O}{\uparrow}}{Cl} \rightarrow O < H{-}O{-}\underset{\underset{O}{\downarrow}}{\overset{\overset{O}{\uparrow}}{Cl}} \rightarrow O$$

(one hydroxide group)

$$H{-}O{-}N{=}O < H{-}O{-}N\begin{smallmatrix}\nearrow O\\ \searrow\!\!\!= O\end{smallmatrix}$$

(one hydroxide group)

$$(H{-}O)_2S{=}O < (H{-}O)_2S({=}O)_2$$

(two hydroxide groups)

The effect of introducing extra oxygen atoms clearly has a pronounced effect on the acidity of these molecules.

11.7 The Lewis theory of acids and bases

According to this theory an acid is 'a substance which can accept a pair of electrons (an electron pair)' whereas a base is 'a substance which can donate a pair of electrons'. The scope of acid-base reactions is thus considerably extended to embrace reactions in which protons are not involved. The reaction between boron trifluoride and ammonia is an example:

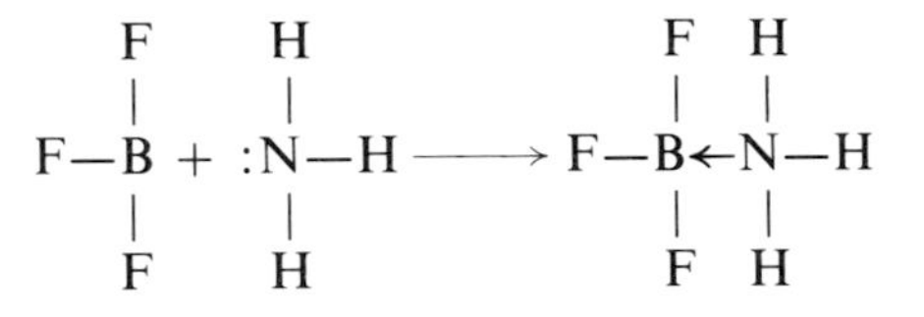

Lewis acids, e.g. boron trifluoride, have important parts to play in catalysing many organic reactions.

Other examples of Lewis acid-base behaviour include the formation of complex ions, e.g.

$$\underset{\text{Acid}}{Cu^{2+}} + \underset{\text{Base}}{4\,{:}NH_3} \longrightarrow [Cu(NH_3)_4]^{2+}$$

and the reaction of sulphur trioxide with metallic oxides, e.g.

$$Ca^{2+}O^{2-}(s) + SO_3(s) \longrightarrow Ca^{2+}SO_4^{\,2-}(s)$$

or

$$\underset{\text{Base}}{O^{2-}(s)} + \underset{\text{Acid}}{SO_3(s)} \longrightarrow SO_4^{\,2-}(s)$$

11.8 Suggested reading

L. Davies, *Theories of Acids and Bases—A review,* School Science Review, No. 152. Vol. 44, 1962.

Ralph G. Pearson, *Hard and Soft Acids and Bases, HSAB, Part I, Fundamental Principles,* Chemical Education, No. 9, Vol. 45, 1968.

11.9 Questions on chapter 11

1 Give and explain a definition of an acid
   (a) making use of the concept of electrons and
   (b) without the use of this concept.
   Are there any substances covered by one of the definitions you give which would not be covered by the other? Which do you consider to be the better definition, and why? (O & C)

2 What is the Brønsted-Lowry definition of (a) an acid and (b) a base? In some reactions water functions as an acid while in others it acts as a base; give one example of each type of behaviour.

3 Discuss the Arrhenius and the Brønsted-Lowry theories of acids and bases and show that the latter theory is more embracing.

4 Write equations in terms of the Brønsted-Lowry concept for each of the following reactions and indicate the conjugate acid-base pairs in each case:
   (a) the interaction of ethanoic acid and water,
   (b) the reaction between ammonia and hydrogen chloride,
   (c) the hydrolysis of a solution of iron (III) chloride,
   (d) the hydrolysis of a solution of potassium carbonate,
   (e) the reaction between an amide ion, NH_2^-, and water.

5 Explain why solutions of copper (II) sulphate, barium chloride and sodium sulphide show respectively an acidic reaction, a neutral reaction and an alkaline reaction to indicators.

6 Distinguish between alkalis, acids and amphoteric hydroxides. Discuss the following statement, giving illustrative examples:
   'Alkalis, oxyacids and amphoteric hydroxides contain one or more –OH groups in association with an element E which may be united with other atoms; in the presence of a polar solvent such as water, dissociation of E—O—H can take place in two ways'.

7 Explain why substances such as lead (II) hydroxide and zinc hydroxide can function as acids and bases under different conditions.

8 Sodium hydroxide is a strong base, silicic acid, $SiO(OH)_2$ is a weak acid, and perchloric acid, ClO_3OH, is a strong acid in aqueous solution; what explanation can you give for this behaviour?

9 In each pair given below, which acid of the two would you expect to be the stronger?
   (a) carbonic acid and nitric acid,
   (b) sulphurous acid and sulphuric acid,
   (c) sulphuric acid and hydrogen sulphate ion, HSO_4^-,
   (d) bromic acid, $HBrO_3$, and chloric acid, $HClO_3$.
   Give reasons in each case.

10 In each pair given below, which base of the two would you expect to be the stronger?
   (a) nitrate ion or nitrite ion,
   (b) carbonate ion or hydrogen carbonate ion, HCO_3^-,
   (c) bromate ion, BrO_3^-, or chlorate ion, ClO_3^-,
   (d) phosphate ion, PO_4^{3-}, or sulphate ion, SO_4^{2-}.
   Give reasons in each case.

11 What are meant by Lewis acids and bases? From your definitions show that the interaction of the following can be considered to be acid-base reactions in the Lewis sense.
   (a) $BF_3(g) + NH_3(g) \longrightarrow BF_3.NH_3(s)$
   (b) $Ca^{2+}O^{2-}(s) + SO_3(s) \longrightarrow Ca^{2+}SO_4^{2-}(s)$
   (c) $Ag^+(aq) + 2CN^-(aq) \longrightarrow Ag(CN)_2^-(aq)$

Chapter 12

Occurrence and extraction of metals

12.1 Introduction

Most metals except the very unreactive ones such as silver, gold and platinum are found as minerals (compounds of the metal) mixed with earthy material (gangue). The mixture as such is called an ore. Since the source of metals and the techniques used to extract them are determined by their nature, the Periodic Table should provide a good starting point for discussing the principles involved. The following scheme was originated by Hulme.

12.2 The principal sources of metals

Metal sources may be classified broadly into five main types as shown in the shortened version of the Periodic Table shown in Table 12A.

Table 12A Classification of metal sources into five main types

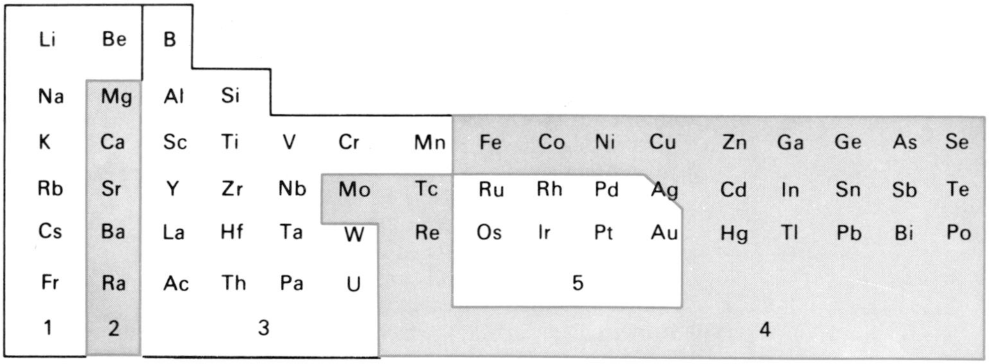

Li	Be	B													
Na	Mg	Al	Si												
K	Ca	Sc	Ti	V	Cr	Mn	Fe	Co	Ni	Cu	Zn	Ga	Ge	As	Se
Rb	Sr	Y	Zr	Nb	Mo	Tc	Ru	Rh	Pd	Ag	Cd	In	Sn	Sb	Te
Cs	Ba	La	Hf	Ta	W	Re	Os	Ir	Pt	Au	Hg	Tl	Pb	Bi	Po
Fr	Ra	Ac	Th	Pa	U			5							
1	2		3								4				

Boron, silicon, arsenic, tellurium and polonium are generally considered to be non-metallic.

Type 1 metals
These are very electropositive metals found as soluble salts, e.g. chlorides, carbonates, sulphates, etc., in the Stassfurt deposits in North Germany, and as constituents of alumino-silicate rocks. The metals are extracted by electrolytic techniques.

Type 2 metals
These are the electropositive alkaline-earth metals, found principally as insoluble carbonates and sulphates (magnesium sulphate is soluble). They are extracted by electrolysis.

Type 3 metals

The main sources of these metals are oxides and mixed oxides. A variety of methods are used to extract these metals, including electrolysis and chemical reduction with carbon, carbon monoxide and more reactive metals.

Type 4 metals

These metals occur principally as sulphides, and less frequently as oxides. The sulphides are almost always converted to oxides—prior to reduction with carbon, carbon monoxide and in some instances hydrogen—or sulphates for subsequent electrolytic treatment.

Type 5 metals

These are unreactive metals and occur in the free state, or as easily reducible compounds.

The borderlines between these five types must not be interpreted too rigidly; for instance copper, zinc and lead occur as carbonates, the only workable sources of iron are its oxides and carbonate, and mercury does occur in the free state and is easily obtained from its sulphide by simple thermal treatment.

12.3 An outline of the stages involved in metal extraction

In order to keep transport costs to a minimum, the ore is concentrated after mining to eliminate as much of the worthless material as possible. One of the most important methods for concentrating sulphide ores is known as the ore flotation process.

In this process the finely pulverised ore is mixed with water, to which one or more chemical 'frothing' agents are added. When air is blown into the mixture a froth is produced and the earthy material is 'wetted' and sinks. The sulphide ore particles, however, rise to the surface in the froth, where they can be skimmed off the surface. After the addition of acid to break up the froth, the concentrated ore is filtered and dried.

Bauxite, hydrated aluminium oxide ($Al_2O_3.2H_2O$), is purified prior to electrolytic reduction by treatment with sodium hydroxide solution which dissolves the aluminium oxide. The addition of small quantities of pure alumina precipitates aluminium hydroxide, which is dried and dehydrated to give pure aluminium oxide. Pure aluminium is now extracted electrolytically.

Roasting of the ore

In this process the concentrated ore is heated in a controlled amount of air. The purpose of this operation might be:

(a) To convert the sulphide ore into its oxide prior to reduction of the oxide to the metal itself:

e.g.

$$\underset{\text{zinc blende}}{2ZnS(s)} + 3O_2(g) \longrightarrow 2ZnO(s) + 2SO_2(g)$$

At this stage impurities such as arsenic are frequently driven off.

(b) To convert the sulphide ore partially into its oxide, which is then reduced to the metal by further reaction with the sulphide ore:

e.g.

$$\underset{\text{galena}}{2PbS(s)} + 3O_2(g) \longrightarrow 2PbO(s) + 2SO_2(g)$$

$$PbS(s) + 2PbO(s) \longrightarrow 3Pb(s) + SO_2(g)$$

Sintering
This involves heating the material until partial fusion occurs and larger, more easily handled material is obtained. Zinc oxide and powdered coke are converted into briquettes by a sintering process prior to smelting.

Smelting
This involves the reduction of the ore to the molten metal at a high temperature. Substances called fluxes are added, their function being to combine with the gangue to form a liquid slag which floats on the surface of the molten metal:

e.g.
$$Fe_2O_3(s) + 3CO(g) \longrightarrow 2Fe(l) + 3CO_2(g)$$
$$\underset{\text{gangue}}{SiO_2(s)} + \underset{\text{flux}}{CaO(s)} \longrightarrow \underset{\text{slag}}{CaSiO_3(l)}$$

Refining
The purpose of refining metals is to make them as pure as is necessary for the particular job they are required to do. Numerous techniques are available, including the following.

Electrolytic refining
Copper is purified electrolytically by making the impure copper the anode of an electrolytic cell, which contains an electrolyte of copper (II) sulphate solution and a thin strip of pure copper as the cathode. By the appropriate choice of voltage, pure copper is transferred from the anode to the cathode.

Zone-refining
This method is applied on a very small scale to produce metals and some non-metals in an extremely high degree of purity. The method depends upon the principle that an impure molten metal will deposit pure crystals on solidifying. The metal, in the form of a rod, is melted over a very narrow region at one end; this molten region is transferred from one end of the rod to the other by slowly moving a furnace. Impurities collect in the molten region and are swept to one end of the metal. Germanium is purified by this method before use as a semiconductor material in transistors.

12.4 Factors determining the methods used to extract metals

As has been seen in section 12.2, the principal source of a particular metal and the type of method used to extract it from its ore are broadly related to the position of the metal in the Periodic Table. For the alkali and alkaline-earth metals (types 1 and 2) electrolysis of the fused halides is the only feasible method, since chemical reducing agents are unsuitable. On the other hand, zinc and chromium can be extracted by chemical reduction of their oxides or by the electrolysis of aqueous solutions of their salts. The enormous quantities of iron needed for conversion into steel for structural purposes demand a cheap process operating on a large scale and here coke provides the answer. For molybdenum and tungsten which are only required on a small scale the reducing agent employed is hydrogen. The supply of cheap coal and hydroelectricity, the nature of the ore—one ore might be more amenable to one type of treatment and another may require a different solution altogether, the scale of operation, the purity demanded of the final product and the value of any by-products—sulphur dioxide is produced in many roasting processes and is vitally important for the manufacture of sulphuric acid—these are some of the factors that must be carefully balanced in deciding which extraction technique to adopt if alternatives are at hand. In deciding whether a particular method is likely to prove possible in the chemical sense, however, it is only by the application of thermodynamics that an answer can be found.

12.5 The electrode potential series as a guide to the general methods used to extract metals

The standard electrode potential of a metal (p. 131) measures the tendency for it to go into solution as hydrated ions. This process is an oxidation of the metal and becomes increasingly difficult as the electrode potential becomes more positive, thus it is more difficult to oxidise copper (standard electrode potential +0·34 V) to copper (II) ions than to carry out the similar change from sodium (standard electrode potential −2·71 V) to sodium ions. The process of extracting metals from their ores is essentially a reduction of the ores to the metals, so it can be argued that a metal higher up in the electrode potential series should be more difficult to extract than one which occurs lower down. This is certainly true in the general sense but, since extraction of metals seldom occurs in aqueous solution, it is hardly surprising that no precise relationship can be found.

Table 12B lists some metals in order of increasing positive electrode potential, together with their main sources and methods of extraction. Three main types of extraction are employed.

Electrolysis

Metals with high negative electrode potentials are extracted by this method. Except for aluminium, halides are employed, since oxyacid salts would decompose thermally at the high temperatures needed to operate the cells and give infusible metallic residues. Fused salts must be used since any water present would lead to selective discharge of hydrogen. Often halide impurities are added to lower the melting point and thus economise on the electrical power needed to keep the salt molten. Chlorine is a valuable by-product.

Chemical reduction

A variety of reducing agents are employed, e.g. carbon, carbon monoxide and reactive metals such as sodium, aluminium and magnesium. Sulphides are not suitable for reduction by carbon, so it is general practice to convert them into oxides by initial roasting in air. The presence of small traces of oxygen have an adverse effect on the properties of titanium and this metal is obtained by reduction, in the complete absence of air, of very pure titanium (IV) chloride with magnesium.

Thermal reduction

Mercury is obtained from its sulphide by simply heating in air. Platinum is obtained in a similar manner from ammonium hexachloroplatinate (IV) into which the ore is first converted. This metal and also silver and gold—which need special methods for their production—occur as the free metals.

Table 12B Extraction of metals

Electrode Process	Standard Electrode Potential/V	Main Occurrence	Main Method of Extraction	Equation for Extraction
Li, Li^+	−3·04	Spodumene $LiAl(SiO_3)_2$	Electrolysis of fused LiCl with KCl added	↑
K, K^+	−2·92	Carnallite $KCl.MgCl_2.6H_2O$	Reduction of fused KCl with Na vapour	
Ba, Ba^{2+}	−2·90	Witherite $BaCO_3$ Barytes $BaSO_4$	Electrolysis of fused $BaCl_2$	
Sr, Sr^{2+}	−2·89	Strontianite $SrCO_3$ Celestine $SrSO_4$	Electrolysis of fused $SrCl_2$	
Ca, Ca^{2+}	−2·87	Limestone $CaCO_3$ Gypsum $CaSO_4$	Electrolysis of fused $CaCl_2$ and CaF_2	Most involve electrolytic reduction: $M^{n+} + ne^- \longrightarrow M$
Na, Na^+	−2·71	Rock salt NaCl Chile saltpetre $NaNO_3$	Electrolysis of fused NaCl with $CaCl_2$ added	
Mg, Mg^{2+}	−2·37	Carnallite $KCl.MgCl_2.6H_2O$ Magnesite $MgCO_3$	Electrolysis of fused $MgCl_2$ with KCl added	
Be, Be^{2+}	−1·70	Beryl $3BeO.Al_2O_3.6SiO_2$	Electrolysis of fused BeF_2 with NaF added	
Al, Al^{3+}	−1·66	Bauxite $Al_2O_3.2H_2O$ Silicate rocks	Electrolysis of Al_2O_3 in molten Na_3AlF_6	↓
Mn, Mn^{2+}	−1·18	Pyrolusite MnO_2 Hausmannite Mn_3O_4	Reduction of oxide with Al or C	$3Mn_3O_4 + 8Al \longrightarrow 9Mn + 4Al_2O_3$
Ti, Ti^{4+}	−0·95	Ilmenite $TiO_2.FeO$ Rutile TiO_2	Reduction of $TiCl_4$ with Mg or Na	$TiCl_4 + 2Mg \longrightarrow Ti + 2MgCl_2$
Zn, Zn^{2+}	−0·76	Zinc blende ZnS Calamine $ZnCO_3$	Reduction of ZnO with C or electrolysis of $ZnSO_4$	$ZnO + C \longrightarrow Zn + CO$
Cr, Cr^{3+}	−0·74	Chromite $FeO.Cr_2O_3$	Reduction of Cr_2O_3 with Al	$Cr_2O_3 + 2Al \longrightarrow 2Cr + Al_2O_3$
Fe, Fe^{2+}	−0·44	Magnetite Fe_3O_4 Haematite Fe_2O_3	Reduction of oxides with CO	$Fe_2O_3 + 3CO \longrightarrow 2Fe + 3CO_2$
Co, Co^{2+}	−0·28	Smaltite $CoAs_2$	Reduction of Co_3O_4 with Al	$3Co_3O_4 + 8Al \longrightarrow 9Co + 4Al_2O_3$
Ni, Ni^{2+}	−0·25	Millerite NiS	Reduction of NiO with CO	$NiO + 5CO \longrightarrow Ni(CO)_4 + CO_2$ $Ni(CO)_4 \longrightarrow Ni + 4CO$
Sn, Sn^{2+}	−0·14	Cassiterite SnO_2	Reduction of SnO_2 with C	$SnO_2 + 2C \longrightarrow Sn + 2CO$
Pb, Pb^{2+}	−0·13	Galena PbS	Reduction of PbO with C	$PbO + C \longrightarrow Pb + CO$
Bi, Bi^{3+}	+0·32	Bismuth glance Bi_2S_3 Bismuthite Bi_2O_3	Reduction of Bi_2O_3 with C	$Bi_2O_3 + 3C \longrightarrow 2Bi + 3CO$
Cu, Cu^{2+}	+0·34	Copper pyrites $CuFeS_2$ Cuprite Cu_2O	Partial oxidation of sulphide ore	$2Cu_2O + Cu_2S \longrightarrow 6Cu + SO_2$
Ag, Ag^+	+0·80	Argentite Ag_2S Occurs as metal	Special methods involving use of sodium cyanide	$Ag_2S + 4NaCN \longrightarrow 2NaAg(CN)_2 + Na_2S$ $2NaAg(CN)_2 + Zn \longrightarrow 2Ag + Na_2Zn(CN)_4$
Hg, Hg^{2+}	+0·85	Cinnabar HgS	Direct reduction of HgS by heat alone	$HgS + O_2 \longrightarrow Hg + SO_2$
Pt, Pt^{2+}	+1·20	Occurs as metal Sperrylite $PtAs_2$	Thermal decomposition of $(NH_4)_2PtCl_6$	$(NH_4)_2PtCl_6 \longrightarrow Pt + 2NH_4Cl + 2Cl_2$
Au, Au^{3+}	+1·50	Occurs as metal	Special methods involving use of sodium cyanide	Similar to that for silver $2NaAu(CN)_2 + Zn \longrightarrow 2Au + Na_2Zn(CN)_4$

12.6 Free energy changes for reactions involving oxides

The standard electrode potential of a metal provides a very good indication of the ease or difficulty of extracting the metal from its compounds. However, since most metals of industrial importance are obtained by chemical reduction of their oxides, the free energy changes occurring during these processes are of more fundamental importance in summarising the principles involved.

A reaction carried out in a closed system takes place with a decrease in free energy (chapter 8). While metallurgical processes may or may not be carried out under these conditions, nevertheless no great inaccuracy is introduced by assuming closed system conditions to apply; the relevant equation is:

$$\Delta G = \Delta H - T\Delta S$$

It is possible to calculate changes in enthalpy ΔH and in entropy ΔS for chemical reactions from a wide range of experimental data. When corrections are made for variations in ΔH and ΔS with temperature (these are small except for a larger change in ΔS when a liquid boils, since there is a large increase in disorder) it becomes possible to calculate free energy changes at a series of temperatures. These values are conveniently plotted on a graph, the method first used by Ellingham; they refer to reactions involving one mole of oxygen at one atmosphere pressure. Figure 12.1 shows the variation of free energies of formation of aluminium oxide and chromium (III) oxide as a function of temperature.

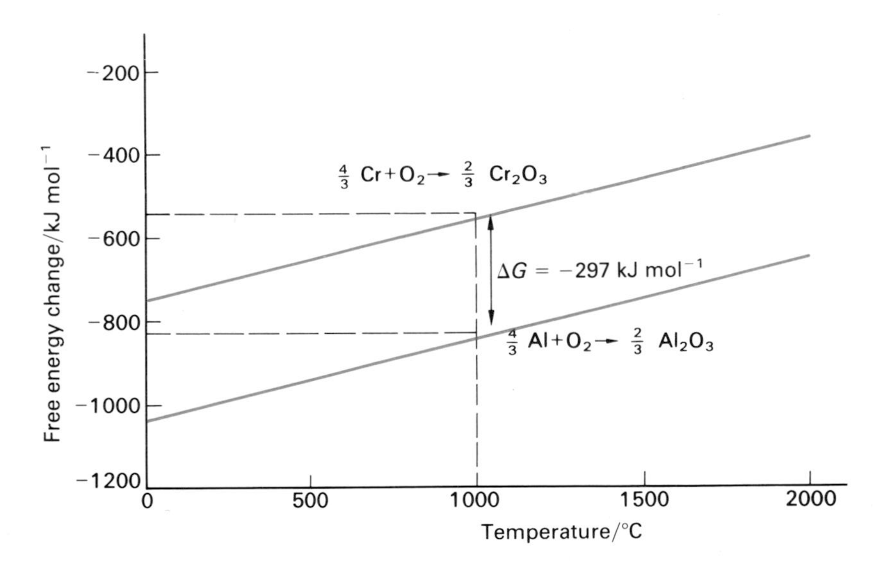

FIG. 12.1. *The free energies of formation of aluminium oxide and chromium* (III) *oxide as a function of temperature*

The free energy of formation of aluminium oxide is more negative than that of chromium (III) oxide at all temperatures, so that aluminium should be capable of reducing the latter to chromium. This is easily seen by considering the two individual reactions:

$$\tfrac{4}{3}\text{Al(s)} + \text{O}_2\text{(g)} \longrightarrow \tfrac{2}{3}\text{Al}_2\text{O}_3\text{(s)} \qquad \Delta G = -827\ \text{kJ mol}^{-1}$$
$$\tfrac{4}{3}\text{Cr(s)} + \text{O}_2\text{(g)} \longrightarrow \tfrac{2}{3}\text{Cr}_2\text{O}_3\text{(s)} \qquad \Delta G = -540\ \text{kJ mol}^{-1}$$

(the free energy changes are for a temperature of 1000°C)

Free energy changes can be treated algebraically just like enthalpies of reaction (p. 101); thus by subtraction:

$$\tfrac{4}{3}Al(s) + \tfrac{2}{3}Cr_2O_3(s) \longrightarrow \tfrac{4}{3}Cr(s) + \tfrac{2}{3}Al_2O_3(s) \qquad \Delta G = -287\,\text{kJ mol}^{-1}$$

Since there is a decrease in free energy, the reaction is thermodynamically possible. The fact that it does not take place at low temperatures is because reactions between solids generally involve high energies of activation.

It is therefore clear that aluminium and magnesium find industrial application as reducing agents because their oxide formation is attended by a large decrease in free energy.

The reducing action of carbon and carbon monoxide can readily be understood in a similar manner. Figure 12.2 shows the variation of free energies of formation of carbon monoxide, carbon dioxide and zinc oxide with temperature. There are three curves for carbon, corresponding to complete oxidation of carbon to carbon dioxide, partial oxidation to carbon monoxide and oxidation of carbon monoxide to the dioxide.

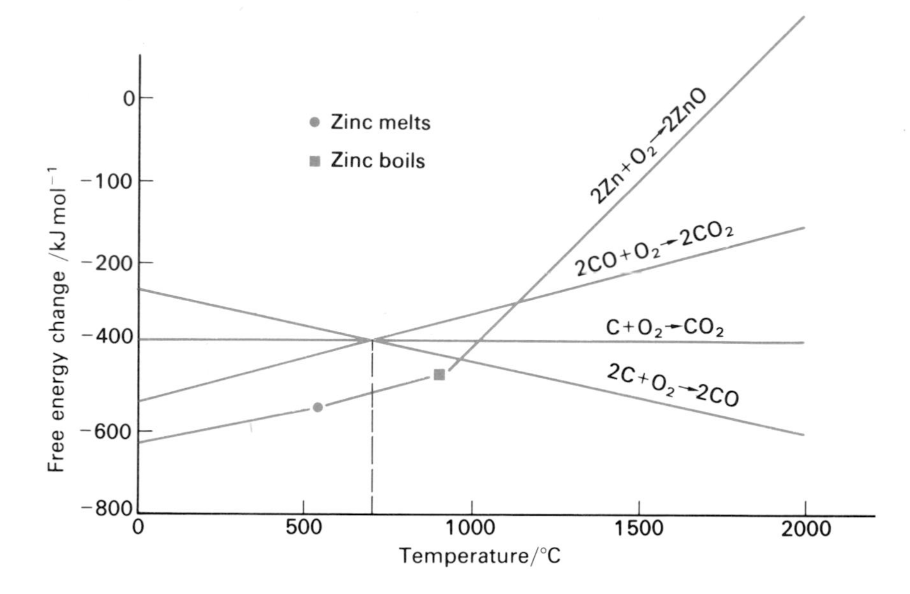

FIG. 12.2. *The variation of the free energies of formation of four systems as a function of temperature*

The three curves pass through a common point at about 710°C; thus the free energies of formation of carbon dioxide from carbon monoxide, and carbon dioxide from carbon are identical at this temperature:

$$2CO(g) + O_2(g) \longrightarrow 2CO_2(g) \quad \Delta G = x\,\text{kJ mol}^{-1}$$
$$C(s) + O_2(g) \longrightarrow CO_2(g) \quad \Delta G = x\,\text{kJ mol}^{-1}$$

Subtracting one equation from the other and rearranging the following result is obtained:

$$CO_2(g) + C(s) \rightleftharpoons 2CO(g) \quad \Delta G = 0$$

i.e. an equilibrium is set up. It is clear from the figure that below a temperature of about 710°C carbon monoxide is a more effective reducing agent than carbon but above this temperature the reverse is true.

All three oxidation curves for the carbon system lie above that for the oxidation of zinc, until a temperature of approximately 1000°C is reached. At this point carbon is thermodynamically capable of reducing zinc oxide to zinc. Since this temperature is greater than the boiling point of zinc (b.p. 907°C) it will be formed as a vapour. The overall equation for the reduction is:

$$ZnO(s) + C(s) \longrightarrow Zn(g) + CO(g)$$

It is interesting to note that the value of carbon as a reducing agent is due to the marked increase in disorder that takes place when carbon (an ordered solid) reacts with one mole of oxygen to give two moles of carbon monoxide. The net effect is an extra mole of gas and hence an increase in disorder (an increase in entropy). It is a fact that in the region of 2000°C carbon is thermodynamically capable of reducing most metallic oxides to metals.

12.7 Suggested reading

D. J. G. Ives, *Principles of the Extraction of Metals,* Royal Institute of Chemistry, Monographs for Teachers, No. 3, 1960.

Alun H. Price, *The extraction of metals from their oxides and sulphides,* School Science Review, No. 219, Vol. 62, 1980.

12.8 Questions on chapter 12

1 Metal sources can be classified broadly into five main types. Give one example of each type, including the main method of extraction for each.

2 Give an account of the principal methods used for the extraction of metals. Indicate, for representative examples, what determines the choice of method in particular instances. (Oxford Schol.)

3 Discuss the statement: 'Free energy changes that occur during reduction of metallic oxides are of more importance than the actual position of a particular metal in the electrode potential series, when discussing the extraction of metals'.

4 Carbon monoxide is a more effective reducing agent for metallic oxides than carbon below a temperature of about 710°C but above this temperature the reverse is true. How do you account for this?

5 In the region of about 2000°C carbon is thermodynamically capable of reducing most metallic oxides to metals but at lower temperatures this is certainly not true. What explanation can you offer?

6 The standard free energies of formation of magnesium oxide and carbon monoxide at temperatures of 1000°C and 2000°C are given below (they refer to the reactions involving one mole of oxygen at one atmosphere pressure). Calculate the free energy change for the reaction:

$$2MgO + C \longrightarrow 2Mg + 2CO$$

at each of the two temperatures and comment on your answers.

$2Mg + O_2 \longrightarrow 2MgO$ $\quad \Delta G^{\ominus}_{1000°C} = -941\ \text{kJ mol}^{-1}$

$\quad \Delta G^{\ominus}_{2000°C} = -314\ \text{kJ mol}^{-1}$

$C + O_2 \longrightarrow 2CO$ $\quad \Delta G^{\ominus}_{1000°C} = -439\ \text{kJ mol}^{-1}$

$\quad \Delta G^{\ominus}_{2000°C} = -628\ \text{kJ mol}^{-1}$

7 The standard free energies of formation of metallic oxides decrease (become less negative) with an increase in temperature whereas the free energy of formation of carbon monoxide increases (becomes more negative) as the temperature rises. What explanation can you give for these observations and what important consequence results?

8 Magnesium and aluminium are frequently used for the extraction of metals such as manganese, chromium and cobalt from their respective oxides. What is the principal reason why these two metals are used as reducing agents?

9 Discuss some of the factors that must be considered when deciding upon the methods and reducing agents employed for extracting metals from their compounds.

10 *Metals are usually extracted from their ores by a reduction process.* Illustrate this statement by reference to magnesium and lead. Why do the methods used for extracting these two metals differ in principle?

Given a mixture of magnesium oxide and lead(II) oxide (lead monoxide), how would you prepare:

(a) magnesium sulphate crystals,

(b) lead(IV) oxide (lead dioxide)? (C)

11 Pure metals are usually isolated from their ores by reduction. Describe the isolation of:

(a) zinc from zinc blende (ZnS),

(b) iron from haematite (Fe_2O_3),

(c) aluminium from bauxite ($Al_2O_3.nH_2O$),

(d) manganese from pyrolusite (MnO_2).

Details of the purification of the metals and concentration of the ores are not required. Comment upon the methods of reduction used and suggest reasons for their choice. (C)

12 The figure below shows the variation of the standard free energy change with temperature for the reaction of certain elements with oxygen according to the equations shown.

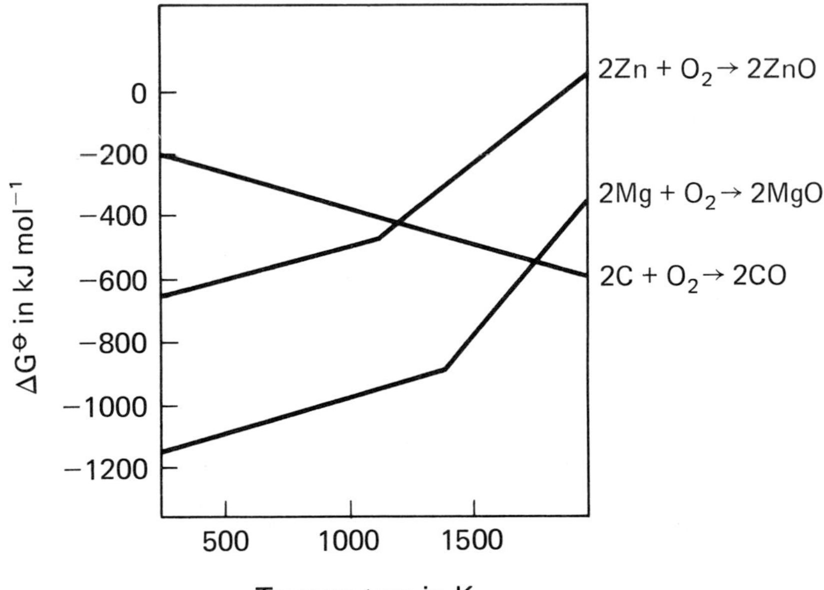

(a) The standard enthalpy changes for all these reactions hardly vary with temperature. Explain qualitatively why the standard free energy change for the reaction $2C + O_2 \rightarrow 2CO$ decreases with a rise in temperature whereas for metals the standard free energy change for the reaction $2M + O_2 \rightarrow 2MO$ increases with a rise in temperature.

(b) Explain carefully how and why the information presented in the figure enables the feasibility to be assessed of the reduction of zinc oxide to zinc by (i) carbon at 800 K, (ii) magnesium at 1000 K and (iii) carbon at 1500 K.

(c) Use the figure to comment on the thermal stability of magnesium oxide, with respect to decomposition into magnesium and oxygen, at 1500 K.

(d) Why is it that several metals are prepared by electrolytic methods? (O)

Chapter 13

Hydrogen

13.1 History and occurrence of hydrogen

Turquet de Mayerne (1655) and Boyle (1672) collected an inflammable gas by reacting iron with sulphuric acid. A century later Cavendish investigated the properties of this gas and called it 'inflammable air', but it was Lavoisier who called it by its present name, hydrogen.

Hydrogen occurs in the free state in some volcanic gases and in the outer atmosphere of the sun; other stars are composed almost entirely of hydrogen. The extremely high temperatures that are commonplace in stars (10^6—10^7 °C) enable nuclear fusion of hydrogen atoms to occur, resulting in a colossal liberation of energy; several reaction schemes have been put forward for this process which ultimately results in the formation of helium:

$$4{}^1_1\text{H} \longrightarrow {}^4_2\text{He} + 2{}^{\;0}_{+1}e \text{ (energy released)}$$

positive electron or positron

The hydrogen liberated during chemical reactions is very readily lost, since the mass of the hydrogen molecule is so small and its speed so high that it can escape from the earth's gravitational field. The main sources of hydrogen are water, and petroleum and natural gas, where it occurs in combination with carbon. The element is an essential ingredient in all living matter, being found in proteins and fats.

13.2 Some general remarks about the chemistry of hydrogen

The hydrogen atom consists of one proton and one electron. Most of the chemistry of hydrogen can be explained in terms of its tendency to acquire the electronic configuration of the noble gas helium; this it can do by gaining an additional electron to give the hydride ion, H^-, by sharing its electron and accepting a part share of an electron from another atom as in the hydrogen molecule H—H, and by accepting a lone pair of electrons, which it does as a proton when it combines with, for example, water and ammonia to give the hydroxonium, H_3O^+, and ammonium, NH_4^+, ions respectively. These three distinct types of behaviour are discussed later in the chapter.

13.3 Preparation of hydrogen

Hydrogen can readily be obtained by the action of certain metals on water (or steam) and dilute acids. Nitric acid and concentrated sulphuric acid must be avoided, since hydrogen appears as water in these instances and oxides of nitrogen and sulphur dioxide are evolved respectively. It is also formed when water is electrolysed in the presence of small amounts of acids, alkalis and salts, steam is passed over heated carbon, and when an ionic hydride, e.g. $Ca^{2+}(H^-)_2$, is hydrolysed.

Action of water on metals

This is essentially a process in which hydrogen ions are reduced.

$$2H^+(aq) + 2e^- \longrightarrow H_2(g)$$

Metals with high negative electrode potentials (Groups 1A and 2A) react with cold water, e.g.

$$2Na(s) + 2H_2O(l) \longrightarrow 2Na^+OH^-(aq) + H_2(g)$$

Magnesium reacts only slowly with water but vigorously with steam; the oxide rather than the hydroxide is formed:

$$Mg(s) + H_2O(g) \longrightarrow Mg^{2+}O^{2-}(s) + H_2(g)$$

Action of dilute acids on metals

This again is a reduction of hydrogen ions. All metals higher than hydrogen in the electrode potential series react with dilute hydrochloric and sulphuric acids to give hydrogen; the more negative the electrode potential of the metal the more vigorous the reaction. Lead does not show much evidence of reaction with dilute hydrochloric or sulphuric acids since an insoluble layer of salt prevents appreciable action.

Hydrogen is normally prepared in the laboratory by reacting zinc with dilute hydrochloric acid (1 vol. of conc. hydrochloric acid to 1 vol. of water). The gas, however, is frequently contaminated with hydrogen sulphide and arsine from traces of zinc sulphide and arsenic in the zinc.

$$Zn(s) + 2H^+(aq) \longrightarrow Zn^{2+}(aq) + H_2(g)$$

Very pure zinc reacts only slowly with these acids and this phenomenon has been attributed to a high energy of activation for one or both of the reactions:

$$2H^+(aq) + 2e^- \longrightarrow 2H(g)$$
$$2H(g) \longrightarrow H_2(g)$$

The reaction can be catalysed by either copper or platinum, which by themselves do not evolve hydrogen from acids, the hydrogen being evolved from the non-zinc surface. Platinum is known to be a very effective catalyst for the recombination of hydrogen atoms.

Action of strong alkalis on zinc and aluminium

Zinc, and more particularly aluminium, react with aqueous solutions of sodium and potassium hydroxides to give hydrogen:

$$Zn(s) + 2OH^-(aq) + 2H_2O(l) \longrightarrow \underset{\text{zincate ion}}{Zn(OH)_4{}^{2-}(aq)} + H_2(g)$$

$$2Al(s) + 2OH^-(aq) + 6H_2O(l) \longrightarrow \underset{\text{aluminate ion}}{2Al(OH)_4{}^-(aq)} + 3H_2(g)$$

Hydrolysis of ionic hydrides

In this type of reaction a hydride ion is oxidised to hydrogen by loss of an electron; it can equally well be regarded as an acid-base reaction in the

Brønsted-Lowry sense (p. 146), e.g.

$$Ca^{2+}(H^-)_2(s) + 2H_2O(l) \longrightarrow Ca^{2+}(OH^-)_2(aq/s) + 2H_2(g)$$

or

$$H^- + H_2O(l) \longrightarrow OH^-(aq) + H_2(g)$$

13.4 Manufacture of hydrogen

Action of steam on coke or hydrocarbons

The first process involving coke is carried out by passing steam over white-hot coke at a temperature of about 1200°C. The reaction is endothermic and the temperature falls to 800°C. A blast of air is now turned on to raise the temperature to 1200°C again; the two reactions are:

$$C(s) + H_2O(g) \longrightarrow \underset{\text{water gas}}{CO(g) + H_2(g)} \qquad \text{(endothermic)}$$

$$2C(s) + (O_2(g) + 4N_2(g)) \longrightarrow \underset{\text{producer gas}}{2CO(g) + 4N_2(g)} \qquad \text{(exothermic)}$$

The water gas is now treated with more steam at a temperature of about 400°C in the presence of a special iron catalyst; the carbon monoxide is oxidised to carbon dioxide and an extra molecule of hydrogen is formed:

$$\underset{\text{water gas}}{CO(g) + H_2(g)} + H_2O(g) \rightleftharpoons CO_2(g) + 2H_2(g) \qquad \text{(shift reaction)}$$

Carbon dioxide is removed by scrubbing with water under pressure or by treatment with a hot solution of potassium carbonate.

The above method is now almost completely obsolete in the UK and has been replaced by a very similar process which uses methane or naphtha (C_4—C_{10} alkanes) in place of coke. The first stage of the process is operated at a temperature of about 900°C and a pressure of 30 atmospheres with a nickel catalyst, when a mixture of carbon monoxide and hydrogen (synthesis gas (p. 225–226)) is formed. The second stage (the shift reaction) is exactly similar to the water gas/steam reaction above, and carbon dioxide is removed by the same technique.

$$C_nH_{2n+2}(g) + nH_2O(g) \longrightarrow \underset{\text{synthesis gas}}{nCO(g) + (2n + 1)H_2(g)}$$

$$nCO(g) + nH_2O(g) \rightleftharpoons nCO_2(g) + nH_2 \qquad \text{(shift reaction)}$$

Other methods of producing hydrogen

Hydrogen is produced as a by-product during the cracking of hydrocarbons, and in the manufacture of chlorine and sodium hydroxide by electrolysing a concentrated solution of sodium chloride.

13.5 Properties of hydrogen

Physical properties

Hydrogen is a colourless gas without taste or smell. It can be liquefied by compression and cooling in liquid nitrogen, followed by sudden expansion. Liquid hydrogen (b.p. −253°C) becomes solid at −259°C.

Chemical properties
Hydrogen burns in air and, under certain conditions, reacts explosively with oxygen and the halogens, e.g.

$$2H_2(g) + O_2(g) \longrightarrow 2H_2O(l)$$
$$H_2(g) + Cl_2(g) \longrightarrow 2HCl(g)$$

It reacts partially with boiling sulphur to give hydrogen sulphide,

$$H_2(g) + S(l) \rightleftharpoons H_2S(g)$$

and with nitrogen at elevated temperature and pressure in the presence of a catalyst to form ammonia (p. 249). With most of the Group 1A and 2A metals it forms ionic hydrides (p. 176), and with transition metals a series of rather ill-defined compounds—interstitial hydrides—is formed (p. 165). At high temperature it will reduce many metallic oxides to metals; this property is used for the extraction of tungsten and molybdenum from their oxides, e.g.

$$WO_3(s) + 3H_2(g) \longrightarrow W(s) + 3H_2O(l)$$

In the presence of finely divided nickel at 150°C or finely divided platinum or palladium at room temperature, hydrogen can reduce the $-C{=}C-$ and $-C{\equiv}C-$ bonds in organic compounds forming saturated compounds, e.g.

$$H_2C{=}CH_2(g) + H_2(g) \longrightarrow H_3C{-}CH_3(g)$$

This technique is useful in determining the number of such groupings in organic compounds.

13.6 Hydrides

Ionic hydrides

Only the Group 1A and 2A metals are sufficiently electropositive to force the hydrogen atom to accept an electron. These hydrides, which are white crystalline solids, are formed by heating the metal in hydrogen at temperatures up to 700°C. The Group 1A hydrides, e.g. Na^+H^-, have the sodium chloride structure, i.e. each cation is surrounded by six equidistant hydride ions and vice versa; the structures of the Group 2A hydrides, e.g. $Ca^{2+}(H^-)_2$, are more complex.

Except for lithium hydride they decompose before their melting points are reached, and the fact that they contain the hydride ion is proved by electrolysing them in fused alkali halides, when hydrogen is evolved at the anode. The hydride ion is a very reactive entity and all ionic hydrides are decomposed by water and air (probably initiated by traces of water vapour in the air):

$$H^- + H_2O(l) \longrightarrow OH^-(aq) + H_2(g)$$

Covalent hydrides

By far the greatest number of hydrides come under this classification and, except for a few, they are gaseous at ordinary temperature. They are formed by the elements from Group 4B to Group 7B in the Periodic Table.

Representative examples are:

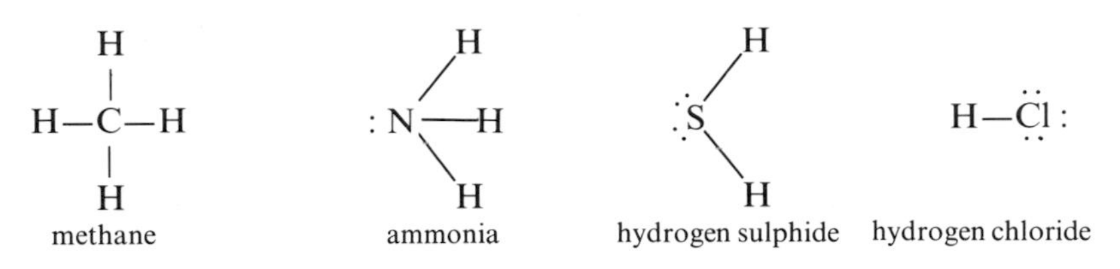

The stability of the hydrides of the elements from a particular periodic group decreases with increasing atomic number (as the element becomes more 'metallic'); thus hydrogen chloride is stable to heat while hydrogen iodide is easily decomposed into its elements. Some are so unstable in the presence of small traces of air, e.g. stannane, SnH_4, that special methods are necessary for their preparation.

In the covalent hydrides, hydrogen is showing its natural tendency to acquire the stable electronic configuration of helium by electron sharing.

Interstitial hydrides

These are ill-defined compounds formed by a number of transition metals in which hydrogen is accommodated in the lattice of the transition element. Some expansion of the metal lattice occurs, since the density of the hydride is less than that of the parent metal. No definite chemical formula can be allocated to these substances, i.e. they are non-stoichiometric. Although the composition can be varied by changes in temperature and pressure, formulae such as $TiH_{1 \cdot 73}$ and $ZrH_{1 \cdot 92}$ have been reported. The uptake of hydrogen is reversible and can in all cases be removed by pumping at a sufficiently high temperature. The use of finely divided palladium for making extremely pure hydrogen depends on this reversibility.

Transition metals form other interstitial compounds, e.g. nitrides and carbides, since nitrogen and carbon, like hydrogen, have small atoms that can penetrate the metal lattice without causing too much distortion.

Covalent hydrides that are electron-deficient

These are compounds which seem to be intermediate in structure between ionic and covalent hydrides. They are polymeric and it is thought that hydrogen atoms act as a bridging unit. Two typical examples are beryllium and magnesium hydrides, the structure of the latter being represented as:

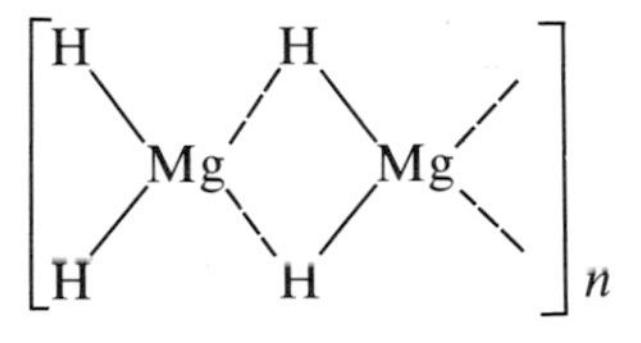

Simple valency theories are inadequate for dealing with structures of this type but more advanced ones are able to provide some explanation of the bonding.

13.7 Active hydrogen

Atomic hydrogen

The hydrogen molecule can be dissociated into atoms using high energy sources, such as a discharge tube containing hydrogen at low pressure, or a high current density arc at high temperature. This dissociation is highly endothermic:

$$H_2(g) \longrightarrow 2H(g) \quad \Delta H^{\ominus}(298\,K) = +435{\cdot}9\,kJ\,mol^{-1}$$

Many metals are able to catalyse the recombination of hydrogen atoms, e.g. platinum and tungsten, which results in the liberation of the same quantity of energy as is needed to effect the dissociation. This effect is used in the atomic hydrogen blowlamp for welding metals. As is to be expected, atomic hydrogen is a very powerful reducing agent, e.g. it will reduce metallic oxides and chlorides to metals, and oxygen to hydrogen peroxide.

Nascent hydrogen

When hydrogen is bubbled into an iron (III) solution no reduction takes place; if the reaction is carried out by adding a mixture of an iron (III) salt in dilute sulphuric acid to zinc an immediate reduction to iron (II) ions takes place. It was at one time thought that an explanation of these observations required the participation of nascent hydrogen (hydrogen at the instant of formation). This notion is now known to be false, since zinc would be expected to be a better reducing agent than hydrogen on account of its higher position in the electrode potential series; indeed, zinc itself effects the reduction without the need for dilute acid:

$$2Fe^{3+}(aq) + Zn(s) \longrightarrow 2Fe^{2+}(aq) + Zn^{2+}(aq)$$

It should be thermodynamically possible for hydrogen gas to reduce iron (III) ions to the iron (II) state, since its standard electrode potential (arbitrarily fixed at zero) is more negative than that for the system $Fe^{3+}(aq)/Fe^{2+}(aq)$ ($E^{\ominus} = +0{\cdot}76\,V$). The fact that it does not do so must be due to a high energy of activation for the reaction, a consequence of the fact that the bond uniting two hydrogen atoms together is a very strong one.

13.8 The proton

The bare proton, H^+, can only be produced when energetic conditions are used, e.g. by applying a high voltage across a discharge tube containing hydrogen at low pressure. In purely chemical systems the proton attracts lone pairs of electrons and assumes the electronic configuration of the noble gas helium as in the hydroxonium and ammonium ions:

$$\left[\begin{array}{c} H{-}\ddot{O}{\rightarrow}H \\ | \\ H \end{array}\right]^{+} \qquad \left[\begin{array}{c} H \\ | \\ H{-}N{\rightarrow}H \\ | \\ H \end{array}\right]^{+}$$

It is interesting to note that solid hydrated chloric (VII) acid, $HClO_4.H_2O$, and ammonium chlorate (VII), $NH_4^+ClO_4^-$, are isomorphous, indicating that the former should be written as $H_3O^+ClO_4^-$.

Some recent results have tended to show that the hydroxonium ion can hydrate further by attracting three more water molecules to give $H^+(H_2O)_4$ or $H_9O_4^+$.

13.9 The isotopes of hydrogen and preparation of some compounds containing deuterium

There are three isotopes of hydrogen of relative masses 1, 2 and 3. They are called ordinary hydrogen, deuterium and tritium respectively and differ in that ordinary hydrogen has no neutrons, deuterium has one and tritium has two neutrons in the nucleus. Tritium is the only one that is radioactive. The ratio of ordinary hydrogen to deuterium in hydrogen compounds is about 6000:1; tritium occurs in even smaller amounts.

On prolonged electrolysis of water, deuterium is concentrated in the residue and it is possible to obtain virtually pure deuterium oxide (D_2O) by this treatment. Its physical properties differ from those of ordinary water; thus it boils at 101·4°C, freezes at 3·8°C and has a specific gravity of 1·10 at 20°C. In chemical properties it behaves almost identically. Deuterium (or heavy hydrogen) can be obtained from deuterium oxide (or heavy water) by reaction with a reactive metal, e.g.

$$2Na(s) + 2D_2O(l) \longrightarrow 2Na^+OD^-(aq) + D_2(g)$$

It is slightly less reactive than ordinary hydrogen but otherwise its properties are almost identical.

Deuterium is used as a tracer for elucidating a wide range of reaction mechanisms, and so are its compounds. Many of these compounds can be readily obtained from deuterium oxide, thus DCl, ND_3 and C_2D_2, the deuterium equivalents of hydrogen chloride, ammonia and ethyne can be obtained as follows:

$$PCl_3(l) + 3D_2O(l) \longrightarrow D_3PO_3(aq) + 3DCl(g)$$
$$(Mg^{2+})_3(N^{3-})_2(s) + 6D_2O(l) \longrightarrow 3Mg^{2+}(OD^-)_2(s) + 2ND_3(g)$$
$$Ca^{2+}C_2{}^{2-}(s) + 2D_2O(l) \longrightarrow Ca^{2+}(OD^-)_2(aq/s) + C_2D_2(g)$$

Deutero-ethyne, C_2D_2, is a useful source of deuterium-containing organic compounds.

Tritium, T, can be obtained by bombardment of deuterium compounds with deuterons:

$$^2_1D + ^2_1D \longrightarrow ^3_1T + ^1_1H$$

It decays by emission of β particles (electrons from the nucleus, 1 neutron $\rightarrow$ 1 proton + 1 electron) to give an isotope of helium:

$$^3_1T \longrightarrow ^3_2He + \beta \text{ particle}$$

13.10 Uses of hydrogen

Until this century only small quantities of hydrogen were required as a fuel in the form of town gas and water gas, for filling balloons and in the oxyhydrogen blowlamp for welding. Nowadays, however, large quantities of the gas are employed in the processes listed below.

(a) Manufacture of ammonia by the Haber process. This is used in turn to manufacture nitric acid, which can then be converted into explosives, dyestuffs and nitrogenous fertilisers.
(b) Manufacture of hydrogen chloride and hydrochloric acid. The hydrogen and chlorine are reacted together; the product is hydrogen chloride, which forms hydrochloric acid with water.
(c) Manufacture of organic chemicals, e.g. methanol, CH_3OH. This compound is obtained by reacting carbon monoxide with hydrogen at 400°C and 300 atmospheres pressure in the presence of zinc oxide and chromium (III) oxide.

(d) Manufacture of margarine. Groundnut oil is reacted with hydrogen in the presence of a nickel catalyst to give a solid fat.
(e) Extraction of some metals from their oxides. Molybdenum and tungsten are extracted by reduction of their oxides with hydrogen.
(f) Liquid hydrogen has been used as a rocket fuel.

13.11 Questions on chapter 13

1 State five reactions of hydrogen which illustrate its chief chemical properties, and give two industrial uses of the gas.
Explain the different types of chemical bonding which occur in the following compounds: (a) sodium hydride, (b) hydrogen chloride,
(c) potassium hydrogen fluoride. (O & C)

2 Discuss the chemistry of hydrogen in relation to its electronic structure. (Oxford Schol.)

3 Metals may react with (a) acids, (b) alkalis or (c) steam, liberating hydrogen in each case. Give one example of each type of reaction.
Account for the types of bonds present in (i) sodium hydride, (ii) water, (iii) the hydrated hydrogen cation.
Explain the meaning of reduction. A solution of iron (III) chloride acidified with hydrochloric acid is unaffected when hydrogen is bubbled through it, but when zinc dissolves in this solution hydrogen is evolved and the iron (III) ions are reduced. Explain this behaviour in terms of electron transfer. (AEB)

4 Outline the chemistry of the hydrides. How far can they be grouped into types? (Oxford Schol.)

5 How is hydrogen obtained on a large scale? What are its principal uses? How and under what conditions does it react with (a) halogens, (b) alkenes, (c) sulphur? (L)

6 How is pure deuterium oxide, D_2O, prepared from ordinary water? Starting with a sample of pure D_2O, how would you prepare, in the laboratory, samples of ND_3, DCl and KOD? (Camb. Schol.)

7 Outline the principal features in the chemistry of ionic and molecular hydrides.
Starting with heavy water, D_2O, as the only source of deuterium, write reaction schemes for the preparation of the following: NaD, ND_3, $CH_2D.CH_2Br$, PD_4I, NaD_2PO_2.
Suggest why the compound $LiAlH_4$ is considerably more soluble than $NaBH_4$ in ether. (O & C)

8 Explain why the chemical properties of isotopes of a given element are closely similar to one another. In what physical properties would you expect differences between isotopes to be shown?
Given a supply of heavy water, D_2O, how would you prepare samples of (a) DI, (b) ND_3, (c) CaD_2, (d) CH_3OD? (Oxford Schol.)

9 Discuss the statement: 'Hydrogen can act as an oxidising agent and as a reducing agent'. Give examples to illustrate your answer.

10 Outline one method for the manufacture of hydrogen from either crude oil or natural gas. State two important uses of hydrogen.
Give explanations and illustrative reactions for the following statements.
(i) The hydrides of the elements Na, P, S, Cl, show increasing acidity with increasing atomic number.
(ii) The hydrides of the elements F, Cl, Br, I, show increasing reducing power with increasing atomic number. (C)

11 (a) Describe in detail the bonding which occurs in the compounds formed between hydrogen and
(i) sodium (in sodium hydride), (ii) carbon (in methane),
(iii) nitrogen (in ammonia).
(b) Describe the reactions, if any, which take place between water and the hydrides of the elements in (a).
(c) Comment upon the significance of the relative values of the following boiling points of the halogen hydrides.

HF	HCl	HBr	HI	
19·5	−85	−67	−36	(°C)

(JMB)

Chapter 14

Group O the noble (inert) gases

14.1 Some physical data of the noble gases

	Atomic Number	Electronic Configuration	First Ionisation Energy/kJ mol^{-1}	van der Waals' Radius/nm	M.p. /°C
He	2	2 $1s^2$	2372	0·120	−271·4 at 30 atm.
Ne	10	2·8 $\ldots 2s^2 2p^6$	2080	0·160	−248·7
Ar	18	2.8.8 $\ldots 3s^2 3p^6$	1519	0·190	−189·2
Kr	36	2.8.18.8 $\ldots 4s^2 4p^6$	1351	0·200	−157
Xe	54	2.8.18.18.8 $\ldots 5s^2 5p^6$	1170	0·220	−111·5
Rn	86	2.8.18.32.18.8 $\ldots 6s^2 6p^6$	1037	—	−71

The noble gases, except for helium, have eight electrons in their outer shell. Since they form chemical compounds less readily than other elements, this closed shell of eight electrons must be an extremely stable one. Theories of valency have been developed according to which it is suggested that elements react to achieve a stable noble gas configuration either by loss and gain of electrons, or by electron sharing (p. 51–54). Although many exceptions are known where electronic configurations other than that of a noble gas are achieved, these simple ideas lead to an understanding of a large range of diverse chemical structures.

14.2 The discovery of argon

In 1785 Cavendish noted that when air was sparked with an excess of oxygen, and the remaining oxygen then removed, a small residue of gas always remained. This observation escaped notice until 1895 when Rayleigh found that the density of nitrogen prepared from ammonia was 1·2505 g dm^{-3} while that of nitrogen prepared from air was 1·2572 g dm^{-3}. By passing nitrogen obtained from the air repeatedly over heated magnesium (which forms magnesium nitride with nitrogen), Ramsay obtained traces of gas which, on spectroscopic examination, showed no signs of nitrogen but contained new lines proved to be characteristic of the new element argon. The gas was contaminated with minute quantities of other noble gases.

14.3 Occurrence and uses of the noble gases

The concentration of these gases in parts per million in the atmosphere are He 5·2 ppm, Ne 18 ppm, Ar 9300 ppm, Kr 1·0 ppm, Xe 0·08 ppm.

Helium

This gas was first discovered by spectroscopic examination of light from the sun; it is formed by nuclear fusion of hydrogen atoms in stellar reactions (p. 161). It occurs, together with natural gas, in various parts of the USA and Canada, some natural gas wells containing as much as 7 per cent of the element. It has probably accumulated as a result of natural radioactive decay processes. Because it is very light and non-inflammable it has been used to inflate the tyres of large aircraft, thus increasing their maximum payload by several hundred pounds. An oxygen/helium mixture is used by divers, since helium is less soluble in the blood than nitrogen, which is notorious for causing decompression sickness, or the 'bends'. It is used to provide inert atmospheres during certain welding operations, and seems to have a bright future in gas-cooled nuclear reactors since it does not become radioactive and is non-corrosive. Liquid helium is used in low temperature research.

Argon

Argon is extracted from the atmosphere, one useful source being Haber process plants, where it tends to accumulate in the recycled gases. It is used in electric lamps to suppress the evaporation of the filament, thus allowing higher temperatures to be used, with a consequent increase in efficiency. Like helium, it is used to provide an inert atmosphere for welding operations.

Neon, krypton and xenon

These noble gases are extracted from liquid air along with argon; they can be separated by fractional distillation and selective absorption on charcoal. Neon is used in shop signs and street lamps; since the characteristic red glow produced when an electrical discharge is passed through neon at low pressure is able to penetrate fog, it finds a use in devices for illuminating airfields. Krypton is used in discharge tubes; there are no significant uses for xenon.

Radon

This gas is radioactive and is not found in the atmosphere. The most stable isotope of radon is ${}^{222}_{86}Rn$, which is obtained during the decay of radium chloride solution:

$${}^{226}_{88}Ra \longrightarrow {}^{222}_{86}Rn + {}^{4}_{2}He$$

It has been used in the treatment of cancer because of its radioactivity.

14.4 Chemical compounds of the noble gases

Following the observation that oxygen reacted with the powerful oxidising agent platinum hexafluoride to give the ionic compound dioxygenyl hexafluoroplatinate, $O_2^+PtF_6^-$, it was reasoned that xenon might react in a similar manner, since the first ionisation energies of molecular oxygen and xenon are almost identical. In 1962 this prediction was verified when it was found that xenon and platinum hexafluoride react together at room temperature to give the solid xenon hexafluoroplatinate, $Xe^+PtF_6^-$. The excitement caused by this discovery led to an active search for other noble gas compounds and in a short time three fluorides of xenon, XeF_2, XeF_4 and XeF_6, had been obtained as white crystalline solids.

$$Xe(g) + 3F_2(g) \xrightarrow[\text{pressure}]{300°C} XeF_6(s)$$
xenon hexafluoride

$$Xe(g) + 2F_2(g) \xrightarrow{400°C} XeF_4(s)$$
xenon tetrafluoride

$$Xe(g) + F_2(g) \xrightarrow[\text{mercury arc}]{\text{light from}} XeF_2(s)$$
xenon difluoride

Other compounds of xenon were obtained by treatment of these fluorides with water; thus partial and complete hydrolysis of xenon hexafluoride gave respectively $XeOF_4$ and XeO_3. Krypton difluoride, KrF_2, has been prepared but is less stable than the xenon analogue, presumably because of the higher ionisation energy of the krypton atom. Radon difluoride, RnF_2, has been prepared and characterised, but no fluorides of helium, neon and argon have yet been isolated.

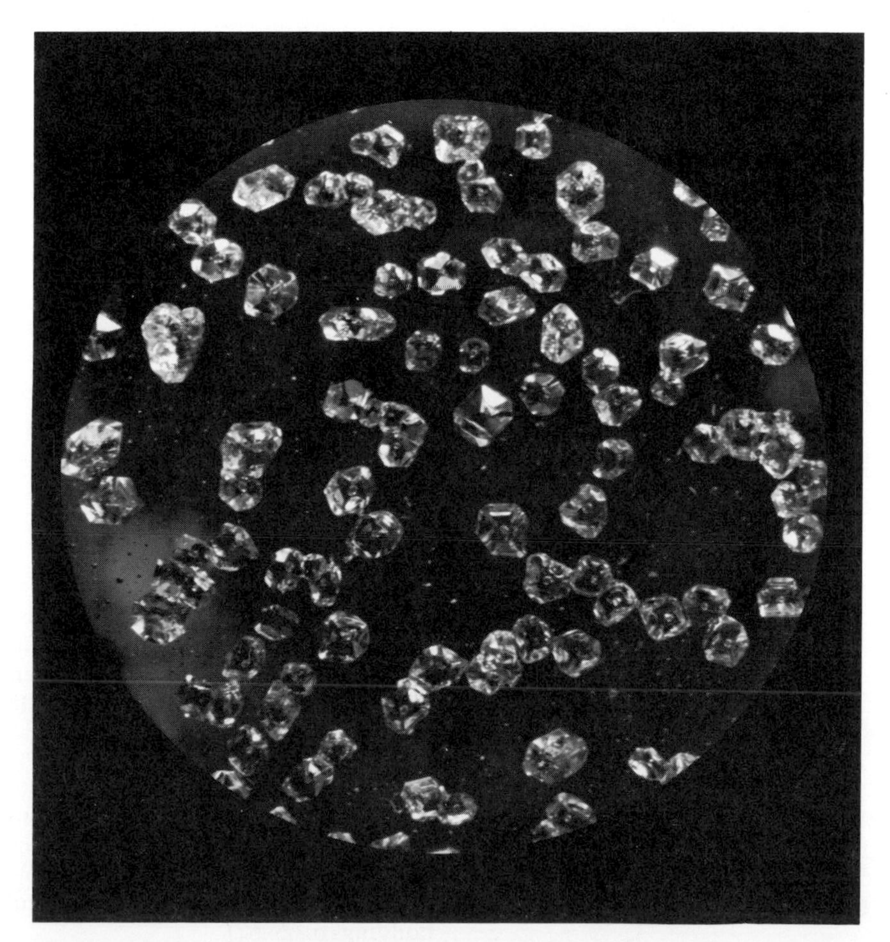

Plate 5. Xenon tetrafluoride crystals in a quartz container (magnification × 50) (By courtesy of Professor Neil Bartlett, University of California)

14.5 Structures of the noble gas compounds

The structures of xenon difluoride (fig. 14.1(a)) and xenon tetrafluoride (fig. 14.1(b)) have been proved to be linear and square planar respectively; there is also evidence that xenon hexafluoride assumes a non-octahedral configuration (fig. 14.1(c)). These structures and also the one for xenon trioxide (fig. 14.1(d)) can be explained by the simple theory of electron pair repulsion developed in chapter 6 (p. 75–79).

FIG. 14.1. *The structures of some xenon compounds*

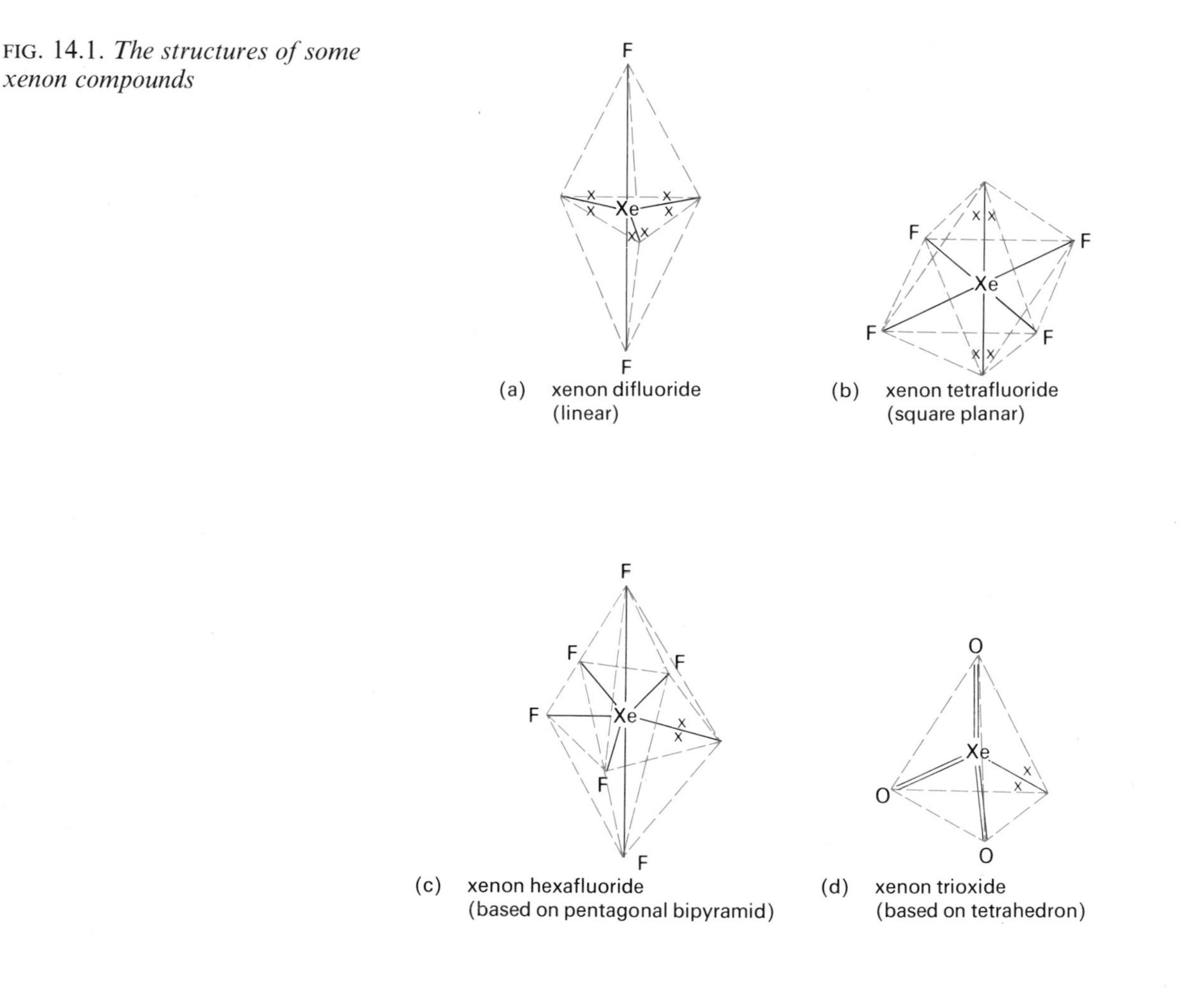

The bonding in these compounds is presumably predominantly covalent and doubtless involves electron promotion from the *p* to the unoccupied *d* levels in the xenon atom. An extremely electronegative entity, e.g. the fluorine atom is necessary to bring this about.

14.6 Hydrates and enclosure compounds of the noble gases

When the noble gases argon, krypton and xenon are compressed with water, hydrates are formed containing six water molecules for each noble gas atom, e.g. $Xe.6H_2O$. These compounds are in reality enclosure compounds (clathrates) in which the noble gas atoms are trapped in a network of water molecules, the water molecules being held together by hydrogen bonding (p. 66–67).

Similar types of clathrates are formed when noble gases at pressures of 10—40 atmospheres are brought in contact with a solution of quinol, $C_6H_4(OH)_2$, in water. On crystallisation, crystals different from those of pure quinol are formed; they contain noble gas atoms trapped in a quinol network and are quite stable. On heating, the hydrogen bonds holding individual quinol molecules together are broken and the noble gas is released.

It must not be thought that clathrates are true chemical compounds, since there is no evidence to suggest that chemical bonds involving the noble gases are present.

14.7 Suggested reading

J. H. Holloway, *Noble Gas Chemistry*, Education in Chemistry, Vol. 10, 140, 1973.
N. K. Jha, *Recent Advances in the Chemistry of Noble Gas Elements*, R.I.C. Reviews, Vol. 4, 147, 1971.
A. G. Sharpe, *Inorganic Chemistry*, Longman, 1981.

14.8 Questions on chapter 14

1. Give an account of the way in which a knowledge of the properties and electronic structures of the noble gases has been used in advancing our understanding of valency and the periodic system of the elements. (O & C)
2. Discuss the following:
   (a) Xenon forms a series of fluorides and oxides.
   (b) Argon has so far resisted attempts to induce it to form chemical compounds.
   (c) The shapes of the compounds of xenon can be predicted by the theory of 'electron pair repulsion'.
3. What is meant by a clathrate? Do the clathrate compounds of the noble gases have chemical bonds between the noble gas and other molecules?
4. Give an account of the isolation and uses of the noble gases.

Chapter 15

Group 1A the alkali metals

15.1 Some physical data of Group 1A elements

	Atomic Number	Electronic Configuration	First Ionisation Energy/kJ mol^{-1}	Standard Electrode Potential/V	Atomic Radius/nm	Ionic Radius/nm	M.p. /°C	B.p. /°C
Li	3	2.1 $1s^2 2s^1$	520	−3·04	0·133	0·060	180	1326
Na	11	2.8.1 $\ldots 2s^2 2p^6 3s^1$	496	−2·71	0·157	0·095	98	883
K	19	2.8.8.1 $\ldots 3s^2 3p^6 4s^1$	418	−2·92	0·203	0·133	64	756
Rb	37	2.8.18.8.1 $\ldots 4s^2 4p^6 5s^1$	403	−2·92	0·216	0·148	39	688
Cs	55	2.8.18.18.8.1 $\ldots 5s^2 5p^6 6s^1$	374	−3·02	0·235	0·169	29	690

15.2 Some general remarks about Group 1A

The elements lithium, sodium, potassium, rubidium, caesium and francium are called the alkali metals. Not much is known about the last-named, since it is radioactive and all its isotopes are exceedingly short-lived; it is formed during the radioactive decay of actinium.

The metals are extremely reactive and electropositive (prone to lose electrons) and exist in combination with other elements or radicals as positive ions, e.g. Na^+Cl^-, $(K^+)_2SO_4^{2-}$; they are therefore never found in the free state in nature. They all adopt the body-centred cubic structure in which each atom is surrounded by eight nearest neighbours with six more atoms only slightly further distant (fig. 15.1).

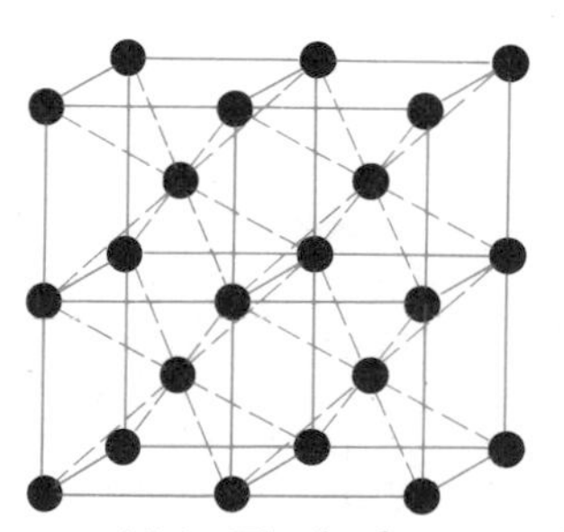

FIG. 15.1. *The body-centred cubic structure of the alkali metals*

All alkali metal atoms have one electron in the outer shell preceded by a closed shell containing eight electrons—except for lithium which has a closed shell of two. In chemical combination this single electron is very readily transferred, giving a unipositive metal ion with the stable electronic configuration of a noble gas, e.g. Na^+(2.8) is isoelectronic with Ne(2.8). Compounds of the alkali metals are therefore generally predominantly ionic and exist as high melting-point solids in which as many ions of opposite charge surround each other as possible (e.g. sodium chloride and caesium chloride structures p. 74). These compounds are generally water soluble and white—unless the anion happens to be coloured.

The alkali metals have lower ionisation energies than any other group of similar elements; even so the removal of the first electron from the caesium atom requires the input of 374 kJ mol^{-1} of energy and for the other alkali metals the energy is somewhat larger still. The reason why the single electron is so readily transferred during chemical combination is because the resulting ions attract each other strongly, with the liberation of energy, and therefore the overall process is highly exothermic (p. 56–57).

The atoms and ions of the alkali metals increase in size with increasing atomic number, since each succeeding member has an extra closed shell of electrons. As the atomic size increases it becomes progressively less difficult to ionise the single valency electron, and the reactivity of the elements increases from lithium to caesium. Lithium is anomalous in having an abnormally high negative electrode potential and this is because the lithium ion, being much smaller than those of the other alkali metals, is more strongly hydrated in aqueous solution (p. 135–136). Lithium is thus a better reducing agent than the other Group 1A metals in aqueous solution but a poorer reducing agent when solvation cannot take place.

All the alkali metal compounds give characteristic flame colorations which are used to identify them—lithium (red), sodium (yellow), potassium (lilac), rubidium (red) and caesium (blue).

Although the alkali metals form predominantly ionic compounds, they can form covalent molecules such as Li_2, Na_2, K_2, etc., which are found to the extent of about 1 per cent in the vapours of these metals. Covalent compounds such as lithium phenyl, C_6H_5Li, exist but they tend to be highly reactive.

15.3 Occurrence, extraction and uses of the alkali metals

Lithium occurs in alumino-silicate rocks, e.g. spodumene, $Li^+Al^{3+}(SiO_3^{2-})_2$ sodium in rock salt, Na^+Cl^-, and Chile saltpetre, $Na^+NO_3^-$, and potassium in carnallite, $K^+Cl^-.Mg^{2+}(Cl^-)_2.6H_2O$, and saltpetre, $K^+NO_3^-$. Compounds of rubidium and caseium are obtained in small amounts during the recrystallisation of naturally occurring salts of the other alkali metals. Francium is produced during the radioactive decay of actinium.

Lithium and sodium are extracted by electrolysis of their fused chlorides, other halides being added to lower the melting point and thus economise on electrical power (see Table 12B, p. 156). Potassium is obtained by reduction of fused potassium chloride with sodium vapour. Rubidium and caesium are obtained by similar methods. Sodium is extracted by the electrolysis of fused sodium chloride to which a little calcium chloride has been added (Downs process). A graphite anode and steel cathode are employed; chlorine is a valuable co-product.

Cathode		**Anode**
Na^+ discharged ←	Na^+Cl^-	→ Cl^- discharged
$2Na^+ + 2e^- \rightarrow 2Na$		$2Cl^- \longrightarrow 2Cl\cdot + 2e^-$
		$2Cl\cdot \longrightarrow Cl_2$

Lithium is used to a slight extent as a component in alloys; in the form of lithium aluminium hydride, $Li^+AlH_4^-$, it finds a use in organic research laboratories, since this compound is a selective reducing agent, e.g. it will reduce a carboxylic acid grouping to an alcohol but generally will leave $C{\equiv}C$ and $C{=}C$ bonds intact. Sodium vapour is used in road lighting. An alloy with potassium is used in a sealed system for withdrawing heat from nuclear reactors, e.g. the Dounreay fast breeder. Over 100 000 tonnes of sodium a year are alloyed with lead for the manufacture of lead tetraethyl, which raised the 'octane rating' of petrol. It reacts at red heat with ammonia and carbon to form sodium cyanide, which is used in the extraction of silver and gold. It has also been used as an alternative to magnesium in the extraction of titanium. In the laboratory it is used as a reducing agent, either alone or as an amalgam.

Potassium is sometimes used as a laboratory reducing agent and caesium, when alloyed with aluminium and barium, is used in the construction of photoelectric cells; but, apart from a few such special uses, these metals and rubidium possess few advantages over sodium, which is cheaper.

15.4 Properties of the alkali metals

Physical properties

The metals are soft and silvery coloured; they are extremely good conductors of heat and electricity. Lithium, sodium and potassium are less dense and rubidium and caesium more dense than water. Because they rapidly tarnish and lose their silvery appearance in air, they are generally stored under oil. Their melting and boiling points decrease along the series from lithium to caesium.

Chemical properties

The metals are very reactive, increasingly so with increasing atomic number; thus lithium reacts quietly with water, sodium and potassium react with increasing vigour, and rubidium and caesium with exceptional violence. In all cases an alkali and hydrogen are produced, e.g.

$$2K(s) + 2H_2O(l) \longrightarrow 2K^+OH^-(aq) + H_2(g)$$

They react with a variety of non-metals when heated to give oxides, sulphides, halides, hydrides, etc., e.g.

$$4Li(s) + O_2(g) \longrightarrow 2(Li^+)_2O^{2-}(s)$$
$$2Na(s) + S(s) \longrightarrow (Na^+)_2S^{2-}(s)$$
$$2K(s) + Cl_2(g) \longrightarrow 2K^+Cl^-(s)$$
$$2Rb(s) + H_2(g) \longrightarrow 2Rb^+H^-(s)$$

Lithium alone reacts with nitrogen to give the nitride $(Li^+)_3N^{3-}$, since both the lithium and nitrogen atoms are very small and the resulting nitride has a very compact structure with a high lattice energy.

The metals burn in a stream of hydrogen chloride and react with ammonia when heated, e.g.

$$2Na(s) + 2HCl(g) \longrightarrow 2Na^+Cl^-(s) + H_2(g)$$
$$2Na(s) + 2NH_3(g) \longrightarrow 2Na^+NH_2^-(s) + H_2(g)$$

15.5 Oxides of the alkali metals

Lithium only forms the monoxide $(Li^+)_2O^{2-}$ when heated in oxygen, sodium forms the monoxide, and the peroxide $(Na^+)_2O_2^{2-}$ if an excess of oxygen is used. The other alkali metals give the superoxide when reacted with oxygen, e.g. $K^+O_2^-$. The lithium ion, by virtue of its small size, is not able to surround itself by sufficient peroxide ions to give a stable crystal lattice and consequently only the monoxide exists. The ions of potassium, rubidium and caesium get progressively larger in this order and are able to form stable structures with the superoxide ion—the largest of the three oxide ions. These three ions are related as follows:

$$O^{2-} \xrightarrow{\frac{1}{2}O_2} O_2^{2-} \xrightarrow{O_2} 2O_2^-$$

All three oxide ions are unstable in the presence of water, e.g.

$$(Li^+)_2O^{2-}(s) + H_2O(l) \longrightarrow 2Li^+OH^-(aq)$$
or
$$O^{2-} + H_2O(l) \longrightarrow 2OH^-(aq)$$

$$(Na^+)_2O_2^{2-}(s) + 2H_2O(l) \longrightarrow 2Na^+OH^-(aq) + H_2O_2(l)$$
or
$$O_2^{2-} + 2H_2O(l) \longrightarrow 2OH^-(aq) + H_2O_2(l)$$

$$2K^+O_2^-(s) + 2H_2O(l) \longrightarrow 2K^+OH^-(aq) + H_2O_2(l) + O_2(g)$$
or
$$2O_2^- + 2H_2O(l) \longrightarrow 2OH^-(aq) + H_2O_2(l) + O_2(g)$$

The above three reactions involve the abstraction of protons from water molecules; the oxide ions thus function as strong bases in the Brønsted-Lowry sense (p. 146).

15.6 The hydroxides of sodium and potassium, M^+OH^-

The hydroxides and other compounds of the alkali metals have similar sets of properties. The descriptive chemistry of these compounds will, therefore, be restricted to those of sodium and potassium.

Manufacture of sodium hydroxide and potassium hydroxide by the Castner-Kellner process and the Gibbs diaphragm cell process

(a) The Castner-Kellner process

In this process a saturated solution of sodium chloride flows through the cell (fig. 15.2) in the same direction as a shallow stream of mercury which constitutes the cathode; the anode consists of a number of titanium

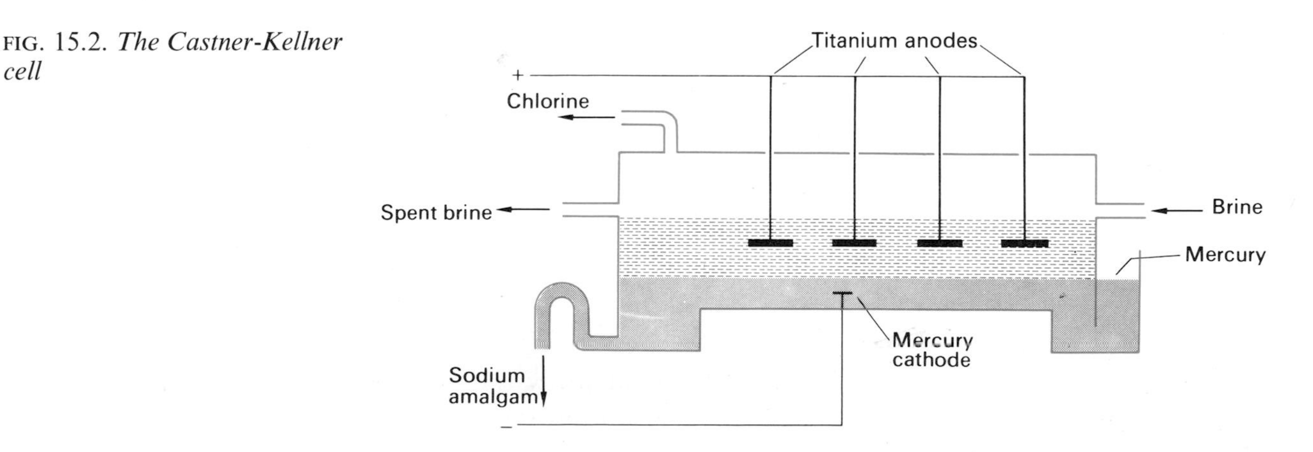

FIG. 15.2. *The Castner-Kellner cell*

blocks. On electrolysis chlorine is discharged at the anode and sodium at the cathode, where it dissolves in the mercury and is removed from the cell. The sodium amalgam is passed through water where the sodium reacts to form 50 per cent sodium hydroxide solution of high purity, the reaction being catalysed by the presence of iron grids. The mercury is then returned to the cell. The products are thus sodium hydroxide, chlorine and hydrogen.

Cathode		**Anode**
Na^+ discharged $\longleftarrow$	Na^+Cl^-	$\longrightarrow$ Cl^- discharged
$2Na^+ + 2e^- \rightarrow 2Na$	$H_2O \rightleftharpoons H^+ + OH^-$	$2Cl^- \rightarrow 2Cl\cdot + 2e^-$
$Na + Hg \rightarrow Na/Hg$		$2Cl\cdot \rightarrow Cl_2$
$2Na/Hg + 2H_2O \longrightarrow 2Na^+OH^- + H_2 + 2Hg$		

Sodium is discharged in preference to hydrogen in the cell, since hydrogen has a high overvoltage at a mercury electrode. This amounts to saying that the discharge of hydrogen ions or the combination of hydrogen atoms to give molecules is difficult to achieve at a mercury surface, i.e. mercury is a poor catalyst for either or both of these processes. Since the sodium dissolves in the mercury which is circulated through the cell, the formation of sodium hydroxide and hydrogen in the electrolytic cell itself is prevented.

(b) The Gibbs diaphragm cell process

In this process a concentrated solution of sodium chloride is electrolysed in a cell containing titanium anodes and steel mesh cathodes, with asbestos diaphragms incorporated in them. Chlorine is evolved at the anodes and hydrogen at the cathodes; sodium hydroxide solution (contaminated with sodium chloride) is tapped off from the cathode compartments.

Although this process produces a less pure solution of sodium hydroxide than that obtained from the Castner-Kellner proccss, it is likely to become increasingly important in the future in view of the health hazards associated with the use of mercury.

Potassium hydroxide is obtained in a similar manner, using a concentrated solution of potassium chloride in place of the brine.

Reactions of sodium and potassium hydroxides

Sodium and potassium hydroxides are white deliquescent solids which are caustic and slimy to touch; they dissolve readily in water with the vigorous evolution of heat. In aqueous solution they are completely dissociated and their reactions are essentially those of the hydroxide ion which is a strong base, i.e. they will neutralise acids and displace ammonia from ammonium salts when heated:

$$Na^+OH^-(aq) + H^+Cl^-(aq) \longrightarrow Na^+Cl^-(aq) + H_2O(l)$$
or
$$OH^-(aq) + H^+(aq) \longrightarrow H_2O(l)$$

$$Na^+OH^-(aq) + NH_4^+Cl^-(aq) \longrightarrow Na^+Cl^-(aq) + H_2O(l) + NH_3(g)$$
or
$$OH^-(aq) + NH_4^+(aq) \longrightarrow H_2O(l) + NH_3(g)$$

In aqueous solution they react with many salts and precipitate the corresponding basic hydroxide, e.g.

$$Cu^{2+}SO_4^{2-}(aq) + 2Na^+OH^-(aq) \longrightarrow Cu(OH)_2(s) + (Na^+)_2SO_4^{2-}(aq)$$
or
$$Cu^{2+}(aq) + 2OH^-(aq) \longrightarrow Cu(OH)_2(s)$$

The hydroxides of the less electropositive metals are amphoteric (p. 149) and dissolve in an excess of alkali, e.g. the hydroxides of aluminium, lead (II), tin (II) and zinc:

$$Al^{3+}(aq) + 3OH^-(aq) \longrightarrow Al(OH)_3(s)$$
$$Al(OH)_3(s) + OH^-(aq) \longrightarrow \underset{\text{aluminate ion}}{Al(OH)_4^-(aq)}$$
$$Zn^{2+}(aq) + 2OH^-(aq) \longrightarrow Zn(OH)_2(s)$$
$$Zn(OH)_2(s) + 2OH^-(aq) \longrightarrow \underset{\text{zincate ion}}{Zn(OH)_4^{2-}(aq)}$$

Sodium and potassium hydroxides react with a variety of non-metals, e.g. the halogens, silicon, sulphur, white phosphorus; sodium and potassium salts are formed in which the non-metal is incorporated into the anion. These individual reactions are discussed in subsequent chapters dealing with non-metals.

Uses of sodium and potassium hydroxides
Sodium hydroxide is used in the laboratory for absorbing carbon dioxide and other acidic gases, in a number of organic reactions involving hydrolysis and in volumetric analysis. Industrially it is used in the manufacture of soap (essentially sodium stearate) and sodium formate (obtained by heating sodium hydroxide with carbon monoxide under pressure). A solution of potassium hydroxide in alcohol is used in organic chemistry; industrially the main use of potassium hydroxide is in the manufacture of soft soap.

15.7 The carbonates of sodium and potassium, $(M^+)_2CO_3^{2-}$

The manufacture of sodium carbonate—the Solvay process
Almost all the 16 million tonnes of sodium carbonate used annually in the UK are manufactured by the ammonia-soda or Solvay process. In theory the process involves the reaction between sodium chloride and calcium carbonate to produce sodium carbonate and calcium chloride, but in practice other chemicals are required, since the reaction as given below does not take place:

$$Ca^2CO_3^{2-}(s) + 2Na^+Cl^-(s) \longrightarrow (Na^+)_2CO_3^{2-}(s) + Ca^{2+}(Cl^-)_2(s)$$

The raw materials are sodium chloride, calcium carbonate, a fuel and ammonia (practically all recovered in the process). The calcium carbonate is strongly heated to give quicklime and carbon dioxide:

$$Ca^{2+}CO_3^{2-}(s) \longrightarrow Ca^{2+}O^{2-}(s) + CO_2(g) \quad (1)$$

The carbon dioxide is now passed up a large tower, fitted with perforated plates, down which a concentrated aqueous solution of sodium chloride saturated with ammonia trickles. The reactions taking place in the tower can be represented by the equations:

$$NH_3(aq) + H_2O(l) \rightleftharpoons NH_3.H_2O(aq) \rightleftharpoons NH_4^+(aq) + OH^-(aq) \quad (2)$$
$$Na^+(aq) + Cl^-(aq) + NH_4^+(aq) + OH^-(aq) + CO_2(g) \rightarrow$$
$$Na^+(aq) + HCO_3^-(aq) + NH_4^+(aq) + Cl^-(aq) \quad (3)$$
$$Na^+(aq) + HCO_3^-(aq) \longrightarrow Na^+HCO_3^-(s) \quad (4)$$

Sodium hydrogen carbonate, which is not very soluble in sodium chloride solution,—the common ion effect—is filtered and heated to produce sodium carbonate:

$$2Na^+HCO_3^-(s) \longrightarrow (Na^+)_2CO_3^{2-}(s) + H_2O(l) + CO_2(g) \quad (5)$$

The sodium carbonate at this stage is contaminated with ammonium salts. If required pure, it is dissolved in water and carbon dioxide is blown through the solution. The precipitate of sodium hydrogen carbonate is filtered and heated to produce pure sodium carbonate; recrystallisation from water produces washing soda, $(Na^+)_2CO_3^-.10H_2O$.

$$(Na^+)_2CO_3^{2-}(aq) + H_2O(l) + CO_2(g) \longrightarrow 2Na^+HCO_3^-(s) \quad (6)$$

precipitated leaving ammonium salts in solution

$$2Na^+HCO_3^-(s) \longrightarrow (Na^+)_2CO_3^{2-}(s) + H_2O(l) + CO_2(g) \quad (7)$$

The Solvay process is a very economical process since:
(a) sodium chloride and calcium carbonate are cheap;
(b) quicklime and ammonium chloride formed in reactions (1) and (3) respectively are reacted together to produce ammonia so that, apart from making up small losses, no additional ammonia is required;
(c) carbon dioxide formed in reactions (5) and (7) is reintroduced into the Solvay tower.

Calcium chloride is the other product for which no very large scale uses have yet been found.

The manufacture of potassium carbonate

Potassium carbonate cannot be made by the Solvay process since potassium hydrogen carbonate is rather too soluble in water to be precipitated. It can, however, be made by passing carbon dioxide into a solution of potassium hydroxide; evaporation and subsequent ignition give the carbonate:

$$K^+OH^-(aq) + CO_2(g) \longrightarrow K^+HCO_3^-(aq)$$

$$2K^+HCO_3^-(aq) \longrightarrow (K^+)_2CO_3^{2-}(s) + H_2O(l) + CO_2(g)$$

Properties and uses of sodium and potassium carbonates

Both carbonates are freely soluble in water and give an alkaline reaction (p. 148). They react with some salts in solution and precipitate the corresponding carbonate, e.g.

$$Ca^{2+}(Cl^-)_2(aq) + (Na^+)_2CO_3^{2-}(aq) \longrightarrow Ca^{2+}CO_3^{2-}(s) + 2Na^+Cl^-(aq)$$

Because of their alkaline reaction in solution, both sodium carbonate and potassium carbonate frequently precipitate basic carbonates, e.g.

$$3Zn^{2+}(aq) + CO_3^{2-}(aq) + 4OH^-(aq) + 2H_2O(l) \longrightarrow ZnCO_3.2Zn(OH)_2.2H_2O(s)$$

In these instances the normal carbonate can usually be obtained by using sodium hydrogen carbonate solution (p. 427).

Sodium carbonate is often used in the laboratory as a volumetric reagent. Industrially it is used in a wide variety of ways, including the manufacture of glass (p. 233), in the preparation of sodium salts, in the treatment of hard water (p. 294), for the manufacture of soap, and in papermaking.

Potassium carbonate, which is more expensive, is used as a drying agent in organic chemistry since it is deliquescent; mixed with sodium carbonate it is used as 'fusion mixture' for getting insoluble salts into solution during analytical procedures.

15.8 The hydrogen carbonates of sodium and potassium, $M^+HCO_3^-$

These can be obtained by passing carbon dioxide through a cold concentrated solution of the corresponding carbonate, e.g.

$$(Na^+)_2CO_3^{2-}(aq) + CO_2(g) + H_2O(l) \longrightarrow 2Na^+HCO_3^-(s)$$

Although the hydrogen carbonates of several metals exist in aqueous solution, only those of the alkali metals can be obtained as solids and even these tend to decompose readily, e.g.

$$2K^+HCO_3^-(s) \longrightarrow (K^+)_2CO_3^{2-}(s) + H_2O(l) + CO_2(g)$$

Sodium hydrogen carbonate is much less soluble in water than potassium hydrogen carbonate and both solutions are alkaline by hydrolysis, but not to the same extent as the corresponding carbonates:

$$HCO_3^-(aq) + H_2O(l) \rightleftharpoons H_2CO_3(aq) + OH^-(aq)$$

Because sodium hydrogen carbonate solution is less alkaline than sodium carbonate solution, the former is often used to precipitate normal carbonates which would otherwise be formed as basic carbonates (p. 223), e.g.

$$Cu^{2+}(aq) + 2HCO_3^-(aq) \longrightarrow CuCO_3(s) + H_2O(s) + CO_2(g)$$

The hydrogen carbonate ions tend to be held together in crystal structures by hydrogen bonding (p. 67). Sodium hydrogen carbonate contains polymeric anions whereas the potassium salt contains the cyclic dimeric anion.

FIG. 15.3. *Structures of the anions in (a)* $Na^+HCO_3^-$ *and (b)* $K^+HCO_3^-$

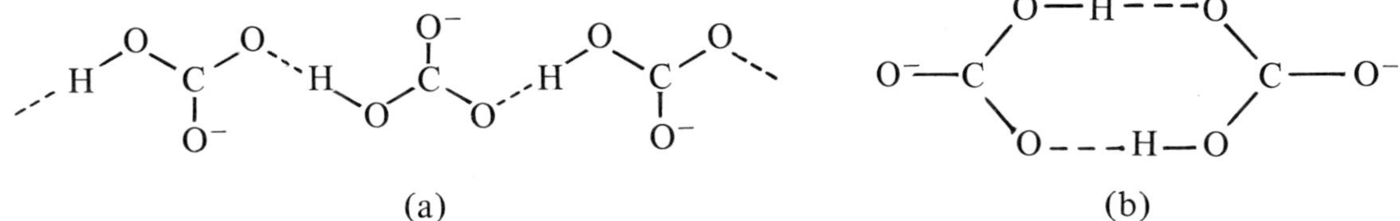

15.9 The halides of sodium and potassium, M^+X^-

The fluorides, M^+F^-

The fluorides can be obtained by reacting aqueous solutions of the carbonate or hydroxide with an aqueous solution of hydrofluoric acid. Polythene vessels must be used since the acid attacks glass.

$$Na^+OH^-(aq) + H^+F^-(aq) \longrightarrow Na^+F^-(aq) + H_2O(l)$$

If an excess of the acid is used, it is possible to isolate the solid hydrogen fluorides $Na^+HF_2^-$ and $K^+HF_2^-$. The hydrogen fluoride ion contains one molecule of hydrogen fluoride, hydrogen bonded to the very electronegative fluoride ion:

$$(F—H---F)^-$$

The hydrogen fluorides decompose on heating to give the normal fluorides and hydrogen fluoride.

The chlorides, M^+Cl^-

Sodium chloride occurs extensively in sea water; as a solid it is found as rock salt, which is coloured owing to the presence of impurities. In Great

Britain, salt deposits occur below ground level and the salt is brought to the surface as a solution in water. Pure sodium chloride can be obtained by passing hydrogen chloride gas through a saturated solution of the impure salt, when it is precipitated—the common ion effect.

The solubility of sodium chloride in water is affected very little by rise in temperature: 35·5 g at 0°C and 39 g at 100°C. When added to ice, it is possible to achieve a temperature of −23°C and an ice-salt mixture is often used as a cheap freezing mixture in the laboratory.

Sodium chloride is a vital raw material in many industrial processes, and the British alkali industry is sited in Cheshire where there are extensive salt deposits. It is used in the manufacture of sodium, sodium hydroxide, chlorine, sodium carbonate, sodium chlorate (I) (p. 341), sodium chlorate (V) (p. 342) and in the preparation of many sodium salts. It is also used to produce the sodium silicate glaze on earthenware vessels. Its more familiar uses include the flavouring and curing of foods.

Potassium chloride is far less abundant than sodium chloride and is found in sea water and in the Stassfurt (Germany) deposits as carnallite, $K^+Cl^-.Mg^{2+}(Cl^-)_2.6H_2O$. It is extracted from the latter source by fractional crystallisation; potassium chloride is less soluble in water at ordinary temperatures than magnesium chloride and can be crystallised out.

Potassium chloride is used to counteract potassium deficiency in soil and in the production of potassium hydroxide.

The bromides, M^+Br^-

Potassium bromide is more important than sodium bromide and is used as a sedative and in the production of silver bromide for use in photography. It can be made by adding bromine to a warm concentrated solution of potassium hydroxide. A mixture of potassium bromide and potassium bromate (V) is formed, which can be crystallised out. On heating with charcoal the bromate (V) is reduced to the bromide:

$$3Br_2(l) + 6OH^-(aq) \longrightarrow 5Br^-(aq) + \underset{\text{bromate (V)}}{BrO_3^-(aq)} + 3H_2O(l)$$

$$2K^+BrO_3^-(s) + 3C(s) \longrightarrow 2K^+Br^-(s) + 3CO_2(g)$$

The iodides, M^+I^-

Potassium iodide is more important than the corresponding sodium salt and its aqueous solution is used in the laboratory for dissolving iodine, which is not very soluble in water, as the polyiodide, $K^+I_3^-$. Sodium iodide also forms a polyiodide.

Potassium iodide can be obtained by a process similar to the one used for making potassium bromide, bromine being replaced by iodine.

The structures of the alkali halides

All the alkali metal halides except the chloride, bromide and iodide of caesium have the rock salt structure (the cations and anions have a co-ordination number of six). The chloride, bromide and iodide of caesium adopt a different structure (the cations and anions have a co-ordination number of eight). The radius ratio criterion which is useful in predicting the co-ordination numbers of ionic compounds has been discussed previously (p. 73).

15.10 The nitrates and nitrites of sodium and potassium

The nitrates, $M^+NO_3^-$

Sodium nitrate occurs as Chile saltpetre, along with sodium iodate (V) from which iodine is extracted, in the desert regions in Chile. It can be purified by recrystallisation from water, being less soluble than sodium iodate (V) at ordinary temperature.

Potassium nitrate occurs as saltpetre; it can also be made by fractional crystallisation of a hot aqueous solution containing sodium nitrate and potassium chloride. The ions present in the solution are Na^+, NO_3^-, K^+ and Cl^-; at about 100°C sodium chloride is the least soluble substance and can be crystallised out. On cooling to about 15°C potassium nitrate is the least soluble substance and crystallises out. It is purified by further recrystallisation from aqueous solution.

Both sodium and potassium nitrates are used as nitrogenous fertilisers. The latter is a constituent in gunpowder—sodium nitrate cannot be used since it is deliquescent; it is also used as a meat preservative.

When heated, the nitrates decompose to give the nitrite and oxygen, e.g.

$$2K^+NO_3^-(s) \longrightarrow 2K^+NO_2^-(s) + O_2(g)$$

The nitrites, $M^+NO_2^-$

The nitrites can be made by the thermal decomposition of the corresponding nitrates (see above); on a large scale a 50:50 mixture of nitrogen dioxide and nitrogen oxide is absorbed in sodium hydroxide solution:

$$2Na^+OH^-(aq) + NO_2(g) + NO(g) \longrightarrow 2Na^+NO_2^-(aq) + H_2O(l)$$

The nitrites can be purified by crystallisation from water.

On treating cold aqueous solutions of the nitrites with dilute hydrochloric acid a blue colour develops and, on warming, a mixture of nitrogen oxide and nitrogen dioxide is evolved:

$$2Na^+NO_2^-(aq) + 2H^+Cl^-(aq) \longrightarrow 2Na^+Cl^-(aq) + H_2O(l) + NO(g) + NO_2(g)$$

The blue colour is presumably due to the presence of N_2O_3 which readily decomposes into NO and NO_2.

Sodium nitrite is used extensively in the production of dyes; with aniline or its derivatives in the presence of hydrochloric acid below about 5°C it reacts to give diazonium salts, e.g. $[C_6H_5N_2]^+Cl^-$, from which the so-called azo-dyes are prepared.

15.11 The sulphates and hydrogen sulphates of sodium and potassium

The sulphates, $(M^+)_2SO_4^{2-}$

The sulphates can be prepared by titrating aqueous solutions of the alkalis with dilute sulphuric acid and allowing the solutions to crystallise, e.g.

$$2Na^+OH^-(aq) + (H^+)_2SO_4^{2-}(aq) \longrightarrow (Na^+)_2SO_4^{2-}(aq) + H_2O(l)$$

Sodium sulphate crystallises as the decahydrate $(Na^+)_2SO_4^{2-}.10H_2O$ which effloresces to give the anhydrous salt; when heated above 32·5°C it also gives the anhydrous salt, and this temperature is the transition point for the system:

$$(Na^+)_2SO_4^{2-}.10H_2O(s) \rightleftharpoons (Na^+)_2SO_4^{2-}(s) + 10H_2O(l)$$

Potassium sulphate crystallises as the anhydrous salt and it is found as such in the Stassfurt deposits. It is used for making up potassium deficiency in soil, as a drying agent and for making potash alum $K^+Al^{3+}(SO_4^{2-})_2.12H_2O$, which is used as a mordant in dyeing.

The hydrogen sulphates, $M^+HSO_4^-$

These can be obtained by crystallising from a warm solution of the sulphate in concentrated sulphuric acid, e.g.

$$(Na^+)_2SO_4^{2-}(s) + H_2SO_4(l) \longrightarrow 2Na^+HSO_4^-(s)$$

The hydrogen sulphates are acidic in water since dissociation of the hydrogen sulphate anion occurs to give hydrogen and sulphate ions:

$$HSO_4^-(aq) \rightleftharpoons H^+(aq) + SO_4^{2-}(aq)$$

When heated, hydrogen sulphates first decompose to give the pyrosulphate and water; the pyrosulphate subsequently decomposes to give the sulphate and sulpur trioxide, e.g.

$$2Na^+HSO_4^-(s) \longrightarrow (Na^+)_2S_2O_7^{2-}(s) + H_2O(l)$$
$$(Na^+)_2S_2O_7^{2-}(s) \longrightarrow (Na^+)_2SO_4^{2-}(s) + SO_3(s)$$

15.12 Other salts of sodium and potassium

Sodium and potassium salts of other oxyacids are discussed elsewhere in the book, e.g. chlorates (I) (p. 341), chlorates (V) (p. 342), phosphates (p. 269) and sulphites (p. 310).

15.13 Hydrolysis of alkali metal salts

The alkali metal halides, nitrates and sulphates are neutral in aqueous solution; but other salts, for instance hydrogen carbonates, carbonates, cyanides and sulphides, show an alkaline reaction in solution. This is because the latter group of anions behave as strong bases and are able to abstract protons from solvent water molecules, thus giving rise to hydroxyl ions:

$$CO_3^{2-}(aq) + H_2O(l) \rightleftharpoons HCO_3^-(aq) + OH^-(aq)$$
$$HCO_3^-(aq) + H_2O(l) \rightleftharpoons H_2CO_3(aq) + OH^-(aq)$$
$$CN^-(aq) + H_2O(l) \rightleftharpoons HCN(aq) + OH^-(aq)$$
$$S^{2-}(aq) + H_2O(l) \rightleftharpoons HS^-(aq) + OH^-(aq)$$
$$HS^-(aq) + H_2O(l) \rightleftharpoons H_2S(aq) + OH^-(aq)$$

When alkali metal cyanides and sulphides are warmed with water volatile hydrogen cyanide and hydrogen sulphide are driven off as gases; the hydrolysis therefore goes to completion and solutions of alkalis remain.

15.14 Summary of trends in Group 1A

All the alkali metals are extremely electropositive, increasingly so with increasing atomic number; they therefore tend to form predominantly electrovalent compounds and these are generally soluble in water. Lithium, because of the small size of its atom and unipositive ion, tends to form compounds in which there is some degree of covalent character, and in many respects behaves like its diagonal neighbour in the Periodic Table—magnesium (see next section).

The alkali metals form hydroxides, which are strong bases, and salts, which are either neutral in solution (those of a strong acid, e.g. the nitrates) or alkaline (those of a weak acid, e.g. the carbonates). Their salts are generally stable to heat; thus the carbonates are not decomposed in the Bunsen flame and strong heating is required to decompose the nitrates to the nitrites and oxygen. Only the hydrogen carbonates of the alkali metals are sufficiently stable to exist as solids. Although many of the alkali metal salts are hydrated in the solid state, there is little tendency for the alkali metal ions to hydrate strongly in aqueous solution. Lithium is exceptional in this respect and is sufficiently hydrated in aqueous solution to have the highest negative electrode potential of this group of elements.

As the atomic and ionic sizes of the alkali metals increase with increasing atomic number there is a gradual increase in chemical reactivity—all the metals are sufficiently reactive to force the hydrogen atom to accept an electron as in the ionic hydrides—and thus a gradual increase in stability of their compounds. As the ionic size increases the complexity of the oxides increases (p. 176); there is also a tendency for the larger ions of potassium, rubidium and caesium to form increasingly stable polyhalides, e.g. $K^+I_3^-$ and $Cs^+I_3^-$. Both tendencies can be traced to an increase in lattice energy that occurs as the alkali metal ion increases in size.

15.15 The anomalies of lithium—the diagonal relationship with magnesium

In general the differences between the first and second members of a particular periodic group are more pronounced than those between the second and third; in this respect lithium is no exception and its abnormally high negative electrode potential has already been noted.

In many aspects of its chemistry lithium shows similarities with magnesium; thus it only forms a monoxide on heating in oxygen, its carbonate is fairly readily decomposed to the oxide and carbon dioxide, and the nitrate on decomposition gives the oxide, nitrogen dioxide and oxygen. Other similarities with magnesium include the insolubilities of its carbonate and phosphate in water, the formation of a carbide and nitride by direct combination of the elements, and a strong tendency for its ions to hydrate in aqueous solution. Its halides, in particular, are soluble in many organic solvents, indicating that they possess appreciable covalent character. In all these respects lithium shows marked differences from the other alkali metals.

Many of the resemblances to magnesium can be explained as being due to these two elements having very similar electropositivities. Since electropositivity increases from top to bottom in any periodic group and decreases from left to right across a particular period, the increase in electropositivity in going down one place in the Periodic Table (magnesium is one place lower down than lithium) is compensated for by the decrease which occurs in moving across a period from left to right (magnesium is one place to the right of lithium).

The appreciable covalent character of several compounds of lithium depends on the fact that its lone valency electron is not completely transferred during chemical reaction; this is readily understood in terms of the small size of the lithium atom.

15.16 Suggested reading

M. Tingle, *The ammonia-soda process*, School Science Review, No. 213, Vol. 60. 1979

15.17 Questions on chapter 15

1 The elements of Group 1A in the Periodic Table are in the order lithium, sodium, potassium, rubidium, caesium, and their atomic numbers are 3, 11, 19, 37 and 55 respectively. Deduce their probable electronic structures. Use these structures to

explain the relative reactivities of the elements, the type of bond they form and the relative stability of their compounds. Thus, predict for lithium and caesium the chemical properties of the elements and of their hydrides, hydroxides, carbonates and chlorides. (S)

2 Outline an electrolytic method for the industrial preparation of sodium hydroxide from brine. How, and under what conditions, does sodium hydroxide react with (i) carbon monoxide, (ii) ammonium alum, (iii) silicon, (iv) phosphorus?

3 Justify the statement that common salt is a very important raw material for chemical industry.

4 'The success of an industrial process depends, amongst other things, upon a detailed knowledge of the chemical reactions taking place, and upon efficient economic operations.'

Justify this statement, without giving details of the industrial plant, by considering the following points, all of which are of importance in the Solvay process used for the preparation of sodium carbonate:

(a) ionic equilibria,
(b) the solubility of gases and solids as a function of temperature,
(c) counter-current flow processes,
(d) the use of by-products.

Explain the following: (i) when a solution of sodium hydrogen carbonate is heated, a gaseous product is observed, (ii) when a solution of sodium hydrogen carbonate is added to a solution of lead nitrate, lead carbonate is precipitated. (JMB)

5 Give a concise account of the extraction of sodium from sodium chloride. How may the following be prepared from the metal:

(a) sodium sulphide,
(b) sodium hydride,
(c) sodium peroxide,
(d) sodamide?

Write equations to show the action of water on these compounds.

6 (a) Write the electronic structures of the Group 1 elements, lithium to caesium, in terms of *s*, *p* and *d* orbitals.

(b) For the elements lithium to caesium, state, and explain qualitatively, the general trends in
(i) first ionization energy,
(ii) atomic radius.

(c) Give reasons for the following:
(i) the large negative standard electrode potential of lithium;
(ii) the differences in the structures of the crystal lattices of sodium chloride and caesium chloride;
(iii) the fact that lithium salts are often hydrated but hydrated caesium salts are rare.

(d) Give **two** chemical properties of lithium or its compounds which are not typical of the other Group 1 elements and their compounds. By means of descriptions or equations show the ways in which these properties are atypical. (AEB)

7 Give a comparative account of the chemistry of the Group 1A metals and their compounds.

8 Lithium shows more differences from sodium than does sodium from potassium. Explain why this should be so, and discuss the diagonal relationship between lithium and magnesium.

9 Define the terms ionization energy and standard electrode potential and discuss the relationship between them. The values for the ionization energies of the alkali metals Li, Na, and K are respectively 5·39, 5·14 and 4·34 eV. The corresponding standard values for the $M^+(aq)/M(s)$ couple are $-3{\cdot}0$ V, $-2{\cdot}7$ V, and $-2{\cdot}9$ V. Comment on these values. (Oxford Schol.)

10 State briefly the important features common to the chemistry of sodium and potassium.

How do sodium and potassium differ in
(a) the natural sources used for the manufacture of their compounds, and
(b) the reagents available for their detection in analysis? (O & C)

Chapter 16

Group 2A the alkaline earth metals

16.1 Some physical data of Group 2A elements

	Atomic Number	Electronic Configuration	Ionisation energy/kJ mol^{-1} First	Second	Standard Electrode Potential/V	Atomic Radius/ nm	Ionic Radius/ nm	M.p. /°C	B.p. /°C
Be	4	2.2 $1s^2 2s^2$	899	1757	−1·85	0·089	0·031	1280	2477
Mg	12	2.8.2 ....$2s^2 2p^6 3s^2$	738	1451	−2·37	0·136	0·065	650	1100
Ca	20	2.8.8.2 ....$3s^2 3p^6 4s^2$	590	1145	−2·87	0·174	0·099	850	1487
Sr	38	2.8.18.8.2 ....$4s^2 4p^6 5s^2$	549	1064	−2·89	0·191	0·113	768	1380
Ba	56	2.8.18.18.8.2 ....$5s^2 5p^6 6s^2$	502	965	−2·90	0·198	0·135	714	1640
Ra	88	2.8.18.32.18.8.2 ....$6s^2 6p^6 7s^2$	508	975	−2·92	—	0·148	700	1140

16.2 Some general remarks about Group 2A

Beryllium differs considerably in its chemistry from the other members of this group, and its compounds when anhydrous show a considerable degree of covalent character—a consequence of the small size of the beryllium atom. Magnesium too forms compounds which show appreciable covalent character. The other four members of this group referred to as 'the alkaline earths' have a very similar chemistry and their compounds are essentially ionic. Radium is radioactive and is formed during the decay of $^{238}_{92}U$ compounds.

All the members of this group are highly reactive and are thus never found in the free state in nature. Beryllium and magnesium have the hexagonal close-packed structure, and strontium the cubic close-packed structure; these two basic structures are two ways in which spheres of equal radius can be close packed and the co-ordination number of each sphere is twelve. Calcium can adopt either of these two structures but barium has the body-centred cubic structure of the alkali metals (p. 174) in which the co-ordination number is eight.

FIG. 16.1. *The close packing of metal atoms*

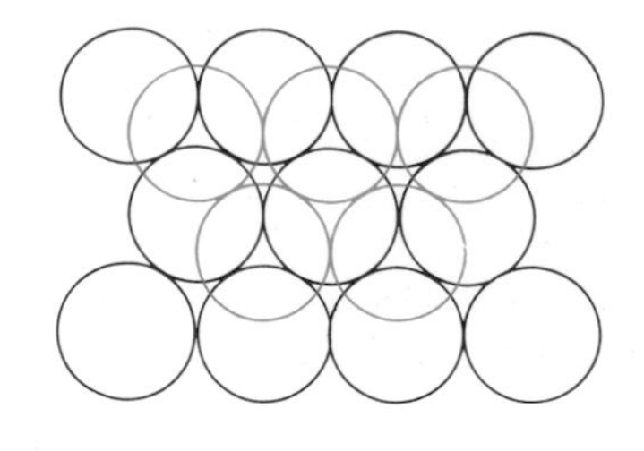

Spheres of equal radius can be packed so that any one sphere is in contact with six others. A second layer can be built up so that each sphere in this layer is in contact with three spheres in the first layer. This second layer is shown by green circles in fig. 16.1. A third layer can be constructed in one of two ways. If the spheres in this layer are arranged to be directly over those in the first layer, the resulting structure is said to be hexagonal close-packed and the basic structure is repeated every two layers (ABABABAB....). If, however, the spheres in this third layer are arranged in an alternative manner the structure is said to be cubic-close packed if the basic structure is repeated every three layers (ABCABCABC......). By making models of these two structures it is easily seen that they represent two different ways of packing spheres together in the most economical manner.

The Group 2A metals have two electrons in the outer shell preceded by a closed shell containing eight electrons—except beryllium which has a closed shell of two. In chemical combination these two outer electrons are transferred, giving a dipositive metal ion with the stable electronic configuration of a noble gas, e.g. Mg^{2+} which is isoelectronic with Na^+(2.8) and Ne(2.8). Compounds formed by the last four members of this group—calcium, strontium, barium and radium—are therefore predominantly ionic and exist as high melting-point solids in which as many ions as possible of opposite charge surround each other (see p. 73). The compounds of magnesium are generally predominantly ionic but, as mentioned above, they do show some covalent character, i.e. an incomplete transfer of the two outer electrons during chemical combination. This trend is even more pronounced with beryllium and its chloride, for example, $BeCl_2$ is considered to be essentially covalent, since it shows very little electrical conductivity in the fused state.

The first ionisation energies of this group of metals are considerably higher than those of the corresponding alkali metals and the energy necessary to form M^{2+} ions—the sum of the first and second ionisation energies—is therefore very much greater than that necessary to form unipositive alkali metal ions. The reason why these metals form ionic compounds is that the assembly of positive and negative ions into a symmetrical crystal structure results in the liberation of a very large amount of energy and the overall formation of an ionic compound is thus exothermic (see p. 56 for the factors that determine the formation of ionic compounds).

The Group 2A metals have higher melting and boiling points and higher densities than corresponding metals in Group 1A; this is because the former have two and the latter only one valency electron per atom for bonding the atoms into a metallic structure.

As the atomic number increases so the ionisation energies decrease and the electrode potentials become more negative; there is therefore an increase in reactivity on passing down this group.

Calcium, strontium and barium compounds give characteristic flame colorations which can be used to identify them—calcium (yellowish-red), strontium (crimson) and barium (apple-green).

This group of metals forms more covalent compounds than those of Group 1A, particularly beryllium and magnesium. The highly reactive Grignard reagents which are covalent, e.g. C_2H_5MgBr, are important reagents in organic chemistry.

16.3 Occurrence, extraction and uses of the Group 2A metals

Beryllium occurs as beryl, $3BeO.Al_2O_3.6SiO_2$, magnesium as magnesite, $Mg^{2+}CO_3^{2-}$, kieserite, $Mg^{2+}SO_4^{2-}.H_2O$ and carnallite, $K^+Cl^-.Mg^{2+}(Cl^-)_2.6H_2O$. It also occurs in sea water, from which it is being extracted on an increasing scale; the central atom in the chlorophyll molecule is magnesium. Calcium, barium and strontium occur as carbonates and sulphates. The presence of magnesium and calcium compounds in natural waters is the principal cause of hardness (p. 293).

The metals are extracted by electrolysis of their fused halides, other halides being added to lower the melting point and in the case of beryllium to make the melt a better electrical conductor (see Table 12B p. 156). Magnesium has more uses than the other Group 2A metals; its extraction from sea water involves the precipitation of magnesium hydroxide by the addition of slaked lime (obtained from oyster shells—principally calcium carbonate), conversion into magnesium chloride by heating in a stream of hydrogen chloride and finally electrolysis of the fused chloride in the presence of other halides. A steel-lined tank as cathode and a graphite rod as anode are used; magnesium is discharged at the cathode and floats on the molten electrolyte. It is protected from atmospheric oxidation by a blanket of town gas.

$$\underset{\text{in sea water}}{Mg^{2+}(aq)} + Ca^{2+}(OH^-)_2(s) \longrightarrow Mg^{2+}(OH^-)_2(s) + Ca^{2+}(aq)$$

$$Mg^{2+}(OH^-)_2(s) + 2HCl(g) \longrightarrow Mg^{2+}(Cl^-)_2(s) + 2H_2O(l)$$

Cathode		**Anode**
Mg^{2+} discharged $\longleftarrow$	$Mg^{2+}(Cl^-)_2$	$\longrightarrow$ Cl^- discharged
$Mg^{2+} + 2e^- \rightarrow Mg$		$2Cl^- \longrightarrow 2Cl\cdot + 2e^-$ $2Cl\cdot \longrightarrow Cl_2$

Beryllium is used for making containers for atomic fuel since it absorbs very few neutrons and does not become radioactive. It is also used as a window material in X-ray apparatus since it is transparent to X-rays. An alloy with copper has a higher tensile strength than any other non-ferrous material.

Magnesium is used extensively in the construction of many light alloys, principally in conjunction with aluminium. It is also used to afford cathodic protection to structures such as ships' hulls and buried pipelines; the greater reactivity of magnesium ensures that this metal corrodes in preference to the main steel structure. Magnesium is also used as a reducing agent in the extraction of some metals (see Table 12B p. 156).

Calcium, strontium and barium metals have no very extensive uses.

16.4 Properties of the Group 2A metals

Physical properties

The metals are very much harder than those of Group 1A; they are good conductors of heat and electricity. When pure they are silvery coloured, but quickly tarnish on exposure to air because an oxide film covers their surface.

Chemical properties

The metals are very reactive, but less so than the alkali metals; the reactivity increases with increasing atomic number. Thus beryllium fails to decompose steam, magnesium burns in steam, and calcium, strontium and barium decompose water with increasing vigour to give the hydroxide and hydrogen, e.g.

$$Ca(s) + 2H_2O(l) \longrightarrow Ca^{2+}(OH^-)_2(aq/s) + H_2(g)$$

At a suitable temperature they combine with a variety of non-metals to give oxides, sulphides, halides and nitrides; calcium, strontium and barium also combine with hydrogen to give hydrides.

$$2Mg(s) + O_2(g) \longrightarrow 2Mg^{2+}O^{2-}(s)$$
$$Ca(s) + S(s) \longrightarrow Ca^{2+}S^{2-}(s)$$
$$Sr(s) + Cl_2(g) \longrightarrow Sr^{2+}(Cl^-)_2(s)$$
$$3Ba(s) + N_2(g) \longrightarrow (Ba^{2+})_3(N^{3-})_2(s)$$
$$Ca(s) + H_2(g) \longrightarrow Ca^{2+}(H^-)_2(s)$$

With dilute hydrochloric and dilute sulphuric acids they give the corresponding salt and hydrogen, e.g.

$$Ca(s) + 2H^+Cl^-(aq) \longrightarrow Ca^{2+}(Cl^-)_2(aq) + H_2(g)$$

16.5 Oxides of the Group 2A metals

The normal oxide $M^{2+}O^{2-}$ is formed on heating the metals in oxygen; strontium and barium also form the peroxide $M^{2+}O_2^{2-}$ on prolonged heating, particularly if pressure is used. The normal oxides are conveniently prepared by decomposition of the carbonates.

Beryllium oxide, BeO, is not attacked by water, presumably because it is covalent. The other oxides are ionic and the oxide and peroxide ions act as bases, abstracting protons from water molecules:

$$O^{2-} + H_2O(l) \longrightarrow 2OH^-(aq)$$
$$O_2^{2-} + 2H_2O(l) \longrightarrow 2OH^-(aq) + H_2O_2(l)$$

Magnesium oxide has a melting point of 2800°C and is used for manufacturing linings for open-hearth steel furnaces. Calcium oxide has a melting point only slightly lower; it is used in many metallurgical operations to produce a slag with impurities in metal ores (p. 154). In the laboratory it is used for drying ammonia and taking out the last traces of water from ethanol. Both oxides are made by decomposing their carbonates, which are abundant in nature.

16.6 The hydroxides of magnesium and calcium, $M^{2+}(OH^-)_2$

The descriptive chemistry of this group of elements is mainly restricted to the compounds of magnesium and calcium, since the compounds of strontium and barium are very similar to those of calcium. The chemistry of beryllium is best considered in relation to that of aluminium, its diagonal neighbour in the Periodic Classification (p. 197).

Manufacture of calcium hydroxide (slaked lime)
Magnesium hydroxide has no significant uses. Calcium hydroxide is manufactured by heating limestone to a temperature of about 1000°C and then adding water to the oxide; the latter reaction—known as slaking—is highly exothermic:

$$Ca^{2+}CO_3^{2-}(s) \longrightarrow \underset{\text{quicklime}}{Ca^{2+}O^{2-}(s)} + CO_2(g)$$

$$Ca^{2+}O^{2-}(s) + H_2O(l) \longrightarrow \underset{\text{slaked lime}}{Ca^{2+}(OH^-)_2(s)}$$

Reactions of magnesium and calcium hydroxides

Magnesium and calcium hydroxides are white solids and both are sufficiently soluble in water to show an alkaline reaction to indicators. A solution of calcium hydroxide is known as lime water.

Magnesium and calcium hydroxides, like the oxides, are strong bases, thus they will neutralise acids and displace ammonia from ammonium salts when heated:

$$Mg^{2+}(OH^-)_2(s) + 2H^+Cl^-(aq) \longrightarrow Mg^{2+}(Cl^-)_2(aq) + 2H_2O(l)$$

or

$$OH^-(aq) + H^+(aq) \longrightarrow H_2O(l)$$

$$Ca^{2+}(OH^-)_2(s) + 2NH_4{}^+Cl^-(s) \longrightarrow Ca^{2+}(Cl^-)_2(s) + 2H_2O(l) + 2NH_3(g)$$

or

$$OH^- + NH_4{}^+ \longrightarrow H_2O(l) + NH_3(g)$$

The solubilities of the Group 2A hydroxides increase with increasing atomic number and barium hydroxide is sufficiently soluble in water to be used as an M/10 solution in volumetric analysis.

Uses of calcium hydroxide

Calcium hydroxide has many uses; the following are some of its more important applications.

(a) Calcium hydroxide displaces ammonia from ammonium salts when the mixture is heated. It is used in recovering ammonia from ammonium chloride in the Solvay process (p. 179). It can be employed in the laboratory preparation of this gas.

(b) It can be used as a cheap alkali in treating 'acid' soil. Thus vegetable acids set free by the decay of organic matter can be neutralised and nitrifying bacteria, which do not function effectively in acid soil, allowed to operate.

(c) Mortar is prepared by mixing 1 part of slaked lime and 3 parts of sand into a paste with water. The mortar first sets and then slowly reacts with atmospheric carbon dioxide to form small crystals of calcium carbonate which effectively bind the grains of sand:

$$Ca^{2+}(OH^-)_2(s) + CO_2(g) \longrightarrow Ca^{2+}CO_3{}^{2-}(s) + H_2O(l)$$

The weathering of mortar is caused by the formation of calcium hydrogen carbonate, which is soluble in water, by the action of rain which is a dilute solution of carbonic acid:

$$Ca^{2+}CO_3{}^{2-}(s) + \underbrace{H_2O(l) + CO_2(g)}_{\text{rain water}} \longrightarrow Ca^{2+}(HCO_3{}^-)_2(aq)$$

(d) Temporary hard water can be softened by the addition of calculated quantities of calcium hydroxide as in the Clark's process (p. 294).

(e) Calcium hydrogen sulphite can be made by passing sulphur dioxide into a suspension of calcium hydroxide in water:

$$Ca^{2+}(OH^-)_2(s) + 2SO_2(g) \longrightarrow Ca^{2+}(HSO_3{}^-)_2(aq)$$

Calcium hydrogen sulphite solution removes the lignin from wood, leaving cellulose (a polysaccharide) which is used in paper manufacture.

(f) Bleaching powder is manufactured by passing chlorine over moist calcium hydroxide. Its structure is complex and it is known to contain Ca^{2+}, OCl^-, Cl^- and OH^- ions and water. It is usually represented by the oversimplified formula $Ca^{2+}O^{2-}.Cl_2$.

$$Ca^{2+}(OH^-)_2(s) + Cl_2(g) \longrightarrow Ca^{2+}O^{2-}.Cl_2(s) + H_2O(l)$$

(g) A suspension of calcium hydroxide in water (milk of lime) is used as whitewash. The solution obtained by filtering this suspension (lime water) is used for detecting carbon dioxide.

16.7 The carbonates of magnesium and calcium, $M^{2+}CO_3^{2-}$

Occurrence of magnesium and calcium carbonates

Magnesium carbonate occurs as magnesite and in association with calcium carbonate as dolomite, $Mg^{2+}CO_3^{2-}.Ca^{2+}CO_3^{2-}$. Calcium carbonate occurs naturally in two crystalline forms: calcite—Iceland spar (which is colourless), marble, limestone and chalk (which is microcrystalline and is composed of the shell remains of dead marine animals) all have this basic structure: aragonite is the second crystalline form and is metastable with respect to calcite; it is found in coral shells.

Properties of magnesium and calcium carbonates

These carbonates are only sparingly soluble in water and on strong heating decompose to give the oxides and carbon dioxide. Like all carbonates they react readily with dilute acids to give salts, carbon dioxide and water.

Uses of magnesium and calcium carbonates

Magnesium carbonate on strong heating decomposes into the oxide and carbon dioxide. Magnesium oxide is used for making linings for open-hearth steel furnaces.

Calcium carbonate is a basic raw material in the Solvay process (p. 179). It is also used for making cement, which is essentially calcium aluminium silicate. A mixture of limestone and clay is heated in long rotating cylindrical kilns, and the product, which is cement, is then ground to a fine powder. It is mixed with water into a paste, either alone or with sand, and in the course of a few hours sets solid, hardening still more in the course of many years. Concrete is made by adding sand and rubble to cement. It is used in the same way as cement and when applied over a steel framework is known as reinforced concrete. The chemical changes responsible for the setting of cement and concrete are highly complex.

Calcium carbonate is used in the glass industry (p. 233), and for making quicklime and slaked lime, whose uses have already been mentioned (p. 190–191).

16.8 The hydrogen carbonates of magnesium and calcium, $M^{2+}(HCO_3^-)_2$

The solid hydrogen carbonates are unknown at room temperature, but they are the cause of temporary hardness in water (p. 293). Rain water attacks any rocks containing calcium and magnesium carbonates and a dilute solution of the hydrogen carbonates is formed, e.g.

$$Ca^{2+}CO_3^{2-}(s) + H_2O(l) + CO_2(g) \longrightarrow Ca^{2+}(HCO_3^-)_2(aq)$$

When water containing the dissolved hydrogen carbonates is boiled a deposit of the carbonates results by reversal of the above equation. The formation of stalactites and stalagmites in limestone caves is due to the decomposition of the hydrogen carbonate as the solution gradually evaporates.

16.9 The halides of magnesium and calcium, $M^{2+}(X^-)_2$

The fluorides, $M^{2+}(F^-)_2$

Magnesium fluoride is virtually insoluble in water; calcium fluoride and the fluorides of the other Group 2A metals are only slightly more soluble. This insolubility is in marked contrast to the ready solubility of the corresponding chlorides, bromides and iodides and is due to the very high lattice energies of these fluorides.

Calcium fluoride is found naturally as fluorspar; it is used for making hydrogen fluoride by reaction with concentrated sulphuric acid:

$$Ca^{2+}(F^-)_2(s) + H_2SO_4(l) \longrightarrow Ca^{2+}SO_4{}^{2-}(s) + 2HF(g)$$

The chlorides, $M^{2+}(Cl^-)_2$

Both the chlorides crystallise from water as hexahydrates,

$$M^{2+}(Cl^-)_2.6H_2O$$

and are very deliquescent. Because it is cheap—it is a by-product of the Solvay process—and deliquescent, anhydrous calcium chloride is often used to dry gases and organic liquids. It cannot, however, be used to dry ammonia and ethanol since it forms complexes having the respective formulae:

$$Ca^{2+}(NH_3)_8(Cl^-)_2 \text{ and } Ca^{2+}(C_2H_5OH)_4(Cl^-)_2$$

When heated, hydrated magnesium chloride is hydrolysed by its water of crystallisation with the evolution of hydrogen chloride and the formation of the oxide:

$$Mg^{2+}(Cl^-)_2(s) + H_2O(l) \longrightarrow Mg^{2+}O^{2-}(s) + 2HCl(g)$$

It is therefore not possible to obtain the anhydrous chloride by evaporation of an aqueous solution of the chloride; it can, however, be obtained if the evaporation is carried out in a stream of hydrogen chloride which has the effect of reversing the reaction above. Hydrated calcium chloride also suffers hydrolysis on heating but to a markedly less extent.

Magnesium and calcium are extracted from their fused chlorides but, since calcium metal has no large scale uses, virtually all the Solvay process calcium chloride is waste product. Use is sometimes made of its deliquescent nature for keeping down dust in coalmines and on secondary roads.

The bromides, $M^{2+}(Br^-)_2$ and iodides, $M^{2+}(I^-)_2$

These are similar to the chlorides; the bromide and iodide of magnesium are appreciably soluble in organic solvents such as alcohols, ketones and ethers, with which they form complexes.

16.10 The nitrates of magnesium and calcium, $M^{2+}(NO_3{}^-)_2$

These can be obtained by reacting the metals, oxides, hydroxides or carbonates with dilute nitric acid and crystallising the resulting solutions. Like all nitrates, except those of the alkali metals, they decompose on heating to give the oxide, nitrogen dioxide and oxygen, e.g.

$$2Ca^{2+}(NO_3{}^-)_2(s) \longrightarrow 2Ca^{2+}O^{2-}(s) + 4NO_2(g) + O_2(g)$$

Calcium nitrate is used as a nitrogenous fertiliser; a mixture of the nitrate and the dihydrogen phosphate, $Ca^{2+}(H_2PO_4{}^-)_2$, called 'nitrophos' is made by reacting rock phosphate $(Ca^{2+})_3(PO_4{}^{3-})_2$ with nitric acid.

16.11 The sulphates of magnesium and calcium, $M^{2+}SO_4^{2-}$

Magnesium sulphate occurs as Epsom salts, $Mg^{2+}SO_4^{2-}.7H_2O$, and in association with the chloride and sulphate of potassium in the Stassfurt deposits. It is used as a purgative and for other medical purposes.

Calcium sulphate occurs as anhydrite, $Ca^{2+}SO_4^{2-}$, and as gypsum, $Ca^{2+}SO_4^{2-}.2H_2O$. Although very much less soluble in water than magnesium sulphate which is freely soluble, both these sulphates are the principal cause of permanent hardness in natural waters (p. 294).

Both magnesium sulphate and calcium sulphate are reduced to sulphides on heating with carbon, e.g.

$$Mg^{2+}SO_4^{2-}(s) + 2C(s) \longrightarrow Mg^{2+}S^{2-}(s) + 2CO_2(g)$$

When heated to a temperature a little in excess of 100°C, gypsum loses three-quarters of its water of crystallisation and becomes Plaster of Paris:

$$2Ca^{2+}SO_4^{2-}.2H_2O(s) \longrightarrow (Ca^{2+}SO_4^{2-})_2.H_2O(s) + 3H_2O(l)$$

On the addition of water, Plaster of Paris evolves heat and quickly sets, expanding somewhat in the process. It is used in making plaster casts, as a surface for walls and in surgery to keep an injured limb rigid.

Anhydrite was formerly used as a source of sulphur dioxide in the manufacture of sulphuric acid (p. 311), but this raw material is now uneconomic because its use involves high energy costs. The process involved heating calcium sulphate with coke at 1000°C when reduction to the sulphide occurred:

$$Ca^{2+}SO_4^{2-}(s) + 2C(s) \longrightarrow Ca^{2+}S^{2-}(s) + 2CO_2(g)$$

The temperature was raised to 1200°C, when excess calcium sulphate reacted with the calcium sulphide:

$$Ca^{2+}S^{2-}(s) + 3Ca^{2+}SO_4^{2-}(s) \longrightarrow 4Ca^{2+}O^{2-}(s) + 4SO_2(g)$$

The temperature was increased further to 1500°C when the sand or shale reacted with the calcium oxide to form calcium silicate, (Portland cement):

$$Ca^{2+}O^{2-}(s) + SiO_2(s) \longrightarrow Ca^{2+}SiO_3^{2-}(s)$$

Thus coke, sand and anhydrite produced sulphur dioxide and Portland cement, and the overall reaction is:

$$2Ca^{2+}SO_4^{2-}(s) + C(s) + 2SiO_2(s) \longrightarrow 2Ca^{2+}SiO_3^{2-}(s) + 2SO_2(g) + CO_2(g)$$

Ammonium sulphate, a nitrogenous fertiliser, is manufactured by reacting powdered calcium sulphate with an ammonia solution containing carbon dioxide (essentially ammonium carbonate). This reaction takes place because calcium carbonate is even less soluble in water than calcium sulphate:

$$Ca^{2+}SO_4^{2-}(s) + (NH_4^+)_2CO_3^{2-}(aq) \longrightarrow Ca^{2+}CO_3^{2-}(s) + (NH_4^+)_2SO_4^{2-}(aq)$$

Calcium carbonate is filtered off and the ammonium sulphate is crystallised from the filtrate.

16.12 Calcium salts of other oxyacids

Calcium hydrogen sulphite, $Ca^{2+}(HSO_3^-)_2$, is used in the manufacture of paper (p. 320), and calcium phosphate (V), $(Ca^{2+})_3(PO_4^{3-})_2$ is the source of phosphatic fertilisers and phosphorus (p. 262).

16.13 The carbides and nitrides of magnesium and calcium

The carbides (acetylides), $M^{2+}(C \equiv C)^{2-}$

The carbides can be made by reacting the oxide with carbon at high temperature; calcium carbide is of considerable industrial importance and is made by reacting quicklime with coke in an electric furnace at a temperature of about 2000°C. On the addition of water, calcium carbide gives ethyne, which is a vital raw material in the organic chemical industry.

$$Ca^{2+}O^{2-}(s) + 3C(s) \longrightarrow Ca^{2+}(C \equiv C)^{2-}(s) + CO(g)$$
$$Ca^{2+}(C \equiv C)^{2-}(s) + 2H_2O(l) \longrightarrow Ca^{2+}(OH^-)_2(aq/s) + HC \equiv CH(g)$$

The nitrides, $(M^{2+})_3(N^{3-})_2$

The nitrides are formed on heating the metals in nitrogen. Addition of water causes hydrolysis and ammonia is liberated.

$$3Mg(s) + N_2(g) \longrightarrow (Mg^{2+})_3(N^{3-})_2(s)$$
$$(Mg^{2+})_3(N^{3-})_2(s) + 6H_2O(l) \longrightarrow 3Mg^{2+}(OH^-)_2(s) + 2NH_3(g)$$

The formation of magnesium nitride was used by Ramsay in obtaining the noble gases from the air. The hydrolysis of the nitride to give ammonia is sometimes used as a laboratory test for nitrogen.

16.14 Hydrolysis of the Group 2A metal salts

The soluble salts of strong acids (e.g. chlorides and nitrates) of strontium and barium show a neutral reaction in aqueous solution, and those of magnesium and calcium are only very slightly acidic. The corresponding salts of beryllium are hydrolysed in solution and invariably show an acid reaction.

This behaviour is very much in line with the progressive increase in ionic size on moving down this particular group. Although the majority of beryllium compounds are essentially covalent, the hydrated ion $[Be(H_2O)_4]^{2+}$ exists in solution and is much smaller than the magnesium ion (p. 187). Because of its small size, the beryllium ion is able to attract water molecules very strongly and thus weaken the hydrogen-oxygen bonds in bonded water molecules; other solvent water molecules are thus able to act as Brønsted-Lowry bases and abstract protons:

$$Be(H_2O)_4{}^{2+}(aq) + H_2O(l) \rightleftharpoons Be(H_2O)_3(OH)^+(aq) + H_3O^+(aq)$$

The hydrogen carbonates and sulphides of these metals, like those of the alkali metals, show an alkaline reaction in aqueous solution and for exactly the same reason (p. 184).

Although not classified as salts, hydrides, carbides and nitrides of these metals are hydrolysed by water, since the hydride, carbide, and nitride ions are strong bases:

$$H^- + H_2O(l) \longrightarrow OH^-(aq) + H_2(g)$$
$$(C \equiv C)^{2-} + 2H_2O(l) \longrightarrow 2OH^-(aq) + HC \equiv CH(g)$$
$$N^{3-} + 3H_2O(l) \longrightarrow 3OH^-(aq) + NH_3(g)$$

16.15 Summary of trends in Group 2A

The elements of this group are very electropositive, although less so than corresponding members of the alkali metal group. Electropositivity increases with increasing atomic number. They therefore form predominantly ionic compounds, although beryllium is exceptional in forming compounds in which the bonding is usually covalent. The reason for this can be traced to the small size of the beryllium atom compared with the sizes of the other Group 2A atoms; the beryllium nucleus is thus able to exert a considerable force of attraction on its outer two valency electrons and thus prevent their transfer during chemical reaction. The chemistry of beryllium is compared with that of aluminium, its diagonal neighbour (see p. 197).

The Group 2A metals form hydroxides which are strong bases, except that of beryllium which is amphoteric; the solubility of these hydroxides increases with increasing atomic number. Their salts are fairly stable to heat but less so than the corresponding salts of the alkali metals. Thus the carbonates can be decomposed into the oxides and carbon dioxide at high temperature, and the nitrates on decomposition give the oxides, nitrogen dioxide and oxygen; the stability to heat increases with increasing atomic number. The hydrogen carbonates exist in solution but any attempt to isolate them as solids leads to decomposition into the carbonates.

The solubilities of the Group 2A salts show a definite trend but in some cases this shows up with increasing atomic number as an increase and in others as a decrease in solubility. The fluorides increase in solubility with increasing atomic number, but the chlorides, carbonates and sulphates decrease in solubility in this order. In order to explain these observations one must consider changes of lattice and hydration energies (p. 60) for comparable salts. The fluorides increase in solubility in the order $Mg < Ca < Sr < Ba$ because the lattice energy decreases more rapidly along the series than does the hydration energy. The reverse is true for the chlorides, carbonates and sulphates because in these compounds the lattice energy of comparable salts decreases less rapidly than hydration energy.

As the atomic and ionic sizes of the Group 2A metals increase with increasing atomic number there is an increase in chemical reactivity, and calcium, strontium and barium are sufficiently reactive to force the hydrogen atom to accept an electron and form ionic hydrides. Ionic carbides and nitrides can be formed by direct combination and in this respect there is no parallel to be found in Group 1A, except for lithium which has a small enough ion to form a stable carbide and nitride lattice.

The Group 2A metals form no oxides containing the superoxide ion, (p. 176) presumably because their ions are too small to form a stable crystal lattice—their ions are very much smaller than those of the corresponding alkali metals (p. 72). The larger ions of strontium and barium are able to form peroxides like the alkali metals and for the same reason, but pressure is required.

FIG. 16.2. *The structure of the calcium tetraethanolate ion*

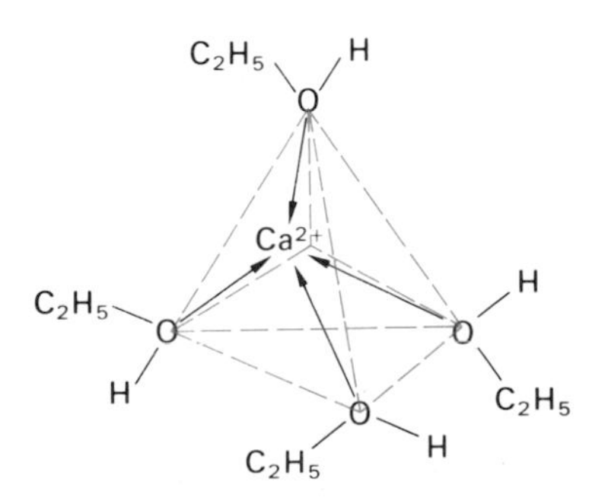

The tendency for the Group 2A metal ions to form complexes decreases with increasing atomic number; thus of the heavier ions only calcium forms a complex with ethanol. The calcium tetraethanolate ion presumably involves dative bonding from the oxygen atoms of ethanol molecules, with the four molecules of ethanol arranged tetrahedrally (theory of electron pair repulsion (fig. 16.2)).

Magnesium, having a smaller ion than calcium, forms complexes more readily and the solubility of its bromide and iodide in alcohols, ketones and ethers is probably due to the formation of datively bonded complexes involving bonds between the magnesium ion and the oxygen of the organic liquids. The complex $MgCl_4^{2-}$ ion is present in the compound $(NH_4^+)_2MgCl_4^{2-}$ and the chloride ions are probably arranged tetrahedrally around the central magnesium ion. Beryllium having the smallest ion of this group of elements naturally tends to form complexes the most readily, since the small positively charged ion is able to exert considerable attraction on other ions or molecules that have lone pairs of electrons. Ammonium fluoroberyllate, $(NH_4^+)_2BeF_4^{2-}$, contains the tetrahedral BeF_4^{2-} ion; the hydrated beryllium ion, $[Be(H_2O)_4]^{2+}$ is also tetrahedral.

The formulae of these complexes are consistent with the theory of octet completion, since four dative bonds would provide eight electrons; their tetrahedral shape is also consistent with the idea that electron pairs tend to arrange themselves in such a way that they are as far from each other as possible.

16.16 The anomalies of beryllium—the diagonal relationship with aluminium

Beryllium shows many resemblances to aluminium in its chemistry and for exactly the same reasons that lithium resembles magnesium (p. 185). The differences between beryllium and magnesium are even more striking than those between lithium and sodium, and its similarities to aluminium are so close that it was at one time thought to have a valency of three. Among its diagonal similarities with aluminium can be listed the following:

(a) Both metals are made passive by nitric acid.

(b) Both metals react with sodium hydroxide solution to evolve hydrogen:

$$Be(s) + 2OH^-(aq) + 2H_2O(l) \longrightarrow \underset{\text{beryllate ion}}{Be(OH)_4^{2-}(aq)} + H_2(g)$$

$$2Al(s) + 2OH^-(aq) + 6H_2O(l) \longrightarrow \underset{\text{aluminate ion}}{2Al(OH)_4^-(aq)} + 3H_2(g)$$

(c) The oxides and hydroxides of beryllium and aluminium are amphoteric (p. 149).

(d) The chlorides are covalent polymeric solids when anhydrous, $(BeCl_2)_x$ and $(AlCl_3)_x$, which readily dissolve in organic solvents. They are readily hydrolysed by water, with the evolution of hydrogen chloride.

(e) Beryllium carbide, Be_2C, and aluminium carbide, Al_4C_3, give methane on treatment with water, unlike the ionic carbides of the Group 2A metals. They are therefore referred to as methides:

$$Be_2C(s) + 4H_2O(l) \longrightarrow 2Be(OH)_2(s) + CH_4(g)$$
$$Al_4C_3(s) + 12H_2O(l) \longrightarrow 4Al(OH)_3(s) + 3CH_4(g)$$

(f) Similar complexes of beryllium and aluminium have similar stabilities, e.g. BeF_4^{2-} and AlF_6^{3-}. The former is tetrahedral and the latter octahedral in shape, in agreement with the simple theory of electron pair repulsion (p. 75–79).

16.17 Questions on chapter 16

1 Some physical properties of the alkaline earth metals (Group II of the Periodic Table) are tabulated on page 197.

Discuss the importance of **each** of these physical properties. Explain any observable trends in their magnitude and show how they may be used to interpret the chemical properties of the alkaline earth metals and their compounds.

The element radium, Ra (atomic number 88), is also an alkaline earth metal. Predict the approximate magnitude of the ionic radius, first ionisation energy and standard electrode potential for this element.

Element	Atomic number	Electronic structure	Ionic radius M^{2+}/nm	First ionisation energy/ kJ mol^{-1}	Standard electrode potential $E^{\ominus}$/V
Beryllium	4	2,2	0·031	900	−1·85
Magnesium	12	2,8,2	0·065	736	−2·38
Calcium	20	2,8,8,2	0·099	590	−2·87
Strontium	38	2,8,18,8,2	0·113	548	−2·89
Barium	56	2,8,18,18,8,2	0·135	502	−2·90

The radioactive decay of the isotope radium-226 can be represented by the equation

$$^{226}_{88}Ra \longrightarrow X + \alpha$$

State, giving reasons, in which group of the Periodic Table element X occurs. (O)

2 Taking the elements beryllium to barium of Group 2A as examples, show how chemical properties vary with atomic number in a group of the Periodic Table.

Confine your answer to a discussion of the properties of:
(a) the elements,
(b) the oxides,
(c) the chlorides,
(d) the sulphates. (O & C)

3 Illustrate the trends to be found in the chemistry of the Group II elements beryllium to barium by writing a comparative account of the chemistry of their hydrides, oxides, carbonates, chlorides and sulphates. (O)

4 Compare and contrast the chemistry of the Group 1A metals with that of the Group 2A metals. Confine your answer to a discussion of the following:
(a) atomic and ionic radii,
(b) ionisation energies,
(c) melting points of the metals,
(d) basic nature of the oxides and hydroxides,
(e) general stability of their compounds,
(f) solubilities of their compounds.
Give explanations for differences whenever possible.

5 (a) How is calcium metal obtained? (Technical details are not required.) Why is this the preferred method?
(b) Describe the chemical properties of (i) calcium oxide, (ii) calcium hydride, in each case giving reasons for the stated properties.
(c) Some calcium compounds are used industrially on a large scale. Select any **one** such compound and describe in outline how and why it is used in industry.
(d) Calcium does not form any stable compounds in which it has an oxidation state of +3. Briefly discuss the reasons for this. (O & C)

6 Give in outline the chemistry of the metal calcium and of four of its compounds, which are of importance in nature or industry. (O & C)

7 From your knowledge of the chemistry of calcium and the general trends in Group 2A predict the chemistry of radium. Confine your answer to a discussion of the following:
(a) reactivity of radium metal (excluding its radioactivity),
(b) the nature and reactivity of its compounds, e.g. oxide, hydride, chloride, carbonate and sulphate,
(c) the solubilities of the carbonate and sulphate in comparison with those of barium.

8 Discuss the diagonal relationship of beryllium with aluminium.

9 Explain the meaning of the terms 'first electron affinity', 'first ionization energy' and 'electronegativity'. Discuss the trends in first ionization energy and electronegativity in Groups IA and IIA of the Periodic Table. In each case, relate your answer to the chemical properties of the elements. Mention and explain three other trends which occur in one or both of these periodic groups. Discuss briefly the relationships which exist between the two groups of elements. (L)

Chapter 17

Group 3B boron, aluminium, gallium, indium and thallium

17.1 Some physical data of Group 3B elements

	Atomic Number	Electronic Configuration	Ionisation energy /kJ mol^{-1} First	Second	Third	Standard Electrode Potential /V $M^{3+} + 3e^- \rightarrow M$	Atomic Radius /nm	Ionic Radius M^{3+}/nm	M.p. /°C	B.p. /°C
B	5	2.3 $1s^2 2s^2 2p^1$	800	2427	3650		0·080	(0·020) Estimated Value	2300	3930
Al	13	2.8.3 ....$2s^2 2p^6 3s^2 3p^1$	578	1816	2744	−1·66	0·125	0·050	660	2470
Ga	31	2.8.18.3 ....$3s^2 3p^6 3d^{10} 4s^2 4p^1$	579	1979	2962	−0·52	0·125	0·062	29·8	2400
In	49	2.8.18.18.3 ....$4s^2 4p^6 4d^{10} 5s^2 5p^1$	558	1820	2705	−0·34	0·150	0·081	157	2000
Tl	81	2.8.18.32.18.3 ....$5s^2 5p^6 5d^{10} 6s^2 6p^1$	589	1970	2880	+0·72	0·155	0·095	304	1460

17.2 Some general remarks about Group 3B

All these elements exhibit a group valency of three, but because of the very large input of energy that is necessary to form the 3-valent ions—the sum of the first three ionisation energies—their compounds when anhydrous are either essentially covalent or contain an appreciable amount of covalent character. Boron never forms a B^{3+} ion since the enormous amount of energy required to remove three electrons from a small atom cannot be repaid with the formation of a stable crystal lattice, even with the most electronegative fluorine atom.

The electronic configurations of the boron and aluminium atoms are similar inasmuch as the penultimate shell has a noble gas configuration, whereas the penultimate shell of the gallium, indium and thallium atoms contains eighteen electrons.

Boron, which is non-metallic, and aluminium, which is clearly metallic, are best considered separately. Gallium, indium and thallium are weakly metallic and can profitably be studied as a group; in many aspects their chemistry is similar to that of aluminium.

Boron

17.3 Occurrence, extraction and uses of boron

Boron occurs principally as borates, e.g. sodium tetraborate (borax), $(Na^+)_2B_4O_7^{2-}.10H_2O$, in which the boron atom is part of an anionic complex.

Boron can be obtained as an amorphous brown powder by treating borax with hydrochloric acid, igniting the boric acid, H_3BO_3, to give the oxide, B_2O_3, and finally reducing the latter with magnesium at a high temperature:

$$B_2O_3(s) + 3Mg(s) \longrightarrow 2B(s) + 3Mg^{2+}O^{2-}(s)$$

It is used in the construction of high impact-resistant steel and, since it absorbs neutrons, in reactor rods for controlling atomic reactions.

A crystalline form of boron can be obtained by the thermal decomposition of boron tri-iodide on a tantalum filament:

$$2BI_3(s) \longrightarrow 2B(s) + 3I_2(s)$$

17.4 Properties of boron

Amorphous boron is a very reactive element combining directly with oxygen, sulphur, nitrogen and the halogens to give respectively an oxide, sulphide, nitride and a halide. The oxide, sulphide and nitride are giant molecules with covalent bonds running completely through the structures.

$$4B(s) + 3O_2(g) \longrightarrow 2B_2O_3(s)$$
$$4B(s) + 6S(s) \longrightarrow 2B_2S_3(s)$$
$$2B(s) + N_2(g) \longrightarrow 2BN(s)$$
$$2B(s) + 3Cl_2(g) \longrightarrow 2BCl_3(g)$$

At red heat it will reduce steam to hydrogen:

$$2B(s) + 3H_2O(l) \longrightarrow B_2O_3(s) + 3H_2(g)$$

17.5 The halides of boron, BX_3

The volatility of the halides decreases with increasing relative molecular mass, thus BF_3 and BCl_3 are gases, BBr_3 is a liquid and BI_3 is a white solid. They are covalent and exist as BX_3 molecules, their structures being planar with the halogen atoms pointing towards the apices of an equilateral triangle to minimise electron repulsion (p. 75).

The formation of three covalent B—X bonds leaves the boron atom two electrons short of a noble gas configuration; the octet is readily completed by the formation of addition compounds with molecules that have lone pairs of electrons, e.g. the boron trichloride-ammonia addition compound which is, as expected, roughly tetrahedral in shape.

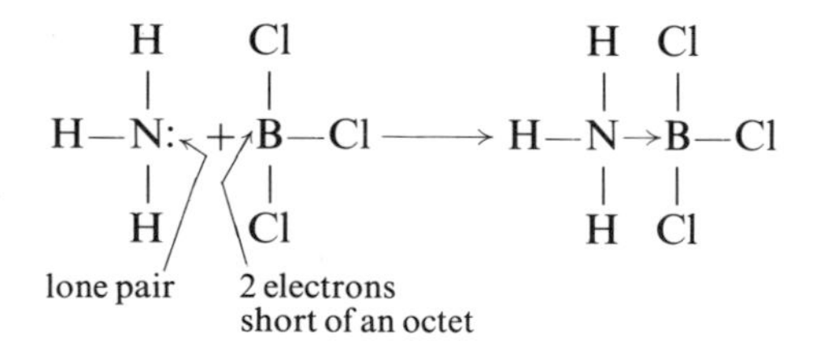

The halides react vigorously with water to give the halogen hydride, except boron trifluoride which gives fluoboric acid, HBF_4, which in solution contains the tetrahedral BF_4^- ion; boric acid is also formed, e.g.

$$BCl_3(g) + 3H_2O(l) \longrightarrow H_3BO_3(aq) + 3HCl(g)$$
$$4BF_3(g) + 3H_2O(l) \longrightarrow 3HBF_4(aq) + H_3BO_3(aq)$$

These hydrolysis reactions are thought to proceed via the formation of a boron trihalide-water complex involving the formation of a dative bond from the oxygen of the water molecule:

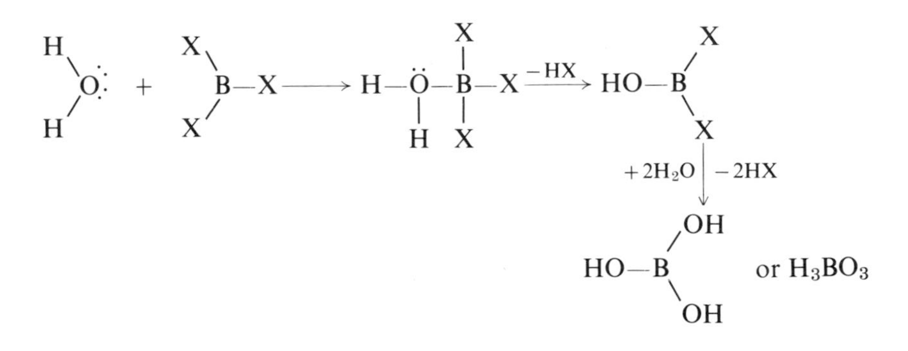

Boron trifluoride is used as a Friedel-Crafts catalyst in organic chemistry, particularly for polymerisation reactions. Its mode of action depends on its ability to act as an electron pair acceptor by completing the octet round the central boron atom.

17.6 Diborane, B_2H_6

This compound, which is an inflammable and very reactive gas, is the simplest hydride of boron. It can be made by the reduction of boron trichloride with lithium aluminium hydride and must be handled in vacuum systems which employ mercury valves, since it attacks tap-grease.

$$4BCl_3(g) + 3Li^+AlH_4^- \longrightarrow 2B_2H_6(g) + 3Li^+Cl^-(s) + 3AlCl_3(s)$$

The structure of this molecule immediately poses problems since the number of valency electrons is twelve, yet at least fourteen are required for conventional single covalent bonded structures similar to ethane, i.e. the molecule is electron deficient. Physical evidence shows that the structure can be considered to involve something resembling boron—boron double bonds, the extra two electrons required for this being provided by two of the six hydrogen atoms which bridge the boron atoms as protons:

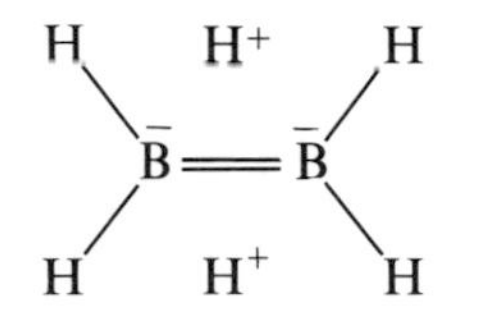

A more sophisticated but more difficult theory envisages the two bridging hydrogen atoms forming bonds with both boron atoms.

Other hydrides of boron are B_4H_{10}, B_5H_9, B_5H_{11} and $B_{10}H_{14}$; they are all electron deficient.

17.7 Boron trioxide, B_2O_3

Boron trioxide can be obtained by burning boron in oxygen or by fusing orthoboric acid:

$$2H_3BO_3(s) \longrightarrow B_2O_3(s) + 3H_2O(l)$$

It is usually obtained as a glassy material whose structure consists of randomly orientated three-dimensional networks of BO_3 groups, each oxygen atom uniting two boron atoms:

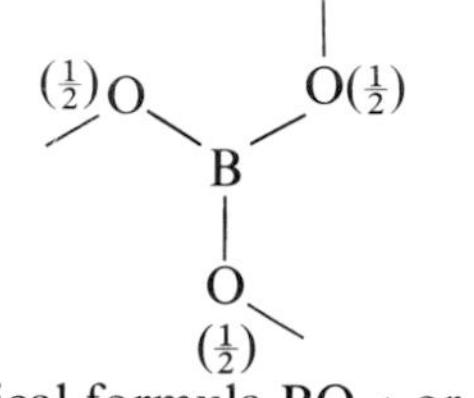

empirical formula $BO_{1\frac{1}{2}}$ or B_2O_3

Boron trioxide is an acidic oxide and slowly reacts with water, forming orthoboric acid. When fused with metallic oxides it forms borate glasses which are often coloured; this is the basis of the borax bead test in qualitative analysis.

17.8 Orthoboric acid, H_3BO_3

Orthoboric acid is formed when the boron halides are hydrolysed or when dilute hydrochloric acid is added to a solution of borax:

$$B_4O_7^{2-}(aq) + 2H^+(aq) + 5H_2O(l) \longrightarrow 4H_3BO_3(aq)$$

It is obtained as a white solid on subsequent crystallisation.

Orthoboric acid is a weak monobasic acid and in aqueous solution the boron atom completes its octet by removing OH^- from water molecules:

$$B(OH)_3(aq) + 2H_2O(l) \longrightarrow B(OH)_4^-(aq) + H_3O^+(aq)$$

It therefore functions as a Lewis acid and not as a proton donor.

When heated, orthoboric acid first forms metaboric acid (empirical formula HBO_2) and then boron trioxide.

The structure of orthoboric acid is based on the planar $B(OH)_3$ unit, individual units being joined together by hydrogen bonds to give a two-dimensional sheet. It is now known that the hydrogen bonds linking the oxygens of different $B(OH)_3$ groups are unsymmetrical, i.e. they are not midway between the oxygen atoms. The layers of two-dimensional sheets are held together by weak van der Waals' forces, and this accounts for the cleavage of the solid into flakes (fig. 17.1).

FIG. 17.1. *The structure of orthoboric acid*

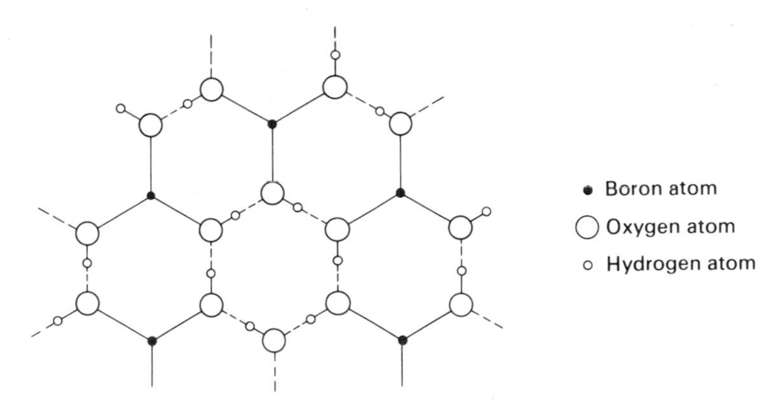

17.9 Borates

Boron, like silicon, has a great affinity for oxygen, and a multitude of structures exist containing rings or chains of alternating boron and oxygen atoms. The simple BO_3^{3-} ion is rather uncommon but does occur in $(Mg^{2+})_3(BO_3^{3-})_2$; the ion, as expected, has a planar structure.

The more complex borates are based on triangular BO_3 units, thus sodium metaborate, $(Na^+)_3B_3O_6^{3-}$, and calcium metaborate, $Ca^{2+}B_2O_4^{2-}$, have borate anions of the structures below:

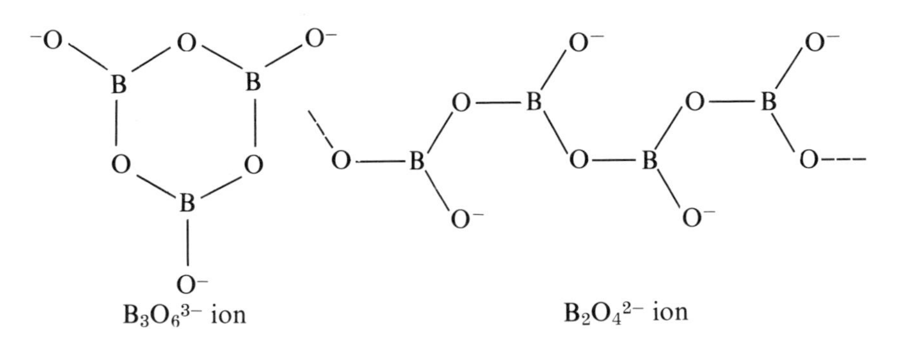

Note that the borate ion in calcium metaborate is polymeric and $B_2O_4^{2-}$ simply represents its empirical formula. The polymeric metaborate chain anions are held together by Ca^{2+} ions between them.

When borax is fused with transition metal salts a coloured glassy material is frequently obtained; these coloured beads are generally metaborates, e.g. copper metaborate has the empirical formula $Cu(BO_2)_2$. During their formation rapid cooling does not allow the complex metaborate anions time to order themselves into a regular repeating crystalline structure and a randomly organised solid—or glass—is the result. A glass can be thought of as a liquid which has been cooled until it is too viscous to flow.

17.10 Borazine ($B_3N_3H_6$) and boron nitride (BN)

When ammonia and diborane, in the ratio of two molecules to one, are reacted together at high temperature the volatile compound borazine is formed; the molecule has a cyclic structure reminiscent of benzene:

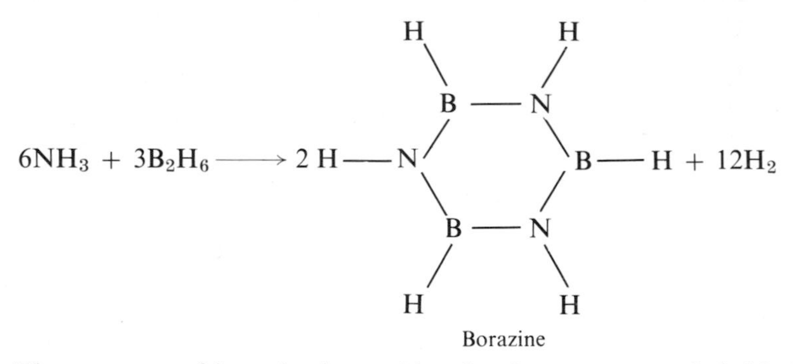

The structure of borazine is considered to be a resonance hybrid of the two structures:

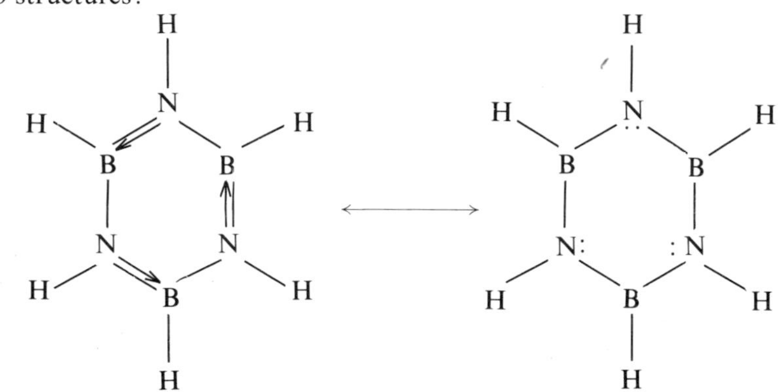

It is isoelectronic with benzene and resembles the latter in some of its physical and chemical properties. Its boiling point is 55°C, some twenty five degrees lower than that of benzene, and it forms addition compounds more readily than benzene, e.g. it will react with hydrogen chloride, the chlorine atom attaching itself to the boron, whereas benzene is unreactive towards this reagent.

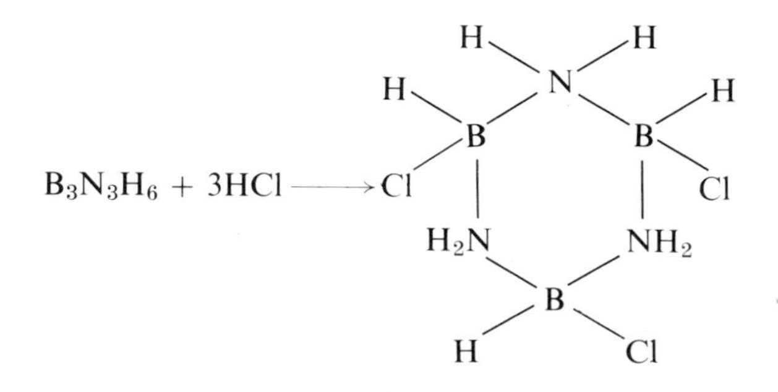

Boron nitride (empirical formula BN) is formed by the direct union of boron and nitrogen at white heat; it has a structure similar to that of graphite (p. 80) and is thus a giant molecule, but differs from graphite in only being a semiconductor of electricity. Whereas the benzene nucleus is the basic repeating structure in the layer structure of graphite, the borazine nucleus is the basic repeating structure in boron nitride.

Another form of boron nitride has been synthesised and has the diamond structure (p. 79), in which boron and nitrogen atoms alternate in the crystal lattice. This form, as expected, is nearly as hard as diamond itself.

17.11 Occurrence, extraction and uses of aluminium

Aluminium

Aluminium is the most abundant metallic element in the earth's crust, occurring in a variety of aluminosilicates such as clay, micas and feldspars. The only ore of aluminium from which it is profitable to extract the metal is bauxite, hydrated aluminium oxide, $Al_2O_3.2H_2O$, which has accumulated through the slow but persistent weathering action on clay.

The bauxite is first freed from silica and iron (III) oxide impurities by dissolving it in sodium hydroxide (iron (III) oxide remains undissolved) and then precipitating aluminium hydroxide by seeding the solution with a little aluminium hydroxide (silica remains dissolved):

$$Al_2O_3(s) + 2OH^-(aq) + 3H_2O(l) \longrightarrow \underset{\text{aluminate ion}}{2Al(OH)_4^-(aq)}$$

$$Al(OH)_4^-(aq) \longrightarrow Al(OH)_3(s) + OH^-(aq) \text{ (takes place on seeding)}$$

Pure aluminium oxide is now obtained by heating the hydroxide:

$$2Al(OH)_3(s) \longrightarrow Al_2O_3(s) + 3H_2O(l)$$

It is dissolved in molten cryolite, $(Na^+)_3AlF_6^{3-}$, and electrolysed at about 900°C, using a number of graphite blocks as anodes and a graphite lined bath as the cathode. Aluminium is discharged at the cathode and collects at the bottom of the molten electrolyte as a liquid, from where it can be tapped off and allowed to solidify. Oxygen is evolved at the anodes, which are slowly burnt away as carbon dioxide.

The extraction of aluminium is only economic where cheap electricity is available from water power, e.g. the Western Highlands of Scotland, Norway and the Canadian Rocky Mountains. A low voltage is used to avoid decomposing the molten cryolite which acts as a solvent; a very high current density is used. One theory assumes that aluminium oxide dissociates into Al^{3+} and AlO_3^{3-}:

$$Al_2O_3 \rightleftharpoons Al^{3+} + AlO_3^{3-}$$

Cathode	**Anode**
Al^{3+} discharged	AlO_3^{3-} discharged
$4Al^{3+} + 12e^- \rightarrow 4Al$	$4AlO_3^{3-} \rightarrow 2Al_2O_3 + 3O_2 + 12e^-$

$(Na^+)_3AlF_6^{3-}$
Molten cryolite solvent

The process, however, is probably much more complex.

Aluminium alloys (Duralumin Al/Mg/Cu and Magnalium Al/Mg) are light and strong and are therefore used in the construction of aircraft and ships. Because of their low moment of inertia they are suitable alloys for the construction of buses, tube trains and piston heads. Aluminium conducts heat well (saucepans) and is made passive by nitric acid which makes it a suitable material for handling this corrosive liquid. Weight for weight it is a better electrical conductor than copper and is used in the National Grid for conveying electrical power. Aluminium foil is used for wrapping chocolates and making milk bottle tops. Its ability to reflect heat and light efficiently accounts for the use of aluminium painted storage tanks, which do not overheat dangerously when exposed to rays from the sun. The great affinity of aluminium for oxygen makes the metal a useful reducing agent for the small scale extraction of metals such as chromium (p. 360).

17.12 Properties of aluminium

Aluminium is a light metal possessing considerable strength, yet is malleable and ductile; it has the cubic close-packed structure (p. 188). It is not as reactive as its high negative electrode potential would imply, because normally there is a very thin oxide layer on its surface. When this oxide layer is removed by rubbing with mercury, the metal reacts rapidly with moisture in the air, forming a 'moss-like growth' of aluminium hydroxide, and becomes very hot in the process.

It combines directly with oxygen, sulphur, nitrogen and the halogens when heated to a sufficiently high temperature. The oxide and fluoride are essentially ionic, the rest are predominantly covalent.

$$4Al(s) + 3O_2(g) \longrightarrow 2(Al^{3+})_2(O^{2-})_3(s)$$
$$4Al(s) + 6S(s) \longrightarrow 2Al_2S_3(s)$$
$$2Al(s) + N_2(g) \longrightarrow 2AlN(s)$$
$$2Al(s) + 3F_2(g) \longrightarrow 2Al^{3+}(F^-)_3(s)$$

Aluminium reacts with moderately concentrated hydrochloric acid to give the chloride and hydrogen. The pure metal is not readily attacked by dilute sulphuric acid, but with the concentrated acid it gives the sulphate and sulphur dioxide. It is made passive by nitric acid and this has been attributed to the formation of an impenetrable oxide layer on its surface. It is attacked by sodium hydroxide solution with the liberation of hydrogen:

$$2Al(s) + 2OH^-(aq) + 6H_2O(l) \longrightarrow 2Al(OH)_4^-(aq) + 3H_2(g)$$

17.13 The halides of aluminium

Aluminium fluoride, $Al^{3+}(F^-)_3$

This compound can be made by the direct combination of aluminium and fluorine. It is the only ionic halide of aluminium and is sparingly soluble in water.

Aluminium chloride, Al_2Cl_6

This is made by passing hydrogen chloride or chlorine over heated aluminium; since it is readily attacked by the moisture in the air it must be prepared under anhydrous conditions, the tube of soda lime being used to keep out moisture as well as to absorb excess chlorine (fig. 17.2).

$$2Al(s) + 3Cl_2(g) \longrightarrow Al_2Cl_6(s)$$
$$2Al(s) + 6HCl(g) \longrightarrow Al_2Cl_6(s) + 3H_2(g)$$

It can also be obtained by passing a stream of chlorine over a mixture of aluminium oxide and charcoal heated to about 1000°C:

$$(Al^{3+})_2(O^{2-})_3(s) + 3C(s) + 3Cl_2(g) \longrightarrow Al_2Cl_6(s) + 3CO(g)$$

FIG. 17.2. *Preparation of anhydrous aluminium chloride*

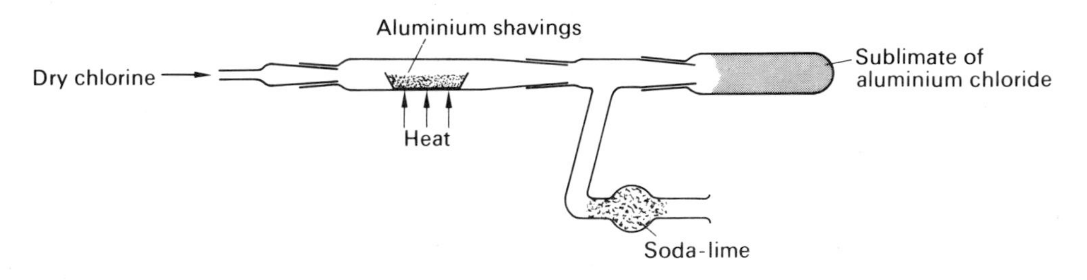

When pure, aluminium chloride is a white solid which sublimes at a temperature of about 180°C. Relative molecular mass determinations in solution, e.g. in benzene, and in the vapour state indicate that the molecule has the formula Al_2Cl_6, unlike the halides of boron which are monomeric. The aluminium atoms complete their octets by dative bonding from two chlorine atoms, the arrangement of the chlorine atoms about each aluminium atom being roughly tetrahedral:

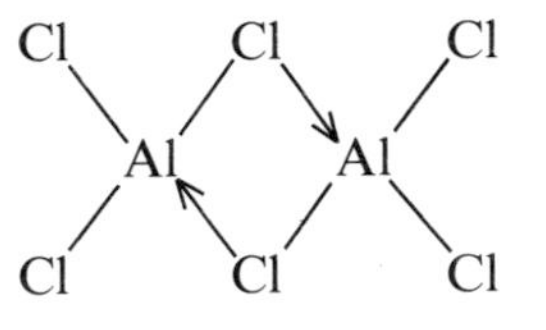

At high temperatures aluminium chloride exists as the monomer $AlCl_3$ in which the aluminium atom has a sextet of electrons; as predicted this molecule is planar and symmetrical:

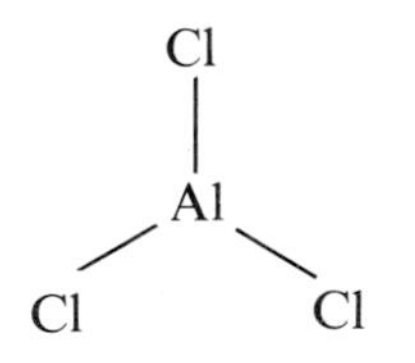

When aluminium chloride is added to water it reacts exothermically to give hydrated aluminium ions, $[Al(H_2O)_6]^{3+}$ and chloride ions, Cl^-. The energy needed to break the Al—Cl covalent bonds is derived from the high enthalpy of hydration of the small highly charged Al^{3+} ion.

$$\underset{\text{covalent}}{Al_2Cl_6(s)} + 12H_2O(l) \longrightarrow \underset{\text{ionic}}{2[Al(H_2O)_6]^{3+}\ (Cl^-)_3(s)}$$

Hydrated aluminium chloride is readily soluble in water, yielding $[Al(H_2O)_6]^{3+}(aq)$ and $Cl^-(aq)$ ions. Unlike anhydrous aluminium chloride it is insoluble in organic solvents and has no catalytic activity (see later).

The dimeric covalent aluminium chloride molecule is readily split in the presence of molecules which contain lone pairs; for instance, with ethers it forms the tetrahedral aluminium chloride-ether complex in which the octet is completed:

$$\underset{\text{an ether}}{2R_2O} + Al_2Cl_6 \longrightarrow \underset{\text{aluminium chloride-ether complex}}{2[R_2O \rightarrow AlCl_3]}$$

Like boron trichloride, it also forms complexes of the same type with halide ions, e.g. $AlCl_4^-$, which explains its action as a Friedel-Crafts catalyst in many organic reactions:

$$\underset{\text{an acid chloride}}{2RCOCl} + Al_2Cl_6 \longrightarrow \underset{\text{ion pair}}{2[RCO^+\ldots\ldots AlCl_4^-]}$$

$$\underset{\text{a carbonium ion}}{RCO^+} + C_6H_6 \longrightarrow (RCOC_6H_6)^+ \longrightarrow \underset{\text{an aromatic ketone}}{RCOC_6H_5} + H^+$$

Aluminium bromide and aluminium iodide, Al_2Br_6 and Al_2I_6

The structures and properties of these are very similar to those of the chloride.

17.14 Aluminium oxide (alumina), $(Al^{3+})_2(O^{2-})_3$

Aluminium oxide occurs naturally as bauxite (p. 205) which is contaminated with silica and iron (III) oxide impurities. When freshly prepared it reacts readily with dilute acids and strong alkalis to give salts. It is thus amphoteric, unlike the oxide of boron which is exclusively acidic (p. 202).

$$(Al^{3+})_2(O^{2-})_3(s) + 6H^+(aq) \longrightarrow 2Al^{3+}(aq) + 3H_2O(l)$$

$$(Al^{3+})_2(O^{2-})_3(s) + 2OH^-(aq) + 3H_2O(l) \longrightarrow \underset{\text{aluminate ion}}{2Al(OH)_4^-(aq)}$$

It is used for dehydrating alcohols to the corresponding alkenes, e.g.

$$\underset{\text{ethanol}}{CH_3{-}CH_2{-}OH(l)} \xrightarrow[300^\circ C]{\text{Alumina}} \underset{\text{ethene}}{H_2C{=}CH_2(g)} + H_2O(l)$$

As a finely divided powder it is frequently used as the absorbing stationary phase in column chromatography, e.g. for separating chlorophylls.

A very hard crystalline form of aluminium oxide called corundum occurs naturally; it is used as an abrasive. It can be made by heating amorphous aluminium oxide to about 2000°C.

Aluminium forms a series of mixed oxides with other metals, some of these occurring naturally as semi-precious stones. These include ruby (Cr^{3+}), and blue sapphire (Co^{2+},Fe^{2+},Ti^{4+}); these are now produced synthetically and find an application as bearings in watches.

17.15 Aluminium hydroxide, $Al(OH)_3$

This is formed as a gelatinous precipitate when ammonium hydroxide solution is added to a solution containing aluminium ions:

$$Al^{3+}(aq) + 3OH^{-}(aq) \longrightarrow Al(OH)_3(s)$$

Unlike the hydroxide of boron which is exclusively an acid, aluminium hydroxide is amphoteric (see p. 149 for the factors that determine the acidic and basic natures of hydroxides).

Gelatinous aluminium hydroxide is converted into a form on standing which is insoluble in acids and strong alkalis. Both forms of the hydroxide readily adsorb dyes and, in the dyeing industry, aluminium hydroxide is precipitated in the fibres of cloth to ensure strong binding of the dye. It is referred to as a mordant (from the Latin, *mordere* = to bite).

17.16 Aluminium sulphate, $(Al^{3+})_2(SO_4^{2-})_3.18H_2O$, and the alums, $M^+Al^{3+}(SO_4^{2-})_2.12H_2O$

Aluminium sulphate, $(Al^{3+})_2(SO_4^{2-})_3.18H_2O$

This is manufactured by reacting aluminium hydroxide with concentrated sulphuric acid. It crystallises from water with eighteen molecules of water of crystallisation. It gives an acid reaction in aqueous solution (p. 223), forming aluminium hydroxide; because of this ready hydrolysis it is used as a mordant in dyeing. It is also used for sizing paper and waterproofing cloth.

The 3-valent aluminium cation is very effective in precipitating negative colloids and aluminium sulphate is used in styptic pencils for clotting blood. It is also used for treating sewage which contains much material in colloidal solution.

The alums, $M^+Al^{3+}(SO_4^{2-})_2.12H_2O$

When a solution containing potassium, aluminium and sulphate ions is allowed to crystallise, transparent octahedral crystals of potash alum, $K^+Al^{3+}(SO_4^{2-})_2.12H_2O$, are obtained. The solid contains $[K(H_2O)_6]^+$, $[Al(H_2O)_6]^{3+}$ and SO_4^{2-} ions and is a double salt, since it gives the characteristic reactions of its constituent ions in solution.

It is possible to prepare a series of alums in which the potassium ion is replaced by the ammonium ion NH_4^+. It is also possible to replace the 3-valent aluminium cation by 3-valent transition metal cations of about the same ionic size, e.g. Ti^{3+}, Cr^{3+}, Mn^{3+}, Fe^{3+} and Co^{3+}. The alums are isomorphous, readily forming overgrowths and solid solutions; a few typical ones are:

$NH_4^+Al^{3+}(SO_4^{2-})_2.12H_2O$	Ammonium alum
$K^+Cr^{3+}(SO_4^{2-})_2.12H_2O$	Chrome (III) alum
$NH_4^+Fe^{3+}(SO_4^{2-})_2.12H_2O$	Ammonium iron (III) alum

Potash alum, like aluminium sulphate, is used as a mordant in dyeing, since it forms aluminium hydroxide by hydrolysis.

17.17 Other compounds of aluminium

Aluminium hydride, AlH_3, and lithium aluminium hydride, $Li^+AlH_4^-$

When lithium hydride is treated with an excess of aluminium chloride in ether solution aluminium hydride is precipitated as a white solid. It is polymeric but its structure is still unknown.

$$3Li^+H^- + AlCl_3 \longrightarrow AlH_3 + 3Li^+Cl^-$$

When lithium hydride is treated with aluminium chloride in ether and an excess of the chloride is avoided, lithium aluminium hydride is formed:

$$4Li^+H^- + AlCl_3 \longrightarrow Li^+AlH_4^- + 3Li^+Cl^-$$

Lithium aluminium hydride is used in organic chemistry for reducing carboxylic acids to alcohols. It will not reduce C=C bonds, and so can be used as a selective reducing agent, e.g. for reducing unsaturated aldehydes to unsaturated alcohols.

$$\underset{\text{propanoic acid}}{CH_3{-}CH_2{-}COOH} \longrightarrow \underset{\text{propan-l-ol}}{CH_3{-}CH_2{-}CH_2OH}$$

$$\underset{\text{but-2-enal}}{CH_3{-}CH{=}CH{-}CHO} \longrightarrow \underset{\text{2-buten-1-ol}}{CH_3CH{=}CH{-}CH_2OH}$$

Aluminium carbide, Al_4C_3 (aluminium methide)
This covalent solid can be made by direct combination of aluminium and carbon at high temperature. It is hydrolysed by water, with the evolution of methane:

$$Al_4C_3(s) + 12H_2O(l) \longrightarrow 4Al(OH)_3(s) + 3CH_4(g)$$

Aluminium sulphide, Al_2S_3
This is made by heating together aluminium powder and finely divided sulphur; the reaction can be exceedingly violent. It is covalent and is readily hydrolysed by water with the evolution of hydrogen sulphide:

$$Al_2S_3(s) + 6H_2O(l) \longrightarrow 2Al(OH)_3(s) + 3H_2S(g)$$

Aluminium triethyl, $(C_2H_5)_3Al$
This compound is made by heating aluminium powder, ethene and hydrogen under high pressure:

$$2Al(s) + 6C_2H_4(g) + 3H_2(g) \longrightarrow 2(C_2H_5)_3Al(l)$$

It is spontaneously inflammable in air and explosively reactive towards water. The liquid is dimeric and two ethyl groups act as bridging units:

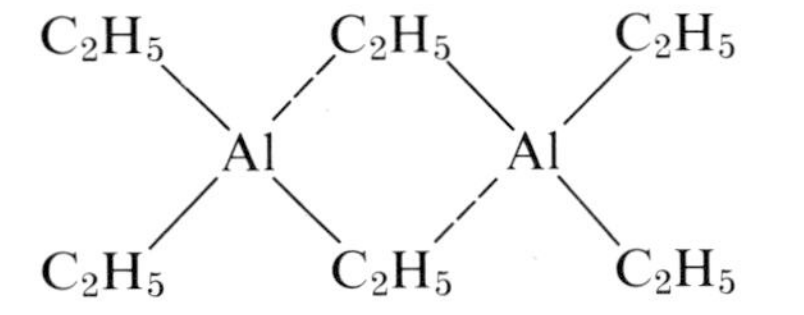

Aluminium triethyl in conjunction with titanium (IV) chloride is the catalyst in one industrial process for converting ethene into polythene. A similar type of catalyst system is used in the polymerisation of propene into polypropene (polypropylene).

17.18 Hydrolysis of aluminium compounds

In solution, aluminium ions are thought to be hydrated by six molecules of water per ion. Because the aluminium ion is highly charged and quite small, the bond uniting the hydrogen and oxygen atoms in co-ordinated water molecules is considerably weakened, and other solvent water molecules are able to act as bases and abstract protons. Hydrated aluminium ions are therefore acidic in aqueous solution:

$$Al(H_2O)_6{}^{3+}(aq) + H_2O(l) \longrightarrow Al(H_2O)_5(OH)^{2+}(aq) + H_3O^+(aq)$$

$$Al(H_2O)_5(OH)^{2+}(aq) + H_2O(l) \longrightarrow Al(H_2O)_4(OH)_2{}^+(aq) + H_3O^+(aq) \text{ etc.}$$

If the solution also contains a strongly basic anion such as $CO_3{}^{2-}$ or S^{2-} further ionisation of hydrated aluminium ions can occur:

$$Al(H_2O)_4(OH)_2{}^+(aq) + CO_3{}^{2-}(aq) \rightleftharpoons \underset{\text{hydrated aluminium hydroxide}}{Al(H_2O)_3(OH)_3(s)} + HCO_3{}^-(aq)$$

These series of reactions show why it is impossible to obtain either aluminium carbonate or aluminium sulphide from aqueous solution. Any attempt to do so results in the precipitation of aluminium hydroxide.

Gallium, indium and thallium

17.19 Comparison of the elements gallium, indium and thallium

Gallium, indium and thallium are obtained by electrolysis of aqueous solutions of their salts. They are soft, white and fairly reactive metals, combining on heating with many non-metals such as the halogens and sulphur. Gallium resembles aluminium in reacting with sodium hydroxide solution; it is a liquid over a remarkable range of temperature (30°—2400°C) for which no adequate explanation is yet possible.

The halides

The trifluorides are high melting-point ionic solids which resemble aluminium fluoride. Gallium (III) chloride is a dimeric covalent solid which is rapidly hydrolysed by water and which dissociates into the monomer on heating; it is thus very similar in behaviour to aluminium chloride. Indium (III) chloride, however, is monomeric and, since it conducts electricity in the fused state, is appreciably ionic. Thallium (III) chloride decomposes on heating to thallium (I) chloride, TlCl, in which the metal exhibits a valency of one.

The oxides

The 3-valent oxides are similar to aluminium oxide but become progressively less difficult to reduce with increasing atomic number. The oxides are more basic in the same order; thus gallium (III) oxide is less amphoteric than aluminium oxide and thallium (III) oxide is exclusively basic.

1-valent compounds

Gallium and indium both form chlorides with the empirical formula XCl_2 in which they appear to be 2-valent. Physical evidence, however, shows that the correct formulation of these chlorides is $X^{I}(X^{III}Cl_4)$, in which both gallium and indium exhibit a valency of one. The tendency towards univalency is even more pronounced in the case of thallium and is due to the reluctance of the two *s* electrons of the heavier elements to be transferred or to participate in covalent bond formation. It is called the inert pair effect.

Some 1-valent compounds of thallium resemble corresponding compounds of the alkali metals; thus the oxide, Tl_2O, and the hydroxide, TlOH, are strong bases. Other compounds are similar to those of silver, thus thallium (I) chloride is only sparingly soluble in water and is sensitive to light.

17.20 Suggested reading

A. G. Massey, *The hydrides of boron*, Education in Chemistry, No. 1, Vol. 11, 1974.
Oxidation State and the Typical Metals, Open University, S25, Unit 6.

17.21 Questions on chapter 17

1 Boron is non-metallic whereas aluminium is metallic. How does this influence the chemistry of the oxides B_2O_3 and Al_2O_3, and the hydroxy-compounds $B(OH)_3$ and $Al(OH_3)$? Would you expect boron to form a sulphate and nitrate?

2 Discuss the structural chemistry of the following compounds of boron:
(a) $(BN)_x$,
(b) BF_3,
(c) $BF_3.NH_3$,
(d) B_2O_3,
(e) H_3BO_3,
(f) borazine, $B_3N_3H_6$.

3 In what respects is the chemistry of boron similar to that of silicon (its diagonal neighbour in the Periodic Table)?

4 Give a comparative account of the chemistry of sodium, magnesium, and aluminium. (C)

5 Give an account of the manufacture of aluminium. Explain applications based on two physical and two chemical characteristics of the metal. Give a brief description of the chloride, hydroxide and sulphate of aluminium. (O & C)

6 Give equations and explain how pure aluminium oxide is separated from a sample containing small quantities of the oxides of iron and silicon. Using sodium chloride (as the source of chlorine) and aluminium, describe briefly, with the aid of a diagram, how you would make anhydrous aluminium chloride.

How and under what conditions does aluminium react with
(a) water,
(b) sodium hydroxide,
(c) iron (III) oxide? (S)

7 Discuss the uses of four important compounds of aluminium.

8 Explain why a solution of aluminium sulphate in water shows an acidic reaction. When solutions of aluminium sulphate and sodium carbonate are mixed a precipitate of aluminium hydroxide, and not aluminium carbonate, is formed. What explanation can you give for this?

9 Aluminium reacts vigorously when heated with sulphur to give aluminium sulphide but the sulphide cannot be obtained by mixing solutions containing aluminium ions and sulphide ions. What explanation can you give for this? How would you expect aluminium sulphide to react with water?

10 The elements in Group 3B are boron, aluminium, gallium, indium and thallium. From your knowledge of the chemistry of aluminium and the trends which occur in chemical properties as a group is descended, describe what you would deduce to be the main features of the chemistry of gallium, indium and thallium.

Why is it valid to make deductions of this kind? (Camb. Schol.)

11 In some of its chemistry thallium resembles aluminium whereas in other ways it resembles a Group 1A metal. Cite evidence in support of this statement.

12 Indium can show valency states of 1 and 3 but never 2, yet a chloride with the empirical formula $InCl_2$ exists. How can you explain this?

13 Explain how aluminium is obtained from its purified oxide. How and under what conditions does aluminium react with
(a) sodium hydroxide,
(b) iron (III) oxide?

Describe how you could prepare dry crystals of potassium aluminium alum starting from aluminium foil, potassium hydroxide and dilute sulphuric acid.

Explain the observation that the addition of sodium hydroxide to aqueous alum leads to the formation of an aluminium hydroxide precipitate which is dispersed by excess alkali. (JMB)

Chapter 18

Group 4B carbon, silicon, germanium, tin and lead

18.1 Some physical data of Group 4B elements

	Atomic Number	Electronic Configuration	Atomic Radius/nm	Ionic Radius/nm M^{2+}	M^{4+}	M.p. /°C	B.p. /°C
C	6	2.4 $1s^2 2s^2 2p^2$	0·077				3850 (subl.)
Si	14	2.8.4 $....2s^2 2p^6 3s^2 3p^2$	0·117		0·041	1410	2360
Ge	32	2.8.18.4 $....3s^2 3p^6 3d^{10} 4s^2 4p^2$	0·122	0·093	0·053	937	2830
Sn	50	2.8.18.18.4 $....4s^2 4p^6 4d^{10} 5s^2 5p^2$	0·140	0·112	0·071	232	2270
Pb	82	2.8.18.32.18.4 $....5s^2 5p^6 5d^{10} 6s^2 6p^2$	0·154	0·120	0·084	327	1744

18.2 Some general remarks about Group 4B

All these elements exhibit a group valency of four, but because an enormous amount of energy is needed to remove four electrons from their atoms, they form compounds which are predominantly covalent. Similarly the gain of four electrons to give the 4-valent anion is energetically impossible.

Germanium, tin and lead form 2-valent compounds in which the two *s* electrons are inert (inert pair effect). The stability of this state relative to the 4-valent state increases steadily from germanium to lead, i.e. 2-valent germanium compounds tend to be strongly reducing and revert to the 4-valent state, while for lead this is the predominant valency state. 2-valent compounds of tin and lead are often predominantly ionic.

Carbon is non-metallic and so too is silicon; germanium has properties of both metals and non-metals (it is a metalloid), while the elements tin and lead are definitely metallic. There is a smooth transition from non-metallic to metallic properties on passing down the series silicon, germanium, tin and lead, but the first member carbon differs considerably from silicon. The chemistry of silicon is very similar to that of boron, its diagonal neighbour in the Periodic Table.

The chemistry of carbon is dominated by its tendency to form chains and rings of carbon atoms in which other atoms, particularly hydrogen, play an important part. The chemistry of silicon is completely different and here the important feature is the formation of silicon-oxygen bonds which are present in the giant molecule silica, and in the polymeric silicate anions (p. 230–232).

The chemistry of carbon and silicon are best considered separately. Germanium, tin and lead can profitably be studied as a group.

18.3 Occurrence of carbon

Carbon

Some of the planets are thought to be surrounded by a layer of carbon dioxide, and it is probable that all the deposits of carbon compounds in the earth originally came from carbon dioxide in the atmosphere. Plants use atmospheric carbon dioxide for growth and animals feed on plants. The decomposition of dead animals and vegetation, followed by geological upheavals which subjected the decaying matter to great temperatures and pressures for long periods of time in the absence of oxygen, almost certainly accounts for coal, oil and natural gas. The carbonate minerals were formed by the sedimentation of the shells of microscopic sea animals, followed by compression of the layers and the redistribution of the oceans. Pure carbon has two allotropes and both of these—diamond and graphite—are found naturally.

The percentage of carbon dioxide in the atmosphere remains constant at a figure of about 0·03 per cent by volume by the operation of a carbon cycle in nature (fig. 18.1).

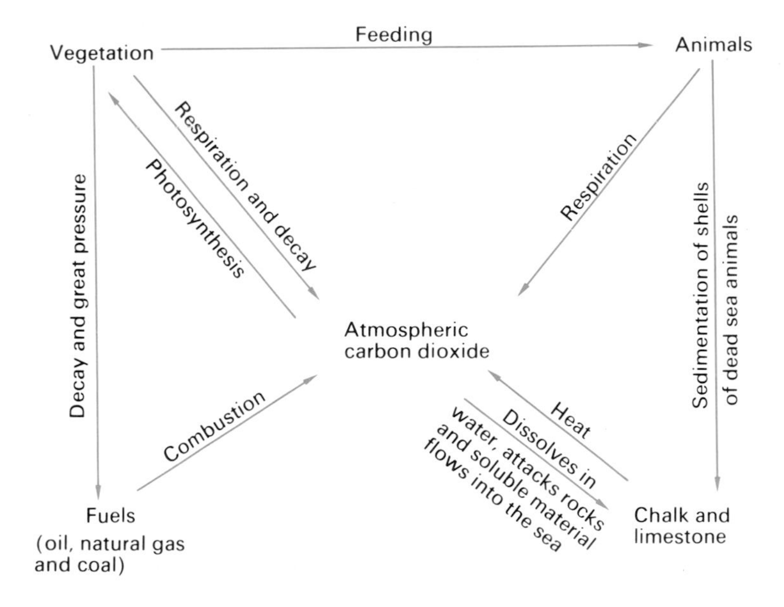

FIG. 18.1. *The carbon cycle*

18.4 The two crystalline forms of carbon—diamond and graphite

Diamond

Diamonds occur naturally in igneous rocks, formed by crystallisation of molten magma from deep within the earth. This igneous rock is blasted and allowed to weather; the light soluble material is then washed away, and the heavy residue is pushed by water under pressure over a bed of grease to which the diamond adheres. Most rock yields about 0·1 g of diamond per tonne, though meteorites containing as much as 1 per cent of diamond have been reported.

Diamond has a density of 3·5 g cm^{-3} and is the hardest naturally occurring substance—a property which explains its use in oilwell drills, as an abrasive for sharpening very hard tools and in dies for the manufacture of tungsten filaments for electric light bulbs. Both these properties of diamond are readily understood in terms of its structure—a three-dimensional array of carbon atoms, each atom being surrounded tetrahedrally by four other carbons. Each diamond crystal is a giant molecule containing very strong bonds (p. 79).

Graphite

Graphite is mined in Mexico, Siberia, Austria and Ceylon, but industrial applications have now outstripped this supply and it is manufactured synthetically (see below).

Its density, 2·3 g cm^{-3}, is less than that of diamond; it is also very soft and slippery and again these properties are readily understood in terms of its structure. Each carbon atom in graphite is surrounded by three more in the same plane, the bonding is covalent and the angle between each covalent bond is 120°. The structure extends in two dimensions and is thus a sheet of carbon atoms arranged in regular hexagons. Individual sheets are held together one above the other only by weak van der Waals' forces, the separation between the sheets being very much greater than the distance between carbon atoms within each sheet (p. 80).

Graphite is less dense than diamond because its structure is less compact. Since the weak forces holding the individual sheets together are easily broken, the sheets are easily moved relative to one another and this accounts for the slippery nature of graphite. Within each sheet every carbon atom is only bonded to three neighbours and thus alternate single and double bonds must be present to complete the octet of each carbon atom. (Since the C—C bond distances in each sheet are identical, resonance structures must be invoked.) The presence of double bonds means that mobile electrons are present, and these can readily be transferred from one hexagon to another by the application of an electric field; graphite, unlike diamond, is therefore a conductor of electricity.

Graphite is used as a lubricant either as a powder or as a dispersion in oil. Mixed with clay it is used in 'lead' pencils. It is chemically very inert and since it conducts electricity it is used as the anode material in many electrolytic processes. Crucibles made of graphite are not attacked by dilute acids or by fused alkalis.

18.5 The manufacture of diamond and graphite

Both diamond and graphite are thermodynamically unstable in the presence of oxygen and, of the two allotropes, graphite is more stable than diamond. These facts can be seen from the equations:

$$\text{C(Diamond)} + O_2(g) \longrightarrow CO_2(g) \quad \Delta H^\ominus(298\,\text{K}) = -395{\cdot}4\,\text{kJ mol}^{-1}$$
$$\text{C(Graphite)} + O_2(g) \longrightarrow CO_2(g) \quad \Delta H^\ominus(298\,\text{K}) = -393{\cdot}5\,\text{kJ mol}^{-1}$$

The reason why diamond and graphite do not react with oxygen at room temperature and diamond does not spontaneously transform itself into graphite is because high activation energies oppose these reactions.

The allotropy of carbon is referred to as **monotropy**, since at ordinary temperatures and pressures one form (graphite) is always thermodynamically more stable than the other (diamond).

Graphite is manufactured by the Acheson process which requires cheap electrical power. A mixture of coke and sand is maintained at a very high temperature for about 24 hours in an electric furnace. Silicon carbide is the probable intermediate which decomposes into graphite and silicon, the latter substance volatilising away.

$$SiO_2(s) + \underset{\text{coke}}{3C(s)} \longrightarrow SiC(s) + 2CO(g)$$

$$SiC(s) \longrightarrow Si(s) + \underset{\text{graphite}}{C(s)}$$

Although graphite is more stable than diamond at ordinary temperatures and pressures, in the region of 3000°C and at extremes of pressure (60—120 thousand atmospheres) an equilibrium between graphite and diamond exists. However, even under such extremes the transformation of graphite into diamond is still extremely slow and a catalyst is required to speed up the rate of attainment of equilibrium. Since 1955 diamonds have been successfully made from graphite by utilising high temperatures and extremes of pressure in the presence of a rhodium catalyst. Only small stones have yet been produced but they are of vital importance in industry.

18.6 Impure forms of carbon

Charcoal is made by heating wood or sugar in a limited supply of oxygen. Sugar charcoal is very pure.

Coke is obtained from coal. It is used in the extraction of several metals from their oxides.

Animal charcoal is obtained by charring bones and consists of finely divided carbon supported on calcium phosphate. It is used for adsorbing colouring matter, e.g. in sugar refining.

Carbon black is a form of carbon obtained by burning hydrocarbons, (e.g. methane), in a deficiency of oxygen. It is used in printer's ink, in shoe polish and is incorporated into rubber to make tyres.

All the above forms of carbon are now known to be microcrystalline forms of graphite. Because they have large surface areas they are chemically more reactive than graphite itself. The large surface area of animal charcoal explains why it is so useful in adsorbing gases and colouring matter.

18.7 Chemical properties of carbon

Carbon in any form will react with oxygen at a sufficiently high temperature to give carbon dioxide; in a deficiency of oxygen, carbon monoxide is formed as well. Charcoal will combine directly with sulphur, some metals and fluorine.

$$C(s) + 2S(s) \longrightarrow CS_2(l)$$
$$Ca(s) + 2C(s) \longrightarrow Ca^{2+}(C \equiv C)^{2-}(s)$$
$$C(s) + 2F_2(g) \longrightarrow CF_4(g)$$

It will reduce steam, forming water gas (p. 163) and many oxides of metals; these reductions are of industrial importance.

It is not attacked by dilute acids, but concentrated nitric acid and sulphuric acid are reduced if warmed with carbon according to the equations:

$$C(s) + 4HNO_3(aq) \longrightarrow 2H_2O(l) + 4NO_2(g) + CO_2(g)$$
$$C(s) + 2H_2SO_4(l) \longrightarrow 2H_2O(l) + 2SO_2(g) + CO_2(g)$$

18.8 The hydrides of carbon

A limitless number of hydrides of carbon exist and owe their existence to the unique property of carbon atoms to catenate (link together) into stable structures. A few typical hydrocarbons are given below; a serious study of these compounds, however, lies outside the scope of this chapter and is dealt with in books devoted to organic chemistry.

	General formula	Examples
The alkanes	C_nH_{2n+2}	CH_4, $H_3C—CH_3$, $H_3C—CH_2—CH_3$
The alkenes	C_nH_{2n}	$H_2C{=}CH_2$, $H_3C—CH{=}CH_2$
The alkynes	C_nH_{2n-2}	$H—C{\equiv}C—H$, $H_3C—C{\equiv}C—H$

Typical ring compounds are based on cyclohexane and benzene:

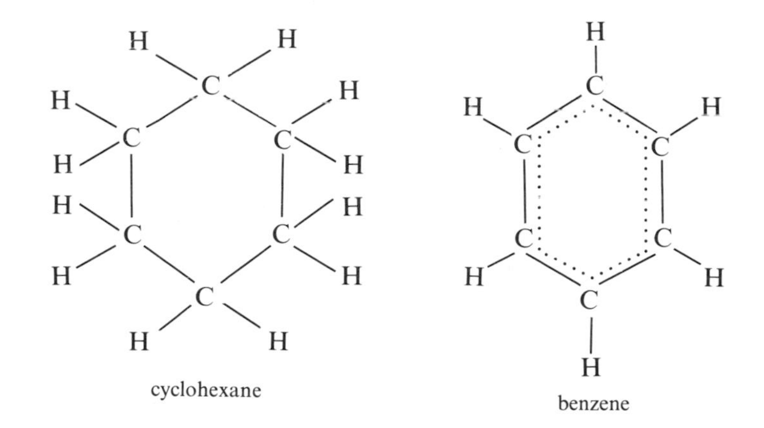

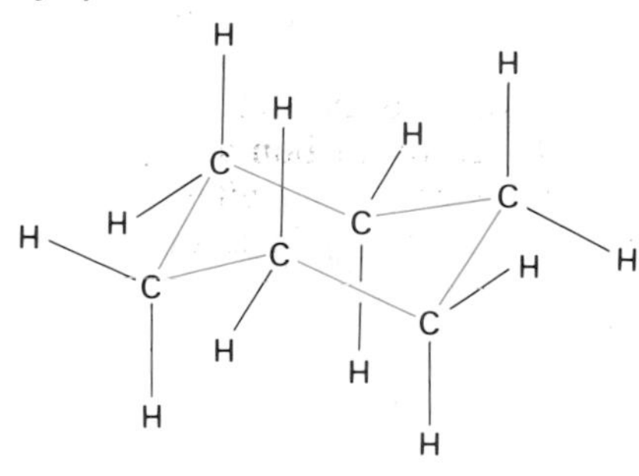

FIG. 18.2. *The puckered structure of cyclohexane*

Cyclohexane rings are puckered in order to maintain approximate tetrahedral valencies for each carbon atom (fig. 18.2). Benzene rings are planar in order to minimise electron pair repulsion (p. 75) and resonance structures are involved (p. 81).

The tendency for the Group 4B elements to catenate decreases rapidly with increasing atomic number, only silicon and germanium forming hydrides corresponding to the alkanes and these being very unstable in the presence of oxygen and/or water. Hydrides containing up to six silicon or germanium atoms in a chain are known with certainty. Tin and lead only form a hydride of the type XH_4, which is extremely unstable.

The striking difference in stability of the hydrides of carbon compared with the hydrides of the other Group 4B elements can be traced to three factors:

(a) the relatively high C—C bond energy

C—C bond energy	346 kJ mol^{-1}	Si—Si bond energy	226 kJ mol^{-1}
Ge—Ge bond energy	186 kJ mol^{-1}	Sn—Sn bond energy	151 kJ mol^{-1}

(b) the relatively high C—H bond energy

C—H bond energy	413 kJ mol^{-1}	Si—H bond energy	326 kJ mol^{-1}
Ge—H bond energy	289 kJ mol^{-1}	Sn—H bond energy	251 kJ mol^{-1}

(c) the carbon atom does not have easily accessible *d* orbitals, i.e. carbon cannot expand its octet (p. 64).

The ready hydrolysis of the hydrides of silicon, for example, in the presence of alkali, is thought to proceed via the formation of a complex, in which the oxygen atom of the hydroxyl radical or water molecule is datively bonded to the silicon atom. This is possible because silicon has *d* orbitals which can be utilised without too large an energy change.

It must be stressed, however, that the stability of hydrocarbons with respect to oxidation at ordinary temperatures is a kinetic rather than a thermodynamic effect, since the oxidation process is attended by a large decrease in free energy once under way. In other words, a high activation energy barrier opposes oxidation and factor (c) above is no doubt of prime importance.

Only carbon of the Group 4B elements is able to multiple bond to itself and thus silicon, germanium, tin and lead do not form hydrides corresponding to the alkenes and alkynes. The reluctance of larger atoms to form multiple bonds is also observed in Groups 5B and 6B; thus the nitrogen molecule contains a triple bond which is not found in the structures of phosphorus, arsenic, antimony and bismuth. Similarly, a double bond unites oxygen atoms in the oxygen molecule, but a double bond is not a feature of the structures of sulphur, selenium, tellurium nor polonium.

18.9 The halides of carbon

Carbon forms halides of the type CX_4 where X = F, Cl, Br or I. The stability to heat decreases with increasing atomic number of the halogen; the boiling points increase in this order. Thus the tetrafluoride is a gas, the tetrachloride a liquid, and the tetrabromide and tetraiodide are solids.

Tetrachloromethane is often used as a solvent and, since it produces a heavy vapour which is non-inflammable, in fire-extinguishers. It is made commercially by reacting carbon disulphide with chlorine in the presence of an iron catalyst:

$$CS_2(l) + 3Cl_2(g) \longrightarrow CCl_4(l) + S_2Cl_2(l)$$

All the tetrahalides of carbon are resistant to hydrolysis, tetrafluoromethane being particularly inert. This is a kinetic effect and is no doubt due to the fact that carbon does not have easily accessible *d* orbitals which can be used for initial bonding of water molecules.

A multitude of compounds containing carbon-fluorine bonds have been synthesised, and a new organic chemistry based on carbon and fluorine (fluorocarbons) has now appeared. The polymer polytetrafluoroethene (PTFE) is made up of $—CF_2—$ units and is particularly resistant to heat and to chemical attack.

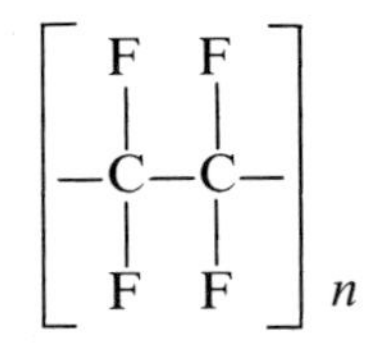

There is no analogous chemistry for the other Group 4B elements since they are reluctant to catenate and, even when they do, these types of compound are susceptible to attack by oxygen and water.

18.10 The oxides of carbon

Carbon forms at least three oxides, and claims that a further two exist have been made.

Carbon dioxide, CO_2

This gas is present in the atmosphere to the extent of about 0·03 per cent by volume; it is due to the operation of a carbon cycle in nature that this figure remains constant (p. 214).

In the laboratory it can conveniently be made by the action of dilute hydrochloric acid on marble chips:

$$CO_3^{2-}(aq) + 2H^+(aq) \longrightarrow CO_2(g) + H_2O(l)$$

Industrially it is produced as a by-product during the manufacture of quicklime and in fermentation processes:

$$Ca^{2+}CO_3^{2-}(s) \longrightarrow Ca^{2+}O^{2-}(s) + CO_2(g)$$

$$\underset{\text{glucose}}{C_6H_{12}O_6(aq)} \xrightarrow{\text{enzymes}} \underset{\text{ethanol}}{2C_2H_5OH(aq)} + 2CO_2(g)$$

It is a colourless, odourless and heavy gas which dissolves in its own volume of water at ordinary temperature and pressure. Like all gases, it dissolves much more readily in water when the pressure is increased and this principle is used in the manufacture of soda water and fizzy drinks.

Carbon dioxide is the acid anhydride of carbonic acid, and a solution of it in water will slowly turn blue litmus wine-red. An equilibrium is set up and, when the solution is boiled, all the carbon dioxide is evolved:

$$CO_2(g) + H_2O(l) \rightleftharpoons H_2CO_3(aq) \rightleftharpoons H^+(aq) + HCO_3^-(aq) \rightleftharpoons 2H^+(aq) + CO_3^{2-}(aq)$$

Carbon dioxide readily reacts with alkalis forming the carbonate and, if the carbon dioxide is in excess, the hydrogen carbonate. This is the basis of the lime water test for this gas:

$$Ca^{2+}(OH^-)_2(aq) + CO_2(g) \longrightarrow Ca^{2+}CO_3^{2-}(s) + H_2O(l) \quad (1)$$

$$Ca^{2+}CO_3^{2-}(s) + H_2O(l) + CO_2(g) \longrightarrow Ca^{2+}(HCO_3^-)_2(aq) \quad (2)$$

Reaction (2) above also accounts for the formation of temporarily hard water (p. 293).

Carbon dioxide is easily liquefied (critical temperature = 31·1°C) and a cylinder of the gas under pressure is a convenient fire extinguisher. When the highly compressed gas is allowed to expand rapidly solid carbon dioxide ('dry ice') is formed. Solid carbon dioxide sublimes at −78°C and, since no messy liquid is produced, it is a convenient means of producing low temperatures.

Unlike the dioxides of the other Group 4B elements, carbon dioxide exists in the form of discrete linear molecules:

$$O=C=O$$

Carbon monoxide, CO

Carbon monoxide, together with carbon dioxide, is formed when carbon or carbonaceous matter is burnt in a deficiency of air or oxygen. It is also produced when carbon dioxide is reduced by red hot carbon; the importance of this reversible reaction in connection with the extraction of metals has been discussed previously (p. 158–159).

$$C(s) + CO_2(g) \rightleftharpoons 2CO(g)$$

A mixture of carbon monoxide and hydrogen (synthesis gas) is formed during the first stage in the industrial production of hydrogen (p. 163).

In the laboratory it can be obtained by dehydrating methanoic acid with concentrated sulphuric acid:

$$HCOOH(l) \xrightarrow{-H_2O} CO(g)$$

If oxalic acid is dehydrated in the same way, carbon dioxide is formed as well:

$$(COOH)_2(s) \xrightarrow{-H_2O} CO(g) + CO_2(g)$$

Carbon monoxide is a colourless, odourless gas which burns in air with a blue flame, forming carbon dioxide. It is exceedingly poisonous, combining with the haemoglobin in the blood more readily than oxygen so that normal respiration is impeded very quickly. Ordinary gas masks are no protection against the gas, since it is not readily absorbed by active charcoal. In the presence of air, a mixture of manganese (IV) oxide and copper (II) oxide catalytically oxidises it to carbon dioxide, and this mixed catalyst is used in the breathing apparatus worn by rescue teams in mine disasters.

Carbon monoxide is a powerful reducing agent, being employed industrially in the extraction of iron and nickel:

$$Fe_2O_3(s) + 3CO(g) \longrightarrow 2Fe(s) + 3CO_2(g)$$
$$NiO(s) + CO(g) \longrightarrow Ni(s) + CO_2(g)$$

It reacts with many transition metals, forming volatile carbonyls (p. 400); the formation of nickel carbonyl followed by its decomposition is the basis of the Mond process (p. 386) for obtaining very pure nickel:

$$Ni(s) + 4CO(g) \xrightarrow{90^{\circ}C} Ni(CO)_4(l) \xrightarrow{180^{\circ}C} Ni(s) + 4CO(g)$$

The transition metal carbonyls are covalent, the carbon atom of the carbon monoxide molecule being bonded to the metal atom. The structure of nickel carbonyl, which is a tetrahedral molecule (to minimise electron pair repulsion, (p. 74)) is given on the opposite page; it is a resonance hybrid of the two forms:

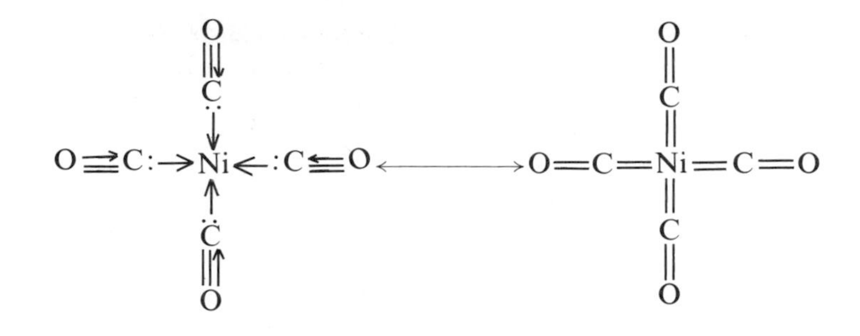

In addition to reacting with oxygen, carbon monoxide combines with sulphur to give carbonyl sulphide and with chlorine in the presence of light to give carbonyl chloride (phosgene), used in the production of polyurethane foam plastics. Carbonyl chloride is an exceedingly poisonous gas.

$$CO(g) + S(s) \longrightarrow \underset{\text{carbonyl sulphide}}{COS(g)}$$

$$CO(g) + Cl_2(g) \longrightarrow \underset{\text{carbonyl chloride}}{COCl_2(g)}$$

Although carbon monoxide is not a true acid anhydride since it does not react with water to produce an acid, it reacts under pressure with fused sodium hydroxide to give sodium methanoate:

$$Na^+OH^-(l) + CO(g) \longrightarrow HCOO^-Na^+(s)$$

Methanoic acid is produced from the sodium salt by the addition of dilute hydrochloric acid.

With hydrogen under pressure and in the presence of a zinc oxide/chromium (III) oxide catalyst it reacts to give methanol; this reaction is of industrial importance.

$$CO(g) + 2H_2(g) \xrightarrow{ZnO/Cr_2O_3} CH_3OH(l)$$

Carbon monoxide is readily absorbed by an ammoniacal solution of copper (I) chloride to give $CuCl.CO.2H_2O$. It reduces an ammoniacal solution of silver nitrate to silver (black) and, in the absence of other gaseous reducing agents, this serves as a test for the gas. It can be estimated by reaction with iodine pentoxide, the iodine which is produced quantitatively being titrated with standard sodium thiosulphate solution.

$$5CO(g) + I_2O_5(s) \longrightarrow I_2(s) + 5CO_2(g)$$

Carbon monoxide is considered to have a structure involving the three resonance forms:

$$:C:\ddot{O}: \longleftrightarrow :C::\ddot{O} \longleftrightarrow :C\vdots O:$$

$$(:\overset{+}{C}-\ddot{O}:^{-}) \qquad (:C=\ddot{O}) \qquad (:\overset{-}{C}\equiv\overset{+}{O}:)$$

Carbon suboxide, C_3O_2
This is an evil-smelling gas and can be made by dehydrating propanedioic (malonic) acid, of which it is the anhydride, with phosphorus pentoxide:

$$3\,CH_2(COOH)_2 + P_4O_{10} \longrightarrow 3C_3O_2 + 4H_3PO_4$$

When heated to about 200°C, it decomposes into carbon dioxide and carbon:

$$C_3O_2(g) \longrightarrow CO_2(g) + 2C(s)$$

The molecule is thought to have a linear structure:

$$O=C=C=C=O$$

18.11 Carbonic acid, carbonates and hydrogen carbonates

Carbonic acid
When carbon dioxide dissolves in water the greater part of it is only loosely hydrated. This hydrated species is in equilibrium with carbonic acid, hydrogen ions, and hydrogen carbonate and carbonate anions:

$$CO_2(g) + H_2O(l) \rightleftharpoons CO_2(aq) \rightleftharpoons H_2CO_3(aq) \rightleftharpoons H^+(aq) + HCO_3^-(aq) \rightleftharpoons 2H^+(aq) + CO_3^{2-}(aq)$$

Although pure carbonic acid cannot be isolated, solid carbonates are plentiful and the Group 1A metals form solid hydrogen carbonates.

Carbonates contain discrete planar ions in which the three carbon-oxygen bond lengths are identical; the ion is thus a resonance hybrid containing equal contributions from the three forms:

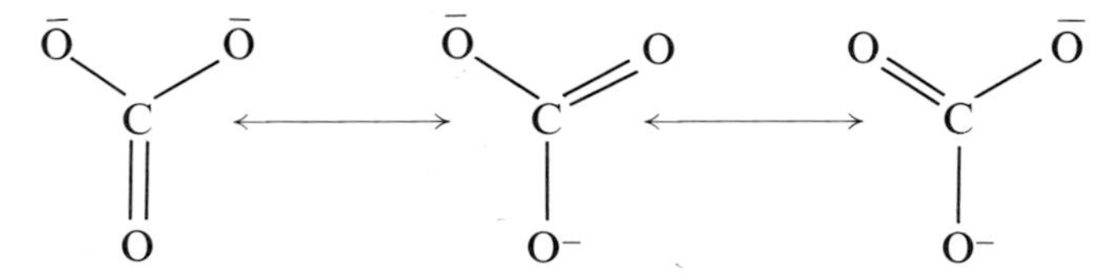

Silicon does not form an analogous discrete silicate anion, SiO_3^{2-}, because of the reluctance of the larger silicon atom to form a multiple bond with oxygen. Silicates contain polymeric silicate anions (p. 231), some of which contain the anion which approximates to the empirical formula SiO_3^{2-}.

Carbonates and hydrogen carbonates
Metallic carbonates are insoluble in water, except those of the Group 1A metals. These last-named soluble carbonates can be obtained by saturating a solution of the alkali with carbon dioxide, and then adding a second identical volume of the alkali:

$$OH^-(aq) + CO_2(g) \longrightarrow HCO_3^-(aq)$$
$$HCO_3^-(aq) + OH^-(aq) \longrightarrow CO_3^{2-}(aq) + H_2O(l)$$

When a solution of an alkali metal carbonate is added to an aqueous solution of a salt it is possible to precipitate a carbonate, a basic carbonate or an hydroxide. The product obtained depends upon the size and charge of the metallic cation.

Precipitation of an hydroxide
The addition of a soluble carbonate to solutions containing aluminium, chromium (III) or iron (III) ions results in the precipitation of the hydroxide. Since the aluminium, chromium (III) and iron (III) ions are highly charged and of small size they are strongly hydrated in solution and show an acid reaction (p. 148), e.g.

$$Al(H_2O)_6^{3+}(aq) + H_2O(l) \rightleftharpoons Al(H_2O)_5(OH)^{2+}(aq) + H_3O^+(aq)$$
$$Al(H_2O)_5(OH)^{2+}(aq) + H_2O(l) \rightleftharpoons Al(H_2O)_4(OH)_2^+(aq) + H_3O^+(aq) \text{ etc.}$$

The addition of a soluble carbonate (the addition of carbonate anions which are strong bases) results in further ionisation, with the precipitation of hydrated aluminium hydroxide:

$$Al(H_2O)_4(OH)_2^+(aq) + CO_3^{2-}(aq) \rightleftharpoons Al(H_2O)_3(OH)_3(s) + HCO_3^-(aq)$$

Precipitation of a basic carbonate
The addition of a soluble carbonate to solutions containing copper (II) or zinc ions results in the precipitation of the basic carbonate, i.e. a mixture of the carbonate and the hydroxide (p. 412). The copper (II) and zinc ions are larger and carry one unit of charge less than the cations discussed above; their solutions are thus less acidic and the ionisation process is more difficult. It is therefore understandable why both carbonate and hydroxide are precipitated.

Precipitation of the carbonate
The larger cations, e.g. those of the Group 2A metals, show little tendency to hydrate in solution and consequently are neutral or only very slightly acidic. The normal carbonate is precipitated in these cases, e.g.

$$Ba^{2+}(aq) + CO_3^{2-}(aq) \longrightarrow Ba^{2+}CO_3^{2-}(s)$$

Metallic ions that give rise to basic carbonates when they interact with carbonate ions can generally be made to produce the normal carbonate on treatment with a soluble hydrogen carbonate (the hydrogen carbonate ion is a weaker base than the carbonate ion because it carries one unit of negative charge less):

$$Cu^{2+}(aq) + 2HCO_3^-(aq) \longrightarrow CuCO_3(s) + H_2O(l) + CO_2(g)$$

Only the hydrogen carbonates of the alkali metals can be isolated as solids and these decompose to give the carbonate at about 70°C (p. 180).

Soluble carbonates and hydrogen carbonates can be estimated volumetrically by titration with standard hydrochloric acid, using methyl orange as indicator:

$$CO_3^{2-}(aq) + 2H^+(aq) \longrightarrow H_2O(l) + CO_2(g)$$
$$HCO_3^-(aq) + H^+(aq) \longrightarrow H_2O(l) + CO_2(g)$$

An M/20 solution of sodium carbonate has a pH of 11·5 and an M/10 solution of sodium hydrogen carbonate a pH of 8·4, thus methyl orange (pH range 2·9—4·0) reacts alkaline (yellow, pH $\geqslant$ 4·0) but turns red in the presence of one drop of hydrochloric acid at the end-point. When an M/20 solution of sodium carbonate is titrated with acid as far as the hydrogen carbonate stage the pH change will be from 11·5 to 8·4; phenolphthalein changes colour over this region, and thus a mixture of carbonate and hydrogen carbonate can be estimated by using both indicators:

$$CO_3^{2-}(aq) + H^+(aq) \longrightarrow HCO_3^-(aq) \text{ (phenolphthalein used)}$$
$$HCO_3^-(aq) + H^+(aq) \longrightarrow H_2O(l) + CO_2(g) \text{ (methyl orange used)}$$

The stability of carbonates to heat correlates well with the position of the particular metal in the electrode potential series, i.e. the Group 1A metallic carbonates are difficult to decompose. All carbonates evolve carbon dioxide when treated with a dilute acid.

18.12 Carbides

The term carbide is generally applied to compounds in which carbon is bonded to elements of lower or approximately the same electronegativity. This definition excludes compounds in which oxygen, sulphur, phosphorus, nitrogen and the halogens are united with carbon. It is also usual to exclude the hydrides of carbon.

Carbides are conveniently classified in terms of the bonding present into salt-like (ionic), interstitial and covalent carbides.

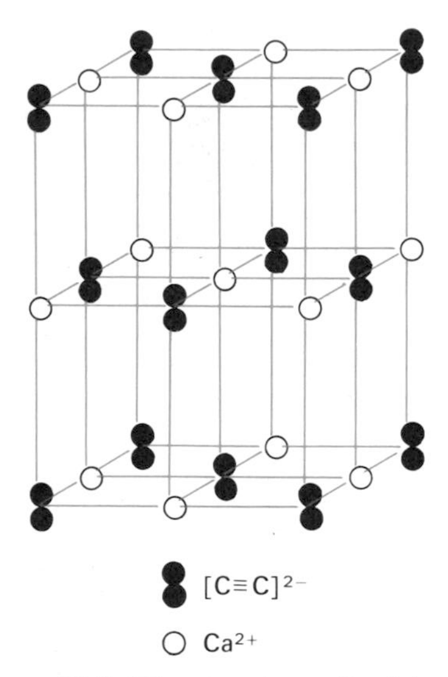

FIG. 18.3. *The structure of calcium carbide*

Salt-like (ionic) carbides

The reactive metals, i.e. those of Groups 1A and 2A, form carbides which are colourless solids and non-conductors of electricity. With the exception of beryllium carbide, they are decomposed by water, with the liberation of ethyne:

$$Ca^{2+}(C \equiv C)^{2-}(s) + 2H_2O(l) \longrightarrow Ca^{2+}(OH^-)_2(aq/s) + H{-}C \equiv C{-}H(g)$$

Many of these carbides have a structure in some ways resembling that of sodium chloride, one cell axis being elongated to accommodate the $(C \equiv C)^{2-}$ ions (tetragonal symmetry). Notice that the co-ordination number of each ion is six, the same as in the structure of sodium chloride, but that the axis along which the acetylide ions lie is elongated (fig. 18.3).

Beryllium and aluminium carbides give methane on hydrolysis with water and, although normally included in this category, are best thought of as a transition from true ionic to essentially covalent carbides. The carbides of the lanthanides are considered to be ionic.

Interstitial carbides

These are formed by many transition metals and are similar to the interstitial hydrides (p. 165), i.e. they are non-stoichiometric. Carbon atoms fit into the spaces between metal atoms in the lattice without too much distortion; thus these carbides have very high melting points and are conductors of electricity. It has been shown that metals with an atomic radius of less than 0·13 nm cannot take up carbon atoms without causing distortion of the metal lattice. In these cases carbides with properties intermediate between ionic and interstitial are formed, e.g. Fe_3C, Mn_3C and Ni_3C.

Covalent carbides

These are giant molecules with covalent bonding running completely through the structure. Typical examples are silicon carbide, SiC, which has the diamond structure (p. 79) and boron carbide, B_4C, whose structure is more complex. Both are extremely hard, boron carbide being the harder of the two, and are used as abrasives.

Plate 6. Production platform in BP's Forties oilfield in the North Sea (By courtesy of British Petroleum Company Ltd.)

18.13 Manufacture and uses of synthesis gas

Manufacture of synthesis gas

Synthesis gas (a mixture of carbon monoxide and hydrogen) is formed during the first stage in the manufacture of hydrogen (p. 163). In the UK methane (present in Natural Gas) or naphtha (typically C_4—C_{10} alkanes, obtained from crude oil) are reacted with steam at a temperature of about 900°C and a pressure of about 30 atmospheres; a nickel catalyst is also employed:

$$CH_4(g) + H_2O(g) \longrightarrow CO(g) + 3H_2(g) \quad (1)$$

$$\underset{\text{(present in naphtha)}}{C_7H_{16}(g)} + 7H_2O(g) \longrightarrow 7CO(g) + 15H_2(g) \quad (2)$$

Where a high H_2/CO ratio is desirable in the synthesis gas, e.g. for the manufacture of hydrogen (p. 163), it is advantageous to use methane rather than naphtha (see equations (1) and (2) above).

Uses of synthesis gas

(a) Manufacture of hydrogen

This process has been discussed earlier in the book (p. 163). Synthesis gas is the principal source of hydrogen for the Haber process (p. 249)

(b) Manufacture of methanol, CH_3OH

This is obtained by passing synthesis gas over a zinc oxide/chromium(III) oxide catalyst at 300°C and 300 atmospheres pressure (p. 221). More recently a catalyst system based on copper has been employed, thereby allowing a lower operating temperature and pressure to be employed (250°C and 70 atmospheres respectively).

Methanol is oxidised to methanal, $H_2C=O$, as the first stage in the production of Bakelite. The formation of ethanoic acid, CH_3COOH, from methanol is outlined below.

(c) Manufacture of higher alcohols

These are produced by reacting synthesis gas with an alkene in the presence of a complex of rhodium (a transition metal) and triphenylphosphine, $(C_6H_5)_3P$. Using propene as the alkene, the first product is the aldehyde butanal, $CH_3CH_2CH_2CHO$, but with excess hydrogen this is converted into butan-1-ol, $CH_3CH_2CH_2CH_2OH$:

$$CH_3CH=CH_2(g) + CO(g) + H_2(g) \longrightarrow CH_3CH_2CH_2CHO(l)$$

$$CH_3CH_2CH_2CHO(l) + H_2(g) \longrightarrow CH_3CH_2CH_2CH_2OH(l)$$

The alcohols produced from longer chain alkenes can be sulphonated and used as detergents (see a standard organic chemistry text-book for details).

(d) Manufacture of ethanoic acid, CH_3COOH

In the processes discussed above, synthesis gas has been used as such. However, the carbon monoxide can be separated from the hydrogen in the mixture, e.g. by liquefaction. Reaction of methanol with carbon monoxide is becoming increasingly important in the production of ethanoic acid:

$$CH_3OH(l) + CO(g) \longrightarrow CH_3COOH(l)$$

For other uses of carbon monoxide see p. 220–221.

Silicon

18.14 Occurrence of silicon and its extraction

Silicon is the second most abundant element occurring in the earth's crust (about 28 per cent by weight) as the oxide, silica, in a variety of forms, e.g. sand, quartz and flint, and as silicates in rocks and clays.

The element is obtained from silica by reduction with carbon in an electric furnace:

$$SiO_2(s) + 2C(s) \longrightarrow Si(s) + 2CO(g)$$

Extremely pure silicon is obtained from 'chemically' pure silicon by the method of zone-refining (p. 154), and is used in the construction of transistors. Single crystals of silicon are used in circuits called **silicon chips.** On such a chip, no larger than 1 cm^2 in area, there can be the equivalent of several thousand transistors. Such devices are used in small computers called microprocessors.

Plate 7. The Ferranti F100-L microprocessor constructed from a silicon chip 6 mm square. This magnified photograph shows the complexity of the circuits on the chip. (By courtesy of Ferranti Limited)

18.15 Properties of silicon

Silicon is a very high melting-point solid with the same structure as diamond (p. 79). The non-existence of an allotrope with the graphite structure clearly shows the inability of silicon atoms to multiple bond with themselves. In the massive form, silicon is chemically rather unreactive but powdered silicon is attacked by the halogens and alkalis:

$$Si(s) + 2Cl_2(g) \longrightarrow SiCl_4(l)$$

$$Si(s) + 2OH^-(aq) + H_2O(l) \longrightarrow \underset{\text{empirical formula}}{SiO_3^{2-}}(aq) + 2H_2(g)$$

It is not attacked by acids except hydrofluoric acid, with which it forms hexafluorosilicic acid:

$$Si(s) + 6HF(aq) \longrightarrow H_2SiF_6(aq) + 2H_2(g)$$

A highly active form of silicon, capable of displacing hydrogen from water, has been prepared.

18.16 The hydrides of silicon (silanes)

A mixture of hydrides of silicon, containing up to six silicon atoms in a chain, can be obtained by treating magnesium silicide with dilute hydrochloric acid:

$$Mg_2Si(s) + H^+(aq) \longrightarrow Mg^{2+}(aq) + SiH_4(g) + H_3Si{-}SiH_3(g), \text{ etc.}$$

The hydrides can be separated by fractional distillation, and individual isomers have been separated by vapour phase chromatography.

The first member of the series, monosilane, is conveniently made by the reduction of silicon tetrachloride with lithium aluminium hydride:

$$SiCl_4(l) + Li^+AlH_4^- \longrightarrow SiH_4(g) + Li^+Cl^-(s) + AlCl_3(s)$$

The hydrides are gases or volatile liquids, the boiling points rising with increasing molecular mass. Their stability to heat decreases with increasing chain length (reluctance of silicon atoms to catenate). They are all spontaneously inflammable in air and are hydrolysed by water—more rapidly in the presence of alkali. This reactivity has been attributed to the availability of *d* orbitals for bonding with attacking species, which are not available to the carbon atom (p. 218).

$$SiH_4(g) + 2O_2(g) \longrightarrow SiO_2(s) + 2H_2O(l)$$

$$SiH_4(g) + 2OH^-(aq) + H_2O(l) \longrightarrow SiO_3^{2-}(aq) + 4H_2(g)$$

18.17 The halides of silicon

Silicon forms a halide of the general formula SiX_4 with fluorine, chlorine, bromine and iodine. With chlorine, a series of chlorides up to Si_6Cl_{14} is known. All the halides, with the exception of silicon tetrafluoride, are hydrolysed completely by water to silicon dioxide

$$SiCl_4(l) + 2H_2O(l) \longrightarrow SiO_2(s) + 4HCl(aq)$$

Silicon tetrafluoride is hydrolysed only partially as below:

$$2SiF_4(g) + 2H_2O(l) \longrightarrow SiO_2(s) + SiF_6^{2-}(aq) + 2HF(aq) + 2H^+(aq)$$

These reactions are thought to proceed by the formation of a dative bond from the oxygen atom of a water molecule to the silicon atom, cf. hydrolysis of the hydrides of silicon. This explains the marked contrast in behaviour of these compounds and the halides of carbon.

18.18 Silicon dioxide (silica), SiO_2

Silicon dioxide occurs in a number of different crystalline forms, but each form contains silicon atoms bonded tetrahedrally to four oxygen atoms by means of single Si—O bonds. Silicon dioxide is thus quite unlike carbon dioxide in its structure, and once again the reluctance of silicon to form multiple bonds (this time with oxygen atoms) is noted.

The three crystalline forms of silicon dioxide are quartz, tridymite and cristobalite, each of which has a high and low temperature modification. Quartz is the most thermodynamically stable of the three forms, but there is no tendency for the other two forms to transform themselves into quartz, since the migration of atoms in solids is a process requiring very high activation energies.

The structure of crystobalite can be visualised in terms of the diamond structure, carbon atoms being replaced by silicon atoms with one oxygen atom midway between two silicons. In quartz and tridymite spiral structures are present and these forms have optical isomers (right- and left-handed spirals). Since all three forms of silicon dioxide have covalent bonds extending in three dimensions, they are giant molecules and high melting-point solids. The empirical formula corresponds to SiO_2 because each oxygen atom is shared equally by two silicon atoms (fig. 18.4).

FIG. 18.4. *In silicon dioxide each silicon atom is surrounded tetrahedrally by oxygen atoms*

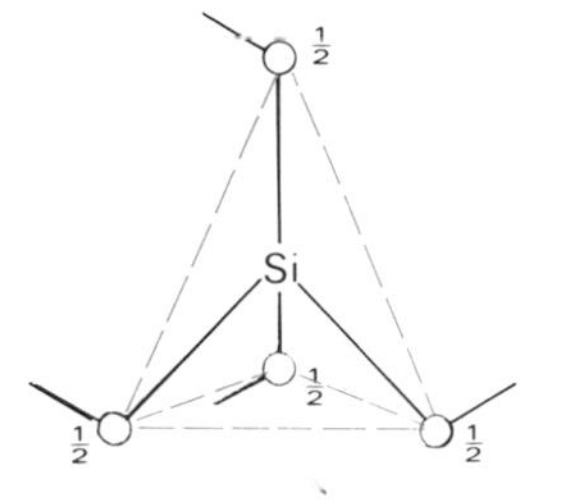

Colourless quartz is called rock crystal; purple quartz is the gem amethyst. It is used in the construction of special lenses and prisms, since it is transparent to both infrared and ultraviolet radiation, and its ability to control radio frequencies is utilised in the construction of electronic equipment, i.e. radar devices. Since its industrial applications have now outstripped its natural production, single crystals of quartz are 'grown' artificially from other forms of silica. Sand is essentially small grains of quartz with impurities; it is formed by the weathering of silicate rocks.

When molten silica is allowed to cool, the substance called silica glass is obtained. It consists of randomly orientated chains, sheets and three-dimensional networks of SiO_4 groups, cf. boron trioxide glass (p. 202). It has a softening point higher than that of ordinary glass and a very low coefficient of expansion. It can, therefore, be heated and cooled rapidly without cracking. It is used in the construction of special chemical apparatus since it is not attacked by acids, except hydrofluoric acid, nor by alkalis unless they are fused. Its main disadvantage is that it is more brittle than ordinary glass and it is expensive.

Silica is a non-volatile acidic oxide (although it does not react with water and forms complex mixtures of silicates only when fused with alkalis or alkali carbonates) and finds numerous applications, e.g. the extraction of phosphorus (p. 262), the manufacture of concrete (p. 192) and the manufacture of glass (p. 233). When heated with coke it forms silicon carbide (empirical formula SiC), one of the hardest man-made substances, which is used as an abrasive.

18.19 Silicic acid and silica gel

When silica is fused with sodium carbonate, carbon dioxide is evolved and it is possible to obtain a water soluble glass (water glass). A number of sodium silicates can be prepared from water glass, including sodium orthosilicate, $(Na^+)_4SiO_4^{4-}$:

$$SiO_2(s) + 2(Na^+)_2CO_3^{2-}(s) \longrightarrow (Na^+)_4SiO_4^{4-}(s) + 2CO_2(g)$$

The orthosilicate anion, SiO_4^{4-}, is a very strong base and is able to abstract protons from water molecules, eventually condensing into polysilicate anions by loss of water. The first step in this condensation reaction is:

$$2HSiO_4^{3-} \longrightarrow \underset{\text{pyrosilicate ion}}{Si_2O_7^{6-}} + H_2O$$

Carried to its logical conclusion, the result is a long chain which is called a metasilicate ion and which has the formula $(SiO_3)_n^{2n-}$. When a solution of a silicate is acidified, a gel is formed which is thought to be a tangled mass of polymeric silicic acid molecules:

```
        H     H     H     H     H
        |     |     |     |     |
        O     O     O     O     O
        |     |     |     |     |
----O—Si—O—Si—O—Si—O—Si—O—Si----
        |     |     |     |     |
        O     O     O     O     O
        |     |     |     |     |
        H     H     H     H     H
```

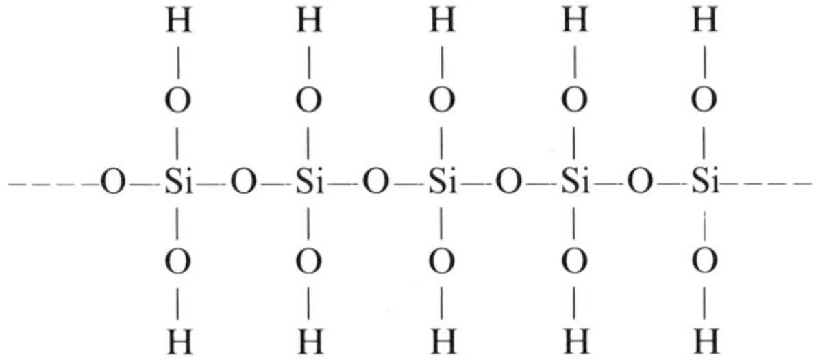

The gel contains large amounts of water held in a cage-like structure of polysilicic acid molecules. When it is heated this water is driven off and the polysilicic acid itself dehydrates, the material hardening and shrinking in the process. The end product is called silica gel—which of course is no longer a gel in the accepted sense of the word. It is a substance with an exceedingly large surface area and is used as a drying agent and as an inert supporting material for many finely divided catalysts.

18.20 Silicates

The tendency of silicon to form single covalent bonds with oxygen atoms has already been noted in the structures of silica and polysilicic acid. These structures and also those of a bewildering variety of silicates are now readily understood in terms of the linking together of tetrahedral SiO_4 units. Pauling considers the silicon-oxygen bond to be about 50 per cent ionic, and it is sometimes convenient to discuss the structures of silicates in terms of Si^{4+} ions tetrahedrally surrounded by four much larger oxygen atoms. Examples of some typical silicates are given below.

Silicates containing discrete SiO_4^{4-} anions (Fig. 18.5(a))

Orthosilicates contain the simple SiO_4^{4-} ion, one example being beryllium orthosilicate, $(Be^{2+})_2SiO_4^{4-}$. As mentioned above, the SiO_4 group is tetrahedral, as would be expected.

Silicates containing $Si_2O_7^{6-}$ anions (one oxygen atom shared) (Fig. 18.5(b))

When one oxygen atom is shared between two tetrahedra, the pyrosilicate anion, $Si_2O_7^{6-}$, is the result. A typical pyrosilicate is $(Sc^{3+})_2Si_2O_7^{6-}$.

Silicates containing extended anions (two oxygen atoms shared) (Fig. 18.5(c) and (d))

When each tetrahedron shares two oxygen atoms, it is possible to have closed ring anions such as $Si_3O_9^{6-}$. Another possibility is the formation of infinite chains, the formula of these anions approximating to $(SiO_3)_n^{2n-}$. Examples of silicates containing these anions are $Be^{2+}Ti^{4+}Si_3O_9^{6-}$, and $Ca^{2+}Mg^{2+}(SiO_3^{2-})_2$.

Silicates containing extended anions (three oxygen atoms shared) (Fig. 18.5(e))

When each tetrahedron shares three oxygen atoms, silicates in the form of extended sheets result. The empirical formula of these polysilicate anions is $SiO_{2\frac{1}{2}}^-$ or $Si_4O_{10}^{4-}$ Anions of this type are found in micas and clays and account for their ready cleavage into thin slices.

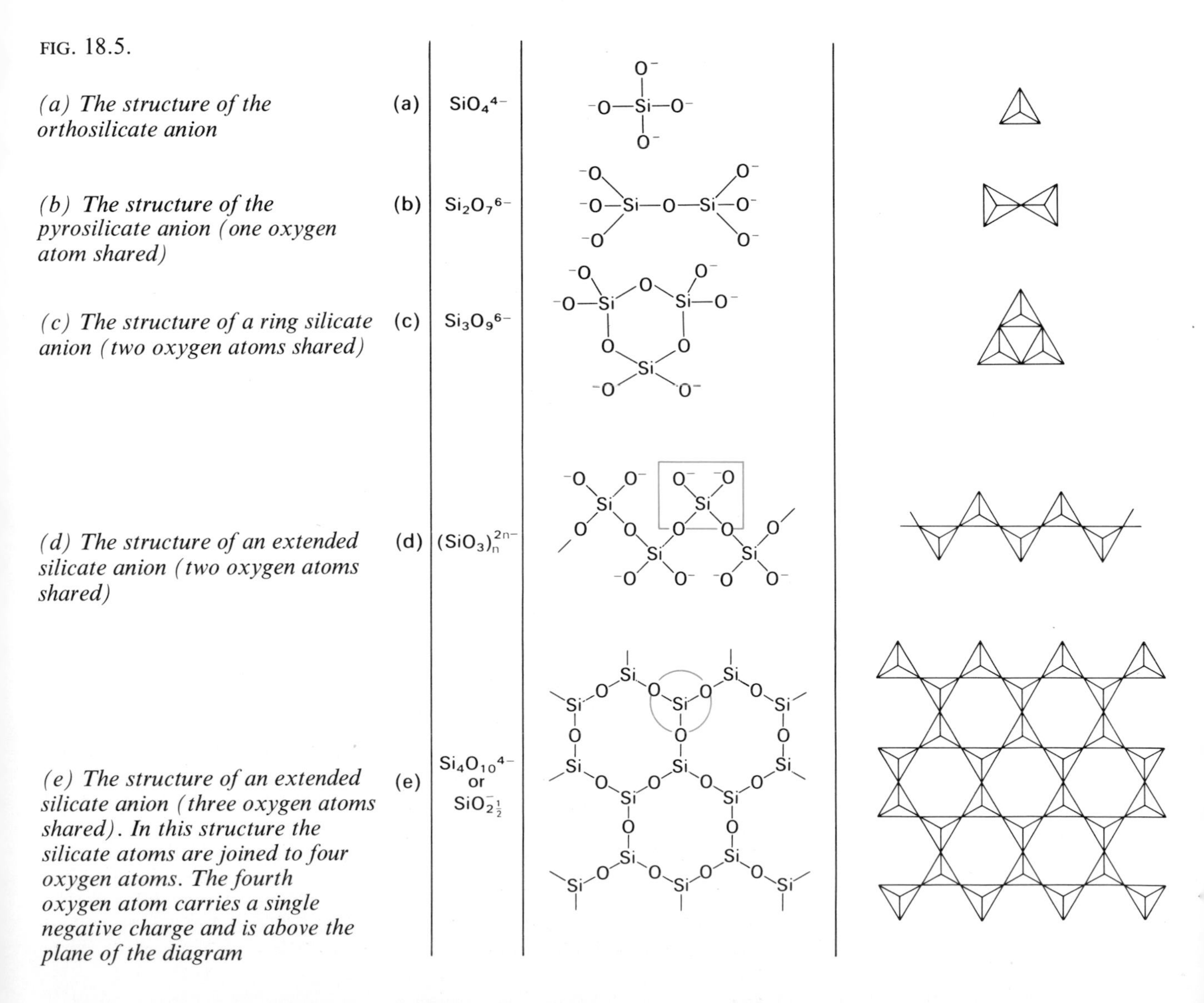

FIG. 18.5.

(a) The structure of the orthosilicate anion

(b) The structure of the pyrosilicate anion (one oxygen atom shared)

(c) The structure of a ring silicate anion (two oxygen atoms shared)

(d) The structure of an extended silicate anion (two oxygen atoms shared)

(e) The structure of an extended silicate anion (three oxygen atoms shared). In this structure the silicate atoms are joined to four oxygen atoms. The fourth oxygen atom carries a single negative charge and is above the plane of the diagram

Plate 8. A piece of amosite ore showing the strand-like structure of asbestos (a silicate) at each end of the rock (By courtesy of Cape Asbestos Fibres Limited, Uxbridge)

Silica and aluminosilicates (four oxygen atoms shared)

When all four oxygen atoms of each tetrahedron are shared a three-dimensional structure results. The different three-dimensional structures that are possible correspond to the different crystalline forms of silica (for example, fig. 18.4).

Some of the Si^{4+} ions can be replaced by Al^{3+} ions and the network must now carry an overall negative charge which is balanced by the incorporation of cations such as Na^+ and Ca^{2+} into the structure. Compounds of this type are called aluminosilicates; one such compound is analcite, which is a zeolite, having the empirical formula $Na^+(AlSi_2O_6)^-.H_2O$. Both natural and synthetic zeolites have a very open structure which allows ion-exchange to take place, and they are used in the treatment of hard water. They also function as 'molecular sieves' since cavities are present in their structures into which molecules can pass. Molecules which only just manage to squeeze into these cavities are strongly adsorbed but very small and very large molecules are not affected. A particular zeolite has been found that will adsorb straight chain alkanes but not branched chain or aromatic substances.

18.21 Silicate glasses

Crystalline silicates are formed by very slow cooling of silicate melts. If the cooling is more rapid there is no time for an orderly structure to form and a glass results. When a glass is heated it gradually softens instead of melting at a sharp temperature, since a gradual breaking of chemical bonds takes place in a disorganised solid.

Soda glass, a mixture of calcium and sodium silicates, is manufactured by fusing the two carbonates with white sand at a temperature of about 1500°C. The non-volatile silicon dioxide displaces carbon dioxide thus:

$$Ca^{2+}CO_3{}^{2-}(s) + SiO_2(s) \longrightarrow Ca^{2+}SiO_3{}^{2-}(s) + CO_2(g)$$

$$(Na^+)_2CO_3{}^{2-}(s) + SiO_2(s) \longrightarrow (Na^+)_2SiO_3{}^{2-}(s) + CO_2(g)$$

The properties of glass can be altered by varying the ingredients; thus lead glass, which has a high refractive index and is used for certain lenses and cut-glass articles, is obtained by replacing most of the calcium with lead. Coloured glasses are made by adding metallic oxides that form coloured silicates; e.g. blue glass (cobalt glass used for eliminating the sodium flame) is manufactured by adding traces of cobalt (II) oxide to the melt. The silicon can also be partially replaced by aluminium, boron and phosphorus. A borosilicate glass, known as 'Pyrex', in which some of the silicon is replaced by boron, is particularly useful for making laboratory apparatus, since it withstands higher temperatures than ordinary soda glass.

18.22 Silicon disulphide, SiS_2

Silicon disulphide is, as might be expected, a giant molecule, unlike carbon disulphide which has the structure S=C=S. It crystallises in the form of long silky needles, a process consistent with its chainlike structure (p. 80). Like carbon disulphide, it can be obtained by direct combination.

18.23 The silicones

These compounds are polymeric, the polymer chain containing alternately linked silicon and oxygen atoms. Alkyl or aryl groups, e.g. CH_3 or C_6H_5, are attached to the polymer backbone by means of covalent bonds to the silicon atoms. A typical silicone has the formula:

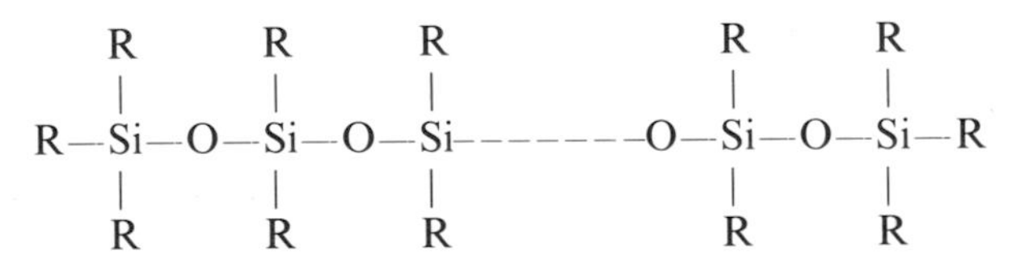

where R is an alkyl or aryl group.

Silicones are obtained by reacting a chloroalkane or a chlorobenzene with silicon in the presence of a copper catalyst and at a temperature of about 300°C. A mixture of alkyl or aryl chlorosilanes results:

$$RCl + Si \xrightarrow{Cu} R_3SiCl + R_2SiCl_2 + RSiCl_3$$

After fractional distillation, the silane derivatives are hydrolysed and the 'hydroxides' immediately condense by intermolecular elimination of water. The final product depends upon the number of hydroxyl groups originally bonded to the silicon atom:

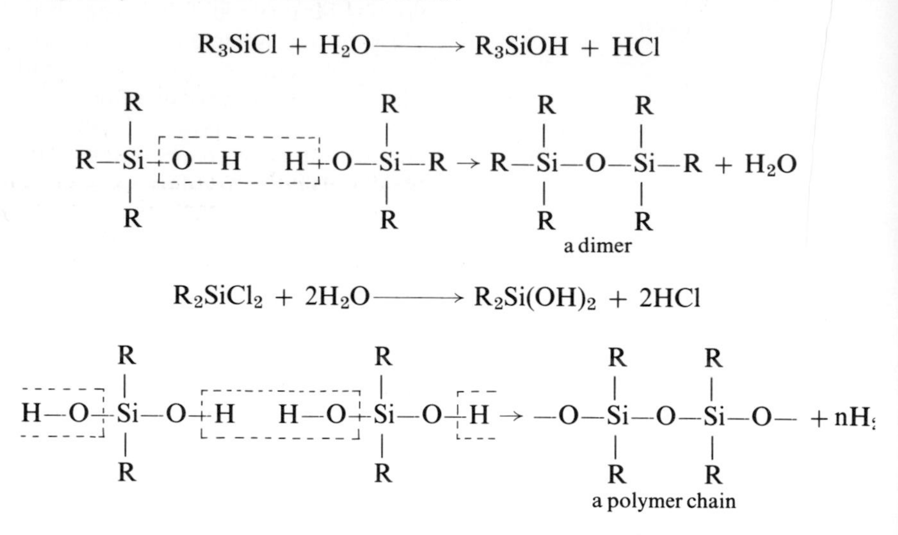

The polymer chain depicted above is terminated by incorporating a small quantity of the monochlorosilane derivative into the hydrolysis mixture.

Hydrolysis of the trichlorosilane derivative gives a two-dimensional structure. By blending a mixture of chlorosilanes before hydrolysis, it is possible to produce polymers of varying chain length, R_3SiOH acting as a chain stopper and $RSi(OH)_3$ as a cross-linking agent.

The hydrocarbon layer along the silicon-oxygen chain makes silicones water-repellent. Silicone fluids are thermally stable and their viscosity alters very little with temperature, and silicone rubbers retain their elasticity at much lower temperatures than ordinary rubber.

Germanium, tin and lead

18.24 Occurrence, extraction and uses of germanium, tin and lead

Germanium

Germanium occurs to the extent of about 6 per cent by weight in the ore germanite, (Cu,Ge,Fe,Zn,Ga)(S,As). When the ore is strongly heated with hydrochloric acid germanium (IV) chloride, $GeCl_4$, distils off. After purification by fractional distillation it is hydrolysed by water to germanium (IV) oxide, GeO_2, from which the metal is obtained by reduction with carbon or hydrogen.

$$GeCl_4(l) + 2H_2O(l) \longrightarrow GeO_2(s) + 4HCl(aq)$$
$$GeO_2(s) + 2H_2(g) \longrightarrow Ge(s) + 2H_2O(l)$$

Very pure germanium is obtained by zone refining (p. 154); it is a semiconductor and like silicon is used in the construction of transistors.

Tin

The only ore of tin of any importance is cassiterite or tinstone, SnO_2, which is mined in Malaysia, Nigeria, Indonesia and Bolivia. The ore is pulverised and washed with water to float away lighter impurities.

After roasting to remove sulphur and arsenic, it is reduced to the metal by heating with anthracite; limestone is added to produce a slag with the impurities.

$$SnO_2(s) + 2C(s) \longrightarrow Sn(l) + 2CO(g)$$

The impure tin is refined by applying just sufficient heat to melt it; the residue usually contains iron, arsenic and lead.

Tin is a constituent of many alloys, including type metal (Sn,Sb,Pb), bronze (Sn,Cu), and solder (Sn,Pb). Large quantities are used in the manufacture of tinplate as food containers. One method of tinplating involves electrolytic deposition of tin on the article to be plated (cathode), using an electrolyte of acidified tin (II) sulphate, $Sn^{2+}SO_4^{2-}$.

Tinplate is only effective provided the layer of tin remains intact; once its surface becomes scratched, exposing the underlying steel, an electrochemical cell is set up and rusting is accelerated (the electrode potential of iron is more negative than that of tin) (fig. 18.6).

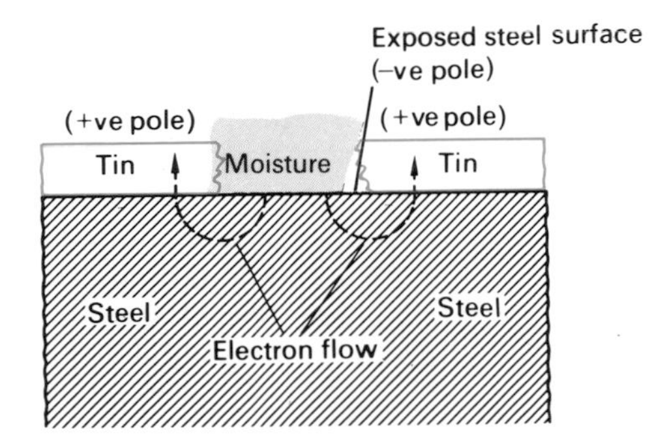

FIG. 18.6. *Rusting is accelerated on a damaged tinplated steel article*

At the exposed steel surface (−ve pole) $Fe(s) \longrightarrow Fe^{2+}(aq) + 2e^-$

At the tin surface (+ve pole) $2H_2O(l) \rightleftharpoons 2H^+(aq) + 2OH^-(aq)$

$$2H^+(aq) + 2e^- \longrightarrow H_2(g)$$

$$Fe^{2+}(aq) + 2OH^-(aq) \longrightarrow Fe(OH)_2(s) \longrightarrow \text{Rust}$$

A coating of zinc on steel prevents rusting since zinc has a more negative electrode potential than iron. Galvanised steel plate, however, cannot be used for making food containers, since poisonous zinc ions would enter the food in preference to the non-poisonous iron (II) ions if the surface of the containers became scratched.

Lead

Lead occurs principally as galena, PbS, often in association with zinc blende, ZnS, in Canada and at Broken Hill in Australia.

The ore containing galena and zinc blende is concentrated into two fractions, one being rich in galena and the other in zinc blende. The galena is roasted to convert it into lead (II) oxide, which is then reduced to the metal either by further reduction with galena or with coke:

$$2PbS(s) + 3O_2(g) \longrightarrow 2PbO(s) + 2SO_2(g)$$
$$2PbO(s) + PbS(s) \longrightarrow 3Pb(l) + SO_2(g)$$
$$PbO(s) + C(s) \longrightarrow Pb(l) + CO(g)$$

A more modern method involves roasting the mixed sulphides to obtain the oxides. The mixed oxides are then fed into a blast furnace and reduced to their respective metals with coke. Molten lead is tapped from the bottom of the furnace; the zinc vapour which emerges from the top of the furnace is condensed by a shower of lead droplets. A solution of lead and zinc leaves the condenser which on cooling releases its zinc. Since zinc is less dense than lead it can be removed continuously as an upper layer.

Mixed sulphides
$$\begin{cases} 2PbS(s) + 3O_2(g) \longrightarrow 2PbO(s) + 2SO_2(g) \\ 2ZnS(s) + 3O_2(g) \longrightarrow 2ZnO(s) + 2SO_2(g) \end{cases}$$

Mixed oxides
$$\begin{cases} PbO(s) + C(s) \longrightarrow Pb(l) + CO(g) & \text{(lead tapped from bottom of the furnace with slag)} \\ ZnO(s) + C(s) \longrightarrow Zn(g) + CO(g) & \text{(zinc vapour from the top of the furnace condensed by molten lead droplets)} \end{cases}$$

Crude lead is further treated by various processes to remove copper, silver, gold, antimony and arsenic.

Lead has been used since Roman times for plumbing, although copper and PVC (polyvinyl chloride) are nowadays generally preferred. It is used for covering underground telephone cables and as a roofing material. Alloys containing lead include type metal (Sn 10 per cent, Sb 20 per cent, Pb 70 per cent) and soft solder (Sn 50 per cent, Pb 50 per cent); lead used in accumulators is hardened by incorporating about 10 per cent of antimony. Lead is used in the manufacture of paints, e.g. basic lead(II) carbonate and trilead tetroxide (Pb_3O_4) are paint ingredients; its use in the production of lead tetraethyl is on the decline, particularly in the USA, with the introduction of tighter pollution controls.

18.25 Properties of germanium, tin and lead

Germanium is hard and brittle, tin is soft and lead is a very soft metal. The principal reactions of germanium and tin are very similar, both elements exhibiting the group valency of 4. Lead, on the other hand, tends to form 2-valent compounds since its two *s* electrons are rather inert (the inert pair effect see p. 63). The principal reactions are summarised below.

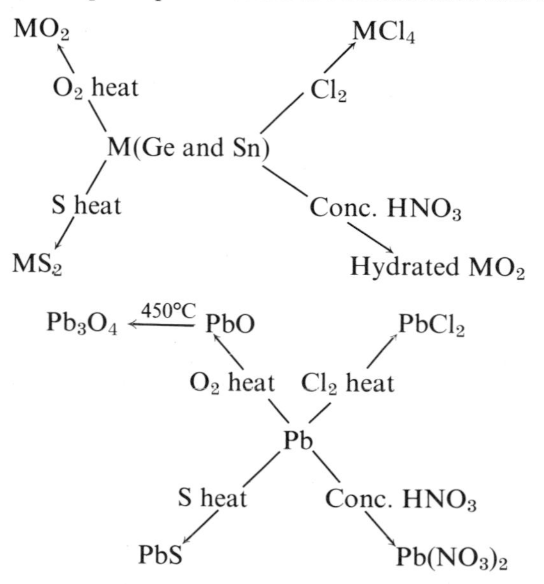

Germanium is not attacked by hydrochloric acid and the dilute acid has little effect on tin and lead. Tin is inert because of the high hydrogen over-voltage at the surface of this metal (p. 162); in the case of lead, the low solubility of lead (II) chloride, $PbCl_2$, limits the reaction.

Hot concentrated alkalis attack the elements, germanium and tin, forming respectively a germanate (IV) and a stannate (IV) in which the elements exhibit a valency of four. Lead forms a plumbate (II) in which it is 2-valent.

$$Ge(s) + 2OH^-(aq) + H_2O(l) \longrightarrow GeO_3^{2-}(aq) + 2H_2(g)$$
$$Sn(s) + 2OH^-(aq) + H_2O(l) \longrightarrow SnO_3^{2-}(aq) + 2H_2(g)$$
$$Pb(s) + 2OH^-(aq) \longrightarrow PbO_2^{2-}(aq) + H_2(g)$$

18.26 The structures of germanium, tin and lead

Germanium has the same type of structure as diamond (p. 79).

Tin exhibits allotropy, three crystalline forms being known with the transition temperatures as shown below.

$$\underset{\text{'grey'}}{\alpha\text{–tin}} \underset{}{\overset{13{\cdot}2°C}{\rightleftharpoons}} \underset{\text{'white'}}{\beta\text{–tin}} \overset{161°C}{\rightleftharpoons} \gamma\text{–tin} \overset{232°C}{\rightleftharpoons} \text{liquid tin}$$

Although β–tin should change into α–tin at a temperature below 13·2°C, the transformation only becomes rapid at about −50°C, unless some α–tin is added to catalyse the reaction. α–Tin has the diamond structure whereas both β–tin and γ–tin are metallic (an approach to close packing of the atoms). Since α–tin has a more open structure than β– and γ–tin its density is considerably less than the densities of the other two allotropes.

This type of allotropy in which two allotropes are equally stable at the transition temperature is referred to as **enantiotropy** (cf. the allotropy of sulphur, p. 300). The allotropy of carbon at ordinary temperatures and pressures is a different phenomenon—monotropy—and has been discussed previously (p. 215).

Lead only exists in one form which is metallic in nature. The changeover from non-metallic to metallic nature with increasing atomic number is thus reflected in the structures of the Group 4B elements.

Compounds of germanium, tin and lead

Unlike carbon and silicon, germanium, tin and lead can exhibit valencies of 2 and 4; in the 2-valent state the two *s* electrons are inert, taking no part in chemical bonding. 2-valent germanium compounds are strongly reducing and revert to the 4-valent state; the same is true of 2-valent tin compounds but to a lesser extent. Lead is predominantly 2-valent and here the 4-valent state tends to be strongly oxidising.

4-valent compounds of these elements are generally predominantly covalent (a few such as SnF_4 contain ionic bonding), whereas many of the 2-valent compounds are essentially ionic.

18.27 The hydrides of germanium, tin and lead

Germanium forms a series of hydrides with formulae analogous to those of the alkanes; the whole series, up to and including the hydride containing six germanium atoms in a chain, is known with certainty. They are rapidly oxidised by oxygen to germanium (IV) oxide and water, e.g.

$$GeH_4(g) + 2O_2(g) \longrightarrow GeO_2(s) + 2H_2O(l)$$

Surprisingly, they are resistant to attack by 30 per cent alkali and in this respect differ from the silanes.

Tin only forms one hydride, stannane, SnH_4, which can be obtained by reduction of tin (IV) chloride with lithium aluminium hydride at low temperature:

$$SnCl_4(l) + Li^+AlH_4^- \longrightarrow SnH_4(g) + Li^+Cl^-(s) + AlCl_3(s)$$

It decomposes slowly at room temperature and rapidly on heating, but resists attack by 15 per cent alkali.

Plumbane, PbH_4, was thought to be formed by treating an alloy of magnesium and lead (containing $^{212}_{82}Pb$ which is redioactive) with dilute acid. The detection of radioactivity in the gas phase was taken as evidence that a volatile compound of lead had been formed. However, some doubt surrounds its actual existence.

18.28 The halides of germanium, tin and lead

The halides of germanium

Germanium forms all four tetrahalides. Germanium (IV) chloride can be made by passing chlorine over the heated metal. It is a fuming colourless liquid, readily hydrolysed by water, with the evolution of hydrogen chloride. Digermanium hexachloride, Ge_2Cl_6, which contains the weak Ge—Ge bond is also known.

Germanium (II) chloride, $GeCl_2$, is a colourless solid and is probably partially ionic. It is obtained by passing germanium (IV) chloride vapour over heated germanium, but it is unstable with respect to the tetrachloride and begins to disproportionate at about 75°C:

$$GeCl_4(g) + Ge(s) \rightleftharpoons 2GeCl_2(s)$$

The halides of tin

Like germanium, tin forms all four tetrahalides. Tin (IV) chloride, $SnCl_4$, is made by passing chlorine over the metal. It is a colourless liquid which fumes in moist air, owing to hydrolysis. It is not completely hydrolysed by water unless the solution is dilute and in the presence of a little water it is possible to obtain the solid hydrate $SnCl_4.5H_2O$. This compound has ionic character and presumably contains the $[Sn(H_2O)_4]^{4+}$ ion.

Tin (IV) fluoride, unlike the chloride, is a white solid and is considered to be essentially ionic $Sn^{4+}(F^-)_4$.

Tin (II) chloride can be obtained as an anhydrous solid on passing hydrogen chloride over heated tin, or as the dihydrate by reacting tin with concentrated hydrochloric acid:

$$Sn(s) + 2HCl(g) \longrightarrow SnCl_2(s) + H_2(g)$$
$$Sn(s) + 2HCl(aq) + 2H_2O(l) \longrightarrow SnCl_2.2H_2O(s) + H_2(g)$$

Although more stable with respect to the tetrachloride than the corresponding dichloride of germanium, it is quite a powerful reducing agent, e.g. it will reduce mercury (II) salts to mercury and iron (III) salts to the iron (II) state:

$$2HgCl_2(aq) + Sn^{2+}(aq) \longrightarrow \underset{\text{white solid}}{Hg_2Cl_2(s)} + Sn^{4+}(aq) + 2Cl^-(aq)$$

$$Hg_2Cl_2(s) + Sn^{2+}(aq) \longrightarrow 2Hg(l) + Sn^{4+}(aq) + 2Cl^-(aq)$$
$$2Fe^{3+}(aq) + Sn^{2+}(aq) \longrightarrow Sn^{4+}(aq) + 2Fe^{2+}(aq)$$

Both the anhydrous chloride and the dihydrate are hydrolysed by water, the solution becoming milky with the formation of a basic salt:

$$SnCl_2(s) + H_2O(l) \rightleftharpoons Sn(OH)Cl(s) + HCl(aq)$$

FIG. 18.7. *The structure of tin (II) chloride*

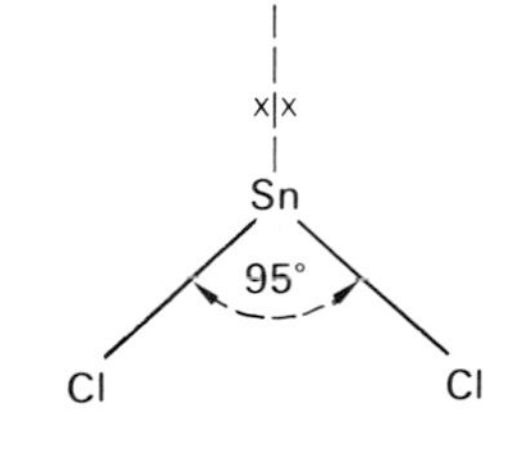

The solution can be kept clear in the presence of hydrochloric acid with which the chloride forms $SnCl_3^-$ and $SnCl_4^{2-}$ complexes. Tin is also added to prevent aerial oxidation.

Anhydrous tin (II) chloride is an angular covalent molecule as expected, since two of the outer electrons of the tin atom are not involved in bonding (fig. 18.7). The bond angle is 95°, which is consistent with the theory that repulsion between lone pair-bonding pair electrons is greater than that between bonding pair-bonding pair electrons (p. 76).

The halides of lead

Lead forms a tetrafluoride and a tetrachloride; the former is a solid and appreciably ionic, $Pb^{4+}(F^-)_4$, while the latter is a covalent liquid, $PbCl_4$, readily hydrolysed by water and easily decomposed into lead (II) chloride, $PbCl_2$, and chlorine. The tetrabromide and tetraiodide do not exist, presumably because bromine and iodine are not sufficiently strong oxidising agents to convert Pb^{II} to Pb^{IV}.

The stable valency state of lead is 2 and all four dihalides can be obtained as sparingly soluble solids by adding halide ions to a soluble lead (II) salt, e.g. the nitrate:

$$Pb^{2+}(aq) + 2X^-(aq) \longrightarrow PbX_2(s) \quad (X^- = F^-, Cl^-, Br^- \text{ or } I^-)$$

The fluoride is ionic, $Pb^{2+}(F^-)_2$, and the chloride, bromide and iodide contain some ionic character which decreases with increasing atomic number of the halogen. Except for the iodide, which is yellow, they are white solids.

Lead (II) chloride and lead (II) iodide are of importance in qualitative analysis and their formation serves as tests for Pb^{2+} in solution. Both are much more soluble in hot water than in cold and the chloride dissolves in concentrated hydrochloric acid with the formation of the $PbCl_4^{2-}$ complex ion:

$$PbCl_2(s) \rightleftharpoons Pb^{2+}(aq) + 2Cl^-(aq)$$
$$+$$
$$4Cl^-(aq) \quad \text{From conc. hydrochloric acid}$$
$$\updownarrow$$
$$PbCl_4^{2-}(aq)$$

18.29 The oxides of germanium, tin and lead

The oxides of germanium

Germanium reacts with oxygen when heated to give germanium (IV) oxide which is a white solid; this oxide can also be obtained by the action of concentrated nitric acid on germanium.

It is less acidic than silicon dioxide, although it reacts with basic oxides to give germanates (IV) which are often isomorphous with the silicates.

$$(Na^+)_2O^{2-}(s) + GeO_2(s) \longrightarrow (Na^+)_2GeO_3^{2-}(s)$$

Germanium (II) oxide, GeO, has been obtained by reduction of Ge^{IV} in solution. It is unstable in the presence of oxygen, reverting to germanium (IV) oxide:

$$2GeO(s) + O_2(g) \longrightarrow 2GeO_2(s)$$

The oxides of tin

Tin (IV) oxide, like germanium (IV) oxide, is obtained by heating the element in oxygen or by treating it with concentrated nitric acid. It is a white solid and occurs naturally as the ore cassiterite from which tin is extracted.

It is an amphoteric oxide reacting with concentrated sulphuric acid to give tin (IV) sulphate and with fused alkalis to give stannates (IV):

$$SnO_2(s) + 2H_2SO_4(l) \longrightarrow Sn(SO_4)_2(s) + 2H_2O(l)$$
$$SnO_2(s) + 2OH^- \longrightarrow SnO_3^{2-} + H_2O(l)$$

(the stannate (IV) ion is known to be $Sn(OH)_6^{2-}$, although it is often written in the dehydrated form, SnO_3^{2-}).

Tin (II) oxide, SnO, is a black solid and can be obtained by heating tin (II) oxalate (the carbon monoxide evolved provides a reducing atmosphere, preventing aerial oxidation to tin (IV) oxide):

$$\begin{matrix} COO^- \\ | \\ COO^- \end{matrix} Sn^{2+}(s) \longrightarrow SnO(s) + CO(g) + CO_2(g)$$

It is amphoteric (more basic than tin (IV) oxide), forming tin (II) salts with acids and stannates (II) with alkalis (air must be excluded since stannates (II) are readily oxidised to stannates (IV)).

$$SnO(s) + 2H^+(aq) \longrightarrow Sn^{2+}(aq) + H_2O(l)$$
$$SnO(s) + 2OH^-(aq) \longrightarrow SnO_2^{2-}(aq) + H_2O(l)$$

(the stannate (II) ion is $Sn(OH)_4^{2-}$, but is often written in the dehydrated form, SnO_2^{2-}).

The oxides of lead

Lead (IV) oxide, PbO_2, is a brown solid obtained when a soluble lead (II) salt is warmed with sodium chlorate (I) which oxidises it:

$$Pb^{2+}(aq) + H_2O(l) + ClO^-(aq) \longrightarrow PbO_2(s) + Cl^-(aq) + 2H^+(aq)$$

It can also be obtained by treating trilead tetroxide, Pb_3O_4, with dilute nitric acid (trilead tetroxide behaves as if it were a mixture of 2PbO and PbO_2).

$$Pb_3O_4(s) + 4HNO_3(aq) \longrightarrow 2Pb(NO_3)_2(aq) + PbO_2(s) + 2H_2O(l)$$

Lead (IV) oxide is a powerful oxidising agent (Pb^{IV} is unstable with respect to Pb^{II}); for instance, it gives lead (II) oxide and oxygen when heated, combines vigorously with sulphur dioxide to give lead (II) sulphate and liberates chlorine from concentrated hydrochloric acid when the mixture is warmed:

$$2PbO_2(s) \longrightarrow 2PbO(s) + O_2(g)$$
$$PbO_2(s) + SO_2(g) \longrightarrow PbSO_4(s)$$
$$PbO_2(s) + 4HCl(aq) \longrightarrow PbCl_2(s) + 2H_2O(l) + Cl_2(g)$$

Lead (IV) oxide is considered to be amphoteric, since it reacts with hydrochloric acid at low temperatures to give lead (IV) chloride (this decomposes readily into lead (II) chloride and chlorine at normal temperatures) and with alkalis to give plumbates (IV) (the plumbate (IV) ion is usually written as PbO_3^{2-} but it is in fact $Pb(OH)_6^{2-}$, cf. the stannate (IV) ion). It forms the positive plate of the lead accumulator (p. 242).

Lead (II) oxide, PbO, exists in two forms, one of which is a red solid and the other a yellow solid. It can be made by the thermal decomposition of lead (II) carbonate or lead (II) nitrate.

Although it is amphoteric, reacting with acids to form lead (II) salts and with alkalis to form plumbates (II) (the plumbate (II) ion is PbO_2^{2-} or $Pb(OH)_4^{2-}$, cf. the stannate (II) ion), its basic properties predominate.

Trilead tetroxide, Pb_3O_4, can be prepared by heating lead (II) oxide in air at 400°C.

$$6PbO(s) + O_2(g) \rightleftharpoons 2Pb_3O_4(s)$$

It decomposes if heated to 500°C into lead (II) oxide and oxygen.

X-ray examination shows that it has a layer structure, each layer of PbO_2 being sandwiched between two layers having the PbO structure. It thus behaves as a mixture of PbO_2 and 2PbO.

It is used in the production of lead (IV) oxide for use in lead accumulators and in the manufacture of red lead paint.

18.30 The sulphides of germanium, tin and lead

Germanium and tin form the monosulphide and disulphide, the former being readily converted to the latter on oxidation (cf. the oxides of these two elements). Lead only forms the monosulphide, PbS. All the sulphides are solids.

Tin (II) sulphide, SnS, is obtained as a dark brown solid when hydrogen sulphide is passed through a solution of a tin (II) salt:

$$Sn^{2+}(aq) + H_2S(g) \longrightarrow SnS(s) + 2H^+(aq)$$

Tin (II) sulphide is oxidised to tin (IV) sulphide, SnS_2, by treatment with ammonium polysulphide (this is usually represented as $(NH_4)_2S_x$ and behaves as a mixture of $(NH_4)_2S$ and sulphur). The tin (IV) sulphide then dissolves to give ammonium trithiostannate (IV).

$$SnS(s) + S \longrightarrow SnS_2(s)$$
$$(NH_4^+)_2S^{2-}(aq) + SnS_2(s) \longrightarrow (NH_4^+)_2SnS_3^{2-}(aq)$$

Lead (II) sulphide, PbS, can be made by passing hydrogen sulphide through a solution containing lead (II) ions:

$$Pb^{2+}(aq) + H_2S(g) \longrightarrow PbS(s) + 2H^+(aq)$$

It is a black solid and is insoluble in ammonium polysulphide.

18.31 The oxysalts of lead

Only the oxysalts of lead are of any importance. Most of the lead (II) salts are insoluble in water, e.g. $PbSO_4$ (white) and $PbCrO_4$ (yellow). Lead (II) nitrate and lead (II) ethanoate, $Pb(CH_3COO)_2.3H_2O$ are, however, freely soluble.

The only stable lead (IV) oxysalt is lead (IV) ethanoate, $Pb(CH_3COO)_4$, which is obtained as a white solid on treating trilead tetroxide with hot ethanoic acid. As expected it is an oxidising agent and is used in organic chemistry for oxidising 1,2-diols to aldehydes or ketones, e.g.

$$\begin{matrix} R-CH-OH \\ | \\ R-CH-OH \end{matrix} \xrightarrow[{[O]}]{Pb(CH_3COO)_4} 2RCH=O + H_2O$$

18.32 Lead tetraethyl, $Pb(C_2H_5)_4$

Lead tetraethyl is a highly poisonous covalent liquid produced by reacting chloroethane vapour with a sodium-lead alloy:

$$4C_2H_5Cl(l) + 4Na/Pb(s) \longrightarrow Pb(C_2H_5)_4(l) + 3Pb(s) + 4Na^+Cl^-(s)$$

It is an antiknock additive used in petrol.

18.33 The lead accumulator

When lead plates are dipped into sulphuric acid (one volume of the concentrated acid diluted to about four times its original volume with distilled water), a layer of lead (II) sulphate is formed and the action then ceases. If direct current is passed through this arrangement, electrons are withdrawn from the positive terminal, oxidation occurs and a deposit of lead (IV) oxide is formed on the plate:

$$PbSO_4(s) + 2H_2O(l) \longrightarrow PbO_2(s) + SO_4^{2-}(aq) + 4H^+(aq) + 2e^-$$

A corresponding reduction occurs at the cathode:

$$PbSO_4(s) + 2e^- \longrightarrow Pb(s) + SO_4^{2-}(aq)$$

The above redox reactions take place when a lead accumulator is charged, lead (IV) oxide being formed on what will become the positive plate and lead on what will become the negative plate; sulphuric acid is also formed. When the accumulator is in use, sulphuric acid is consumed and a check on its specific gravity indicates when it must be recharged; lead (II) sulphate is formed on both plates:

$$PbO_2(s) + SO_4^{2-}(aq) + 4H^+(aq) + 2e^- \longrightarrow PbSO_4 + 2H_2O$$
$$Pb(s) + SO_4^{2-}(aq) \longrightarrow PbSO_4(s) + 2e^-$$

The voltage supplied by the lead accumulator is just in excess of 2 volts.

18.34 Summary of trends in Group 4B

Carbon and silicon are non-metallic, germanium is a metalloid and tin and lead are weakly metallic. This trend is reflected in the properties and structures of the elements; thus diamond is a non-conductor of electricity (graphite, however, will conduct), silicon and germanium are semi-conductors and adopt the diamond structure, whereas tin and lead are conductors. Tin exhibits enantiotropy, one form having the diamond structure and the other two having closer packed atoms, this being a feature of metals; lead, on the other hand, only exists in one form, which is metallic. Carbon, germanium and tin show non-metallic characteristics in forming dioxides on treatment with nitric acid, but lead shows the typical behaviour of a metal and forms the nitrate.

The oxides of carbon are acidic (carbon monoxide will react with fused sodium hydroxide to form sodium methanoate) and are discrete molecules. Silicon dioxide is also acidic, germanium (IV) oxide shows amphoteric tendencies and the oxides of tin and lead are decidedly amphoteric; all these oxides are giant molecules and thus solids, unlike the oxides of carbon. The 4-valent oxides of tin and lead are more acidic than the corresponding 2-valent oxides and the properties of lead (II) oxide are predominantly basic. This behaviour is typical of elements which form more than one oxide.

The outstanding feature of carbon is its ability to form a limitless range of hydrocarbons which are fairly resistant to chemical attack. In the case of silicon there is an extensive chemistry based on silicon-oxygen bonds, e.g. bonds of this type are present in silica, the diverse range of silicates and the silicones.

The 'inert pair' effect is apparent in the chemistry of germanium, tin and lead and accounts for the gradual increase in stability of the 2-valent state with respect to the 4-valent state with increasing atomic number of the Group 4B elements. Thus 2-valent germanium and tin compounds are reducing agents whereas for lead this is the stable valency state. Many of these 2-valent compounds contain appreciable ionic character, which is almost completely lacking in the corresponding 4-valent compounds.

Carbon, unlike the other Group 4B elements, cannot expand its octet and this factor is probably of prime importance in accounting for the stability of CH_4 and CCl_4, for instance, to attack by oxygen and water at room temperature. The tetrahydrides and tetrachlorides of the other Group 4B elements are generally unstable in the presence of oxygen and water, and it is thought that intermediates are formed involving dative bonds from the attacking species to the Group 4B atom (expansion of the octet).

Expansion of the octet also explains why silicon, germanium, tin and lead can form complex anions, e.g. SiF_6^{2-}, $GeCl_6^{2-}$, $SnCl_6^{2-}$, $Sn(OH)_6^{2-}$, $PbCl_6^{2-}$, $Pb(OH)_6^{2-}$, which are octahedral in shape as expected.

18.35 Suggested reading

E. Abel, *Silicon*, Education in Chemistry, Vol. 15, 48, 1978.

D. Kolb and K. E. Kolb, *The Chemistry of Glass*, Journal of Chemical Education, Vol. 56, 604, 1979.

18.36 Questions on chapter 18

1 Describe briefly the preparation of the chlorides of the elements carbon, silicon, tin and lead. Discuss the chemistry of these halides in relation to the position of the four elements in the Periodic Table. (Oxford Schol.)

2 Give a comparative account of the properties of the hydrides, oxides and chlorides of carbon, silicon, tin and lead. To what extent do these properties justify the classification together of these elements? (Camb. Schol.)

3 Discuss the general chemistry of the group of elements carbon, silicon, tin and lead. (O)

4 Describe how carbon monoxide can be prepared and collected in the laboratory.
How and under what conditions does carbon monoxide react with
(a) chlorine,
(b) hydrogen,
(c) sodium hydroxide,
(d) nickel? (C)

5 Explain why CH_4 and CCl_4 are resistant towards hydrolysis whereas the silicon analogues of these compounds are very readily hydrolysed.

6 Discuss the classification of the carbides.

7 (a) Describe with essential experimental details a laboratory preparation of carbon monoxide.
(b) Draw a structural formula to indicate the type of bonding believed to be present in this gas.
(c) How, and under what conditions, does carbon monoxide react with
(i) chlorine,
(ii) iodine (V) oxide (iodine pentoxide),
(iii) sodium hydroxide?
(d) $50{\cdot}0\,cm^3$ of a mixture of carbon monoxide, carbon dioxide and hydrogen were exploded with $25{\cdot}0\,cm^3$ of oxygen. After explosion, the volume measured at the original room temperature and pressure was $37{\cdot}0\,cm^3$. After treatment with potassium hydroxide solution the volume was reduced to $5{\cdot}0\,cm^3$. Calculate the percentage composition by volume of the original mixture. (AEB)

8 Much of the chemistry of carbon can be summed up in the statement: 'carbon atoms can bond to each other to give stable chain and ring structures; the carbon atom can also multiple bond to itself and to other non-metals such as oxygen'. The chemistry of silicon is influenced 'by its reluctance to multiple bond to itself and to oxygen and its ability to form strong single covalent bonds with oxygen'. By discussing appropriate examples, show that these two statements are true.

9 Starting from silica, how could you obtain:
(a) a colloidal dispersion of hydrated silica (silicic acid),
(b) silica gel?
Comment upon the following facts:
(i) scientific instruments which need to be kept dry may have silica gel containing a little anhydrous cobalt (II) chloride placed inside them;
(ii) if silica is fused with sodium carbonate, carbon dioxide is evolved but if carbon dioxide is passed through an aqueous solution of sodium silicate, hydrated silica is precipitated;
(iii) the boiling point of methane is below that of the corresponding hydride of silicon but the boiling point of water is above that of hydrogen sulphide. (JMB)

10 Describe briefly the composition of ordinary glass. Give an account of its manufacture, and comment on its physical state.
Compare the oxides, chlorides and hydrides of carbon and silicon, and discuss the inclusion of these two elements in the same group of the Periodic Table. (L)

11 Give an account of the chemical properties of the element tin and describe four of its principal compounds.
The element germanium (Mendeléeff's eka-silicon) lies in Group 4B of the Periodic Table, below carbon and silicon and above tin and lead. What properties would you predict for this element, for its oxide GeO_2 and for its chloride $GeCl_4$? (O & C)

12 Compare the chemistry of tin with that of lead in relation to the stability and properties of
(a) their chlorides,
(b) their oxides, and
(c) their sulphides. (C)

13 A substance **X** is dissolved in hot concentrated hydrochloric acid, and on treating the solution with sodium hydroxide solution, a white precipitate formed which dissolved in excess, giving a solution with strongly reducing properties.
On heating **X** with sulphur, a brown powder **Y** was formed which dissolved in hot concentrated hydrochloric acid. **Y** dissolved on warming with yellow ammonium sulphide, and on adding hydrochloric acid, a yellow precipitate was formed.
When **X** was strongly heated in air, a white powder **Z** was obtained which could be dissolved only in concentrated sulphuric acid. When **Z** was fused with sodium hydroxide, and extracted with hot water, white crystals were obtained.
Identify **X**, **Y** and **Z** and account for the reactions described, and discuss briefly the chemical properties of **X** shown in them. (S)

14 Give the name and formula for the chief ore of lead and describe how the metal is obtained from it. How may a sample of lead (IV) oxide be obtained from trilead tetroxide? Explain what happens when lead (IV) oxide is treated with
- (a) warm concentrated hydrochloric acid,
- (b) cold concentrated hydrochloric acid followed by ammonium chloride solution,
- (c) sulphur dioxide. (AEB)

15 Describe carefully how you would obtain from metallic tin:
- (a) two different anhydrous chlorides of tin;
- (b) a solution of sodium hexahydrostannate (IV) (sodium stannate (IV)).

What explanation can you offer for the following experimental facts?
- (i) If hydrogen sulphide is passed through a dilute acidified solution of tin (II) chloride a brown precipitate is obtained. The precipitate, after removal and washing, is found to form a clear solution with yellow ammonium sulphide solution. On the addition of dilute hydrochloric acid to the clear solution a yellow precipitate is formed and hydrogen sulphide is given off.
- (ii) An acidified solution of tin (II) chloride becomes cloudy on dilution with water. (JMB)

16 Discuss the main features of the structural chemistry of the silicates in terms of the linking together of tetrahedral SiO_4 structural units. Mica and asbestos are silicates, the former being readily split into thin slices and the latter being fibrous; show how the structures of these two substances account for these properties.

17 Discuss the preparation, properties, structures and uses of the silicones. How is the length of a silicone chain controlled and the cross-linking together of silicone chains achieved?

Illustrate the transition from non-metallic to metallic character of carbon, silicon and tin by describing the structure of, and giving one chemical property of, any **one** type of compound of these elements.

One of the products of the reaction between magnesium silicide (Mg_2Si) and sulphuric acid is a gas, **X**.

A 0·620 g sample of *X* occupied 224 cm^3. When hydrolysed, this sample yielded 1568 cm^3 of hydrogen and a residue (SiO_2) which, after it had been strongly heated, weighed 1·200 g. (All gas volumes corrected to s.t.p.)

What is the molecular formula of **X**?

Write an equation for the hydrolysis of **X**. (O & C)

19 Tabulate two differences and two similarities in the chemistry of carbon and silicon.

When tin (IV) chloride was allowed to react with an excess of ethyl magnesium bromide (C_2H_5MgBr) a liquid **A** was isolated, the vapour density of which was 117; 0·1935g of **A** gave 0·1240g of SnO_2 on repeated evaporation with nitric acid. **A** contained no magnesium and gave no reaction with silver nitrate solution.

On heating 1·41g of **A** with 0·52g of tin (IV) chloride in a sealed tube for some time, 1·93g of a liquid **B** were obtained, 0·2240g of which gave 0·1332g of silver chloride when treated with silver nitrate; 0·1865g of **B** gave 0·1164g of SnO_2 on treatment with nitric acid. The vapour density of **B** was 121.
- (a) What is the formula of **A**?
- (b) What is the formula of **B**?
- (c) Write an equation for the reaction of **A** with tin (IV) chloride to give **B**.
- (d) What would you expect to be the product of the reaction of **B** with silver acetate? (O & C)

20 This question concerns the group of elements carbon, silicon, tin and lead.
- (a) Discuss:
  - (i) the variation of the acid/base character of the oxides MO and MO_2 with increase in atomic number
  - (ii) the difference in acid/base character of the oxides MO and MO_2 for any one element
  - (iii) the reactivity of the tetrachlorides with water
  - (iv) the relative thermal stability of the dichlorides and tetrachlorides.
- (b) How could:
  - (i) carbon monoxide be made from carbon,
  - (ii) lead monoxide be made from lead? (L)

Chapter 19

Group 5B nitrogen, phosphorus, arsenic, antimony and bismuth

19.1 Some physical data of Group 5B elements

	Atomic Number	Electronic Configuration	Atomic Radius/nm	Ionic Radius/nm M^{3+}	M.p. /°C	B.p. /°C
N	7	2.5 $1s^22s^22p^3$	0·074		−210	−196
P	15	2.8.5 $\ldots 2s^22p^63s^23p^3$	0·110		44·1 (white)	280 (white)
As	33	2.8.18.5 $\ldots 3s^23p^63d^{10}4s^24p^3$	0·121	0·069		613 (sublimation)
Sb	51	2.8.18.18.5 $\ldots 4s^24p^64d^{10}5s^25p^3$	0·141	0·090	630	1380
Bi	83	2.8.18.32.18.5 $\ldots 5s^25p^65d^{10}6s^26p^3$	0·152	0·120	271	1560

19.2 Some general remarks about Group 5B

In theory the Group 5B elements can complete the octet in chemical combination by gaining three electrons to form the 3-valent anion, by forming three covalent bonds, or by losing five electrons; the last possibility is ruled out on energetic grounds. Only nitrogen (and possibly phosphorus to a slight extent) forms the 3-valent ion and reactive metals are required for it to be possible; the N^{3-} ion is present in ionic nitrides, e.g. $(Li^+)_3N^{3-}$ and $(Ca^{2+})_3(N^{3-})_2$. The majority of compounds formed by this group of elements are covalent.

Antimony and bismuth can form the 3-valent cation X^{3+} (the inert-pair effect), the Sb^{3+} ion being present in $(Sb^{3+})_2(SO_4{}^{2-})_3$ and the Bi^{3+} ion in $Bi^{3+}(F^-)_3$ and $Bi^{3+}(NO_3{}^-)_3.5H_2O$.

Because phosphorus, arsenic, antimony and bismuth have vacant *d* orbitals they are able to form 5-covalent compounds which are not possible for nitrogen, e.g. in the formation of PCl_5, one of the 3*s* electrons of the phosphorus atom is promoted to the 3*d* level, giving five unpaired electrons for valency purposes (p. 265).

Nitrogen and phosphorus are non-metallic; metallic properties first become apparent with arsenic and become progressively more important for antimony and bismuth. Of these elements only nitrogen is able to

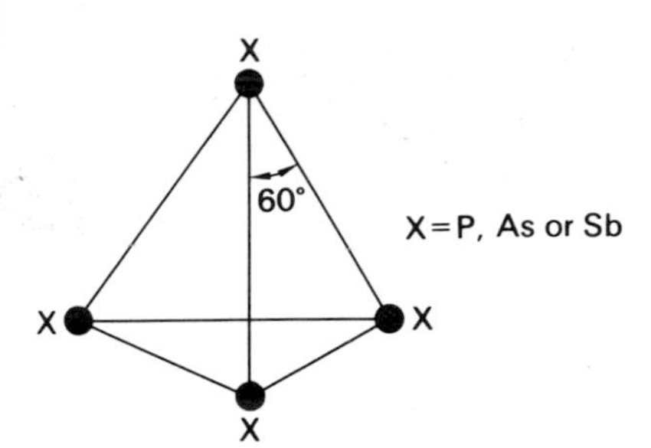

FIG. 19.1. *The structure of one allotropic form of phosphorus, arsenic and antimony*

multiple bond with itself, the triple bond being present in the nitrogen molecule, N≡N. Phosphorus, arsenic and antimony are allotropic, the less dense allotrope containing X_4 tetrahedra with X—X—X bond angles of 60° which introduce considerable strain (fig. 19.1).

The denser allotropic forms of phosphorus, arsenic and antimony are more stable and more 'metallic' (closer packing of the atoms) (p. 263). Bismuth also adopts a metallic structure.

There is little resemblance between the chemistry of nitrogen and phosphorus and these two elements are considered separately; arsenic, antimony and bismuth are then treated together.

Nitrogen

19.3 Occurrence and preparation of nitrogen

Nitrogen occurs in the atmosphere to the extent of about 78 per cent by volume. It is an essential ingredient in all living matter (together with carbon, hydrogen and oxygen), and its presence can be shown by the fact that ammonia is evolved when vegetable or animal products are heated. The oxidation of ammonia with copper (II) oxide is a convenient way of obtaining nitrogen in the laboratory.

When a stream of dry ammonia is passed over heated copper (II) oxide the ammonia is oxidised and the products are nitrogen, water and copper:

$$2NH_3(g) + 3CuO(s) \longrightarrow N_2(g) + 3H_2O(l) + 3Cu(s)$$

Industrially, nitrogen is obtained from the atmosphere. One process involves compressing air to about 10 atmospheres after filtration to remove dust; the air is now freed from carbon dioxide and moisture and further compressed to about 200 atmospheres. After cooling to room temperature sudden expansion results in a further cooling of the gas. This cooled air is now recycled and a further drop in temperature takes place on expansion. Eventually the air becomes so cold that it liquefies; fractional distillation gives gaseous nitrogen, b.p. −196°C, (which is usually contaminated with argon and a little oxygen), and liquid oxygen, b.p. −183°C.

19.4 The nitrogen cycle

Nitrogen compounds—proteins in particular—are required for healthy development of plants and animals; plants obtain nitrogen compounds from the soil, which they convert into proteins, while animals obtain their protein materials by eating plants. Decay of plants and death of animals result in nitrogen compounds being released into the soil, some of which are broken down to nitrogen which then finds its way into the atmosphere. Nitrogen compounds must be introduced into soil to make good deficiencies, sodium nitrate (Chile saltpetre) and ammonium salts constituting nitrogenous fertilisers. The atmosphere acts as a nitrogen reservoir, the percentage of nitrogen in it remaining constant by the operation of a nitrogen cycle in nature (fig. 19.2) (cf. the carbon cycle p. 214).

FIG. 19.2. *The nitrogen cycle*

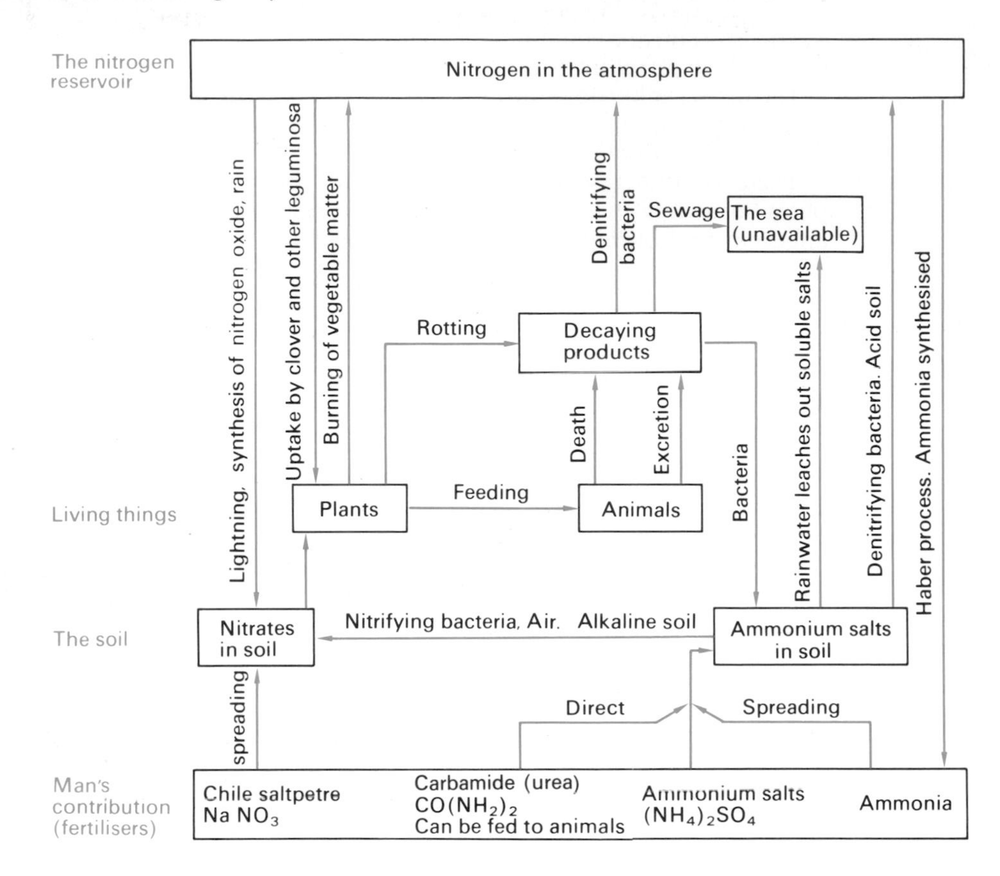

19.5 The properties of nitrogen

Nitrogen is a colourless and odourless diatomic gas. It liquefies at $-196°C$ and freezes at $-210°C$. Chemically it is rather inert and this is due in part to the very large $N \equiv N$ bond energy (946 kJ mol^{-1}). Because it is inert it is often used to provide an inert atmosphere when handling easily oxidisable chemicals; the same property also accounts for it being used as a 'carrier gas' in vapour phase chromatography. Liquid nitrogen is used as a cooling agent.

Although nitrogen is rather inert, it will combine with hydrogen under pressure in the presence of a catalyst to form ammonia and with oxygen, when sparked, to give nitrogen oxide. The former reaction is of immense industrial importance (p. 249); the latter reaction takes place in the atmosphere during lightning flashes.

$$N_2(g) + 3H_2(g) \rightleftharpoons 2NH_3(g)$$
$$N_2(g) + O_2(g) \rightleftharpoons 2NO(g)$$

It will also combine with some other non-metals at high temperature to give nitrides, e.g. boron nitride and silicon nitride which are giant covalent molecules.

Lithium and the Group 2A metals combine directly with nitrogen under the right conditions to give ionic nitrides (p. 262). Some transition metals also combine with the gas at red heat, if they are finely divided, to form interstitial nitrides (p. 262).

19.6 The hydrides of nitrogen

Ammonia, NH_3

Laboratory preparation

Ammonia can be made in the laboratory by heating an ammonium salt with an alkali, e.g.

$$Ca^{2+}(OH^-)_2(s) + 2NH_4{}^+Cl^-(s) \longrightarrow Ca^{2+}(Cl^-)_2(s) + 2H_2O(l) + 2NH_3(g)$$

or

$$OH^- + NH_4{}^+ \longrightarrow H_2O(l) + NH_3(g)$$

It is also obtained when an ionic nitride is hydrolysed with water:

$$N^{3-} + 3H_2O(l) \longrightarrow 3OH^-(aq) + NH_3(g)$$

The gas can be dried by passing it over calcium oxide.

Industrial production—the Haber process

In this process nitrogen obtained from the air and hydrogen obtained from synthesis gas, which in turn is obtained from methane or naphtha (p. 225–226), are reacted together at a high pressure (at least 250 atmospheres) in the presence of a finely divided iron catalyst at 500°C. Under these conditions about 15 per cent of the gases are converted into ammonia. The gases are cooled while still under pressure, and the ammonia is removed as a liquid; the unreacted nitrogen and hydrogen are recycled for further conversion into ammonia.

$$N_2(g) + 3H_2(g) \rightleftharpoons 2NH_3(g) \quad \Delta H^\ominus(298\,K) = -92{\cdot}0\,kJ\,mol^{-1}$$

By Le Chatelier's Principle an increase in pressure should increase the yield of ammonia, since there is a decrease in the number of moles of gas when ammonia is formed.

Pressure	**Yield of ammonia** at 550°C with a catalyst
1 atmosphere	Negligible
100 atmospheres	7%
1000 atmospheres	41%

In addition to increasing the yield, the reaction should reach equilibrium faster at higher and higher pressures, since the number of gaseous collisions will increase. However, with increasing pressure the cost of industrial plant becomes more expensive, e.g. pipes have to be thicker to withstand the pressure, and a compromise is reached between high yield at high cost and lower yield at lower cost.

Since the synthesis of ammonia is an exothermic process, a decrease in temperature will lead to a higher percentage conversion of nitrogen and hydrogen into ammonia. However, a decrease in temperature will decrease the rate of reaction and equilibrium conditions will take a longer time to achieve. A compromise has again to be reached between high yield in a long time and a lower yield in a shorter time.

Temperature	**Yield of ammonia** at 250 atmospheres with a catalyst
1000°C	Negligible
550°C	15%
200°C	88%

Ammonia, produced by the Haber process, accounts for practically all the ammonia used industrially. It is used in the manufacture of nitric acid, urea, and nitrogenous fertilisers, e.g. ammonium nitrate. Liquid ammonia is a refrigerant; it is used to cool the supporting legs of the Canadian-Alaskan pipeline which would otherwise become warm, by heat generated by the flowing oil, and thus cause damage to the permafrost.

Properties of ammonia

Ammonia is an easily liquefiable gas with a characteristic smell. It is readily soluble in water and the solution turns red litmus blue. A solution of ammonia in water is commonly called ammonium hydroxide, although NH_4OH molecules are not present. The solution, which contains ammonia hydrogen bonded to water, is best represented as an equilibrium between $NH_3(aq)$, $NH_4^+(aq)$ and $OH^-(aq)$ ions:

$$NH_3(aq.) + H_2O(l) \rightleftharpoons NH_4^+(aq) + OH^-(aq)$$

$$K = \frac{[NH_4^+(aq)][OH^-(aq)]}{[NH_3(aq)]} = 1{\cdot}81 \times 10^{-5}\,\text{mol dm}^{-3}\ (\text{at } 298\,\text{K})$$

The low value for the dissociation constant means that an aqueous solution of ammonia is a weak alkali; a buffer solution of aqueous ammonia and ammonium chloride is used in qualitative analysis for the selective precipitation of metallic hydroxides.

The ammonia molecule is pyramidal in shape and contains a lone pair of electrons which can bond to a proton to form the tetrahedral ammonium ion, NH_4^+, i.e. ammonia neutralises acids to form ammonium salts:

$$NH_3 + H^+ \longrightarrow NH_4^+$$

The formation of a white smoke of ammonium chloride, which is formed when ammonia is brought in contact with hydrochloric acid fumes, is often used as a test for ammonia.

Ammonium salts generally resemble the corresponding alkali metal salts in solubility and structure; they are, however, thermally unstable, e.g.

$$\begin{aligned} NH_4^+Cl^-(s) &\rightleftharpoons NH_3(g) + HCl(g) \\ NH_4^+NO_2^-(s) &\longrightarrow N_2(g) + 2H_2O(l) \\ NH_4^+NO_3^-(s) &\longrightarrow N_2O(g) + 2H_2O(l) \end{aligned}$$

Ammonium salts can be estimated quantitatively by heating with an excess of standard sodium hydroxide until all the ammonia has been expelled, followed by titration with standard acid.

Many transition metal ions form complexes with ammonia, e.g. $[Ni(NH_3)_6]^{2+}$ and $[Co(NH_3)_6]^{3+}$ which are both octahedral in shape, with the lone pair on the ammonia molecule pointing towards the metal cation.

The Groups 1A and 2A metals are sufficiently electropositive to react with ammonia on heating to give ionic amides, e.g.

$$2Na(s) + 2NH_3(g) \longrightarrow 2Na^+NH_2^-(s) + H_2(g)$$

The amide ion is a very strong base and these compounds are hydrolysed by water with the liberation of ammonia:

$$NH_2^- + H_2O(l) \longrightarrow OH^-(aq) + NH_3(g)$$

Ammonia is a reducing agent, chlorine and sodium chlorate (I) being strong enough oxidising agents to react with the gas at room temperature to produce nitrogen:

$$2NH_3(g) + 3Cl_2(g) \longrightarrow N_2(g) + 6HCl(g) \quad (NH_4^+Cl^- \text{ formed with excess } NH_3)$$
$$2NH_3(g) + 3OCl^-(aq) \longrightarrow N_2(g) + 3Cl^-(aq) + 3H_2O(l)$$

Nitrogen is also formed when ammonia is passed over heated copper (II) oxide (p. 247).

Although it shows no reaction towards air, it burns with a pale green flame in oxygen; nitrogen and steam are produced:

$$4NH_3(g) + 3O_2(g) \longrightarrow 2N_2(g) + 6H_2O(g)$$

The presence of a platinum catalyst almost completely eliminates the above reaction and allows the otherwise slow oxidation to nitrogen oxide to become dominant. This is the first stage of the industrial method for making nitric acid (p. 257).

$$4NH_3(g) + 5O_2(g) \longrightarrow 4NO(g) + 6H_2O(g)$$

Liquid ammonia as a solvent

Liquid ammonia is ionised to a slight extent:

$$2NH_3 \rightleftharpoons NH_4^+ + NH_2^-$$
$$\text{cf.} \quad 2H_2O \rightleftharpoons H_3O^+ + OH^-$$

Whereas in water acids are substances that provide hydroxonium ions and alkalis are substances that provide hydroxyl ions, in liquid ammonia acids give rise to ammonium ions, e.g. ammonium salts, and bases give rise to amide ions, e.g. ionic amides. Thus in liquid ammonia an ammono-acid neutralises an ammono-base to give a salt and ammonia (an exact parallel with aqueous systems), e.g.

$$\underset{\text{ammono-acid}}{NH_4^+Cl^-} + \underset{\text{ammono-base}}{Na^+NH_2^-} \longrightarrow \underset{\text{salt}}{Na^+Cl^-} + \underset{\text{ammonia}}{2NH_3}$$

or

$$NH_4^+ + NH_2^- \longrightarrow 2NH_3$$

Ionic nitrides such as lithium nitride function as bases in liquid ammonia:

$$3NH_4^+Cl^- + (Li^+)_3N^{3-} \longrightarrow 3Li^+Cl^- + 4NH_3$$

or

$$3NH_4^+ + N^{3-} \longrightarrow 4NH_3$$

More parallels with aqueous systems exist; for instance hydrolysis and ammonolysis are essentially very similar processes:

$$SiCl_4 \xrightarrow{\text{hydrolysis}} Si(OH)_4 \xrightarrow{\text{heat}} SiO_2 + 2H_2O$$
$$3SiCl_4 \xrightarrow{\text{ammonolysis}} 3Si(NH_2)_4 \xrightarrow{\text{heat}} Si_3N_4 + 8NH_3$$

Hydrazine, N_2H_4

Hydrazine is a colourless fuming covalent liquid and is made by the oxidation of ammonia with sodium chlorate (I) in the presence of gelatine (this is thought to complex with heavy metal ions and so prevent them from catalysing the oxidation of ammonia to nitrogen). The reaction proceeds in two stages:

$$NH_3(g) + OCl^-(aq) \longrightarrow OH^-(aq) + NH_2Cl(g)$$
$$NH_2Cl(g) + NH_3(g) + OH^-(aq) \longrightarrow N_2H_4(l) + Cl^-(aq) + H_2O(l)$$

Hydrazine is an endothermic compound which burns in air to give nitrogen and steam with a vigorous evolution of heat. This fact accounts for the interest shown in it as a potential rocket fuel. It is a weaker base than ammonia but, because it contains two lone pairs of electrons (one lone pair on each nitrogen atom), it is bifunctional; it thus forms two chlorides, for example $(NH_2{-}NH_3^+)Cl^-$ and $(NH_3^+{-}NH_3^+)(Cl^-)_2$.

Aldehydes and ketones condense with hydrazine to give hydrazones, e.g.

$$\mathrm{R{-}\overset{\displaystyle H}{\overset{|}{C}}{=}O + H_2N{-}NH_2 \longrightarrow R{-}\overset{\displaystyle H}{\overset{|}{C}}{=}N{-}NH_2 + H_2O}$$

a hydrazone

A derivative of hydrazine, 2,4-dinitrophenylhydrazine, forms yellow crystalline condensation products with aldehydes and ketones in the same manner, and a determination of their melting points affords a useful method of determining the identity of these compounds.

Hydrazoic acid, HN_3, and azides

Hydrazoic acid is a colourless, highly explosive liquid. It can be made by treating sodamide with dinitrogen oxide at about 190°C; sodium azide is formed, which reacts with dilute sulphuric acid to give hydrazoic acid:

$$2Na^+NH_2^-(s) + N_2O(g) \longrightarrow Na^+N_3^-(s) + Na^+OH^-(s) + NH_3(g)$$
$$2Na^+N_3^-(s) + (H^+)_2SO_4^{2-}(aq) \longrightarrow 2HN_3(l) + (Na^+)_2SO_4^{2-}(aq)$$

Hydrazoic acid is covalent and its molecule is asymmetrical:

$$\mathrm{H{-}N{=}\overset{+}{N}{=}\overset{-}{\ddot{N}}}$$

Covalent azides such as lead azide, $Pb(N_3)_2$, are explosive and are used as detonators, whereas ionic azides, e.g. sodium azide, $Na^+N_3^-$, decompose non-explosively when heated into the metal and nitrogen. The azide ion is linear as expected.

19.7 The halides of nitrogen

These are of no great importance, but it is of interest to note that nitrogen trifluoride, NF_3, is a covalent gas resistant to hydrolysis, whereas the trichloride, NCl_3, is a covalent and highly explosive yellow oil readily decomposed by water into ammonia and chloric (I) acid. This hydrolysis is thought to proceed by a stepwise replacement of chlorine, the lone pair on the nitrogen atom bonding to a hydrogen atom of a water molecule:

$$\mathrm{Cl{-}\underset{\underset{\displaystyle Cl}{|}}{\overset{\overset{\displaystyle Cl}{|}}{N}}{:} + H{-}\overset{\overset{\displaystyle H}{|}}{O} \longrightarrow Cl{-}\underset{\underset{\displaystyle Cl}{|}}{\overset{\overset{\displaystyle Cl}{|}}{N}}{:}{-}{-}H{-}\overset{\overset{\displaystyle H}{|}}{O} \longrightarrow Cl{-}\underset{\underset{\displaystyle Cl}{|}}{\ddot{N}}{-}H + HClO}$$

$$NHCl_2 + H_2O \longrightarrow NH_2Cl + HClO$$
$$NH_2Cl + H_2O \longrightarrow NH_3 + HClO$$

19.8 The oxides of nitrogen

The best known oxides of nitrogen are dinitrogen oxide, N_2O, nitrogen oxide, NO, dinitrogen trioxide, N_2O_3, nitrogen dioxide, NO_2, and dinitrogen pentoxide, N_2O_5. They are all endothermic compounds, i.e. they liberate heat when they decompose into their elements, and this is due to the high stability of the nitrogen molecule.

Dinitrogen oxide, N_2O

This gas can be prepared by the cautious decomposition of ammonium nitrate; it is safer, however, to heat a mixture of sodium nitrate and ammonium sulphate, the ammonium nitrate decomposing as fast as it is formed:

$$2Na^+NO_3^-(s) + (NH_4^+)_2SO_4^{2-}(s) \longrightarrow 2NH_4^+NO_3^-(s) + (Na^+)_2SO_4^{2-}(s)$$

$$NH_4^+NO_3^-(s) \longrightarrow N_2O(g) + 2H_2O(l)$$

It can be freed from slight traces of nitrogen oxide impurity by passing through a solution of iron (II) sulphate (p. 380).

Dinitrogen oxide has a sweet taste and is decomposed into its elements with a glowing splint which is thus rekindled. Its main use is as an anaesthetic.

The molecule has a linear structure and is a resonance hybrid of the two forms:

$$:\overset{\cdot\cdot}{\underset{}{N}}{=}\overset{+}{N}{=}\overset{\cdot\cdot}{\underset{\cdot\cdot}{O}} \longleftrightarrow :N{\equiv}\overset{+}{N}{-}\overset{\cdot\cdot}{\underset{\cdot\cdot}{O}}:$$

Nitrogen oxide, NO

Nitrogen oxide can be prepared by the action of 50 per cent nitric acid on copper:

$$3Cu(s) + 8HNO_3(aq) \longrightarrow 3Cu^{2+}(aq) + 6NO_3^-(aq) + 4H_2O(l) + 2NO(g)$$

The main impurity is nitrogen dioxide, and the gas can be purified by passing it through a concentrated solution of iron (II) sulphate with which it reacts to form a dark brown solution containing the addition compound $FeSO_4.NO$. The action of heat on this compound, in the absence of air, produces reasonably pure nitrogen oxide.

Small quantities of the gas are formed in the atmosphere during lightning flashes; it is also produced during the catalytic oxidation of ammonia, and is a reactive intermediate in the industrial production of nitric acid (p. 257).

Nitrogen oxide is a colourless gas and virtually insoluble in water. It acts as a reducing agent, combining immediately with oxygen to form the brown gas nitrogen dioxide, and with chlorine in the presence of activated charcoal to give nitrosyl chloride, NOCl:

$$2NO(g) + O_2(g) \longrightarrow 2NO_2(g)$$

$$2NO(g) + Cl_2(g) \longrightarrow 2NOCl(g)$$

It is oxidised quantitatively by an acidified solution of potassium manganate (VII) and this reaction can be used to estimate the gas:

$$3MnO_4^-(aq) + 4H^+(aq) + 5NO(g) \longrightarrow 3Mn^{2+}(aq) + 5NO_3^-(aq) + 2H_2O(l)$$

When heated to about 1000°C nitrogen oxide decomposes into nitrogen and oxygen; thus strongly burning substances such as magnesium are oxidised by it, but it will not rekindle a glowing splint.

$$2Mg(s) + 2NO(g) \longrightarrow 2Mg^{2+}O^{2-}(s) + N_2(g)$$

In the presence of water it will oxidise sulphur dioxide to sulphuric acid, being itself reduced to dinitrogen oxide:

$$SO_2(g) + H_2O(l) + 2NO(g) \longrightarrow H_2SO_4(aq) + N_2O(g)$$

Nitrogen oxide is unusual in that its molecule contains an odd number of electrons, so the octet cannot be completed round the nitrogen atom. The structure of the molecule is a resonance hybrid of the two forms:

$$:\dot{N}=\ddot{O}: \longleftrightarrow :\overset{-}{\ddot{N}}=\overset{+}{\dot{O}}:$$

Only in the liquid and solid has the tendency to dimerisation been observed and loose units, of the type below, been shown to exist (cf. the dimerisation of nitrogen dioxide (p. 255)):

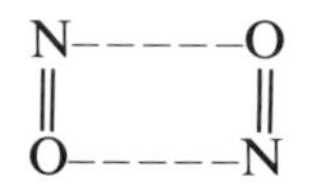

The loss of one electron from the nitrogen oxide molecule should produce a positive ion isoelectronic with the nitrogen molecule. This entity, the nitrosonium ion NO^+, does indeed exist in compounds such as $NO^+ClO_4^-$ and $NO^+HSO_4^-$.

Dinitrogen trioxide, N_2O_3

This oxide only exists in the pure condition as a pale blue solid at low temperatures. The liquid, which is also pale blue, and the vapour contain dinitrogen trioxide in equilibrium with nitrogen oxide and nitrogen dioxide, the dissociation being about 90 per cent complete at room temperature:

$$N_2O_3(g) \rightleftharpoons NO(g) + NO_2(g)$$

A convenient way of making the unstable liquid is to condense equal volumes of nitrogen oxide and nitrogen dioxide at −20°C. When this same mixture is dissolved in aqueous sodium hydroxide almost pure sodium nitrite is formed, dinitrogen trioxide is thus the acid anhydride of the unstable nitrous acid.

$$NO(g) + NO_2(g) + 2Na^+OH^-(aq) \longrightarrow 2Na^+NO_2^-(aq) + H_2O(l)$$

The structure of the oxide is thought to involve resonance between forms such as:

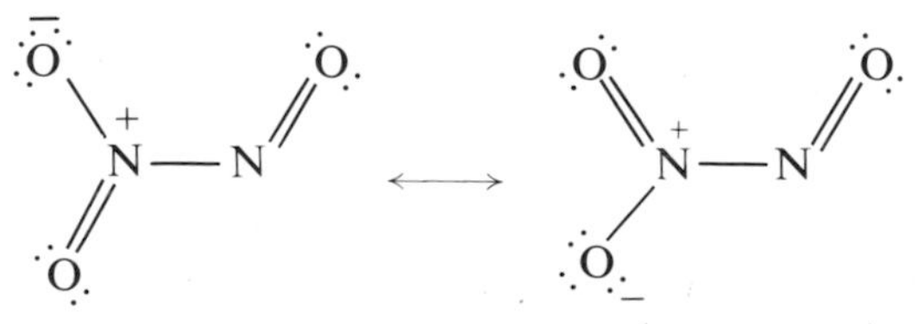

Nitrogen dioxide, NO_2 (dinitrogen tetroxide, N_2O_4)

Nitrogen dioxide exists in equilibrium with the dimer N_2O_4, the extent of the dimerisation decreasing with increasing temperature. At low temperature, in the solid state, the oxide exists almost entirely as the pale yellow N_2O_4; as the temperature is increased the colour darkens as NO_2, which is black, is produced. Dissociation into NO_2 is about 90 per cent complete at 100°C and 100 per cent complete at 150°C, when the colour is black. Further increase in temperature results in a loss of colour as the nitrogen dioxide decomposes into nitrogen oxide and oxygen.

$$N_2O_4(g) \underset{\text{cool}}{\overset{\text{heat}}{\rightleftharpoons}} 2NO_2(g) \underset{\text{cool}}{\overset{\text{heat}}{\rightleftharpoons}} 2NO(g) + O_2(g)$$

Nitrogen dioxide can be prepared by the action of heat on most metallic nitrates, but lead nitrate is usually used since it does not contain any water of crystallisation. Oxygen is evolved as well but the two gases can be separated by passing them through a freezing mixture of ice and salt; nitrogen dioxide condenses as a pale yellow liquid, b.p. 22°C.

$$2Pb(NO_3)_2(s) \longrightarrow 2PbO(s) + 4NO_2(g) + O_2(g)$$

Nitrogen dioxide is a dense poisonous gas and very soluble in water, forming a mixture of nitrous and nitric acids, of which it is the mixed anhydride:

$$2NO_2(g) + H_2O(l) \longrightarrow HNO_2(aq) + HNO_3(aq)$$

Nitrous acid then decomposes gradually into nitric acid and nitrogen oxide:

$$3HNO_2(aq) \longrightarrow HNO_3(aq) + 2NO(g) + H_2O(l)$$

With alkalis it forms a mixture of the nitrite and nitrate:

$$2NO_2(g) + 2OH^-(aq) \longrightarrow NO_2^-(aq) + NO_3^-(aq) + H_2O(l)$$

Because nitrogen dioxide decomposes into nitrogen oxide and oxygen at about 150°C it will oxidise strongly-burning substances such as magnesium, phosphorus and even charcoal. At room temperature it will oxidise hydrogen sulphide to sulphur, sulphurous acid to sulphuric acid and iodide ions to iodine, and is itself reduced to nitrogen oxide:

$$H_2S(g) + NO_2(g) \longrightarrow NO(g) + H_2O(l) + S(s)$$
$$SO_2(aq) + H_2O(l) + NO_2(g) \longrightarrow H_2SO_4(aq) + NO(g)$$
$$2I^-(aq) + H_2O(l) + NO_2(g) \longrightarrow I_2(s) + 2OH^-(aq) + NO(g)$$

Nitrogen dioxide, like nitrogen oxide, has a molecule containing an odd number of electrons, its structure being angular and a resonance hybrid of the two forms:

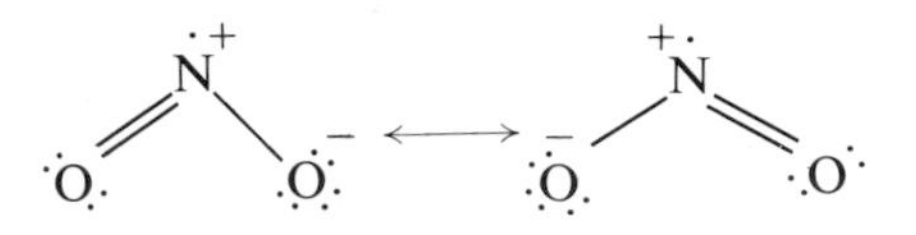

The dimer N_2O_4 has a planar structure, the odd electron on each of the nitrogen atoms of the NO_2 molecules pairing to form a weak N—N bond:

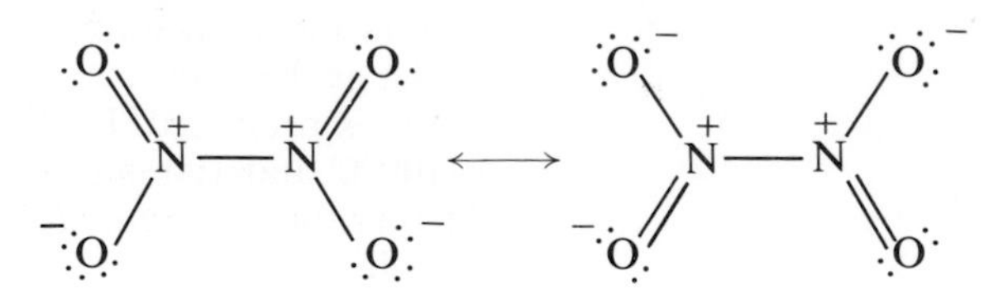

The molecule is a resonance hybrid since all four N—O bonds are equivalent.

Liquid dinitrogen tetroxide will ionise slightly into NO^+ and NO_3^- ions:

$$N_2O_4 \rightleftharpoons NO^+ + NO_3^-$$

This ionisation is increased by adding substances capable of removing either the NO^+ or NO_3^- ions. Thus zinc nitrate, for example, will dissolve in the liquid to form the complex anion, $[Zn(NO_3)_4]^{2-}$, which is present in the nitrosonium salt:

$$(NO^+)_2[Zn(NO_3)_4]^{2-}$$

Dinitrogen pentoxide, N_2O_5

This is a colourless solid and can be made by dehydrating nitric acid—of which it is the acid anhydride—with phosphorus (V) oxide

$$4HNO_3(l) + P_4O_{10}(s) \longrightarrow 4HPO_3(s) + 2N_2O_5(s)$$

It is thought that the structure of the gaseous molecule is a resonance hybrid of forms such as:

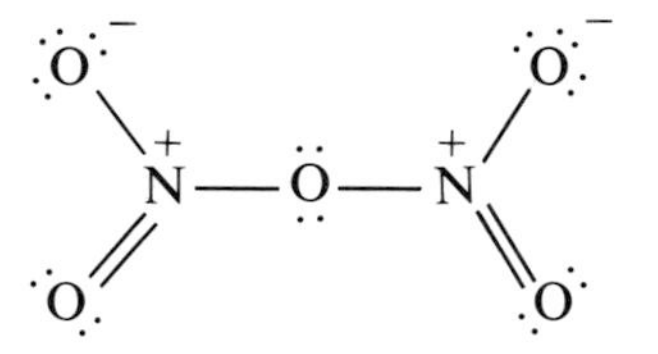

The solid is ionic, $NO_2^+NO_3^-$, (nitronium nitrate). The nitronium ion NO_2^+ is isoelectronic with carbon dioxide and has a similar linear structure (p. 219); the nitrate ion is planar and symmetrical (p. 261).

19.9 The oxyacids of nitrogen

Nitrous acid, HNO_2, and nitrites

An aqueous solution of nitrous acid can be obtained by adding dilute hydrochloric acid to a cold dilute solution of sodium nitrite:

$$NO_2^-(aq) + H^+(aq) \longrightarrow HNO_2(aq)$$

It is a weak and unstable acid, any attempt to concentrate it resulting in its decomposition:

$$3HNO_2(aq) \longrightarrow HNO_3(aq) + H_2O(l) + 2NO(g)$$

Nitrites of the alkali metals are fairly stable and can be obtained by the thermal decomposition of the nitrate, e.g.

$$2Na^+NO_3^-(s) \longrightarrow 2Na^+NO_2^-(s) + O_2(g)$$

Nitrous acid and acidified solutions of nitrites (which produce nitrous acid) are oxidising agents; for instance, iodides are oxidised to iodine and iron (II) ions converted into iron (III) ions:

$$2I^-(aq) + 4H^+(aq) + 2NO_2^-(aq) \longrightarrow I_2(s) + 2H_2O(l) + 2NO(g)$$
$$Fe^{2+}(aq) + 2H^+(aq) + NO_2^-(aq) \longrightarrow Fe^{3+}(aq) + H_2O(l) + NO(g)$$

Strong oxidising agents, such as acidified potassium manganate (VII), force nitrous acid and nitrites to behave as reducing agents, and this is a convenient way of estimating nitrites:

$$2MnO_4^-(aq) + 5NO_2^-(aq) + 6H^+(aq) \longrightarrow 2Mn^{2+}(aq) + 5NO_3^-(aq) + 3H_2O(l)$$

Ionic nitrites evolve the brown gas nitrogen dioxide when treated with dilute acid, and this can be used to distinguish them from nitrates which do not react.

Sodium nitrite, in particular, is important in the production of diazonium salts (made by treating an aromatic amine with sodium nitrite and dilute hydrochloric acid below about 5°C) which couple with phenols to produce dyes.

The nitrite ion is angular and a resonance hybrid:

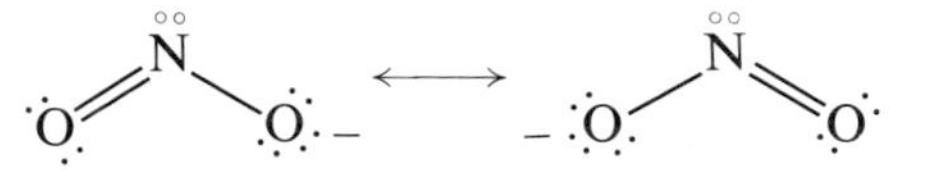

Nitric acid, HNO_3

Preparation and manufacture of nitric acid

This can be prepared by treating sodium nitrate with concentrated sulphuric acid and warming the mixture to distil over the nitric acid, which boils at a lower temperature than sulphuric acid:

$$NO_3^- + H_2SO_4(l) \longrightarrow HSO_4^- + HNO_3(l)$$

Industrially it is made by the catalytic oxidation of ammonia. A mixture of ammonia and air is passed over a platinum-rhodium catalyst at about 850°C, when nitrogen oxide and steam are formed:

$$4NH_3(g) + 5O_2(g) \longrightarrow 4NO(g) + 6H_2O(g)$$

The mixture is cooled when the nitrogen oxide reacts with air to form nitrogen dioxide, which is then reacted with water and more air to produce nitric acid:

$$2NO(g) + O_2(g) \longrightarrow 2NO_2(g)$$
$$4NO_2(g) + 2H_2O(l) + O_2(g) \longrightarrow 4HNO_3(aq)$$

Any surplus nitrogen dioxide is reacted with sodium hydroxide solution, forming sodium nitrite and sodium nitrate.

Acid up to 75 per cent strength can be obtained by this process. If such a solution is boiled, nitric acid distils until a solution of 68 per cent strength is left. If dilute solutions of nitric acid are boiled, water distils off until a solution of acid of the same strength remains. This mixture distils unchanged at a temperature of 120·5°C and is a 'constant boiling' or azeotropic mixture.

Pure nitric acid can be obtained from this 68 per cent acid by adding phosphorus (V) oxide to combine with the water, followed by distillation.

Physical properties of nitric acid

Pure nitric acid is a colourless liquid, b.p. 86°C. Its vapour consists of molecules which are planar; resonance is involved:

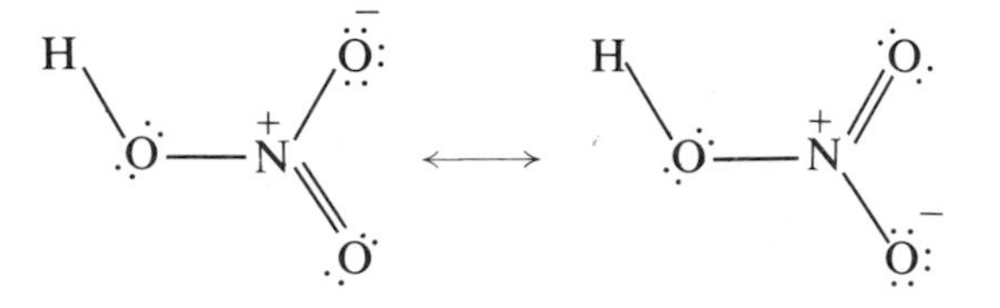

The pure liquid, however, is conducting and some ionisation occurs:

$$2HNO_3 \rightleftharpoons NO_2^+ + NO_3^- + H_2O$$
$$\text{cf.}\quad 2H_2O \rightleftharpoons H_3O^+ + OH^-$$

Ionisation of nitric acid is much enhanced in the presence of a strong proton acceptor, and in aqueous solution it behaves as a strong acid:

$$HNO_3(l) + H_2O(l) \rightleftharpoons H_3O^+(aq) + NO_3^-(aq)$$

Chemical properties of nitric acid

In aqueous solution, nitric acid shows many of the reactions of a strong acid, e.g. it displaces carbon dioxide from a carbonate and neutralises a base to form a salt and water. With the possible exception of magnesium under certain conditions, it will not react with metals to evolve hydrogen, because it is an oxidising agent.

Concentrated nitric acid decomposes photochemically and gradually assumes a yellow colour due to the presence of dissolved nitrogen dioxide:

$$4HNO_3(aq) \longrightarrow 4NO_2(g) + 2H_2O(l) + O_2(g)$$

The same reaction occurs, but at a much faster rate, when the acid is heated.

Concentrated nitric acid in the presence of concentrated sulphuric acid reacts with many organic compounds. For instance, glycerol (propane-1,2,3-triol) reacts with this mixture to give glyceryl trinitrate (nitroglycerine); the temperature must be kept below 30°C or the reaction will become uncontrollable:

$$\begin{array}{l} CH_2\text{—}OH \\ | \\ CH\text{—}OH \\ | \\ CH_2\text{—}OH \\ \text{Glycerol} \end{array} + 3HNO_3 \longrightarrow \begin{array}{l} CH_2\text{—}O\text{—}NO_2 \\ | \\ CH\text{—}O\text{—}NO_2 \\ | \\ CH_2\text{—}O\text{—}NO_2 \\ \text{Glyceryl trinitrate} \end{array} + 3H_2O$$

Glyceryl trinitrate is an explosive which detonates violently with slight shock. It decomposes exothermically, being transformed from a liquid to a disordered mixture of gases, i.e. there is a large decrease in free energy (p. 112). Dynamite is made by absorbing glyceryl trinitrate in Kieselguhr; it retains the explosive qualities of glyceryl trinitrate but is much safer to handle.

Benzene and its derivatives are nitrated by the same reagents, the reacting species being the nitronium ion, NO_2^+. It is thought that the NO_2^+ ion first attacks the benzene ring, with the formation of a positively charged complex; this then loses a proton to give the nitro compound, e.g.

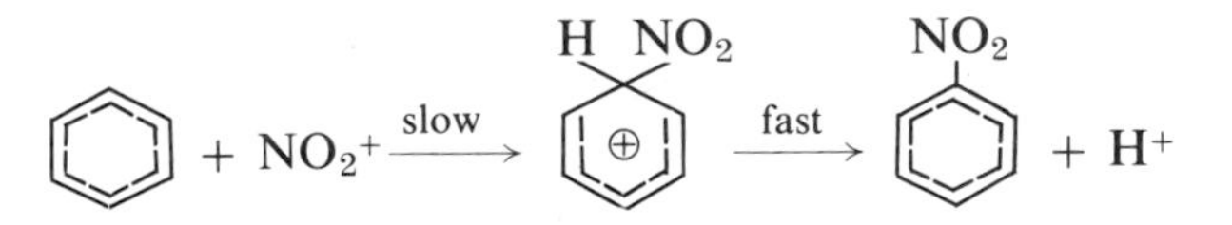

The purpose of the sulphuric acid is to increase the NO_2^+ concentration beyond that already present in the pure nitric acid; the following two ionisations have been postulated:

$$2HNO_3 \rightleftharpoons NO_2^+ + NO_3^- + H_2O \quad (1)$$
$$HNO_3 + H_2SO_4 \rightleftharpoons NO_2^+ + HSO_4^- + H_2O \quad (2)$$

If these equations are essentially correct the addition of nitrate ions, NO_3^-, should reduce the concentration of NO_2^+ by reversal of equation (1) above (the mass action effect). This is proved to be true, since nitration in the presence of nitrate ions is retarded. The existence of the NO_2^+ ion has been conclusively established in salts such as $NO_2^+ClO_4^-$.

The usual reduction products of nitric acid are nitrogen oxide (for the dilute acid) and nitrogen dioxide (for the more concentrated acid), although others are possible, such as dinitrogen oxide and ammonia. These redox reactions are discussed below.

(a) Oxidation of non-metals by nitric acid

Non-metals such as sulphur and phosphorus give their highest oxides when treated with concentrated nitric acid; these then react with the water produced to give acids:

$$S(s) + 6HNO_3(aq) \longrightarrow H_2SO_4(aq) + 6NO_2(g) + 2H_2O(l)$$
$$P(s) + 5HNO_3(aq) \longrightarrow H_3PO_4(aq) + 5NO_2(g) + H_2O(l)$$

The same type of reaction often occurs with weak metals; thus tin is oxidised to tin (IV) oxide:

$$Sn(s) + 4HNO_3(aq) \longrightarrow SnO_2(s) + 4NO_2(g) + 2H_2O(l)$$

(b) Oxidation of metals by nitric acid

Most metals react with nitric acid to form nitrates; nitrogen oxide tends to predominate when the acid is dilute, and nitrogen dioxide if the concentration of the acid is increased, e.g.

$$3Cu(s) + 8HNO_3(aq) \longrightarrow 3Cu^{2+}(aq) + 6NO_3^-(aq) + 4H_2O(l) + 2NO(g) \quad \text{(50 per cent acid)}$$

or
$$\begin{cases} 3Cu(s) \longrightarrow 3Cu^{2+}(aq) + 6e^- & \text{(oxidation)} \\ 8HNO_3(aq) + 6e^- \longrightarrow 6NO_3^-(aq) + 4H_2O(l) + 2NO(g) & \text{(reduction)} \end{cases}$$

$$Cu(s) + 4HNO_3(aq) \longrightarrow Cu^{2+}(aq) + 2NO_3^-(aq) + 2H_2O(l) + 2NO_2(g) \quad \text{(conc. acid)}$$

or
$$\begin{cases} Cu(s) \longrightarrow Cu^{2+}(aq) + 2e^- & \text{(oxidation)} \\ 4HNO_3(aq) + 2e^- \longrightarrow 2NO_3^-(aq) + 2H_2O(l) + 2NO_2(g) & \text{(reduction)} \end{cases}$$

In very rare instances, using very dilute nitric acid and a fairly electropositive metal, the reduction of nitric acid can proceed further to give dinitrogen oxide or hydrogen, e.g.

$$4Zn(s) + 10HNO_3(aq) \longrightarrow 4Zn^{2+}(aq) + 8NO_3^-(aq) + 5H_2O(l) + N_2O(g)$$
$$Mg(s) + 2HNO_3(aq) \longrightarrow Mg^{2+}(aq) + 2NO_3^-(aq) + H_2(g)$$

By employing Devarda's alloy (45 per cent Al, 5 per cent Zn, 50 per cent Cu) in alkaline solution, nitric acid and nitrates can be converted quantitatively into ammonia:

$$4Zn(s) + NO_3^-(aq) + 7OH^-(aq) + 6H_2O(l) \longrightarrow 4Zn(OH)_4^{2-}(aq) + NH_3(g)$$

Nitric acid of any concentration has no effect on very unreactive metals such as gold and platinum. The concentrated acid has no effect either on aluminium, iron or chromium; this is unexpected. It is thought that the surface of these three metals is first attacked, a thin impervious oxide film being formed which stops further action; the phenomenon is referred to as passivity.

(c) Oxidation of some compounds by nitric acid

Compounds of metals that exhibit variable valency can often be oxidised from a lower to a higher valency state with nitric acid; for instance an acidified solution of an iron (II) salt is converted into the iron (III) state by dilute nitric acid:

$$3Fe^{2+}(aq) + 4H^+(aq) + NO_3^-(aq) \longrightarrow 3Fe^{3+}(aq) + 2H_2O(l) + NO(g)$$

Other compounds oxidised by nitric acid include hydrogen sulphide and hydrogen chloride:

$$3H_2S(g) + 2HNO_3(aq) \longrightarrow 4H_2O(l) + 2NO(g) + 3S(s)$$
$$3HCl(g) + HNO_3(aq) \longrightarrow Cl_2(g) + NOCl(g) + 2H_2O(l)$$

The former reaction is a nuisance in qualitative analysis, since the formation of sulphur is easily mistaken for a metallic sulphide and is difficult to remove; nitric acid should be avoided here whenever possible. The latter reaction is the basis of the solvent action of aqua regia

(3 volumes of concentrated hydrochloric acid and 1 volume of concentrated nitric acid). Aqua regia is capable of dissolving gold and platinum, which are resistant to attack by nitric acid. It is thought that the metals are first oxidised by the nitric acid (the reaction being catalysed by the nitrosyl chloride NOCl) and the cations are removed as fast as they are formed by complexing with chloride ions, so preventing the reverse reaction, e.g.

$$Au(s) + 6HNO_3(aq) \rightleftharpoons Au^{3+}(aq) + 3NO_3^+(aq) + 3H_2O(l) + 3NO_2(g)$$

$$+$$

$$4Cl^-(aq) \text{ (from the conc. hydrochloric acid)}$$

$$\upharpoonleft\downharpoonright$$

$$AuCl_4^-(aq)$$

Uses of nitric acid

Nitric acid is used in the manufacture of nitrogenous fertilisers, e.g. ammonium nitrate and sodium nitrate. Reaction with organic compounds leads to the production of explosives, e.g. glyceryl trinitrate and trinitrotoluene, and intermediates used in the manufacture of dyes.

19.10 Nitrates

Metallic nitrates can be prepared by reacting the metal or its oxide, hydroxide or carbonate with nitric acid; they are all soluble in water. Their stability to heat generally accords well with the position of the particular metal in the electrode potential series. Sodium nitrate and potassium nitrate decompose on heating into the nitrite and oxygen, e.g.

$$2Na^+NO_3^-(s) \longrightarrow 2Na^+NO_2^-(s) + O_2(g)$$

The majority of other nitrates give nitrogen dioxide, oxygen and the metallic oxide, unless the latter is unstable to heat, when it decomposes into the metal and oxygen, e.g.

$$2Mg^{2+}(NO_3^-)_2(s) \longrightarrow 2Mg^{2+}O^{2-}(s) + 4NO_2(g) + O_2(g)$$

$$2Ag^+NO_3^-(s) \longrightarrow 2Ag(s) + 2NO_2(g) + O_2(g)$$

Ammonium nitrate gives dinitrogen oxide and water when cautiously heated:

$$NH_4^+NO_3^-(s) \longrightarrow N_2O(g) + 2H_2O(l)$$

Nitrates can be detected by the brown ring test with iron (II) sulphate solution and cold concentrated sulphuric acid. In the absence of ammonium ions, they can also be detected by the production of ammonia when they are reacted with Devarda's alloy in alkaline solution (p. 260).

The nitrate ion is planar and symmetrical, being a resonance hybrid of the three forms:

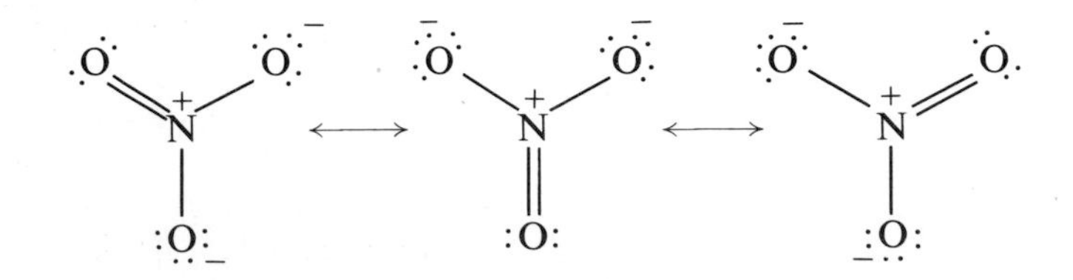

19.11 Nitrides

Like the carbides, nitrides can be classified as ionic, covalent and interstitial. Ionic nitrides are formed by lithium and the Group 2A elements, either by direct synthesis or, more often, by heating the metal in ammonia to form the amide which is then decomposed:

$$6Li(s) + N_2(g) \longrightarrow 2(Li^+)_3N^{3-}(s)$$

$$Ba(s) + 2NH_3(g) \longrightarrow Ba^{2+}(NH_2^-)_2(s) + H_2(g)$$
$$3Ba^{2+}(NH_2^-)_2(s) \longrightarrow (Ba^{2+})_3(N^{3-})_2(s) + 4NH_3(g)$$

They are high melting-point solids and readily hydrolysed by water to the hydroxide and ammonia:

$$N^{3-} + 3H_2O(l) \longrightarrow NH_3(g) + 3OH^-(aq)$$

Covalent nitrides are formed by a multitude of elements which are non-metallic. Examples include silicon nitride, Si_3N_4, and boron nitride, BN, which are giant molecules. Boron nitride exists in the 'graphite' and 'diamond' forms (p. 79–80).

Nitrogen atoms are able to fit into the spaces between transition metal atoms, without causing too much distortion of the basic metal lattice, forming non-stoichiometric interstitial nitrides. Like the corresponding carbides, these nitrides are extremely hard, very inert, and possess electrical conductivity. They are generally prepared by heating the finely divided metal in a stream of nitrogen or ammonia at a temperature in excess of 1000°C.

Phosphorus

19.12 Occurrence of phosphorus and its extraction

Calcium phosphate (V) occurs as apatite, $Ca^{2+}(F^-)_2.3(Ca^{2+})_3(PO_4^{3-})_2$, and rock phosphate, $(Ca^{2+})_3(PO_4^{3-})_2$, which is formed by the weathering of apatite. Large deposits occur in the USA, North Africa and in Russia. Bones contain about 58 per cent of calcium phosphate (V) in addition to calcium carbonate, fat and organic matter containing nitrogen; phosphorus compounds are also present in animal and vegetable tissues.

Phosphorus is extracted by heating a mixture of phosphate rock, silica and coke in an electric furnace to a temperature of about 1500°C (fig. 19.3). The more volatile phosphorus (V) oxide, P_4O_{10}, is first displaced from calcium phosphate (V) by the involatile silica, SiO_2, and subsequently reduced to white phosphorus with coke.

The phosphorus distils over as a vapour in a stream of carbon monoxide and is condensed under water. The slag of calcium silicate is periodically tapped off from the furnace.

$$2(Ca^{2+})_3(PO_4^{3-})_2(s) + 6SiO_2(s) \longrightarrow 6Ca^{2+}SiO_3^{2-}(s) + P_4O_{10}(s)$$
$$P_4O_{10}(s) + 10C(s) \longrightarrow P_4(g) + 10CO(g)$$

White phosphorus is stored and transported under water, since it readily inflames in air once it becomes dry.

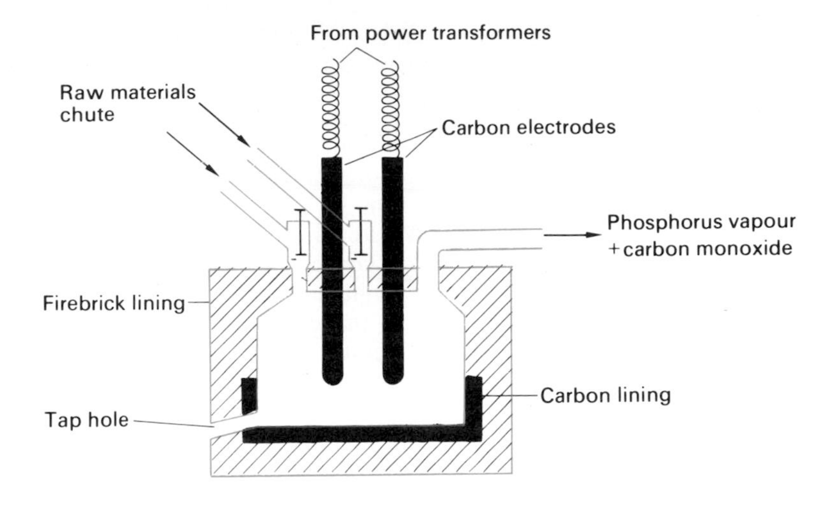

FIG. 19.3. *Electric furnace for extracting phosphorus*

19.13 The allotropy of phosphorus

The allotropy of phosphorus is rather complex but, essentially, there are three allotropic forms known as white, red and black phosphorus.

White phosphorus is formed as a soft, waxy solid whenever phosphorus vapour is condensed; structurally it contains tetrahedral P_4 units held together by van der Waals' forces (fig. 19.1). Since the P—P—P bond angles are 60° in each of these P_4 units there is a considerable amount of strain, and this makes itself felt in the high chemical reactivity of this allotrope.

White phosphorus very slowly changes into the red variety in the course of many years; this change can be accelerated by raising the temperature, and commercially this allotrope is made from white phosphorus by heating in the absence of air to 270°C for several days. Its structure is not known with certainty but it is certainly macromolecular; it is denser than white phosphorus.

The third allotrope, black phosphorus, can be obtained by subjecting white phosphorus to high pressures at 200°C in the absence of air. It too is macromolecular, with each phosphorus atom surrounded by three more atoms. It is an electrical conductor resembling graphite in this respect and also in its flakiness. Its density is higher than that of red phosphorus.

Only white and red phosphorus are normally encountered in the laboratory, their more important differences being shown in Table 19A.

Table 19A The important differences between white and red phosphorus

White phosphorus	Red phosphorus
Yellow waxy solid	Red brittle powder
Density = 1·8 g cm^{-3}	Density = 2·2 g cm^{-3}
Melting point = 44°C	Melting point = 590°C (under pressure)
Boiling point = 280°C	Sublimes at 400°C (1 atm pressure)
Toxic	Non-toxic
Soluble in carbon disulphide	Insoluble in carbon disulphide
Ignites in air at 35°C	Ignites in air at 260°C
Ignites in chlorine	Does not ignite in chlorine unless heated
Reacts with hot sodium hydroxide solution to produce phosphine	Does not react with hot sodium hydroxide solution

White phosphorus has much lower melting and boiling points than red phosphorus, because very little energy is required to sever the weak bonds between the individual P_4 units in the former allotrope. Similarly, white phosphorus is easily dispersed as P_4 in carbon disulphide, but this solvent is, of course, ineffectual in breaking down the strong macromolecular structure of red phosphorus. The toxicity of the white allotrope is presumably due to its very much higher vapour pressure.

White phosphorus is less dense than the red variety because its structure is less compact. The more open structure of the white form also accounts for its higher chemical reactivity, which is further enhanced by the fact that considerable strain is present in the P_4 units (fig. 19.1); this is relieved by chemical reaction.

19.14 The hydrides of phosphorus

Phosphine, PH_3, the phosphorus analogue of ammonia, can be obtained by heating a concentrated solution of sodium hydroxide with white phosphorus:

$$P_4(s) + 3OH^-(aq) + 3H_2O(l) \longrightarrow PH_3(g) + \underset{\text{phosphinate}}{3H_2PO_2^-(aq)}$$

It is evolved as a very poisonous gas with an unpleasant smell. As made by the above method it is spontaneously inflammable, forming phosphorus (V) oxide on contact with air; this is due to the presence of P_2H_4 or phosphorus vapour impurities. When the gas is pure it is not spontaneously inflammable, but it will readily oxidise to phosphorus (V) oxide when ignited:

$$4PH_3(g) + 8O_2(g) \longrightarrow P_4O_{10}(s) + 6H_2O(l)$$

Phosphine can also be obtained by hydrolysis of calcium phosphide (cf. the hydrolysis of ionic nitrides (p. 262)):

$$P^{3-} + 3H_2O(l) \longrightarrow PH_3(g) + 3OH^-(aq)$$

The phosphine molecule is pyramidal like the molecule of ammonia but, unlike ammonia, it is only slightly soluble in water. This is probably due to the fact that the phosphorus atom is not sufficiently electronegative to hydrogen bond to the hydrogen of a water molecule. The presence of hydrogen bonding in ammonia and its absence in phosphine also explains why phosphine, which is a heavier molecule than ammonia, has a lower boiling point (p. 67).

The lone pair of electrons of the phosphine molecule can bond to a proton to form the phosphonium ion, PH_4^+, which is tetrahedral like the ammonium ion, e.g.

$$PH_3(g) + HI(g) \longrightarrow \underset{\text{phosphonium iodide}}{PH_4^+I^-(s)}$$

Phosphonium iodide, which is one of the most stable salts containing the phosphonium ion, is less thermally stable than ammonium salts because phosphine is less basic than ammonia, and decomposition occurs at about 60°C:

$$PH_4^+I^-(s) \longrightarrow PH_3(g) + HI(g)$$

Unlike ammonium salts it is also decomposed by water and this is a convenient method of making pure phosphine:

$$PH_4^+I^-(s) + H_2O(l) \longrightarrow PH_3(g) + H_3O^+(aq) + I^-(aq)$$

Phosphine is a much more powerful reducing agent than ammonia, converting silver and copper (II) salts in solution to their phosphides, which subsequently react to give the free metals, e.g.

$$6Ag^+(aq) + PH_3(g) + 3H_2O(l) \longrightarrow 6Ag(s) + H_3PO_3(aq) + 6H^+(aq)$$

Diphosphine, P_2H_4, the phosphorus analogue of hydrazine, is formed in trace amounts during the preparation of phosphine. As mentioned above it is spontaneously inflammable; it is also devoid of basic properties.

19.15 The halides of phosphorus

Whereas nitrogen is restricted to a covalency of 3, phosphorus can also form compounds in which it is 5-covalent. This is possible since phosphorus has *d* orbitals available, and promotion of one 3*s* electron to the 3*d* level results in five unpaired electrons:

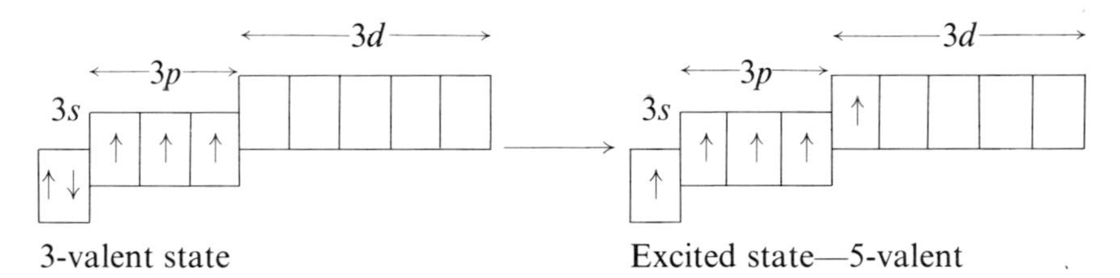

The trihalides, PX_3

All the four trihalides exist and, except for the trifluoride, can be made by direct synthesis. The trifluoride is a gas, the trichloride and tribromide are liquids, and the tri-iodide is a solid. Their molecules have the expected pyramidal structure.

The trichloride, which is the most important of these trihalides, is obtained by passing chlorine over white phosphorus. The phosphorus burns with a pale green flame and phosphorus trichloride distils and is condensed as a colourless liquid. Since it is attacked by air and water, it is necessary to displace the air from the apparatus with a stream of carbon dioxide and to include a soda-lime drying tube.

$$P_4(s) + 6Cl_2(g) \longrightarrow 4PCl_3(l)$$

Phosphorus trichloride is readily hydrolysed by water to phosphonic acid, H_3PO_3, and hydrogen chloride:

$$PCl_3(l) + 3H_2O(l) \longrightarrow H_3PO_3(aq) + 3HCl(g)$$

It is thought that the reaction takes place in stages, with the formation of complexes in which the oxygen atom of a water molecule is attached to the phosphorus atom (expansion of the octet can occur since the phosphorus atom has *d* orbitals available):

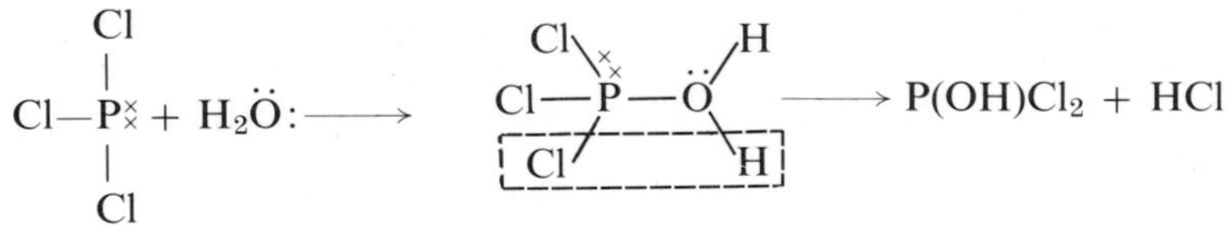

$$PCl_3 + H_2\ddot{O}: \longrightarrow Cl_3P\text{—}\ddot{O}H_2 \longrightarrow P(OH)Cl_2 + HCl$$

$$P(OH)Cl_2 + H_2O \longrightarrow P(OH)_2Cl + HCl$$
$$P(OH)_2Cl + H_2O \longrightarrow P(OH)_3 + HCl$$

(This should be compared with the hydrolysis of silicon tetrachloride where *d* orbitals are also available, and contrasted with the hydrolysis of nitrogen trichloride where they are not.)

Phosphorus trichloride reacts with many compounds containing the —OH group, and it is used in organic chemistry for the preparation of acid chlorides and alkyl chlorides, e.g.

$$3CH_3COOH(l) + PCl_3(l) \longrightarrow 3CH_3COCl + H_3PO_3$$
$$3C_2H_5OH(l) + PCl_3(l) \longrightarrow 3C_2H_5Cl + H_3PO_3$$

It readily combines with oxygen and chlorine (reversibly), the phosphorus atom increasing its covalency from three to five:

$$2PCl_3(l) + O_2(g) \longrightarrow \underset{\text{phosphorus trichloride oxide}}{2POCl_3(l)}$$
$$PCl_3(l) + Cl_2(g) \rightleftharpoons PCl_5(s)$$

The pentahalides, PX_5

Except for the pentaiodide, all the pentahalides have been prepared. Phosphorus pentafluoride is a gas; the pentachloride and pentabromide are solids.

Phosphorus pentachloride is prepared by passing chlorine through a flask into which phosphorus trichloride is dripping. Since it dissociates into the trichloride and chlorine very readily, the experiment is conducted in an ice-cooled apparatus.

$$PCl_3(l) + Cl_2(g) \rightleftharpoons PCl_5(s)$$

Like the trichloride it is attacked by compounds containing the hydroxyl group, e.g.

$$PCl_5(s) + H_2O(l) \longrightarrow POCl_3(l) + 2HCl(g)$$
$$POCl_3(l) + 3H_2O(l) \longrightarrow H_3PO_4(aq) + 3HCl(g)$$
$$CH_3COOH(l) + PCl_5(s) \longrightarrow CH_3COCl(l) + POCl_3(l) + HCl(g)$$

In the vapour state the phosphorus pentachloride molecule has a trigonal bipyramidal structure (fig. 19.4(a)); in the solid state it is ionic, having the structure $(PCl_4^+)(PCl_6^-)$ (fig. 19.4(b)).

FIG. 19.4.
(a) The structure of phosphorus pentachloride vapour
(b) The structure of solid phosphorus pentachloride

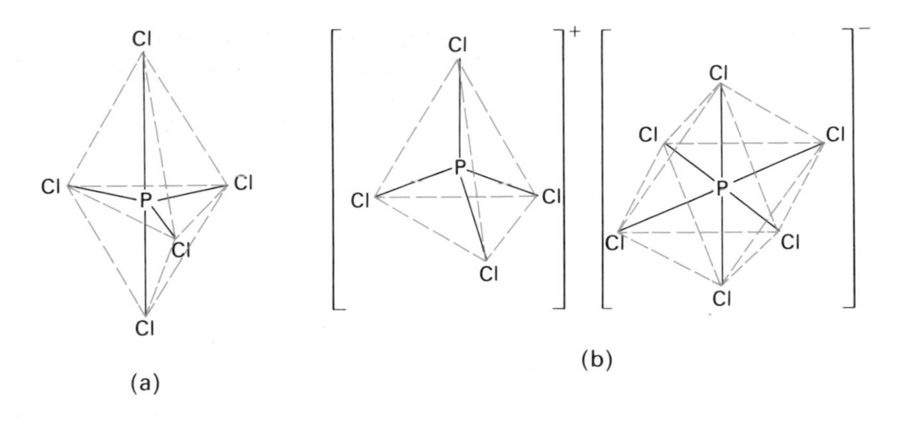

Phosphorus pentabromide is also ionic in the solid state, but its structure is $(PBr_4^+)Br^-$. This has been partially attributed to the impossibility of packing six bromines round a central phosphorous atom to give a stable PBr_6^- ion—a steric effect. Steric hindrance is also a possible explanation of why PI_5 does not exist—the iodine atom being larger than the bromine atom.

19.16 The oxides of phosphorus, P_4O_6 and P_4O_{10}

Phosphorus (III) oxide, P_4O_6

Phosphorus (III) oxide, once thought to be P_2O_3 but now known to be the dimer, is obtained as a white solid (m.p. 23·8°C) by burning phosphorus in a limited amount of air:

$$P_4(s) + 3O_2(g) \longrightarrow P_4O_6(s)$$

FIG. 19.5. *The structure of phosphorus (III) oxide*

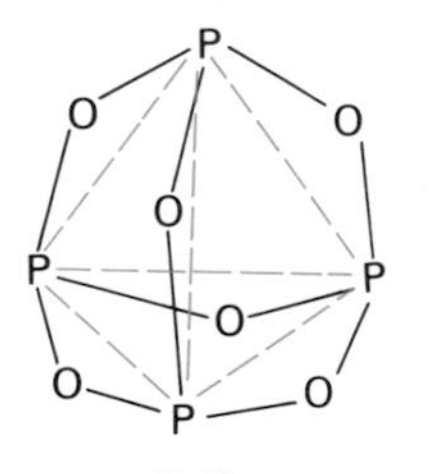

It can be freed from small amounts of phosphorus (V) oxide, which are also formed, by making use of its higher volatility.

Phosphorus (III) oxide reacts readily with oxygen when heated, to give phosphorus (V) oxide, and produces phosphonic acid with water:

$$P_4O_6(s) + 2O_2(g) \longrightarrow P_4O_{10}(s)$$
$$P_4O_6(s) + 6H_2O(l) \longrightarrow 4H_3PO_3(aq)$$

Its relative molecular mass in solution and in the vapour state corresponds with the formula P_4O_6. This structural unit is based on the tetrahedral P_4 molecule, each oxygen atom bridging two phosphorus atoms (fig. 19.5).

Phosphorus (V) oxide, P_4O_{10}

Phosphorus (V) oxide is formed as a white solid when phosphorus is burnt in an excess of air:

$$P_4(s) + 5O_2(g) \longrightarrow P_4O_{10}(s)$$

It reacts with water, forming polytrioxophosphoric acid (empirical formula HPO_3) which on boiling gives phosphoric (V) acid, H_3PO_4:

$$P_4O_{10}(s) + 2H_2O(l) \longrightarrow 4HPO_3(aq)$$

$$HPO_3(aq) + H_2O(aq) \longrightarrow H_3PO_4(aq)$$

FIG. 19.6. *The structure of phosphorus (V) oxide*

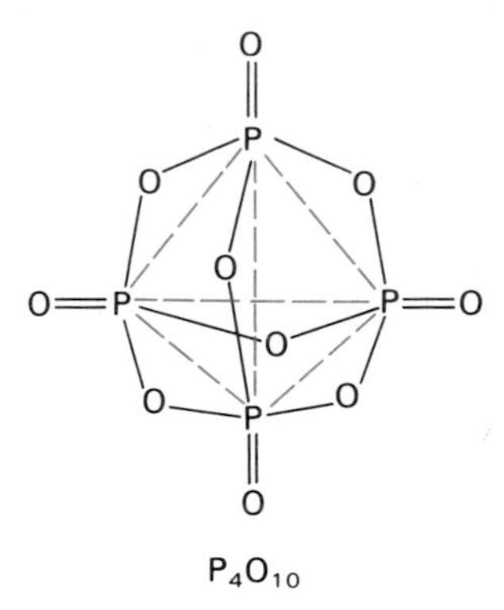

Because of its high affinity for water it is often used as a drying agent. It will also remove the elements of water from compounds and has been used for dehydrating nitric acid to dinitrogen pentoxide, and for converting acid amides to acid nitriles, e.g.

$$2HNO_3 \xrightarrow{-H_2O} N_2O_5$$

$$\underset{\text{ethanamide}}{CH_3CONH_2} \xrightarrow{-H_2O} \underset{\text{ethanonitrile}}{CH_3C{\equiv}N}$$

Phosphorus (V) oxide exists in a number of polymorphic forms, at least three of which are crystalline. One of these crystalline forms is composed of P_4O_{10} units which are structurally related to those of phosphorus (III) oxide, except that each phosphorus atom is bonded to an extra oxygen. Because these bonds are shorter than the other P—O bonds it is thought that double bonding is involved, i.e. phosphorus expands its octet instead of datively bonding to oxygen by means of its lone pair (fig. 19.6):

19.17 The oxyacids of phosphorus

The main oxyacids of phosphorus are the phosphonic acids and the phosphoric (V) acids, of which phosphorus (III) oxide and phosphorus (V) oxide respectively are the acid anhydrides. Other oxyacids of phosphorus exist, but they are not directly based on these oxides and will not be considered here.

Phosphonic acid, H_3PO_3

This is the most important acid based on phosphorus (III) oxide, and can be obtained from the latter by the addition of water:

$$P_4O_6(s) + 6H_2O(l) \longrightarrow 4H_3PO_3(aq)$$

It can also be obtained by hydrolysis of phosphorus trichloride:

$$PCl_3(l) + 3H_2O(l) \longrightarrow H_3PO_3(aq) + 3HCl(g)$$

The pure acid is a deliquescent solid and in aqueous solution it is moderately strong. Since it is dibasic its structure is thought to contain only two hydroxyl groups, although its formation by hydrolysis of phosphorus trichloride suggests three. It is thought that an equilibrium exists of the type shown below, which is very largely displaced to the right:

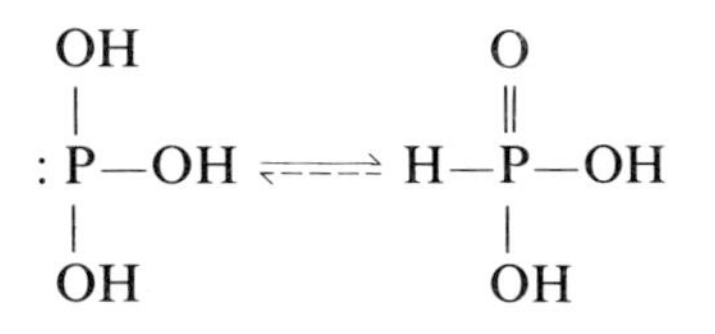

The phosphorus atom is therefore essentially 5-covalent.

The alkali metal salts of phosphonic acid, e.g. $Na^+H_2PO_3^-$ and $(Na^+)_2HPO_3^{2-}$, are soluble in water; the majority of other salts are insoluble. The acid and its salts are strong reducing agents, the free acid decomposing into phosphine—which is a powerful reducing agent itself—and phosphoric (V) acid when heated:

$$4H_3PO_3(s) \longrightarrow PH_3(g) + 3H_3PO_4(s)$$

The phosphoric acids and phosphates

A variety of phosphoric acids based on phosphorus (V) oxide exists; in many respects the structural chemistry of these acids and their anions is similar to that of the silicates (p. 230–231).

Phosphoric (V) acid, H_3PO_4, and phosphates (V)

Phosphoric (V) acid is manufactured by heating rock phosphate with sulphuric acid; a purer product is obtained by reacting phosphorus (V) oxide with water and then boiling.

$$(Ca^{2+})_3(PO_4^{3-})_2(s) + 3H_2SO_4(aq) \longrightarrow 3Ca^{2+}SO_4^{2-}(s) + 2H_3PO_4(aq)$$
$$P_4O_{10}(s) + 6H_2O(l) \longrightarrow 4H_3PO_4(aq)$$

Phosphoric (V) acid is a deliquescent crystalline solid, although it is usually encountered as a concentrated and very viscous aqueous solution; this high viscosity is due to the presence of hydrogen bonding which links individual molecules of the acid into very large aggregates.

The acid is tribasic, and relatively weak, neutralisation with alkalis taking place in stages. When an M/10 solution of sodium hydroxide is added to an M/10 solution of the acid, methyl orange changes colour at the stage corresponding to conversion into sodium dihydrogen phosphate (V), $Na^+H_2PO_4^-$:

$$Na^+OH^-(aq) + H_3PO_4(aq) \longrightarrow H_2O(l) + Na^+H_2PO_4^-(aq) \text{ (pH 4·4)}$$

If phenolphthalein is used as indicator, twice as much alkali is needed before the indicator changes colour and this corresponds to the formation of disodium hydrogen phosphate (V), $(Na^+)_2HPO_4^{2-}$:

$$2Na^+OH^-(aq) + H_3PO_4(aq) \longrightarrow 2H_2O(l) + (Na^+)_2HPO_4^{2-}(aq) \text{ (pH 9·6)}$$

The third stage leading to the formation of trisodium phosphate (V), $(Na^+)_3PO_4^{3-}$, cannot be realised in practice since this salt is extensively hydrolysed in aqueous solution, i.e. the phosphate (V) ion, PO_4^{3-}, is a strong base:

$$(Na^+)_3PO_4^{3-}(aq) + H_2O(l) \rightleftharpoons Na^+OH^-(aq) + (Na^+)_2HPO_4^{2-}(aq)$$

or

$$PO_4^{3-}(aq) + H_2O(l) \rightleftharpoons OH^-(aq) + HPO_4^{2-}(aq)$$

Most phosphates (V) are insoluble in water but they usually dissolve in the presence of a strong acid such as hydrochloric acid. This is due to the formation of weakly dissociated phosphoric (V) acid which allows the equilibrium below to move over to the right, e.g.

$$(Ca^{2+})_3(PO_4^{3-})_2(s) \rightleftharpoons 3Ca^{2+}(aq) + 2PO_4^{3-}(aq)$$

$$+$$
$$6H^+(aq) \text{ from hydrochloric acid}$$
$$\updownarrow$$
$$2H_3PO_4(aq)$$

FIG. 19.7.
(a) The structure of phosphoric (V) acid
(b) The structure of the dihydrogen phosphate (V) anion
(c) The structure of the hydrogen phosphate (V) anion
(d) The structure of the phosphate (V) anion

Phosphoric (V) acid has an approximately tetrahedrally shaped molecule (fig. 19.7(a)); the phosphate (V) anions also adopt a tetrahedral configuration (fig. 19.7(b), (c) and (d)) and in these instances resonance is involved.

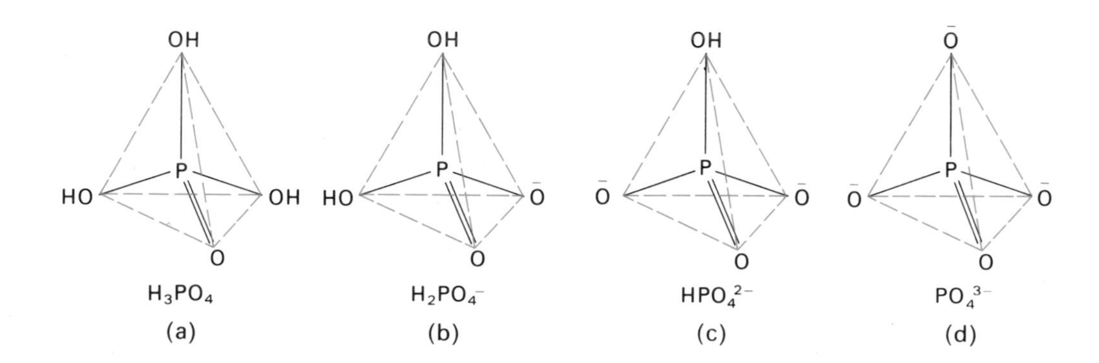

Disphosphoric (V) acid, $H_4P_2O_7$, diphosphates (V) and other chain polyphosphates (V)

Diphosphoric (V) acid can be obtained by heating phosphoric (V) acid to 220°C, when two molecules condense, with the elimination of water:

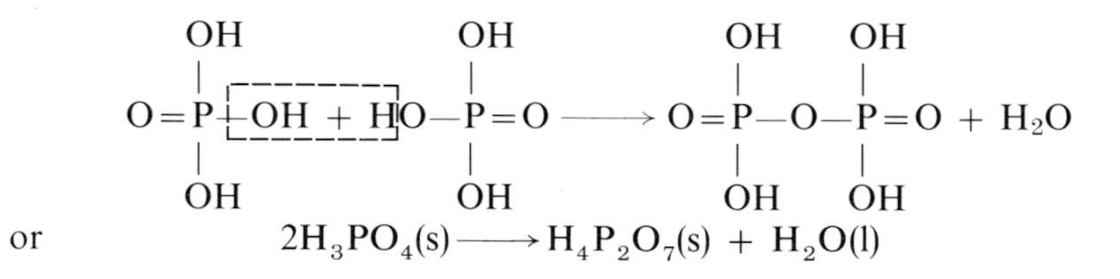

or $$2H_3PO_4(s) \longrightarrow H_4P_2O_7(s) + H_2O(l)$$

Although diphosphoric (V) acid is tetrabasic only two sodium salts are known:

$$(Na^+)_4P_2O_7^{4-} \text{ and } (Na^+)_2H_2P_2O_7^{2-}$$

The tetrasodium salt is obtained when disodium hydrogen phosphate (V) is heated at 500°C:

$$2(Na^+)_2HPO_4^{2-}(s) \longrightarrow (Na^+)_4P_2O_7^{4-}(s) + H_2O(l)$$

or $$O{=}P(O^-)_2{-}OH + HO{-}P(O^-)_2{=}O \longrightarrow O{=}P(O^-)_2{-}O{-}P(O^-)_2{=}O + H_2O$$

FIG. 19.8.
(a) The structure of diphosphoric (V) acid
(b) The structure of the diphosphate (V) anion
(c) The structure of the triphosphate (V) anion

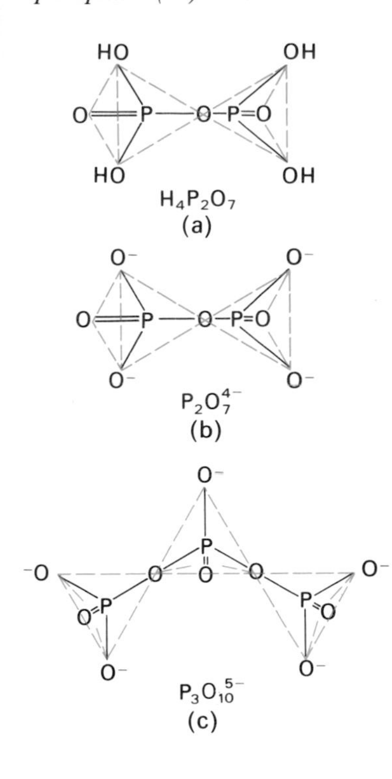

The disodium salt is conveniently obtained by heating sodium dihydrogen phosphate (V) at 200°C:

$$2Na^+H_2PO_4^-(s) \longrightarrow (Na^+)_2H_2P_2O_7^{2-}(s) + H_2O(l)$$

or

$$O{=}\underset{O^-}{\overset{OH}{P}}{-}OH + HO{-}\underset{O^-}{\overset{OH}{P}}{=}O \longrightarrow O{=}\underset{O^-}{\overset{OH}{P}}{-}O{-}\underset{O^-}{\overset{OH}{P}}{=}O + H_2O$$

Diphosphoric (V) acid (fig. 19.8(a)) and its anions (fig. 19.8(b)) have a structure based on tetrahedral PO_4 units linked together by an oxygen atom common to both.

The linking together of more PO_4 tetrahedral units gives rise to linear (or chain) polyphosphates (V), e.g. the tripolyphosphate (V) anion $P_3O_{10}^{5-}$ (V) contains three PO_4 units (fig. 19.8(c)):
Chain polyphosphates (V) are able to form water soluble complexes with many metals and are used in the softening of hard water. They complex with calcium and magnesium ions and so prevent them forming a scum with soap.

Polytrioxophosphoric (V) acid, HPO_3, and polytrioxophosphates (V)
Polytrioxophosphoric (V) acid is obtained by the dehydration of phosphoric (V) acid at 316°C. It is a glassy polymeric solid and a mixture of several acids, with the empirical formula HPO_3.

$$H_3PO_4(s) \longrightarrow HPO_3(s) + H_2O(l)$$

Although the structures of the various acids are still uncertain, mixtures of polytrioxophosphates (V) have been separated by chromatographic methods. These ions are built up from PO_4 units and are cyclic as opposed to linear. The $P_3O_9^{3-}$ ion, is shown below:

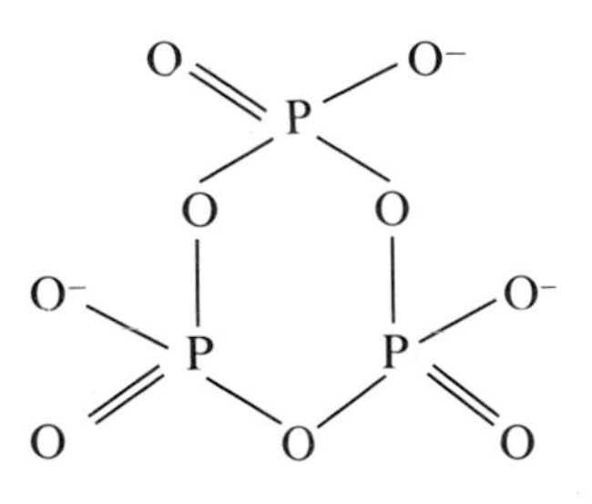

Sodium polytrioxophosphate (V), empirical formula $Na^+PO_3^-$, is formed when sodium dihydrogen phosphate (V) is heated:

$$Na^+H_2PO_4^-(s) \longrightarrow Na^+PO_3^-(s) + H_2O(l)$$

It is a mixture, containing ions of differing ring size.

Tests for phosphates (V)

Phosphates (V) give a yellow precipitate of ammonium phosphomolybdate when warmed to about 60°C with nitric acid and ammonium molybdate solution. This same test is also given by condensed phosphates (V), since hydrolysis to phosphates (V) takes place on warming; precipitation of ammonium phosphomolybdate in these cases, however, is a much slower process.

Phosphates (V) precipitate magnesium ammonium phosphate (V) when added to an ammoniacal solution of magnesium chloride. If this precipitate is washed, filtered and ignited it forms magnesium diphosphate (V). This can be weighed and thus phosphates (V) can be estimated quantitatively.

$$2Mg^{2+}(NH_4^+)PO_4^{3-}(s) \longrightarrow (Mg^{2+})_2P_2O_7^{4-}(s) + 2NH_3(g) + H_2O(l)$$

Some important phosphates

All naturally occurring phosphates are phosphates (V), the most abundant of these being rock phosphate, $(Ca^{2+})_3(PO_4^{3-})_2$, the source of phosphorus and other phosphates. Rock phosphate is mostly consumed by the fertiliser industry in the manufacture of 'superphosphate of lime', 'triple superphosphate' and 'nitrophos'—a combined phosphatic and nitrogenous fertiliser.

$$(Ca^{2+})_3(PO_4^{3-})_2(s) + 2H_2SO_4(aq) \longrightarrow \underbrace{Ca^{2+}(H_2PO_4^-)_2(aq) + 2Ca^{2+}SO_4^{2-}(s)}_{\text{'Superphosphate of lime'}}$$

$$(Ca^{2+})_3(PO_4^{3-})_2(s) + 4H_3PO_4(aq) \longrightarrow \underbrace{3Ca^{2+}(H_2PO_4^-)_2(aq)}_{\text{'Triple superphosphate'}}$$

$$(Ca^{2+})_3(PO_4^{3-})_2(s) + 4HNO_3(aq) \longrightarrow \underbrace{Ca^{2+}(H_2PO_4^-)_2(aq) + 2Ca^{2+}(NO_3^-)_2(aq)}_{\text{'Nitrophos'}}$$

Other phosphatic fertilisers are ammonium dihydrogen phosphate (V) and diammonium hydrogen phosphate (V), which also counteract nitrogen deficiency.

Commercially important sodium phosphates include disodium hydrogen diphosphate (V)—an ingredient in baking powder—and disodium hydrogen phosphate (V), tetrasodium diphosphate (V) and sodium polytrioxophosphate (V)—used in the processing of cheese. Condensed phosphates of sodium are used in the softening of water (p. 295).

Organic phosphates are intimately bound up with life processes, being involved in photosynthesis and protein synthesis. The latter process is still poorly understood but it is thought that deoxyribonucleic acid (DNA) provides the information as to how the amino acids are linked together into protein chains, another organic phosphate called ribonucleic acid (RNA) carrying this information to the protein 'factories'.

19.18 Polymers containing phosphorus and nitrogen

When a mixture of phosphorus pentachloride and ammonium chloride is fused, a mixture of compounds with the empirical formula $PNCl_2$ is obtained. This same mixture of compounds can also be obtained by refluxing in an inert organic solvent.

$$nPCl_5 + nNH_4^+Cl^- \longrightarrow (PNCl_2)_n + 4nHCl$$

The lower members of this mixture have been separated by chromatographic methods and have cyclic structures, e.g. $(PNCl_2)_3$ which has the structure shown below:

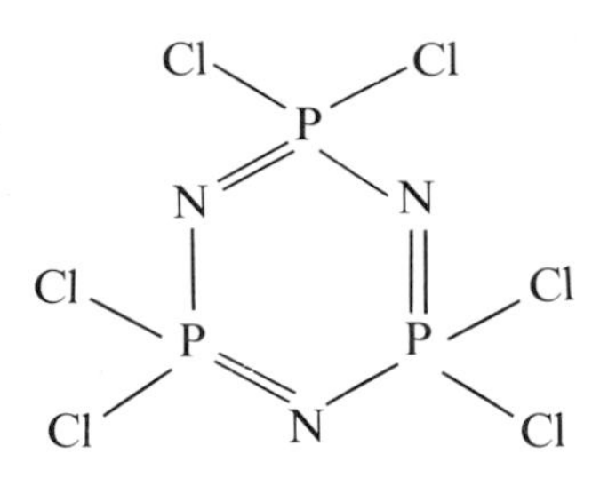

The highest members are linear polymers, the two free ends of the chain being 'stopped' by phosphorus pentachloride fragments (depicted by X and Y in the diagram below):

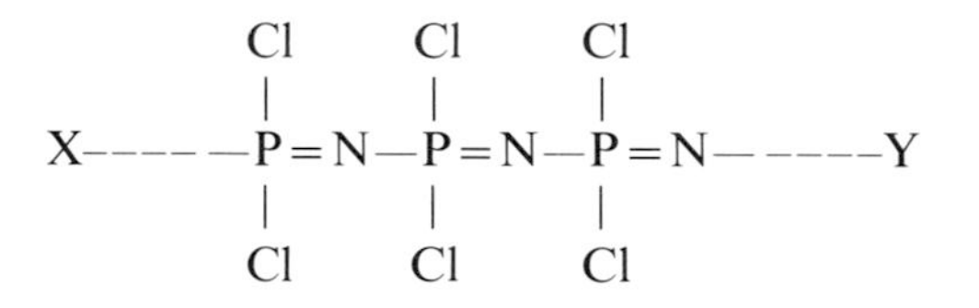

The linear polymers are rubberlike; unfortunately they are not very stable.

19.19 Uses of phosphorus and its compounds

The annual production of phosphorus in this country is about 24 000 tonnes, of which 90 per cent is converted into phosphoric (V) acid. Another important application of phosphorus is for the preparation of phosphorus chlorides, thiophosphoryl chloride, $PSCl_3$, and phosphorus pentasulphide, P_2S_5, which are used in the manufacture of insecticides, plasticisers and oil additives. Smaller quantities are used in the production of the alloy phosphor-bronze and in matches.

Phosphoric (V) acid is used in the rustproofing of steel and in the soft drink industry but its greatest outlet is in the preparation of phosphates (p. 269). When purity of product is the prime consideration, i.e. in the foodstuffs industry, phosphoric (V) acid is produced from phosphorus via phosphorus (V) oxide, otherwise the cheaper route from rock phosphate and sulphuric acid is employed.

Arsenic, antimony and bismuth

19.20 Occurrence, extraction and uses of arsenic, antimony and bismuth

All three elements occur as sulphides of the general formula M_2S_3, although other ores are known. They can be extracted by roasting in air to form the oxides which are subsequently reduced with carbon:

$$2M_2S_3(s) + 9O_2(g) \longrightarrow 2M_2O_3(s) + 6SO_2(g)$$
$$M_2O_3(s) + 3C(s) \longrightarrow 2M(s) + 3CO(g)$$

The uses of arsenic are dependent on its toxicity, e.g. arsenical insecticides; antimony is used for hardening metals such as tin and lead, e.g. the lead plates in accumulators, and bismuth is used in fusible alloys, e.g. Wood's metal which has a melting point of about 70°C (an alloy of bismuth, lead, tin and cadmium).

19.21 Properties of arsenic, antimony and bismuth

The elements combine on heating with oxygen, sulphur and the halogens:

$$4M(s) + 3O_2(g) \longrightarrow M_4O_6(s)\ (2Bi_2O_3)$$
$$2M(s) + 3S(s) \longrightarrow M_2S_3(s)$$
$$2M(s) + 3X_2 \longrightarrow 2MX_3\ (X_2 \text{ is a halogen})$$

Arsenic also forms some pentafluoride with fluorine and antimony some pentahalide with fluorine and chlorine.

Concentrated sulphuric acid attacks the elements, sulphur dioxide being liberated; antimony and bismuth form the sulphates, whereas arsenic, being less metallic, gives the oxide As_4O_6. The elements are also attacked by nitric acid with the evolution of oxides of nitrogen, but in this instance only bismuth forms the salt; arsenic and antimony form oxides.

Arsenic alone is attacked by fused alkali to give the arsenate (III):

$$2As(s) + 6OH^- \longrightarrow 2AsO_3^{3-}(s) + 3H_2(g)$$

19.22 The structures of arsenic, antimony and bismuth

Arsenic and antimony exhibit allotropy, an unstable form of each element, structurally similar to that of white phosphorus (p. 263), being formed by rapid condensation of their vapours. This unstable form is readily transformed into a denser allotrope which is metallic and similar in structure to that of black phosphorus (p. 263). Bismuth does not exhibit allotropy and adopts this latter structure.

As in the case of germanium, tin and lead in Group 4B, the change over from non-metallic to metallic nature with increasing atomic number is reflected in the structures of this group of elements.

Compounds of arsenic, antimony and bismuth

Like phosphorus, arsenic, antimony and bismuth exhibit covalencies of three and five although the stability of the higher valency state decreases with increasing atomic number, i.e. as the elements become more metallic. In addition, antimony and bismuth can display an electrovalency of 3 (the inert pair effect, p. 63).

19.23 The hydrides of arsenic, antimony and bismuth

Arsenic, antimony and bismuth form gaseous hydrides of the general formula MH_3, which are increasingly unstable in this order.

Arsine, AsH_3, and stibine, SbH_3, are conveniently obtained by reduction of the corresponding trichloride with lithium aluminium hydride, e.g.

$$4AsCl_3(l) + 3Li^+AlH_4^- \longrightarrow 4AsH_3(g) + 3Li^+Cl^-(s) + 3AlCl_3(s)$$

They can also be obtained by reduction of arsenic and antimony compounds with zinc and dilute sulphuric acid.

In this procedure, arsenic-free zinc is added to dilute sulphuric acid; hydrogen is evolved and is passed through a horizontal glass tube. The arsenic-containing compound is then added to the acid and reduction to arsine occurs:

$$AsO_4^{3-}(aq) + 4Zn(s) + 11H^+(aq) \longrightarrow AsH_3(g) + 4Zn^{2+}(aq) + 4H_2O(l)$$

The horizontal glass tube is heated and decomposition of arsine into arsenic and hydrogen occurs, the arsenic appearing as a shiny black mirror just beyond the heated area. Under carefully controlled conditions the intensity of the arsenic deposit can be used to estimate the amount of arsenic in compounds. Antimony compounds give a similar black mirror of antimony under the same conditions, but the two elements can be differentiated since arsenic is soluble in sodium chlorate (I) solution (oxidation to arsenate (V) occurs), whereas antimony is not:

$$2As(s) + 5OCl^-(aq) + 3H_2O(l) \longrightarrow 2H_3AsO_4(aq) + 5Cl^-(aq)$$

Like phosphine, arsine and stibine are powerful reducing agents; for example, they will reduce silver and copper (II) compounds in solution to their respective metals, e.g.

$$6Ag^+(aq) + AsH_3(g) + 3H_2O(l) \longrightarrow 6Ag(s) + H_3AsO_3(aq) + 6H^+(aq)$$

Arsine, which is exceedingly poisonous, and stibine, which is only slightly less so, have pyramidal molecules like ammonia and phosphine; the H–M–H bond angle becomes smaller with increasing relative molecular mass:

NH_3	PH_3	AsH_3	SbH_3
106°45′	93°50′	91°35′	91°30′

This has been explained as being due to a decrease in electronegativity of the central atom with increasing atomic number, resulting in the lone pair of electrons being less firmly bound. As a consequence of this, the repulsion between the lone pair-bonding pair electrons becomes more pronounced and so decreases the bond angle.

It was noted earlier in the chapter (p. 264) that phosphine was less basic than ammonia; this trend continues with arsine and stibine, which are even less basic than phosphine, e.g. they cannot bond to a proton to form ions analogous to PH_4^+.

Bismuthine, BiH_3, is extremely unstable and was first detected by treating a bismuth-magnesium alloy, containing radioactive bismuth, with dilute acid. The detection of radioactivity in the gas phase showed that a volatile hydride of bismuth existed.

19.24 The halides of arsenic, antimony and bismuth

The trihalides

All the trihalides exist as solids, except the trifluoride and trichloride of arsenic which are liquids. Like the hydrides, they are essentially covalent and have pyramidal molecules, except for BiF_3 which is thought to contain an appreciable amount of ionic bonding.

Arsenic trihalides are hydrolysed by water with the formation of hydrated arsenic (III) oxide, As_4O_6, e.g.

$$4AsCl_3(l) + 6H_2O(l) \longrightarrow As_4O_6(aq) + 12HCl(aq)$$

This hydrolysis is probably similar to that of the phosphorus trihalides and takes place in stages (p. 266).

The trichlorides of antimony and bismuth are incompletely hydrolysed by water, forming insoluble oxychlorides which contain respectively the antimonyl, SbO^+, and bismuthyl, BiO^+, cations. This behaviour is consistent with the increase in metallic character that occurs lower down in a periodic group.

$$SbCl_3(s) + H_2O(l) \rightleftharpoons SbOCl(s) + 2HCl(aq)$$
$$BiCl_3(s) + H_2O(l) \rightleftharpoons BiOCl(s) + 2HCl(aq)$$

The addition of concentrated hydrochloric acid forces the above two equilibria to the left and, as a result of this, the solution becomes clear. The dissolution of the oxychlorides is probably partially due to the formation of chloride complexes with $SbCl_3$ and $BiCl_3$, which reinforce the simple mass action effect.

The pentahalides

The pentahalides known to exist are AsF_5 (liquid), SbF_5 (liquid), $SbCl_5$ (liquid) and BiF_5 (solid). Both antimony pentachloride and bismuth pentafluoride are powerful oxidising agents and easily decompose into the trihalide and the halogen:

$$SbCl_5(l) \longrightarrow SbCl_3(s) + Cl_2(g)$$
$$BiF_5(s) \longrightarrow BiF_3(s) + F_2(g)$$

The instability of the 5-valent with respect to the 3-valent state accounts for the non-existence of pentabromides and pentaiodides, e.g. neither bromine nor iodine is a sufficiently powerful oxidising agent to convert Sb^{III} to Sb^{V}. However, the remarkable stability of antimony pentafluoride and the marked instability of arsenic pentachloride (decomposition into $AsCl_3$ and Cl_2 occurs above $-50°C$) are not easily explained.

19.25 The oxides of arsenic, antimony and bismuth

The 3-valent oxides

Arsenic and antimony (III) oxides are obtained when the respective metals are heated in oxygen:

$$4As(s) + 3O_2(g) \longrightarrow As_4O_6(s)$$
$$4Sb(s) + 3O_2(g) \longrightarrow Sb_4O_6(s)$$

Arsenic (III) oxide exists in two polymorphic forms, the one normally encountered being based on As_4O_6 units which are structurally similar to those of phosphorus (III) oxide (p. 267); in the vapour state the same As_4O_6 molecules are present. Antimony (III) oxide has a similar structure both in the solid state and as a vapour; above about 570°C, however, a polymeric form also exists.

Arsenic (III) oxide is an acidic oxide, reacting with sodium hydroxide solution to give an arsenate (III):

$$As_4O_6(s) + 4OH^-(aq) \longrightarrow 4AsO_2^-(aq) + 2H_2O(l)$$

It is slightly soluble in water, forming what is thought to be the hydrated oxide $As_2O_3(aq.)$ rather than a definite acid.

Antimony (III) oxide is amphoteric, dissolving in sodium hydroxide solution to form an antimonate (III) and in concentrated sulphuric acid to give a sulphate:

$$Sb_4O_6(s) + 4OH^-(aq) \longrightarrow 4SbO_2^-(aq) + 2H_2O(l)$$

$$Sb_4O_6(s) + 6H_2SO_4(l) \longrightarrow 2Sb_2(SO_4)_3(aq) + 6H_2O(l)$$

Bismuth (III) oxide is exclusively basic and exists in several forms, one of which is ionic; it is therefore represented by the formula Bi_2O_3; it can be made by the standard methods used for obtaining metallic oxides, e.g. by heating the nitrate and carbonate.

Bismuth hydroxide, like the oxide, is basic and can be precipitated by adding sodium hydroxide solution to a solution of a bismuth salt:

$$Bi^{3+}(aq) + 3OH^-(aq) \longrightarrow Bi(OH)_3(s)$$

The 5-valent oxides

Arsenic and antimony (V) oxides can be obtained by oxidation of the elements with concentrated nitric acid, e.g.

$$2Sb(s) + 10HNO_3(aq) \longrightarrow Sb_2O_5(s) + 5H_2O(l) + 10NO_2(g)$$

Since their structures are unknown they are represented by the respective empirical formulae, As_2O_5 and Sb_2O_5.

Both oxides are unstable with respect to the 3-valent state, decomposition into the 3-valent oxide and oxygen occurring on gentle heating, e.g.

$$2As_2O_5(s) \longrightarrow As_4O_6(s) + 2O_2(g)$$

They are acidic oxides, reacting with alkalis to give arsenates (V), AsO_4^{3-}, and antimonates (V), $Sb(OH)_6^-$, respectively.

It is possible that bismuth (V) oxide exists, although it has never been obtained in a pure form. In view of the instability of the 5-valent oxides of arsenic and antimony to mild heating and the increased metallic character of bismuth, it would be expected to be very unstable with respect to bismuth (III) oxide.

19.26 The oxyacids and oxysalts of arsenic, antimony and bismuth

The acids and their anions in valency state (III)

There is some doubt about the nature of arsenic (III) acid and it is thought that the action of water on arsenic (III) oxide gives the hydrated oxide. However, the arsenate (III) ions AsO_3^{3-} and AsO_2^{2-} (empirical formula) are formed on treating arsenic (III) oxide with hot alkali. Arsenates (III) are reducing agents, being readily oxidised to arsenates (V) by dichromate (VI) and manganate (VII) ions and even by iodine, e.g.

$$AsO_3^{3-}(aq) + I_2(aq) + 2OH^-(aq) \rightleftharpoons AsO_4^{3-}(aq) + 2I^-(aq) + H_2O(l)$$

The above reaction is reversible but can be forced completely to the right by adding sodium hydrogen carbonate solution, which produces hydroxyl ions by hydrolysis. Sodium hydroxide and sodium carbonate cannot be used since the iodate (I) would be formed (p. 334). The reaction can also be forced completely to the left by adding sodium thiosulphate solution which removes the iodine. The reaction, which is quantitative, can therefore be used for estimating either arsenates (III) or arsenates (V) volumetrically.

Antimony (III) acid does not exist but antimonates (III) containing the SbO_2^- ion (empirical formula) are known.

There is no lower oxyacid of bismuth since the hydroxide, $Bi(OH)_3$, is basic.

The acids and their anions in valency state (V)

Arsenic (V) acid, H_3AsO_4, is obtained when arsenic of arsenic (III) oxide is oxidised with concentrated nitric acid; it can be crystallised as a white hydrated solid, $H_3AsO_4.\frac{1}{2}H_2O$.

$$As_4O_6(s) + 4HNO_3(aq) + 4H_2O(l) \longrightarrow 4H_3AsO_4(aq) + 2NO(g) + 2NO_2(g)$$

It is a tribasic acid, forming salts containing the AsO_4^{3-} anion, which are generally isomorphous with the corresponding phosphates (V).

Although arsenic forms no acids corresponding to the diphosphoric (V) and polytrioxophosphoric (V) acids, salts containing condensed arsenate (V) anions are known; they are more readily hydrolysed than the corresponding condensed phosphates (V), giving the $AsO_4^{3-}(aq)$ ion.

Antimony (V) acid has never been isolated in the pure form, although solid antimonates (V) such as $K^+[Sb(OH)_6]^-$ are known (note that the co-ordination number of antimony in antimonates (V) is six, whereas phosphorus and arsenic both have a co-ordination number of four in phosphates (V) and arsenates (V)).

There is no higher acid of bismuth corresponding to arsenic (V) acid and although sodium bismuthate (V), $Na^+BiO_3^-$, exists it has never been obtained in a pure form. The bismuthate (V) ion is a powerful oxidising agent, e.g. it will oxidise Mn^{2+} to MnO_4^-.

19.27 The sulphides of arsenic, antimony and bismuth

Four sulphides of arsenic, As_4S_3, As_4S_4, As_2S_3 and As_2S_5, can be obtained by heating arsenic and sulphur together in the correct proportions. Arsenic (III) sulphide, As_2S_3 can also be obtained by passing hydrogen sulphide through a solution of an arsenate (III) containing hydrochloric acid:

$$2AsO_3^{3-}(aq) + 6H^+(aq) + 3H_2S(g) \longrightarrow As_2S_3(s) + 6H_2O(l)$$

Arsenic (V) sulphide, As_2S_5, can be obtained in a similar manner from an arsenate (V):

$$2AsO_4^{3-}(aq) + 6H^+(aq) + 5H_2S(g) \longrightarrow As_2S_5(s) + 8H_2O(l)$$

Both arsenic (III) sulphide and arsenic (V) sulphide are yellow solids.

Both sulphides of arsenic dissolve in sodium hydroxide and sodium sulphide solutions, the relevant equations being:

$$2As_2S_3(s) + 4OH^-(aq) \longrightarrow AsO_2^-(aq) + 3AsS_2^-(aq) + 2H_2O(l)$$

$$4As_2S_5(s) + 24OH^-(aq) \longrightarrow 3AsO_4^{3-}(aq) + 5AsS_4^{3-}(aq) + 12H_2O(l)$$

$$As_2S_3(s) + S^{2-}(aq) \longrightarrow 2AsS_2^-(aq)$$
$$As_2S_5(s) + 3S^{2-}(aq) \longrightarrow 2AsS_4^{3-}(aq)$$

The above reactions are analogous to the action of sodium hydroxide solution on the oxides of arsenic and are not unexpected, since oxygen and sulphur are both members of the same periodic group. The sulphides of arsenic are therefore showing 'acidic' tendencies in the above reactions.

Antimony (III) sulphide, Sb_2S_3, and antimony (V) sulphide, Sb_2S_5, can be obtained by direct combination, or by passing hydrogen sulphide through solutions containing respectively 3-valent and 5-valent antimony compounds. Antimony (III) sulphide is precipitated as an orange solid, although a grey form also exists.

Like the corresponding sulphides of arsenic, they are soluble in sodium sulphide solution, forming thioantimonate (III) and thioantimonate (V) anions, i.e. they are 'acidic'.

Bismuth only forms the 3-valent sulphide, Bi_2S_3, which is precipitated as a dark brown solid when hydrogen sulphide is passed through a solution containing bismuth ions:

$$2Bi^{3+}(aq) + 3H_2S(g) \longrightarrow Bi_2S_3(s) + 6H^+(aq)$$

It is not soluble in sodium sulphide solution since it is devoid of 'acidic' properties.

19.28 Salts containing the antimony and bismuth cations

There is no evidence that arsenic forms a cation but the more metallic antimony and bismuth can form respectively the Sb^{3+} and Bi^{3+} cations in which a pair of electrons remains inert (the inert pair effect p. 63). These cations are present in $Sb_2(SO_4)_3$, $Bi_2(SO_4)_3$, $Bi(NO_3)_3$ and BiF_3; bismuth forms more salts than antimony because of its greater 'metallic character'.

Compounds containing either the antimony or bismuth cation are hydrolysed by water, forming the oxycation:

$$Sb^{3+}(aq) + H_2O(l) \rightleftharpoons SbO^+(aq) + 2H^+(aq)$$
$$Bi^{3+}(aq) + H_2O(l) \rightleftharpoons BiO^+(aq) + 2H^+(aq)$$

19.29 Summary of trends in Group 5B

Nitrogen is a non-metal and so too is phosphorus, although one allotrope (black phosphorus) shows metallic conduction. Arsenic and antimony are weakly metallic in that their stable allotrope has the same structure as black phosphorus. Bismuth does not display allotropy and it adopts the black phosphorus structure.

Only nitrogen of the Group 5B elements is able to multiple bond to itself, and the nitrogen molecule is very stable; the high stability of this molecule is the prime reason why the oxides of nitrogen are endothermic compounds.

The increase in 'metallic character' that occurs with increasing atomic number makes itself apparent in a number of ways:

(a) Nitrogen (and possibly phosphorus) in combination with a very reactive metal such as calcium forms the X^{3-} ion.

(b) The hydrides XH_3 decrease in thermal stability with increasing relative molecular mass, i.e. they become better reducing agents. They also become less basic in the same order, and only nitrogen and phosphorus can form the XH_4^+ cation.

(c) The oxides of nitrogen (except dinitrogen oxide and nitrogen oxide) are acidic; so too are the oxides of phosphorus and arsenic. Antimony (III) oxide is amphoteric, antimony (V) oxide is acidic and bismuth (III) oxide is exclusively basic.

Two quite general observations emerge from this, namely that:

(i) the oxides become less 'acidic' and more 'basic' with increasing 'metallic character'.

(ii) for a particular element, the higher oxide is the more 'acidic', e.g. Sb_4O_6 is amphoteric while Sb_2O_5 is acidic.

(d) Only antimony and bismuth can form cations of the type X^{3+} (inert pair effect), and of the two, bismuth does so more readily than antimony. The formation of cations is a characteristic property of metals.

Nitrogen, unlike the other Group 5B elements, cannot expand its octet (*d* orbitals cannot be utilised without too great an energy change) and is thus restricted to a covalency of three, e.g. the highest chloride of nitrogen is NCl_3 while that of phosphorus is PCl_5. The fact that nitrogen cannot utilise *d* orbitals, whereas phosphorus can, accounts for the different modes of hydrolysis of the trichlorides (p. 252 and p. 266); nitrogen trichloride acts as an electron donor, whereas phosphorus trichloride acts as an electron acceptor.

Expansion of the octet also explains why phosphorus, arsenic, antimony and bismuth can form complex anions, e.g. PCl_6^-, AsF_6^-, $SbCl_6^-$ and $BiCl_5^{2-}$.

19.30 Suggested reading

J. A. Martin et al., *The Synthesis of nitric acid from ammonia*, Education in Chemistry, No. 3, Vol. 14, 1977.

G. E. G. Mattingly, *Inorganic Fertilisers*, Education in Chemistry, No. 2, Vol. 16, 1979.

19.31 Questions on chapter 19

1 Discuss the inclusion of nitrogen and phosphorus in the same group of the Periodic Classification by comparing the physical and chemical properties of: (a) these two elements; (b) their hydrides, and (c) their chlorides. (L)

2 Compare and contrast the chemical behaviour of nitrogen and phosphorus by reference to the properties of:

(a) the elements,
(b) their hydrides,
(c) their chlorides,
(d) nitric acid and orthophosphoric acid (phosphoric (V) acid).

Give reasons for any similarities or differences in behaviour that you quote. (O & C)

3 Give an account of the reactions of the oxides of nitrogen. Comment on any unusual features of their electronic or molecular structure.

Sulphur (1g) dissolved completely in excess of liquid ammonia to give a gas and also a solid containing only nitrogen and sulphur. The gas turned lead acetate paper black and when collected was found to occupy a volume of 418 cm^3 at s.t.p. Deduce the empirical formula of the solid and write an equation for the reaction. Suggest some properties that the solid substance might exhibit. (O & C)

4 Give electronic formulae for ammonia, nitrous oxide (dinitrogen oxide) and nitric acid molecules, carefully distinguishing between any different bonds present.

Describe briefly how you would prepare reasonably pure specimens of two of these compounds.

Give an example of

(a) ammonia acting as a reducing agent,
(b) nitrous oxide (dinitrogen oxide) acting as an oxidising agent,
(c) nitric acid acting as a nitrating agent. (AEB)

5 Stating the appropriate conditions, describe what happens when ammonia reacts with

(a) oxygen,
(b) silver nitrate,
(c) sodium,
(d) carbon dioxide,
(e) chlorine.

A solution containing ammonium hydroxide, a very weak base, can be prepared by dissolving ammonia in water. Describe the equilibria which occur in solution and account for the very small dissociation constant of ammonium hydroxide. (AEB)

6 A mixture of sodium nitrite and sodium nitrate was dissolved in water and the volume made up to 250 cm^3. In a titration, 31·35 cm^3 of this solution decolorised 25·00 cm^3 of a potassium permanganate solution containing 3·16 g/dm^3, which had been acidified with dilute sulphuric acid. The mixture resulting from the reduction of 25·00 cm^3 of the original solution with an excess of potassium hydroxide and Devarda's alloy was steam-distilled and the distillate collected in 50·00 cm^3 of 0·1M sulphuric acid. The excess of sulphuric acid required 32·50 cm^3 of 0·2M sodium hydroxide solution for neutralisation. Calculate the weights of sodium nitrite and sodium nitrate present in the original solution. (L)

7 Briefly compare and contrast the properties of

(a) the hydrides,
(b) the oxyacids, of nitrogen and phosphorus.

The action of nitrogen dioxide on phosphorus trichloride yields a liquid **Y**, $P_2O_3Cl_4$, which is hydrolysed by water to a mixture of orthophosphoric acid (phosphoric (V) acid) and hydrochloric acid. What does this suggest concerning the structure of **Y**? (Camb. Schol.)

8 (a) Name six different products which may be obtained by the reduction of nitric acid or a nitrate ion. In each case give the equation for the reaction in a way which indicates the number of electrons gained in the reduction.

Under what conditions may some of these reaction products be obtained by the reaction between nitric acid and metals? You should attempt to state the conditions for as many as possible of the six products you name.

(b) Hydroxylamine may be estimated volumetrically by adding to it an excess of iron (III) ion, boiling, and titrating the resultant mixture with cerium (IV) sulphate solution. The reaction is:

$$2NH_2OH + 4Fe^{3+} \longrightarrow N_2O + H_2O + 4Fe^{2+} + 4H^+$$

Cerium (IV) sulphate is an oxidising agent able to perform the oxidation iron (II) to iron (III), and in the process is itself reduced according to the equation:

$$Ce^{4+} + e \longrightarrow Ce^{3+}$$

In an estimation of this type, a standard cerium (IV) sulphate solution was made by dissolving 50·30g of the double salt, $Ce(SO_4)_2.2(NH_4)_2SO_4.2H_2O$ to make 1 litre of solution. The formula weight of this double salt is 632 and for the purposes of this exercise the sample may be regarded as pure. 10 cm^3 of an aqueous solution of hydroxylamine **A** was diluted to 1 dm^3, 50 cm^3 of this diluted solution taken, boiled with an excess of iron (III) solution, and then titrated to equivalence with the standard cerium (IV) solution. If 60·2 cm^3 of the standard cerium (IV) solution was required, what was the concentration, in g per litre, of the original hydroxylamine solution **A**? (JMB)

9 Give an account of the preparation and properties of the oxides of nitrogen. Discuss their structures. (Camb. Schol.)

10 (a) The Haber process for the manufacture of ammonia can be represented by the following equation:

$$N_2(g) + 3H_2(g) \rightleftharpoons 2NH_3(g) \qquad \Delta H^\ominus = -92\,kJ.$$

(i) From chemical theory, show how the most suitable conditions for the production of ammonia can be predicted.
(ii) State the conditions used in practice and give reasons why these are chosen.

(b) Describe how ammonia is converted into nitric acid on a large scale.

(AEB)

11 Some reactions of nitrogen and its compounds are given below.

Stating clearly your reasons, identify the compounds **A** to **G** and describe the stereochemistry and the electronic structures of the *nitrogen*-containing ions or molecules in these compounds.

When magnesium is heated in nitrogen, a pale grey compound, **A**, is produced. When **A** is hydrolysed, a colourless gas **B** is produced which dissolves in water to give an alkaline solution. The reaction of **B** with sodium chlorate (I) leads to the formation of a colourless liquid **C** with empirical formula NH_2. The reaction of **C** with sulphuric acid in a 1:1 molecular ratio produces a salt **D**, $N_2H_6SO_4$, which contains one cation and one anion per formula unit. An aqueous solution of **D** reacts with nitrous acid to give a solution which, when neutralized with ammonia, produces a salt **E**, with empirical formula NH, which contains one anion and one cation per 'molecular' unit.

The gas **B** reacts with heated sodium to give a solid **F** and hydrogen. The reaction of **F** with N_2O in a 1:1 molecular ratio gives a solid **G**, which contains the same anion as **E**, and water. **G** decomposes when heated to give sodium and nitrogen.

(C)

12 Describe the preparation, properties and reactions of the chlorides and oxides of phosphorus. Explain why phosphorus forms two chlorides, but nitrogen only one.

(L)

13 What is meant by the 'strength' of an acid and of a base?

Give one example in each case of an inorganic compound which, when dissolved in water, forms

(a) a strong acid, (c) a strong base,
(b) a weak acid, (d) a weak base.

How would you account for the wide variation in the acid/base strengths of these compounds in terms of the fundamental properties of their constituent atoms?

Liquid ammonia dissociates slightly into NH_4^+ ions and NH_2^- ions. By analogy with water, state which of the following compounds you would expect to be acidic or basic in liquid ammonia;

(a) NH_4Cl, (d) $Zn(NH_2)_2$,
(b) KNH_2, (e) NaOH.
(c) HCl,

(O & C)

14 When white phosphorus is heated carefully with sodium hydroxide solution, a colourless, poisonous, inflammable gas is evolved. This gas contains 91·02% P and 8·98% H. (H = 1; P = 31.)

(a) Write the formula for the white phosphorus molecule.
(b) Draw a diagram to illustrate the geometry of the white phosphorus molecule.
(c) Deduce the empirical formula of the gas evolved in the statement above.
(d) This gas is very weakly basic, and reacts with hydrogen iodide to form a solid compound. Write the equation for this reaction.
(e) By analogy with ammonium iodide (or otherwise), write the equation for the reaction between the solid formed in (d) with sodium hydroxide solution.
(f) The gas formed from the original white phosphorus is spontaneously inflammable in air. Write the equation for this combustion, naming the products.
(g) Phosphorus and nitrogen form the hydrides P_2H_4 and N_2H_4 (hydrazine).
 (i) Construct a bond diagram to show the structure of hydrazine. (*The structure contains the N-N bond.*)
 (ii) Comment on the geometry of the hydrazine molecule and estimate bond angles on the the diagram.
 (iii) Explain why NCl_3 is the highest chloride of nitrogen but phosphorus can form PCl_5.
(h) 1·07 g of a nitrogen compound was boiled with an excess of sodium hydroxide solution and all the nitrogen was released as ammonia gas. This ammonia was

found to neutralize 200 cm^3 0·100M hydrochloric acid. Calculate the percentage by mass of nitrogen in the compound. (N = 14.)

(S)

15 (a) Outline a method for the laboratory preparation of **three** of the following: phosphine; phosphorus trichloride; phosphorus pentachloride, phosphorus (V) oxide. You should confine your answer in each case to naming the reactants, stating the essential conditions, indicating how the product is isolated and providing an equation for the reaction.

(b) State two important physical differences between ammonia and phosphine. How may these be accounted for? (O)

16 'Metallic character increases as one proceeds down a periodic group'. Discuss this statement with reference to the Group 5B elements.

17 Compare and contrast the structural chemistry of the borates, silicates and phosphates.

18 One of the products, **F**, of the reaction between phosphorus pentachloride, PCl_5, and ammonium chloride, NH_4Cl, was found to have the molecular formula $P_3N_3Cl_6$.

In compound **F**, all the phosphorus and all the nitrogen atoms occupy identical positions, that is, they are equivalent.

F reacts with dimethylamine in ether solution at −78°C in the molecular ratio 1:4 respectively to give, as the principal products, **G** and **H**, both of which have the molecular formula $P_3N_3Cl_4(NMe_2)_2$. **F** reacts with dimethylamine in ether solution under reflux (35°C) in the molecular ratio 1:8 respectively to give **I**, $P_3N_3Cl_2(NMe_2)_4$, as the principal product. **F** reacts with excess dimethylamine in chloroform solution under reflux (62°C) to give **J**, $P_3N_9Me_{12}$.

When **F** is treated with sodium azide, NaN_3, in acetone, the product **K** contained P 24·03% and N 75·97%.

(i) Suggest possible structures for the compounds **F**, **G**, **H**, **I** and **J**.

(ii) How many compounds of the type $P_3N_3Cl_2(NMe_2)_4$ are possible theoretically? Suggest possible structures for them.

(iii) Comment upon the different conditions by which **F** is converted to **G** and **H**, to **I**, and to **J** respectively.

(iv) Calculate the empirical formula of compound **K**, and attempt to write an equation for its formation from **F** and sodium azide.

[N = 14·0; P = 31·0.] (C)

19 Give a brief account of the chemistry of the principal oxides of nitrogen.

Dinitrogen oxide (nitrous oxide, N_2O) has a linear structure containing the skeleton N—N—O. Starting with a sample of ^{15}N-labelled potassium nitrate, $K^{15}NO_3$, how may samples of dinitrogen oxide be prepared in which (a) the terminal nitrogen atom, (b) the central nitrogen atom, (c) both nitrogen atoms are labelled with ^{15}N?

(O & C)

20 (a) Two possible ways of fixing atmospheric nitrogen may be represented by the equations:

$$\tfrac{1}{2}N_2(g) + \tfrac{3}{2}H_2O(g) \longrightarrow NH_3(g) + \tfrac{3}{4}O_2(g)$$

$$\tfrac{1}{2}N_2(g) + \tfrac{1}{2}H_2O(g) + \tfrac{5}{4}O_2(g) \longrightarrow HNO_3(l).$$

Calculate the value of $\Delta H^{\ominus}$ for each of these processes at 298 K.

Why do you think that the process represented by the second equation does not provide a suitable method of fixing nitrogen?

Comment on the ways in which the synthesis of ammonia from nitrogen and hydrogen is controlled to give the maximum economic yield of ammonia.

(b) Discuss the principles underlying the uses of ion-exchange and osmosis in the purification of water. (O & C)

Chapter 20

Group 6B oxygen, sulphur, selenium, tellurium and polonium

20.1 Some physical data of Group 6B elements

	Atomic Number	Electronic Configuration	Atomic Radius/nm	Ionic Radius/nm X^{2-}	M.p. /°C	B.p. /°C
O	8	2.6 $1s^2 2s^2 2p^4$	0·074	0·140	−218	−183
S	16	2.8.6 ....$2s^2 2p^6 3s^2 3p^4$	0·104	0·184	119*	445
Se	34	2.8.18.6 ....$3s^2 3p^6 3d^{10} 4s^2 4p^4$	0·117	0·198	217†	685
Te	52	2.8.18.18.6 ....$4s^2 4p^6 4d^{10} 5s^2 5p^4$	0·137	0·221	450	990
Po	84	2.8.18.32.18.6 ....$5s^2 5p^6 5d^{10} 6s^2 6p^4$	0·140		254	960

* For monoclinic sulphur
† For grey selenium

20.2 Some general remarks about Group 6B

The Group 6B elements show the usual gradation from non-metallic to metallic properties with increasing atomic number that occurs in any periodic group. Oxygen and sulphur are non-metals, selenium and tellurium are semiconductors and polonium is metallic.

These elements can enter into chemical combination and complete their octets by gaining two electrons to form the 2-valent ion, e.g. O^{2-}, S^{2-}, except for polonium which is too metallic, and by forming two covalent bonds, e.g. the hydrides H_2O, H_2S, H_2Se, H_2Te and H_2Po.

The two heavier members of this group can form the 4-valent cation X^{4+} (the inert pair effect), e.g. there is evidence of the presence of Te^{4+} ions in the dioxide TeO_2 and of Po^{4+} ions in the dioxide, PoO_2, and sulphate, $Po(SO_4)_2$.

Because sulphur, selenium, tellurium and polonium have vacant *d* orbitals that can be utilised without too great an energy change, they are able to form covalent compounds in which the octet of electrons is expanded; for instance, the valencies of sulphur in H_2S, SCl_4 and SF_6 are two, four and six respectively. Oxygen, in common with other first row members of the Periodic Table, cannot expand its octet.

Oxygen exists in the form of discrete molecules, a double bond uniting two oxygen atoms together, $O{=}O$. The atoms of the other Group 6B elements do not multiple bond to themselves and sulphur, in particular, shows a strong tendency to catenate, puckered S_8 rings being present in rhombic and monoclinic sulphur (the two main allotropic forms of this

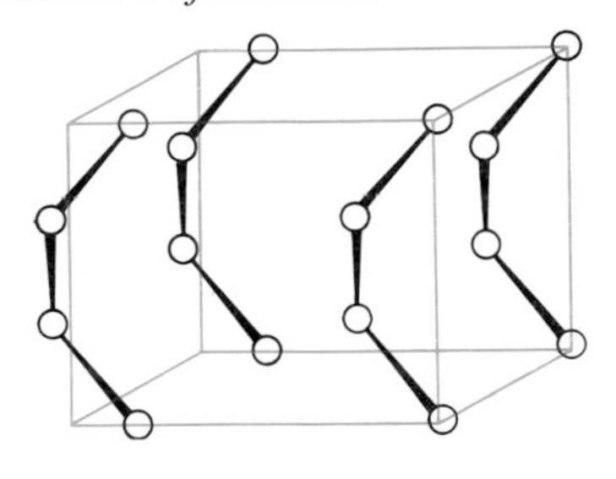

FIG. 20.1. *The spiral chain structure of selenium*

element). There are two forms of selenium corresponding in structure to rhombic and monoclinic sulphur in which Se_8 rings are present. These forms, however, are readily converted into a 'metallic' form of the element called grey selenium which contains infinite spiral chains of selenium atoms (fig. 20.1). As far as is known, there is only one form of tellurium which has the same structure as grey selenium. Polonium is truly metallic, exhibiting a co-ordination number of six in both its allotropic forms.

Since the chemistry of oxygen and sulphur have little in common, these two elements are treated separately, and compared in section 20.32. A brief account of the chemistry of selenium, tellurium and polonium, in so far as it illustrates group trends, is included.

Oxygen

20.3 Occurrence and preparation of oxygen

Oxygen occurs in the atmosphere to the extent of about 21 per cent by volume (23 per cent by weight). This percentage remains constant by the operation of the highly complex process termed photosynthesis, whereby green plants exposed to sunlight utilise carbon dioxide (one product of respiration and combustion) and water to build up carbohydrate molecules, at the same time releasing oxygen. The element is present in the earth's crust and in water to the extent of about 50 per cent and 89 per cent by weight respectively. It is an essential ingredient in all living matter (together with carbon, hydrogen and nitrogen) and is of prime importance in respiration and combustion processes. Although only slightly soluble in water, enough oxygen dissolves to support marine life.

Oxygen can be obtained in the laboratory in a variety of ways:

By the thermal decomposition of the oxides of metals low in the electrode potential series

The oxides of most metals are thermally stable. However, the oxides of silver and mercury are thermally unstable:

$$2Ag_2O(s) \longrightarrow 4Ag(s) + O_2(g)$$
$$2HgO(s) \longrightarrow 2Hg(l) + O_2(g)$$

The latter reaction is of historical interest, being the method used by Priestley to obtain oxygen in 1774.

By the thermal decomposition of higher oxides

So-called higher oxides such as Pb_3O_4 and PbO_2 decompose thermally into the lower oxide and oxygen:

$$2Pb_3O_4(s) \longrightarrow 6PbO(s) + O_2(g)$$
$$2PbO_2(s) \longrightarrow 2PbO(s) + O_2(g)$$

By the decomposition of peroxides

Hydrogen peroxide is readily decomposed into water and oxygen by catalysts such as finely divided metals and manganese dioxide:

$$2H_2O_2(aq) \longrightarrow 2H_2O(l) + O_2(g)$$

Solid peroxides such as sodium peroxide evolve oxygen on the addition of water:

$$2(Na^+)_2O_2^{2-}(s) + 2H_2O(l) \longrightarrow 4Na^+OH^-(aq) + O_2(g)$$

By the thermal decomposition of salts containing anions rich in oxygen

Examples of salts that decompose thermally to give oxygen include nitrates, manganates (VII) and chlorates (V):

$$2K^+NO_3^-(s) \longrightarrow 2K^+NO_2^-(s) + O_2(g)$$
$$2K^+MnO_4^-(s) \longrightarrow (K^+)_2MnO_4^{2-}(s) + MnO_2(s) + O_2(g)$$
$$2K^+ClO_3^-(s) \longrightarrow 2K^+Cl^-(s) + 3O_2(g)$$

By electrolysis

Electrolysis of aqueous solutions of alkalis and acids results in the discharge of hydrogen at the cathode and oxygen at the anode. Pure water also decomposes in the same way but, since pure water is a very poor conductor of electricity, hydrogen and oxygen are only discharged slowly.

Industrially, oxygen is obtained from the atmosphere by first removing carbon dioxide and water vapour. The remaining gases are then liquefied and fractionally distilled to give nitrogen and oxygen (p. 247).

20.4 The properties of oxygen

Oxygen is a colourless and odourless diatomic gas. It liquefies at −183°C and freezes at −218°C. Chemically it is very reactive, forming compounds with all other elements except the noble gases and, apart from the halogens and some unreactive metals, these can be made to combine directly with oxygen under the right conditions. The action of a silent electrical discharge upon oxygen gives ozone in concentrations up to 10 per cent.

In addition to its importance in normal respiration and combustion processes, oxygen is used in oxyacetylene welding and cutting, in the manufacture of many metals—particularly steel—and in hospitals, high altitude flying and in mountaineering. The combustion of fuels, e.g. kerosene in liquid oxygen, provides the tremendous thrust required to put satellites into space.

20.5 Ozone, O_3—an allotrope of oxygen

Occurrence

Ozone is too reactive to remain for long in the atmosphere at sea level, but at a height of about 20 kilometres it is formed from atmospheric oxygen by the energy of sunlight. This ozone layer protects the earth's surface from an excessive concentration of ultraviolet radiation. Very small concentrations of ozone are formed in the vicinity of electrical machinery.

Preparation

When a slow dry stream of oxygen is passed through a silent electrical discharge up to 10 per cent conversion to ozone occurs (fig. 20.2):

$$3O_2 \longrightarrow 2O_3 \qquad \Delta H^\ominus(298\,K) = +284\,kJ\,mol^{-1}$$

Since ozone is an endothermic compound it is necessary to use a silent electrical discharge, otherwise sparking would generate heat and decompose it.

FIG. 20·2. *Laboratory preparation of ozone*

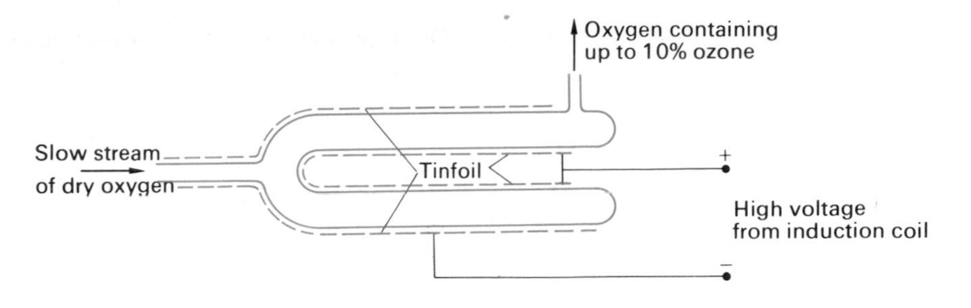

If concentrations of ozone greater than 10 per cent are required a battery of ozonisers can be used, and pure ozone (b.p. −112°C) condensed out in a vessel surrounded by liquid oxygen.

Properties

Ozone has a characteristic smell and in small concentrations it is harmless. However, if the concentration rises to above about 100 parts per million breathing becomes uncomfortable and it causes headaches. When pure it is a pale blue gas; liquid ozone is a darker blue and solid ozone is violet-black.

Ozone is thermodynamically unstable with respect to oxygen, since its decomposition into oxygen results in the liberation of heat (ΔH is negative) and an increase in entropy (ΔS is positive since the triatomic molecule is broken down into the simpler diatomic oxygen). These two effects re-inforce each other, resulting in a large negative free energy change (ΔG) for its conversion into oxygen:

$$\Delta G = \Delta H - T\Delta S$$

It is not really surprising, therefore, that high concentrations of ozone can be dangerously explosive.

Ozone is a much more powerful oxidising agent than molecular oxygen; for instance, it will oxidise lead sulphide to lead sulphate, iodide ions to iodine and iron (II) ions to the iron (III) state:

$$PbS(s) + 4O_3(g) \longrightarrow PbSO_4(s) + 4O_2(g)$$
$$2I^-(aq) + H_2O(l) + O_3(g) \longrightarrow 2OH^-(aq) + I_2(s) + O_2(g)$$
$$2Fe^{2+}(aq) + 2H^+(aq) + O_3(g) \longrightarrow 2Fe^{3+}(aq) + H_2O(l) + O_2(g)$$

The iodine liberated when ozone reacts with an excess of potassium iodide solution (acidified to remove the hydroxyl ions that are formed) can be titrated with a standard solution of sodium thiosulphate. This is a quantitative method of estimating the gas.

Organic compounds containing carbon-carbon double bonds react with ozone to give ozonides, which can be hydrolysed to give aldehydes and ketones. This reaction has been used to determine the position of carbon-carbon double bonds in organic molecules, e.g.

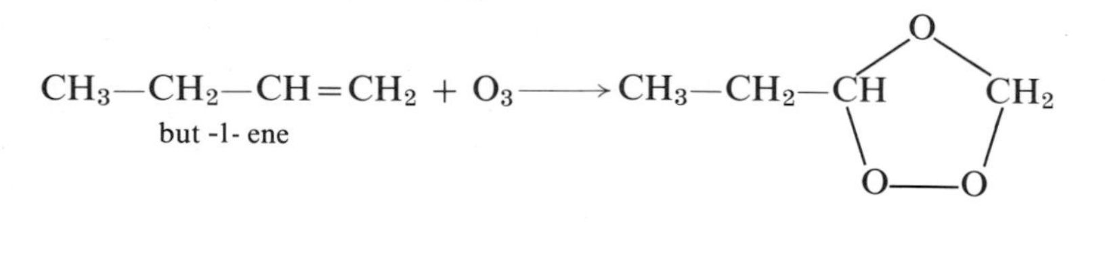

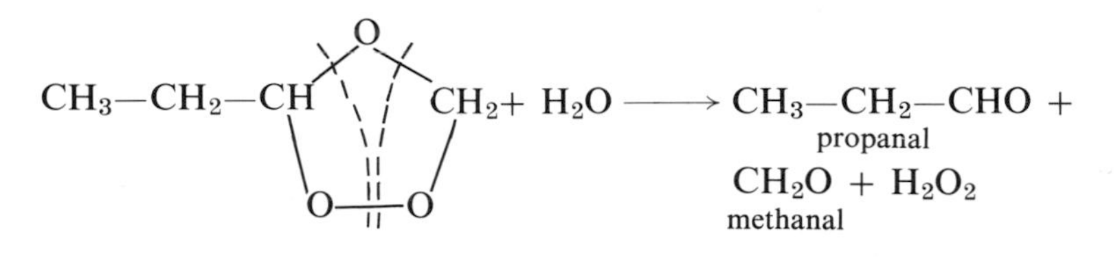

Laboratory experiments have shown that nitrogen oxide combines very rapidly with ozone and there is thus the possibility that nitrogen oxide emission from the exhaust systems of supersonic aircraft, which normally operate in the stratosphere, might be slowly depleting the concentration of the ozone layer in the upper atmosphere (p. 286).

Another threat to this ozone layer is probably posed by the use of two fluorochloromethanes, i.e. $CFCl_3$ and CF_2Cl_2, which are used in aerosol sprays and as refrigerants. It is ironic that the non-toxic properties of these two compounds, which make them so useful, mean that they are not destroyed in the lower atmosphere. Laboratory experiments have shown that short wavelength ultra-violet radiation of the type present in the stratosphere, and which the ozone layer effectively filters out, decomposes these compounds forming chlorine atoms, which then rapidly react with ozone.

Ozone 'tails' mercury, partially converting it into the oxide so that it sticks to the walls of the vessel containing it. This serves as a conclusive test for the presence of ozone in the presence of excess oxygen.

It is used in the treatment of domestic water supplies, and in the production of some important pharmaceuticals.

The structure of ozone

The two oxygen-oxygen bond lengths in the ozone molecule are identical and intermediate in value between those for an oxygen-oxygen double bond and an oxygen-oxygen single bond. The molecule is angular as expected (minimisation of electron pair repulsion p. 75) with a bond angle of about 117°. It is a resonance hybrid of two main forms:

20.6 Oxides

Classification of oxides

Any realistic attempt to classify the diverse range of oxides requires a knowledge of their structures and their mutual reactions.

(a) Structures of oxides

The oxides of most metals are essentially ionic and consist of a regular array of positive and negative ions, their arrangement being dictated by the radius ratio criterion and the necessity for maintaining overall electrical neutrality (p. 74). For instance, magnesium oxide adopts the sodium chloride structure (p. 75) in which each magnesium ion is surrounded symmetrically by six oxide ions and vice versa, and sodium monoxide, $(Na^+)_2O^{2-}$, has a structure in which the co-ordination numbers of the sodium and oxide ions are four and eight respectively.

The oxides of non-metals and weak metals tend to be covalent and structures ranging from those of discrete molecules, e.g. carbon dioxide (O=C=O), right through to three-dimensional giant molecules, e.g. silicon dioxide (empirical formula $(SiO_2)_n$) are observed.

Irrespective of whether an oxide is ionic or covalent, there are four well-defined types of structures which oxides can exhibit:

Normal oxides, involving bonds only between the element and oxygen. Typical examples are $Mg^{2+}O^{2-}$, O=C=O, and the giant molecule silicon dioxide:

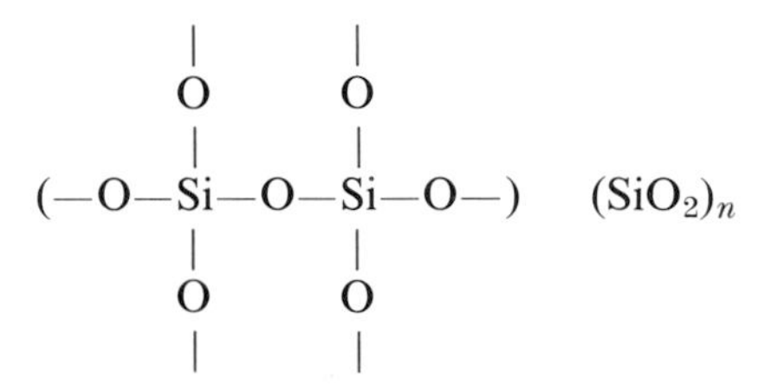

Peroxides, involving bonds between atoms of oxygen in addition to bonds between the element and oxygen. Typical examples are $Ba^{2+}(O—O)^{2-}$ and H—O—O—H.

Suboxides, involving bonds between atoms of the element in addition to bonds between the element and oxygen. A typical example is O=C=C=C=O.

Mixed oxides, which behave as though they contained separate quantities of two distinct oxides of the same element. A typical example is Pb_3O_4 ($2PbO:PbO_2$).

(b) Mutual reactions of oxides

There is a strong tendency for a metallic oxide to react with a non-metallic oxide to form a salt; this tendency can be given a quantitative meaning by employing tabulated values of standard free energy changes (quoted at 298 K and 1 atmosphere pressure).

The standard free energy changes for reactions involving a selection of metallic oxides and a common non-metallic oxide are given below:

Reaction	$\Delta G^\ominus$(298 K)/kJ mol^{-1}
$(K^+)_2O^{2-}(s) + CO_2(g) \longrightarrow (K^+)_2CO_3^{2-}(s)$	−356
$Ba^{2+}O^{2-}(s) + CO_2(g) \longrightarrow Ba^{2+}CO_3^{2-}(s)$	−218
$Ca^{2+}O^{2-}(s) + CO_2(g) \longrightarrow Ca^{2+}CO_3^{2-}(s)$	−130
$Mg^{2+}O^{2-}(s) + CO_2(g) \longrightarrow Mg^{2+}CO_3^{2-}(s)$	−67

From the above figures it is apparent that the basic tendencies of the metallic oxides decrease in the order:

$$(K^+)_2O^{2-} > Ba^{2+}O^{2-} > Ca^{2+}O^{2-} > Mg^{2+}O^{2-}$$

For a number of reactions involving a common metallic oxide and a variety of non-metallic oxides, the following free energy changes are obtained:

Reaction	$\Delta G^\ominus(298\ K)/kJ\ mol^{-1}$
$Ca^{2+}O^{2-}(s) + SO_3(s) \longrightarrow Ca^{2+}SO_4^{2-}(s)$	−347
$Ca^{2+}O^{2-}(s) + N_2O_5(s) \longrightarrow Ca^{2+}(NO_3^-)_2(s)$	−272
$Ca^{2+}O^{2-}(s) + CO_2(g) \longrightarrow Ca^{2+}CO_3^{2-}(s)$	−130
$Ca^{2+}O^{2-}(s) + SiO_2(s) \longrightarrow Ca^{2+}SiO_3^{2-}(s)$	−92
$Ca^{2+}O^{2-}(s) + H_2O(l) \longrightarrow Ca^{2+}(OH^-)_2(s)$	−59

From these figures the following order of decreasing acidic tendencies of the non-metallic oxides is observed:

$$SO_3 > N_2O_5 > CO_2 > SiO_2 > H_2O$$

Metallic and non-metallic oxides can be arranged in a series according to their basic and acidic tendencies. Towards the middle of this table these tendencies will begin to merge and it is not, therefore, surprising that some oxides can show both basic and acidic properties under different conditions, e.g. zinc oxide. Such oxides are said to be amphoteric.

$$Zn^{2+}O^{2-}(s) + (Na^+)_2O^{2-}(s) \longrightarrow (Na^+)_2ZnO_2^{2-}(s)$$

(zinc oxide showing acidic properties)

$$Zn^{2+}O^{2-}(s) + CO_2(g) \longrightarrow Zn^{2+}CO_3^{2-}(s)$$

(zinc oxide showing basic properties)

Using this system of classification, water would be regarded as an amphoteric oxide.

It is possible to arrive at five possible classifications of oxides on the basis of their acid-base properties:

Very basic oxides

Oxides which react readily with amphoteric oxides and very readily with acidic oxides, e.g. potassium monoxide, sodium monoxide and calcium oxide.

Moderately basic oxides

Oxides which scarcely react with amphoteric oxides but readily react with acidic oxides, e.g. magnesium oxide, iron (II) oxide, iron (III) oxide and copper (II) oxide.

Amphoteric oxides

Oxides which react with very basic oxides and acidic oxides, e.g. aluminium oxide, zinc oxide and water.

Acidic oxides

Oxides which react with amphoteric oxides and with basic oxides of both types, e.g. sulphur trioxide, dinitrogen pentoxide, carbon dioxide and silicon dioxide.

Neutral oxides

Oxides which do not react with any others, e.g. dinitrogen oxide and nitrogen oxide.

The boundaries between these five classes are not rigid and not all the predicted reactions occur easily; nevertheless this classification eliminates the need for such terms as 'higher oxide', which convey little meaning.

Using structure and acid-base properties to describe oxides, magnesium oxide is a 'normal basic oxide', barium peroxide is a 'very basic peroxide', aluminium oxide is a 'normal amphoteric oxide' and sulphur trioxide is a 'normal acidic oxide'. These names are precise and convey information about the structure and reactions of each oxide.

Preparation of oxides

The oxides of many non-metals and the reactive metals can be obtained by direct combination with oxygen; other methods that can be used to obtain the oxides of metals include the thermal decomposition of hydroxides, carbonates and nitrates, e.g.

$$Cu(OH)_2(s) \longrightarrow CuO(s) + H_2O(l)$$
$$Ca^{2+}CO_3^{2-}(s) \longrightarrow Ca^{2+}O^{2-}(s) + CO_2(g)$$
$$2Mg^{2+}(NO_3^-)_2(s) \longrightarrow 2Mg^{2+}O^{2-}(s) + 4NO_2(g) + O_2(g)$$

The oxides of some non-metals and weak metals can be obtained by oxidation of the element with nitric acid, e.g.

$$C(s) + 4HNO_3(aq) \longrightarrow CO_2(g) + 4NO_2(g) + 2H_2O(l)$$
$$Sn(s) + 4HNO_3(aq) \longrightarrow SnO_2(s) + 4NO_2(g) + 2H_2O(l)$$

20.7 Hydroxides

Hydroxides, like oxides, can be classified as basic, amphoteric and acidic; basic hydroxides which are soluble in water are called alkalis (see p. 148 for the factors which determine the acid-base properties of hydroxides in aqueous solution).

The hydroxides of the Group 1A and 2A metals can be obtained by the addition of water to the respective oxides, the oxide ion being a strong base, e.g.

$$(Na^+)_2O^{2-}(s) + H_2O(l) \longrightarrow 2Na^+OH^-(aq)$$

or

$$O^{2-} + H_2O(l) \longrightarrow 2OH^-(aq)$$

The very low solubility of other ionic oxides in water, e.g. zinc oxide, prevents any appreciable hydroxide formation. These hydroxides can be obtained by adding an aqueous solution of an alkali to an aqueous solution of a salt (an excess of the alkali must be avoided in many cases since amphoteric hydroxides will dissolve), e.g.

$$Zn^{2+}(aq) + 2OH^-(aq) \longrightarrow Zn(OH)_2(s)$$

$$Zn(OH)_2(s) + 2OH^-(aq) \longrightarrow Zn(OH)_4^{2-}(aq)$$

Many acidic hydroxides can be obtained by the addition of water to the respective non-metallic oxide, e.g.

$$SO_3(s) + H_2O(l) \longrightarrow H_2SO_4(aq) \quad ((HO)_2SO_2)$$
$$N_2O_5(s) + H_2O(l) \longrightarrow 2HNO_3(aq) \quad (HONO_2)$$

20.8 Water

FIG. 20.3. *The structure of the water molecule*

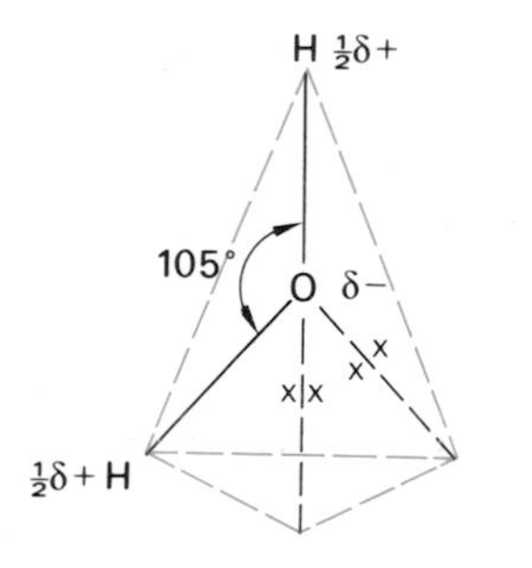

FIG. 20.4. *The open structure of ice*

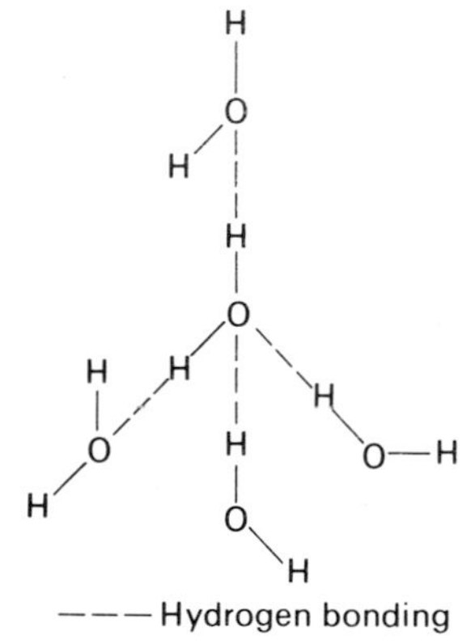

Water, the most abundant liquid, has some very unusual properties; for instance it has a higher density than ice (it has its maximum density at 4°C), and its melting and boiling points are abnormally high by comparison with those of the other Group 6B hydrides (p. 67). These properties are readily understandable in terms of the high electronegativity of the oxygen atom (only the fluorine atom has a higher electronegativity value than oxygen).

The water molecule has an angular structure (fig. 20.3) the two bonding pairs and two lone pairs of electrons being arranged approximately tetrahedrally round the oxygen atom (to minimise repulsion between electron pairs p. 75). The molecule is also extensively polarised because of the high electronegativity of the oxygen atom, partial positive charges residing on each hydrogen atom and a negative charge on the oxygen.

At very low temperatures (−183°C) each water molecule is hydrogen bonded to four more water molecules and an open three-dimensional structure is the result (fig. 20.4). At higher temperatures some of the hydrogen bonds break but even at 0°C liquid water still retains much of the open structure of ice. Water is denser than ice since a closer packing of water molecules takes place when hydrogen bonds are broken. Hydrogen bonding also explains why water has abnormally high melting and boiling points, i.e. water is an associated liquid but the other Group 6B hydrides are not, since sulphur, selenium and tellurium are insufficiently electronegative to form hydrogen bonds.

Water is very stable towards heat, a consequence of the very high O—H bond energy (average O—H bond energy is about 460 kJ mol^{-1}; nevertheless, at about 1000°C dissociation into hydrogen and oxygen occurs to a very slight extent. Reactive metals such as potassium, sodium and calcium react with water, liberating hydrogen (p. 162), and some other metals, e.g. magnesium and zinc, react with steam (p. 162). The reaction between steam and coke or hydrocarbons is the basis of the industrial production of hydrogen (p. 163).

Pure water is a very poor conductor of electricity and at 298 K one dm^3 of water contains only 10^{-7} moles of H_3O^+(aq) ions and an equal number of OH^-(aq) ions, i.e. the pH of pure water is 7. Despite the fact that water is such a poor conductor of electricity it will allow other substances, which by themselves are virtually non-conductors when pure, to ionise in it. This is due to the fact that the water molecule contains lone pairs of electrons and is extensively polarised; both these factors make the water molecule a strong proton acceptor, e.g.

$$\underset{\text{Acid (1)}}{HNO_3(l)} + \underset{\text{Base (2)}}{H_2O(l)} \rightleftharpoons \underset{\text{Base (1)}}{NO_3^-(aq)} + \underset{\text{Acid (2)}}{H_3O^+(aq)}$$

In the presence of much stronger proton acceptors than the water molecule, e.g. the carbonate ion, water is forced to act as a Brønsted-Lowry acid (p. 147). This explains why the carbonates of the Group 1A metals show an alkaline reaction in solution, e.g.

$$\underset{\text{Acid (1)}}{H_2O(l)} + \underset{\text{Base (2)}}{CO_3^{2-}(aq)} \rightleftharpoons \underset{\text{Base (1)}}{OH^-(aq)} + \underset{\text{Acid (2)}}{HCO_3^-(aq)}$$

The formation of an alkaline solution with the Group 1A metal carbonates is referred to as salt hydrolysis; salt hydrolysis also occurs when transition metal salts are dissolved in water, but in these instances an acidic solution results, e.g.

$$\underset{\text{Acid (1)}}{Fe(H_2O)_6^{3+}(aq)} + \underset{\text{Base (2)}}{H_2O(l)} \rightleftharpoons \underset{\text{Base (1)}}{Fe(H_2O)_5(OH)^{2+}(aq)} + \underset{\text{Acid (2)}}{H_3O^+(aq)}$$

In the latter case, water molecules function as bases and detach protons from other water molecules, which are strongly bound to the small, highly charged transition metal cation (p. 148). The hydration of ions in aqueous solution (particularly cations), and the evolution of energy that occurs in this hydration, is one of the factors that accounts for water being such a good solvent for many electrovalent compounds (p. 60).

Many covalent compounds are also hydrolysed by water, particularly halides; for instance, silicon tetrachloride and phosphorus trichloride are hydrolysed to silicic and phosphonic acids respectively with the evolution of hydrogen chloride. It is thought that these reactions proceed via the formation of an intermediate compound in which water molecules are datively bonded to the silicon and phosphorus atoms (p. 266).

20.9 Hardness of water

Water is described as being 'hard' if it forms an insoluble scum before it forms a lather with soap. The hardness of natural water is generally caused by the hydrogen carbonates and sulphates of calcium and magnesium but, in fact, any soluble salts that form a scum with soap would cause hardness.

Soap is a mixture of the sodium salts of long chain fatty acids and includes sodium octadecanoate (stearate), $C_{17}H_{35}COO^-Na^+$ which reacts with the calcium and magnesium ions to form a scum (precipitate):

$$Ca^{2+}(aq) + 2C_{17}H_{35}COO^-(aq) \longrightarrow (C_{17}H_{35}COO^-)_2Ca^{2+}(s)$$

$$Mg^{2+}(aq) + 2C_{17}H_{35}COO^-(aq) \longrightarrow (C_{17}H_{35}COO^-)_2Mg^{2+}(s)$$

Soap will not produce a lather with water until all the calcium and magnesium ions have been precipitated as octadecanoates; hard water, therefore, wastes soap.

Water is described as being temporarily hard if it contains calcium and magnesium hydrogen carbonates, since this type of hardness is easily removed, and permanently hard if it contains calcium and magnesium sulphates. Both types of hardness frequently occur together in hard water.

Temporary hardness in water

Rain water dissolves small quantities of carbon dioxide from the atmosphere and is thus a very dilute solution of carbonic acid. This water attacks calcium and magnesium carbonates in any rocks over which it flows and the soluble hydrogen carbonates are formed:

$$Ca^{2+}CO_3^{2-}(s) + H_2O(l) + CO_2(g) \longrightarrow Ca^{2+}(HCO_3^-)_2(aq)$$

Temporary hardness in water is easily removed by boiling, as the hydrogen carbonates decompose readily and the insoluble carbonate is precipitated:

$$Ca^{2+}(HCO_3^-)_2(aq) \xrightarrow{\text{boil}} Ca^{2+}CO_3^{2-}(s) + H_2O(l) + CO_2(g)$$

It can also be removed by Clark's process, which involves the addition of slaked lime, $Ca^{2+}(OH^-)_2$:

$$Ca^{2+}(HCO_3^-)_2(aq) + Ca^{2+}(OH^-)_2(aq) \longrightarrow 2Ca^{2+}CO_3^{2-}(s) + 2H_2O(l)$$

It is essential to add only the correct amount of slaked lime because any excess will cause artificial hardness.

Permanent hardness in water

Permanent hardness is introduced when water passes over rocks containing the sulphates of calcium and magnesium. Neither boiling nor the addition of slaked lime will remove this type of hardness; it can, however, be removed, together with temporary hardness, by the addition of washing soda, $(Na^+)_2CO_3^{2-}$, or by the Permutit process.

(a) *Addition of washing soda (sodium carbonate)*

The calcium and magnesium ions in the water react with the carbonate ions from the sodium carbonate and a precipitate results:

$$Ca^{2+}(aq) + CO_3^{2-}(aq) \longrightarrow Ca^{2+}CO_3^{2-}(s)$$

$$Mg^{2+}(aq) + CO_3^{2-}(aq) \longrightarrow Mg^{2+}CO_3^{2-}(s)$$

The water now contains dissolved sodium salts but these have no effect upon the soap.

(b) *The ion exchange process*

Permutit is a complex chemical compound called sodium aluminium silicate (abbreviated here to $Na^+AlSilicate^-$) which is insoluble in water. When a slow stream of hard water is passed through this material, the calcium and magnesium ions exchange with the sodium ions in the Permutit thus:

$$Ca^{2+}(aq) + 2Na^+AlSilicate^-(s) \longrightarrow Ca^{2+}(AlSilicate^-)_2(s) + 2Na^+(aq)$$
$$Mg^{2+}(aq) + 2Na^+AlSilicate^-(s) \longrightarrow Mg^{2+}(AlSilicate^-)_2(s) + 2Na^+(aq)$$

Eventually the Permutit is converted into a mixture of calcium and magnesium aluminium silicates and has to be regenerated; this is achieved by passing a concentrated solution of sodium chloride through it:

$$Ca^{2+}(AlSilicate^-)_2(s) + 2Na^+(aq) \longrightarrow 2Na^+AlSilicate^-(s) + Ca^{2+}(aq)$$

The calcium chloride is then completely washed out of the Permutit before re-use.

(c) *The Calgon process*

The sodium salts of chain polyphosphates (V) (p. 272) are able to complex with calcium and magnesium ions and so prevent them reacting with soap to form a scum. One such commercial preparation is called Calgon which complexes with calcium and magnesium ions in hard water, at the same time releasing an equivalent amount of sodium ions into the water.

The estimation of hardness in water

The hardness of water is best determined by titration with ethylenediamine tetra-acetic acid (abbreviated to EDTA) in the presence of special organic dyes which function as indicators. The theory of the method is outlined below.

Ethylenediamine tetra-acetic acid (EDTA) has the formula

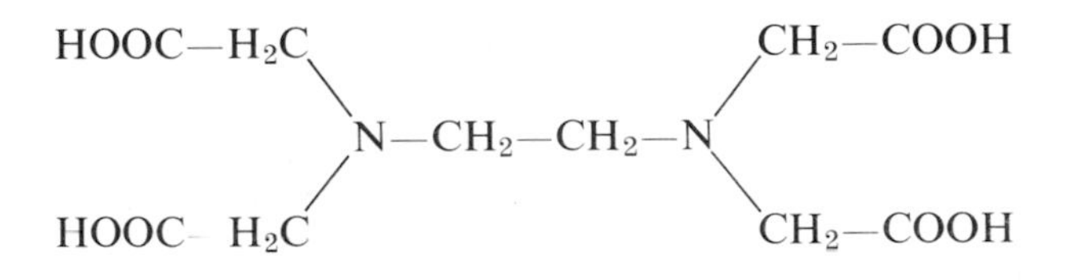

and is a tetrabasic acid, but since it is not very soluble in water it is usual to use the disodium salt. Under appropriate conditions the quadruply charged anion shown below can be obtained; this anion forms 1:1 complexes of great stability with many metallic cations (1 mole of the anion reacts with 1 mole of the cation), co-ordinating onto the cations by means of lone pairs of electrons (marked with an asterisk below).

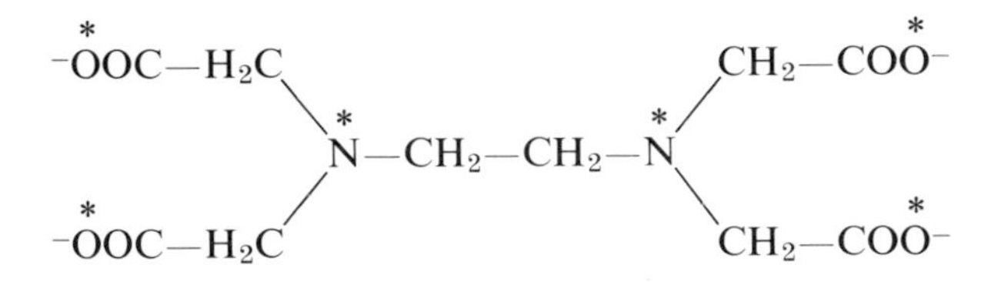

In order to obtain the EDTA completely in the form of the above anion, it is necessary to add $OH^-(aq)$ ions which remove the $H^+(aq)$ ions as unionised water molecules, and thus allow the following four equilibria to swing over completely to the right (it is customary to denote EDTA as H_4Y when discussing these reactions).

$$H_4Y(aq) \rightleftharpoons H_3Y^-(aq) + H^+(aq)$$
$$H_3Y^-(aq) \rightleftharpoons H_2Y^{2-}(aq) + H^+(aq)$$
$$H_2Y^{2-}(aq) \rightleftharpoons HY^{3-}(aq) + H^+(aq)$$
$$HY^{3-}(aq) \rightleftharpoons Y^{4-}(aq) + H^+(aq)$$

A method of determining the stage when all the cations in solution are complexed with the EDTA anion is necessary, and this is provided by special organic dyes. The basic requirement here is that the dye must complex with the cations, but not as strongly as the EDTA; furthermore, it must show one colour when uncomplexed and another colour when complexed with only a minute amount of cations. We are now in a position to discuss the EDTA titration.

A measured quantity of hard water is pipetted out and an alkaline buffer solution added (to ensure that when the EDTA is added, it is all in the form of $Y^{4-}(aq)$). The indicator is now added (Eriochrome Black T) which complexes with the calcium and magnesium ions producing a red solution. EDTA is now added from a burette and gradually displaces the Eriochrome indicator in the complex. At the end-point, when all the Eriochrome has been displaced, the solution turns blue. The titration reading is a measure of the total concentration of calcium and magnesium ions in solution.

If the whole experiment is now repeated with another indicator (Murexide), which only complexes with the calcium ions, the titration reading now is a measure of the calcium ions in solution. By difference the magnesium ion concentration can be determined.

20.10 Preparation of pure water

The removal of hardness in water by the addition of washing soda, Calgon, or by the Permutit process produces water that contains dissolved sodium salts. While this is no disadvantage for most purposes, there are occasions when water free from dissolved solids is required. Two methods are available for producing pure water.

(a) By distillation. This method tends to be rather expensive but is the only means available for removing all solid solutes.

(b) By the use of ion exchange resins. This method is generally less expensive than distillation and can be employed to remove dissolved salts.

Purification of water by ion exchange resins

The principle here is the same as the one applied in the Permutit process but in this case the water must be passed through two complex compounds instead of one. These two complex compounds, called resins, are essentially organic polymers; one contains acidic groupings along the length of the polymer chain, whereas the other contains basic groupings. They are called respectively cation exchangers and anion exchangers. The layout of a typical laboratory apparatus is shown in fig. 20.5. The hard water passes through a column containing the cation exchanger, when the calcium and magnesium ions are exchanged for hydrogen ions, and then through an anion exchanger where the hydrogen carbonate and sulphate ions are replaced by hydroxyl ions. If we consider the removal of calcium sulphate, and represent the cation and anion exchangers by the abbreviations H^+R^- and $R_1^+OH^-$ respectively, the equations are:

FIG. 20.5. *Laboratory ion exchange column*

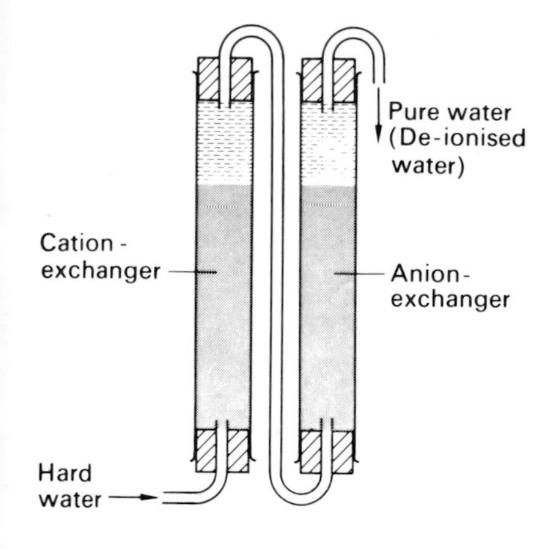

$$Ca^{2+}SO_4^{2-}(aq) + 2H^+R^-(s) \longrightarrow \underset{\text{passes out of first column as very dilute solution}}{(H^+)_2SO_4^{2-}(aq)} + Ca^{2+}(R^-)_2(s)$$

$$\underset{\text{from first column}}{(H^+)_2SO_4^{2-}(aq)} + 2R_1^+OH^-(s) \longrightarrow (R_1^+)_2SO_4^{2-}(s) + \underset{\text{pure water}}{2H_2O(l)}$$

The water that passes out of the second column is therefore freed from dissolved salts. The spent resins are regenerated eventually by running dilute sulphuric acid and sodium carbonate solution through the respective columns. Modern ion exchange columns now incorporate both cation and anion exchangers in a single unit.

20.11 Hydrogen peroxide, H_2O_2

Laboratory preparation of hydrogen peroxide

Hydrogen peroxide can be obtained by adding barium peroxide to dilute sulphuric acid at 0°C. A dilute solution is obtained and can be filtered to remove the barium sulphate:

$$Ba^{2+}(O{-}O)^{2-}(s) + (H^+)_2SO_4^{\ 2-}(aq) \longrightarrow Ba^{2+}SO_4^{\ 2-}(s) + H_2O_2(aq)$$

Industrial production of hydrogen peroxide

Hydrogen peroxide is manufactured by catalytically reducing 2-butylanthraquinone to 2-butylanthraquinol, which is then oxidised with oxygen-enriched air to hydrogen peroxide. 2-Butylanthraquinone is reformed and thus functions as a catalyst. The hydrogen peroxide is then extracted with water to give about a 20 per cent solution.

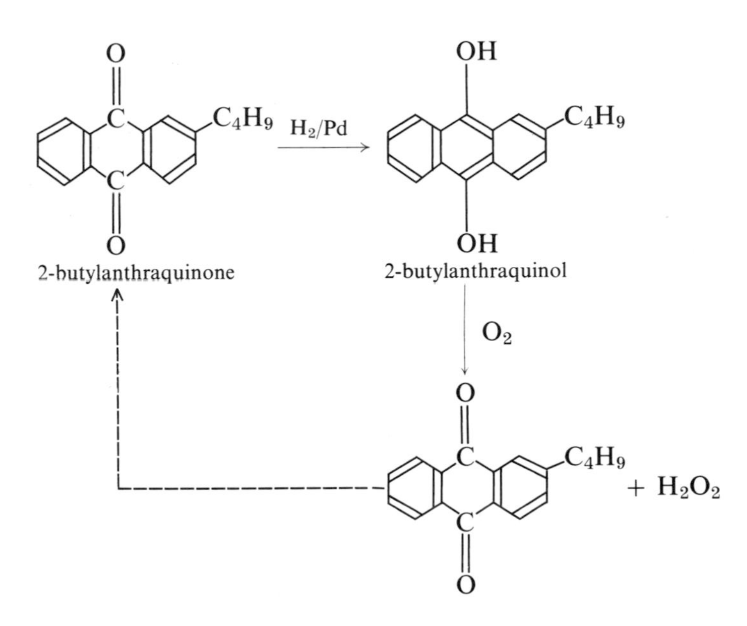

A second industrial method of manufacture involves the oxidation of propan-2-ol with oxygen under slight pressure:

$$\underset{\text{propan-2-ol}}{(CH_3)_2CHOH(l)} + O_2(g) \longrightarrow \underset{\text{propanone}}{(CH_3)_2C=O(l)} + H_2O_2(l)$$

Water is now added and the acetone (propanone) recovered by distillation. Once again a dilute solution of hydrogen peroxide is formed.

An aqueous solution of hydrogen peroxide can be concentrated by distillation under reduced pressure; the pure substance is very easily decomposed in the presence of light, and finely divided metals can cause explosive decomposition. It is stored either in smooth dark glass bottles or in smooth aluminium containers.

Physical properties of hydrogen peroxide

Pure hydrogen peroxide is a pale blue viscous liquid which boils at 150°C with decomposition, but it can be distilled under reduced pressure; it freezes at −0·9°C. The liquid is an associated substance like water, owing to the presence of hydrogen bonding.

Chemical properties of hydrogen peroxide

Hydrogen peroxide is thermodynamically unstable with respect to oxygen and water, since its decomposition results in the liberation of heat (ΔH is negative) and an increase in entropy (ΔS is positive since 2 moles of hydrogen peroxide form 2 moles of water and 1 mole of oxygen). These two effects reinforce one another, resulting in a large negative free energy change:

$$2H_2O_2(l) \longrightarrow 2H_2O(l) + O_2(g) \qquad \Delta G^{\ominus}(298\,K) = -238{\cdot}4\,kJ\,mol^{-1}$$
$$\Delta H^{\ominus}(298\,K) = -196{\cdot}6\,kJ\,mol^{-1}$$

Its decomposition into water and oxygen is catalysed by many finely divided solids such as manganese (IV) oxide and metals, and also by traces of alkali. Its inherent instability is principally accounted for by the fact that O—O single bonds are not very strong (cf. the instability of ozone p. 287).

Hydrogen peroxide is generally used in the laboratory as an aqueous solution, a '20 volume' solution containing about 6 per cent of hydrogen peroxide and so called because 1 cm³ of the liquid gives about 20 cm³ of oxygen when it decomposes at s.t.p. Even a dilute solution is powerfully oxidising, e.g. lead sulphide is oxidised to lead sulphate, iron (II) ions are converted into iron (III) ions, and iodide ions are oxidised to iodine in acid solution:

$$PbS(s) + 4H_2O_2(aq) \longrightarrow PbSO_4(s) + 4H_2O(l)$$
$$2Fe^{2+}(aq) + H_2O_2(aq) + 2H^+(aq) \longrightarrow 2Fe^{3+}(aq) + 2H_2O(l)$$
$$2I^-(aq) + H_2O_2(aq) + 2H^+(aq) \longrightarrow I_2(aq) + 2H_2O(aq)$$

Powerful oxidising agents such as silver oxide and an acidified solution of potassium manganate (VII) force hydrogen peroxide to assume the role of a reducing agent:

$$Ag_2O(s) + H_2O_2(aq) \longrightarrow 2Ag(s) + H_2O(l) + O_2(g)$$
$$2MnO_4^-(aq) + 5H_2O_2(aq) + 6H^+(aq) \longrightarrow 2Mn^{2+}(aq) + 8H_2O(l) + 5O_2(g)$$

The last reaction is a convenient volumetric method of estimating the concentration of an aqueous solution of hydrogen peroxide. Incidentally, by using isotopically labelled hydrogen peroxide, it has definitely been established that the molecular oxygen originates from this source and not from the potassium manganate (VII).

When an acidified solution of potassium dichromate (VI) is treated with hydrogen peroxide, chromium peroxide, CrO_5, is formed which can be extracted with ether, producing a blue solution. This reaction serves as a qualitative test for hydrogen peroxide.

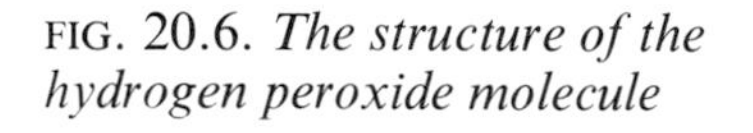

FIG. 20.6. *The structure of the hydrogen peroxide molecule*

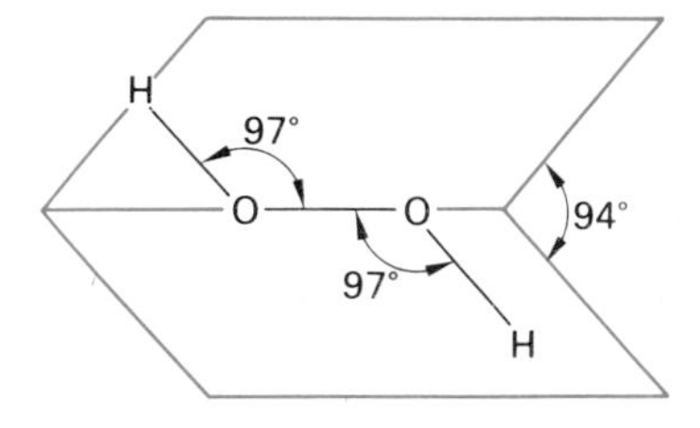

The structure of hydrogen peroxide

The hydrogen peroxide molecule contains two O—H groups, which do not lie in the same plane, joined together through the two oxygen atoms. The structure shown (fig. 20.6) is the one that reduces to a minimum the repulsion between the lone pairs on the oxygen atoms.

Uses of hydrogen peroxide

The main use of hydrogen peroxide is as a bleaching agent. Other uses include the preparation of peroxides, particularly organic peroxides which are used for initiating polymerisation reactions, and the preparation of antiseptics. Liquid hydrogen peroxide in conjunction with liquid hydrazine has been used for rocket propulsion.

Sulphur

20.12 Occurrence of sulphur

Sulphur occurs in combination with many metals as sulphides, if they are insoluble in water, e.g. the sulphides of zinc, lead, copper and mercury. Sulphur dioxide is recovered as a by-product during the extraction of these metals from their sulphide ores. Another valuable source of sulphur dioxide is iron pyrites, FeS_2.

Many sulphate minerals occur, including anhydrite, $Ca^{2+}SO_4^{2-}$, gypsum, $Ca^{2+}SO_4^{2-}.2H_2O$, and Epsom salts, $Mg^{2+}SO_4^{2-}.7H_2O$. Other important sources of sulphur are crude oil and natural gas, from which it is extracted as hydrogen sulphide.

Sulphur occurs in the free state in Japan, and underground in Texas and Louisiana, where it was discovered by Frasch in 1867.

In order to extract this underground sulphur (the Frasch process), three concentric pipes are sunk deep into the ground (fig. 20.7). Superheated water at 170°C is forced down the outer pipe into the sulphur which is melted; compressed air blown through the inner pipe now forces the sulphur as a liquid to the surface, where it is allowed to solidify. Sulphur of about 99·5 per cent purity is obtained by this process.

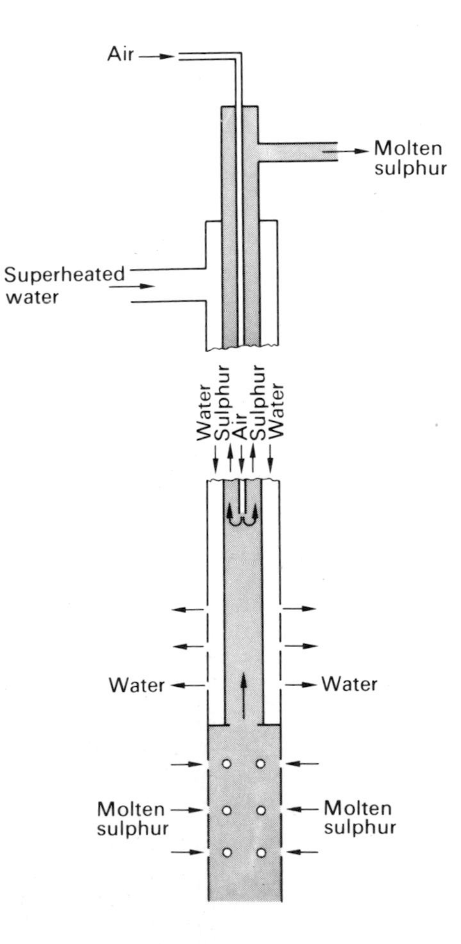

FIG. 20.7. *The Frasch process for extracting sulphur*

20.13 The allotropy of sulphur

Unlike oxygen which is a discrete molecule, two atoms being united by a double bond, sulphur atoms show a marked reluctance to double bond with themselves and the two main allotropes of sulphur contain S_8 molecules, in which single bonds unite sulphur atoms into a puckered octagonal ring (fig. 20.8). The high relative molecular mass of these S_8 structural units explains why sulphur, unlike oxygen, is a solid. The bond angles of 105° are consistent with the simple theory of electron pair repulsion (p. 75).

FIG. 20.8. *The structure of the* S_8 *molecule*

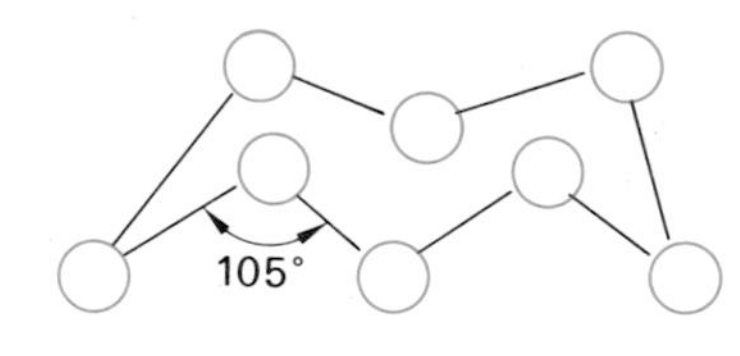

Rhombic sulphur

This is the form of sulphur normally encountered and consists of S_8 structural units packed together to give crystals whose shape is shown in fig. 20.9(a). Fairly large crystals can be obtained by allowing a solution of powdered roll sulphur in carbon disulphide to evaporate slowly; they are yellow, transparent and have a density of 2·06 g cm^{-3}.

FIG. 20.9. *The crystal shapes of (a) rhombic sulphur, (b) monoclinic sulphur*

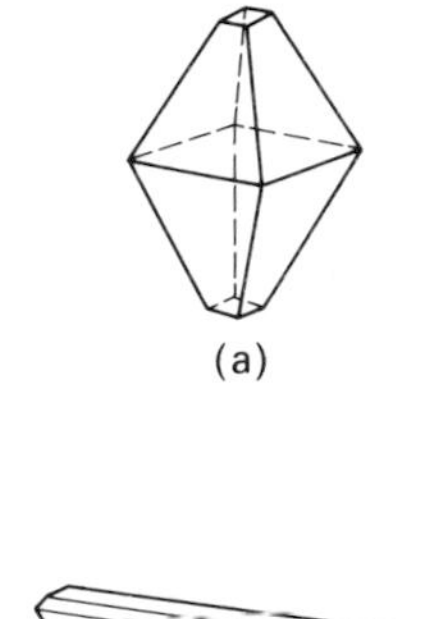

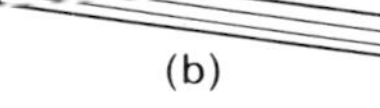

Monoclinic sulphur

This form of sulphur is formed when molten sulphur is allowed to crystallise above 95·6°C. Like rhombic sulphur it consists of S_8 structural units, but these are arranged differently in the crystal lattice (fig. 20.9(b)). The temperature of 95·6°C is the transition temperature for sulphur; below this temperature rhombic sulphur is the more stable allotrope and above it, monoclinic sulphur is the more stable of the two forms. This type of allotropy, in which a definite transition point exists where two forms become equally stable, is called **enantiotropy** (cf. the allotropy of tin p. 237).

$$\text{Rhombic sulphur} \overset{95{\cdot}6°\text{C}}{\rightleftharpoons} \text{Monoclinic sulphur}$$

Crystals of monoclinic sulphur are amber-yellow in colour and have a density of 1·96 g cm^{-3}. As they gradually change over into rhombic sulphur below 95·6°C, each crystal retains its overall shape but changes into a mass of small rhombic crystals.

Amorphous sulphur

A number of forms of sulphur which possess no regular crystalline form can be obtained when sulphur is liberated in chemical reactions, e.g. by the action of dilute hydrochloric acid on a solution of sodium thiosulphate:

$$S_2O_3^{2-}(aq) + 2H^+(aq) \longrightarrow H_2O(l) + SO_2(g) + S(s)$$

Plastic sulphur

This is obtained, as an amber-brown soft and elastic solid, by pouring nearly boiling sulphur into cold water. It consists of a completely random arrangement of chains of sulphur atoms which, when stretched, align themselves parallel to each other. On standing, it slowly changes over into rhombic sulphur, as the chains of sulphur atoms break and reform the S_8 cyclic units.

The action of heat on sulphur

Both rhombic and monoclinic sulphur melt to a yellow liquid. Owing to the conversion of rhombic to monoclinic sulphur, and also to possible variations in the percentage of allotropes of liquid sulphur formed, the melting points are not sharp; rhombic sulphur melts at approximately 113°C and monoclinic sulphur at approximately 119°C. As the temperature rises the colour of the liquid darkens until it is nearly black, and it becomes viscous. At about 200°C the viscosity begins to fall and at its boiling point of 445°C the liquid is again mobile. When sulphur vapour comes in contact with a cool surface it sublimes to give a pale yellow solid.

There is still some doubt concerning a complete explanation of these observations, but a recent theory runs as follows: as the sulphur melts the S_8 rings begin to open and it is possible that other ring systems containing possibly six and four sulphur atoms form. It is known, however, that sulphur chains begin to form and reach their maximum chain length at 200°C, corresponding to the maximum viscosity of liquid sulphur. The decrease in viscosity of liquid sulphur that occurs above 200°C is explained as being due to the breakdown of these long chains and the re-formation of S_8 rings. Sulphur vapour contains S_8 rings, together with smaller fragments such as S_6, S_4 and S_2. At very high temperatures atomic sulphur is formed.

20.14 Chemical properties of sulphur

Sulphur combines with most metals when heated, finely divided metals such as magnesium and aluminium (which are high up in the electrode potential series) reacting with considerable vigour, e.g.

$$Mg(s) + S(s) \longrightarrow Mg^{2+}S^{2-}(s)$$

Non-metals that combine with sulphur directly include fluorine, chlorine, oxygen and carbon; hydrogen combines reversibly to a slight extent when passed through molten sulphur near its boiling point.

$$\begin{aligned} S(s) + 3F_2(g) &\longrightarrow SF_6(g) \\ 2S(s) + Cl_2(g) &\longrightarrow S_2Cl_2(l) \\ S(s) + O_2(g) &\longrightarrow SO_2(g) \\ C(s) + 2S(s) &\longrightarrow CS_2(l) \\ H_2(g) + S(l) &\rightleftharpoons H_2S(g) \end{aligned}$$

Sulphur is oxidised by concentrated nitric and sulphuric acids; with hot concentrated solutions of alkalis a sulphide and a sulphite are formed, which react with more sulphur to form polysulphides and a thiosulphate respectively.

$$\begin{aligned} S(s) + 6HNO_3(aq) &\longrightarrow 2H_2O(l) + H_2SO_4(aq) + 6NO_2(g) \\ S(s) + 2H_2SO_4(l) &\longrightarrow 2H_2O(aq) + 3SO_2(g) \end{aligned}$$

$$3S(s) + 6OH^-(aq) \longrightarrow 2S^{2-}(aq) + SO_3^{2-}(aq) + 3H_2O(l)$$

followed by:

$$S^{2-}(aq) + nS(s) \longrightarrow \underset{\text{polysulphide ion}}{S_{n+1}^{2-}(aq)} \qquad n = 1 \text{ to } 8$$

$$SO_3^{2-}(aq) + S(s) \longrightarrow \underset{\text{thiosulphate ion}}{S_2O_3^{2-}(aq)}$$

20.15 Hydrogen sulphide, H_2S

Occurrence and preparation of hydrogen sulphide

Hydrogen sulphide occurs in great quantities in the natural gas deposits of France and Canada and these sources are of major importance to the sulphuric acid industry. Very small quantities of the gas are released from bad eggs.

It is usually prepared in the laboratory by he action of dilute hydrochloric acid on iron (II) sulphide in a Kipp's apparatus:

$$FeS(s) + 2H^+(aq) + 2Cl^-(aq) \longrightarrow Fe^{2+}(aq) + 2Cl^-(aq) + H_2S(g)$$

FIG. 20.10. *The structure of the hydrogen sulphide molecule*

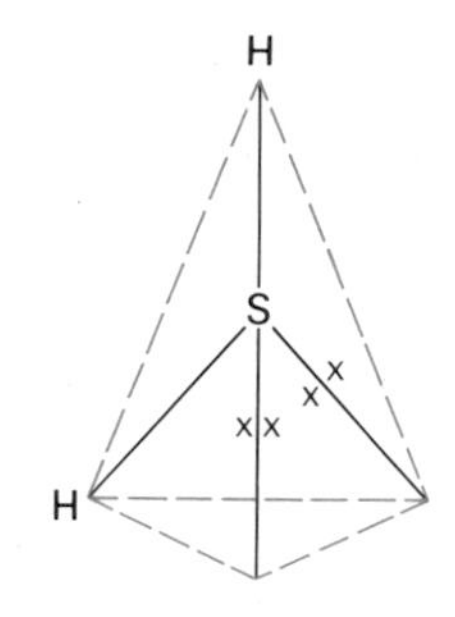

Since iron (II) sulphide always contains uncombined iron, the gas is contaminated with free hydrogen. A purer gas can be obtained by warming antimony (III) sulphide with concentrated hydrochloric acid:

$$Sb_2S_3(s) + 6HCl(aq) \longrightarrow 2SbCl_3(aq) + 3H_2S(g)$$

Physical properties of hydrogen sulphide

Unlike water, hydrogen sulphide is a gas, since the electronegativity of the sulphur atom is insufficient to allow it to participate in hydrogen bonding (p. 66). However, as would be expected, its molecule has an angular structure with the two lone pair and two bonding pair electrons taking up an approximately tetrahedral position (fig. 20.10).

It is an extremely poisonous substance (as little as 1 part per 1000 parts of air is fatal), but fortunately its smell becomes intolerable long before the fatal concentration is reached.

Chemical properties of hydrogen sulphide

These are most conveniently discussed under the following three headings.

(a) Acid properties

Hydrogen sulphide has no effect on litmus paper when dry, but once it becomes wet it will turn blue litmus paper a wine-red colour, showing that it functions as a weak acid in solution. Ionisation in water takes place in two stages, the second occurring only to a very minute extent:

$$H_2O(l) + H_2S(g) \rightleftharpoons H_3O^+(aq) + HS^-(aq)$$
$$H_2O(l) + HS^-(aq) \rightleftharpoons H_3O^+(aq) + S^{2-}(aq)$$

In the presence of hydroxyl ions, a much stronger proton acceptor than water molecules, the ionisation of hydrogen sulphide is much more extensive; thus the reaction with sodium hydroxide solution produces the hydrosulphide and sulphide, i.e. Na^+HS^- and $(Na^+)_2S^{2-}$.

$$OH^-(aq) + H_2S(g) \rightleftharpoons H_2O(l) + HS^-(aq)$$
$$OH^-(aq) + HS^-(aq) \rightleftharpoons H_2O(l) + S^{2-}(aq)$$

Even so, the reactions do not go to completion, i.e. the alkali metal hydrosulphides and sulphides show an alkaline reaction in solution, thus demonstrating the presence of $OH^-(aq)$ ions.

(b) Reactions involving precipitation

Hydrogen sulphide is used in qualitative analysis for precipitating the sulphides of many metals; in practice the precipitation is done under controlled conditions so that two groups of sulphides can be distinguished (analysis Groups II and IV which are in no way related to the Groups 2 and 4 of the periodic classification).

Group II —the sulphides of mercury (II), lead, bismuth, copper (II), cadmium, arsenic, antimony and tin.
Group IV—the sulphides of nickel, cobalt, manganese and zinc.

The sulphides of Group II have smaller solubility products than those of Group IV, i.e. they are less soluble in water. In order to ensure that the Group II sulphides precipitate but the Group IV sulphides do not, the sulphide ion concentration must be decreased to an even smaller value than it already is in a saturated solution of hydrogen sulphide. This is achieved by carrying out the precipitation in the presence of acid, the additional $H_3O^+(aq)$ ions shifting the two equilibria (shown below) to the left and thus reducing $[S^{2-}(aq)]$.

$$H_2O(l) + H_2S(g) \rightleftharpoons H_3O^+(aq) + HS^-(aq)$$
$$H_2O(l) + HS^-(aq) \rightleftharpoons H_3O^+(aq) + S^{2-}(aq)$$
$$\longleftarrow$$

Both equilibria shifted to the left by the extra H_3O^+ ions

The points made above are best considered by using an example.

EXAMPLE

'Show that both copper (II) sulphide and nickel sulphide precipitate when hydrogen sulphide is passed through molar solutions of copper (II) and nickel salts, but that only copper (II) sulphide precipitates when the solution is made 0·3 molar with respect to hydrochloric acid'. The solubility products of CuS and NiS are approximately 10^{-36} and 10^{-21} mol² dm⁻⁶ respectively.

(a) In a neutral solution the sulphide ion concentration is approximately 10^{-15} mol dm^{-3}.

$[Cu^{2+}(aq)][S^{2-}(aq)] = 10^{-36}\ mol^2\ dm^{-6}$ $\quad [Cu^{2+}(aq)] = 10^{-36}/10^{-15} = 10^{-21}\ mol\ dm^{-3}$

$[Ni^{2+}(aq)][S^{2-}(aq)] = 10^{-21}\ mol^2\ dm^{-6}$ $\quad [Ni^{2+}(aq)] = 10^{-21}/10^{-15} = 10^{-6}\ mol\ dm^{-3}$

The minimum concentration of $Cu^{2+}(aq)$ and $Ni^{2+}(aq)$ necessary for precipitation to occur (10^{-21} and 10^{-6} mol dm^{-3} respectively) is considerably less than the concentrations of these ions present (molar solutions), therefore both sulphides precipitate.

(b) In 0·3 molar solution of hydrochloric acid a saturated solution of hydrogen sulphide has a sulphide ion concentration of approximately 10^{-22} mol dm^{-3}.

$[Cu^{2+}(aq)][S^{2-}(aq)] = 10^{-36}\ mol^2\ dm^{-6}$ $\quad [Cu^{2+}(aq)] = 10^{-36}/10^{-22} = 10^{-14}\ mol\ dm^{-3}$

$[Ni^{2+}(aq)][S^{2-}(aq)] = 10^{-21}\ mol^2\ dm^{-6}$ $\quad [Ni^{2+}(aq)] = 10^{-21}/10^{-22} = 10\ mol\ dm^{-3}$

The minimum concentrations of $Cu^{2+}(aq)$ and $Ni^{2+}(aq)$ necessary for precipitation to occur (10^{-14} and 10 mol dm^{-3}) are now very much larger, and only in the case of copper (II) sulphide can precipitation occur, i.e. the minimum concentration of $Ni^{2+}(aq)$ ions needed for precipitation to occur is ten times higher than the concentration present.

The above example shows clearly that the acidity of the solution can influence the precipitation of metallic sulphides. However, the problem is seldom as simple as this, and often one has to consider the possibility of complex ion formation. Even more serious is the fact that widely divergent values of solubility products can be found in the literature.

To ensure the complete precipitation of the Group IV sulphides, the solution is made alkaline (generally by adding aqueous ammonia solution in the presence of ammonium chloride). The presence of the OH^- ensures a greater sulphide ion concentration than that present in a neutral solution of hydrogen sulphide (the OH^- ions combine with the H_3O^+ ions, formed when the hydrogen sulphide ionises, to form water, and more hydrogen sulphide ionises as a result).

$$H_2O(l) + H_2S(g) \rightleftharpoons H_3O^+(aq) + HS^-(aq)$$
$$H_2O(l) + HS^-(aq) \rightleftharpoons H_3O^+(aq) + S^{2-}(aq)$$
$$+$$
$$OH^-(aq)$$
$$\upharpoonleft\downharpoonright$$
$$2H_2O(aq)$$
$$\longrightarrow$$

Both equilibria shifted this way by the addition of OH^- ions

The formation of a black precipitate of lead sulphide, formed when hydrogen sulphide comes in contact with a filter paper strip dipped in lead ethanoate solution, is a convenient test for the gas:

$$Pb^{2+}(aq) + H_2S(g) \longrightarrow PbS(s) + 2H^+(aq)$$

(c) Hydrogen sulphide as a reducing agent

Hydrogen sulphide is a reducing agent because it can readily give up its hydrogen and deposit sulphur. It burns in air to form sulphur which is oxidised to sulphur dioxide:

$$2H_2S(g) + 3O_2(g) \longrightarrow 2SO_2(g) + 2H_2O(l)$$

It reacts with dissolved oxygen in water, so that a saturated solution of hydrogen sulphide will slowly go cloudy after standing for some days:

$$2H_2S(aq) + O_2(g) \longrightarrow 2H_2O(l) + 2S(s)$$

Other reducing reactions of hydrogen sulphide include the conversion of moist chlorine to hydrogen chloride, moist sulphur dioxide to sulphur, sulphuric acid to sulphur dioxide, iron (III) ions to iron (II) ions, and acidified dichromate (VI) and manganate (VII) solutions to chromium (III) and manganese (II) ions respectively:

$$\text{(moist) } Cl_2(g) + H_2S(g) \longrightarrow 2HCl(g) + S(s)$$
$$\text{(moist) } SO_2(g) + 2H_2S(g) \longrightarrow 2H_2O(l) + 3S(s)$$
$$H_2SO_4(l) + H_2S(g) \longrightarrow SO_2(g) + 2H_2O(l) + S(s)$$
$$2Fe^{3+}(aq) + H_2S(g) \longrightarrow 2Fe^{2+}(aq) + 2H^+(aq) + S(s)$$
$$Cr_2O_7^{2-}(aq) + 8H^+(aq) + 3H_2S(g) \longrightarrow 2Cr^{3+}(aq) + 7H_2O(l) + 3S(s)$$
$$2MnO_4^-(aq) + 6H^+(aq) + 5H_2S(g) \longrightarrow 2Mn^{2+}(aq) + 8H_2O(l) + 5S(s)$$

Note that in the reactions above the hydrogen sulphide is always oxidised to sulphur.

20.16 Other hydrides of sulphur

Sulphur forms a number of hydrides which contain catenated sulphur atoms such as H_2S_2, H_2S_3, H_2S_4, etc. These are all yellow oils which readily decompose into hydrogen sulphide and free sulphur; they can be represented by the general formula:

$$H—(S)_n—H$$

The first member of the series ($n = 2$) is the sulphur analogue of hydrogen peroxide and decomposes in a similar manner:

$$2H_2S_2(l) \longrightarrow 2H_2S(g) + 2S(s)$$
$$\text{cf. } 2H_2O_2(l) \longrightarrow 2H_2O(l) + O_2(g)$$

20.17 Metallic sulphides

Metallic sulphides are less ionic than the corresponding oxides, and only the sulphides of Groups 1A and 2A appear to be essentially ionic. Like the corresponding oxides, these ionic sulphides are soluble in water and give a similar alkaline reaction:

$$S^{2-} + H_2O(l) \rightleftharpoons HS^-(aq) + OH^-(aq)$$
$$\text{cf. } O^{2-} + H_2O(l) \longrightarrow 2OH^-(aq)$$

The hydrolysis of the sulphide ion is only partially complete in the cold, but on boiling it goes to completion, since hydrogen sulphide is evolved; the loss of hydrogen sulphide from the system thus allows further hydrolysis of the hydrosulphide ion to take place unimpeded:

$$HS^-(aq) + H_2O(l) \longrightarrow H_2S(g) + OH^-(aq) \text{ (goes to completion on boiling)}$$

Aqueous solutions of these ionic sulphides react with sulphur on heating to form a number of polysulphides which contain chain polysulphide anions:

$$(K^+)_2S^{2-}(aq) + nS(s) \longrightarrow (K^+)_2S_{n+1}{}^{2-}(aq)$$

These anions, as expected, possess an approximately tetrahedral inter-sulphur bond angle (electron pair repulsion p. 75); thus the structure of the $S_3{}^{2-}$ ion ($n = 2$) is as shown below:

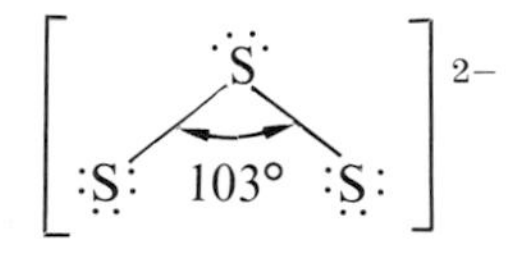

By far the greatest proportion of metallic sulphides are covalent and these are insoluble in water; many occur naturally, e.g. ZnS, PbS, etc., and are of importance for the metals they contain (p. 153). The sulphides of transition elements behave very much like alloys and many show metallic properties, e.g. metallic lustre and conductivity. Generally they cannot be allotted definite formulae, i.e. they are non-stoichiometric; for example iron (II) sulphide has a slightly variable composition, and one specimen has been described as having the composition $Fe_{0.858}S$, in which it is deficient in iron.

Many transition elements also form disulphides in which discrete S_2 groups are present; they too are covalent solids. Iron pyrites, FeS_2, is an important source of sulphur (p. 299).

20.18 The halides of sulphur

The fluorides of sulphur

Sulphur forms three covalent fluorides with formulae SF_4, SF_6 and S_2F_{10}, the first two being gases and the other one a very volatile liquid. Sulphur hexafluoride, SF_6, is made by the direct combination of sulphur and fluorine, traces of the other two fluorides being formed at the same time.

Sulphur tetrafluoride, SF_4, contains 4-valent sulphur and is immediately hydrolysed by water to sulphur dioxide and hydrogen fluoride:

$$SF_4(g) + 2H_2O(l) \longrightarrow SO_2(g) + 4HF(aq)$$

FIG. 20.11. *The structure of (a) sulphur tetrafluoride, (b) sulphur hexafluoride, (c) disulphur decafluoride*

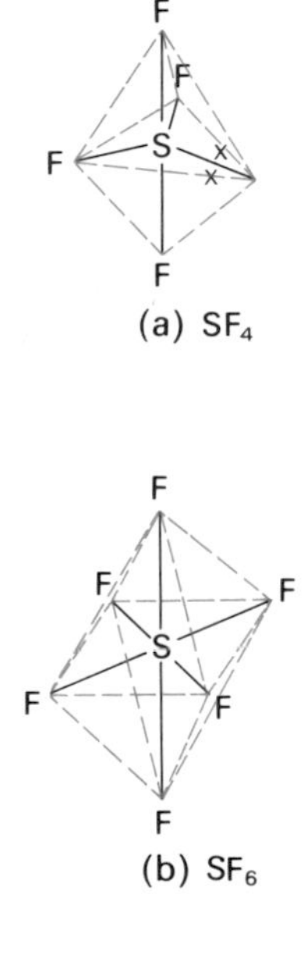

(a) SF_4

(b) SF_6

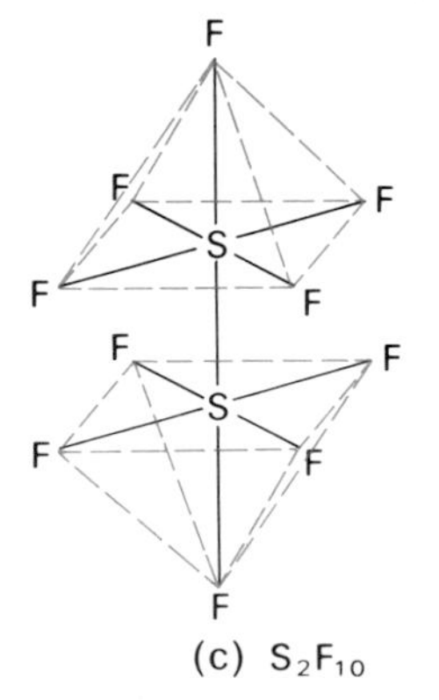

(c) S_2F_{10}

Sulphur hexafluoride, SF_6, contains 6-valent sulphur and is chemically inert towards fused alkali and oxygen. This extreme stability is probably due to the fact that sulphur is exhibiting its maximum valency; the screening of the central sulphur atom by the six fluorine atoms is probably of equal importance. It is used as an insulating material in high voltage transformers.

Disulphur decafluoride, S_2F_{10}, is exceedingly poisonous (no explanation of its toxic properties has yet been advanced). It is chemically rather inert, but not to the same extent as sulphur hexafluoride.

The structures of the three fluorides of sulphur are very much in line with those predicted by electron pair repulsion (fig. 20.11), although the structure of SF_4 cannot be predicted specifically (two possible theoretical structures are possible).

The chlorides of sulphur

The known chlorides of sulphur are disulphur dichloride, S_2Cl_2, sulphur dichloride, SCl_2, and sulphur tetrachloride, SCl_4, the latter being very unstable and decomposing at about $-31°C$. Disulphur dichloride is formed when chlorine is passed over molten sulphur, and can be distilled off as a red, repulsive-smelling liquid:

$$2S(l) + Cl_2(g) \longrightarrow S_2Cl_2(l)$$

It is used for cross-linking the hydrocarbon chains in rubber and so hardening it, a process called vulcanisation. Its molecule is similar in shape to that of hydrogen peroxide (p. 298).

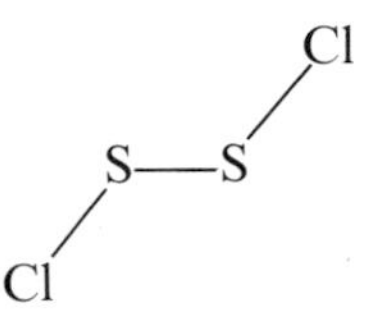

Sulphur dichloride is prepared by reacting chlorine with disulphur dichloride at 0°C. It is a red liquid which very readily dissociates into chlorine and disulphur dichloride:

$$S_2Cl_2(l) + Cl_2(g) \underset{\text{above } 0°C}{\overset{\text{at } 0°C}{\rightleftharpoons}} 2SCl_2(l)$$

It is interesting to note that sulphur displays valencies of 2, 4 and 6 in its halides and that the very electronegative fluorine atom is required to bring out the highest valency. This is possible because sulphur has easily accessible *d* orbitals available, and successive electron promotion from the 3*p* and 3*s* to the 3*d* levels results in four and six unpaired electrons:

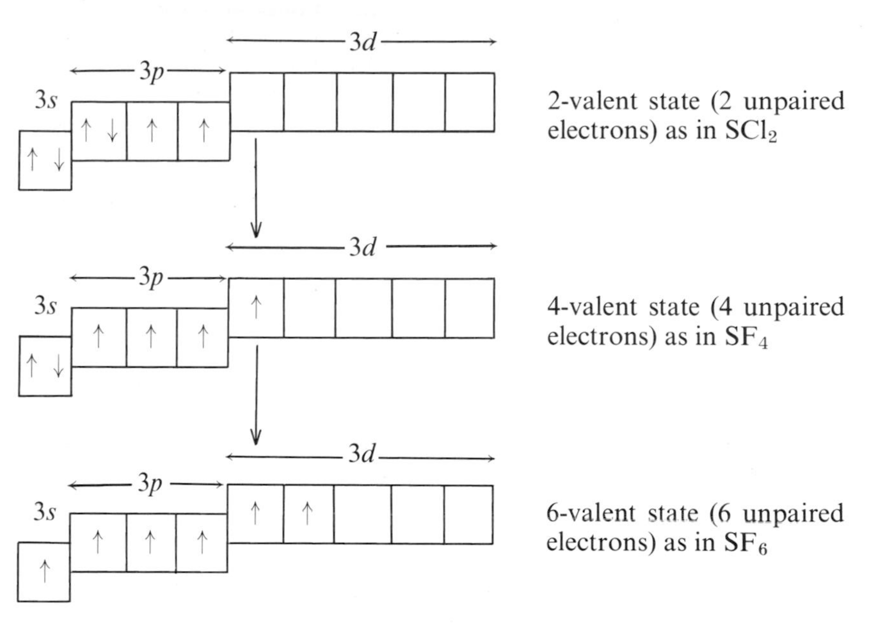

Unlike sulphur, oxygen is restricted to a covalency of 2 since the oxygen atom does not possess easily accessible *d* orbitals, cf. nitrogen has a covalency of 3, but phosphorus shows covalencies of 3 and 5 (p. 265).

20.19 The oxides of sulphur

Sulphur forms several oxides, but the only ones we shall be concerned with here are sulphur dioxide, SO_2, and sulphur trioxide, SO_3.

Sulphur dioxide, SO_2

Preparation

Sulphur dioxide is formed, together with a little sulphur trioxide, when sulphur is burnt in air or oxygen:

$$S(s) + O_2(g) \longrightarrow SO_2(g)$$

In the laboratory it is readily generated by reacting a sulphite with dilute sulphuric acid:

$$SO_3^{2-}(aq) + 2H^+(aq) \longrightarrow H_2O(l) + SO_2(g)$$

It can also be obtained by heating concentrated sulphuric acid with copper, the following equation only partially summarising the reaction (there are significant side reactions):

$$Cu(s) + 2H_2SO_4(l) \longrightarrow Cu^{2+} + SO_4^{2-} + 2H_2O(l) + SO_2(g)$$

Industrially it is produced as a by-product during the roasting of sulphide ores and by other methods (p. 311).

Physical properties
Sulphur dioxide is a colourless dense gas with a choking smell. It is very soluble in water—1 volume of water at 0°C dissolves about 80 volumes of the gas. It boils at −10°C, liquefies under three atmospheres at 20°C and is used as a fumigant.

The sulphur dioxide molecule is angular and, because the sulphur-oxygen bond lengths (which are both equal) are shorter than S—O bonds, it is thought that both bonds are double, i.e. the sulphur atom expands its octet and is 4-valent:

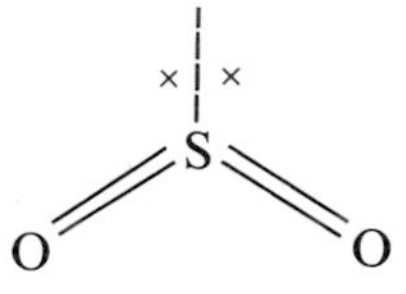

Chemical properties
Sulphur dioxide dissolves readily in water, forming a solution of sulphurous acid, but any attempt to isolate this acid by evaporation results in sulphur dioxide being recovered:

$$SO_2(g) + H_2O(l) \rightleftharpoons H_2SO_3(aq) \rightleftharpoons H^+(aq) + HSO_3^-(aq) \rightleftharpoons 2H^+(aq) + SO_3^{2-}(aq)$$

It reacts very readily with sodium hydroxide solution, forming sodium sulphite, which then reacts with more sulphur dioxide to form the hydrogen sulphite:

$$2Na^+OH^-(aq) + SO_2(g) \longrightarrow (Na^+)_2SO_3^{2-}(aq) + H_2O(l)$$
$$(Na^+)_2SO_3^{2-}(aq) + H_2O(l) + SO_2(g) \longrightarrow 2Na^+HSO_3^-(aq)$$

In its reaction with water and alkalis, the behaviour of sulphur dioxide is very similar to that of carbon dioxide (p. 219).

Sulphur dioxide reacts with chlorine in the presence of charcoal (which acts as a catalyst) to give sulphur dichloride dioxide, SO_2Cl_2, and is oxidised catalytically to sulphur trioxide by oxygen in the presence of vanadium (V) oxide:

$$SO_2(g) + Cl_2(g) \longrightarrow SO_2Cl_2(l)$$
$$2SO_2(g) + O_2(g) \longrightarrow 2SO_3(g)$$

When moist, sulphur dioxide behaves as a reducing agent, e.g. it converts iron (III) ions to iron (II) ions and decolorises acidified potassium manganate (VII) solution, the latter reaction being a convenient test for the gas. However, these reactions are best regarded as redox reactions involving the sulphite ion, and are discussed in section 20.20.

Sulphur trioxide, SO_3

Preparation
Sulphur trioxide is evolved as a white smoke when some metallic sulphates are strongly heated, e.g.

$$(Fe^{3+})_2(SO_4^{2-})_3(s) \longrightarrow (Fe^{3+})_2(O^{2-})_3(s) + 3SO_3(s)$$

It is best prepared by passing a mixture of sulphur dioxide and oxygen (bubbled through concentrated sulphuric acid to dry the mixed gases) over heated platinised asbestos. The sulphur trioxide formed is condensed as

FIG. 20.12. *The cyclic structure of sulphur trioxide*

a white solid in a receiver cooled in ice, and fitted with a sulphuric acid trap to prevent the entry of moisture:

$$2SO_2(g) + O_2(g) \xrightarrow[400°C]{\text{Pt catalyst}} 2SO_3(g)$$

Physical properties

Sulphur trioxide exists in several polymorphic forms, one form having a melting point of 17°C and a trimeric cyclic structure, in which four oxygen atoms are arranged approximately tetrahedrally around each sulphur atom (fig. 20.12):

This form gradually changes over into a linear polymerised structure in the presence of moisture (actually two slightly different forms are known). The strand-like form of the solid accounts very neatly for its fibrous needle-like appearance (fig. 20.13):

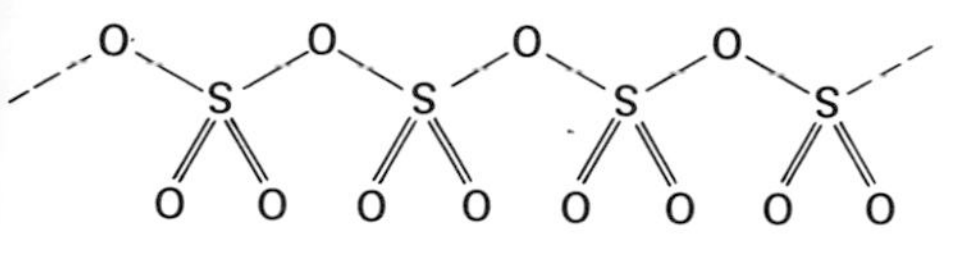

FIG. 20.13. *The linear polymerised structure of sulphur trioxide*

On heating, the structure of sulphur trioxide is broken down and discrete SO_3 molecules are present in the vapour; these have a symmetrical planar structure (electron pair repulsion):

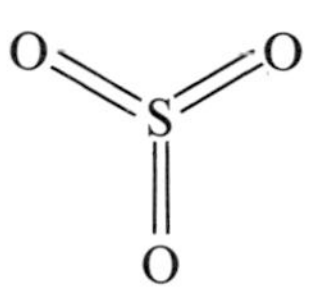

Note that sulphur shows a valency of 6 in all the structural modifications of sulphur trioxide.

Chemical properties

Sulphur trioxide is a powerful acidic oxide; thus it fumes in moist air and reacts explosively with water to form sulphuric acid:

$$SO_3(s) + H_2O(l) \longrightarrow H_2SO_4(l)$$

It continues to dissolve in the sulphuric acid to form pyrosulphuric acid (fuming sulphuric acid or oleum), $H_2S_2O_7$.

$$H_2SO_4(l) + SO_3(s) \longrightarrow H_2S_2O_7(l)$$

With basic oxides it reacts exothermically to form sulphates, e.g.

$$Ca^{2+}O^{2-}(s) + SO_3(s) \longrightarrow Ca^{2+}SO_4{}^{2-}(s)$$

In some of its reactions it functions as an oxidising agent; for example, it will oxidise hydrogen bromide to free bromine:

$$2HBr(g) + SO_3(s) \longrightarrow H_2O(l) + Br_2(l) + SO_2(g)$$

20.20 Sulphurous acid and its salts

An aqueous solution of sulphurous acid, H_2SO_3, is obtained when sulphur dioxide is passed into water; the solution contains $H^+(aq)$, $HSO_3^-(aq)$ and $SO_3^{2-}(aq)$ ions, together with free sulphur dioxide (p. 308).

Although pure sulphurous acid does not exist, sulphites of the alkali metals, e.g. $(Na^+)_2SO_3^{2-}$, can be obtained as solids (see p. 308 for preparation). Hydrogen sulphites, e.g. $Na^+HSO_3^-$ also exist in solution, but when attempts are made to isolate them, two hydrogen sulphite ions condense with the elimination of water, and a pyrosulphite is deposited:

$$2Na^+HSO_3^-(aq) \rightleftharpoons \underset{\text{sodium pyrosulphite}}{(Na^+)_2S_2O_5^{2-}(aq)} + H_2O(l)$$

or

$$2HSO_3^-(aq) \rightleftharpoons S_2O_5^{2-}(aq) + H_2O(l)$$

The above reaction is reversible, since pyrosulphites give the reactions of hydrogen sulphites in aqueous solution.

Moist sulphur dioxide (or sulphurous acid) and sulphites behave as reducing agents, for instance, on standing in the air a solution of the acid is slowly oxidised to a solution of sulphuric acid, and sulphites often give a test for the sulphate ion (p. 316). These reactions are best formulated as an oxidation of the sulphite ion to the sulphate ion:

$$2SO_3^{2-}(aq) + O_2(g) \longrightarrow 2SO_4^{2-}(aq)$$

Other reducing reactions of the sulphite ion include the conversion of chlorine to the chloride ion, iron (III) ions to iron (II) ions, and dichromate (VI) and manganate (VII) solutions to chromium (III) and manganese (II) ions respectively:

$$Cl_2(g) + SO_3^{2-}(aq) + H_2O(l) \longrightarrow 2Cl^-(aq) + SO_4^{2-}(aq) + 2H^+(aq)$$
$$2Fe^{3+}(aq) + SO_3^{2-}(aq) + H_2O(l) \longrightarrow 2Fe^{2+}(aq) + SO_4^{2-}(aq) + 2H^+(aq)$$
$$Cr_2O_7^{2-}(aq) + 3SO_3^{2-}(aq) + 8H^+(aq) \longrightarrow 2Cr^{3+}(aq) + 3SO_4^{2-}(aq) + 4H_2O(l)$$
$$2MnO_4^-(aq) + 5SO_3^{2-}(aq) + 6H^+(aq) \longrightarrow 2Mn^{2+}(aq) + 5SO_4^{2-}(aq) + 3H_2O(l)$$

Note that the reducing action of the sulphite ion is very similar to that of hydrogen sulphide (p. 304), except that the sulphate ion is formed. Hydrogen sulphide, however, is a stronger reducing agent since it reacts with sulphites, forcing them to assume the role of oxidising agent:

$$SO_3^{2-}(aq) + 2H^+(aq) + 2H_2S(g) \longrightarrow 3H_2O(l) + 3S(s)$$

The sulphite ion has a pyramidal structure, the lone pair of electrons on the sulphur atom occupying the fourth tetrahedral position (fig. 20.14).

FIG. 20.14. *The structure of the sulphite ion*

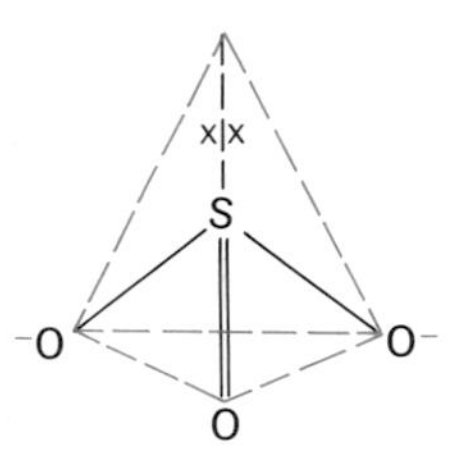

20.21 Sulphuric acid, H_2SO_4

Sulphuric acid is used industrially on an enormous scale and in 1980 it was estimated that world production exceeded 100 million tonnes.

Manufacture of sulphuric acid

Essentially the manufacture of this acid involves the conversion of sulphur dioxide into sulphur trioxide which is then dissolved in water to form sulphuric acid.

(a) Sources of sulphur dioxide

The percentages of sulphuric acid manufactured from different sources of sulphur in Great Britain are as follows:
Imported sulphur – mostly from Poland, Mexico and the USA (Frasch process) and from France and Canada, where it is extracted from 'sour' natural gas (methane contaminated with hydrogen sulphide). About 90% of the UK's requirements are satisfied by these sources of sulphur.

The roasting of sulphide ores, generally carried out for metal production, produces sulphur dioxide. In the UK this source of sulphur dioxide accounts for only about 5% of the sulphuric acid produced, and is obtained entirely from the sulphide ores of lead and zinc, e.g.

$$2ZnS(s) + 3O_2(g) \longrightarrow 2ZnO(s) + 2SO_2(g)$$

Desulphurization of imported crude oil at refineries in the UK accounts for about 5% of the sulphur dioxide converted into sulphuric acid. North Sea oil contains only insignificant amounts of sulphur compounds and it is uneconomic to obtain sulphur dioxide from this source.

In order to produce sulphur dioxide from sulphur, the liquid element is sprayed into a furnace at about 1000°C where it is burnt in air, which has previously been dried by passage through concentrated sulphuric acid. About 10% of the emerging gases is sulphur dioxide.

Sulphur dioxide is converted into sulphuric acid by the 'lead chamber' process and by the 'Contact process'. The former process is no longer used in the UK, so only the latter process is described here.

(b) The Contact process

A mixture of sulphur dioxide and air (approximately 8·5% SO_2 and 12·5% O_2 by volume) is passed over a vanadium (V) oxide catalyst at a temperature of about 430°C. The reaction is exothermic and the temperature rises to about 600°C; the mixture is cooled to about 430°C by passage through a heat exchanger (the heat extracted being used to heat the initial sulphur dioxide/air mixture), and at this stage the conversion of sulphur dioxide to sulphur trioxide is about 66% complete. The mixture is then passed through three more converters, at each stage the emerging gas stream being cooled to 430° as described above before passing from one converter to another. The final gas stream (overall conversion of sulphur dioxide to sulphur trioxide being about 98% complete) is then cooled, and the sulphur trioxide absorbed in 98% sulphuric acid to give 100% sulphuric acid or, if the absorption is carried further, fuming sulphuric acid or oleum, $H_2S_2O_7$. Dilution of the fuming sulphuric acid with water gives sulphuric acid.

$$2SO_2(g) + O_2(g) \rightleftharpoons 2SO_3(g) \qquad \Delta H^{\ominus}(298\,K) = -196\,kJ\,mol^{-1}$$

$$SO_3(g) + H_2SO_4(l) \longrightarrow H_2S_2O_7(l)$$

$$H_2S_2O_7(l) + H_2O(l) \longrightarrow 2H_2SO_4(l)$$

In order to reduce pollution, due to the discharge of unchanged sulphur dioxide into the atmosphere, modern Contact plants are designed to absorb the sulphur trioxide from the gas stream in 98% sulphuric acid after it emerges from the third converter. The unchanged sulphur dioxide (with air) is then allowed to pass through the fourth converter and the extra amount of sulphur trioxide similarly absorbed. In this modified process only about 0·05% of the sulphur dioxide remains unconverted and is allowed to pass into the atmosphere.

Since the catalytic oxidation of sulphur dioxide to sulphur trioxide is an exothermic reaction, a high yield of the trioxide will be favoured by a reasonably low temperature, but the temperature must not be too low or the rate of reaction will become too slow. A temperature of about 430°C is the optimum one.

The reaction takes place with a decrease in volume, hence an increase in pressure should increase the equilibrium yield of sulphur trioxide and also the rate of reaction. Since both yield and reaction rate are quite satisfactory at a pressure slightly greater than atmospheric, the extra expense of making pressure equipment is not justified.

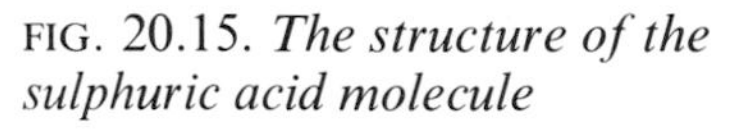
FIG. 20.15. *The structure of the sulphuric acid molecule*

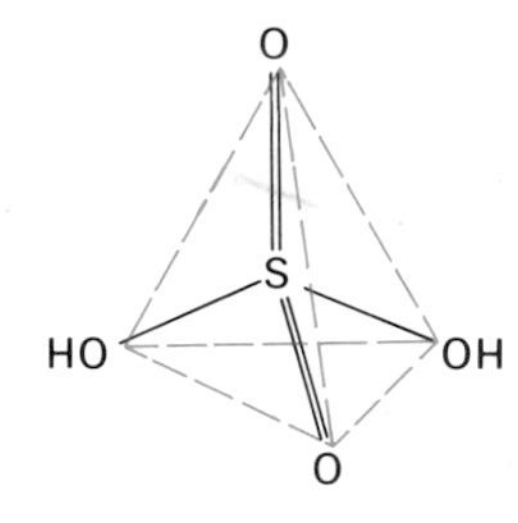

Physical properties of sulphuric acid

Pure sulphuric acid is a viscous liquid which has a density of 1·85 g cm^{-3} and a freezing point of 10·5°C. In the absence of water it will not turn blue litmus red nor will it react with metals to form hydrogen. It decomposes on boiling to form sulphur trioxide and steam, and a constant boiling point mixture is formed, containing 98·3 per cent of acid.

Structure of sulphuric acid

Pure sulphuric acid is covalent, its molecule having an approximately tetrahedral structure and containing 6-valent sulphur (fig. 20.15).

Its high boiling point (270°C) and viscosity are presumably due to the presence of hydrogen bonding which link the molecules into larger aggregates:

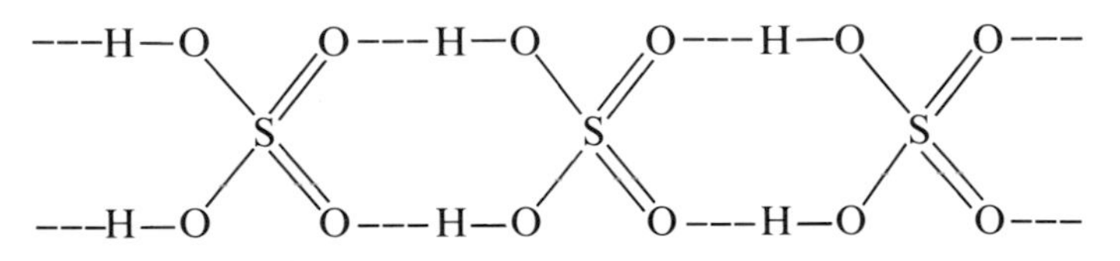

Chemical properties of sulphuric acid

These are most conveniently discussed under several headings.

(a) Acid properties

Concentrated sulphuric acid reacts violently with water to give a solution which exhibits the properties of a strong acid, e.g. it displaces carbon dioxide from a carbonate, neutralises a base to form a salt and water, and reacts with metals high up in the electrode potential series to form a salt and hydrogen.

The ionisation of sulphuric acid in water takes place in two stages, i.e. it is dibasic; the first ionisation is virtually complete, whereas the second takes place to the extent of about 10 per cent:

$$H_2SO_4(l) + H_2O(l) \rightleftharpoons H_3O^+(aq) + HSO_4^-(aq) \quad \text{(virtually complete)}$$
$$HSO_4^-(aq) + H_2O(l) \rightleftharpoons H_3O^+(aq) + SO_4^{2-}(aq) \quad \text{(about 10\% complete)}$$

An aqueous solution of the acid therefore contains $H_3O^+(aq)$ ions and very many more $HSO_4^-(aq)$ than $SO_4^{2-}(aq)$ ions. In the presence of $OH^-(aq)$ ions, which are stronger proton acceptors than the neutral water molecules, more extensive ionisation of the $HSO_4^-(aq)$ ions takes place and $SO_4^{2-}(aq)$ ions predominate:

$$HSO_4^-(aq) + OH^-(aq) \rightleftharpoons H_2O(l) + SO_4^{2-}(aq) \quad \text{(equilibrium to the right)}$$

This explains why the interaction of sodium hydroxide solution and dilute sulphuric acid produces sodium sulphate, and why special conditions are needed to crystallise out the hydrogen sulphate (p. 183).

(b) Dehydrating properties

Concentrated sulphuric acid has such an affinity for water that it will remove it from mixtures and compounds with the evolution of much heat (the acid should always be diluted by pouring it into water while stirring, and not the other way round). Gases are dried in the laboratory by passing them through a wash bottle containing concentrated sulphuric acid, except of course the gases that chemically combine with the acid.

Many organic compounds lose the elements of water when treated with the concentrated acid; thus glucose gives carbon, methanoic acid produces carbon monoxide and oxalic acid gives an equimolecular mixture of carbon monoxide and carbon dioxide:

$$C_6H_{12}O_6(s) - 6H_2O \longrightarrow 6C(s)$$

$$HCOOH(l) - H_2O \longrightarrow CO(g)$$

$$\begin{matrix} COOH \\ | \\ COOH \end{matrix}(s) - H_2O \longrightarrow CO(g) + CO_2(g)$$

The dehydration of organic compounds is of industrial importance; thus ethoxyethane is made by dehydrating ethanol and the insecticide DDT is made by dehydrating a mixture of chlorobenzene and trichloroethanal (chloral):

$$2C_2H_5OH \xrightarrow{-H_2O} \underset{\text{ethoxyethane}}{C_2H_5{-}O{-}C_2H_5}$$

$$2C_6H_5Cl + CCl_3CHO \xrightarrow{-H_2O} \underset{\text{DDT}}{ClC_6H_4{-}\overset{\displaystyle CCl_3 \atop |}{\underset{\displaystyle | \atop H}{C}}{-}C_6H_4Cl}$$

(c) Oxidising properties

Hot concentrated sulphuric acid functions as an oxidising agent, although less effectively than concentrated nitric acid. A number of reduction products of sulphuric acid can be formed, their relative proportions in any one particular reaction depending upon such factors as concentration of the acid, strength of the reducing agent present, and temperature. These possible reduction products are shown schematically below (underlined):

$$H_2SO_4 \xrightarrow{-[O]} (H_2SO_3) \longrightarrow H_2O + \underline{SO_2}$$

$$H_2SO_4 \xrightarrow{-4[O]} \underline{H_2S}$$

which may be followed by oxidation of the hydrogen sulphide to sulphur:

$$3H_2S(g) + H_2SO_4(l) \longrightarrow 4H_2O(l) + 4S(s)$$

Non-metals such as carbon and sulphur reduce sulphuric acid when heated to sulphur dioxide, they themselves being oxidised to their dioxides:

$$C(s) + 2H_2SO_4(l) \longrightarrow 2H_2O(l) + CO_2(g) + 2SO_2(g)$$
$$S(s) + 2H_2SO_4(l) \longrightarrow 2H_2O(l) + 3SO_2(g)$$

With metals, it is possible to obtain sulphur dioxide, hydrogen sulphide and sulphur in addition to the metallic sulphate; further reaction between the metallic sulphate and hydrogen sulphide can give rise to significant amounts of the metallic sulphide. These reactions are, therefore, quite complex; however, in general it can be said that sulphur dioxide is the main reduction product when metals low down in the electrode potential series are used, e.g. copper, but that increasing amounts of hydrogen sulphide are formed with more reactive metals, e.g. zinc. The following equations, therefore, do no more than indicate the main reactions that take place:

$$Cu(s) + 2H_2SO_4(l) \longrightarrow CuSO_4(aq) + 2H_2O(l) + 2SO_2(g)$$
$$Zn(s) + \underset{98\%\ \text{acid}}{2H_2SO_4(l)} \longrightarrow ZnSO_4(aq) + 2H_2O(l) + 2SO_2(g)$$

$$4Zn(s) + \underset{90\%\ \text{acid}}{5H_2SO_4(aq)} \longrightarrow 4ZnSO_4(aq) + 4H_2O(l) + H_2S(g)$$

Compounds oxidised by concentrated sulphuric acid include hydrogen bromide and hydrogen iodide:

$$2HBr(g) + H_2SO_4(l) \longrightarrow 2H_2O(l) + SO_2(g) + Br_2(l)$$
$$8HI(g) + H_2SO_4(l) \longrightarrow 4H_2O(l) + H_2S(g) + 4I_2(s)$$

These reactions explain why the free halogen is always liberated, together with the halogen hydride, when bromides and iodides are treated with concentrated sulphuric acid.

(d) Displacement reactions

When concentrated sulphuric acid is added to a nitrate there is little sign of any reaction in the cold; nevertheless an equilibrium reaction is set up:

$$NO_3^- + H_2SO_4(l) \rightleftharpoons HSO_4^- + HNO_3(l)$$

On warming, the equilibrium moves to the right, since nitric acid, which has a lower boiling point than sulphuric acid, is removed from the system as a vapour.

Other displacement reactions involving sulphuric acid are similar and take place because the substance that is displaced has a lower boiling point than sulphuric acid, e.g.

$$Cl^- + H_2SO_4(l) \rightleftharpoons HCl(g) + HSO_4^-$$

Hydrogen chloride is readily displaced from this system since it is a gas.

In some instances the substance displaced is unstable and breaks down to give volatile products which are lost from the system, thus enabling the reaction to proceed, e.g.

$$SO_3^{2-} + 2H_2SO_4(l) \rightleftharpoons 2HSO_4^- + H_2SO_3 \rightleftharpoons H_2O(l) + SO_2(g)$$

In the above reaction the loss of sulphur dioxide as a gas allows the reaction to proceed to completion.

These reactions, which depend on the high boiling point of concentrated sulphuric acid, explain why this acid is useful in analysis for detecting the presence of certain acid radicals.

(e) Reactions with organic compounds

Alkenes undergo addition reactions with concentrated sulphuric acid; if the addition products are diluted with water and heated, alcohols are produced. This type of reaction is of industrial importance, e.g.

$$\underset{\text{but-1-ene}}{CH_3{-}CH_2{-}CH{=}CH_2} + H_2SO_4 \longrightarrow CH_3{-}CH_2{-}\underset{\displaystyle OSO_3H}{\underset{|}{CH}}{-}CH_3$$

$$CH_3{-}CH_2{-}\underset{\displaystyle OSO_3H}{\underset{|}{CH}}{-}CH_3 + H_2O \longrightarrow \underset{\text{butan-2-ol}}{CH_3{-}CH_2{-}\underset{\displaystyle OH}{\underset{|}{CH}}{-}CH_3} + H_2SO_4$$

Sulphuric acid also reacts with long chain hydrocarbons with the elimination of water, e.g.

$$C_{18}H_{30} + H_2SO_4 \longrightarrow \underset{\text{a sulphonic acid}}{C_{18}H_{29}SO_3H} + H_2O$$

The sodium salts of sulphonic acids are used in detergent preparations, e.g. sodium benzene dodecyl sulphonate, $C_{18}H_{29}SO_3^-Na^+$, is the main surface active constituent of Tide, Daz, Surf and Omo.

The importance of dehydration reactions involving sulphuric acid and organic compounds has been mentioned earlier, e.g. the preparation of ethoxyethane and DDT (p. 313).

20.22 Uses of sulphuric acid

Sulphuric acid is used on an enormous scale; in Great Britain alone over 4 million tonnes of the concentrated acid are manufactured. About one third of the total output is used in the manufacture of fertilisers, e.g. 'superphosphate' and ammonium sulphate. Other uses include:

Paint manufacture—titanium (IV) oxide and lithopone (p. 357)
Synthetic fibres—rayon and plastics
Acid production—hydrochloric and hydrofluoric acids
Metallurgy—pickling steel and sulphates for electrolysis
Detergents—mostly alkyl-aryl sulphonates
Petroleum refining—mostly extraction of alkenes
Organic chemicals—dyestuffs and explosives and drugs.

When used in the manufacture of dyestuffs and explosives its purpose is to enhance the nitrating property of nitric acid (p. 259).

20.23 Sulphates

Metallic sulphates can be obtained by reacting the metal (if above hydrogen in the electrode potential series), or its oxide, hydroxide or carbonate with dilute sulphuric acid. The Group 1A metals also form hydrogen sulphates which can be isolated as solids, but certain conditions are required for their isolation (p. 183). In general, metallic sulphates are soluble in water and crystallise with water of crystallisation. Notable exceptions are barium sulphate and lead sulphate, which are virtually insoluble, and calcium sulphate, which is only sparingly soluble.

Sulphates are more thermally stable than the corresponding nitrates, and only those of metals reasonably low down in the electrode potential series decompose on strong heating; the metallic oxide and oxides of sulphur are formed, e.g.

$$(Fe^{3+})_2(SO_4{}^{2-})_3(s) \longrightarrow (Fe^{3+})_2(O^{2-})_3(s) + 3SO_3(s)$$

The hydrogen sulphates of the alkali metals decompose on heating to give the pyrosulphate, which then decomposes into the sulphate, evolving sulphur trioxide, e.g.

$$2Na^+HSO_4{}^-(s) \longrightarrow (Na^+)_2S_2O_7{}^{2-}(s) + H_2O(l)$$
$$(Na^+)_2S_2O_7{}^{2-}(s) \longrightarrow (Na^+)_2SO_4{}^{2-}(s) + SO_3(s)$$

Solutions of sulphates (and hydrogen sulphates) can be detected by reaction with an aqueous solution of barium chloride, a white precipitate of barium sulphate being formed which is insoluble in dilute hydrochloric acid:

$$Ba^{2+}(aq) + SO_4{}^{2-}(aq) \longrightarrow Ba^{2+}SO_4{}^{2-}(s)$$

Sulphites generally give the same reaction, since they are easily oxidised to sulphates and old specimens often contain appreciable amounts of sulphate ions. However, sulphites but not sulphates evolve sulphur dioxide on treatment with dilute hydrochloric acid, so no confusion should arise:

$$\underset{\text{sulphite}}{SO_3{}^{2-}} + 2H^+(aq) \longrightarrow H_2O(l) + SO_2(g)$$

Like the molecule of sulphuric acid, the sulphate ion has a tetrahedral configuration; since all the sulphur-oxygen bond lengths are identical, it is considered that resonance is involved (fig. 20.16):

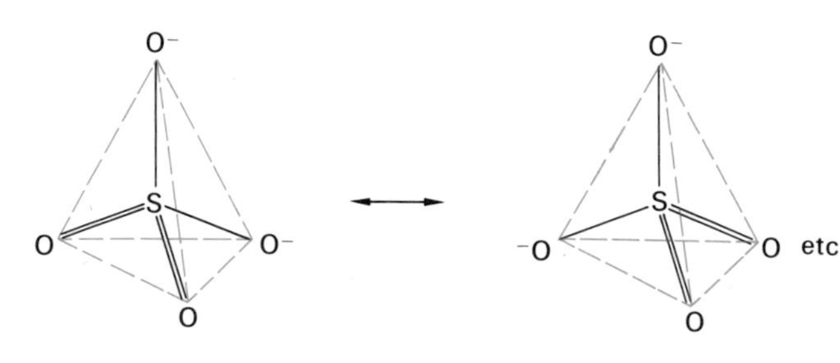

FIG. 20.16. *The structure of the sulphate ion*

20.24 The chloro derivatives of sulphurous and sulphuric acids

Sulphur dichloride oxide, $SOCl_2$—a derivative of sulphurous acid
Sulphur dichloride oxide (thionyl chloride) is made by reacting sulphur dioxide (or a sulphite) with phosphorus pentachloride, the liquid product then being separated by distillation:

$$SO_2(g) + PCl_5(s) \longrightarrow SOCl_2(l) + POCl_3(l)$$

It is hydrolysed by water, like most covalent halides and, on warming, sulphur dioxide and hydrogen chloride are expelled:

$$SOCl_2(l) + H_2O(l) \longrightarrow SO_2(g) + 2HCl(g)$$

It is used mainly in organic chemistry for the replacement of an hydroxyl group by the chlorine atom, e.g. the conversion of carboxylic acids to acid chlorides and alcohols to chloroalkanes:

$$CH_3COOH(l) + SOCl_2(l) \longrightarrow CH_3COCl(l) + SO_2(g) + HCl(g)$$
$$C_2H_5OH(l) + SOCl_2(l) \longrightarrow C_2H_5Cl(g) + SO_2(g) + HCl(g)$$

The sulphur dichloride oxide molecule is pyramidal with the lone pair electrons occupying the fourth tetrahedral position (fig. 20.17).

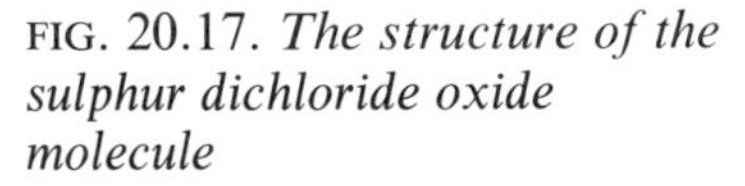
FIG. 20.17. *The structure of the sulphur dichloride oxide molecule*

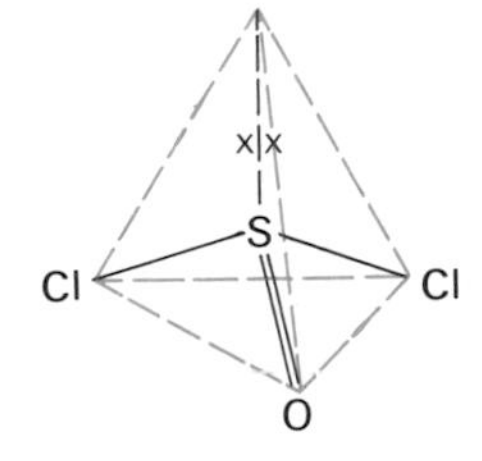

Chlorosulphonic acid, $HOSO_2Cl$—a derivative of sulphuric acid
This is obtained by reacting hydrogen chloride with fuming sulphuric acid, the water produced being absorbed by more sulphuric acid:

$$H_2SO_4(l) + HCl(g) \longrightarrow HOSO_2Cl(l) + H_2O(l)$$

Chlorosulphonic acid is a colourless liquid which fumes in moist air owing to hydrolysis; with much water, it is readily converted back again into sulphuric acid:

$$HOSO_2Cl(l) + H_2O(l) \longrightarrow H_2SO_4(aq) + HCl(g)$$

Its molecule has a tetrahedral shape like that of sulphuric acid, the only difference being the replacement of one OH group by the chlorine atom.

Sulphur dichloride dioxide, SO_2Cl_2—a derivative of sulphuric acid
When concentrated sulphuric acid is reacted with phosphorus pentachloride both OH groups are replaced by chlorine atoms to produce sulphur dichloride dioxide:

$$H_2SO_4(l) + 2PCl_5(s) \longrightarrow SO_2Cl_2(l) + 2POCl_3(l) + 2HCl(g)$$

It can also be made by the direct combination of sulphur dioxide and chlorine in the presence of charcoal:

$$SO_2(g) + Cl_2(g) \longrightarrow SO_2Cl_2(l)$$

FIG. 20.18. *The structure of the sulphur dichloride dioxide molecule*

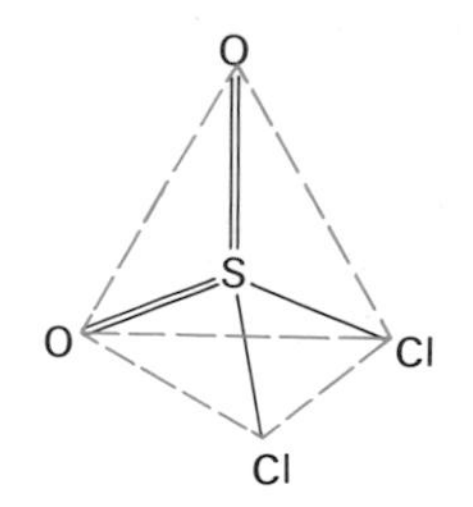

Sulphur dichloride dioxide is a colourless liquid but its boiling point (69°C) is much lower than that of chlorosulphonic acid, since the possibility of hydrogen bonding is now ruled out. Like chlorosulphonic acid, it is readily hydrolysed by water:

$$SO_2Cl_2(l) + 2H_2O(l) \longrightarrow H_2SO_4(aq) + 2HCl(g)$$

The molecule of sulphur dichloride dioxide is tetrahedral as expected (fig. 20.18):

20.25 Oxyacids of sulphur containing linked sulphur atoms

Thiosulphuric acid, $H_2S_2O_3$

This acid has never been isolated, but its salts are well known. They contain the thiosulphate ion, $S_2O_3^{2-}$, which can be regarded as being derived from the sulphate ion by replacement of one oxygen atom by sulphur; its shape is similar to that of the sulphate ion.

The best known thiosulphate is the sodium salt, $(Na^+)_2S_2O_3^{2-}.5H_2O$, which is used in photography for 'fixing' the negative (p. 417). It is obtained by boiling a solution of sodium sulphite with sulphur, followed by filtration and crystallisation:

$$SO_3^{2-}(aq) + S(s) \longrightarrow S_2O_3^{2-}(aq)$$

The thiosulphate ion is unstable in the presence of acid and breaks down to give the sulphite (which subsequently reacts to give sulphur dioxide and water) and free sulphur:

$$S_2O_3^{2-}(aq) + 2H^+(aq) \longrightarrow H_2O(l) + SO_2(g) + S(s)$$

It also functions as a reducing agent; thus chlorine is reduced to chloride ions, and this reaction is utilised for removing excess chlorine from bleached fabrics:

$$S_2O_3^{2-}(aq) + 4Cl_2(g) + 5H_2O(l) \longrightarrow 2SO_4^{2-}(aq) + 10H^+(aq) + 8Cl^-(aq)$$

The milder oxidising agent iodine behaves rather differently, though it too is converted into its ions:

$$2S_2O_3^{2-}(aq) + I_2(aq) \longrightarrow 2I^-(aq) + \underset{\text{tetrathionate ion}}{S_4O_6^{2-}(aq)}$$

This reaction is used in volumetric analysis for estimating iodine solutions and indirectly a variety of oxidising agents, e.g. copper (II) ions:

$$2Cu^{2+}(aq) + 2I^-(aq) \longrightarrow 2Cu^+ + I_2(aq)$$
$$2S_2O_3^{2-}(aq) + I_2(aq) \longrightarrow 2I^-(aq) + S_4O_6^{2-}(aq)$$
$$\text{or} \quad \underset{\text{2 moles}}{2Cu^{2+}(aq)} \equiv \underset{\text{1 mole}}{I_2(aq)} \equiv \underset{\text{2 moles}}{2S_2O_3^{2-}(aq)}$$

In the above example a known volume of the copper (II) solution is added to an excess of potassium iodide solution; the liberated iodine is then titrated with sodium thiosulphate solution, using starch as indicator.

The polythionic acids, $H_2S_nO_6$, (n = 2, 3, 4, 5 or 6)
These are obtained, together with much free sulphur, when hydrogen sulphide is bubbled through a solution of sulphurous acid, i.e. the reaction between moist sulphur dioxide and hydrogen sulphide is more complex than hitherto described (p. 310). The acids themselves are readily decomposed into sulphur and sulphur dioxide.

Neither the acids nor their salts are of any great importance other than once again illustrating the marked tendency for the sulphur atom to catenate, the structure of the anions having been established as containing zigzag sulphur chains.

20.26 Oxyacids of sulphur containing linked oxygen atoms

Peroxomonosulphuric acid, H_2SO_5
This acid is formed when hydrogen peroxide reacts with chlorosulphonic acid using a 1 : 1 mole ratio:

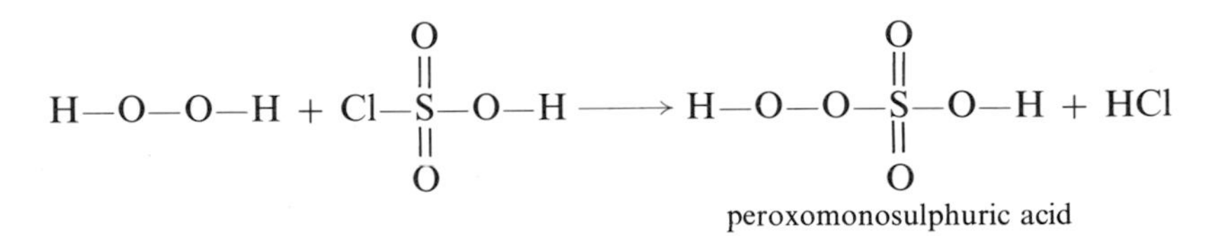

Like hydrogen peroxide it is an oxidising agent; thus it converts iron (II) ions into iron (III) ions. Although its structure is unknown, it is likely to resemble that of hydrogen peroxide, the SO_2OH group replacing one hydrogen atom; the four oxygen atoms are presumably arranged tetrahedrally around the central sulphur atom.

Peroxodisulphuric acid, $H_2S_2O_8$
This acid can be obtained by reacting 1 mole of hydrogen peroxide with 2 moles of chlorosulphonic acid:

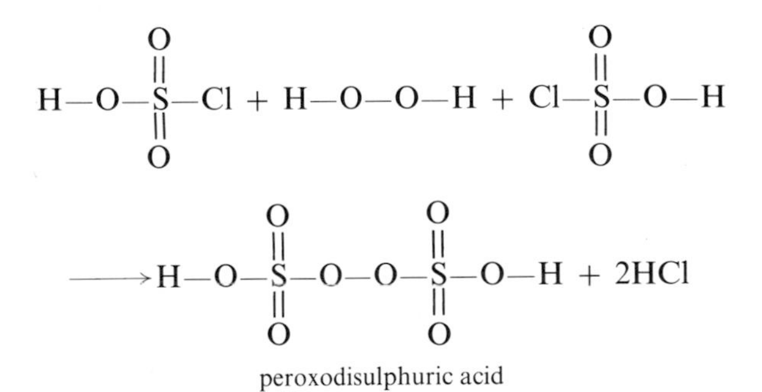

Its salts are better known than the acid itself, e.g. potassium peroxodisulphate (potassium persulphate), $(K^+)_2S_2O_8^{2-}$, which is made by electrolysing a saturated solution of potassium hydrogen sulphate using a platinum wire anode and a high current density at 0°C.

Cathode ⟵	$K^+HSO_4^-$	⟶ **Anode**
H^+ discharged	$H_2O \rightleftharpoons H^+ + OH^-$	HSO_4^- ions discharged
$2H^+ + 2e^- \rightarrow H_2$		$2HSO_4^- \rightarrow 2H^+ + S_2O_8^{2-} + 2e^-$

Sparingly soluble potassium peroxodisulphate forms at the anode and can be crystallised out.

Peroxodisulphuric acid and its anion are extremely powerful oxidising agents; thus iron (II) ions are oxidised to iron (III) ions and iodide ions to iodine:

$$2Fe^{2+}(aq) + S_2O_8^{2-}(aq) \longrightarrow 2Fe^{3+}(aq) + 2SO_4^{2-}(aq)$$
$$2I^-(aq) + S_2O_8^{2-}(aq) \longrightarrow I_2(aq) + 2SO_4^{2-}(aq)$$

In the presence of silver ions, which act as catalysts, metallic copper is oxidised to copper (II) ions, and manganese (II) ions to manganate (VII) ions:

$$Cu(s) + S_2O_8^{2-}(aq) \longrightarrow Cu^{2+}(aq) + 2SO_4^{2-}(aq)$$
$$2Mn^{2+}(aq) + 5S_2O_8^{2-}(aq) + 8H_2O(l) \longrightarrow 2MnO_4^-(aq) + 16H^+(aq) + 10SO_4^{2-}(aq)$$

Note that in each case the reduction product is the sulphate ion.

By X-ray analysis it has been shown that the peroxodisulphate ion has a structure in which four oxygen atoms are arranged in an approximately tetrahedral fashion about each sulphur atom (fig. 20.19):

FIG. 20.19. *The structure of the peroxodisulphate ion*

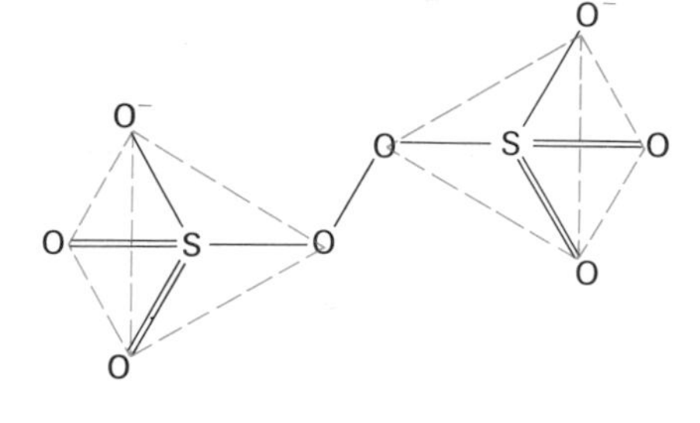

20.27 Uses of sulphur and its compounds

The most important use of sulphur is in the manufacture of sulphur dioxide, most of which is converted into sulphuric acid. Sulphur is also used in vulcanising rubber, in the manufacture of gunpowder, for spraying plants, fruit, etc., since it is toxic to lower forms of life, and for making sodium thiosulphate (used in photography) and carbon disulphide (a solvent used in the manufacture of rayon).

Sulphur dioxide, in addition to being used in the industrial production of sulphuric acid, is used as a bleaching agent. When neutralised with calcium hydroxide solution, the solution of calcium hydrogen sulphite produced is used for bleaching woodpulp to make paper.

The widespread uses of sulphuric acid have already been listed (p. 315).

Selenium, tellurium and polonium

20.28 The elements, selenium, tellurium and polonium

Selenium and tellurium are recovered from the anode 'sludge' remaining after the electrolytic refining of copper. Selenium exists in three different allotropic forms, the two non-metallic forms containing Se_8 structural units similar to the S_8 rings present in solid sulphur. The 'metallic' form of selenium contains infinite spiral chains of selenium atoms (see fig. 20.1, p. 285); this form of the element is weakly conducting, but its conductivity is enhanced in the presence of light and this explains the use of selenium in photoelectric cells. Tellurium has the same structure as 'metallic' selenium and is a slightly better conductor of electricity.

Polonium is obtained by neutron-irradiation of bismuth and is itself radioactive; its structure is truly metallic (closer packing of the atoms) and its metallic properties are similar to those of lead. It is only available in milligramme quantities, but such is the sophistication of modern chemical techniques that a significant amount of chemistry has been done on this very small scale.

20.29 The hydrides of selenium, tellurium and polonium

Selenium, tellurium and polonium form gaseous hydrides of the type H_2X where X is Se, Te or Po. With hydrogen sulphide they form part of a well-graded series; for instance their boiling points increase with increasing relative molecular mass and they become increasingly thermally unstable in the same order.

Selenides, tellurides and polonides exist and are generally covalent solids like the corresponding sulphides. Selenides and tellurides of the Group 1A metals, like the sulphides, are ionic and contain respectively the Se^{2-} and Te^{2-} ions. Polyselenides and polytellurides of the alkali metals are formed in the same way as polysulphides (p. 305) and have similar formulae, e.g. Se_4^{2-}, Te_4^{2-} are comparable with S_4^{2-}.

20.30 The halides of selenium, tellurium and polonium

Like sulphur, selenium and tellurium form halides in which they show valencies of 2, 4 and 6. Halides of polonium containing 2- and 4-valent polonium have been obtained, but so far no 6-valent halide has been prepared. Typical examples of these halides are shown below.

$SeCl_2$ (gas)	$SeCl_4$ (solid)	SeF_4 (liquid)	SeF_6 (gas)
$TeCl_2$ (solid)	$TeCl_4$ (solid)	TeF_4 (solid)	TeF_6 (gas)
$PoCl_2$ (solid)	$PoCl_4$ (solid)		

They are all covalent and appear to have similar structures to the halides of sulphur (p. 306). Whereas SF_6 is remarkably inert towards water and alkalis, SeF_6 is hydrolysed slowly and TeF_6 more rapidly still. Unlike the tetrachlorides of sulphur, those of selenium, tellurium and polonium form chloride complexes, e.g. $SeCl_6^{2-}$, $TeCl_6^{2-}$ and $PoCl_6^{2-}$.

20.31 The oxides and oxyacids of selenium and tellurium

Selenium forms a solid dioxide, SeO_2, which has a chainlike structure; it dissolves in water to form selenious acid, H_2SeO_3. This acid forms hydrogen selenites and selenites but, unlike sulphurous acid, it is an oxidising agent. The trioxide of selenium, SeO_3, has not been obtained in a pure form; it dissolves in water to form selenic acid, H_2SeO_4, which is a strong oxidising agent. Hydrogen selenates and selenates are known, the latter often being isomorphous with the corresponding sulphates.

Tellurium dioxide, TeO_2, contains Te^{4+} ions and, although practically insoluble in water, reacts with alkalis to form hydrogen tellurites and tellurites. It also reacts with nitric acid, forming a basic nitrate, so it is amphoteric; this is in line with the more 'metallic' character of tellurium. Tellurium trioxide, TeO_3, is also insoluble in water and dissolves in alkalis to form hydrogen tellurates and tellurates; the formula of the parent acid is H_6TeO_6, $[Te(OH)_6]$, and is thus not analogous to the formulae of sulphuric and selenic acids.

Polonium dioxide, PoO_2, contains the Po^{4+} ion and is more basic than tellurium dioxide, forming a sulphate and nitrate. The higher oxide of polonium is unknown.

20.32 A comparison of the chemistry of oxygen and sulphur

Oxygen and sulphur are non-metallic but, apart from this, there are few points of similarity. Oxygen exists in the form of discrete molecules, a double bond being present in the molecule, whereas S_8 molecules contain —S—S— bonds in the two crystalline allotropes of sulphur. The reluctance of oxygen to form —O—O— bonds is shown in the instability of ozone with respect to molecular oxygen and in the powerful oxidising properties of hydrogen peroxide, peroxides, and peroxodisulphuric acid and its salts.

Both elements form the 2-valent anion in combination with reactive metals and both elements also show a covalency of 2. However, sulphur can, in addition, show covalencies of 4 and 6 since it alone can utilise *d* orbitals.

Oxygen forms the hydrides H_2O and H_2O_2, the latter being thermodynamically unstable with respect to water and molecular oxygen. Sulphur similarly forms the sulphides H_2S and H_2S_2, but whereas water is a liquid H_2S is a gas, since sulphur is not sufficiently electronegative to participate to any extent in hydrogen bonding. The higher hydride of sulphur, like hydrogen peroxide, is unstable with respect to the lower hydride and the element.

Both oxygen and sulphur form a wide range of oxides and sulphides but whereas the oxides of metals can be considered to be essentially ionic only the sulphides of Groups 1A and 2A show appreciable ionic character. The alkali metal oxides and sulphides behave similarly towards hydrolysis; thus sodium oxide, for example, forms the hydroxide while the corresponding sulphide forms the hydrosulphide and hydroxide (p. 305). The sulphide ion alone shows a tendency towards catenation; thus warm aqueous solutions of alkali metal sulphides dissolve sulphur with the formation of polysulphide ions.

20.33 Suggested reading

E. Emmett, *Vanadium catalyses in sulphuric acid manufacture*, Education in Chemistry, No. 3, Vol. 12, 1975.
Non-metals and Semi-metals, Open University, S25, Unit 10.
Water, A Unilever Educational Booklet, 1976.

20.34 Questions on chapter 20

1 Give four different types of reaction by which oxygen may be prepared.
Discuss the variation in the chemical properties of the oxides formed by the elements of the second short period (sodium to argon). (O & C)

2 (a) Describe in outline how oxygen (dioxygen) may be converted into
(i) a sample of oxygen containing trioxygen (ozone) and
(ii) an aqueous solution of hydrogen peroxide.
(b) Suggest how the concentrations of trioxygen in sample (i) and of hydrogen peroxide in sample (ii) could be determined.
(c) Trioxygen is said to react readily with hydrogen peroxide according to the equation

$$H_2O_2 + O_3 = H_2O + 2O_2.$$

Suggest a method that would enable you to show that this equation is correct, assuming that an aqueous solution of hydrogen peroxide of known concentration and a sample of oxygen containing a known amount of trioxygen are available. (O & C)

3 Describe one method for preparing a sample of ozonised oxygen. The equilibrium between oxygen and ozone is represented by

$$3O_2 \rightleftharpoons 2O_3 \text{ and } \Delta H = +284\,\text{kJ mol}^{-1}$$

Explain why only a negligible quantity of ozone is formed when an electric spark is passed through oxygen.

The times for the same volume of pure oxygen and pure ozone to effuse through a small hole under the same conditions were 44 and 54 sec respectively. What information about the composition of ozone do these data provide? (AEB)

4 Discuss the principles involved in the use of ethylenediamine tetra-acetic acid (EDTA) for determining the concentrations of Ca^{2+} and Mg^{2+} in hard water.

5 How may an aqueous solution of hydrogen peroxide be prepared in the laboratory?
How does hydrogen peroxide react with
(a) lead sulphide,
(b) acidified potassium iodide,
(c) acidified potassium permanganate?
200 cm^3 of oxygen (measured at s.t.p.) were obtained by the decomposition of 10 cm^3 of a solution of hydrogen peroxide. Calculate the concentration of hydrogen peroxide in grammes per dm^3. (O & C)

6 Describe the properties of the element sulphur. Make a comparison of the properties and reactions of
(a) sulphur dioxide and sulphur trioxide,
(b) sulphurous acid and sulphuric acid. (O & C)

7 Give an account of the element sulphur including a description of its main allotropic forms. Rhombic and monoclinic sulphur behave similarly in chemical reactions, but oxygen and ozone and red and white phosphorus behave differently. What explanation can you suggest for these facts? (O & C)

8 'Sulphur is in many ways chemically similar to oxygen'. Discuss this statement in relation to the properties of the following pairs:
(a) hydrogen sulphide and water,
(b) potassium thiosulphate and potassium sulphate,
(c) carbon disulphide and carbon dioxide.
25 cm^3 of an acidified solution of bromine were added to an excess of an acidified solution of potassium iodide. This solution was then equivalent to 25 cm^3 of a solution of sodium thiosulphate. When the bromine solution reacted directly with the same amount of the thiosulphate solution it was found that 200 cm^3 of the bromine solution were required for complete reaction. Obtain an equation for the reaction of bromine with thiosulphate. (Camb. Schol.)

9 Describe how you would prepare (using sulphur as the only source of that element) specimens of
(a) disulphur dichloride, S_2Cl_2,
(b) sodium thiosulphate.
How, and under what conditions, does sulphuric acid react with (i) benzene, (ii) sulphur, (iii) iron? (C)

10 Give a short account of the reactions of sulphur dioxide and of the sulphites.
A is one of the substances formed when sulphur dioxide and phosphorus pentachloride react together. It is a fuming liquid which dissolves in water to give a pungent-smelling liquid which is acid to litmus. 0·595g of **A** were dissolved in water and the solution made up to 1 dm^3. Dilute nitric acid and a slight excess of silver nitrate solution were added to 100 cm^3 of this solution and the white precipitate was collected, dried and weighed. The weight of precipitate was 1·43g. Dilute hydrochloric acid followed by barium chloride solution were added to a second 100 cm^3 portion of the solution. No precipitate formed, but on further addition of hydrogen peroxide solution to the mixture a white precipitate was thrown down. After collecting and drying, this precipitate weighed 1·17g. Deduce the empirical formula of the substance **A**. (O & C)

11 Sulphuryl chloride (SO_2Cl_2) reacts with ammonia gas to form two products, **A** and **B**.
Substance **A** liberates ammonia when warmed with sodium hydroxide solution and gives a white precipitate when added to a solution of silver nitrate in dilute nitric acid. This precipitate is soluble in ammonium hydroxide solution
Substance **B** contains no chlorine and has a molecular weight of about 100. It loses ammonia when heated and forms another substance **C**, which was found to have a molecular weight of about 240.
All of the hydrogen in **C** can be replaced by silver to form a silver salt **D**. When

0·3100g of **D** was added to hydrochloric acid, **C** was re-formed and 0·2375g of silver chloride was precipitated.

(a) Calculate the percentage of silver in **D**.
(b) Deduce the molecular formulae of **B** and **C**.
(c) Write equations for the reaction between sulphuryl chloride and ammonia and for the thermal decomposition of **B**.
(d) Suggest a structural formula for **C**. (O & C)

12 An element **X** burns in air to give a compound XO_2 which reacts with PCl_5 to give $XOCl_2$.XO_2 reacts extremely slowly with oxygen at room temperature to give XO_3 though the reaction is thermodynamically favourable. However, in the presence of a catalyst the reaction proceeds readily.

XO_2 dissolves in strong alkali to give the anion, XO_3^{2-}, which can be reduced by zinc in the presence of excess XO_2 to give $X_2O_4^{2-}$. However, aqueous solutions of $X_2O_4^{2-}$ decompose on standing to give $X_2O_3^{2-}$ and HXO_3^-. When 0·1741g of the sodium salt of $X_2O_4^{2-}$ are added to an excess of ammoniacal silver nitrate solution, 0·2158g of silver are precipitated. Calculate the atomic weight of **X** and discuss the reactions. (Oxford Schol.)

13 Mention two large scale methods (other than the burning of sulphur) by which sulphur dioxide is produced.

Outline the conversion of sulphur dioxide to concentrated sulphuric acid by the contact process, paying particular attention to the physico-chemical principles involved.

Describe and explain one method in each case by which it can be shown that sulphuric acid is

(a) dibasic,
(b) a dehydrating agent.

Under what conditions does sulphuric acid react with (i) magnesium, (ii) copper? State the nature of the reaction in each case and explain it in terms of electron transfer. (JMB)

14 What are the principal differences between sulphurous and sulphuric acids?

Sulphur dioxide and chlorine react in the presence of a catalyst to give a compound of the formula SO_2Cl_2, which is completely decomposed by water. How would you attempt to show that the decomposition is represented by the equation:

$$SO_2Cl_2 + 2H_2O \longrightarrow 4H^+ + SO_4^{2-} + 2Cl^- \qquad \text{(C)}$$

15 Discuss

(a) the similarities between oxygen and sulphur which justify classifying them in the same group of the Periodic Table,
(b) the differences which exist between the chlorides and hydrides of nitrogen and phosphorus in spite of their being in the same group. (S)

16 'Sulphur atoms show a marked tendency to catenate but this phenomenon is much less marked in the chemistry of oxygen'. Discuss, with appropriate examples, the validity of this statement.

17 Discuss the chemistry of four oxyacids of sulphur.

18 Give a concise comparative account of the similarities in the chemistry of the elements selenium, tellurium and polonium. By citing examples, discuss whether sulphur has more in common with these three elements than with oxygen.

19 (a) Explain how changes in temperature, pressure and the presence of a catalyst effect (i) the rate of reaction, (ii) the position of equilibrium with reference to the following system.

$$2SO_2(g) + O_2(g) \rightleftharpoons 2SO_3(g) \qquad \Delta H = -188\,\text{kJ mol}^{-1}$$

(b) The Contact Process for the manufacture of concentrated sulphuric acid involves the above reaction.

(i) Give an equation to represent the large scale production of sulphur dioxide from a source other than sulphur.
(ii) State the operating conditions for the oxidation of the sulphur dioxide.
(iii) Describe with the aid of equations the conversion of the sulphur trioxide into concentrated sulphuric acid.

(c) How and under what conditions does sulphuric acid react with (i) ethanedioic acid (oxalic acid), (ii) ethanol? (AEB)

Chapter 21

Group 7B the halogens (fluorine, chlorine, bromine, iodine and astatine)

21.1 Some physical data of Group 7B elements

	Atomic Number	Electronic Configuration	Atomic Radius/nm	Ionic Radius/nm X^-	M.p. /°C	B.p. /°C
F	9	2.7 ....$1s^2 2s^2 2p^5$	0·072	0·136	−220	−188
Cl	17	2.8.7 ....$2s^2 2p^6 3s^2 3p^5$	0·099	0·181	−101	−34·7
Br	35	2.8.18.7 ....$3s^2 3p^6 3d^{10} 4s^2 4p^5$	0·114	0·195	−7·2	58·8
I	53	2.8.18.18.7 ....$4s^2 4p^6 4d^{10} 5s^2 5p^5$	0·133	0·216	114	184
At	85	2.8.18.32.18.7 ....$5s^2 5p^6 5d^{10} 6s^2 6p^5$				

21.2 Some general remarks about Group 7B

All members of Group 7B are non-metallic, although there is the usual increase in 'metallic' character with increasing atomic number, e.g. dipyridine iodine nitrate can be written as $[I(pyridine)_2]^+NO_3^-$, containing the I^+ ion as part of a complex. Fluorine and chlorine are gases, bromine is a volatile liquid, and iodine is a dark shiny coloured solid. Astatine is radioactive and very short-lived; what little chemistry that has been carried out with this element has employed tracer techniques.

These elements can enter into chemical combination and complete their octets by gaining one electron to form the 1-valent ion, e.g. F^-, Cl^-, etc., and by forming one covalent bond, e.g. the elements themselves F_2, Cl_2, Br_2, I_2 and their hydrides HF, HCl, HBr and HI.

Because chlorine, bromine and iodine have easily accessible *d* orbitals available, they are able to form covalent compounds in which the octet of electrons is expanded; for instance, iodine shows valencies of 1, 3, 5 and 7 respectively in the compounds ICl, ICl_3, IF_5 and IF_7. Like nitrogen and oxygen (the first members of Group 5B and 6B respectively), fluorine cannot expand its octet and is thus restricted to a covalency of 1.

The molecules of the halogens are diatomic with only weak van der Waals' forces operating between the individual molecules; however, in the case of iodine these forces are sufficiently strong to bind the iodine molecules into a three dimensional lattice (fig. 21.1). This structure is easily broken down on heating, and in fact iodine sublimes at one atmosphere pressure if warmed gently.

FIG. 21.1. *The structure of iodine*

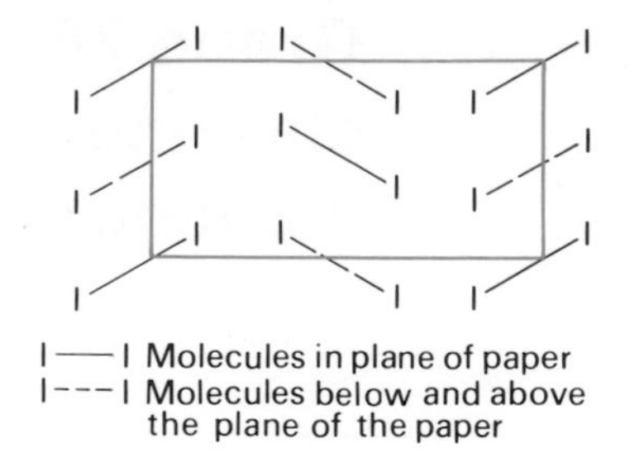

There is sufficient difference between fluorine and chlorine in chemical behaviour to warrant a separate treatment of the former element. Chlorine, bromine and iodine are treated as a group, and a short account of astatine, in so far as it illustrates group trends, is included.

Fluorine

21.3 Occurrence and preparation of fluorine

The most important naturally occurring compound of fluorine is fluorspar (calcium fluoride), $Ca^{2+}(F^-)_2$; cryolite, $(Na^+)_3AlF_6^{3-}$, is of importance in the extraction of aluminium (p. 205).

The only possible method of preparing fluorine is by electrolysis, and even then the chemicals used must be free from traces of water which react rapidly with the free halogen. The electrolyte used is hydrogen fluoride to which some potassium fluoride has been added to increase the electrical conductivity. One type of cell which operates at about 250°C employs an equimolar mixture HF—KF; carbon is used as the anode and a steel vessel serves as the cathode. The overall process involves discharge of fluorine at the anode and hydrogen at the cathode; hydrogen fluoride is added to the cell periodically to maintain the correct composition of the mixture.

Fluorine discharged at the anode: $$2F^- \longrightarrow F_2 + 2e^-$$
Hydrogen discharged at the cathode: $$2H^+ + 2e^- \longrightarrow H_2$$

The fluorine is contaminated with hydrogen fluoride which is removed by passing it over sodium fluoride pellets:

$$Na^+F^-(s) + HF(g) \longrightarrow \underset{\text{sodium hydrogen fluoride}}{Na^+HF_2^-(s)}$$

21.4 Properties of fluorine

Fluorine is a pale yellow gas with an irritating smell; it is extremely poisonous. It liquefies at −188°C and freezes at −220°C.

It is the most chemically reactive non-metal, combining directly with all other non-metals except nitrogen, oxygen and the noble gases (except krypton and xenon, p. 171). Most metals catch fire in fluorine and even gold and platinum are attacked if heated in a stream of the gas. Elements generally attain their highest valency state in combination with fluorine, e.g. the highest fluoride of sulphur is SF_6 but the highest chloride is SCl_4, and silver forms a 2-valent fluoride but only a 1-valent chloride.

Fluorine is the most electronegative of elements (a Pauling electronegativity value of 4·0 compared with 3·0 for chlorine) and will displace all other halogens from their ionic halides, e.g.

$$2Na^+Cl^-(s) + F_2(g) \longrightarrow 2Na^+F^-(s) + Cl_2(g)$$

or

$$2Cl^- + F_2(g) \longrightarrow 2F^- + Cl_2(g)$$

It displaces oxygen from silicon dioxide and water (some hydrogen peroxide and ozone are formed as well); with cold dilute solutions of alkalis it gives oxygen difluoride, F_2O, and with warm concentrated alkalis oxygen is liberated:

$$SiO_2(s) + 2F_2(g) \longrightarrow SiF_4(g) + O_2(g)$$
$$2H_2O(l) + 2F_2(g) \longrightarrow 4H^+F^-(aq) + O_2(g)$$
$$\text{(cold dilute) } 2OH^-(aq) + 2F_2(g) \longrightarrow F_2O(g) + 2F^-(aq) + H_2O(l)$$
$$\text{(warm conc.) } 4OH^-(aq) + 2F_2(g) \longrightarrow O_2(g) + 4F^-(aq) + 2H_2O(l)$$

21.5 Why is fluorine such a reactive element?

It is convenient to discuss the extreme chemical reactivity of fluorine under two headings, namely the formation of covalent fluorides, and the formation of ionic fluorides.

The formation of covalent fluorides

Fluorine has an abnormally low bond energy by comparison with the other halogens (F_2, 158; Cl_2, 242; Br_2, 193; I_2, 151 kJ mol^{-1}). This low value has been partially ascribed to repulsion between the non-bonding electrons on the fluorine atoms (the other halogens have longer bond lengths and this repulsive force is of little importance). By contrast with this low bond energy, covalent bonds formed between fluorine and other elements are generally very strong, e.g. C—F, 484; C—Cl, 338; C—Br, 276; C—I, 238 kJ mol^{-1}.

These two effects reinforce each other and, as a result, the formation of tetrafluoromethane, for example, is much more exothermic than, say, the formation of tetrachloromethane. The enthalpy changes involved in forming one mole of tetrafluoromethane and one mole of tetrachloromethane (in the gaseous state) from carbon and the halogen are set out below:

$$C(s) + 2F_2(g) \longrightarrow CF_4(g)$$

(a) Conversion of solid carbon to the gaseous state (the enthalpy of atomisation, S is absorbed)

$$C(s) \longrightarrow C(g) \qquad S = +x \text{ kJ mol}^{-1}$$

(b) Conversion of fluorine molecules into fluorine atoms (twice the bond energy, $2D$, is absorbed)

$$2F_2(g) \longrightarrow 4F(g) \qquad 2D = +316 \text{ kJ mol}^{-1}$$

(c) Combination of gaseous carbon and atomic fluorine to give 1 mole of tetrafluoromethane (four times the C—F bond energy, $4D_{C-F}$ is evolved)

$$C(g) + 4F(g) \longrightarrow CF_4(g) \qquad 4D_{C-F} = -1936 \text{ kJ mol}^{-1}$$

The overall enthalpy of formation of tetrafluoromethane from solid carbon and fluorine molecules is thus:

$$-1936 + 316 + x = (-1620 + x)\,\text{kJ mol}^{-1}$$

Using an exactly similar calculation, the enthalpy of formation of gaseous tetrachloromethane from solid carbon and chlorine molecules is:

$$-1352 + 484 + x = (-868 + x)\,\text{kJ mol}^{-1}$$

The enthalpy of formation of tetrafluoromethane is therefore more exothermic than the enthalpy of formation of tetrachloromethane to the extent of 752 kJ mol^{-1} (in practice it is slightly less than this value, since tetrachloromethane is a liquid and heat is evolved when the gaseous form of the substance condenses to a liquid).

The reason why many elements attain their highest valency state in combination with fluorine can be directly attributed to the low fluorine bond energy and the high element-fluorine bond energy.

The formation of ionic fluorides

When considering the enthalpy changes involved in the formation of ionic fluorides, the low bond energy of the fluorine molecule is again of importance. The energy changes involved in the formation of sodium fluoride and sodium chloride (both of which have a 6:6 structure, i.e. six cations surround each anion and vice versa) are set out below:

$$\text{Na(s)} + \tfrac{1}{2}\text{F}_2\text{(g)} \longrightarrow \text{Na}^+\text{F}^-\text{(s)}$$

(a) Conversion of solid sodium to gaseous atoms (the enthalpy of atomisation S, is absorbed)

$$\text{Na(s)} \longrightarrow \text{Na(g)} \qquad S = +109\,\text{kJ mol}^{-1}$$

(b) Removal of the outer electron of the sodium atom to give a sodium ion (the ionisation energy, I, is absorbed)

$$\text{Na(g)} \longrightarrow \text{Na}^+\text{(g)} + e^- \qquad I = +496\,\text{kJ mol}^{-1}$$

(c) Conversion of fluorine molecules into fluorine atoms (half the bond energy, $D/2$, is absorbed)

$$\tfrac{1}{2}\text{F}_2\text{(g)} \longrightarrow \text{F(g)} \qquad D/2 = +79\,\text{kJ mol}^{-1}$$

(d) Addition of an electron to a fluorine atom to give a fluoride ion (the electron affinity, E, is evolved)

$$\text{F(g)} + e^- \longrightarrow \text{F}^-\text{(g)} \qquad E = -333\,\text{kJ mol}^{-1}$$

(e) The bringing together of the sodium and fluoride ions to form solid sodium fluoride (the lattice energy, U, is evolved)

$$\text{Na}^+\text{(g)} + \text{F}^-\text{(g)} \longrightarrow \text{Na}^+\text{F}^-\text{(s)} \qquad U = -920\,\text{kJ mol}^{-1}$$

The overall enthalpy of formation of sodium fluoride is thus:

$$(109 + 496 + 79 - 333 - 920) = -569\,\text{kJ mol}^{-1}$$

A similar calculation shows that the enthalpy of formation of sodium chloride is $-411\,\text{kJ mol}^{-1}$ (p. 57), i.e. the formation of sodium fluoride is more exothermic than the formation of sodium chloride to the extent of about $158\,\text{kJ mol}^{-1}$. The most important factor here is the appreciable difference in the lattice energies of sodium fluoride and sodium chloride. The lattice energy of the fluoride is more negative, since the smaller fluoride ion can bind sodium ions more firmly than the large chloride ion.

Since the lattice energy of a fluoride is always greater than that of a chloride for comparable crystal structures, it is not really surprising that some chlorides are forced to assume covalent structures on energetic grounds, whereas the corresponding fluoride is essentially ionic, e.g. tin (IV) chloride is a volatile covalent liquid while the fluoride is an ionic solid.

21.6 Hydrogen fluoride, HF

Hydrogen fluoride can be made by the action of concentrated sulphuric acid on calcium fluoride:

$$Ca^{2+}(F^-)_2(s) + H_2SO_4(l) \longrightarrow Ca^{2+}SO_4^{2-}(s) + 2HF(g)$$

When dry it is known as anhydrous hydrofluoric acid and is available as a liquid which is usually stored in mild steel cylinders. Its boiling point (19·5°C) is considerably higher than that of hydrogen chloride (−84°C) and this is because the individual HF units are associated into larger aggregates by means of hydrogen bonding. Extreme care should be observed when handling this substance since its vapour is highly toxic and it causes painful burns if it comes in contact with the skin (a face shield and thick rubber gloves must be worn).

A solution of hydrogen fluoride in water behaves as a weak acid, in marked contrast with aqueous solutions of the other hydrogen halides which are strongly acidic. In order to trace the reason for this weakness (the dissociation constant, K, for an aqueous solution of hydrogen fluoride at 298 K is 7×10^{-4}) it is necessary to consider a number of individual energy terms; when this is done, it turns out that the very high H—F bond energy is the principal reason why ionisation into $H_3O^+(aq)$ and $F^-(aq)$ ions in water is not very extensive:

$$HF(l) + H_2O(l) \rightleftharpoons H_3O^+(aq) + F^-(aq)$$ (equilibrium well over to the left)

An aqueous solution of hydrofluoric acid attacks most metals with the formation of the metallic fluoride; these salts can also be obtained by the action of the acid on the metallic oxide, hydroxide or carbonate.

Moist hydrogen fluoride and aqueous solutions of the acid attack silica and glass, so the solution is stored in polythene containers.

$$SiO_2(s) + 4HF(aq) \longrightarrow SiF_4(g) + 2H_2O(l)$$
$$SiF_4(g) + 2HF(aq) \longrightarrow \underset{\text{fluorosilicic acid}}{(H^+)_2SiF_6^{2-}(aq)}$$

Unlike the other hydrogen halides it forms acid salts, e.g. potassium hydrogen fluoride, $K^+HF_2^-$, which can be made by passing hydrogen fluoride over potassium fluoride.

21.7 The fluorides

Fluorides are more ionic than the corresponding chlorides, typical ionic fluorides being those of Groups 1A and 2A (see p. 329 for a discussion of the importance of lattice energy in connection with ionic fluorides).

Most ionic fluorides are soluble in water, except those with very high lattice energies, and this explains why calcium fluoride is insoluble in water, unlike the other halides of calcium. The ready solubility of silver fluoride, in marked contrast to the insolubility of other silver halides, can be ascribed to the ionic nature of the fluoride (the other halides are appreciably covalent).

Several transition metals form fluorides in which they exert their highest valency, e.g. UF_6 and OsF_8. These are covalent compounds and exist as gases or volatile liquids, for it is energetically impossible to form, say, U^{6+} and Os^{8+} ions.

The fluorides of non-metals are covalent but the element-fluorine bond is generally appreciably polarised; often these compounds contain the non-metal in its highest valency state, e.g. SF_6 and IF_7. An extensive organic chemistry based on carbon-fluorine bonds is now being built up. Polytetrafluoroethene (empirical formula CF_2) is a plastic remarkable for its resistance to chemical attack; this extreme stability has been attributed to the great strength of the C—F bond and to the shielding of the carbon by the fluorine atoms (a steric effect). It seems likely that the shielding effect is of more importance since the free energy change for the hydrolysis of tetrafluoromethane has been calculated to be negative, i.e. this hydrolysis is thermodynamically possible:

$$CF_4(g) + 2H_2O(l) \longrightarrow CO_2(g) + 4HF(g) \quad \Delta G^{\ominus}(298\,K) = -344\ kJ\,mol^{-1}$$

21.8 The oxides of fluorine

Oxygen difluoride, F_2O

This gas is prepared by the action of fluorine on cold dilute sodium hydroxide solution:

$$2OH^-(aq) + 2F_2(g) \longrightarrow F_2O(g) + 2F^-(aq) + H_2O(l)$$

It oxidises metals and some non-metals to a mixture of the fluoride and oxide.

It is covalent with an angular molecule, similar in shape to that of hydrogen sulphide (electron pair repulsion).

Oxygen monofluoride, F_2O_2

Oxygen monofluoride is obtained by the action of a silent electrical discharge on a fluorine/oxygen mixture at low temperature and pressure. It decomposes into its constituent elements above −95°C.

Chlorine, bromine and iodine

21.9 Occurrence and extraction of chlorine, bromine and iodine

The main source of chlorine is sodium chloride, which occurs in sea water and in salt deposits, e.g. the Dead Sea and Stassfurt deposits and the rock salt deposits in Cheshire. Electrolysis of an aqueous solution of sodium chloride produces sodium hydroxide, chlorine and hydrogen (see p. 177–178 for a description of the Castner-Kellner process and the Gibbs diaphragm cell process.

Large quantities of chlorine are used in the production of organic chlorides and often appreciable quantities of hydrogen chloride are formed in these processes. This 'by-product' hydrogen chloride is catalytically oxidised back again to chlorine and this process is now of major industrial importance; the catalyst used is copper mixed with rare earth chlorides and supported on silica.

$$4HCl(g) + O_2(g) \longrightarrow 2H_2O(l) + 2Cl_2(g)$$

Bromides, e.g. those of potassium, sodium and magnesium, occur in sea water and in salt deposits, but they are not nearly so abundant as chlorides.

Most of the world's supply of bromine is obtained from sea water (bromide content about 1 part in 15 000). The sea water is first acidified (to prevent appreciable hydrolysis of chlorine and bromine) and then treated with chlorine. The bromide ions are oxidised to bromine which is then expelled as a vapour by passing air through the liquid mixture. The air, containing a very small concentration of bromine, is mixed with sulphur dioxide and water is then added; the bromine is reduced back again to bromide ions and the air recycled to pick up more bromine, which is reduced to bromide and added to the aqueous solution. By this means a solution containing a high concentration of bromide ions is obtained. Treatment of this solution with chlorine results in the displacement of bromine in high concentration, which can be distilled off on heating the solution. Dilute sulphuric acid is produced in the process and is used to treat the sea water prior to addition of chlorine.

$$\underset{\text{in sea water}}{2Br^-(aq)} + Cl_2(g) \longrightarrow 2Cl^-(aq) + \underset{\text{low concentration}}{Br_2(g)}$$

$$Br_2(g) + SO_2(g) + 2H_2O(l) \longrightarrow 4H^+(aq) + SO_4^{2-}(aq) + \underset{\text{high concentration}}{2Br^-(aq)}$$

$$\underset{\text{high concentration}}{2Br^-(aq)} + Cl_2(g) \longrightarrow 2Cl^-(aq) + \underset{\text{high concentration}}{Br_2(g)}$$

Iodides occur only in minute concentrations in sea water, but certain species of seaweed can use these minute amounts and concentrate the iodine in themselves as organic compounds. This source, together with the brine wells in California, supply some of the world's iodine, but the main source of this element is sodium iodate (V), $Na^+IO_3^-$, which occurs in the sodium nitrate deposits in Chile.

The mother liquor containing sodium iodate (V), i.e. the liquid remaining after crystallisation of the sodium nitrate, is treated with sodium hydrogen sulphite, when the iodate (V) is reduced to iodine:

$$2IO_3^-(aq) + 5HSO_3^-(aq) \longrightarrow 3HSO_4^-(aq) + 2SO_4^{2-}(aq) + I_2(s) + H_2O(l)$$

The reaction involves three stages; first of all the iodate (V) is slowly reduced to iodide:

$$IO_3^-(aq) + 3HSO_3^-(aq) \longrightarrow I^-(aq) + 3HSO_4^-(aq) \text{ (slow reaction)}$$

The iodide then rapidly reacts with unchanged iodate (V) in the presence of hydrogen ions:

$$IO_3^-(aq) + 5I^-(aq) + 6H^+(aq) \longrightarrow 3I_2(s) + 3H_2O(l) \text{ (fast reaction)}$$

A reaction occurs between the iodine and unchanged hydrogen sulphite, and iodine is only precipitated once all the hydrogen sulphite has been consumed:

$$I_2(s) + HSO_3^-(aq) + H_2O(l) \longrightarrow 2I^-(aq) + HSO_4^-(aq) + 2H^+(aq)$$

Crude iodine may be contaminated with iodine monochloride, ICl, and iodine monobromide, IBr. It is purified by sublimation in the presence of potassium iodide which removes these impurities:

$$K^+I^-(s) + ICl(l) \longrightarrow K^+Cl^-(s) + I_2(g)$$

21.10 Laboratory preparation of chlorine, bromine and iodine

Chlorine can be obtained in the laboratory by oxidising concentrated hydrochloric acid with either manganese (IV) oxide (heat required) or potassium manganate (VII) (no heat required):

$$Mn^{4+}(O^{2-})_2(s) + 4H^+Cl^-(aq) \longrightarrow Mn^{2+}(Cl^-)_2(aq) + 2H_2O(l) + Cl_2(g)$$

$$2K^+MnO_4^-(s) + 16H^+Cl^-(aq) \longrightarrow 2K^+Cl^-(aq) + 2Mn^{2+}(Cl^-)_2(aq) + 8H_2O(l) + 5Cl_2(g)$$

Bleaching powder (see p. 191) also evolves chlorine when treated with dilute acids.

Chlorine from any of these sources can be dried by passing it through concentrated sulphuric acid and collected by downward delivery. When it is obtained by oxidation of concentrated hydrochloric acid it is passed through water to dissolve any hydrogen chloride fumes before being dried.

Hydrobromic and hydriodic acids, the bromine and iodine equivalents of hydrochloric acid, are unstable in the presence of air and are not often used as laboratory reagents; therefore the laboratory preparations of bromine and iodine, although similar, are not quite analogous to the methods used for obtaining chlorine.

Bromine is obtained by heating a mixture of potassium bromide, manganese (IV) oxide and concentrated sulphuric acid. The reaction is carried out in a 'Quickfit' distillation apparatus and the bromine vapour is condensed in a receiver cooled under running water. The reaction takes place in two stages:

$$K^+Br^-(s) + H_2SO_4(l) \longrightarrow K^+HSO_4^-(s) + HBr(g)$$

$$Mn^{4+}(O^{2-})_2(s) + 4HBr(g) \longrightarrow Mn^{2+}(Br^-)_2(s) + 2H_2O(l) + Br_2(g)$$

Iodine is obtained by a similar reaction using a mixture of potassium iodide, manganese (IV) oxide and concentrated sulphuric acid.

21.11 Properties of chlorine, bromine and iodine

Physical properties

Chlorine is a greenish-yellow poisonous gas with an extremely irritating smell. At room temperature under a pressure of about 7 atmospheres it liquefies to a yellow liquid. Bromine is a dark red liquid with an unpleasant and poisonous vapour, and iodine is a dark shiny solid which produces a purple vapour on heating. Although the vapour pressure of iodine is low at room temperature, solid iodine has a distinctive smell; in high concentrations the vapour of iodine is unpleasant and poisonous.

Chlorine and bromine are moderately soluble and iodine sparingly soluble in water (there is some chemical reaction with the water, p. 333–334).

All three halogens are much more soluble in organic solvents such as tetrachloromethane. Iodine imparts a brown colour to water, ethanol and aqueous potassium iodide solution; when dissolved in solvents such as benzene and tetrachloromethane a purple solution is obtained. The brown solutions have been attributed to some chemical interaction between the iodine molecules and the solvent, e.g. the lone pairs of electrons on the oxygen atom in the water and ethanol molecules and on the iodide ion can form a co-ordinate bond with iodine molecules:

$$C_2H_5\text{—}\underset{\underset{\displaystyle H}{|}}{\ddot{O}}: + I\text{—}I \rightleftharpoons C_2H_5\text{—}\underset{\underset{\displaystyle H}{|}}{\ddot{O}}\rightarrow I\text{—}I$$

$$:\ddot{\underset{..}{I}}:^- + I\text{—}I \rightleftharpoons [:\ddot{\underset{..}{I}}\rightarrow I-I]^-$$

The reason why iodine is much more soluble in aqueous potassium iodide than in water itself is because the I_3^- complex ion is very stable.

The purple colour of solutions of iodine in benzene and tetrachloromethane (similar to the colour of iodine vapour) is due to the separation of the dissolved iodine into I_2 molecules, i.e. the weak van der Waals' forces which are present in solid iodine (fig, 21.1 p. 326) are broken; otherwise there is little interaction with these non-polar solvents.

Iodine forms a deep blue colour with starch solution and this is an extremely sensitive test for traces of the element. It is thought that iodine molecules reside inside the spiral structure of starch molecules.

Chemical properties

(a) Reactions with elements

Chlorine, bromine and iodine combine with many metals and non-metals; in general the combination with chlorine is the most vigorous and with iodine the least vigorous, although there are some notable exceptions to this rule, e.g. potassium explodes with bromine and iodine. Exceptions such as these can be attributed to a greater halogen concentration in liquid bromine and solid iodine as compared with gaseous chlorine. Carbon, nitrogen and oxygen do not combine with any of the halogens directly.

Metals that form more than one chloride generally form the higher one in combination with chlorine, e.g. iron forms iron (III) chloride and not iron (II) chloride, and tin forms the tin (IV) compound. Non-metals often give a mixture of chlorides, e.g. phosphorus gives both phosphorus trichloride and phosphorus pentachloride, the latter predominating if an excess of chlorine is used.

(b) Reactions with water and alkalis

Chlorine and bromine are moderately soluble in water (bromine more so than chlorine), while iodine is only very sparingly soluble. Although hydrolysis of the halogens takes place in solution, this is only extensive in the case of chlorine, i.e. aqueous solutions of bromine and iodine are essentially solutions containing respectively Br_2 and I_2 molecules.

It is possible to discuss the dissolution of chlorine in water in terms of two equilibrium reactions:

$$Cl_2(g) + H_2O(l) \rightleftharpoons Cl_2(aq)$$

$$Cl_2(aq) + 2H_2O(l) \rightleftharpoons H_3O^+(aq) + Cl^-(aq) + \underset{\text{chloric (I) acid}}{HOCl(aq)}$$

The dissociation constant for the second reaction, K, is given by the usual expression:

$$K = \frac{[H_3O^+(aq)][Cl^-(aq)][HOCl(aq)]}{[Cl_2(aq)]} = 4{\cdot}2 \times 10^{-4}\ \text{mol}^2\ \text{dm}^{-6} \text{ at } 298\ \text{K}$$

For bromine and iodine the numerical values of K are respectively $7{\cdot}2 \times 10^{-9}$ and $2{\cdot}0 \times 10^{-13}$, thus bromic (I) acid, HOBr, and iodic (I) acid, HOI, are formed in negligible amounts.

Chlorine dissolves to a much greater extent in the presence of $OH^-(aq)$ ions, i.e. in alkaline solution, which remove the $H_3O^+(aq)$ ions forming water molecules, thereby allowing the hydrolysis reaction to proceed very much further to the right:

$$\begin{array}{ccccccc} Cl_2(aq) + 2H_2O(l) \rightleftharpoons & H_3O^+(aq) & + & Cl^-(aq) & + & HOCl(aq) & \\ & \Big\downarrow OH^-(aq) & & & & \Big\downarrow OH^-(aq) & \\ & 2H_2O(l) & & & & H_2O(l) + ClO^-(aq) & \end{array}$$

A similar reaction takes place with solutions of bromine and iodine, although the bromate (I), BrO^-, and iodate (I), IO^-, ions rapidly disproportionate (undergo self-oxidation-reduction) to give the halide and halate (V) ions:

$$3BrO^-(aq) \longrightarrow 2Br^-(aq) + \underset{\text{bromate (V) ion}}{BrO_3^-(aq)}$$

$$3IO^-(aq) \longrightarrow 2I^-(aq) + \underset{\text{iodate (V) ion}}{IO_3^-(aq)}$$

The chlorate (I) ion also undergoes disproportionation into the chloride and chlorate (V) ions, but this reaction is significant only above about 75°C.

The action of alkalis on the three halogens can be summarised as follows:

Chlorine reacts with cold dilute sodium hydroxide solution to give sodium chloride and sodium chlorate (1):

$$2OH^-(aq) + Cl_2(g) \longrightarrow Cl^-(aq) + ClO^-(aq) + H_2O(l)$$

Chlorine reacts with warm concentrated sodium hydroxide solution to give sodium chloride and sodium chlorate (V):

$$6OH^-(aq) + 3Cl_2(g) \longrightarrow 5Cl^-(aq) + \underset{\text{chlorate (V) ion}}{ClO_3^-(aq)} + 3H_2O(l)$$

Bromine and iodine react with sodium hydroxide solution to give the halide and halate (V):

$$6OH^-(aq) + 3Br_2(l) \longrightarrow 5Br^-(aq) + \underset{\text{bromate (V) ion}}{BrO_3^-(aq)} + 3H_2O(l)$$

$$6OH^-(aq) + 3I_2(s) \longrightarrow 5I^-(aq) + \underset{\text{iodate (V) ion}}{IO_3^-(aq)} + 3H_2O(l)$$

(c) Oxidising reactions

Chlorine is a powerful oxidising agent and will remove hydrogen from hydrocarbons depositing carbon, e.g. it reacts explosively with ethyne,

$$H{-}C{\equiv}C{-}H(g) + Cl_2(g) \longrightarrow 2C(s) + 2HCl(g)$$

Under more controlled conditions it can be used to produce chlorinated hydrocarbons, e.g. the reaction with ethyne can be controlled to give 1,2-dichloroethene and 1,1,2,2-tetrachlorethane, the latter compound being known as Westron, which is an important industrial solvent for fats and oils.

$$H{-}C{\equiv}C{-}H \xrightarrow{Cl_2} \underset{\text{1,2-dichloroethene}}{H{-}C(Cl){=}C(Cl){-}H} \xrightarrow{Cl_2} \underset{\substack{\text{1,1,2,2-tetrachloroethane}\\ \text{(Westron)}}}{H{-}CCl_2{-}CCl_2{-}H}$$

Bromine reacts with hydrocarbons in a similar but less vigorous manner; iodine shows little reaction under the same conditions.

The decrease in oxidising power with increasing atomic number of the halogens is convincingly demonstrated by a series of displacement reactions; thus chlorine will displace bromine and iodine from aqueous solutions of bromides and iodides respectively and bromine will displace iodine from an aqueous solution of an iodide, e.g.

$$2Br^-(aq) + Cl_2(g) \longrightarrow Br_2(l) + 2Cl^-(aq)$$

or

$$2Br^-(aq) \longrightarrow Br_2(l) + 2e^- \quad \text{(oxidation)}$$

$$Cl_2(g) + 2e^- \longrightarrow 2Cl^-(aq) \quad \text{(reduction)}$$

$$2I^-(aq) + Br_2(l) \longrightarrow I_2(s) + 2Br^-(aq)$$

Chlorine (or its aqueous solution) and aqueous solutions of bromine and iodine oxidise hydrogen sulphide to sulphur:

$$H_2S(g) + X_2 \longrightarrow 2HX(g) + S(s) \qquad (X_2 = Cl_2, Br_2 \text{ or } I_2)$$

They will also oxidise sulphur dioxide in solution to sulphuric acid (essentially oxidation of the sulphite ion to the sulphate ion):

$$SO_3^{2-}(aq) + H_2O(l) + X_2 \longrightarrow 2H^+(aq) + 2X^-(aq) + SO_4^{2-}(aq)$$

Chlorine and bromine oxidise the thiosulphate ion, $S_2O_3^{2-}$, to the sulphate ion (p. 318), whereas the milder oxidising agent iodine converts it quantitatively into the tetrathionate, $S_4O_6^{2-}$, and this reaction can be used to estimate iodine solutions (p. 318).

21.12 Uses of chlorine, bromine and iodine

Chlorine is used extensively for the manufacture of organic chemicals such as tetrachloromethane and trichloroethene, which are used as solvents, and vinyl chloride, which is the monomer used in the production of the plastic PVC. Other organic compounds containing chlorine include Dettol, TCP and DDT. Another large scale use of chlorine is as a bleaching agent for cotton, rayon and woodpulp; it is also used for sterilising water supplies.

Bromine is used mainly for the manufacture of 1,2-dibromoethane, which is added to petrol to remove lead as volatile lead bromide and so prevent it from fouling the sparking plugs (the lead arises from lead tetraethyl, an antiknock additive in petrol). Another use is for the production of silver bromide, which forms the basis of photographic films.

Iodine is converted into silver iodide which is also used in the manufacture of photographic films. Other uses include the manufacture of alkyl iodides and iodoform, the preparation of iodised salt and tincture of iodine (a dilute solution of iodine in ethanol which is used medicinally), and the preparation of dyes used in colour photography.

21.13 The hydrides of chlorine, bromine and iodine

Preparation

(a) Direct synthesis

All three halogens combine under appropriate conditions with hydrogen to give the halogen hydride:

$$H_2(g) + X_2(g) \longrightarrow 2HX(g)$$

Chlorine explodes violently with hydrogen when the mixture is irradiated with ultraviolet light, but a stream of hydrogen can be burnt safely in chlorine:

$$H_2(g) + Cl_2(g) \longrightarrow 2HCl(g)$$

This method is used industrially to manufacture hydrogen chloride, and hence hydrochloric acid.

Hydrogen and bromine combine smoothly at about 300°C, in the presence of a platinum catalyst to give hydrogen bromide, HBr, but under the same conditions the reaction between hydrogen and iodine vapour is reversible:

$$H_2(g) + Br_2(g) \rightarrow 2HBr(g)$$
$$H_2(g) + I_2(g) \rightleftharpoons 2HI(g)$$

(b) Reaction of an ionic halide with concentrated sulphuric acid

Hydrogen chloride can be obtained by the action of concentrated sulphuric acid on an ionic chloride, e.g. sodium chloride, (displacement of a volatile acid by a less volatile one, p. 314).

The reaction takes place in two stages, the first one occurring at room temperature and the second on heating strongly:

$$Na^+Cl^-(s) + H_2SO_4(l) \longrightarrow Na^+HSO_4^-(s) + HCl(g) \text{ (in the cold)}$$
$$Na^+HSO_4^-(s) + Na^+Cl^-(s) \longrightarrow (Na^+)_2SO_4^{2-}(s) + HCl(g) \text{ (on heating)}$$

Similar methods cannot be usefully employed to obtain hydrogen bromide and hydrogen iodide, since these two hydrides are reducing agents and are readily oxidised by concentrated sulphuric acid to the free halogens (see under the reducing properties of the halogen hydrides).

(c) The action of water on phosphorus trihalides

Phosphorus trichloride is readily hydrolysed by water to give phosphonic acid and hydrogen chloride:

$$PCl_3(l) + 3H_2O(l) \longrightarrow H_3PO_3(aq) + 3HCl(g)$$

This method is not used for preparing hydrogen chloride in the laboratory, since more convenient methods are available. However, a similar method is used for making hydrogen bromide and hydrogen iodide.

Hydrogen bromide is obtained by dropping bromine onto a mixture of red phosphorus and a little water; phosphorus tribromide is formed initially and is then hydrolysed:

$$2P(s) + 3Br_2(l) \longrightarrow 2PBr_3(l)$$
$$PBr_3(l) + 3H_2O(aq) \longrightarrow H_3PO_3(aq) + 3HBr(g)$$

Any unchanged bromine vapour is removed by passing the mixed gases over damp red phosphorus supported on glass wool, and the hydrogen bromide is collected by downward delivery.

Since iodine is a solid a slight variation on this method is used for obtaining hydrogen iodide. Water is dropped onto a mixture of red phosphorus and iodine, and any iodine vapour is removed by passing over damp red phosphorus as before. The gas is collected by downward delivery.

$$2P(s) + 3I_2(s) \longrightarrow 2PI_3(s)$$
$$PI_3(s) + 3H_2O(aq) \longrightarrow H_3PO_3(aq) + 3HI(g)$$

Industrial production of hydrogen chloride and hydrochloric acid

Hydrogen chloride is obtained industrially by burning hydrogen in chlorine; it is also available in considerable quantities as a by-product formed during many organic chlorination reactions, e.g. the chlorination of hydrocarbons. A small amount is still obtained by reacting sodium chloride with concentrated sulphuric acid (p. 314), since the sodium sulphate produced in this process is used in the manufacture of glass.

A solution of hydrogen chloride in water is hydrochloric acid.

Physical properties

The hydrides of chlorine, bromine and iodine are colourless covalent gases with sharp odours; their boiling points increase with increasing relative molecular mass. They dissolve in non-ionising solvents such as toluene and trichloromethane, preserving their covalent character, e.g. a solution of hydrogen chloride in toluene is a non-conductor of electricity and will not affect blue litmus paper.

Chemical properties

(a) Reducing properties

Hydrogen chloride shows little tendency to decompose into its constituent elements when heated; strong heating of hydrogen bromide produces a brown colour of bromine vapour, while copious violet fumes of iodine are obtained when a red-hot steel needle is plunged into a gas jar of hydrogen iodide. The stability of the halogen hydrides to thermal decomposition therefore decreases in the order:

$$HCl > HBr > HI$$

and this is due to the progressive decrease in H—X bond energy.

The removal of hydrogen from hydrogen chloride (or concentrated hydrochloric acid) can be effected by strong oxidising agents such as manganese (IV) oxide and potassium manganate (VII) (p. 368), and these oxidising agents also remove hydrogen from hydrogen bromide and hydrogen iodide (or their concentrated acid solutions in water).

Concentrated sulphuric acid does not attack hydrogen chloride but both hydrogen bromide and hydrogen iodide are oxidised to the free halogen. Hydrogen bromide reduces the sulphuric acid to sulphur dioxide, while the stronger reducing agent hydrogen iodide converts it mainly into hydrogen sulphide:

$$2HBr(g) + H_2SO_4(l) \longrightarrow SO_2(g) + 2H_2O(l) + Br_2(l)$$
$$8HI(g) + H_2SO_4(l) \longrightarrow H_2S(g) + 4H_2O(l) + 4I_2(s)$$

(b) The action of water on the halogen hydrides, and the properties of the acidic solutions so formed.

The hydrides of chlorine, bromine and iodine are extremely soluble in water, the solutions so formed being strongly acidic, e.g. M/10 solutions of the halogen hydrides in water are virtually completely ionised into $H_3O^+(aq)$ and $X^-(aq)$ ions.

$$HX(aq) + H_2O(l) \rightleftharpoons H_3O^+(aq) + X^-(aq)$$

These solutions possess the usual properties of acids, e.g. they will react with metals with negative electrode potentials to form a salt and hydrogen, and with metallic oxides, hydroxides and carbonates.

A concentrated solution of hydrochloric acid evolves gaseous hydrogen chloride when distilled until the composition of the liquid has fallen to 20 per cent HCl:80 per cent H_2O, when this liquid distils unchanged. Similarly, a dilute solution of hydrochloric acid evolves steam on distillation until the composition rises again to 20 per cent HCl:80 per cent H_2O, when this mixture distils unchanged. The liquid mixture that distils unchanged is called a maximum constant boiling point mixture. The same phenomenon occurs with hydrobromic and hydriodic acids.

Solutions of hydrobromic and hydriodic acids, like hydrogen bromide and hydrogen iodide, possess reducing properties, e.g. they are unstable in the presence of air and slowly turn dark owing to the formation of the free halogen:

$$4H^+(aq) + 4Br^-(aq) + O_2(g) \longrightarrow 2Br_2(aq) + 2H_2O(l)$$
$$4H^+(aq) + 4I^-(aq) + O_2(g) \longrightarrow 2I_2(aq) + 2H_2O(l)$$

Concentrated hydriodic acid in conjunction with red phosphorus is used for reducing alcohols, aldehydes and ketones to the corresponding alkanes.

A convenient way of making aqueous solutions of hydrobromic and hydriodic acids is to pass hydrogen sulphide through a mixture of the halogen and water:

$$Br_2(l) + 2H_2O(l) + H_2S(g) \longrightarrow 2H_3O^+(aq) + 2Br^-(aq) + S(s)$$
$$I_2(s) + 2H_2O(l) + H_2S(g) \longrightarrow 2H_3O^+(aq) + 2I^-(aq) + S(s)$$

21.14 Tests for the chloride, bromide and iodide ions

(a) *Action of concentrated sulphuric acid*
Ionic halides evolve a sharp-smelling gas which fumes in moist air (the halogen hydride) when treated with a little concentrated sulphuric acid. A brown colour of bromine is also apparent from a bromide and a purple colour of iodine from an iodide.

(b) *Reaction with silver nitrate solution*
An aqueous solution of a chloride forms a white precipitate of silver chloride when mixed with silver nitrate solution:

$$Cl^-(aq) + Ag^+(aq) \longrightarrow AgCl(s)$$

Under similar conditions a bromide forms a cream precipitate of silver bromide, and an iodide a yellow precipitate of silver iodide.

Silver chloride readily dissolves in dilute ammonia solution; silver bromide is only sparingly soluble, and silver iodide is almost insoluble in ammonia solution (p. 415).

(c) *Displacement reactions*
An aqueous solution of a bromide is oxidised to free bromine when treated with chlorine water (p. 335). If a small quantity of tetrachloromethane is added to the mixture, bromine is extracted into the organic layer and the brown colour of bromine is easily visible.

A similar reaction with an aqueous solution of an iodide produces a purple colour of free iodine in the organic layer.

21.15 The oxides of chlorine, bromine and iodine

The oxides of chlorine
The known oxides of chlorine are chlorine monoxide, Cl_2O, (an orange coloured gas), chlorine dioxide, ClO_2, (a yellow gas), chlorine hexoxide, Cl_2O_6, (a red liquid), and chlorine heptoxide, Cl_2O_7, (a colourless liquid). All four oxides are unstable and dangerously explosive. They react with alkalis as follows:

$$Cl_2O(g) + 2OH^-(aq) \longrightarrow \underset{\text{chlorate (I)}}{2OCl^-(aq)} + H_2O(l)$$

$$2ClO_2(g) + 2OH^-(aq) \longrightarrow \underset{\text{chlorate (III)}}{ClO_2^-(aq)} + \underset{\text{chlorate (V)}}{ClO_3^-(aq)} + H_2O(l)$$

$$Cl_2O_6(l) + 2OH^-(aq) \longrightarrow \underset{\text{chlorate (V)}}{ClO_3^-(aq)} + \underset{\text{chlorate (VII)}}{ClO_4^-(aq)} + H_2O(l)$$

$$Cl_2O_7(l) + 2OH^-(aq) \longrightarrow \underset{\text{chlorate (VII)}}{2ClO_4^-(aq)} + H_2O(l)$$

The molecules of chlorine monoxide and chlorine dioxide are angular, the latter oxide containing an odd number of electrons, cf. the molecules NO and NO_2 (p. 254–255).

Solid chlorine hexoxide has an ionic structure $ClO_2^+ClO_4^-$. Chlorine heptoxide is $O_3Cl—O—ClO_3$ with a ClOCl bond angle of 119°.

The oxides of bromine

These have not been studied very extensively since they readily decompose into bromine and oxygen at low temperatures, e.g. at $-40°C$ or even lower. Oxides with formulae Br_2O, Br_3O_8, BrO_2 and Br_2O_7 have been reported.

The oxides of iodine

The best known oxide is iodine pentoxide, I_2O_5, which is thermally stable at temperatures up to about 300°C. It is a white crystalline solid and can be obtained by dehydrating iodic (V) acid, at 200°C, of which it is the acid anhydride:

$$2HIO_3(s) \longrightarrow I_2O_5(s) + H_2O(l)$$

Iodine pentoxide is an oxidising agent and liberates iodine when reacted with hydrogen sulphide and carbon monoxide:

$$I_2O_5(s) + 5H_2S(g) \longrightarrow 5H_2O(l) + 5S(s) + I_2(s)$$
$$I_2O_5(s) + 5CO(g) \longrightarrow 5CO_2(g) + I_2(s)$$

The reaction with carbon monoxide is quantitative and can be used for estimating this gas, since the liberated iodine can be titrated with a standard solution of sodium thiosulphate (p. 318).

21.16 The oxyacids of chlorine, bromine and iodine

No clear-cut generalisations cover all the halogen oxyacids and their salts, but for aqueous solutions of the oxyacids of chlorine, namely,

HOCl	HOClO	$HOClO_2$	$HOClO_3$
Chloric (I) acid	Chloric (III) acid	Chloric (V) acid	Chloric (VII) acid

increasing oxygen content leads to:
(a) An increase in thermal stability.
(b) An increase in strength of the acid.
(c) A decrease in oxidising power.

Salts of the oxyacids of chlorine are more thermally stable than the acid itself, and in aqueous solution an increase in oxygen content leads to a decrease in oxidising ability (the salts are less effective as oxidising agents than the parent acids).

The halic (I) acids, HOX

Chloric (I) acid, HOCl, is formed when chlorine is passed into water. A similar reaction with bromine and iodine produces decreasingly smaller concentrations of bromic (I) and iodic (I) acids (HOBr and HOI) respectively:

$$X_2 + 2H_2O(aq) \rightleftharpoons H_3O^+(aq) + X^-(aq) + HOX(aq)$$

Addition of mercury (II) oxide allows the above equilibrium reaction to move further to the right, since the halide ions, $X^-(aq)$, are removed as covalent mercury (II) halide (actually as the basic salt $HgO.HgX_2$):

$$2HgO(s) + 2H_3O^+(aq) + 2X^-(aq) \rightleftharpoons HgO.HgX_2(s) + 3H_2O(l)$$

Shaking a mixture of mercury (II) oxide, water and the halogen is a convenient method of obtaining a solution of these acids.

None of these acids exists as a pure substance, and in aqueous solution they are very weak, progressively so with increasing size of the halogen. They are increasingly thermally unstable in the same order.

The most important salt of the halic (I) acids is sodium chlorate (I), Na^+OCl^-, which is made by electrolysing a cold solution of sodium chloride, and allowing the products of electrolysis (sodium hydroxide solution and chlorine) to mix:

$$2OH^-(aq) + Cl_2(g) \longrightarrow Cl^-(aq) + OCl^-(aq) + H_2O(l)$$

It is sold under a variety of trade names like Parazone, Domestos and Milton as a dilute aqueous solution (which also contains sodium chloride) for use as a mild disinfectant and bleaching agent.

Sodium chlorate (I) is alkaline in solution because hydrolysis takes place, i.e. the OCl^- ion is a strong base:

$$OCl^-(aq) + H_2O(l) \rightleftharpoons HOCl(aq) + OH^-(aq)$$

Both chloric (I) acid and the chlorate (I) ion are strong oxidising agents, the former more powerful than the latter. Thus in acid solution iodide ions are oxidised to iodine, and iron (II) ions to iron (III) ions:

$$2I^-(aq) + H^+(aq) + HOCl(aq) \longrightarrow I_2(aq) + H_2O(l) + Cl^-(aq)$$
$$2Fe^{2+}(aq) + H^+(aq) + HOCl(aq) \longrightarrow 2Fe^{3+}(aq) + H_2O(l) + Cl^-(aq)$$

Chloric (I) acid and aqueous chlorates (I) decompose in the presence of sunlight into oxygen and chloride ions:

$$2OCl^-(aq) \longrightarrow 2Cl^-(aq) + O_2(g)$$

On warming to about 75°C, disproportionation into the chloride and chlorate (V) takes place:

$$3OCl^-(aq) \longrightarrow 2Cl^-(aq) + ClO_3^-(aq)$$

Bleaching powder, which is a mixture of several substances (p. 191), contains calcium chlorate (I). It is made by passing chlorine over moist slaked lime; as its name implies, it is used for bleaching fabrics.

Chloric (III) acid, HOClO

Chloric (III) acid is a stronger acid than chloric (I) acid and is known only in solution; it is of little importance. Sodium chlorate (III), $Na^+ClO_2^-$, is used industrially as a textile bleaching agent.

The bromine and iodine analogues of chloric (III) acid are unknown.

The halic (V) acids, $HOXO_2$

Aqueous solutions of chloric (V) acid, $HClO_3$, bromic (V) acid, $HBrO_3$, and iodic (V) acid, HIO_3, can be obtained by reacting dilute sulphuric acid with barium chlorate (V), barium bromate (V) and barium iodate (V) respectively; the insoluble barium sulphate can be separated from the aqueous solution by filtration:

$$Ba^{2+}(XO_3^-)_2(aq) + (H^+)_2SO_4^{2-}(aq) \longrightarrow Ba^{2+}SO_4^{2-}(s) + 2H^+XO_3^-(aq)$$

Neither chloric (V) nor bromic (V) acids can be isolated in the pure state since decomposition takes place if too much water is evaporated from dilute solutions; nevertheless careful evaporation can give solutions containing about 40 per cent of the acids. Iodic (V) acid, on the other hand, is more thermally stable and can be isolated as a pure solid.

The halic (V) acids dissociate extensively in aqueous solution, i.e. they are strong acids:

$$HXO_3(aq) + H_2O(l) \longrightarrow H_3O^+(aq) + XO_3^-(aq)$$

They also behave as powerful oxidising agents, particularly in the concentrated condition; thus a concentrated solution of chloric (V) acid and solid iodic (V) acid ignite many organic compounds, e.g. sugar. Typical oxidation reactions are:

$$ClO_3^-(aq) + 3SO_3^{2-}(aq) \longrightarrow Cl^-(aq) + 3SO_4^{2-}(aq)$$
$$2H^+(aq) + 2BrO_3^-(aq) + 5H_2S(g) \longrightarrow Br_2(aq) + 6H_2O(l) + 5S(s)$$
$$6H^+(aq) + IO_3^-(aq) + 5I^-(aq) \longrightarrow 3I_2(aq) + 3H_2O(l)$$

The alkali metal halates (V) are obtained when the halogen is reacted with a hot concentrated solution of the alkali:

$$3X_2 + 6OH^-(aq) \longrightarrow XO_3^-(aq) + 5X^-(aq) + 3H_2O(l)$$

Like the parent acids, they are oxidising agents.

Sodium chlorate (V), $Na^+ClO_3^-$, is manufactured by electrolysing a concentrated solution of sodium chloride at a temperature of about 70°C. The products of electrolysis are sodium hydroxide solution and chlorine which are allowed to mix freely. The sodium chlorate (V) is separated from the sodium chloride by fractional crystallisation. Potassium chlorate (V) is made in a similar manner from potassium chloride solution.

Sodium chlorate (V) is used as a weedkiller and potassium chlorate (V), which is much less soluble in water, is the oxidising agent in matches and some fireworks.

Potassium bromate (V), $K^+BrO_3^-$, reacts with bromide ions in acid solution to give a quantitative yield of bromine, and this reaction is sometimes made use of in volumetric analysis.

$$BrO_3^-(aq) + 5Br^-(aq) + 6H^+(aq) \longrightarrow 3Br_2(aq) + 3H_2O(l)$$

Potassium iodate (V), $K^+IO_3^-$, is mainly used in volumetric analysis as a primary standard, because it can be obtained in a high degree of purity and is stable in aqueous solution. In acid solution it reacts with an excess of potassium iodide to give a quantitative yield of iodine:

$$IO_3^-(aq) + 5I^-(aq) + 6H^+(aq) \longrightarrow 3I_2(aq) + 3H_2O(l)$$

The liberated iodine can be titrated with sodium thiosulphate solution (which is not a primary standard) and hence the latter can be standardised.

Potassium chlorate (V) decomposes when heated to give the chloride and oxygen, the reaction proceeding via the chlorate (VII):

$$4K^+ClO_3^-(s) \longrightarrow 3K^+ClO_4^-(s) + K^+Cl^-(s) \text{ (at 400°C)}$$
$$3K^+ClO_4^-(s) \longrightarrow 3K^+Cl^-(s) + 6O_2(g) \text{ (at higher temperature)}$$

Potassium bromate (V) and potassium iodate (V) also give the halide and oxygen, but the halates (VII) cannot be isolated:

$$2K^+BrO_3^-(s) \longrightarrow 2K^+Br^-(s) + 3O_2(g)$$
$$2K^+IO_3^-(s) \longrightarrow 2K^+I^-(s) + 3O_2(g)$$

Potassium iodate (V) alone reacts with the parent acid to give acid salts, e.g. $K^+IO_3^-.HIO_3$.

The halic (VII) acids

Chloric (VII) acid, $HClO_4$, can be prepared by vacuum distillation of a mixture of potassium chlorate (VII) and concentrated sulphuric acid; it is a liquid which begins to decompose at its boiling point of 90°C, hence care is needed in its preparation. In contact with organic matter it is dangerously explosive.

Chloric (VII) acid is a very strong acid in aqueous solution; indeed the monohydrate is a solid, $HClO_4.H_2O$, which is isomorphous with ammonium chlorate (VII) and is more correctly formulated as $H_3O^+ClO_4^-$.

Alkali metal chlorates (VII) can be obtained by controlled heating of the corresponding chlorate (V) (p. 342), or by electrolysis of aqueous solutions of the corresponding chlorate (V) when oxidation to the chlorate (VII) occurs at the anode. The oxidising ability of chlorates (VII) is utilised in the manufacture of explosives and detonators.

The bromate (VII) ion, BrO_4^-, has been obtained by oxidation of an alkaline solution of a bromate (V) with fluorine. The free acid can be obtained from this solution by the use of an 'acidic' ion-exchange resin.

It is a powerful oxidising agent.

Unlike chloric (VII) acid, iodic (VII) acid exists in a number of forms, the most common one having the formula, H_5IO_6, in which six oxygen atoms are arranged octahedrally round a central iodine atom (presumably the chlorine atom is too small to accommodate more than four oxygen atoms as in the ClO_4^- ion).

The acid H_5IO_6 is fairly weak in aqueous solution; the pure substance, which is a solid, can be converted into $H_4I_2O_9$ and HIO_4 by controlled heating in vacuo. The molecule of the former acid contains five oxygen atoms round each iodine atom which are themselves linked together via an oxygen bridge; the molecule of the latter acid is tetrahedral like that of chloric (VII) acid. Salts corresponding to the iodic (VII) acids are well known and, like the acids themselves, are powerful oxidising agents.

21.17 The interhalogen compounds

The known interhalogen compounds are listed below; they are covalent substances and polarised in such a way that a fractional negative charge resides on the lighter halogen atom(s). All the compounds listed can be obtained by direct synthesis under the right conditions.

ClF (gas)	ClF_3 (gas)	ClF_5 (gas)	
BrF (gas) BrCl (gas)	BrF_3 (liquid)	BrF_5 (liquid)	
ICl (liquid) IBr (solid)	ICl_3 (solid)	IF_5 (liquid)	IF_7 (gas)

The interhalogen compounds are oxidising agents; the halogen fluorides, except iodine heptafluoride which is a mild fluorinating agent, are among the most reactive substances known, e.g. bromine trifluoride will react vigorously even with asbestos.

Bromine trifluoride conducts electricity to a significant extent and this observation indicates slight ionisation into BrF_2^+ and BrF_4^- ions:

$$2BrF_3 \rightleftharpoons BrF_2^+ + BrF_4^-$$

Substances dissolved in liquid bromine trifluoride which furnish BrF_2^+ ions are regarded as acids and substances which provide BrF_4^- ions are regarded as bases in this non-aqueous solvent, cf. acid-base neutralisation in liquid ammonia (p. 251). The example below illustrates an acid-base neutralisation reaction in liquid bromine trifluoride:

$$\underset{\text{acid}}{BrF_2^+SbF_6^-} + \underset{\text{base}}{Ag^+BrF_4^-} \longrightarrow \underset{\text{salt}}{Ag^+SbF_6^-} + \underset{\text{solvent}}{2BrF_3}$$

The structures of the interhalogen compounds are interesting: thus the bromine atom in bromine trifluoride contains ten outer electrons, i.e. three bonding and two non-bonding pairs, which by the theory of electron-pair repulsion are placed at the apices of a trigonal bipyramid (fig. 21.2). It is not possible to predict the structure specifically in this case (alternatives are available), but experiment shows that the molecule is T-shaped and is as represented. The structures of chlorine trifluoride and iodine trifluoride are similar.

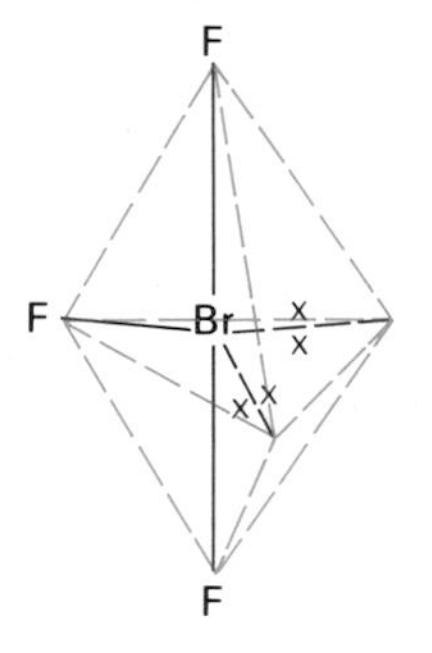

FIG. 21.2. *The structure of bromine trifluoride*

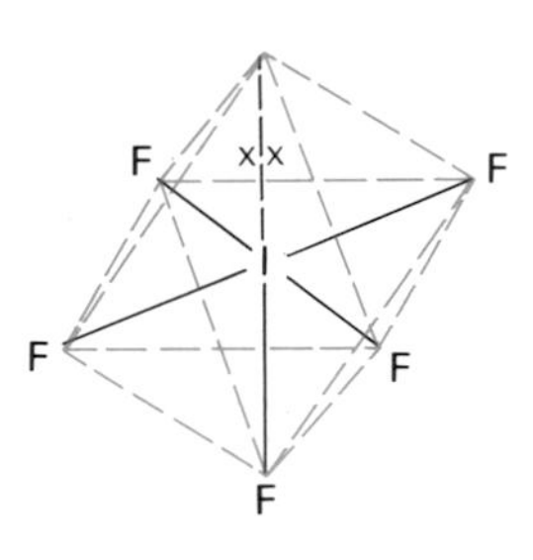

FIG. 21.3. *The structure of iodine pentafluoride*

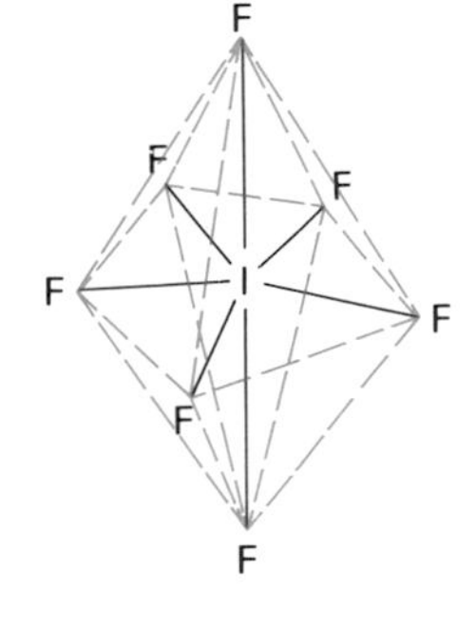

FIG. 21.4. *The structure of iodine heptafluoride*

In iodine pentafluoride the iodine atom is surrounded by twelve valency electrons, i.e. one non-bonding and five bonding pairs, which are arranged at the apices of a regular octahedron (fig. 21.3). The structure of bromine pentafluoride is thought to be similar. The fourteen valency electrons in iodine heptafluoride are arranged in pairs at the apices of a pentagonal bipyramid (fig. 21.4).

Astatine

21.18 Astatine—the heaviest halogen

Astatine was first obtained by the bombardment of bismuth with α-particles:

$$^{209}_{83}\text{Bi} + ^{4}_{2}\text{He} \longrightarrow ^{211}_{85}\text{At} + 2^{1}_{0}\text{n}$$

Astatine (atomic mass 211) has a half-life of only 7·5 hours and, although several more isotopes are now known, none of them has a half-life much longer than this.

From tracer studies, the element has been found to be appreciably volatile like iodine and, also like iodine, to form a mainly physical solution in water from which it can be extracted with benzene. The At^- ion can be obtained by reduction of the element; mild oxidation converts the element into what is thought to be the AtO^- ion and more powerful oxidation into the AtO_3^- ion. There is also some evidence for the existence of an At^+ complex with pyridine, cf. the similar complex formed by iodine (p. 325).

21.19 Summary of trends in Group 7B

All members of this group are non-metallic though there is some evidence for 'metallic' character in the case of iodine and astatine, i.e. the formation of complex cations. Fluorine is exceptionally reactive and this is due to the low bond energy of the fluorine molecule and to the very strong bonds that fluorine forms with most other elements; for these reasons fluorine often forces another element in combination with it to display its highest valency, e.g. SF_6. The reason why some metallic fluorides are ionic while the corresponding chlorides are covalent is mainly due to the much greater lattice energy of an ionic fluoride, i.e. the smaller fluoride ion can bind cations more strongly than the larger chloride ion, e.g. aluminium fluoride is ionic but aluminium chloride is covalent.

Other important differences between fluorine and the rest of the halogens are listed below:

(a) Fluorine oxidises water with the liberation of oxygen, while the other halogens dissolve to give mainly physical solutions (bromine and iodine) or some acid formation in addition (chlorine). However, the action of fluorine on water at 273 K produces the acid HOF which has strong oxidising properties.

(b) Fluorine reacts with alkaline solutions to give oxygen difluoride in the cold and oxygen if the solution is warmed. The other halogens react with alkaline solutions to give an oxysalt, the nature of which depends upon the experimental conditions (p. 334).

(c) Fluorine is restricted to a covalency of one since easily accessible *d* orbitals are not available, whereas the other halogens, particularly iodine, can show valencies of 1, 3, 5 and 7.

(d) Hydrogen fluoride is a liquid owing to the presence of hydrogen bonding, whereas the other halogen hydrides are gases. An aqueous solution of hydrogen fluoride is only a weak acid and this is primarily due to the very high H—F bond energy; the other halogen hydrides function as strong acids in aqueous solution.

Chlorine, bromine and iodine form part of a well-graded series, and can be obtained by oxidation of the corresponding halogen hydride. Chlorine displaces both bromine and iodine from aqueous solutions of bromides and iodides respectively, and bromine will displace iodine from an aqueous iodide; chlorine is therefore the best and iodine the feeblest oxidising agent of these three elements.

The hydrides of chlorine, bromine and iodine can be made by direct synthesis (the reaction between hydrogen and iodine is reversible) and they are all strongly fuming gases; in water they form strongly acidic solutions. With increasing size of the halogen atom the hydrides become increasingly more powerful reducing agents, e.g. hydrogen chloride will not reduce concentrated sulphuric acid, hydrogen bromide reduces it to sulphur dioxide, whereas hydrogen iodide converts it mainly into hydrogen sulphide. The reducing actions of aqueous solutions of the hydrides are similar.

Chlorine and bromine in particular form a variety of oxides which are explosive, e.g. the oxides of bromine decompose below −40°C. The best known oxide of iodine, I_2O_5, however, is a solid and stable up to about 300°C.

Chlorine, bromine and iodine react with alkalis in a similar manner to give the halide and either the halate (I), OX^-, or the halate (V), XO_3^-. The tendency for the halate (I) ion to disproportionate into the halate (V) and halide ions increases with increasing size of the halogen.

The oxyacids and the oxysalts of chlorine, bromine and iodine exhibit varying degrees of oxidising power. For the oxyacids of chlorine in particular, an increase in oxygen content leads to an increase in thermal stability, an increase in acid strength and a decrease in oxidising ability; the oxysalts of chlorine are more thermally stable than the parent acids and an increase in oxygen content also leads to a decrease in oxidising power.

The presence of chloride, bromide and iodide ions can be detected by the formation of a precipitate when mixed with silver ions in aqueous solution; silver chloride is white, silver bromide is cream and silver iodide is yellow.

21.20 Suggested reading

Inorganic Chemicals, ICI Educational Publications, 1978.
G. D. Twigg, *Chlorine*, School Science Review, No. 162, Vol. 47, 1966.
The Halogens and the Noble Gases, Open University, S25, Unit 11.
E. H. Appleman, *Non-existent Compounds*, Account of Chemical Research, Vol. 4, 113, 1973.

21.21 Questions on chapter 21

1 Give a concise comparative account of the chemistry of the halogens (excluding astatine), with particular reference to general variations in properties with increasing atomic number. Indicate in what important respects fluorine and iodine are exceptional members of the group. (W)

2 Give a concise description of the element fluorine and of three of its more important compounds. Indicate briefly how the properties of these compounds differ from those of the corresponding compounds of the other halogens. (O & C)

3 Discuss the trends which occur in Group VII of the Periodic Table, the halogens. Illustrate your answer with suitable examples, paying attention to the abnormal properties of fluorine. (L)

4 Give an account of the chemistry of the fluorides.

5 'The halogens form a well-defined family of elements as indicated by their positions in the Periodic Table'. Discuss this statement with reference to chlorine, bromine, iodine and their more important compounds. (L)

6 Give examples of reactions in which chlorine displaces four different elements from their compounds.

How does chlorine react with
(a) sodium hydroxide,
(b) potassium hexacyanoferrate (II)?

Explain why carbon tetrachloride is a volatile, non-polar liquid and why it is more stable than its corresponding silicon analogue. (AEB)

7 Compare and contrast the properties of the elements chlorine and iodine by reference to their reactions with the following:

(a) potassium iodide,
(b) hydrogen,
(c) sodium hydroxide,
(d) sodium thiosulphate,
(e) concentrated nitric acid.

How do these reactions bear upon the position of these elements in the Periodic Table? (C)

8 The elements silicon, phosphorus, sulphur and chlorine occur in the second short period of the Periodic Table. With reference to the properties of the elements and their simple compounds, discuss the gradation in properties which they exhibit, and show how far these properties are consistent with the positions of the elements in the Periodic Table. (L)

9 For either chlorine or iodine:

(a) Outline a method of preparation of a salt which is formed by the oxyacid of the halogen in its highest oxidation state.
(b) Write equations for the reactions of the selected halogen with:
 (i) hot sodium hydroxide solution,
 (ii) sodium thiosulphate solution,
 (iii) potassium iodide solution,
 (iv) hydrogen sulphide solution.

State the names of the products in each case. (O & C)

10 Describe the reactions involved in the extraction of pure iodine from crude Chile saltpetre.

Describe the preparation and collection of hydrogen iodide in the school laboratory.

By only writing equations and stating conditions, outline the reactions which occur between iodine and

(a) sodium thiosulphate,
(b) potassium hydroxide,
(c) nitric acid,
(d) potassium iodide. (S)

11 Describe, with essential experimental detail, the preparation of bromine in the laboratory.

Discuss the general resemblance between chlorine, bromine and iodine by comparing:

(a) their reactions with hydrogen,
(b) their reactions with potassium iodide solution,
(c) their compounds with silver.

Name the products formed when concentrated sulphuric acid is warmed with (i) potassium chloride, (ii) potassium bromide, (iii) potassium iodide. (JMB)

12 (a) Give the electronic configurations of (i) chlorine, (ii) oxygen.

(b) Using your answers in (a) and by following the usual bonding principles, draw diagrams to show the electronic structures ('dot and cross' diagrams) of
 (i) the Cl^- ion,
 (ii) the ClO^- ion,
 (iii) the ClO_2^- ion,
 (iv) the ClO_4^- ion.

(c) Using the concept of electron repulsion, draw diagrams to show the shape of
 (i) the ClO_2^- ion,
 (ii) the ClO_4^- ion,
 commenting on the diagrams and estimating bond angles.

(d) Name the products and write equations for the reactions of chlorine with
 (i) cold sodium hydroxide solution,
 (ii) hot sodium hydroxide solution.
 Mention *one* commercial use for the product in (i).

(e) Sodium chlorate (V) ($NaClO_3$) disproportionates on heating to its melting-point. Name the likely products of this change and write an equation.

(f) Explain and compare what happens if you heat sodium chlorate (V) above its melting-point in the absence of and in the presence of manganese-(IV) oxide (manganese dioxide). (S)

13 For the separate reactions of the ions, Cl^-, Br^-, I^- with (a) concentrated sulphuric acid and (b) aqueous silver nitrate followed by aqueous nitric acid and then aqueous ammonia,

(i) state what is seen and

(ii) write equations for the reactions which occur.

Account for the differences in terms of oxidation/reduction, and for the similarities and differences in (b) in terms of bonding and the formation of complex ions.

(You may find the following estimates of ionic character in the silver halides useful: silver chloride 26 per cent, silver bromide 19 per cent and silver iodide 9 per cent.) (AEB)

14 Write a short account of the more important oxides and oxyacids of either nitrogen or a halogen. You may restrict your answer to five compounds, at least two of which should be oxides. (O & C)

15 Write a comparative account of either the hydrides or the chlorides of the elements. (Oxford Schol.)

16 Describe how, starting from calcium fluoride, you would prepare

(a) an aqueous solution of hydrogen fluoride,

(b) anhydrous hydrogen fluoride,

(c) fluorine.

Discuss the separate reactions of fluorine and chlorine with (i) alkalis, (ii) sulphur, (iii) carbon. (Camb. Schol.)

17 From your knowledge of the chemistry of iodine, predict what you would think to be the main physical and chemical properties of astatine and its compounds.

18 Give an account of the interhalogen compounds, including some discussion of the geometry of their molecules.

19 Briefly describe the preparation of (*a*) chlorine, and (*b*) iodine.

Suggest a reason why the same method is not used for the preparation of both halogens.

When an excess of liquid chlorine is added to iodine a reaction occurs. When the unreacted chlorine is allowed to evaporate an orange solid of formula ICl_x remains. One mole of ICl_x reacts with excess potassium iodide solution to liberate two moles of iodine (I_2).

Write an equation for the reaction between ICl_x and iodide ions

What is the oxidation state of the iodine in ICl_x? (O & C)

20 (a) Astatine is a *naturally radioactive* member of the halogens having a *half life* of 54·0 seconds. It is considered to decay according to the following sequence:

$$^{216}_{85}At \rightarrow ^{212}_{83}Bi \rightarrow ^{212}_{84}Po \rightarrow ^{208}_{82}Pb$$

(i) Explain the terms in italics.

(ii) Name the particles emitted in each stage and indicate their nature and chief properties.

(iii) Give the number of protons, electrons and neutrons in the isotope of lead obtained.

(b) Show by means of an equation how astatine could possibly be obtained from an aqueous solution of potassium astatide.

(c) Give the name of an astatide which might be expected to be insoluble in water.

(d) If astatoethane (ethyl astatide) was treated with aqueous alkali, would you expect the hydrolysis to be faster or slower than that with bromoethane (ethyl bromide)?

Give reasons for your answer.

What would be the products of such a hydrolysis? (AEB)

Chapter 22

The first transition series (scandium, titanium, vanadium, chromium, manganese, iron, cobalt and nickel)

22.1 Some physical data of first transition series elements

	Atomic Number	Electronic Configuration	Atomic Radius/nm	Ionic Radius /nm M^{3+}	M.p./ °C (approx.)	Density at 298 K/g cm^{-3}	Standard Redox Potentials /V	
							$M^{2+}(aq) + 2e^- \rightarrow M(s)$	$M^{3+}(aq) + 3e^- \rightarrow M(s)$
Sc*	21	2.8.8(1).2 ...$3s^23p^63d^14s^2$	0·144	0·081		2·99		—2·10
Ti	22	2.8.8(2).2 ...$3s^23p^63d^24s^2$	0·132	0·076	1680	4·54	—1·63	—1·21
V	23	2.8.8(3).2 ...$3s^23p^63d^34s^2$	0·122	0·074	1917	6·11	—1·18	—0·85
Cr	24	2.8.8(5).1 ...$3s^23p^63d^54s^1$	0·117	0·069	1890	7·19	—0·91	—0·74
Mn	25	2.8.8(5).2 ...$3s^23p^63d^54s^2$	0·117	0·066	1247	7·42	—1·18	—0·28
Fe	26	2.8.8(6).2 ...$3s^23p^63d^64s^2$	0·116	0·064	1535	7·87	—0·44	—0·04
Co	27	2.8.8(7).2 ...$3s^23p^63d^74s^2$	0·116	0·063	1490	8·90	—0·28	+0·40
Ni	28	2.8.8(8).2 ...$3s^23p^63d^84s^2$	0·115	0·062	1452	8·90	—0·25	
Cu*	29	2.8.8(10).1 ...$3s^23p^63d^{10}4s^1$	0·117		1083	8·94	+0·34	

* The element copper is not a transition element but Cu^{2+} has transitional characteristics. Sc^{3+} is also non-transitional. See Section 22.2 for the definition of a transition element.

22.2 Some general remarks about first transition series elements

A transition element may be defined as one that has a partially filled *d* shell, i.e. the atom of the element has between one and nine *d* electrons; this definition is extended to include the oxidation (valency) states of these elements which also have partially filled *d* shells. There are three transition series, and two inner transition series (the lanthanides and actinides) which are referred to briefly in chapters 26 and 27. In this chapter, we shall only be concerned with the first transition series which runs from scandium to copper; the element zinc, which immediately follows copper in the Periodic Table, has a full complement of *d* electrons and so has the zinc ion, so that this element is non-transitional. The outer electronic configurations of the atoms and some ions of the first transition series are given in Table 22A.

Table 22A The outer electronic configurations of the atoms and some ions of the first transition series

Sc	$3p^63d^14s^2$	Sc^{3+}	$3p^6$ (NT)		
Ti	$3p^63d^24s^2$	Ti^{2+}	$3p^63d^2$	Ti^{3+}	$3p^63d^1$
V	$3p^63d^34s^2$	V^{2+}	$3p^63d^3$	V^{3+}	$3p^63d^2$
Cr	$3p^63d^54s^1$	Cr^{2+}	$3p^63d^4$	Cr^{3+}	$3p^63d^3$
Mn	$3p^63d^54s^2$	Mn^{2+}	$3p^63d^5$	Mn^{3+}	$3p^63d^4$
Fe	$3p^63d^64s^2$	Fe^{2+}	$3p^63d^6$	Fe^{3+}	$3p^63d^5$
Co	$3p^63d^74s^2$	Co^{2+}	$3p^63d^7$	Co^{3+}	$3p^63d^6$
Ni	$3p^63d^84s^2$	Ni^{2+}	$3p^63d^8$		
Cu	$3p^63d^{10}4s^1$	Cu^{2+}	$3p^63d^9$	Cu^+	$3p^63d^{10}$ (NT)
Zn	$3p^63d^{10}4s^2$ (NT)			Zn^{2+}	$3p^63d^{10}$ (NT)

(NT)—non-transitional)

There are five separate 3*d* orbitals, each of which can accommodate two electrons with opposite spins. These five energy levels are degenerate, i.e. have the same energy in the free atoms and ions. In practice each of these seperate energy levels must be singly occupied by an electron before electron pairing takes place (Hund's rule, see p. 43), e.g. in the manganese (II) ion, Mn^{2+}, each of the five 3*d* levels are singly occupied while in the Fe^{2+} ion four of the five 3*d* levels are singly occupied and the fifth holds two electrons with opposite spins.

When all five 3*d* levels are either singly or doubly filled a degree of stability is conferred on the atom or ion, and this explains why the electronic configurations of the atoms of chromium and copper are respectively $3d^54s^1$ and $3d^{10}4s^1$. It also explains why $Fe^{2+}(3d^6)$ is easily oxidised to $Fe^{3+}(3d^5)$ and why $Mn^{2+}(3d^5)$ is not readily oxidised to $Mn^{3+}(3d^4)$.

22.3 Features of transition elements and their ions

The transition elements are the most industrially important metals and this is due mainly to the presence of strong inter-atomic bonding which results in them having high melting points and good mechanical properties. Their ions are usually coloured and they form a wide variety of complexes. Other features include variable oxidation (valency) states, catalytic activity, paramagnetism, the formation of isomorphous compounds and a large range of alloys, and the existence of interstitial compounds. Before reviewing each of these features in turn, the concept of oxidation number which is widely used in transition metal chemistry will be discussed.

Consider manganese (VII) oxide, Mn_2O_7, which is an explosive covalent compound. If electron transfer took place completely from the less to the more electronegative atom, i.e. from manganese to oxygen, the manganese atom would carry a charge of +7 units, since two electrons would be transferred to each oxygen atom (a total of fourteen altogether). In this example the oxidation number of manganese is +7 and is sometimes written as Mn^{VII}. The oxidation number of an element in a compound is simply obtained by considering how many electrons would be gained or lost by each atom in the compound assuming the compound to be ionic (regardless of whether it is ionic or covalent). It is, therefore, an artificial concept, albeit a very useful one.

Metallic properties

The metals are dense and have high melting and boiling points. The high density is accounted for by the relatively small atomic radius, e.g. in the first transition series the largest atom is that of scandium (atomic radius 0·144 nm) which is appreciably smaller than the atom of calcium (atomic radius 0·174 nm), the preceding element in the Periodic Table. The high melting and boiling points are due to strong inter-atomic bonding which involves the participation of both the 4*s* and 3*d* electrons.

Along a particular transition series there is little variation in atomic radii, e.g. from vanadium to nickel the atomic radii contract slightly from 0·122 nm to 0·115 nm, and this partially explains why these elements are used in the production of alloy steels.

Coloured ions

The majority of transition metal ions are coloured, the colours and number of 3*d* electrons present in some hydrated ions being given in Table 22B.

Table 22B The colours of some transition metal ions

Ion	Colour	Configuration	Ion	Colour	Configuration
Sc^{3+} (aq)	Colourless	$3d^0$	Fe^{3+} (aq)	Pale violet	$3d^5$
Ti^{3+} (aq)	Purple	$3d^1$	Fe^{2+} (aq)	Green	$3d^6$
V^{3+} (aq)	Green	$3d^2$	Co^{2+} (aq)	Pink	$3d^7$
Cr^{3+} (aq)	Violet	$3d^3$	Ni^{2+} (aq)	Green	$3d^8$
Mn^{3+} (aq)	Violet	$3d^4$	Cu^{2+} (aq)	Blue	$3d^9$
Mn^{2+} (aq)	Pale pink	$3d^5$	Zn^{2+} (aq)	Colourless	$3d^{10}$

Notice that the hydrated Sc^{3+} and Zn^{2+} ions are non-transitional since they have respectively none and ten 3*d* electrons; these ions are colourless.

The colour of a particular transition metal ion depends upon the nature of the ligands (either neutral molecules such as water which contain lone pairs or negative ions such as the chloride ion) bonded to the ion. The pale blue hydrated copper (II) ion changes to dark blue in the presence of sufficient ammonia and to green if sufficient chloride ions are added; copper (II) chloride solution is therefore either blue or green, depending upon the relative concentrations of water molecules and chloride ions:

$$\underset{\text{pale blue}}{Cu(H_2O)_6^{\ 2+}(aq)} + 4Cl^-(aq) \rightleftharpoons \underset{\text{green}}{CuCl_4^{\ 2-}(aq)} + 6H_2O(l)$$

In practice an aqueous solution of copper (II) chloride will contain a variety of complex ions containing both water molecules and chloride ions, but in dilute solution the hydrated complex will predominate and in large concentrations of chloride ions the chloride complex will be the predominating species.

Similarly hydrated cobalt (II) ions are pink but in the presence of sufficient chloride ions a blue complex is formed, $CoCl_4^{2-}(aq)$.

The colour of a transition metal ion is associated with:

(a) An incomplete *d* level (between 1 and 9 *d* electrons).
(b) The nature of the ligands surrounding the ion.

A complete theory of colour is very complex but, put simply, it is due to the movement of electrons from one *d* level to another. Since the five separate 3*d* orbitals are orientated differently in space an electron (or electrons) which is close to a ligand will be repelled and hence the energy of such orbitals will be raised relative to the others. The degeneracy of the 3*d* levels is therefore destroyed; this is represented pictorially for the copper (II) ion (fig. 22.1):

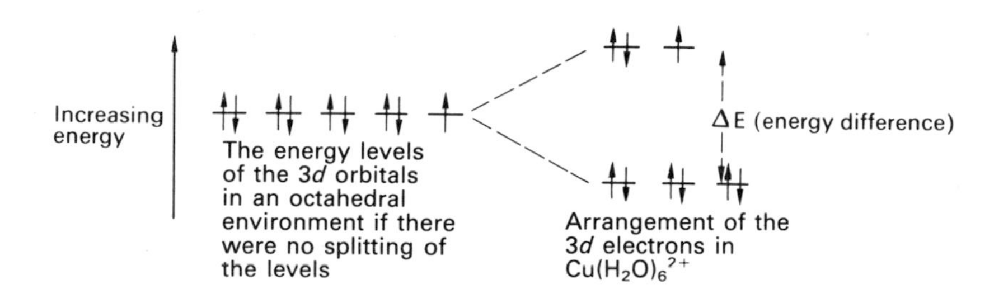

FIG. 22.1. *The splitting of the 3d orbitals of the Cu^{2+} ion in an octahedral environment of water ligands*

It will be seen from the above diagram that two of the 3*d* levels are raised in energy relative to the other three, the energy difference being ΔE where, by the Planck equation:

$$\Delta E = h\nu$$

Radiation of frequency given by the above equation will raise an electron from the lower 3*d* level to the upper one, and in the case of transition metal ions this radiation is part of the visible spectrum. Hydrated copper (II) ions are blue because red light of the appropriate frequency is absorbed (white light minus red gives blue). Since the degree of splitting of the 3*d* levels depends upon the particular ligands themselves, the variation in colour of ions of a particular transition metal is explained.

Complex ions

A complex ion is one that contains a central ion linked to other atoms, ions or molecules which are called ligands. If the ligands are easily removed the complex is said to be unstable and if they are difficult to remove the complex is a stable one. Among the transition metals the manganate (VII) and chromate (VI) anions, respectively MnO_4^- and CrO_4^{2-}, are considered to be stable complexes and can be represented formally as:

$$Mn^{+7}(O^{-2})_4 \quad \text{and} \quad Cr^{+6}(O^{-2})_4$$

Hydrated transition metal ions are unstable complexes.

The majority of transition metal ion complexes contain six ligands surrounding the central ion octahedrally (electron pair repulsion). Others contain four which are either arranged tetrahedrally or less frequently, at the corners of a square. The bonding between the ligand and the transition metal ion can either be predominantly electrostatic or predominantly covalent or in many cases intermediate between the two extremes. Typical complexes include the following:

Table 22C Typical transition metal ion complexes

Octahedral Complexes	Tetrahedral Complexes	Square Planar
$[Fe(CN)_6]^{4-}$ hexacyanoferrate (II) $[Fe(CN)_6]^{3-}$ hexacyanoferrate (III) $[Ni(NH_3)_6]^{2+}$ $[Co(NH_3)_6]^{2+}$	MnO_4^- CrO_4^{2-} $[Cu(CN)_4]^{3-}$	$[Ni(CN)_4]^{2-}$ $[Cu(NH_3)_4]^{2+}$

Variable oxidation state

The variable oxidation states displayed by transition elements are primarily due to the fact that successive ionisation energies of a transition metal atom increase gradually. For other metals, which only have one oxidation state, there is a noticeable break in the values of successive ionisation energies, e.g. for the aluminium atom the successive ionisation energies show a distinct break for the removal of the fourth electron and Al^{4+} cannot be produced by chemical reaction.

$$Al(g) \xrightarrow{578} Al^+(g) \xrightarrow{1820} Al^{2+}(g) \xrightarrow{2740} Al^{3+}(g) \xrightarrow{11\,600} Al^{4+}(g)$$

energy/kJ mol^{-1}

The ionisation energies of a transition metal atom such as vanadium, however, increase more gradually without any distinct breaks.

$$V(g) \xrightarrow{648} V^+(g) \xrightarrow{1370} V^{2+}(g) \xrightarrow{2870} V^{3+}(g) \xrightarrow{4600} V^{4+}(g)$$

energy/kJ mol^{-1}

Except for scandium, which has exclusively an oxidation state of +3, the first transition series elements show an oxidation state of +2 when both 4*s* electrons are involved in bonding; for oxidation states greater than +2, 3*d* electrons are used in addition to both the 4*s* electrons.

The oxidation states shown by the first transition series are given in Table 22D (the most usual oxidation state is in bold type).

Table 22D The oxidation states shown by the first transition series

Sc	Ti	V	Cr	Mn	Fe	Co	Ni	Cu	(Zn)
3	**4**	5	6	7	6	4	4	**2**	**2**
	3	**4**	**3**	6	**3**	3	**2**	1	
	2	3	2	4	2	**2**			
		2		3					
				2					

In general, the lower oxidation states are reducing and the higher oxidation states are oxidising, e.g. Cr^{2+} is rapidly oxidised to Cr^{3+} in the presence of air, and potassium dichromate (VI), $K_2Cr_2O_7$, which can be formally represented as $(K^+)_2(Cr^{+6})_2(O^{-2})_7$, is a powerful oxidising agent readily reduced to Cr^{3+}.

The lower oxidation states of these metals give predominantly basic oxides and the higher oxidation states give predominantly acidic oxides; intermediate oxides are generally amphoteric (Table 22E).

Table 22E The acid-base properties of the oxides of manganese

Metal Oxide	Oxidation State of Metal	Nature of Oxide	Salts
MnO	+2	Basic	Mn^{2+} salts
Mn_2O_3	+3	Basic	Few Mn^{3+} salts (e.g. MnF_3)
MnO_2	+4	Amphoteric	$Mn(SO_4)_2$ and K_2MnO_3
MnO_3	+6	Acidic	K_2MnO_4
Mn_2O_7	+7	Acidic	$KMnO_4$

Catalytic activity

The catalytic activity of transition metals and their compounds is associated with their variable oxidation states. Typical catalysts are vanadium (V) oxide (Contact process), finely divided iron (Haber process), nickel (catalytic hydrogenation), and palladium (II) chloride and a copper (II) salt for the production of ethanal from ethene and water (Wacker process). Haemoglobin, a large molecule containing Fe(II), is the catalyst used in respiration processes.

Catalysis at a solid surface involves the formation of bonds between reactant molecules and the surface catalyst atoms (the first transition series elements have 3*d* electrons in addition to the 4*s* electrons which can be utilised in bonding); this has the effect of increasing the concentration of the reactants at the catalyst surface and also of weakening the bonds in the reactant molecules (the activation energy is lowered).

Transition metal ions function as catalysts by changing their oxidation states, e.g. iron (III) ions catalyse the reaction between iodide and peroxodisulphate ions:

$$2I^-(aq) + S_2O_8^{2-}(aq) \longrightarrow I_2(aq) + 2SO_4^{2-}(aq)$$

A plausible, though possibly oversimplified, explanation of this catalysis might be:

$$2Fe^{3+}(aq) + 2I^-(aq) \longrightarrow 2Fe^{2+}(aq) + I_2(aq)$$
$$2Fe^{2+}(aq) + S_2O_8^{2-}(aq) \longrightarrow 2Fe^{3+}(aq) + 2SO_4^{2-}(aq)$$

It is known that both the above reactions can take place, and it would be expected that two reactions between ions of opposite charge would be faster than one reaction between ions of the same type of charge.

Paramagnetism

Most substances are weakly repelled by a strong magnetic field (diamagnetic) while some others are weakly attracted by it (paramagnetic). If the force of attraction is very large the substance is said to be ferromagnetic, e.g. iron, cobalt and nickel.

Paramagnetism arises because electrons can be regarded as spinning on their axes and, just as an electric current flowing through a wire generates a magnetic moment, so too does a spinning electron; electrons that occupy the same orbital, i.e. have opposite spins, have, of course, zero magnetic moment since the two contributions act in the opposite sense. The fact that the magnetic moments of transition metal ions (Table 22F describes these in arbitrary units) agree fairly closely with those calculated assuming the truth of Hund's rule is evidence for the rule's validity.

Table 22F The paramagnetic moments (arbitrary units) of some transition metal ions

Ion	Occupation of the 3*d* orbitals	Number of unpaired 3*d* electrons	Paramagnetism (arbitrary units)
Sc^{3+}	0	0	0
Ti^{3+}	1	1	1
V^{3+}	1, 1	2	2
Cr^{3+}	1, 1, 1	3	3
Mn^{3+}	1, 1, 1, 1	4	4
Mn^{2+}, Fe^{3+}	1, 1, 1, 1, 1	5	5
Fe^{2+}	2, 1, 1, 1, 1	4	4
Co^{2+}	2, 2, 1, 1, 1	3	3
Ni^{2+}	2, 2, 2, 1, 1	2	2
Cu^{2+}	2, 2, 2, 2, 1	1	1
(Zn^{2+})	2, 2, 2, 2, 2	0	0

The formation of isomorphous compounds

The best known examples of isomorphous salts are the alums of formula $K^+M^{3+}(SO_4^{2-})_2.12H_2O$ where M^{3+} can be Al^{3+}, Ti^{3+}, V^{3+}, Cr^{3+}, Mn^{3+}, Fe^{3+} or Co^{3+}. Undoubtably similarity of ionic size of the transition metal cations in the +3 oxidation state is one reason why these compounds are isomorphous, although Al^{3+} (atomic radius 0·050 nm) is significantly smaller than say Ti^{3+} (atomic radius 0·076 nm).

The formation of interstitial compounds

Transition metals form interstitial hydrides (p. 165) in which the hydrogen is accommodated in the lattice of the transition element. Some expansion of the lattice occurs, since the density of the hydride is less than that of the parent metal. No definite chemical formula can be allotted to these substances, since they are non-stoichiometric. The uptake of hydrogen is reversible and can in all cases be removed by heating in vacuo to a sufficiently high temperature.

Other interstitial compounds include nitrides (p. 262) and carbides (p. 224) which are very similar to interstitial hydrides in general structure.

22.4 Scandium

Scandium

Scandium is of no significance. Unlike the other elements in the first transition series, it never exhibits an oxidation state of +2, having an oxidation state exclusively of +3 and using both its 4*s* and its one 3*d* electrons for bonding. The $Sc^{3+}(aq)$ ion is colourless since no 3*d* electrons are present.

22.5 Occurrence and extraction of titanium

Titanium

Titanium is the seventh most abundant metal in the earth's crust and occurs principally as rutile (titanium (IV) oxide), TiO_2, and ilmenite, $FeO.TiO_2$.

The metal is obtained by first converting the dioxide into titanium (IV) chloride (a covalent liquid) by mixing it with carbon and heating in a stream of chlorine:

$$TiO_2(s) + C(s) + 2Cl_2(g) \longrightarrow TiCl_4(l) + CO_2(g)$$

After fractional distillation, the tetrachloride is reduced to metallic titanium by heating either with magnesium or sodium under a blanket of argon (traces of air make the metal brittle):

$$TiCl_4(l) + 2Mg(s) \longrightarrow Ti(s) + 2MgCl_2(s)$$

22.6 Properties and uses of titanium

Titanium has a high melting point and a density of 4·54 g cm^{-3} (about 60 per cent that of steel and about as strong). It is unattacked by air and water at ordinary temperatures, but at high temperatures it burns in oxygen and nitrogen and will decompose steam on really strong heating.

On heating in chlorine it forms the tetrachloride; it is also attacked by hot concentrated hydrochloric acid and cold concentrated sulphuric acid to give Ti^{3+} salts and hydrogen.

Technological difficulties delayed production of the metal until about 1950 since traces of air and carbon render the hot metal brittle. Its great strength, comparatively low density and its resistance to corrosion make it an excellent material for the construction of supersonic aircraft, space vehicles and nuclear reactors.

Plate 9. The metal titanium combines the properties of low density with high strength and corrosion resistance. The photograph shows a variety of aerospace fasteners manufactured from this metal (By courtesy of T. J. Brooks (Leicester) Limited)

22.7 Compounds of titanium

Titanium has oxidation states of +4 (the most stable), +3 (reducing) and +2 (strongly reducing and unimportant).

Titanium (IV) compounds

Titanium (IV) chloride, $TiCl_4$, is made either by passing chlorine over heated titanium or over a heated mixture of titanium (IV) oxide and carbon:

$$Ti(s) + 2Cl_2(g) \longrightarrow TiCl_4(l)$$

It is a covalent liquid which fumes in moist air, and is hydrolysed by water to give titanium (IV) oxide and hydrogen chloride:

$$TiCl_4(l) + 2H_2O(l) \longrightarrow TiO_2(s) + 4HCl(g)$$

With chloride ions it forms the $TiCl_6^{2-}$ complex ion.

Titanium (IV) chloride is used in conjunction with aluminium triethyl (p. 210) as the Ziegler catalyst for polymerising ethene to polythene and propene to polypropene.

Titanium (IV) oxide, TiO_2, is a white ionic solid and exists in three different polymorphic forms. The rutile structure contains Ti^{4+} ions surrounded octahedrally by six O^{2-} ions (each O^{2-} ion is surrounded symmetrically by three Ti^{4+} ions).

Titanium (IV) oxide dissolves in hot concentrated sulphuric acid to give titanyl sulphate, $TiOSO_4$, and it also forms a titanate, e.g. K_2TiO_3, so it is an amphoteric oxide. Owing to its brilliant whiteness and inertness, titanium (IV) oxide is used as a white pigment.

Titanium (III) compounds

Solutions containing the $Ti(H_2O)_6^{3+}(aq)$ ion can be obtained by reduction of titanium (IV) compounds in acid solution with zinc, e.g. a solution of titanium (IV) chloride in hydrochloric acid is reduced to titanium (III) chloride, $TiCl_3$, with zinc. The hydrated titanium (III) ion, which is purple, is unstable in the presence of air and readily reverts to the +4 oxidation state.

Titanium (II) compounds

These are less important compounds. Titanium (II) chloride, $TiCl_2$, can be obtained by reduction of the tetrachloride with titanium metal:

$$TiCl_4(l) + Ti(s) \longrightarrow 2TiCl_2(s)$$

Like other titanium (II) compounds, it is a strong reducing agent. It is rapidly oxidised by air, acids, and even water, so the Ti^{2+} ion cannot exist in aqueous solution.

Vanadium

22.8 Occurrence and extraction of vanadium

Vanadium can be obtained from the ore carnotite, which contains the elements potassium, uranium, vanadium and oxygen and which is of greater importance as a source of uranium. By a series of complex chemical reactions vanadium (V) oxide, V_2O_5, is first obtained. This is then converted into ferrovanadium, an alloy with iron, by reduction with aluminium in the presence of steel chippings:

$$3V_2O_5(s) + 10Al(s) \longrightarrow \underset{\text{as an alloy with Fe}}{6V(s)} + 5Al_2O_3(s)$$

Since the principal use of the metal is as an alloying ingredient in steel, pure vanadium is seldom extracted.

22.9 Properties and uses of vanadium

Vanadium has a slightly higher boiling point than titanium and a density of 6·11 g cm^{-3}. It resembles titanium in being strong and resistant to corrosion at ordinary temperatures, and in combining with oxygen and nitrogen at elevated temperatures. On heating with chlorine it forms the tetrachloride, VCl_4.

Cold dilute acids do not attack the metal, but it dissolves slowly in hot dilute nitric acid and hot concentrated sulphuric and hydrochloric acids.

Vanadium is used mainly as an alloying ingredient in steel, e.g. for the construction of high-speed tools and exhaust valves.

22.10 Compounds of vanadium

Vanadium has oxidation states of +5, +4, +3 and +2. These are all readily demonstrated by shaking a solution of ammonium vanadate in dilute sulphuric acid with zinc amalgam. The colour of the solution changes from pale yellow to blue, to green, and finally to lavendar (Table 22G):

Table 22G The different oxidation states of vanadium in aqueous solution

Ion	VO_3^- Metavanadate	VO^{2+} Vanadium (IV)	V^{3+} Vanadium (III)	V^{2+} Vanadium (II)
Oxidation state	+5	+4	+3	+2
Colour	pale yellow	blue	green	lavender

Notice that the V^{4+}(aq) cation does not exist as such in aqueous solution since it is highly charged and has a small ionic radius. Ions such as this are able to exert a profound polarising effect on neighbouring water molecules and, in extreme cases, detach O^{2-} ions, i.e. VO^{2+} can be formally regarded as $V^{4+}(O^{2-})$.

Vanadium (V) compounds

Vanadium (V) oxide, V_2O_5, is obtained by the action of heat on ammonium metavanadate, NH_4VO_3, (p. 359):

$$2NH_4VO_3(s) \longrightarrow V_2O_5(s) + 2NH_3(g) + H_2O(l)$$

It is an orange coloured solid and its main use is as a catalyst in the Contact process.

When vanadium (V) oxide is dissolved in strongly alkaline solution the orthovanadate anion, VO_4^{3-}(aq), is obtained, i.e. sodium orthovanadate, Na_3VO_4, can be obtained by treating vanadium (V) oxide with a concentrated solution of sodium hydroxide:

$$V_2O_5(s) + 6OH^-(aq) \longrightarrow 2VO_4^{3-}(aq) + 3H_2O(l)$$

Polyvanadate anions (condensed anions containing vanadium and oxygen) can be obtained by lowering the pH of the solution, but the nature of many of the species present still remains obscure. However, experimental evidence is available which indicates the formation of the anion $V_3O_9^{3-}$(aq):

$$3VO_4^{3-}(aq) + 6H^+(aq) \rightleftharpoons V_3O_9^{3-}(aq) + 3H_2O(l)$$

Ammonium metavanadate is precipitated by adding ammonium chloride to a solution of sodium orthovanadate. Ammonium chloride provides NH_4^+ ions which can combine with OH^- ions to form weakly dissociated aqueous ammonia, i.e. the pH will be lowered. Polyvanadate ions will be formed which have the empirical formula VO_3^-: thus the formation of ammonium metavanadate, NH_4VO_3, is explained.

In strongly acidic solutions vanadium (V) oxide forms the VO_2^+(aq) cation, and can thus act as a basic oxide:

$$V_2O_5(s) + 2H^+(aq) \longrightarrow 2VO_2^+(aq) + H_2O(l)$$

If the solution is made less acidic a number of complex vanadates, the nature of which is still rather obscure, are formed.

Vanadium (IV) compounds

Vanadium (IV) oxide, VO_2, which is amphoteric, can be obtained as a dark blue solid by reduction of vanadium (V) oxide with sulphur dioxide. It is readily oxidised back to vanadium (V) oxide by heating in air.

As mentioned previously V^{4+}(aq) ions do not occur in solution since the charge/ionic size is too large; instead the blue hydrated VO^{2+}(aq) ions are present. A solution of vanadyl sulphate, $VOSO_4$, can be obtained by the mild reduction of ammonium metavanadate, dissolved in dilute sulphuric acid, with sulphur dioxide:

$$2VO_3^-(aq) + 8H^+(aq) + 2e^- \longrightarrow 2VO^{2+}(aq) + 4H_2O(l) \quad \text{(reduction)}$$
$$SO_2(g) + 2H_2O(l) \longrightarrow SO_4^{2-}(aq) + 4H^+(aq) + 2e^- \quad \text{(oxidation)}$$
$$\text{or } 2VO_3^-(aq) + 4H^+(aq) + SO_2(g) \longrightarrow 2VO^{2+}(aq) + SO_4^{2-}(aq) + 2H_2O(l)$$

Vanadium (III) compounds

Vanadium (III) oxide, V_2O_3, is a high melting-point black solid obtained when hydrogen is passed over heated vanadium (V) oxide:

$$V_2O_5(s) + 2H_2(g) \longrightarrow V_2O_3(s) + 2H_2O(l)$$

This oxide is exclusively basic and dissolves in acids to give the green $V(H_2O)_6^{3+}$(aq) ion.

One of the best known V(III) salts is the sulphate, $V_2(SO_4)_3$, which when crystallised from aqueous solution with an equivalent amount of potassium sulphate forms an alum (p. 209).

Vanadium (III) compounds revert to vanadium (IV) compounds in the presence of air, i.e. they are reducing agents. Thus the trioxide combines slowly in the cold with oxygen to give the dioxide, and aqueous solutions containing V^{3+}(aq) ions are oxidised by air to give VO^{2+}(aq) ions.

Vanadium (II) compounds

A black oxide, VO, has been reported which is exclusively basic; it dissolves in acids to give $V(H_2O)_6^{2+}$(aq) ions which are lavender coloured. Solutions containing the same ions can also be obtained by shaking a solution of ammonium metavanadate in dilute sulphuric acid with zinc amalgam (p. 358):

$$2VO_3^-(aq) + 12H^+(aq) + 6e^- \longrightarrow 2V^{2+}(aq) + 6H_2O(l) \quad \text{(reduction)}$$
$$3Zn(s) \longrightarrow 3Zn^{2+}(aq) + 6e^- \quad \text{(oxidation)}$$
or $$2VO_3^-(aq) + 12H^+(aq) + 3Zn(s) \longrightarrow 2V^{2+}(aq) + 6H_2O(l) + 3Zn^{2+}(aq)$$

Solutions containing V^{2+}(aq) ions are strongly reducing and react with water, evolving hydrogen; thus such solutions can only be preserved in the presence of a reducing agent such as zinc amalgam and dilute acid.

Chromium

22.11 Occurrence and extraction of chromium

The main ore of chromium is chromite, $FeCr_2O_4$, ($FeO.Cr_2O_3$). If this compound is heated directly with carbon, ferrochrome, an alloy of chromium and iron is obtained, which is used in the manufacture of stainless steel:

$$FeCr_2O_4(s) + 4C(s) \longrightarrow Fe(s) + 2Cr(s) + 4CO(g)$$

The process of extracting pure chromium from the ore is rather more involved. Chromite is heated strongly with sodium carbonate in the presence of air; sodium chromate (VI), Na_2CrO_4, is obtained, which is dissolved in hot water and concentrated:

$$4FeCr_2O_4(s) + 8Na_2CO_3(s) + 7O_2(g) \longrightarrow 8Na_2CrO_4(s) + 2Fe_2O_3(s) + 8CO_2(g)$$

The sodium chromate (VI) is now converted into sodium dichromate (VI), $Na_2Cr_2O_7$, by acidifying the solution with sulphuric acid; sodium sulphate can be crystallised out first on concentration, followed by the sodium dichromate (VI):

$$2Na_2CrO_4(aq) + H_2SO_4(aq) \longrightarrow Na_2Cr_2O_7(aq) + Na_2SO_4(aq) + H_2O(l)$$

The sodium dichromate (VI) is reduced by carbon to chromium (III) oxide, Cr_2O_3, which is then freed from sodium carbonate formed during the reaction; reduction of this oxide with aluminium produces metallic chromium.

$$Na_2Cr_2O_7(s) + 2C(s) \longrightarrow Cr_2O_3(s) + Na_2CO_3(s) + CO(g)$$
$$Cr_2O_3(s) + 2Al(s) \longrightarrow Al_2O_3(s) + 2Cr(s)$$

Sodium dichromate (VI), produced in the initial stages of this process, is the most industrially important compound of chromium, and conversion into the metal is only one of its uses.

22.12 Properties and uses of chromium

Chromium has the second highest melting point of the first transition series of elements (m.p. 1890°C); its density is 7·19 g cm^{-2}. It is a hard white metal and extremely resistant to chemical attack at room temperature; however, it combines with non-metals such as oxygen and chlorine on heating and will decompose steam at red heat. Chromium (III) compounds are generally formed under these conditions.

Dilute hydrochloric and sulphuric acids slowly attack the metal with the liberation of hydrogen and the formation of chromium (II), $Cr^{2+}(aq)$, ions which readily oxidise to chromium (III) ions, $Cr^{3+}(aq)$, in the presence of air. It is not attacked by nitric acid and this phenomenon, which is called passivity, is thought to be due to the formation of a thin impervious film of oxide on its surface, cf. aluminium and iron. However, the nature of passivity is still very obscure.

Chromium is used in the production of alloy steels which are extremely hard, e.g. they are used in the manufacture of ball bearings. It is also used for plating steel articles, e.g. motor car fittings.

22.13 Compounds of chromium

Chromium has oxidation states of +6 (oxidising), +3 (the most stable) and +2 (reducing). As in the case of vanadium, these are readily demonstrated by shaking an acidified solution of potassium dichromate (VI) with zinc amalgam; the colour changes are orange to green and then to blue (Table 22H):

Table 22H The different oxidation states of chromium in aqueous solution

Ion	$Cr_2O_7^{2-}$ Dichromate (VI)	Cr^{3+} Chromium (III)	Cr^{2+} Chromium (II)
Oxidation state	+6	+3	+2
Colour	orange	green	blue

Chromium (VI) compounds

Chromium (VI) oxide, CrO_3

This is prepared by adding concentrated sulphuric acid to a cold concentrated solution of potassium dichromate (VI); red crystals of chromium (VI) oxide separate from the solution.

Chromium (VI) oxide is very soluble in water, forming 'chromic (VI) acid' which is used for cleaning laboratory glassware. Although chromic (VI) acid has never been isolated, it seems likely that a solution of chromium (VI) oxide in water contains H_2CrO_4 (which is the parent acid of chromates (VI)) and $H_2Cr_2O_7$ (from which dichromates (VI) are derived), i.e. chromium (VI) oxide is an acidic oxide.

Chromium (VI) oxide is powerfully oxidising, thus it immediately inflames ethanol and on heating it forms chromium (III) oxide with the liberation of oxygen:

$$4CrO_3(s) \longrightarrow 2Cr_2O_3(s) + 3O_2(g)$$

Chromates (VI) and dichromates (VI)

Sodium chromate (VI), Na_2CrO_4, and sodium dichromate (VI), $Na_2Cr_2O_7$, are prepared industrially from chromite (p. 360).

The chromate (VI) ion, CrO_4^{2-}, can only exist in solution under alkaline conditions, e.g. potassium and sodium chromates (VI) are stable in aqueous solution, since salt hydrolysis (p. 148) ensures that the solution is alkaline. These solutions are yellow in colour (the colour of the chromate (VI) ion); insoluble chromates (VI) are often yellow if the cation is colourless, e.g. $PbCrO_4$, although silver chromate (VI), Ag_2CrO_4, is red. Chromates (VI) are generally isomorphous with the corresponding sulphates.

In the laboratory, sodium chromate (VI) may be made by the oxidation of a chromium (III) salt with sodium peroxide in aqueous solution:

$$2Cr^{3+}(aq) + 16OH^-(aq) \longrightarrow 2CrO_4^{2-}(aq) + 8H_2O(l) + 6e^- \quad \text{(oxidation)}$$

$$\underset{\text{peroxide ion}}{3O_2^{2-}(aq)} + 6H_2O(l) + 6e^- \longrightarrow 12OH^-(aq) \quad \text{(reduction)}$$

or $$2Cr^{3+}(aq) + 4OH^-(aq) + 3O_2^{2-}(aq) \longrightarrow 2CrO_4^{2-}(aq) + 2H_2O(l)$$

The sodium peroxide provides sufficient OH^- ions in aqueous solution so it is not necessary to add further alkali.

The addition of acid to an aqueous solution of a chromate (VI) results in the elimination of water and the formation of a condensed anion—the dichromate (VI) anion $Cr_2O_7^{2-}$. The colour of the solution changes from yellow (the colour of the chromate (VI) ion) to orange (the colour of the dichromate (VI) ion). The reaction is easily reversed by adding alkali, i.e. OH^- ions:

$$\underset{\text{yellow}}{2CrO_4^{2-}(aq)} + 2H^+(aq) \rightleftharpoons \underset{\text{orange}}{Cr_2O_7^{2-}(aq)} + H_2O(l)$$

Sodium and potassium dichromates (VI) are common laboratory reagents. The former, which is more soluble in water than potassium dichromate (VI), is used in the presence of acid as an oxidising agent, thus a solution of it acidified with sulphuric acid is used for oxidising alcohols to aldehydes or ketones, and aldehydes to fatty acids. Pieces of filter paper impregnated with an acidified solution of sodium dichromate (VI) can be used to test for gaseous reducing agents, e.g. sulphur dioxide and hydrogen sulphide:

$$Cr_2O_7^{2-}(aq) + 8H^+(aq) + 3H_2S(g) \longrightarrow 2Cr^{3+}(aq) + 7H_2O(l) + 3S(s)$$

The orange spot is turned green owing to the formation of chromium (III) ions.

Sodium dichromate (VI) cannot be used as a primary standard in volumetric analysis, since the solid is too deliquescent. However, potassium dichromate (VI) is an ideal primary standard since it does not hydrate, and as its solubility in water increases rapidly with increasing temperature, it is readily obtained pure by recrystallisation. It is made by mixing hot concentrated solutions of sodium dichromate (VI) and potassium chloride, filtering off the precipitated sodium chloride, and cooling the resulting solution, when the less soluble potassium dichromate (VI) crystallises from solution.

Potassium dichromate (VI) is used in volumetric analysis to estimate reducing agents, e.g. Fe^{2+} ions in acid solution. Unlike a solution of potassium manganate (VII) it can be used in the presence of hydrochloric acid, i.e. dichromate (VI) ions do not oxidise Cl^- ions to chlorine, but the stronger oxidising manganate (VII) ions produce chlorine under these conditions.

When a solution of potassium dichromate (VI) is added to an acidified solution of an iron (II) salt the orange dichromate (VI) ions are reduced to the green chromium (III) ions. The indicator used is barium diphenylamine sulphonate which is less readily oxidised than iron (II) ions, but which is oxidised to give a blue colour once all the iron (II) ions have been converted to the iron (III) state. Since the iron (III) ions would affect the indicator and thus give rise to an inaccurate end-point, they are removed by adding some phosphoric (V) acid (before titration) with which they form a fairly stable complex. The relevant redox equations are:

$$Cr_2O_7^{2-}(aq) + 14H^+(aq) + 6e^- \longrightarrow 2Cr^{3+}(aq) + 7H_2O(l) \quad \text{(reduction)}$$

$$6Fe^{2+}(aq) \longrightarrow 6Fe^{3+}(aq) + 6e^- \quad \text{(oxidation)}$$

Industrially sodium dichromate (VI) is used in the tanning of leather, as a source of other chromium compounds, e.g. chromium (VI) oxide, in the preparation of the electrolyte used in chrome plating, as a source of pure chromium, and as a general oxidising agent.

Chromium (VI) dichloride dioxide, CrO_2Cl_2
When concentrated sulphuric acid is added to a mixture of solid potassium dichromate (VI) and sodium chloride, chromium (VI) dichloride dioxide is evolved as a dark red vapour on heating gently. It condenses to a dark red covalent liquid which is immediately hydrolysed by solutions of alkalis to give the chromate (VI):

$$CrO_2Cl_2(l) + 4OH^-(aq) \longrightarrow CrO_4^{2-}(aq) + 2Cl^-(aq) + 2H_2O(l)$$

Since bromides and iodides do not give analogous compounds this is a specific test for chloride ions.

Potassium chlorochromate (VI), $K(CrO_3Cl)$
This is an orange-coloured ionic solid and can be prepared by boiling a solution of potassium dichromate (VI) with concentrated hydrochloric acid. Potassium chlorochromate (VI) separates as a solid on cooling.

$$Cr_2O_7^{2-}(aq) + 2HCl(aq) \longrightarrow \underset{\text{chlorochromate (VI)}}{2CrO_3Cl^-(aq)} + H_2O(l)$$

It reacts with solution of alkalis to give the chromate (VI) and chloride ions:

$$CrO_3Cl^-(aq) + 2OH^-(aq) \longrightarrow CrO_4^{2-}(aq) + Cl^-(aq) + H_2O(l)$$

Similarities between chromium (VI) and sulphur (VI) compounds
Apart from the fact that chromium (VI) compounds are powerful oxidising agents and highly coloured, there is a certain resemblance to sulphur (VI) compounds. For instance sulphur trioxide, SO_3, gives rise to sulphates and pyrosulphates containing respectively the SO_4^{2-} and $S_2O_7^{2-}$ ions, cf. CrO_4^{2-} and $Cr_2O_7^{2-}$. Other similarities include the formation of sulphur dichloride dioxide, SO_2Cl_2, and chromium (VI) dichloride dioxide, CrO_2Cl_2, both of which are hydrolysed by alkali to give the SO_4^{2-} and CrO_4^{2-} ions respectively. Similarly the chlorosulphate, SO_3Cl^-, and chlorochromate (VI), CrO_3Cl^-, ions are hydrolysed by alkali to the sulphate and chromate (VI) ions.

Except for chromium (VI) oxide, whose structure is uncertain, sulphur (VI) and chromium (VI) compounds have similar structures (see pp. 316, 318, 317 for the structures of SO_4^{2-}, SO_2Cl_2 and $HOSO_2Cl$—the acid from which the SO_3Cl^- ion is derived).

Chromium (III) compounds

Chromium (III) oxide, Cr_2O_3
Chromium (III) oxide may be obtained by heating chromium in a stream of oxygen, or by heating ammonium dichromate (VI):

$$(NH_4)_2Cr_2O_7(s) \longrightarrow Cr_2O_3(s) + 4H_2O(l) + N_2(g)$$

Industrially it is obtained by reducing sodium dichromate (VI) with carbon (p. 360).

Chromium (III) oxide is a dark green solid having a similar ionic structure to that of aluminium oxide. It is amphoteric, dissolving in acids to give chromium (III) ions and in concentrated solutions of alkalis to give chromates (III):

$$Cr_2O_3(s) + 6H^+(aq) \longrightarrow 2Cr^{3+}(aq) + 3H_2O(l)$$
$$Cr_2O_3(s) + 6OH^-(aq) + 3H_2O(l) \longrightarrow 2Cr(OH)_6{}^{3-}(aq)$$

Chromium (III) hydroxide, $Cr(OH)_3$
Chromium (III) hydroxide, usually represented by the formula $Cr(OH)_3$, although doubtless of more complex structure, is obtained as a green precipitate when alkali is added to a solution containing chromium (III) ions:

$$Cr^{3+}(aq) + 3OH^-(aq) \longrightarrow Cr(OH)_3(s)$$

Like chromium (III) oxide it dissolves in acids and concentrated alkalis, i.e. it is amphoteric.

Chromium (III) chloride, $CrCl_3$
Anhydrous chromium (III) chloride is a reddish-violet solid obtained when chlorine is passed over strongly heated chromium. The substance is insoluble in water unless either a trace of chromium (II) chloride, $CrCl_2$, is present, or a reducing agent such as tin (II) chloride is added to reduce some chromium (III) to chromium (II). Since the function of chromium (II) ions is catalytic, it would appear that a high energy of activation barrier opposes the dissolution of pure solid chromium (III) chloride.

When crystallised from aqueous solution a dark green hexahydrate is obtained. When a solution of this hydrated chromium (III) chloride is treated with a solution of silver nitrate, only one third of the chlorine precipitates out as silver chloride, i.e. only one third of the chlorine in this compound is ionic; this experimental behaviour is indicative of the formula:

$$[Cr(H_2O)_4Cl_2]^+Cl^-.2H_2O$$

It is possible to form two isomers of the above compound. In one all the chlorine can be precipitated as silver chloride, while in the other only two thirds of the chlorine is converted to the insoluble chloride. These two compounds must have the respective formulae:

$[Cr(H_2O)_6]^{3+}(Cl^-)_3$ grey-blue $[Cr(H_2O)_5Cl]^{2+}(Cl^-)_2.H_2O$ light green

Notice that the successive replacement of neutral water molecules, in the octahedral chromium (III) complex by chloride ions, reduces the charge carried by the complex one unit at a time.

Chromium (III) sulphate, $Cr_2(SO_4)_3.xH_2O$
A number of different hydrates of chromium (III) sulphate can be obtained, with colours ranging from violet to green. One such hydrate gives no precipitate of barium sulphate when treated with a solution containing barium ions, i.e. the sulphate ions are part of the chromium (III) complex; while another gives the theoretical yield of barium sulphate, showing that the sulphate ions are not part of a complex.

FIG. 22.2. *The structure of the hexammine chromium (III) ion*

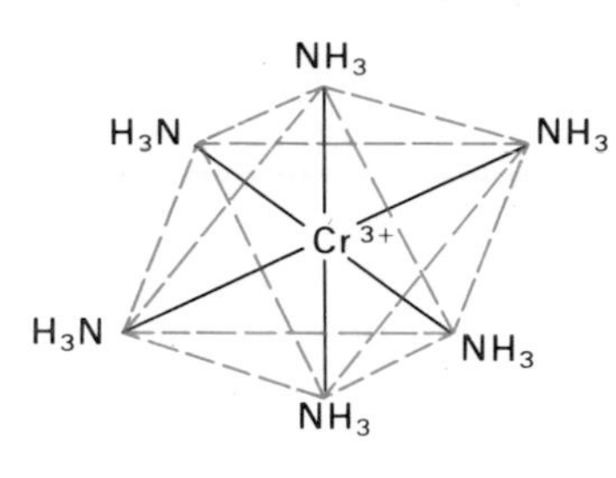

Chromium (III) sulphate forms an alum of formula $K^+Cr^{3+}(SO_4^{2-})_2.12H_2O$ when crystallised from aqueous solution in the presence of an equivalent amount of potassium sulphate. A convenient way of making chrome (III) alum involves the reduction of a solution of potassium dichromate (VI), acidified with sulphuric acid, using sulphur dioxide:

$$K_2Cr_2O_7(aq) + H_2SO_4(aq) + 3SO_2(g) \longrightarrow K_2SO_4(aq) + Cr_2(SO_4)_3(aq) + H_2O(l)$$

On crystallisation, purple octahedral crystals of chrome (III) alum separate from the solution.

Complexes containing chromium (III)

Chromium (III) forms a great number of complexes in which it is surrounded octahedrally by neutral ligands such as water and ammonia (fig. 22.2) or by charged ions such as Cl^-. Some co-ordinating substances occupy two of the octahedral positions, i.e. they are bidendate, e.g. diaminoethane (usually known as ethylenediamine) $H_2N—CH_2—CH_2—NH_2$

FIG. 22.3. *The optical isomers of tris (ethylenediamino) chromium (III) ion*

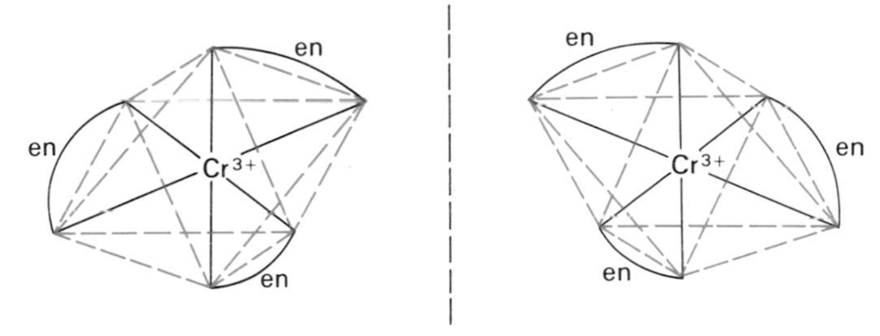

(fig. 22.3) Notice that the ethylenediamine complex can exist in two forms, neither of which has either a centre of symmetry or a plane of symmetry; one form is the mirror image of the other, i.e. they are optical isomers. Preparation of optically active complexes results in the formation of racemic mixtures, many of which have been resolved.

Complexes can be formed which contain more than one type of ligand bonded to the central chromium (III) ion, e.g. the hydrates of chromium (III) chloride discussed previously (p. 364).

There are two forms of the complex $[Cr(NH_3)_4(Cl)_2]^+$ which bears one unit of positive charge, since two Cl^- ions neutralise two of the three charges on Cr^{3+}. One form contains adjacent chlorine atoms, while the other form contains chlorine atoms lying along the diagonals of a square; they are named 'cis' and 'trans' respectively and are geometrical isomers. Since both the cis- and trans-isomers have a planc of symmetry, they are not optically active (fig. 22.4).

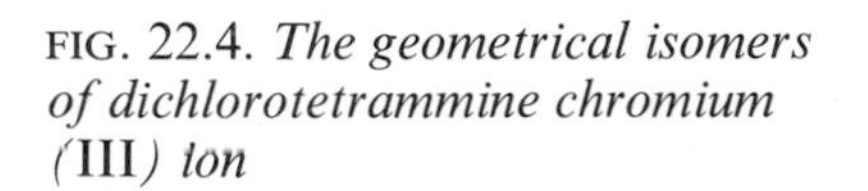
FIG. 22.4. *The geometrical isomers of dichlorotetrammine chromium (III) ion*

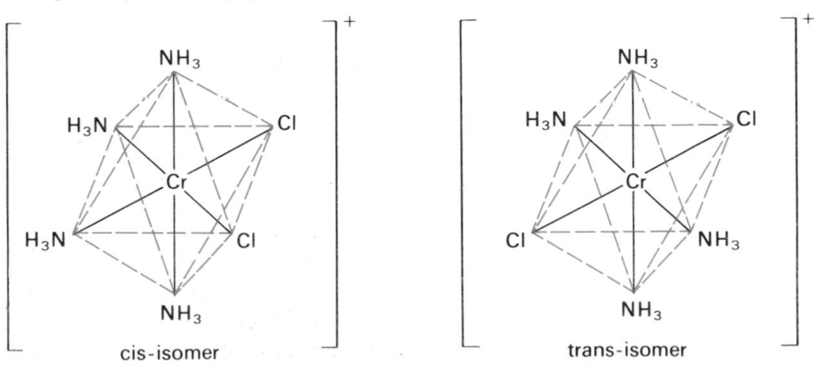

Chromium (II) compounds

This is the least stable oxidation state of chromium, being powerfully reducing.

Chromium (II) chloride, $CrCl_2$, can be made by passing dry hydrogen chloride over heated chromium; it is a white solid which dissolves in water to give the blue $Cr^{2+}(aq)$ ions. Solutions containing $Cr^{2+}(aq)$ ions can be obtained by reduction of either an acidified solution of a dichromate (VI) (p. 361) or an acidified solution of a chromium (III) salt using zinc amalgam:

$$2Cr^{3+}(aq) + Zn(s) \longrightarrow 2Cr^{2+}(aq) + Zn^{2+}(aq)$$

Solution containing $Cr^{2+}(aq)$ ions are readily oxidised by air to $Cr^{3+}(aq)$, and can only be preserved in the presence of an inert atmosphere.

Manganese

22.14 Occurrence and extraction of manganese

Typical ores of manganese are pyrolusite, MnO_2, and hausmannite, Mn_3O_4. The former is the more important.

Pure manganese is obtained by the reduction of Mn_3O_4 with aluminium, followed by distillation of the manganese in vacuo:

$$3Mn_3O_4(s) + 8Al(s) \longrightarrow 4Al_2O_3(s) + 9Mn(s)$$

Reduction of MnO_2 with aluminium is dangerously explosive, and this oxide is first converted into Mn_3O_4 by heating strongly in air before reduction is attempted:

$$3MnO_2(s) \longrightarrow Mn_3O_4(s) + O_2(g)$$

Most manganese is used in the form of alloys with iron, and these are obtained by reducing mixed manganese and iron ores with coke in an electric furnace.

22.15 Properties and uses of manganese

Manganese has a comparatively low melting point (1247°C) and a density of $7{\cdot}42\ g\ cm^{-3}$. It is a hard, brittle metal and, although not readily attacked by air at ordinary temperature, it is fairly reactive, e.g. it combines with a variety of non-metals on heating (these include oxygen, nitrogen, chlorine and sulphur) and will liberate hydrogen from warm water. Dilute acids such as hydrochloric acid and sulphuric acid readily dissolve the metal, forming manganese (II), $Mn^{2+}(aq)$, ions and liberating hydrogen.

Pure manganese has few uses since it is brittle and is attacked by water, but industrially it is very important, in the form of alloys with iron. Ferromanganese (Mn, 80 per cent; Fe, 20 per cent) is used in the manufacture of alloy steels, the presence of manganese conferring hardness, toughness and resistance to wear. Spiegeleisen (Mn, 25 per cent; Fe, 70 per cent; C, 5 per cent) is added to molten steel for the purpose of removing combined oxygen and sulphur, thereby improving its mechanical properties.

22.16 Compounds of manganese

Manganese shows important oxidation states of +7 (powerfully oxidising), +6, +4, +3 and +2 (the most stable). Reduction of manganese (VII) to manganese (II) is readily achieved with a variety of reducing agents, e.g. shaking an acidified solution of potassium manganate (VII) with zinc amalgam:

$$\underset{\substack{\text{manganate (VII)}\\ \text{purple}\\ +7}}{2MnO_4^-(aq)} + 16H^+(aq) + 5Zn(s) \longrightarrow \underset{\substack{\text{pale pink}\\ +2}}{2Mn^{2+}(aq)} + 5Zn^{2+}(aq) + 8H_2O(l)$$

Manganese (VII) compounds

Manganese (VII) oxide, Mn_2O_7
This oxide has been obtained by adding potassium manganate (VII) to well-cooled concentrated sulphuric acid. It is a dark coloured covalent liquid and **no attempt should be made to prepare it, since it decomposes explosively into manganese (IV) oxide and oxygen:**

$$2Mn_2O_7(l) \longrightarrow 4MnO_2(s) + 3O_2(g)$$

Manganese (VII) oxide dissolves in water to give a purple solution which contains 'manganic (VII) acid', and although the pure acid, $HMnO_4$, has never been isolated, the potassium salt is well known.

Potassium manganate (VII), $KMnO_4$
This is the best known compound containing manganese (VII), and can be prepared in the laboratory by fusing manganese (IV) oxide and potassium hydroxide in the presence of an oxidising agent such as potassium chlorate (V). Potassium manganate (VI), K_2MnO_4, a green solid is obtained which is then crushed, dissolved in water and the resulting mixture boiled. Under these conditions potassium manganate (VI) begins to disproportionate, since it is only stable in strongly alkaline solution, and potassium manganate (VII) and manganese (IV) oxide are formed. To assist this disproportionation, a stream of carbon dioxide is bubbled through the solution and this removes the hydroxyl ions as hydrogen carbonate ions and thus allows the reaction to go to completion (see second equation below):

$$3MnO_2(s) + 6OH^- + ClO_3^- \longrightarrow \underset{\text{manganate (VI) green}}{3MnO_4^{2-}} + 3H_2O(l) + Cl^-$$

$$\underset{\text{manganate (VI)}}{3MnO_4^{2-}(aq)} + 2H_2O(l) \rightleftharpoons \underset{\text{manganate (VII)}}{2MnO_4^-(aq)} + MnO_2(s) + 4OH^-(aq)$$

$$4OH^-(aq) + 4CO_2(g) \longrightarrow 4HCO_3^-(aq)$$

After completion of the reaction, the mixture can be freed from solid manganese (IV) oxide by filtration through glass wool, and the resulting solution evaporated until crystallisation of potassium manganate (VII) begins.

Industrially, potassium manganate (VII) is obtained from the potassium manganate (VI) by oxidation with chlorine; this reaction avoids the formation of unwanted manganese (IV) oxide:

$$2MnO_4^{2-}(aq) + Cl_2(g) \longrightarrow 2MnO_4^-(aq) + 2Cl^-(aq)$$

Powerful oxidising agents such as sodium bismuthate (V) oxidise a solution of a manganese (II) salt to the manganate (VII) ion, MnO_4^-. This reaction is carried out in the cold in the presence of dilute nitric acid, and is often used as a confirmatory test for manganese (II) ions, since the manganate (VII) ion is intensely coloured even in very dilute solution:

$$2Mn^{2+}(aq) + 5BiO_3^-(aq) + 14H^+(aq) \longrightarrow 2MnO_4^-(aq) + 5Bi^{3+}(aq) + 7H_2O(l)$$

Potassium manganate (VII) is a powerful oxidising agent, e.g. it will oxidise hydrochloric acid to chlorine and hydrogen peroxide to oxygen. Strips of filter paper impregnated with potassium manganate (VII) solution and acidified with dilute sulphuric acid are decolorised in the presence of some gaseous reducing agents; this is a convenient test for sulphur dioxide and hydrogen sulphide, e.g.

$$2MnO_4^-(aq) + 6H^+(aq) + 5H_2S(g) \longrightarrow 2Mn^{2+}(aq) + 8H_2O(l) + 5S(s)$$

Under alkaline conditions potassium manganate (VII) is reduced first to potassium manganate (VI) and then to manganese (IV) oxide, e.g. alkenes give diols (glycols) when treated with an alkaline solution of potassium manganate (VII), the purple solution turning green as potassium manganate (VI) is formed; a brown residue of manganese (IV) oxide is eventually formed.

Potassium manganate (VII) is used as a volumetric reagent but, unlike potassium dichromate (VI), it is not a primary standard since it cannot be obtained in a high degree of purity; samples of the substance contain some manganese (IV) oxide and aqueous solutions slowly deposit brown manganese (IV) oxide on standing. It is standardised against a standard solution of iron (II) ammonium sulphate acidified with dilute sulphuric acid. Potassium manganate (VII) solution is run into the acidified iron (II) salt from a burette and the purple colour is discharged as virtually colourless manganese (II) ions are formed; at the end-point the solution turns pink, the colour of potassium manganate (VII) in extremely dilute solution, thus it acts as its own indicator. The relevant partial equations are:

$$MnO_4^-(aq) + 8H^+(aq) + 5e^- \longrightarrow Mn^{2+}(aq) + 4H_2O \quad \text{(reduction)}$$
$$5Fe^{2+}(aq) \longrightarrow 5Fe^{3+}(aq) + 5e^- \quad \text{(oxidation)}$$

A standard solution of potassium manganate (VII) can be used for estimating iron (II) salts and oxalates, previously acidified with dilute sulphuric acid before titration, e.g.

$$2MnO_4^-(aq) + 16H^+(aq) + 10e^- \longrightarrow 2Mn^{2+}(aq) + 8H_2O(l) \quad \text{(reduction)}$$
$$5\left|\begin{matrix}COO^-\\ \\COO^-\end{matrix}\right.(aq) \longrightarrow 10CO_2(g) + 10e^- \quad \text{(oxidation)}$$

Similarities between manganese (VII) and chlorine (VII) compounds

Apart from the fact that manganese (VII) compounds are highly coloured, there is a certain amount of resemblance to chlorine (VII) compounds. For example, manganese (VII) oxide, Mn_2O_7, and chlorine heptoxide, Cl_2O_7, are explosive covalent liquids. The manganate (VII) ion, MnO_4^-, and the chlorate (VII) ion, ClO_4^-, are oxidising agents, the former more powerful than the latter; corresponding manganates (VII) and chlorates (VII) are isomorphous.

Manganese (VI) compounds

Neither manganese (VI) oxide, MnO_3, nor manganic (VI) acid, H_2MnO_4, has been prepared and the only pure compounds so far obtained containing manganese in the +6 oxidation state are sodium and potassium manganates (VI); both these compounds are dark green solids, the colour being due to the MnO_4^{2-} ion.

Potassium manganate (VI), K_2MnO_4, can be prepared by fusing a mixture of manganese (IV) oxide, potassium hydroxide and potassium chlorate (V) (p. 367). The manganate (VI) ion, MnO_4^{2-}, readily disproportionates into the manganate (VII) ion, MnO_4^-, and manganese (IV) oxide in neutral or acid solution, but it can be preserved in the presence of alkali (a mass action effect):

$$3MnO_4^{2-}(aq) + 2H_2O(l) \rightleftharpoons 2MnO_4^-(aq) + MnO_2(s) + 4OH^-(aq)$$

(addition of $OH^-(aq)$ ions drives the equilibrium to the left)

Manganese (IV) compounds

The only compound of manganese (IV) of any importance is the dioxide, MnO_2, which occurs naturally as the ore pyrolusite; it can be made by the action of heat on manganese (II) nitrate:

$$Mn(NO_3)_2(s) \longrightarrow MnO_2(s) + 2NO_2(g)$$

Manganese (IV) oxide is a black insoluble solid and is considered to have an essentially ionic structure. It is a powerful oxidising agent, releasing chlorine from concentrated hydrochloric acid, at the same time being reduced to manganese (II):

$$MnO_2(s) + 4HCl(aq) \longrightarrow MnCl_2(aq) + 2H_2O(l) + Cl_2(g)$$

The dioxide exhibits feeble amphoteric properties, e.g. an unstable sulphate, $Mn(SO_4)_2$, has been obtained and so-called 'manganates (IV)' of very doubtful composition have been obtained by fusion of manganese (IV) oxide with Group 2A metallic oxides.

Manganese (III) compounds

Manganese (III) oxide, Mn_2O_3, occurs naturally as a brown solid and can be made by heating manganese (IV) oxide to red heat in air; it is exclusively basic. At about 1000°C manganese (IV) oxide is converted into trimanganese tetroxide, Mn_3O_4, which contains both manganese (III) and manganese (II); it is better represented by the formula $Mn^{II}Mn^{III}{}_2O_4$.

The best known salt containing manganese in the +3 oxidation state is the fluoride, MnF_3, which crystallises as a dihydrate when manganese (III) oxide is reacted with hydrofluoric acid.

In aqueous solution the $Mn^{3+}(aq)$ ion is a strong oxidising agent and readily reverts to $Mn^{2+}(aq)$; it can however be stabilised by complex formation, e.g. $Mn(CN)_6^{4-}(aq)$ which contains manganese (II) is readily oxidised to $Mn(CN)_6^{3-}(aq)$ which contains manganese (III).

Manganese (II) compounds

This is the most stable oxidation state of manganese; the presence of five singly occupied 3*d* orbitals in the Mn^{2+} ion is often cited as an explanation of this stability (p. 350).

Manganese (II) oxide, MnO, is a greyish-green solid and can be prepared by heating manganese (II) oxalate (carbon monoxide produced in this thermal decomposition provides a reducing atmosphere, so preventing possible aerial oxidation to higher oxides):

$$MnC_2O_4(s) \longrightarrow MnO(s) + CO(g) + CO_2(g)$$

It reacts with acids, forming salts, but is not attacked by alkalis; thus it is exclusively a basic oxide.

Manganese (II) hydroxide, $Mn(OH)_2$, precipitates as a white solid when sodium hydroxide is added to an aqueous solution of a manganese (II) salt:

$$Mn^{2+}(aq) + 2OH^-(aq) \longrightarrow Mn(OH)_2(s)$$

However, it is readily oxidised on standing in the air, eventually passing over into hydrated manganese (III) oxide, $Mn_2O_3.xH_2O$, which is brown.

Manganese (II) sulphide, MnS, precipitates as a pale pink solid when hydrogen sulphide is passed through an aqueous solution of a manganese (II) salt in the presence of aqueous ammonia and ammonium chloride (Group IV of qualitative analysis schemes). The solubility product of the sulphide is too high for it to precipitate in acid solution (p. 303).

Soluble salts containing the Mn^{2+} ion, e.g. the chloride, nitrate and sulphate, can be made by reacting the oxide, hydroxide or carbonate of manganese (II) with dilute acids in the usual way, followed by partial evaporation and crystallisation. In solution these salts contain the $Mn(H_2O)_6{}^{2+}(aq)$ ion which is pale pink. Insoluble manganese (II) salts include the carbonate and phosphate (V).

Iron

22.17 Occurrence and extraction of iron

Iron, which is the second most abundant metal occurring in the earth's crust, is extracted from its oxides, haematite, Fe_2O_3, and magnetite, Fe_3O_4, and also from the carbonate siderite, $FeCO_3$. Iron pyrites, FeS_2, is not considered to be an important ore of iron. Iron occurs in meteors; there is also reason to believe that the centre of the earth is composed of a solid inner core of iron and nickel at a pressure of 3·5 million atmospheres and a temperature of 3000-4000°C, surrounded by a molten layer of the two metals at a pressure of 1·5 million atmospheres.

The extraction of iron is carried out in a blast furnace which can vary in size and can be between 25 and 60 metres in height and up to 10 metres in diameter (fig. 22.5). It is constructed from steel with the inner regions lined with firebricks. A charge of iron ore, limestone and coke in the correct proportions is fed into the top of the furnace through a cone and hopper arrangement. Preheated air at a temperature of about 600°C is injected into the furnace through a number of pipes called tuyères (pronounced 'tweers'); the tuyères are fed from a 'bustle' pipe encircling the blast furnace. The blast furnace is provided with two tap holes which are plugged with clay; molten iron is tapped from the lower one and molten slag from the other. The production of iron is a continuous process and, depending upon its size, a blast furnace can produce from 1000 to 1800 tonnes of iron every twenty-four hours.

The energy and reducing agent required for the smelting of iron are obtained by the combustion of coke, the temperature of the charge increasing steadily as it falls through the ascending combustion gases:

$$2C(s) + O_2(g) \longrightarrow 2CO(g)$$

FIG. 22.5. *The manufacture of iron*

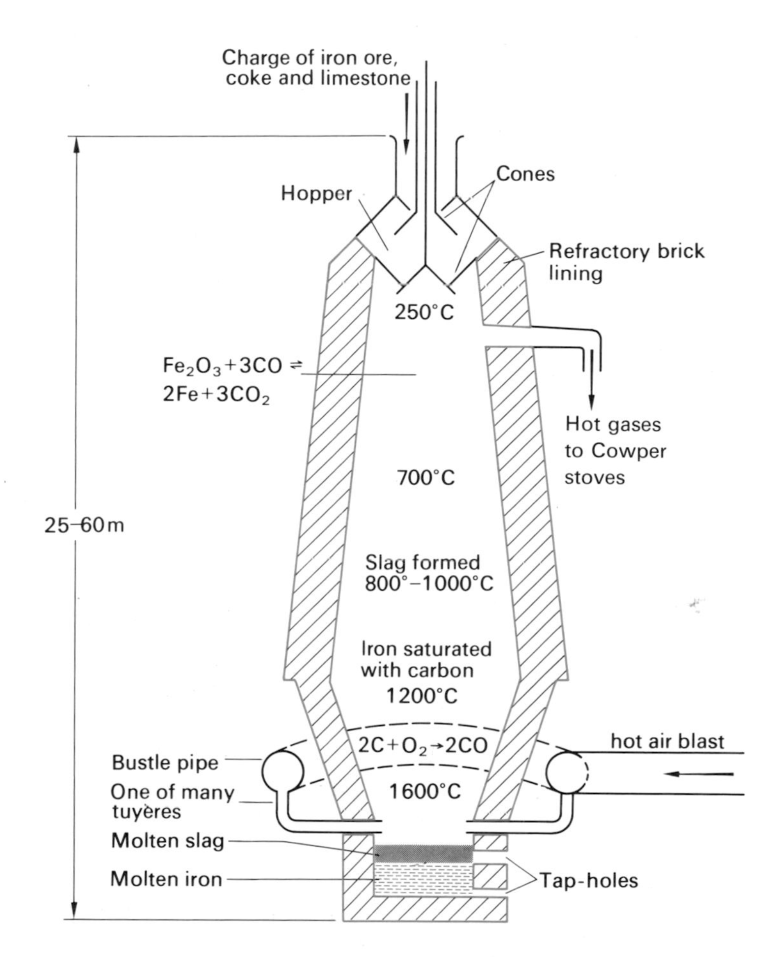

At a temperature of about 700°C the iron ore is reduced to spongy iron by the carbon monoxide:

$$Fe_2O_3(s) + 3CO(g) \rightleftharpoons 2Fe(s) + 3CO_2(g)$$

Since this reaction is reversible a high CO/CO_2 pressure ratio favours the reduction to iron.

The limestone decomposes at about 800°C and the calcium oxide reacts with sandy impurities to form a slag of calcium silicate. More carbon monoxide is produced by the reduction of carbon dioxide:

$$CaCO_3(s) \longrightarrow CaO(s) + CO_2(g)$$

$$\underset{\text{}}{CaO(s)} + \underset{\text{impurity}}{SiO_2(s)} \longrightarrow \underset{\text{slag}}{CaSiO_3(l)}$$

$$C(s) + CO_2(g) \longrightarrow 2CO(g)$$

The reduction of the iron oxide is completed by the coke at a temperature in the region of 1200°C and cementite, Fe_3C, and graphite enter the iron. Other reactions also occur at high temperatures, for instance silica is reduced to silicon and this enters the iron as ferrosilicon:

$$SiO_2(s) + 2C(s) \longrightarrow Si(s) + 2CO(g)$$

A similar reaction occurs with any phosphates present and some Fe_3P is retained by the iron. Other impurities present in the iron include iron (II) sulphide and manganese, which form an alloy.

The molten metal is either run out into moulds of sand, when it is known as pig-iron, or more generally conveyed directly in the liquid form to steelmaking plants. The slag is tapped from the furnace as a liquid and can be used in concrete or blown into a 'woolly' material and used for insulation.

The hot gases emerging from the top of the furnace contain appreciable amounts of carbon monoxide and are burnt in Cowper stoves to preheat the air for the blast.

Cast iron

Iron castings are made by igniting a mixture of pig-iron, scrap iron and coke in cupola furnaces by a blast of hot air. The molten iron is poured into moulds to make articles such as manhole covers, guttering, machinery frames and drainpipes. Cast iron expands slightly on solidifying and therefore faithfully reproduces the shape of the mould. It is extremely hard, but unfortunately is very brittle and will fracture if struck by a sharp blow. The impurities in cast iron lower the melting point from 1535°C for pure iron to approximately 1200°C.

Wrought iron

This is made by heating impure iron with haematite so that impurities are oxidised. Carbon is converted to carbon monoxide and silicon and manganese to a slag.

$$Fe_2O_3(s) + 3C(s) \longrightarrow 2Fe(l) + 3CO(g)$$

As the impurities are removed, the melting point of the iron rises to about 1500°C and the pasty mass is removed from the furnace as balls and worked under a hammer to squeeze out the slag. Wrought iron is tough, malleable and ductile; it can be worked by a blacksmith into chains, railway carriage couplings and ornamental gates, etc.

22.18 Steel

Manufacture of steel

Impure iron is too brittle for most purposes and is converted into steel by removing practically all the sulphur, silicon and phosphorus impurities. The carbon content of impure iron, which may be as high as 4·5 per cent, is reduced to between 0·1 and 1·5 per cent depending on the type of steel required. Several steelmaking processes are available.

Siemens-Martin Open-Hearth Process

In this process a mixture of impure molten iron, scrap steel and limestone is heated in an open-hearth furnace by producer gas. The furnace is lined with a mixture of calcium and magnesium oxides, made by strongly heating dolomite ($CaCO_3.MgCO_3$). The impurities are converted into their oxides by the haematite; carbon is converted into carbon monoxide which escapes and phosphorus and silicon into phosphorus (V) oxide and silicon dioxide respectively, which then combine with the lining to form a slag:

$$Fe_2O_3(s) + 3C(s) \longrightarrow 2Fe(l) + 3CO(g)$$

Oxides of phosphorus and silicon + lining → phosphate and silicate slag

Towards the end of the process, which lasts about 10 hours and produces up to 250 tonnes of steel, spiegeleisen (an alloy of manganese, iron and carbon) is added, together with other alloying metals, depending on the type of steel required.

Heat is conserved during the operation of this type of furnace by passing the producer gas and air through separate chequerwork brick chambers on each side of the furnace. Thus while one set of chambers is heated up by the waste gases the other set is used to preheat the producer gas and air. The flow of gases is reversed about four times an hour.

Modern developments of the open-hearth process include the use of fuel oil instead of producer gas, and enrichment of the air supply with oxygen to provide more efficient combustion. Pure oxygen is also injected into the molten mixture to speed up the oxidation of impurities. By this procedure it has been found possible to halve the processing time.

Very Low Nitrogen Bessemer Process

The Bessemer converter is charged with up to 60 tonnes of molten iron at a temperature of about 1200°C. A blast of oxygen diluted with either steam or carbon dioxide is blown through the converter (fig. 22.6(a));

FIG. 22.6. *Bessemer converters*

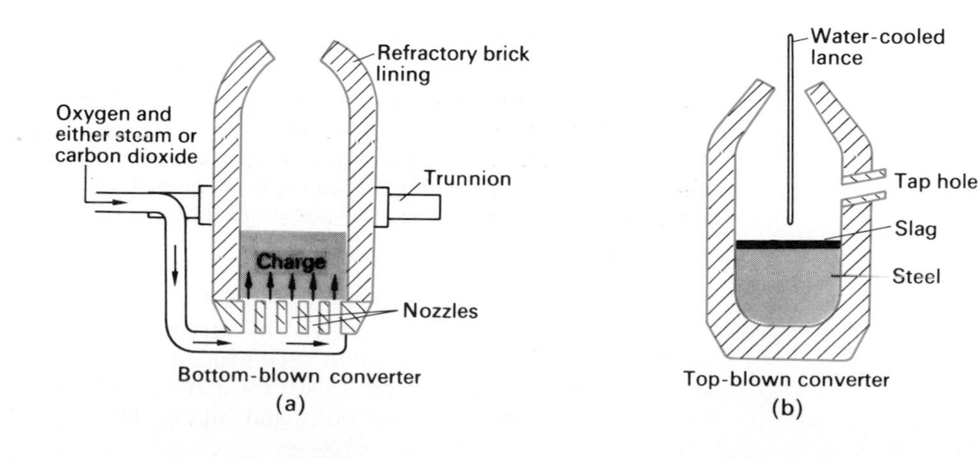

oxygen reacts with the impurities and raises the temperature to about 1900°C. Carbon is oxidised to carbon monoxide which burns at the top of the converter, and the oxides of silicon and manganese form a slag. After several minutes the flame at the top of the converter sinks, indicating that all the carbon has been removed. The molten metal is now run out, after tapping off the slag, and a controlled amount of spiegeleisen and other ingredients added.

Oxygen is used rather than air (original process) since nitrogen is retained by steel and makes it brittle. The oxygen is diluted with steam or carbon dioxide, since otherwise the nozzles would very quickly burn away and a large volume of iron oxide fumes would be formed by oxidation of the iron; carbon dioxide is preferable to steam, since hydrogen in the latter tends to make the steel brittle.

In the top-blown converter (fig. 22.6 (b)) oxygen is injected into the metal from a water-cooled lance suspended just above the slag level. So much heat is given out that scrap metal has to be added to keep the temperature within working limits.

Electrical Process

This is essentially the same as the open-hearth process, except that electrical power is used to heat up the charge, thereby allowing closer control of temperature to be achieved. It is an expensive small scale process used to produce high quality alloy steels.

The structure of steel

A complete discussion of the structure of steel would be very complex and the following is a short and somewhat oversimplified account.

Below about 910°C the structure of iron is body-centred cubic, i.e. each atom of iron is surrounded by eight nearest neighbours situated at the corners of a cube; above this temperature the structure of iron changes to that of a face-centred lattice, i.e. each atom of iron is surrounded by twelve nearest neighbours (p. 188). The presence of carbon in liquid iron lowers the temperature at which one structure changes over to the other; furthermore, whereas the body-centred structure of iron can accommodate virtually no carbon in solid solution, the face-centred lattice can hold up to 1·7 per cent carbon in solid solution. When liquid iron containing about 0·3 per cent of carbon is cooled it crystallises in the face-centred arrangement but changes over into the body-centred structure at about 810°C. The carbon separates as a distinct phase in the form of cementite, Fe_3C, and the final solid contains regions of iron and regions consisting of a sandwich structure of iron alternating with cementite, such a sandwich structure being known as pearlite. The two distinct regions have an important part to play in determining the properties of different steels. The regions of iron confer ductility and the regions of pearlite confer hardness and strength. In steel containing about 0·3 per cent of carbon the regions of iron predominate over the regions of pearlite and the result is a ductile steel; as the carbon content increases so too does the relative proportion of pearlite to iron with the result that a harder less ductile steel is produced.

Properties of steel

The properties of steel can be altered by varying the amount of carbon added during manufacture (Table 22I). Mild steel contains between 0·1 and 0·4 per cent of carbon and is used for making such articles as motor car bodies, tinplate, nuts and bolts and piping. Hard steel containing between 0·5 and 1·5 per cent carbon, can be further hardened and tempered by heat treatment and is used for tools. There is a vast range of alloy steels; a few examples are given in the table below.

Table 22I Some alloy steels

Name of Alloy	Percentage Composition (Approx.)	Use	Special Features
Stainless steel	73Fe, 18Cr, 8Ni + C	Cutlery	Resists corrosion
Tungsten steel	94Fe, 5W + C	High speed cutting tools	Hardness
Invar	64Fe, 36Ni	Watches	Small coefficient of expansion
Manganese steel	86Fe, 13Mn + C	Rock drills	Toughness
Permalloy	78Ni, 21Fe, + C	Electro-magnets	Strongly magnetised by electric current Loses magnetism when current is switched off

Hardening and tempering of steel

Hard steel can be further hardened by heating it to red heat (850°C) followed by plunging it into cold water. This process, known as 'quenching', makes the steel extremely hard but brittle. This phenomenon can be explained as follows: above 850°C the structure is face-centred cubic and the carbon is present in solid solution. Rapid cooling does not allow

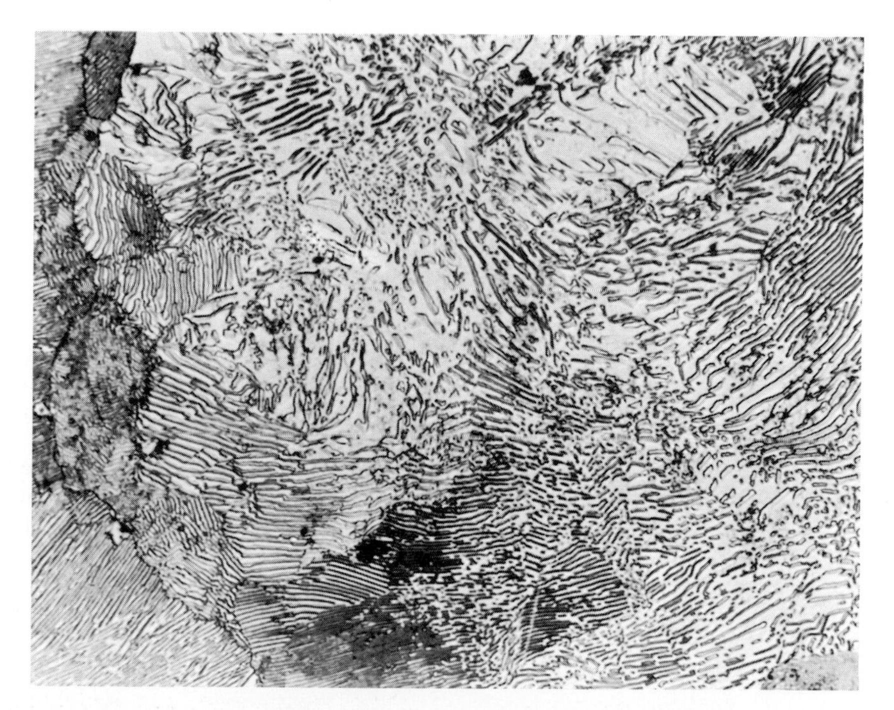

Plate 10. A photograph of an annealed specimen of 0·7% carbon steel showing pearlite structure under magnification (By courtesy of British Steel Corporation (South Wales Group)

sufficient time for the carbon to separate as cementite, Fe_3C, and instead carbon is deposited; the presence of carbon atoms makes it impossible for the iron to adopt a true body-centred structure and a distorted modification which is hard but brittle is the result.

If the steel is now heated to a temperature between 230° and 300°C, and allowed to cool slowly, it retains most of its hardness but the brittleness disappears; this process is known as 'tempering' or 'annealing'. During tempering the surface of the steel takes on certain colours, being yellow at 230°C and blue at 300°C. These colours are due to the formation of a very thin oxide film on the surface of the steel, the thickness of the film determining the final colour. Watch springs are blue, since tempering at 300°C removes all brittleness and produces an extremely tough and springy steel.

During 'tempering' microscopically small areas of cementite are formed and this helps to relieve the strain present in 'quenched steel' without significantly altering the hardness of the steel.

22.19 Properties and uses of iron

Pure iron is a silvery coloured metal with a melting point of 1535°C. It is easily magnetised when placed inside a coil carrying an electric current, but loses its magnetism when the current is switched off. A number of non-metals combine with it on heating, e.g. oxygen, the halogens, nitrogen, sulphur and carbon; iron filings burn in oxygen with a shower of bright sparks forming iron (III) oxide, Fe_2O_3, but in the massive form iron is coated with a layer of magnetic oxide of iron, Fe_3O_4. Pure iron does not corrode significantly at ordinary temperature when exposed either to the action of dry air or air-free water; however the combined action of air and water results in the formation of rust, essentially hydrated iron (III) oxide.

At red heat iron is attacked by steam with the formation of magnetic oxide of iron and hydrogen:

$$3Fe(s) + 4H_2O(g) \rightleftharpoons Fe_3O_4(s) + 4H_2(g)$$

Dilute non-oxidising acids such as sulphuric and hydrochloric acids attack iron, with the formation of iron (II) ions, $Fe^{2+}(aq)$, and hydrogen; in the presence of air the iron (II) ions are slowly oxidised to iron (III) ions, $Fe^{3+}(aq)$. Iron (III) ions are also formed when iron is warmed with dilute nitric acid; concentrated nitric acid renders the metal passive (p. 260).

The uses of cast iron, wrought iron and steel have been dealt with in section 22.18.

22.20 The rusting of iron and steel

If an electrolyte is in contact with two metals which are themselves in contact, the three essential elements of an electrochemical cell are present and corrosion takes place. For instance, rusting takes place at the exposed surfaces of a scratched tinplate, the electrolyte being the moisture in the air (p. 235).

Pure iron out of contact with any other metal will still corrode by an electrolytic mechanism, for it is found that iron exposed to a higher concentration of oxygen will accept electrons from iron where the oxygen concentration is less.

Lower oxygen concentration (−ve pole) $2Fe(s) \longrightarrow 2Fe^{2+}(aq) + 4e^-$

Higher oxygen concentration (+ve pole) $O_2(g) + 2H_2O(l) + 4e^- \longrightarrow 4OH^-(aq)$

At some point on the surface of the iron (between the regions acting as negative and positive poles) iron (II) hydroxide is formed by diffusion of the iron (II) and hydroxyl ions; this is then oxidised to rust (hydrated iron (III) oxide). Rusting is accelerated if the moisture contains a strong electrolyte, since more current can flow through the tiny electrochemical cells, e.g. sea spray and the sodium chloride used to thaw ice on roads in the winter can play havoc with unprotected car chassis.

The fact that rust is formed some way between the positive and negative poles explains why paint must be entirely stripped off slightly corroded areas, for it may well be that the iron under the paint has corroded appreciably.

Protection against rusting can be afforded by connecting a metal with a higher negative electrode potential than iron to the structure, e.g. underground steel pipes have been protected from corrosion by connecting easily renewable magnesium rods along them at specific points. Since magnesium has a higher negative electrode potential than iron it will pass into solution as ions (and thus corrode), leaving the steel structure intact.

22.21 Compounds of iron

Iron shows oxidation states of +3 (the most stable) and +2 (reducing). There is also an unstable oxidation state of +6 which is powerfully oxidising. The presence of five singly occupied 3*d* orbitals in the Fe^{3+} ion and only four singly occupied 3*d* orbitals in the Fe^{2+} ion is often cited as an explanation of the easy oxidation of Fe^{2+} to Fe^{3+}.

Iron (VI) compounds

The ferrate (VI) ion, FeO_4^{2-}, contains iron in the +6 oxidation state, and potassium ferrate (VI), K_2FeO_4, can be obtained by fusing a mixture of iron filings and potassium nitrate (oxidising agent); when the cooled mass is poured into water a dark red solution is obtained, the colour being due to the $FeO_4^{2-}(aq)$ ion. Barium ferrate (VI), $BaFeO_4$, which is a purple solid, can be precipitated by adding barium chloride solution.

The ferrate (VI) ion is tetrahedral and a very powerful oxidising agent, e.g. it will oxidise chromium (III) ions to chromate (VI) ions; corresponding ferrates (VI) and sulphates are isomorphous.

Neither the oxide, FeO_3, nor the parent acid of the ferrates (VI), H_2FeO_4, is known.

Iron (III) compounds

Iron (III) oxide, Fe_2O_3

Iron (III) oxide occurs naturally as haematite; it can be prepared as a rust-coloured solid by heating iron (III) hydroxide or iron (II) sulphate:

$$2FeSO_4(s) \longrightarrow Fe_2O_3(s) + SO_2(g) + SO_3(g)$$

Note that iron (II) sulphate is oxidised to iron (III) oxide in the above reaction.

Naturally occurring iron (III) oxide has a similar structure to that of aluminium oxide, i.e. it is ionic.

Iron (III) hydroxide, $Fe_2O_3.xH_2O$

Iron (III) hydroxide, which is best represented as iron (III) oxide associated with an indefinite number of molecules of water, is precipitated as a rust-coloured gelatinous solid when aqueous solutions containing hydroxyl and iron (III) ions are mixed. It reacts readily with dilute acids to give iron (III) salts but since it also reacts with concentrated solutions of alkalis to give ferrates (III), e.g. $NaFeO_2$, it must be regarded as amphoteric.

Colloidal iron (III) hydroxide is obtained as a rust-coloured 'solution' when a few drops of iron (III) chloride solution are added to a large volume of boiling water.

Tri-iron tetroxide (magnetic oxide of iron), Fe_3O_4

This oxide occurs naturally as the ore magnetite which is magnetic. It is a black solid and is formed when steam is passed over heated iron (p. 376). When reacted with acids it gives a mixture of iron (II) and iron (III) salts (ratio 1 : 2) and this is consistent with a structure containing Fe^{2+} and Fe^{3+} ions (ratio 1 : 2) with enough oxide ions to maintain electrical neutrality.

Uses of iron (III) oxides

Haematite, Fe_2O_3, and magnetite, Fe_3O_4, are major sources of iron and steel. In addition, iron (III) oxide is used as an abrasive (jeweller's rouge) and as a pigment (yellow, red and purple forms can be obtained by a variety of methods).

Iron (III) halides

Iron (III) fluoride, chloride and bromide can be made by heating iron in the presence of the halogen; pure iron (III) iodide cannot be isolated.

Iron (III) chloride, $FeCl_3$, and iron (III) bromide, $FeBr_3$, are dark red covalent solids which in the vapour state exist as dimerised molecules Fe_2Cl_6 and Fe_2Br_6 respectively; these structures are similar to that of aluminium chloride (p. 207). Both compounds dissociate on heating to give the iron (II) halide and the halogen, and both dissolve very readily in water e.g. Fe_2Cl_6 dissolves to produce a yellow solution containing a variety of complex cations, including $Fe(H_2O)_5Cl^{2+}(aq)$, and the $Cl^-(aq)$ ion, crystallisation from solution gives the hexahydrate, e.g. $FeCl_3.6H_2O$.

Anhydrous iron (III) chloride like aluminium chloride is used as a Friedel-Crafts catalyst and this action depends upon the formation of the $FeCl_4^-$ complex ion (see p. 208).

It is not really surprising that iron (III) iodide cannot be obtained in the pure state, since a solution containing iron (III) ions readily oxidises iodide ions to iodine.

$$2Fe^{3+}(aq) + 2I^-(aq) \longrightarrow 2Fe^{2+}(aq) + I_2(aq)$$

The instability of FeI_3 with respect to FeI_2 and iodine means that iodine is not a sufficiently powerful oxidising agent to convert Fe^{II} to Fe^{III}; this behaviour is very much in line with the more extensive dissociation of iron (III) bromide into iron (II) bromide and bromine on heating as compared with the corresponding chloride.

Iron (III) sulphate, $Fe_2(SO_4)_3$

Iron (III) sulphate can be obtained by heating an aqueous acidified solution of iron (II) sulphate with hydrogen peroxide (oxidising agent):

$$2FeSO_4(aq) + H_2SO_4(aq) + H_2O_2(aq) \longrightarrow Fe_2(SO_4)_3(aq) + 2H_2O(l)$$

On crystallising the solution, the hydrate $Fe_2(SO_4)_3.9H_2O$ separates out.

Iron (III) sulphate forms alums (p. 209), the best known being ammonium iron (III) sulphate, $NH_4^+Fe^{3+}(SO_4^{2-})_2.12H_2O$.

Qualitative tests for iron (III) ions

When aqueous solutions containing iron (III) ions and hydroxyl ions are mixed a rust-coloured gelatinous precipitate of iron (III) hydroxide is formed.

The addition of potassium hexacyanoferrate (II) solution to an aqueous solution containing iron (III) ions produces a dark blue precipitate (Prussian blue):

$$\underset{\text{hexacyanoferrate (II)}}{K^+(aq) + Fe^{II}(CN)_6{}^{4-}(aq)} + Fe^{3+}(aq) \longrightarrow \underset{\text{Prussian blue}}{K^IFe^{III}[Fe^{II}(CN)_6]^{4-}(s)}$$

The addition of a solution of potassium hexacyanoferrate (III), $(K^+)_3[Fe^{III}(CN)_6]^{3-}$, to a solution containing iron (III) ions produces a brown or green coloration.

The addition of potassium thiocyanate solution to a solution containing iron (III) ions produces an intense red coloration, the colour being due to the presence of a variety of ions including $Fe(H_2O)_5(SCN)^{2+}(aq)$:

$$Fe^{3+}(aq) + \underset{\text{thiocyanate}}{SCN^-(aq)} \longrightarrow Fe(H_2O)_5(SCN)^{2+}(aq)$$

This is an extremely sensitive test for iron (III) ions and the presence of trace amounts of these ions in an iron (II) solution produces a red coloration. The test must therefore be performed under controlled conditions, i.e. equal volumes of potassium thiocyanate solution are added to equal volumes of solutions containing iron (II) ions and suspected iron (III) ions respectively; a much more intense coloration is produced with the iron (III) solution.

The qualitative tests for iron (II) ions are given on p. 382.

Iron (II) compounds

Iron (II) oxide, FeO

Iron (II) oxide can be obtained as a black solid by heating iron (II) oxalate (the carbon monoxide provides a reducing atmosphere and prevents aerial oxidation to iron (III) oxide):

$$FeC_2O_4(s) \longrightarrow FeO(s) + CO(g) + CO_2(g)$$

The oxide as made by the above method is pyrophoric, becoming incandescent when sprinkled into the air. It has the sodium chloride structure, indicating that it is ionic, but the crystal lattice is somewhat deficient in iron (II) ions, and it is non-stoichiometric. If heated in an inert atmosphere to a high temperature and allowed to cool slowly it tends to disproportionate into iron and magnetic oxide of iron:

$$4FeO(s) \longrightarrow Fe(s) + Fe_3O_4(s)$$

Iron (II) hydroxide, $Fe(OH)_2$

Iron (II) hydroxide when pure is a white solid, but it is normally obtained as a green gelatinous precipitate by adding sodium hydroxide solution to an aqueous solution containing iron (II) ions:

$$Fe^{2+}(aq) + 2OH^-(aq) \longrightarrow Fe(OH)_2(s)$$

The green colour is due to the presence of some iron (III) hydroxide produced either by aerial oxidation or by there being some iron (III) ions in the iron (II) solution. Prolonged standing in air results in complete conversion into iron (III) hydroxide, essentially hydrated iron (III) oxide (p. 377).

Iron (II) hydroxide dissolves readily in dilute acids to produce the green hydrated iron (II) ion, $Fe(H_2O)_6{}^{2+}(aq)$; since it also reacts with concentrated sodium hydroxide solution, it must be regarded as being somewhat amphoteric.

Iron (II) halides

Anhydrous iron (II) chloride, $FeCl_2$, can be made by passing a stream of dry hydrogen chloride over the heated metal, the anhydrous fluoride is made in a similar manner using a stream of dry hydrogen fluoride. Note that the action of either chlorine or fluorine on the heated metal produces the iron (III) halide (p. 333).

Both these halides dissolve in water from which the hydrates $FeF_2.8H_2O$ and $FeCl_2.6H_2O$ can be crystallised; the hydrated salts can also be made by the action of dilute hydrofluoric and dilute hydrochloric acid on the metal.

Iron (II) fluoride and iron (II) chloride, whether anhydrous or hydrated, are predominantly ionic solids.

Iron (II) bromide is obtained by heating iron in the presence of bromine vapour; an excess of metal is used to prevent the formation of any iron (III) bromide. Iron (II) iodide is made by a similar method from iron and iodine but, since iron (III) iodide does not exist, it is not necessary in this case to use an excess of the metal.

Iron (II) sulphate, $FeSO_4$

The action of dilute sulphuric acid on iron produces a solution of iron (II) sulphate and hydrogen is liberated. When the solution is crystallised the green heptahydrate, $FeSO_4.7H_2O$, is deposited.

Since iron (II) salts readily undergo aerial oxidation to the iron (III) state, the preparation is best carried out by maintaining a reducing atmosphere above the solution.

Careful heating of the heptahydrate produces the white anhydrous salt, but further heating produces iron (III) oxide (not iron (II) oxide since this is readily oxidised), sulphur dioxide and sulphur trioxide:

$$FeSO_4.7H_2O(s) \longrightarrow FeSO_4(s) + 7H_2O(l)$$
$$2FeSO_4(s) \longrightarrow Fe_2O_3(s) + SO_2(g) + SO_3(g)$$

A solution of iron (II) sulphate contains the pale green $Fe(H_2O)_6{}^{2+}(aq)$ ions; in the presence of nitrogen oxide one molecule of water is displaced from the complex and replaced by NO to give:

$$Fe(H_2O)_5(NO)^{2+}(aq)$$

This complex ion is dark brown, and the above reaction is the basis of the 'brown ring' test for ionic nitrates (p. 261).

Iron (II) sulphate is one of the cheapest industrial chemicals and is used to make iron (III) oxide (used as a pigment), Prussian blue (a pigment) and inks.

Iron (II) ammonium sulphate, $FeSO_4(NH_4)_2SO_4.6H_2O$
This double salt is obtained by crystallising a solution containing equivalent amounts of iron (II) sulphate and ammonium sulphate. Unlike hydrated iron (II) sulphate it neither effloresces nor oxidises on exposure to the atmosphere. It is used as a primary volumetric standard, especially for standardising potassium manganate (VII) solutions (p. 368).

Iron (II) oxalate, FeC_2O_4
This is a lemon-coloured solid and it is of significance that its aqueous solution is also yellow. Since $Fe(H_2O)_6^{2+}$(aq) ions are pale green, the yellow colour must be due to the presence of an iron (II) oxalate complex which is stable in aqueous solution. The following equilibrium is suggested:

$$\underset{\text{pale green}}{Fe(H_2O)_6^{2+}(aq)} + 2C_2O_4^{2-}(aq) \rightleftharpoons \underset{\text{yellow}}{Fe(C_2O_4)_2(H_2O)_2^{2-}(aq)} + 4H_2O(l)$$

(Equilibrium displaced largely to the right)

Notice that the oxalate ion is bidentate, i.e. each oxalate ion occupies two of the octahedral positions in the complex. Iron (II) oxalate is more correctly formulated as $Fe^{II}[Fe^{II}(C_2O_4)_2]$.

Iron (II) sulphide, FeS
This solid is obtained by heating a mixture of iron filings and sulphur:

$$Fe(s) + S(s) \longrightarrow FeS(s)$$

Samples of iron (II) sulphide having slightly variable compositions, but all deficient in iron, have been reported, i.e. the sulphide is non-stoichiometric. The formula of one specimen has been reported as $Fe_{0 \cdot 858}S$.

The action of dilute acids on iron (II) sulphide produces hydrogen sulphide, and this is a convenient way of making the gas in the laboratory:

$$FeS(s) + 2H^+(aq) \longrightarrow Fe^{2+}(aq) + H_2S(g)$$

Iron (II) carbonate, $FeCO_3$
Iron (II) carbonate occurs naturally as the ore siderite; it precipitates as a white solid when solutions containing iron (II) ions and carbonate ions are mixed:

$$Fe^{2+}(aq) + CO_3^{2-}(aq) \longrightarrow FeCO_3(s)$$

Since it oxidises on standing in the air, eventually forming iron (III) hydroxide (essentially hydrated iron (III) oxide), it must be obtained using solutions made up from 'boiled out' distilled water; air must also be excluded.

It is of significance that iron (III) carbonate does not exist and this is because hydrated iron (III) ions are more acidic than hydrated iron (II) ions (the smaller iron (III) ions polarise co-ordinating water molecules to a greater extent than the larger iron (II) ions which are also less heavily charged). A strong base such as CO_3^{2-} abstracts protons very readily from hydrated iron (III) ions, eventually forming iron (III) hydroxide (hydrated iron (III) oxide):

$$Fe(H_2O)_6^{3+}(aq) + 3CO_3^{2-}(aq) \rightleftharpoons Fe(H_2O)_3(OH)_3(s) + 3HCO_3^-(aq)$$

(equilibrium displaced largely to the right)

$$2Fe(H_2O)_3(OH)_3(s) \longrightarrow \underset{\text{hydrated iron (III) oxide}}{Fe_2O_3(s) + 9H_2O(l)}$$

Aluminium carbonate does not exist for a similar reason (p. 223).

Qualitative tests for iron (II) ions

An aqueous solution containing iron (II) ions produces a green gelatinous precipitate of iron (II) hydroxide when a solution of an alkali is added:

$$Fe^{2+}(aq) + 2OH^-(aq) \longrightarrow Fe(OH)_2(s)$$

The addition of potassium hexacyanoferrate (II) solution to an aqueous solution containing iron (II) ions produces a white precipitate which rapidly turns blue.

The addition of potassium hexacyanoferrate (III) solution to an aqueous solution containing iron (II) ions produces a dark blue precipitate (Turnbull's blue). Turnbull's blue has been shown to be identical with Prussian blue (p. 379), and it is likely that iron (II) ions first reduce the hexacyanoferrate (III) to hexacyanoferrate (II):

$$Fe^{2+}(aq) + Fe^{III}(CN)_6^{3-}(aq) \longrightarrow Fe^{3+}(aq) + Fe^{II}(CN)_6^{4-}(aq)$$

followed by:

$$K^+(aq) + Fe^{II}(CN)_6^{4-}(aq) + Fe^{3+}(aq) \longrightarrow \underset{\text{Turnbull's blue or Prussian blue}}{K^IFe^{III}[Fe^{II}(CN)_6]^{4-}(s)}$$

The addition of potassium thiocyanate solution to a solution containing iron (II) ions does not produce any coloration if iron (III) ions are completely absent. However, in view of the ready oxidation of Fe^{2+} to Fe^{3+} and the extreme sensitivity of the thiocyanate test for iron (III) ions, iron (II) solutions often give some coloration with this reagent.

Cobalt

22.22 Occurrence and extraction of cobalt

The main ores of cobalt are smaltite, $CoAs_2$, and cobaltite, CoAsS; they occur together with ores of nickel, and indeed the two metals resemble each other very closely.

The extraction of cobalt involves a number of complex stages which ultimately result in the formation of tricobalt tetroxide, Co_3O_4. This oxide is then reduced to the metal by heating either with carbon or with aluminium:

$$3Co_3O_4(s) + 8Al(s) \longrightarrow 9Co(s) + 4Al_2O_3(s)$$

Pure cobalt may be obtained by electrolysis of an aqueous solution containing cobalt (II) sulphate, $CoSO_4$, and ammonium sulphate.

22.23 Properties and uses of cobalt

Cobalt has a melting point of 1490°C and a density of 8·90 g cm^{-3}. It is a hard bluish-white metal and is fairly unreactive, e.g. it is not attacked by air or water at ordinary temperatures. Oxygen attacks the metal at high temperatures forming cobalt (II) oxide, CoO, and it combines directly with carbon and sulphur on heating. Dilute acids such as hydrochloric acid and sulphuric acid slowly react with the metal with the liberation of hydrogen and the formation of cobalt (II), $Co^{2+}(aq)$ ions. Concentrated nitric acid renders the metal passive (p. 260) but it reacts with dilute nitric acid, forming cobalt (II) ions and oxides of nitrogen.

Stellite, an alloy of cobalt, chromium and tungsten, is extremely hard, even at red heat, and is used for making high-speed cutting tools and valves for internal combustion engines. An aluminium/cobalt/nickel alloy (Alnico) is used for the manufacture of extremely strong permanent magnets; two other alloys called constantan and nichrome (both of which contain cobalt and nickel) are used for electrical heating elements.

22.24 Compounds of cobalt

Cobalt has oxidation states of +3 (uncommon except when present in a complex) and +2 (the most stable). Although Co^{III} is strongly oxidising and thus easily reduced to Co^{II}, the presence of complexing agents such as NH_3 and CN^- considerably stabilise the higher oxidation state.

Cobalt (III) compounds

The best known examples of 'simple' cobalt (III) compounds are the fluoride, CoF_3, and sulphate, $Co_2(SO_4)_3.18H_2O$. Cobalt (III) fluoride is made by passing fluorine over cobalt at a temperature of about 350°C. It is a brown solid which reacts with water, liberating oxygen. Cobalt (III) sulphate can be obtained by the action of ozone on a solution of cobalt (II) sulphate in fairly concentrated sulphuric acid; it is a blue solid and, like the fluoride, liberates oxygen in contact with water.

It is significant that cobalt (III) fluoride and sulphate require strong oxidising agents for their preparation, i.e. fluorine and ozone respectively. The fact that they exist at all points to their structures being complex.

When present as an integral part of a complex, compounds containing cobalt in the +3 oxidation state are considerably more stable, i.e. they are not such powerful oxidising agents. Many Co^{III} complexes exist and they are all octahedral in shape. Typical ones are $[Co(NH_3)_6]^{3+}$, $[Co(CN)_6]^{3-}$ and $[Co(NO_2)_6]^{3-}$. The $[Co(NO_2)_6]^{3-}$ anion is present in sodium hexanitritocobaltate(III), $Na_3[Co(NO_2)_6]$. Since this salt (but not the corresponding potassium compound) is soluble in water, a solution of this reagent is used as a means of distinguishing between potassium and sodium salts, which otherwise have very similar chemical properties. The $[Co(CN)_6]^{3-}$ ion is of interest in that it is stable in the presence of water unlike $[Co(CN)_5]^{3-}$ which contains Co^{II}; indeed $[Co(CN)_5]^{3-}$ attacks water, liberating hydrogen, and is itself oxidised to $[Co(CN)_6]^{3-}$.

Cobalt (II) compounds

This is the most stable oxidation state of cobalt in solution and in the absence of strong complexing agents such as NH_3 and CN^-, e.g. Co^{II} complexes containing NH_3, and CN^- ligands are readily oxidised to Co^{III} complexes. In solution cobalt (II) salts are pink, the colour being due to the presencee of $Co(H_2O)_6{}^{2+}(aq)$ ions.

Cobalt (II) oxide, CoO, is an olive-green solid and can be prepared by heating cobalt (II) carbonate or cobalt (II) nitrate in the absence of air (in the presence of air it can take up oxygen, forming Co_3O_4 which contains both Co^{II} and Co^{III}):

$$CoCO_3(s) \longrightarrow CoO(s) + CO_2(g)$$
$$2Co(NO_3)_2(s) \longrightarrow 2CoO(s) + 4NO_2(g) + O_2(g)$$

Cobalt (II) hydroxide, $Co(OH)_2$, can be precipitated as a blue solid by adding sodium hydroxide solution to a solution of a cobalt (II) salt:

$$Co^{2+}(aq) + 2OH^-(aq) \longrightarrow Co(OH)_2(s)$$

On standing the colour gradually changes from blue to pink.

Cobalt (II) hydroxide dissolves readily in dilute acids, forming the pink cobalt (II) ion; it also reacts with a hot concentrated solution of potassium hydroxide and so must be considered to show weak amphoteric properties.

Cobalt (II) sulphide, CoS, precipitates as a black solid when hydrogen sulphide is passed through an aqueous solution of a cobalt (II) salt in the presence of aqueous ammonia and ammonium chloride (Group IV of qualitative analysis schemes). The solubility product of the sulphide is too high for it to precipitate in acid solution (p. 303).

Soluble salts containing the Co^{2+} ion, e.g. the chloride, nitrate and sulphate, can be made by reacting the oxide, hydroxide or carbonate of cobalt (II) with dilute acids in the usual way, followed by partial evaporation and crystallisation.

When a solution of a cobalt (II) salt is treated with hydrochloric acid the pink solution deepens in colour and eventually becomes deep blue in the presence of sufficient $Cl^-(aq)$ ions. If a warm solution of a cobalt (II) salt is treated with just sufficient concentrated hydrochloric acid to produce a blue colour, then the original pink colour can be restored on cooling. These results are consistent with an equilibrium of the following type:

$$\underset{\text{pink}}{Co(H_2O)_6{}^{2+}(aq)} + 4Cl^-(aq) \rightleftharpoons \underset{\text{blue}}{CoCl_4{}^{2-}(aq)} + 6H_2O(l)$$

(a) Equilibrium shifts to the right on adding $Cl^-(aq)$ ions.
(b) Equilibrium shifts to the right on raising the temperature.

It is known that the blue $[CoCl_4]^{2-}$ ion is a tetrahedral complex.

Insoluble cobalt (II) salts include the carbonate and oxalate which can be obtained by double decomposition reactions.

Isomerism associated with some cobalt (III) complexes

Cobalt (III), like chromium (III), forms a diverse range of complexes many of which exhibit geometrical and optical isomerism; indeed comparable complexes of both these metals in this oxidation state show many features in common.

An interesting series of complexes are those derived from $CoCl_3$ and ammonia (Table 22J):

Table 22J Complexes derived from $CoCl_3$ and ammonia

Empirical Formula	Colour of Solid	% Ionic Chlorine	Formula of the Complex
(i) $CoCl_3(NH_3)_6$	Orange-yellow	100	$[Co(NH_3)_6]^{3+}\ 3Cl^-$
(ii) $CoCl_3(NH_3)_5$	Violet	67	$[Co(NH_3)_5Cl]^{2+}\ 2Cl^-$
(iii) $CoCl_3(NH_3)_4$	Violet	33	$[Co(NH_3)_4(Cl)_2]^+\ Cl^-$
(iv) $CoCl_3(NH_3)_4$	Green	33	$[Co(NH_3)_4(Cl)_2]^+\ Cl^-$

The formulae of the complexes were determined by reaction with silver nitrate solution; only the chloride which is not incorporated into the complex cation precipitates as silver chloride. Conductivity measurements in aqueous solution allow the number of ions produced to be calculated, i.e. four from compound (i), three from compounds (ii), and two from compounds (iii) and (iv). There is little doubt, therefore, that the formulae shown in Table 22J are correct; the shapes of (and arrangement of the ligands in) these complexes are shown below:

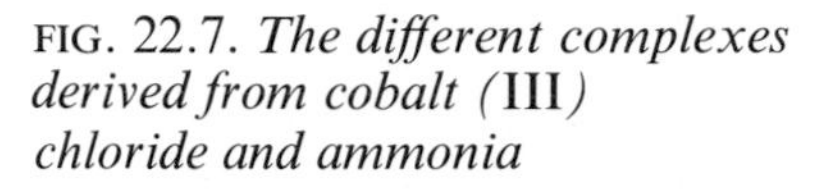

FIG. 22.7. *The different complexes derived from cobalt (III) chloride and ammonia*

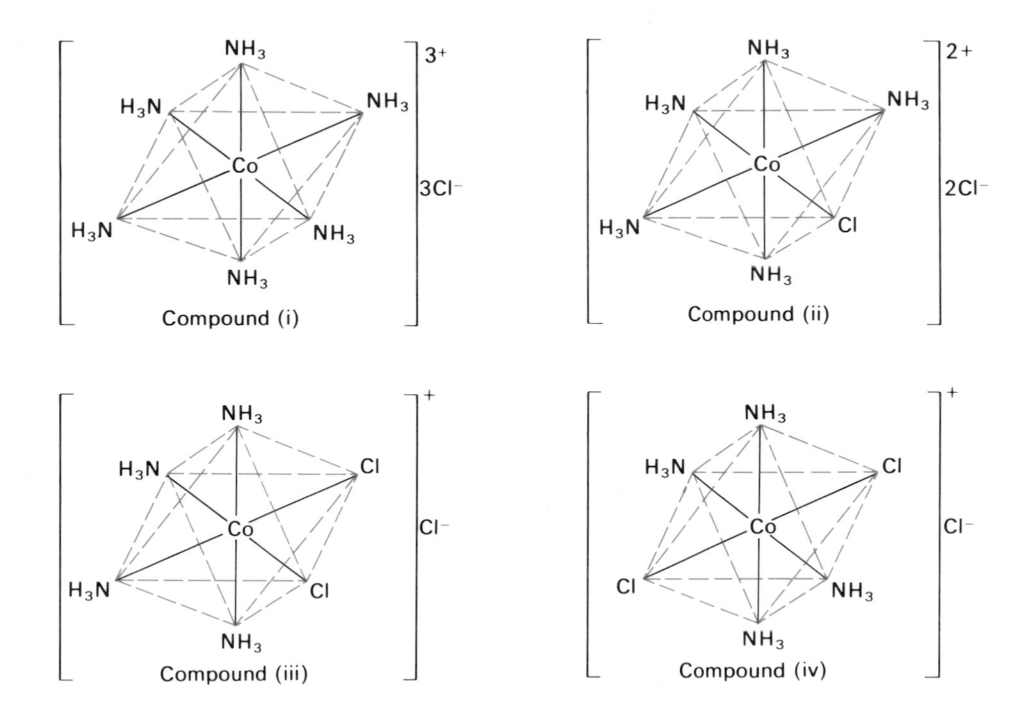

Notice that compounds (iii) and (iv) are respectively the cis and trans isomers (cf. similar complexes of chromium (III)) (fig. 22.4).

The cobalt (III) complex containing three ethylenediamine molecules is potentially optically active, like the corresponding complex of chromium (III) (p. 365), and can be resolved into its *d*- and *l*-forms (fig. 22.8).

FIG. 22.8. *The optical isomers of tris (ethylenediamino) cobalt (III) ion*

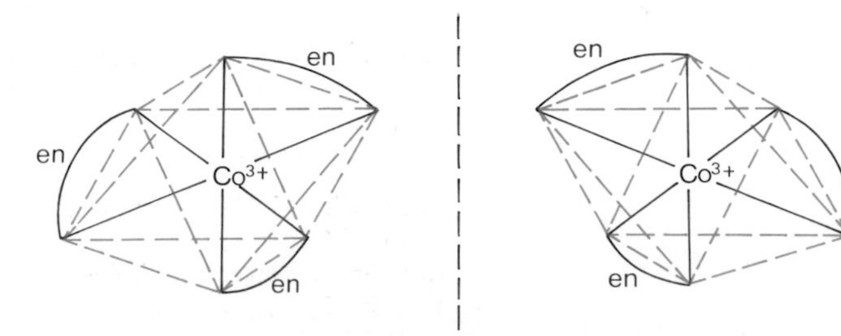

Nickel

22.25 Occurrence and extraction of nickel

Ores of nickel are frequently associated with those of cobalt, typical ones being millerite, NiS, pentlandite (a mixed sulphide of iron and nickel), and garnierite, $(Ni,Mg)SiO_3.xH_2O$.

The extraction of nickel involves a number of rather complex stages which ultimately result in the conversion of the ores into nickel (II) oxide, NiO. The oxide is then reduced to nickel by heating with carbon:

$$NiO(s) + C(s) \longrightarrow Ni(s) + CO(g)$$

Nickel of purity 99·99 per cent can be obtained from nickel (II) oxide by reduction with synthesis gas (a mixture of carbon monoxide and hydrogen) at a temperature of about 350°C. The impure nickel which is formed is then heated to about 60°C in a stream of carbon monoxide; volatile nickel carbonyl, $Ni(CO)_4$, distils leaving impurities behind and is decomposed to give very pure nickel by heating to about 200°C.

$$NiO(s) + CO(g) \longrightarrow \underset{\text{impure}}{Ni(s)} + CO_2(g)$$

$$\underset{\text{impure}}{Ni(s)} + 4CO(g) \xrightarrow{60°C} Ni(CO)_4(g)$$

$$Ni(CO)_4(g) \xrightarrow{200°C} \underset{\text{pure}}{Ni(s)} + 4CO(g)$$

This method of producing extremely pure nickel is known as the Mond process.

22.26 Properties and uses of nickel

The properties of nickel are very similar to those of cobalt; it melts at 1452°C and has a density of 8·90 g cm^{-3}. It is unreactive towards air or water at ordinary temperatures, but combines directly with oxygen on heating to give nickel (II) oxide, NiO. Dilute mineral acids only slowly attack the metal with the formation of nickel (II), $Ni^{2+}(aq)$, ions. However, concentrated nitric acid renders it passive. Since it is not attacked by fused sodium hydroxide, nickel crucibles are convenient receptacles in which to carry out alkali fusions.

Finely divided nickel, which can be made by reduction of nickel (II) oxide with hydrogen, readily absorbs hydrogen and is used as a catalyst in the hydrogenation of alkenes, e.g. unsaturated inedible oils which contain carbon-carbon double bonds can be hydrogenated to saturated solids, which are edible fats.

Alloys containing nickel and cobalt have been mentioned in section 22.23; other alloys containing nickel include monel metal (nickel and copper) which is used in the construction of chemical plant, and cupro-nickel (also nickel and copper), which is used in 'silver' coinage.

22.27 Compounds of nickel

Although a few compounds are known which contain Ni^{III} (e.g. an impure oxide, $Ni_2O_3.2H_2O$) and Ni^{IV} (e.g. an impure oxide, NiO_2, and a complex fluoride, K_2NiF_6), the only important oxidation state of nickel is +2.

Nickel (II) compounds

Nickel (II) oxide, NiO, is a green solid and can be obtained by heating nickel (II) carbonate or nickel (II) nitrate:

$$NiCO_3(s) \longrightarrow NiO(s) + CO_2(g)$$
$$2Ni(NO_3)_2(s) \longrightarrow 2NiO(s) + 4NO_2(g) + O_2(g)$$

It is a typical basic oxide, dissolving in dilute acids to give the green $Ni(H_2O)_6{}^{2+}(aq)$ ions.

Nickel (II) hydroxide, $Ni(OH)_2$, can be precipitated as a green gelatinous solid by adding sodium hydroxide solution to a solution of a nickel (II) salt:

$$Ni^{2+}(aq) + 2OH^-(aq) \longrightarrow Ni(OH)_2(s)$$

It reacts readily with dilute acids but is not attacked by strong alkalis, since it is exclusively basic. The fact that it is soluble in aqueous ammonia, forming a blue solution, is due to the formation of complex ions, i.e. $Ni(H_2O)_2(NH_3)_4{}^{2+}(aq)$ and $Ni(NH_3)_6{}^{2+}(aq)$.

$$\begin{array}{ccl} Ni(OH)_2(s) \rightleftharpoons & Ni^{2+}(aq) & + \; 2OH^-(aq) \\ & + & \\ & 6NH_3(aq) & \\ & \updownarrow & \\ & Ni(NH_3)_6{}^{2+}(aq) & \end{array}$$

The removal of $Ni^{2+}(aq)$ ions as complex ions containing ammonia allows the nickel (II) hydroxide to pass into solution.

Nickel (II) sulphide, NiS, precipitates as a black solid when hydrogen sulphide is passed through an aqueous solution of a nickel (II) salt in the presence of aqueous ammonia and ammonium chloride (Group IV of qualitative analysis schemes). The solubility product of the sulphide is too high for it to precipitate in acid solution (p. 303).

Soluble salts containing the Ni^{2+} ion, e.g. the chloride, $NiCl_2.6H_2O$, the nitrate, $Ni(NO_3)_2.6H_2O$, and the sulphate, $NiSO_4.7H_2O$, can be made by reacting nickel (II) oxide, nickel (II) hydroxide or nickel (II) carbonate with dilute acids in the usual way, followed by partial evaporation and crystallisation. Nickel (II) sulphate forms a double salt when crystallised in the presence of an equivalent amount of ammonium sulphate; this double salt, $(NH_4)_2SO_4,NiSO_4.6H_2O$, is used in aqueous solution as the electrolyte in nickel plating.

Insoluble salts of nickel include the carbonate and phosphate (V) which may be made by double decomposition reactions; in order to avoid the formation of a basic carbonate, nickel (II) carbonate is obtained by adding sodium hydrogen carbonate solution to a solution containing $Ni^{2+}(aq)$ ions (p. 181).

Complexes of nickel

Nickel (II) complexes are fairly numerous and the majority of them are octahedral, e.g. the green $[Ni(H_2O)_6]^{2+}$ and the blue $[Ni(NH_3)_6]^{2+}$. A well-known planar nickel complex is $[Ni(CN)_4]^{2-}$ and another is nickel dimethylglyoximate, which precipitates as a red solid when dimethylglyoxime is added to a solution of a nickel salt made just alkaline by the addition of ammonia; the formation of the latter complex is used as a test for Ni^{2+} ions.

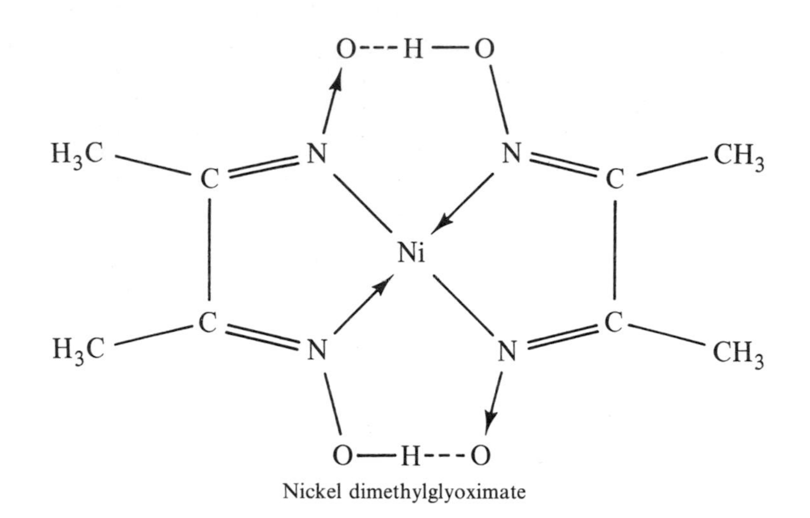

Nickel dimethylglyoximate

If dimethylglyoxime is represented as DMG, the equation for its formation can be represented as:

$$2DMG + Ni^{2+}(aq) + 2OH^-(aq) \longrightarrow Ni(DMG)_2(s) + 2H_2O(l)$$

During the reaction two protons (one from each dimethylglyoxime molecule) are neutralised by the ammonia solution. The complex, which is thus uncharged, is held in the planar configuration by two hydrogen bonds (shown by dotted lines).

22.28 Suggested reading

J. V. Huxley, *The production of cobalt*, Education in Chemistry, No. 9, Vol. 11, 1974.
J. V. Huxley, *The production of nickel*, Education in Chemistry, No. 2, Vol. 15, 1978.

22.29 Questions on chapter 22

1. What do you understand by the phrase *First transition series of elements*? Give the names and symbols of as many as possible of these elements, and discuss, with illustrative examples, *three* properties which are general among the elements and/or their compounds.
Outline a brief explanation for the properties you describe.
Describe the special features of the chemistry of any ONE of these elements. (S)
2. Explain the term 'transition element'. Give an account of the characteristic properties of transition elements, illustrating your answer by reference to manganese.
In what respects do you consider that copper behaves as a transition element? Give reasons for your answer. (L)
3. Transition elements are characterised by variable oxidation state, coloured ions, catalytic activity and the formation of complexes. Illustrate this statement by detailed reference to three transition elements selected from the series titanium to copper exclusively.
Would you classify zinc as a transition element? Give your reasons briefly, defining the term 'transition element' and relating your answer to the earlier discussion. (S)

4 Explain the meaning of the terms 'transition metal' and 'complex ion', illustrating your answers by reference to one of the following elements: chromium, manganese, iron or copper.

Discuss two examples of the use of complex ion formation in qualitative analysis. (O & C)

5 For either chromium or manganese or nickel or copper describe briefly:

(a) how you would prepare a compound of the metal in its lowest oxidation state,
(b) how you would prepare a salt of the metal in its highest oxidation state,
(c) one use of the salt described in (b). (O & C)

6 Give an account of the chemistry of nickel with special reference to

(a) its extraction from natural sources,
(b) its transitional character,
(c) its detection in analysis,
(d) its industrial uses. (O & C)

7 Outline the physico-chemical principles underlying the operation of the blast furnace, commenting on the Ellingham diagrams for carbon and iron.
Give an account of the uses of iron, emphasizing the advantages and disadvantages in each case.
Give the structure of iron (III) chloride. Describe, giving essential experimental details, a method for obtaining a pure sample of anhydrous iron (III) chloride from iron. (L)

8 Give the formulae for, and explain the constitutions of four of the following: hexammine nickel bromide; potassium hexacyanoferrate (II); sodium hexanitrito cobaltate (III); the 'brown-ring' compound (nitrosyl iron (II) sulphate); chrome (III) alum. (O & C)

9 Give an account of the chemistry of iron illustrating particularly

(a) its transitional character;
(b) the use of its compounds in analytical chemistry. (O & C)

10 What do you understand by the term 'complex ion'? Illustrate the importance of complex ion formation in systematic qualitative analysis. (Oxford Schol.)

11 Give a short account of the effect of complex formation on the chemical properties of metal ions in solution, with special reference to one of the elements chromium, manganese, nickel or copper.

When a solution of iron (II) ions was oxidised to one of iron (III) ions with an acidified solution of potassium permanganate (manganate (VII)) 20 cm^3 of the permanganate solution were required to reach the equivalence point. Repeating the titration, but in the presence of a large excess of fluoride ions, however, 25 cm^3 were required. How do you account for this result? (O & C)

12 (a) Justify the classification of either manganese or chromium as a transition element, illustrating your answer by reference to suitable compounds of the element you choose.
(b) Describe the preparation of a crystalline sample of either potassium permanganate (manganate (VII)) from manganese (IV) oxide or potassium dichromate from chromium (III) oxide. (C)

13 Explain the chemistry of the production of steel from cast iron. For each of the following processes give one experimental method:

(a) the oxidation of iron (II) ions to iron (III) ions,
(b) the reduction of iron (III) ions to iron (II) ions,
(c) the preparation of a sample of iron (III) oxide starting from iron.

In (a) and (b) describe how you would demonstrate the success of each process. (JMB)

14 Explain the following facts and in each case give **one** appropriate example.

(a) Transition metals are usually hard solids with high melting points.
(b) Transition metals commonly show a number of relatively stable oxidation states.
(c) Transition metals form a large number of coordination compounds.
(d) The aquo-cations of transition metals are often acidic in solution. (O & C)

15 Outline the preparation of (a) chlorosulphonic acid,
(b) chromyl chloride (chromium (VI) dichloride dioxide).

When a sample of potassium chlorochromate ($KCrClO_3$) was heated, gas was evolved and a solid residue remained. When the gas was shaken with aqueous potassium iodide, iodine was liberated and half the gas disappeared. The remainder proved to be oxygen. When the solid residue was extracted with water, chromium (III) oxide was left behind, and the solution was of a reddish colour and contained chloride. When the solution was treated with dilute sulphuric acid and potassium iodide, three times as much iodine as was liberated by the gas was set free.

Write an equation to describe how the salt decomposed when it was heated.

(Camb. Schol)

16 The element vanadium (V) occurs in the Periodic Table as indicated below:

			C	N	O	F
			Si	P	S	Cl
K	Ca	Sc	Ti	V	Cr	Mn

Discuss the properties you would expect it to have. (Oxford Schol.)

17 If you were given a transition metal how would you attempt to investigate its main oxidation states?

18 Sulphur and chromium have been included in the same group of the Periodic Table; how far can this be justified? (Camb. Schol.)

19 A solution containing vanadium (+ 5 oxidation state) contains the equivalent of 2·55 g of vanadium metal per dm^3 of solution. 25 cm^3 of this solution were reduced by sulphur dioxide and re-oxidised to its original condition by potassium permanganate (manganate (VII)), of which 12·5 cm^3 of an M/50 solution were required. Deduce the oxidation state of the reduced vanadium solution.

Another 25 cm^3 portion of the original vanadium solution, after reduction with zinc and dilute sulphuric acid, required 37·5 cm^3 of the potassium permanganate solution for complete oxidation; while a third portion similarly treated with zinc and dilute sulphuric acid, but through which air was passed prior to titration for about 5 min, required 25 cm^3 of potassium permanganate for complete oxidation. Deduce what you can from these results.

20 (a) A transition metal cation, M^{3+}, forms a complex ion with the formula $[M(NH_3)_4Cl_2]^+$ with the ligands arranged octahedrally round the cation. Draw diagrams to illustrate the number of geometrical isomers. Would any of them be expected to be optically active? Give reasons.

(b) Another transition metal cation forms a complex ion of the formula $[M(NH_3)_3Cl_3]^+$ with a similar arrangement of ligands.

(i) What is the oxidation state of the transition metal?

(ii) Draw the possible geometrical isomers and indicate, with reasons, if any of them would be expected to be optically active.

(c) Pt^{2+} forms the complex $[Pt(NH_3)_2Cl_2]$ which exists in two different forms. What possible arrangement of ligands round the central platinum ion does this rule out? How would you expect the ligands to be arranged round Pt^{2+}?

21 When heated with potassium carbonate in air, a metallic ore (a complex oxide which contains iron in addition to the element **A**) gives a yellow solid **B** which is soluble in water: the iron is removed as an insoluble iron oxide, Fe_3O_4. Using a 336 g sample of the ore, the dry weight of iron oxide obtained was 116 g. If **B** is dissolved in water and the solution acidified, an orange solution of compound **C** is obtained.

A solid sample of **C**, when treated with potassium chloride and concentrated sulphuric acid, gives a red-brown liquid **D**. When hydrogen peroxide is added to an acidified solution of **C**, a blue colour develops in the solution. This blue colour can be extracted into diethyl ether. Addition of pyridine, C_5H_5N, to the blue ethereal solution permits the isolation of a compound **E** which is monomeric and whose molecular formula is given by $C_5H_5NO_5$**A**. Quantitative conversion of all the element **A** in a 336 g sample of the metallic ore into the compound **E** would give 633 g of **E**.

When **C** is heated with ammonium chloride, an oxide **F** of the element **A** is formed. **D** is immediately hydrolysed by water to produce hydrogen chloride and a solution from which **B** can be isolated after the addition of potassium chloride.

(i) Identify the element **A** and the compounds **B**, **C**, **D** and **F**.

(ii) Write equations for the reactions involved in the conversions **B** → **C**, **C** → **D**, **C** → **F** and **D** → **B**.

(iii) Deduce the formula of the ore.

[H = 1·0; C = 12·0; N = 14·0; O = 16·0; Fe = 56·0; Atomic weight of element *A* = 52·0.] (C)

Chapter 23

Some more features of the first transition series

23.1 Introduction

In this chapter the following topics will be discussed in turn:

(a) The causes of acidity of transition metal cations in aqueous solution, the formation of oxycations, e.g. $VO^{2+}(aq)$, and the formation of oxyanions, e.g. $CrO_4^{2-}(aq)$ and $MnO_4^-(aq)$.

(b) The use of redox potential data in interpreting the stability (and instability) of some oxidation states of transition metals in aqueous solution.

(c) A more complete account of the origin of the colours associated with transition metal complexes.

(d) A short account of the carbonyls.

23.2 The causes of acidity of transition metal cations in aqueous solution and some other related effects

The transition metal salts of strong acids, e.g. the chloride, sulphate and nitrate, all show an acid reaction in aqueous solution and for salts containing the same concentration of transition metal ion in solution there is an increase in acidity with increasing charge on the transition metal ion, e.g. $Fe(H_2O)_6^{3+}(aq)$ is more acidic than $Fe(H_2O)_6^{2+}(aq)$. For ions carrying the same charge there is an increase in acidity in the order of increasing atomic number of the transition metal. An increase in acidity occurs when:

(a) the ionic radius of the cation decreases,

(b) the charge on the cation increases.

In solution, cations tend to be surrounded by molecules of water orientated so that the negative end of the water molecules, i.e. the oxygen atoms, are pointing towards the cation. If the cation is small and reasonably heavily charged, the water molecules become sufficiently polarised for other water molecules to function as bases and abstract protons. A number of equilibria exist in solution, two of which are shown below for the hydrated iron (III) ion:

$$Fe(H_2O)_6^{3+}(aq) + H_2O(l) \rightleftharpoons Fe(H_2O)_5(OH)^{2+}(aq) + H_3O^+(aq)$$
$$Fe(H_2O)_5(OH)^{2+}(aq) + H_2O(l) \rightleftharpoons Fe(H_2O)_4(OH)_2^+(aq) + H_3O^+(aq)$$

The reason why simple $Ti^{4+}(aq)$ and $V^{4+}(aq)$ ions do not occur in aqueous solution (they occur as the oxycations $TiO^{2+}(aq)$ and $VO^{2+}(aq)$ is due to the large charge/size ratio of the Ti^{4+} and V^{4+} ions, i.e. they bind firmly to a water molecule and two protons are readily transferred to other water molecules, e.g.

$$Ti(H_2O)_x^{4+}(aq) + 2H_2O(l) \rightleftharpoons TiO(H_2O)_{x-1}^{2+}(aq) + 2H_3O^+(aq)$$

In a similar way it is possible to extend this reasoning to explain the formation of anions containing a transition metal. For instance the $Cr(H_2O)_6^{3+}(aq)$ ion is acidic and can be oxidised to the $CrO_4^{2-}(aq)$ ion in aqueous solution by the addition of sodium peroxide (an oxidising agent and also a source of hydroxyl ions in solution). Successive oxidation of Cr^{3+} would result in the formation of the hypothetical Cr^{6+} ion which, by virtue of its high charge and small size, would exert a profound polarising effect on nearby water molecules. It is not difficult to see why it should be possible to remove protons readily from these water molecules and the presence of OH^- ions would certainly help this process. Under such extreme conditions both protons are abstracted from such water molecules, leaving O^{2-} ions; of course Cr^{6+} ions would not exist as such in the presence of O^{2-} ions, and a redistribution of charge would result in the formation of chromium-oxygen covalent bonds. The fact that the chromate ion is CrO_4^{2-} (which can be represented formally as $Cr^{6+}(O^{2-})_4$) and not, say, CrO_5^{4-} ($Cr^{6+}(O^{2-})_5$) can be partially explained as being due to the fact that the chromium atom is not large enough to accommodate more than four oxygens (a steric effect). In a similar way it is possible to visualise the formation of other oxyanions containing transition metals, e.g. manganates (VI) and manganates (VII).

23.3 The use of standard redox potentials in explaining the stability and instability of some oxidation states

In this section use will be made of tabulated values of standard redox potentials to explain some of the features of the solution chemistry of the transition metals. The discussion is restricted to points mentioned in the previous chapter, which are summarised below:

(a) Except for vanadium, which is fairly resistant to attack by hydrochloric acid, the first transition series elements react with this acid to give hydrated ions and hydrogen.

(b) The stable oxidation states of the transition metals in solution and in the presence of oxygen are TiO^{2+}(aq)(Ti(IV)), VO^{2+}(aq)(V(IV)), Cr^{3+}(aq), Mn^{2+}(aq), Fe^{3+}(aq), Co^{2+}(aq) and Ni^{2+}(aq). It will be seen that, except for iron, there is a tendency for the lower oxidation states to become increasingly stable with increasing atomic number. Oxidation states higher than the stable ones given above (if they can be obtained in solution) tend to be oxidising, i.e. they liberate oxygen from water; conversely, lower oxidation states tend to be reducing, and they liberate hydrogen from water.

(c) Higher oxidation states which are normally oxidising in solution can be stabilised by complex formation, e.g. Co^{3+} is stabilised when complexed with ammonia.

(d) The manganate (VI) ion MnO_4^{2-}(aq) is stable in the presence of OH^-(aq) ions but disproportionates in neutral or acid solution into the manganate (VII) ion MnO_4^-(aq) and manganese (IV) oxide.

The standard electrode potential of a metal is the potential difference established when the metal in contact with a 1 M solution of its ions is joined by a salt bridge to a standard hydrogen electrode (a system containing M.H^+(aq) ions into which pure hydrogen gas at one atmosphere pressure is passing, equilibrium being established in the presence of a platinum electrode coated with platinum black); the standard electrode potential of the hydrogen electrode is arbitrarily taken as zero.

If electrons flow from the metal to the hydrogen electrode through the external circuit, the metal is allotted a negative electrode potential (see chapter 10 for a fuller discussion); if electron flow takes place in the other direction, the metal is said to have a positive electrode potential.

Since transition metals exhibit variable oxidation states, the electrode potentials will depend upon the oxidation states of the ions which are in contact with the metals, e.g. for the systems Fe^{2+}(aq)/Fe(s) and

Fe^{3+}(aq)/Fe(s) the respective standard electrode (redox) potentials are −0·44 V and −0·04 V. Redox potentials for systems containing metal ions in two different oxidation states can readily be obtained, e.g. the standard redox potential for the system Fe^{3}(aq)/Fe^{2+}(aq) is +0·76 V.

Values of redox potentials for the systems M^{2+}(aq)/M(s) and M^{3+}(aq)/M^{2+} (aq) are shown in fig. 23.1. The horizontal lines drawn across the graph at $E^{\ominus} = 0$ and $E^{\ominus} = +1{\cdot}23$ V correspond to redox potentials for the systems:

$H^+(aq)/\frac{1}{2}H_2(g)$	$H^+(aq) + e^- \longrightarrow \frac{1}{2}H_2(g)$	$E^{\ominus} = 0$ V
$\frac{1}{2}O_2(g) + 2H^+(aq)/H_2O(l)$	$\frac{1}{2}O_2(g) + 2H^+(aq) + 2e^- \longrightarrow H_2O(l)$	$E^{\ominus} = +1{\cdot}23$ V

The last system comprises a 1 M solution of hydrogen ions into which pure oxygen at one atmosphere pressure is passed.

FIG. 23.1. *The redox potentials for some aqueous systems*

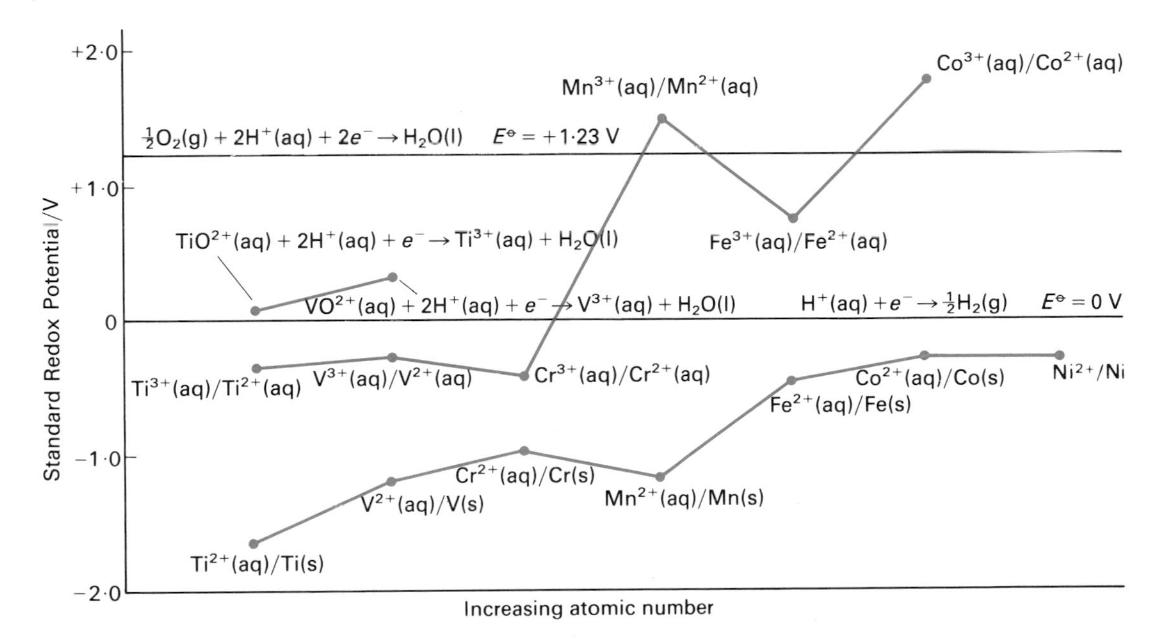

Consider a cell comprising the systems Ti^{2+}(aq)/Ti(s) and $H^+(aq)/\frac{1}{2}H_2(g)$. Since the standard redox potential of titanium is negative, electron flow would take place from the titanium to the hydrogen electrode through the external circuit, i.e. titanium is thermodynamically capable of reducing hydrogen ions to hydrogen and this is reflected in the positive sign attached to the e.m.f. of this cell (see chapter 10):

$$Ti^{2+}(aq) + 2e^- \longrightarrow Ti(s) \qquad E^{\ominus} = -1{\cdot}63\text{ V}$$
$$H^+(aq) + e^- \longrightarrow \tfrac{1}{2}H_2(g) \qquad E^{\ominus} = 0\text{ V}$$
$$\text{or} \quad Ti(s) + 2H^+(aq) \longrightarrow Ti^{2+}(aq) + H_2(g) \qquad E^{\ominus}_{total} = +1{\cdot}63\text{ V}$$

In a similar way it can be seen that Ti^{2+}(aq) ions are potentially capable of reducing hydrogen ions to hydrogen:

$$Ti^{3+}(aq) + e^- \longrightarrow Ti^{2+}(aq) \qquad E^{\ominus} = -0{\cdot}37\text{ V}$$
$$H^+(aq) + e^- \longrightarrow \tfrac{1}{2}H_2(g) \qquad E^{\ominus} = 0\text{ V}$$
$$\text{or} \quad Ti^{2+}(aq) + H^+(aq) \longrightarrow Ti^{3+}(aq) + \tfrac{1}{2}H_2(g) \qquad E^{\ominus}_{total} = +0{\cdot}37\text{ V}$$

From this discussion it can be seen that titanium should, on thermodynamic grounds, dissolve in hydrochloric acid to give $Ti^{3+}(aq)$ ions and hydrogen and, furthermore, that $Ti^{2+}(aq)$ ions should not exist in acid solution. This is indeed the case. Similarly any system with a negative redox potential, i.e. one that lies below the horizontal line drawn on the graph at $E^\circ = 0$ should be capable of liberating hydrogen from hydrochloric acid. The table below shows the predicted behaviour of the transition metals and their ions in the presence of hydrochloric acid.

Table 23A The action of hydrochloric acid on transition metals and their ions

Predicted Action of Hydrochloric Acid	
(a) On Metals	(b) On Ions
$Ti(s) + H^+(aq) \rightarrow Ti^{3+}(aq) + H_2(g)$	$Ti^{2+}(aq) + H^+(aq) \rightarrow Ti^{3+}(aq) + H_2(g)$
$V(s) + H^+(aq) \rightarrow V^{3+}(aq) + H_2(g)$	$V^{2+}(aq) + H^+(aq) \rightarrow V^{3+}(aq) + H_2(g)$
$Cr(s) + H^+(aq) \rightarrow Cr^{3+}(aq) + H_2(g)$	$Cr^{2+}(aq) + H^+(aq) \rightarrow Cr^{3+}(aq) + H_2(g)$
$Mn(s) + H^+(aq) \rightarrow Mn^{2+}(aq) + H_2(g)$	$Mn^{2+}(aq)$ stable
$Fe(s) + H^+(aq) \rightarrow Fe^{2+}(aq) + H_2(g)$	$Fe^{2+}(aq)$ stable
$Co(s) + H^+(aq) \rightarrow Co^{2+}(aq) + H_2(g)$	$Co^{2+}(aq)$ stable
$Ni(s) + H^+(aq) \rightarrow Ni^{2+}(aq) + H_2(g)$	$Ni^{2+}(aq)$ stable

In all these examples, this in fact happens, except that vanadium metal is resistant to attack by hydrochloric acid. This illustrates the important point that redox potential data only allow one to predict whether a reaction is possible in the thermodynamic sense; it is not possible to predict how fast a reaction will take place and, in the case of vanadium, attack by acid can be considered to take place extremely slowly, as a high energy barrier opposes the reaction (see chapter 9).

Since the action of acids on metals proceeds in the presence of oxygen (air) unless special precautions are taken to exclude it, we must now examine whether oxygen can have an effect on these reactions. The line drawn horizontally on the graph at +1·23 V corresponds to the redox potential for the system:

$$\tfrac{1}{2}O_2(g) + 2H^+(aq) + 2e^- \longrightarrow H_2O(l) \qquad E^\circ = +1{\cdot}23\text{ V}$$

Clearly any system which lies below this value is potentially capable of reducing oxygen to water, i.e. oxygen is potentially capable of oxidising the system. This explains why Cr^{2+} and Fe^{2+} ions are readily oxidised to Cr^{3+} and Fe^{3+} in solution, e.g.

$$\begin{aligned} Fe^{3+}(aq) + e^- &\longrightarrow Fe^{2+}(aq) & E^\circ &= +0{\cdot}76\text{ V} \\ \tfrac{1}{2}O_2(g) + 2H^+(aq) + 2e^- &\longrightarrow H_2O(l) & E^\circ &= +1{\cdot}23\text{ V} \\ \text{or}\quad 2Fe^{2+}(aq) + 2H^+(aq) + \tfrac{1}{2}O_2(g) &\longrightarrow 2Fe^{3+}(aq) + H_2O(l) & E^\circ_{total} &= +0{\cdot}47\text{ V} \end{aligned}$$

It also explains why $Ti^{3+}(aq)$ and $V^{3+}(aq)$ ions, which are stable in acid solution in the absence of air, oxidise to $TiO^{2+}(aq)$ and $VO^{2+}(aq)$ respectively if air is present (Fig. 23.1).

Since the redox potentials of the systems $Mn^{3+}(aq)/Mn^{2+}(aq)$ and $Co^{3+}(aq)/Co^{2+}(aq)$ lie above the value +1·23 V, neither $Mn^{3+}(aq)$ nor $Co^{3+}(aq)$ can be formed by the action of hydrochloric acid on the respective metals in the presence of air. In fact $Mn^{3+}(aq)$ and $Co^{3+}(aq)$ ions are potentially capable of liberating oxygen from water; MnF_3 and CoF_3 can be obtained and they do indeed liberate oxygen from water, with the formation of Mn^{2+}(aq) and Co^{2+}(aq) ions respectively.

The redox potentials so far considered relate to hydrated transition metal cations and it is found that these potentials change if the water molecules are displaced by other ligands which form stable complexes with these ions. For instance, the standard redox potential of the system $[Co(NH_3)_6]^{3+}/[Co(NH_3)_6]^{2+}$ is $+0{\cdot}10$ V compared with the value of $+1{\cdot}82$ V for the system $Co^{3+}(aq)/Co^{2+}(aq)$. Since the value $+0{\cdot}10$ V is less positive than the value $+1{\cdot}23$ V, oxygen should be potentially capable of oxidising $[Co(NH_3)_6]^{2+}$ to $[Co(NH_3)_6]^{3+}$ in solution and this is in fact found to be so:

$$Co(NH_3)_6{}^{3+}(aq) + e^- \longrightarrow Co(NH_3)_6{}^{2+}(aq) \qquad E^\ominus = +0{\cdot}10\ V$$
$$\tfrac{1}{2}O_2(g) + 2H^+(aq) + 2e^- \longrightarrow H_2O(l) \qquad E^\ominus = +1{\cdot}23\ V$$

or

$$2Co(NH_3)_6{}^{2+}(aq) + \tfrac{1}{2}O_2(g) + 2H^+(aq) \longrightarrow 2Co(NH_3)_6{}^{3+}(aq) + H_2O(l)$$
$$E^\ominus_{total} = +1{\cdot}13\ V$$

Complex ion formation is a general method used to stabilise oxidation states of elements which would otherwise be unstable.

As a final example of the use of redox potentials consider the following systems:

$$MnO_4{}^-(aq) + e^- \longrightarrow MnO_4{}^{2-}(aq) \qquad E^\ominus = +0{\cdot}564\ V$$
$$MnO_4{}^{2-}(aq) + 4H^+(aq) + 2e^- \longrightarrow MnO_2(s) + 2H_2O(l) \quad E^\ominus = +2{\cdot}260\ V$$

or

$$3MnO_4{}^{2-}(aq) + 4H^+(aq) \longrightarrow 2MnO_4{}^-(aq) + MnO_2(s) + 2H_2O(l)$$
$$E^\ominus_{total} = +1{\cdot}696\ V$$

A positive value for the e.m.f. of the above cell means that in 1 M acid solution manganate (VI) ions disproportionate into manganate (VII) ions and manganese (IV) oxide; this prediction is borne out in practice and manganate (VI) ions can only be prevented from disproportionating in aqueous solution by the addition of alkali.

23.4 Colour and degree of paramagnetism associated with transition metal complexes—an explanation based on crystal field theory

The colour of a transition metal ion is associated with
(a) an incomplete *d* level (between 1 and 9 *d* electrons),
(b) the nature of the ligands surrounding the ion.

The degree of paramagnetism shown by transition metal ions is dependent on the number of unpaired *d* electrons (p. 355); for instance the hydrated Mn^{2+} ion (5 unpaired 3*d* electrons) and the hydrated Fe^{2+} ion (4 unpaired 3*d* electrons) have numerical values of 5 and 4 respectively for their paramagnetic moments (in arbitrary units). However, in other complexes the degree of paramagnetism and the number of 3*d* electrons do not correlate at all, unless it is assumed that in such cases electron pairing occurs before all five 3*d* orbitals are singly occupied (a violation of Hund's rule of maximum multiplicity). For example, the $[Fe(CN)_6]^{4-}$ ion which contains Fe^{II} has no paramagnetic moment, and it must be assumed that the six 3*d* electrons pair and occupy only three of the five 3*d* orbitals.

An explanation of colour and paramagnetism is readily afforded by crystal field theory. This theory, first introduced by Bethe and van Vleck, was extended, mainly by Orgel, to cover transition metal chemistry. The theory starts with the assumption that, as far as transition metal complexes are concerned, the central transition metal ion can be regarded as a point charge, i.e. a charge concentrated into a very small volume, and that the surrounding ligands, be they anions or neutral molecules containing lone pairs of electrons, can likewise be represented as point charges. The bonding between the central metal ion and the surrounding ligands is assumed to be ionic.

Consider the case of a transition metal ion surrounded octahedrally by six ligands. Crystal field theory considers what effect the approach of these six ligands (along the x, y and z axes) have on the $3d$ orbitals of the transition metal ion. The orientations in space of the five $3d$ orbitals (which in the free ion are degenerate) are shown, the approaching ligands being denoted by the letter L (fig. 23.2).

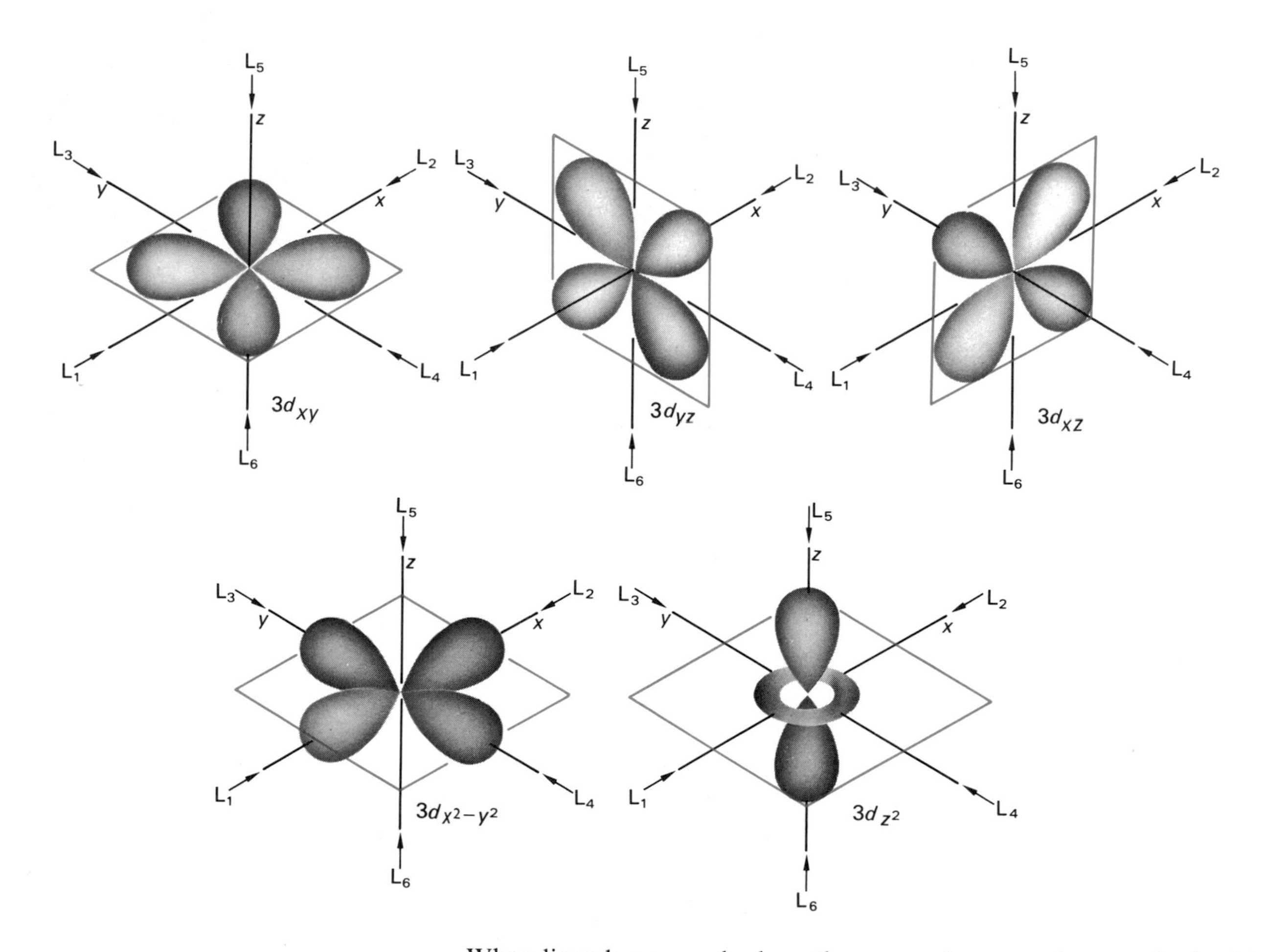

FIG. 23.2. *The splitting of the 3d levels when a transition metal ion is surrounded octahedrally by six ligands*

When ligands approach along the x, y and z axes, electrons in the $3d$ orbitals will be repelled but as can be seen from the above diagrams the effect will be greater for the $3d_{z^2}$ and $3d_{x^2-y^2}$ orbitals since these two orbitals have lobes lying along the line of approaching ligands. The net result is that the energy of the $3d_{z^2}$ and the $3d_{x^2-y^2}$ orbitals is raised relative to the energy of the $3d_{xy}$, $3d_{xz}$ and $3d_{yz}$ orbitals, i.e. the degeneracy of the $3d$ orbitals is now destroyed since their relative orientations in space preclude the possibility of equal interaction with the surrounding ligands; the splitting of the $3d$ levels is shown in fig. 23.3.

FIG. 23.3. *The splitting of the 3d levels in an octahedral environment of ligands*

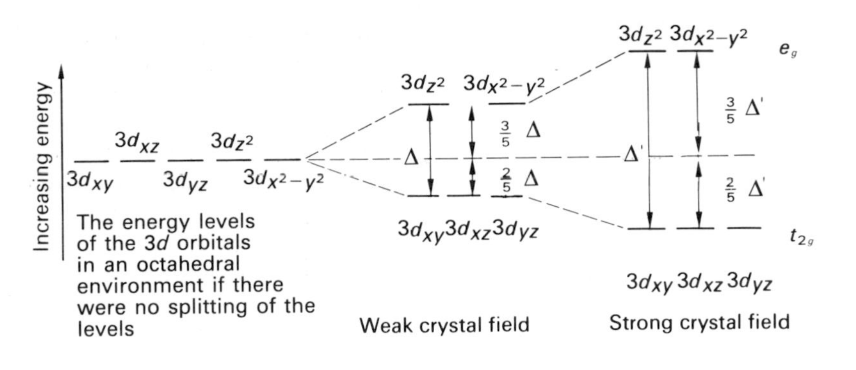

The 3*d* levels are split into an upper group of two (doubly degenerate and labelled e_g) and a lower group of three (trebly degenerate and labelled t_{2g}); the splitting of the levels is represented by the symbol Δ. If we reckon the zero of energy as the state of affairs that would obtain if each of the five 3*d* orbitals had interacted equally with the six ligands, then each of the upper two orbitals is raised by 3/5Δ (or 6/5Δ collectively) while each of the lower three orbitals is lowered by 2/5Δ (or 6/5Δ collectively). As the diagram shows, the degree of splitting depends upon the strength of the crystal field.

If we now consider the example of a transition metal ion with only one 3*d* electron surrounded octahedrally by six ligands, e.g. $[Ti(H_2O)_6]^{3+}$, then this single 3*d* electron will normally occupy one of the three degenerate lower levels (t_{2g}). In order to transfer this electron into an upper level (e_g) radiation of the appropriate frequency must be supplied. Transition metal ions are coloured because radiation in the visible spectrum is of the right frequency to promote this electronic transition; and in particular $[Ti(H_2O)_6]^{3+}$ ions are purple because green light (wavelength about 500 nm) is absorbed, i.e. white light minus green light gives purple light. The relationship between Δ and the frequency of light absorbed is given by the usual expression:

$$\Delta = h\nu$$

where h is Planck's constant and ν is the frequency of the light absorbed.

Similar considerations apply to complexes in which the central transition metal ion has more than one 3*d* electron although needless to say the presence of more than one electron in the 3*d* orbitals leads to slight complications; for instance, hydrated copper (II) ions are blue since light towards the red end of the visible spectrum is absorbed in bringing about the necessary electronic transition.

It is found experimentally that for a given transition series (in this case the first transition series) the value of Δ depends upon (a) the charge carried by the central transition metal ion, (b) the nature of the ligands, and (c) the transition metal ion itself. In general, for a given ligand, the crystal field splitting is greater for M^{3+} octahedral complexes than for M^{2+} octahedral complexes; while for transition metal ions carrying the same charge, the value of Δ increases in the order,

$I^- < Br^- < Cl^- < OH^- < F^- < H_2O < NH_3 <$ ethylenediamine $< NO_2^- < CN^-$,

where the above ions and neutral molecules are the ligands which may surround the transition metal ion. This order is known as **the spectrochemical series.**

Since small changes in the values of Δ can significantly affect the colour of light absorbed by transition metal ions, it is not surprising that transition metal ions can show a wide range of colours in different environments.

In order to explain why the same transition metal ion can often display two widely different degrees of paramagnetism in different environments it is necessary to consider the spectrochemical series. For instance, the CN^- ion produces a greater crystal field splitting than the water molecule; if the value of Δ is sufficiently large, electrons which would occupy each of the five separate 3*d* orbitals before beginning to pair (Hund's rule), crowd into the lower three degenerate levels (which can hold a maximum of six electrons) before the upper two degenerate levels begin to fill. The state of affairs for the $[Fe(H_2O)_6]^{2+}$ ion (which has a paramagnetic moment associated with four unpaired 3*d* electrons) and the $[Fe(CN)_6]^{4-}$ ion (which has zero paramagnetic moment indicating that all six 3*d* electrons are paired) is shown diagrammatically (fig. 23.4).

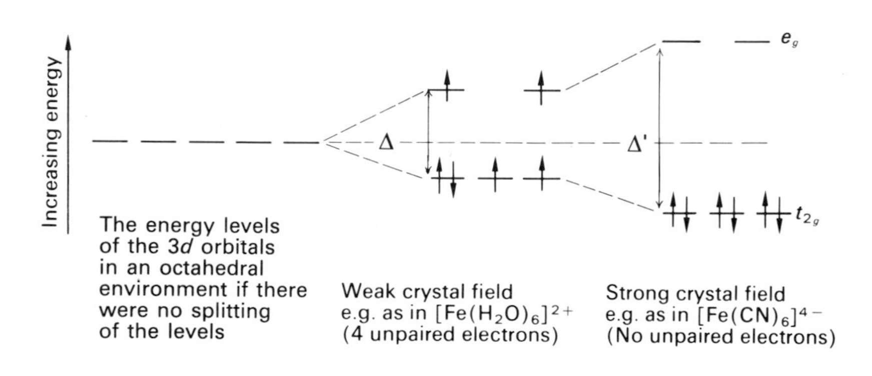

FIG. 23.4. *The occupation of the 3d levels for (a) a weak crystal field, (b) a strong crystal field*

Major anomalies in the values of paramagnetic moments associated with octahedral transition metal complexes can thus be explained in terms of the strength of the crystal field. Table 23b shows how the number of unpaired 3*d* electrons changes when a weak crystal field is replaced by a stronger one.

Transition metal ions that contain between four and seven 3*d* electrons have a different arrangement of these electrons, depending upon whether the crystal field is a weak or a strong one. Such transition metal ions have a larger number of unpaired electrons in a weak field than in a strong field and complexes are referred to as 'high spin' and 'low spin' respectively. Transition metal complexes containing less than four and more than seven 3*d* electrons have the same arrangement of 3*d* electrons irrespective of whether the crystal field is weak or strong.

Tetrahedral complexes, which are less frequent than the octahedral ones dealt with above, can also be discussed in terms of crystal field theory. A tetrahedral environment of ligands leads to an inversion of the 3*d* levels, i.e. the trebly degenerate t_{2g} orbitals are now of higher energy than the doubly degenerate e_g orbitals. Planar complexes can be regarded as distorted octahedral complexes and are discussed briefly in the next section.

Crystal field theory is able to account for the colour and paramagnetism of transition metal complexes. However, the basic assumption on which the theory is based, i.e. that the bonding between the transition metal ion

Table 23B The arrangement of 3*d* electrons in octahedral complexes

Number of 3*d* electrons	Weak field (High spin) t_{2g}	Weak field (High spin) e_g	Number of unpaired 3*d* electrons	Strong field (Low spin) t_{2g}	Strong field (Low spin) e_g	Number of unpaired 3*d* electrons
$3d^1$	↑		1	↑		1
$3d^2$	↑ ↑		2	↑ ↑		2
$3d^3$	↑ ↑ ↑		3	↑ ↑ ↑		3
$3d^4$	↑ ↑ ↑	↑	4	↑↓ ↑ ↑		2
$3d^5$	↑ ↑ ↑	↑ ↑	5	↑↓ ↑↓ ↑		1
$3d^6$	↑↓ ↑ ↑	↑ ↑	4	↑↓ ↑↓ ↑↓		0
$3d^7$	↑↓ ↑↓ ↑	↑ ↑	3	↑↓ ↑↓ ↑↓	↑	1
$3d^8$	↑↓ ↑↓ ↑↓	↑ ↑	2	↑↓ ↑↓ ↑↓	↑ ↑	2
$3d^9$	↑↓ ↑↓ ↑↓	↑↓ ↑	1	↑↓ ↑↓ ↑↓	↑↓ ↑	1

and the ligands is essentially ionic, is often far from true. Molecular orbital theory is able to explain colour and paramagnetism and in addition gives a satisfactory account of the nature of the bonding. Unfortunately the theory, when applied to transition metal complexes, is rather involved and will not be pursued here.

23.5 Planar complexes as distorted octahedral ones

Application of crystal field theory shows that an octahedral arrangement of ligands interacts more strongly with e_g orbitals than with t_{2g} orbitals. It might be expected, therefore, that an asymmetric occupation of the two e_g orbitals would be reflected in a distorted octahedral arrangement of ligands in a transition metal complex. The two possible distributions of the electrons in the e_g orbitals are shown below for a transition metal ion having nine 3*d* electrons (three electrons in the e_g orbitals).

↑↓ d_{z^2} ↑ $d_{x^2-y^2}$ ↑ d_{z^2} ↑↓ $d_{x^2-y^2}$

Consider the arrangement of two electrons in the d_{z^2} orbital and one electron in the $d_{x^2-y^2}$ orbital. The two electrons in the former orbital would be expected to interact more strongly with the ligands directed along the z axis than the one electron in the latter with ligands lying along the x and y axes. Thus the ligands in the xy plane would be drawn closer to the central transition metal ion than the ligands directed along the z axis, i.e. four short and two long bonds would be formed. If the occupation of the two e_g orbitals had been reversed it is easy to see that four long and two short bonds would have resulted. Many octahedrally co-ordinated compounds of copper (II) (nine 3*d* electrons) do indeed have four short bonds and two long ones, e.g. CuF_2 contains Cu^{2+} ions surrounded by six

F^- ions, four F^- ions being at a distance of 0·193 nm and the other two at a distance of 0·227 nm, so the configuration $(d_{z^2})^2(d_{x^2-y^2})^1$ seems to be preferred to the alternative $(d_{z^2})^1(d_{x^2-y^2})^2$.

Consider now the case of a transition metal ion containing eight $3d$ electrons in an octahedral environment. If the two electrons could be forced to pair in the d_{z^2} orbital, leaving the $d_{x^2-y^2}$ orbital vacant, then one would expect a further shortening of the metal-ligand bonds lying in the xy plane. In the extreme case one could regard the octahedral arrangement as distorting sufficiently for the arrangement of ligands to be regarded as essentially planar. Nickel (II), Ni^{2+}, ions have eight $3d$ electrons and in the presence of CN^- ions (which produce a large crystal field) the planar complex $Ni(CN)_4{}^{2-}$ is formed, thus indicating that the two electrons in the e_g orbitals can be forced into the d_{z^2} orbital. Hydrated nickel (II) ions, $Ni(H_2O)_6{}^{2+}$, adopt the octahedral configuration, since water molecules do not produce a sufficient crystal field to cause a significant distortion of this configuration.

23.6 Carbonyls of transition elements

Many transition metals form volatile compounds with carbon monoxide. Typical examples are:

mononuclear carbonyls	$Ni(CO)_4$	$Fe(CO)_5$	$Cr(CO)_6$
binuclear carbonyls	$Mn_2(CO)_{10}$	$Fe_2(CO)_9$	$Co_2(CO)_8$

Nickel carbonyl can be obtained by passing carbon monoxide over nickel at a temperature of about 60°C, whereas iron pentacarbonyl can only be obtained at elevated temperatures and pressures:

$$Ni(s) + 4CO(g) \longrightarrow Ni(CO)_4(l)$$
$$Fe(s) + 5CO(g) \longrightarrow Fe(CO)_5(l)$$

Other carbonyls are obtained by a variety of indirect methods.

The structures of most of the carbonyls are in agreement with those predicted on the basis of electron pair repulsion and are given on the next page.

The mononuclear carbonyls of nickel, iron and chromium, and the binuclear carbonyl of manganese have symmetrical structures (fig. 23.5). The binuclear carbonyls of iron, $Fe_2(CO)_9$, and cobalt, $Co_2(CO)_8$, contain bridging carbon monoxide units and are less symmetrical. The bonding between the carbon monoxide and the transition metal atom is essentially covalent and in the mononuclear carbonyls the M—C—O bonds are linear; since these bonds are somewhat shorter than expected for single bonds it is thought that the transition metal uses $3d$ electrons in 'back-bonding' to the carbon atom. The bonding can thus be represented as a resonance hybrid of the forms shown below:

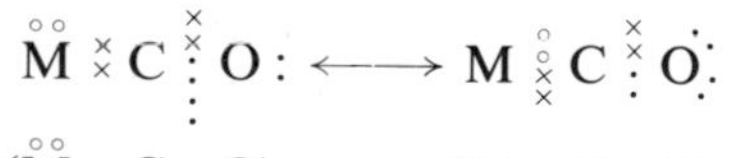

$$\overset{\circ\circ}{M} \; {}^{\times}_{\times} C \, \vdots \, O : \longleftrightarrow M \; {}^{\circ}_{\times} C \, {}^{\times}_{\cdot} \, O \!:$$

$$(\overset{\circ\circ}{M}{-}C{\equiv}O) \qquad (M{=}C{=}O)$$

∘ electron from metal atom
× electron from carbon atom
· electron from oxygen atom

FIG. 23.5. *The structures of some transition metal carbonyls*

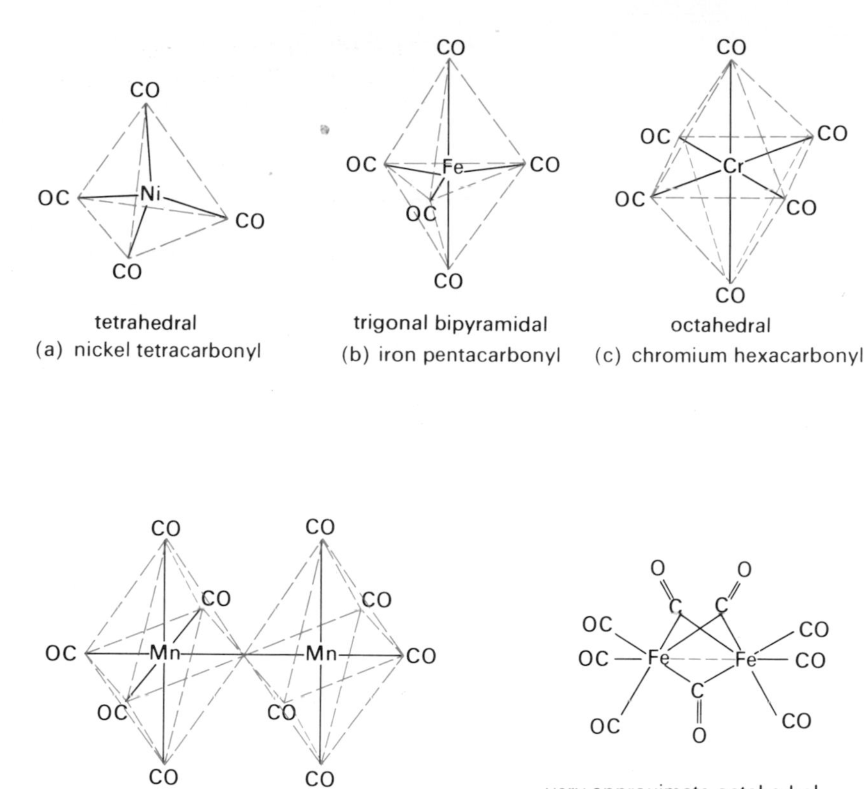

tetrahedral
(a) nickel tetracarbonyl

trigonal bipyramidal
(b) iron pentacarbonyl

octahedral
(c) chromium hexacarbonyl

octahedrons sharing corner

(d) dimeric manganese pentacarbonyl

very approximate octahedral arrangement about each Fe atom

(e) iron enneacarbonyl

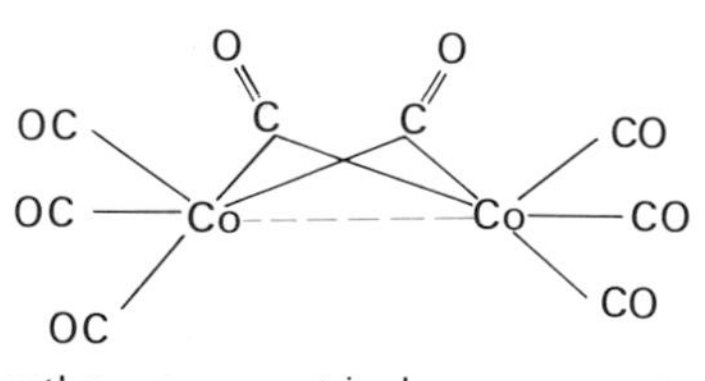

rather unsymmetrical arrangement

(f) dimeric cobalt tetracarbonyl

Nickel carbonyl and iron pentacarbonyl are liquids and highly toxic; the rest of the carbonyls listed in this section are solids. Nickel carbonyl is used in the Mond process for obtaining extremely pure nickel.

23.7 Suggested reading

G. A. Rodley and P. G. Hodgson, *Understanding colour of transition metal compounds*, Education in Chemistry, No. 5, Vol. 14, 1977.

23.8 Questions on chapter 23

1 Explain why aqueous solutions of the chlorides, sulphates and nitrates of transition metals show an acid reaction.

A transition metal forms hydrated ions $M(H_2O)_6^{2+}$ and $M(H_2O)_6^{3+}$.

Two separate solutions, the first 1 M with respect to the M^{2+} cations and the second 1 M with respect to the M^{3+} cations are made up; which solution would you expect to be the more acidic and why?

2 How do you explain the existence of oxycations such as TiO^{2+} and VO^{2+} in aqueous solution rather than hydrated Ti^{4+} and V^{4+} cations?

3 Use the standard redox potentials given below to answer the following questions.

$$Fe^{2+}(aq) + 2e^- \longrightarrow Fe(s) \quad E^\ominus = -0{\cdot}44\ V$$
$$Fe^{3+}(aq) + e^- \longrightarrow Fe^{2+}(aq) \quad E^\ominus = +0{\cdot}76\ V$$
$$H^+(aq) + e^- \longrightarrow \tfrac{1}{2}H_2(g) \quad E^\ominus = 0\ V$$
$$Ti^{2+}(aq) + 2e^- \longrightarrow Ti(s) \quad E^\ominus = -1{\cdot}63\ V$$
$$Ti^{3+}(aq) + e^- \longrightarrow Ti^{2+}(aq) \quad E^\ominus = -0{\cdot}37\ V$$
$$\tfrac{1}{2}O_2(g) + 2H^+(aq) + 2e^- \longrightarrow H_2O(l) \quad E^\ominus = +1{\cdot}23\ V$$

(a) What result would you predict for the action of dilute hydrochloric acid on
   (i) iron,
   (ii) iron (II) ions, both in the complete absence of air? Explain your reasons fully.

(b) What result would you predict for the action of dilute hydrochloric acid on
   (i) titanium,
   (ii) titanium (III) ions, both in the complete absence of air? Explain your reasons fully.

(c) What differences (if any) would you predict if air is not excluded? Give reasons.

4 Using the appropriate standard redox potentials given in question 3 above and the following,

$$Mn^{2+}(aq) + 2e^- \longrightarrow Mn(s) \quad E^\ominus = -1{\cdot}18\ V$$
$$Mn^{3+}(aq) + e^- \longrightarrow Mn^{2+}(aq) \quad E^\ominus = +1{\cdot}51\ V$$

predict what would happen if dilute hydrochloric acid is added to manganese
   (a) in the absence of air,
   (b) in the presence of air. Give your reasons. How would you cxpcct Mn^{3+} ions to react with water? Give reasons.

5 Give an account of the origin of the colour of transition metal salts in solution. Copper (II) ions are light blue but the colour darkens when ammonia is added to an aqueous solution of a copper (II) salt and then turns green if an excess of concentrated hydrochloric acid is added. Account for these observations in a qualitative sense.

6 Write down the number of 3*d* electrons in each of the following ions: Ti^{2+}, V^{3+}, Cr^{3+}, Mn^{2+}, Fe^{2+}, Fe^{3+} and Co^{2+}. Indicate how you would expect the five 3*d* orbitals to be occupied for these hydrated ions (octahedral).

7 The complex ion $[Fe(H_2O)_6]^{2+}$ has a paramagnetic moment associated with four unpaired electrons while the $[Fe(CN)_6]^{4-}$ complex ion has no paramagnetic moment. Similarly CoF_6^{3-} is paramagnetic while $[Co(CN)_6]^{3-}$ has zero paramagnetic moment. How do you account for these facts?

8 (a) The following are standard redox potentials in volts in 1 M acid solution for the reactions

$$M^{n+} + xe^- \rightarrow M^{(n-x)+} \text{ (symbolised as } M^{n+}/M^{(n-x)+}\text{)},$$

where, for example, the process

$$Na^+ + e^- \rightarrow Na \text{ (symbolised as } Na^+/Na\text{)}$$

is defined as having a large *negative* potential:

Cr^{2+}/Cr $-0{\cdot}9$ V, Mn^{2+}/Mn $-1{\cdot}2$ V,
Cr^{3+}/Cr^{2+} $-0{\cdot}4$ V, Mn^{3+}/Mn^{2+} $+1{\cdot}5$ V,
Fe^{2+}/Fe $-0{\cdot}4$ V,
Fe^{3+}/Fe^{2+} $+0{\cdot}8$ V.

Use this data to comment upon:
   (i) the stability in acid solution of Fe^{3+} towards reducing agents as compared to that of either Cr^{3+} or Mn^{3+};
   (ii) the ease with which metallic iron can be oxidised to iron(II) (ferrous) ions compared to the similar process for **either** metallic chromium **or** metallic manganese;
   (iii) the result of treating a solution containing **either** chromium(II) (chromous) **or** manganese(II) (manganous) ions with a solution containing iron(III) (ferric) ions.

Chapter 24

Group 1B copper, silver and gold

24.1 Some physical data of Group 1B elements

	Atomic Number	Electronic Configuration	Atomic Radius/nm	Ionic Radius/nm M^+	M.p. /°C	B.p. /°C	Density /g cm^{-3}	Standard Redox Potential/V	
								$M^+(aq) + e^- \rightarrow M(s)$	
Cu	29	2.8.18.1 ....$3s^23p^63d^{10}4s^1$	0·117	0·096	1083	2595	8·94	+0·52	+0·34 $Cu^{2+}(aq) + 2e^- \rightarrow Cu(s)$
Ag	47	2.8.18.18.1 ....$4s^24p^64d^{10}5s^1$	0·134	0·126	961	2212	10·5	+0·80	
Au	79	2.8.18.32.18.1 ....$5s^25p^65d^{10}6s^1$	0·134	0·137	1063	2966	19·3	+1·68	+1·42 $Au^{3+}(aq) + 3e^- \rightarrow Au(s)$

24.2 Some general remarks about Group 1B

The atoms of these three elements have one electron in the outer shell like the atoms of the Group 1A metals but, whereas the atoms of the latter group of elements have eight electrons in the penultimate shell, the atoms of copper, silver and gold have penultimate shells containing eighteen electrons:

Group 1A		Group 1B	
Potassium	2.8.8.1	Copper	2.8.18.1
Rubidium	2.8.18.8.1	Silver	2.8.18.18.1
Caesium	2.8.18.18.8.1	Gold	2.8.18.32.18.1

The fact that the atom of a Group 1B metal has a much larger atomic number but the same number of electronic shells as its Group 1A counterpart has important consequences and leads to these two groups of metals having a completely different chemistry. For instance, the nucleus of the copper atom carries a charge of 29+ and has a much greater influence on its surrounding electrons than does the nucleus of the potassium atom with a charge of 19+; consequently the copper atom has a smaller atomic radius and a larger first ionisation energy than the potassium atom. The densities of comparable members of the two groups are widely different

(a) because the Group 1B atoms are heavier and (b) because the atomic radii of Group 1B metals are smaller; the much higher melting points of the Group 1B metals can be attributed to (i) their heavier atoms, and (ii) to the fact that *d* electrons participate in inter-atomic bonding in addition to the single outer electron. The important differences in properties between Groups 1A and 1B are given in Table 24A:

Table 24A Some differences in properties between Group 1A and Group 1B

	K	Rb	Cs	Cu	Ag	Au
Atomic Radius/nm (M^+)	0·203	0·216	0·235	0·117	0·134	0·134
Ionisation Energy /kJ mol^{-1}	418	403	374	745	732	891
Standard Electrode Potential/V	−2·92	−2·99	−3·02	+0·52	+0·80	+1·68
Melting point/°C	63·5	39·0	28·5	1083	961	1063
Density/g cm^{-3}	0·86	1·53	1·90	8·94	10·5	19·3

The outstanding feature of copper, silver and gold is their resistance to chemical attack and this is reflected in their positive electrode potentials, whereas the Group 1A metals are noted for their high chemical reactivity, i.e. they have high negative electrode potentials. It is instructive at this stage to enquire into the reasons for these two widely divergent groups of electrode potentials.

Consider the standard electrode potentials of rubidium and silver

$$Rb^+(aq) + e^- \longrightarrow Rb(s) \qquad E^\ominus = -2{\cdot}99\,V$$
$$Ag^+(aq) + e^- \longrightarrow Ag(s) \qquad E^\ominus = +0{\cdot}80\,V$$

Rubidium, with a negative standard electrode potential, has a greater tendency to form hydrated ions than silver, with a positive standard electrode potential. The process of a metal passing into solution in the form of its ions can be considered to take place in three hypothetical stages, each one of which involving an energy change. (A more rigorous treatment would consider free energy changes (p. 113).

Metal (solid)	⟶	Metal (gaseous)	Enthalpy of atomisation *S* absorbed
Metal (gaseous)	⟶	Metal ion + e^-	Ionisation energy *I* absorbed
Metal ion + water	⟶	Metal ion (aq)	Enthalpy of hydration $\Delta H_h^\ominus$ evolved

The overall energy change is obtained by applying Hess's law to the three hypothetical equations. For rubidium we have:

$$Rb(s) \longrightarrow Rb(g) \qquad S = +86\,kJ\,mol^{-1}$$
$$Rb(g) \longrightarrow Rb^+(g) + e^- \qquad I = +403\,kJ\,mol^{-1}$$
$$Rb^+(g) + water \longrightarrow Rb^+(aq) \qquad \Delta H_h^\ominus = -289\,kJ\,mol^{-1}$$

The overall process is endothermic to the extent of 200 kJ mol^{-1}, thus

$$Rb(s) + water \longrightarrow Rb^+(aq) + e^- \qquad \Delta H^\ominus_{total} = +200\,kJ\,mol^{-1}$$

For silver the corresponding changes are:

$$\begin{aligned} Ag(s) &\longrightarrow Ag(g) & S &= +289\,kJ\,mol^{-1} \\ Ag(g) &\longrightarrow Ag^+(g) + e^- & I &= +732\,kJ\,mol^{-1} \\ Ag^+(g) + water &\longrightarrow Ag^+(aq) & \Delta H^\ominus_h &= -464\,kJ\,mol^{-1} \end{aligned}$$

The overall process is endothermic to the extent of 557 kJ mol^{-1}, i.e.

$$Ag(s) + water \longrightarrow Ag^+(aq) + e^- \qquad \Delta H^\ominus_{total} = +557\,kJ\,mol^{-1}$$

The process of rubidium metal passing into solution as its hydrated ions is less endothermic to the extent of about 357 kJ mol^{-1} than the similar process for silver and this large difference, neglecting entropy effects, accounts for the widely different electrode potentials of these two metals. The smaller enthalpy of atomisation of rubidium is counterbalanced by the larger enthalpy of hydration of the silver ion and it can be seen that the large difference in ionisation energy between these two metals is of prime importance. Similar considerations apply to the other pairs of metals, i.e. potassium/copper and caesium/gold.

Copper, silver and gold are known as the 'coinage metals' and are resistant to chemical attack (gold in particular is chemically very unreactive). All three metals exhibit an oxidation state of +1, and in this state, their compounds are largely covalent or at least possess a considerable degree of covalent character, although the principal oxidation states for copper, silver and gold are +2, +1 and +3 respectively. In the higher oxidation states the Group 1B metals utilise their *d* electrons in addition to their single *s* electrons; thus their ions contain an incomplete *d* level and are typically transitional (p. 350). Unlike the Group 1A metals, copper, silver and gold form numerous complexes in all their oxidation states.

Since there are no smooth gradations in properties along the series copper, silver and gold, these three metals are discussed separately.

Copper

24.3 Occurrence and extraction of copper

Although copper does not occur abundantly in nature, many copper-containing ores are known. It is principally extracted from copper pyrites, $CuFeS_2$, copper glance, Cu_2S, and cuprite, Cu_2O; it is also mined as the free element in Northern Michigan, in the USA. The extraction of copper from copper pyrites will be considered.

The pulverised ore is concentrated by the ore flotation process (p. 153) and roasted in a limited supply of air to convert the iron into iron (II) oxide:

$$2CuFeS_2(s) + 4O_2(g) \longrightarrow Cu_2S(s) + 3SO_2(g) + 2FeO(s)$$

After the addition of silica, SiO_2, the mixture is heated in the absence of air to convert the iron (II) oxide into a slag of iron (II) silicate, $FeSiO_3$, which is poured away. The copper (I) sulphide is now reduced to copper by heating in a controlled amount of air:

$$Cu_2S(s) + O_2(g) \longrightarrow 2Cu(s) + SO_2(g)$$

Copper is refined electrolytically to give a product of about 99·95 per cent purity. The impure copper is made the anode of an electrolytic cell which contains pure strips of copper as the cathode and an electrolyte of copper (II) sulphate solution. During electrolysis copper is transferred from the anode to the cathode; an anode sludge containing silver and gold is produced during this process, thus helping to make the process economically feasible.

24.4 Properties and uses of copper

Copper has a melting point of 1083°C and a density of $8{\cdot}94\,g\,cm^{-3}$. It is an attractive golden coloured metal, being very malleable and ductile, and its electrical and thermal conductivities are second only to those of silver. The metal is slowly attacked by moist air and its surface gradually becomes covered with an attractive green layer of basic copper carbonate. At about 300°C it is attacked by air or oxygen and a black coating of copper (II) oxide forms on its surface; at a temperature of about 1000°C copper (I) oxide is formed instead. Copper is also attacked by sulphur vapour, with the formation of copper (I) sulphide, and by the halogens which form the copper (II) halide (except iodine which forms copper (I) iodide).

The metal is not attacked by water or steam and dilute non-oxidising acids such as dilute hydrochloric and dilute sulphuric acid are without effect in the absence of an oxidising agent. Boiling concentrated hydrochloric acid attacks the metal, with the evolution of hydrogen, a surprising result in view of the positive electrode potential of copper; this phenomenon is due to the formation of the $CuCl_2^-(aq)$ ion which drives the equilibrium reaction to the right:

$$2Cu(s) + 2H^+(aq) \rightleftharpoons 2Cu^+(aq) + H_2(g)$$

$$+$$

$$4Cl^-(aq) \text{ (from the hydrochloric acid)}$$

$$\upharpoonleft\downharpoonright$$

$$2CuCl_2^-(aq)$$

Hot concentrated sulphuric acid attacks the metal and so too does dilute and concentrated nitric acid; the equations below do no more than indicate the main reactions since there are significant side reactions:

$$Cu(s) + \underset{\text{conc.}}{2H_2SO_4(l)} \longrightarrow CuSO_4(s) + 2H_2O(l) + SO_2(g)$$

$$3Cu(s) + \underset{\text{dil.}}{8HNO_3(aq)} \longrightarrow 3Cu(NO_3)_2(aq) + 4H_2O(l) + 2NO(g)$$

$$Cu(s) + \underset{\text{conc.}}{4HNO_3(aq)} \longrightarrow Cu(NO_3)_2(aq) + 2H_2O(l) + 2NO_2(g)$$

Copper is used for the windings of dynamos and for conveying electrical power; its resistance to chemical attack and its high thermal conductivity make it a useful metal for the construction of condensers for chemical plants and car radiators. Brass, an alloy of copper and zinc, is used for making cartridge containers, headlamp reflectors and the working parts of watches and clocks. Bronze, an alloy of copper and tin, is used for fabricating bearings and ships' fittings. Phosphor bronze, which contains some phosphorus, is used for watch springs and galvanometer suspensions. Finely divided copper is used as an industrial catalyst, in the oxidation of methanol to methanal.

24.5 Compounds of copper

Copper exhibits oxidation states of +2 (the most common) and +1 (only stable in aqueous solution if part of a stable complex ion). A few compounds containing copper (III) are known, e.g. K_3CuF_6, but this oxidation state of the metal is unimportant.

Copper (I) compounds

In aqueous solution the hydrated copper (I) ion is unstable and disproportionates into the copper (II) ion and copper, i.e. undergoes self oxidation-reduction; this is indicated by the standard redox potentials for the system $Cu^+(aq)/Cu(s)$ and $Cu^{2+}(aq)/Cu^+(aq)$ which are given below:

$$Cu^+(aq) + e^- \longrightarrow Cu(s) \qquad E^\ominus = +0{\cdot}52\ V$$

$$Cu^{2+}(aq) + e^- \longrightarrow Cu^+(aq) \qquad E^\ominus = +0{\cdot}16\ V$$

or

$$2Cu^+(aq) \longrightarrow Cu(s) + Cu^{2+}(aq) \qquad E^\ominus_{total} = +0{\cdot}36\ V$$

The positive e.m.f. of the above cell reaction implies that hydrated copper (I) ions are thermodynamically unstable in solution with respect to copper and hydrated copper (II) ions. The value of the equilibrium constant for this disproportionation reaction has been estimated to be in the order of $10^6\ dm^3\ mol^{-1}$ at 298 K.

$$2Cu^+(aq) \rightleftharpoons Cu(s) + Cu^{2+}(aq)$$

$$K = [Cu^{2+}(aq)]/[Cu^+(aq)]^2 = 10^6\ dm^3\ mol^{-1}$$

i.e. the concentration of $Cu^+(aq)$ ions in solution is extremely minute. The equilibrium can be shifted to the left by adding anions which precipitate out an insoluble copper (I) compound, e.g. I^- ions precipitate insoluble CuI, or by adding a substance which forms a more stable complex ion with Cu^+ than with Cu^{2+}, e.g. ammonia.

Although the chemistry of copper (I) is largely that of its water insoluble compounds and of its stable complexes, other copper (I) compounds are perfectly stable in the absence of moisture, e.g. copper (I) sulphate, Cu_2SO_4.

Copper (I) oxide, Cu_2O

Copper (I) oxide is obtained as a red solid by the reduction of an alkaline solution of copper (II) sulphate. Since the addition of alkali to a solution of a copper (II) salt would result in the precipitation of copper (II) hydroxide, the copper (II) ions are complexed with tartrate ions; under these conditions the $Cu^{2+}(aq)$ ions are present in such low concentration that the solubility product of copper (II) hydroxide is not exceeded.

The experimental procedure is as follows: a solution of copper (II) sulphate is added to an alkaline solution of sodium potassium tartrate, when a deep blue solution is obtained, the colour being due to the presence of a copper (II)-tartrate complex. The solution is warmed with a solution of glucose (reducing agent) when a red deposit of copper (I) oxide is obtained.

Copper (I) oxide reacts with dilute sulphuric acid on warming to give a solution of copper (II) sulphate and a deposit of copper, i.e. disproportionation occurs:

$$Cu_2O(s) + 2H^+(aq) \longrightarrow Cu^{2+}(aq) + Cu(s) + H_2O(l)$$

It dissolves in concentrated hydrochloric acid with the formation of the $CuCl_2^-(aq)$ complex ion.

Copper (I) oxide is a covalent solid, each oxygen atom being surrounded by four copper atoms and each copper atom lying midway between two oxygen atoms (4:2 co-ordination).

Copper (I) chloride, CuCl

This is a white solid which is insoluble in water. It can be obtained by boiling a solution containing copper (II) chloride, excess of copper turnings and concentrated hydrochloric acid. Copper (I) is present in this solution as the $CuCl_2^-(aq)$ complex ion:

$$\begin{array}{rl} Cu(s) + Cu^{2+}(aq) \rightleftharpoons & 2Cu^+(aq) \\ & + \\ & 4Cl^-(aq) \\ & \updownarrow \\ & 2CuCl_2^-(aq) \end{array}$$

On pouring the solution into air-free distilled water, copper (I) chloride precipitates as a white solid. It must be rapidly washed, dried and sealed in the absence of air, since a combination of air and moisture oxidises it to copper (II) chloride.

Copper (I) chloride is essentially covalent and its structure is similar to that of diamond, i.e. each copper atom is surrounded tetrahedrally by four chlorine atoms and each chlorine atom is surrounded tetrahedrally by four copper atoms; in the vapour phase dimeric and trimeric species are present, i.e. $(CuCl)_2$ and $(CuCl)_3$. It is soluble in water in the presence of entities such as Cl^-, $S_2O_3^{2-}$ and NH_3, with which it forms complex ions, e.g. $Cu(NH_3)_2^+(aq)$.

Copper (I) chloride is used in conjunction with ammonium chloride as a catalyst in the dimerisation of ethyne to but-1-ene-3-yne (vinyl acetylene), which is used in the production of synthetic rubber. In the laboratory a mixture of copper (I) chloride and hydrochloric acid is used for converting benzene diazonium chloride into chlorobenzene (Sandmeyer reaction):

$$C_6H_5N_2^+Cl^-(aq) \xrightarrow{CuCl/HCl} C_6H_5Cl(l) + N_2(g)$$

An ammoniacal solution of copper (I) chloride absorbs carbon monoxide and precipitates copper (I) carbide when ethyne is bubbled through it; the latter reaction can be used for purifying ethyne since the carbide readily evolves ethyne when treated with dilute acid.

Other copper (I) halides

Copper (I) fluoride has never been prepared, presumably because fluorine is such a good oxidising agent that it would immediately oxidise it to copper (II) fluoride.

Copper (I) bromide, CuBr, can be made by a method similar to that used for making copper (I) chloride. It resembles the chloride in most respects.

Copper (I) iodide, CuI, precipitates as a white solid when an aqueous solution of a copper (II) salt is treated with an aqueous solution of potassium iodide, so that copper (II) iodide is unstable with respect to copper (I) iodide and iodine:

$$2Cu^{2+}(aq) + 4I^{-}(aq) \longrightarrow 2CuI(s) + I_2(aq)$$

The above reaction can be used to estimate the concentration of a copper (II) salt, since the quantitative formation of iodine can be determined by titration with standard sodium thiosulphate solution (p. 318).

Copper (I) sulphate, Cu_2SO_4

Copper (I) sulphate is obtained as a white solid when copper (I) oxide is heated with anhydrous dimethyl sulphate:

$$Cu_2O(s) + (CH_3)_2SO_4(l) \longrightarrow Cu_2SO_4(s) + (CH_3)_2O(l)$$

In the absence of moisture it is stable but, on dissolving, it disproportionates rapidly into copper (II) sulphate and copper:

$$Cu_2SO_4(s) + H_2O(l) \longrightarrow CuSO_4(aq) + Cu(s)$$

Copper (II) compounds

This is the most common oxidation state of copper and in aqueous solution copper (II) salts are blue, the colour being due to the presence of $Cu(H_2O)_6{}^{2+}(aq)$ ions.

Copper (II) oxide, CuO

Copper (II) oxide may be obtained as a black solid by heating either copper (II) carbonate (actually a basic salt) or copper (II) nitrate:

$$CuCO_3(s) \longrightarrow CuO(s) + CO_2(g)$$
$$2Cu(NO_3)_2(s) \longrightarrow 2CuO(s) + 4NO_2(g) + O_2(g)$$

On heating to a temperature of about 800°C it decomposes into copper (I) oxide and oxygen:

$$4CuO(s) \longrightarrow 2Cu_2O(s) + O_2(g)$$

It reacts readily with dilute mineral acids, on warming, with the formation of copper (II) salts; and is also easily reduced to copper on heating in a stream of hydrogen:

$$CuO(s) + 2H^{+}(aq) \longrightarrow Cu^{2+}(aq) + H_2O(l)$$
$$CuO(s) + H_2(g) \longrightarrow Cu(s) + H_2O(l)$$

The detection of carbon and hydrogen in organic compounds can be achieved by heating with dry copper (II) oxide, when carbon dioxide and water are formed.

Copper (II) hydroxide, $Cu(OH)_2$

Copper (II) hydroxide is precipitated as a blue-green gelatinous solid when an aqueous solution of a copper (II) salt is made alkaline with sodium hydroxide solution. It can be filtered and dried at 100°C to the composition $Cu(OH)_2$; however, heating the unfiltered suspension to about 80°C results in decomposition into copper (II) oxide and water.

Copper (II) hydroxide is readily soluble in an aqueous solution of ammonia with the formation of the intensely blue $Cu(NH_3)_4{}^{2+}(aq)$ ion:

$$\begin{array}{ccc} Cu(OH)_2(s) \rightleftharpoons & Cu^{2+}(aq) & + \ 2OH^-(aq) \\ & + & \\ & 4NH_3(aq) & \\ & \updownarrow & \\ & Cu(NH_3)_4{}^{2+}(aq) & \end{array}$$

It reacts readily with dilute acids to give copper (II) salts but, since the freshly precipitated solid is also slightly soluble in sodium hydroxide solution, it must be considered to be somewhat amphoteric:

$$Cu(OH)_2(s) + 2H^+(aq) \longrightarrow Cu^{2+}(aq) + 2H_2O(l)$$
$$Cu(OH)_2(s) + 2OH^-(aq) \rightleftharpoons Cu(OH)_4{}^{2-}(aq)$$

Copper (II) sulphide, CuS

Copper (II) sulphide precipitates as a black solid when hydrogen sulphide is passed into an aqueous solution of a copper (II) salt, even in the presence of a large concentration of hydrogen ions, since it has a very low solubility product:

$$Cu^{2+}(aq) + H_2S(g) \longrightarrow CuS(s) + 2H^+(aq)$$

Copper (II) chloride, $CuCl_2$

The anhydrous solid is dark brown in colour and can be obtained by passing chlorine over heated copper. It is predominantly covalent and adopts a layer structure in which each copper atom is surrounded by four chlorine atoms at a distance of 0·230 nm and two more at a distance of 0·295 nm (fig. 24.1).

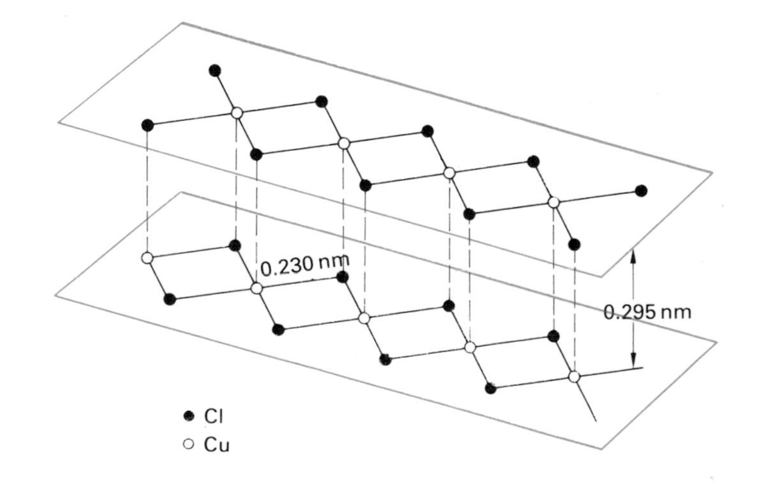

FIG. 24.1. *The layer structure of anhydrous copper (II) chloride (two layers only are shown)*

Copper (II) chloride is very soluble in water; a concentrated aqueous solution is dark brown, the colour being due to the presence of complex ions such as $CuCl_4^{2-}(aq)$, but with dilution the colour changes to green and then to blue. These colour changes are due to the successive replacement of chloride ions in the complexes by water molecules, the final colour being that of the $Cu(H_2O)_6^{2+}(aq)$ ion. The dihydrate, $CuCl_2.2H_2O$, which is a green solid, can be obtained by crystallising the solution.

Other copper (II) halides

Copper (II) fluoride, CuF_2, is a colourless solid and unlike the chloride is considered to have an ionic structure.

Copper (II) bromide, $CuBr_2$, is a black crystalline solid which resembles the corresponding chloride structurally and in other respects. In aqueous solution a number of complex species are present but on dilution the blue colour of the hydrated copper (II) ion predominates (see above).

Any attempt to prepare copper (II) iodide results in the formation of copper (I) iodide and iodine (p. 409). Thus iodine is not sufficiently powerful as an oxidising agent to convert copper (I) to copper (II).

Copper (II) sulphate, $CuSO_4.5H_2O$

Copper (II) sulphate may be prepared by reacting either copper (II) oxide or copper (II) carbonate with dilute sulphuric acid; the solution is heated to obtain a saturated solution and the blue solid pentahydrate separates on cooling (a few drops of dilute sulphuric acid are generally added before heating in order to prevent hydrolysis). On an industrial scale, copper (II) sulphate is obtained by forcing air through a hot mixture of copper and dilute sulphuric acid:

$$2Cu(s) + 4H^+(aq) + O_2(g) \longrightarrow 2Cu^{2+}(aq) + 2H_2O(l)$$

Copper (II) sulphate pentahydrate loses four of its molecules of water of crystallisation on heating to about 100°C; the fifth molecule of water is lost at a temperature of about 250°C:

$$CuSO_4.5H_2O(s) \xrightleftharpoons{100°C} CuSO_4.H_2O(s) + 4H_2O(l)$$

$$\xrightleftharpoons{250°C} CuSO_4(s) + 5H_2O(l)$$

The anhydrous salt decomposes into copper (II) oxide and sulphur trioxide on really strong heating:

$$CuSO_4(s) \longrightarrow CuO(s) + SO_3(g)$$

In the solid pentahydrate each copper (II) ion is surrounded by four water molecules arranged at the corners of a square; the fifth and sixth octahedral positions are occupied by oxygen atoms from sulphate anions and the fifth water molecule is held in position by hydrogen bonding

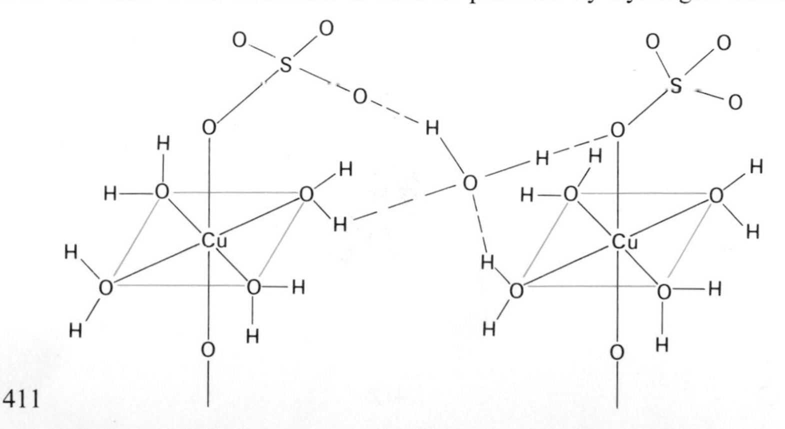

FIG. 24.2. *The partial structure of copper (*II*) sulphate pentahydrate*

Copper (II) sulphate has many industrial uses, including the preparation of Bordeaux mixture (a fungicide) and the preparation of other copper compounds. It is also used in electroplating solutions, in textile dyeing, and as a timber preservative.

Copper (II) nitrate, $Cu(NO_3)_2.3H_2O$
Copper (II) nitrate is a blue deliquescent solid which may be obtained by reacting either copper (II) oxide or copper (II) carbonate with dilute nitric acid and crystallising the resulting solution. Other hydrates containing six and nine molecules of water of crystallisation are known. On heating, it readily decomposes into copper (II) oxide, nitrogen dioxide and oxygen.

Anhydrous copper (II) nitrate can be obtained by reacting copper with a solution of nitrogen dioxide in ethyl ethanoate followed by isolation of the solid product and decomposition of the solvate at 90°C. It sublimes on heating to about 200°C and this seems to suggest that the anhydrous solid is appreciably covalent.

Copper (II) carbonate, $aCuCO_3.bCu(OH)_2$
The pure substance is unknown and any method of preparation, e.g. the addition of either sodium carbonate or sodium hydrogen carbonate solutions to a solution of a copper (II) salt, results in the formation of a green basic carbonate. The ore malachite is $CuCO_3.Cu(OH)_2$; another basic carbonate has the composition $2CuCO_3.Cu(OH)_2$.

Basic copper (II) carbonate is insoluble in water but dissolves in ammonia solution, since the dark blue $Cu(NH_3)_4^{2+}(aq)$ ion is formed and the reaction (see below) moves to the right:

$$CuCO_3(s) \rightleftharpoons Cu^{2+}(aq) + CO_3^{2-}(aq)$$
$$+$$
$$4NH_3(aq)$$
$$\upharpoonleft\downharpoonright$$
$$Cu(NH_3)_4^{2+}(aq)$$

Silver

24.6 Occurrence and extraction of silver

Silver occurs as silver sulphide (silver glance), Ag_2S, and silver chloride (horn silver), AgCl; it also occurs as the free element. Significant amounts of silver are also obtained during the extraction of lead from its ores and the electrolytic refining of copper.

One method of extraction of the metal involves the treatment of the pulverised ore with an aerated solution of sodium cyanide; the silver is taken into solution as the $Ag(CN)_2^-(aq)$ complex. Addition of metallic zinc, which forms a stable cyanide complex, results in the precipitation of silver:

$$2Ag_2S(s) + 8CN^-(aq) + O_2(g) + 2H_2O(l) \longrightarrow 4Ag(CN)_2^-(aq) + 2S(s) + 4OH^-(aq)$$

$$2Ag(CN)_2^-(aq) + Zn(s) \longrightarrow 2Ag(s) + Zn(CN)_4^{2-}(aq)$$

The pure metal is obtained by electrolysis of a solution containing both silver nitrate and nitric acid; the impure silver is the anode and a pure strip of silver serves as the cathode.

24.7 Properties and uses of silver

Silver has a melting point of 961°C and a density of $10{\cdot}5\,g\,cm^{-3}$. It is a white lustrous metal and is very malleable and ductile; its thermal and electrical conductivities exceed those of copper and, indeed, it is the best conductor known. The metal is resistant to attack by air and moisture, although the presence of hydrogen sulphide results in the familiar black stain of silver sulphide. Steam and dilute non-oxidising acids are without effect on the metal; however, it is attacked by hot concentrated sulphuric acid and cold dilute nitric acid, with the formation of silver (I), $Ag^+(aq)$, ions:

$$\underset{\text{conc.}}{2Ag(s) + 2H_2SO_4(l)} \longrightarrow Ag_2SO_4(s) + SO_2(g) + 2H_2O(l)$$

$$\underset{\text{dil.}}{3Ag(s) + 4HNO_3(aq)} \longrightarrow 3AgNO_3(aq) + 2H_2O(l) + NO(g)$$

Concentrated nitric acid produces mainly nitrogen dioxide.

Silver is still used in coinage as an alloy with copper (not in Britain). Large quantities are also used for the manufacture of tableware and jewellery. Silver mirrors are made by the reduction of the complex ion, $Ag(NH_3)_2{}^+(aq)$ with methanal (the silver mirror test for aldehydes is the same basic reaction). The deposition of silver on cheaper articles (usually alloys containing copper) is carried out to produce silverplated cutlery. In this process the article to be plated is the anode and a pure strip of silver is the cathode in an electrolytic cell containing a solution of sodium dicyanoargentate (I), $Na^+[Ag(CN)_2]^-$. The presence of only a very low concentration of silver ions in solution favours the formation of a strong film of silver on the article to be plated:

$$Ag(CN)_2{}^-(aq) \rightleftharpoons Ag^+(aq) + 2CN^-(aq) \text{ (equilibrium over to left)}$$
$$Ag^+(aq) + e^- \longrightarrow Ag(s) \text{ (occurs at the cathode)}$$

24.8 Compounds of silver

The important oxidation state of silver is +1 although compounds of Ag(II) (e.g. AgF_2) and Ag(III) ($K^+[AgF_4]^-$) are known.

Compounds containing either silver (III) or silver (II) are powerful oxidising agents. The instability of $Ag^{2+}(aq)$ with respect to $Ag^+(aq)$ in aqueous solution is indicated by the standard redox potentials for the systems $Ag^{2+}(aq)/Ag^+(aq)$ and $Ag^+(aq)/Ag(s)$ which are given below:

$$Ag^{2+}(aq) + e^- \longrightarrow Ag^+(aq) \qquad E^\ominus = +1{\cdot}98\,V$$
$$Ag^+(aq) + e^- \longrightarrow Ag(s) \qquad E^\ominus = +0{\cdot}80\,V$$
$$\text{or} \quad Ag^{2+}(aq) + Ag(s) \longrightarrow 2Ag^+(aq) \qquad E^\ominus_{total} = +1{\cdot}18\,V$$

The positive e.m.f. of the above cell reaction implies that $Ag^{2+}(aq)$ ions are thermodynamically capable of oxidising silver to $Ag^+(aq)$ ions. Notice that the disproportionation of $Ag^+(aq)$ into $Ag^{2+}(aq)$ and Ag cannot take place since the e.m.f. of this cell reaction is negative (cf. the disproportionation of $Cu^+(aq)$ into $Cu^{2+}(aq)$ and Cu (p. 407)).

Silver (I) compounds

Since the compounds of silver of any importance are those that contain silver (I), it is usual to refer to silver (I) compounds simply as silver compounds.

Silver oxide, Ag_2O

The addition of sodium hydroxide solution to a solution of a silver salt results in the precipitation of brown silver oxide. It is possible that the hydroxide is formed initially and then decomposes into the oxide and water:

$$2Ag^+(aq) + 2OH^-(aq) \longrightarrow 2AgOH$$
$$2AgOH \longrightarrow Ag_2O(s) + H_2O(l)$$

Silver oxide turns moist red litmus blue, i.e. in the presence of water it produces some OH^- ions. The moist oxide is used in organic chemistry for converting alkyl halides into the corresponding alcohol:

$$2RI + Ag_2O(s) + H_2O(l) \longrightarrow 2ROH + 2AgI(s)$$

The use of dry silver oxide results in the formation of an ether:

$$2RI + Ag_2O(s) \longrightarrow R{-}O{-}R + 2AgI(s)$$

where R = alkyl group.

Silver oxide is readily soluble in ammonia solution, forming the $Ag(NH_3)_2^+(aq)$ complex ion:

$$Ag_2O(s) + H_2O(l) + 4NH_3(aq) \longrightarrow 2Ag(NH_3)_2^+(aq) + 2OH^-(aq)$$

This complex ion is readily reduced to silver by aldehydes and 'reducing' sugars, e.g. glucose, and this reaction is used in testing for aldehyde groupings. The solution containing the complex should be washed away immediately after use, since explosions have occurred with this reagent. It has been suggested that explosive silver nitride, Ag_3N, is formed on standing.

Silver oxide is thermally unstable and gentle heating is sufficient to cause decomposition into silver and oxygen:

$$2Ag_2O(s) \longrightarrow 4Ag(s) + O_2(g)$$

Silver sulphide, Ag_2S

The familiar black stain that forms on the surface of silver articles which have been exposed to an atmosphere containing hydrogen sulphide is silver sulphide. It precipitates as a black solid when hydrogen sulphide is bubbled through a solution of a silver salt:

$$2Ag^+(aq) + H_2S(g) \longrightarrow Ag_2S(s) + 2H^+(aq)$$

Since it is exceedingly insoluble in water, it can be precipitated from solutions containing only a minute concentration of silver ions, e.g. from solutions in which the silver ions are bound in the form of stable complexes (see p. 417).

Silver halides, AgX

Silver fluoride, AgF, is a white, water-soluble solid and is considered to have an essentially ionic structure. Unlike the fluoride, the other silver halides are only sparingly soluble in water and their solubilities decrease in the order,

$$AgCl > AgBr > AgI$$

These last three halides are appreciably covalent, their covalent character increasing with increasing atomic number of the halogen atom. They can all be precipitated by mixing solutions containing silver ions and halide ions and this is a convenient test for the halide ion in solution (p. 339):

$$Ag^{+}(aq) + Cl^{-}(aq) \longrightarrow \underset{\text{white}}{AgCl(s)}$$

$$Ag^{+}(aq) + Br^{-}(aq) \longrightarrow \underset{\text{cream}}{AgBr(s)}$$

$$Ag^{+}(aq) + I^{-}(aq) \longrightarrow \underset{\text{yellow}}{AgI(s)}$$

Silver chloride is soluble in ordinary bench ammonia solution, whereas silver bromide dissolves if the concentration of the ammonia solution is increased; silver iodide is virtually insoluble in ammonia solution of any concentration. These observations show the decreasing solubilities of these three halides in water. Thus the addition of ammonia results in silver ions passing into solution as the $Ag(NH_3)_2{}^{+}(aq)$ complex ion:

$$Ag^{+}(aq) + 2NH_3(aq) \rightleftharpoons Ag(NH_3)_2{}^{+}(aq)$$

The fact that silver iodide does not dissolve in ammonia solution means that there are sufficient $Ag^{+}(aq)$ ions in solution for the solubility product of silver iodide to be exceeded. Since silver chloride dissolves under these conditions, the solubility product of silver chloride cannot be exceeded and it must be larger than that of silver iodide. Similarly, the solubility product of silver bromide must be intermediate in value between those of silver chloride and silver iodide.

Silver chloride, bromide and iodide are sensitive to light and decompose slowly, with the formation of a black deposit of silver, e.g.

$$2AgBr(s) \longrightarrow 2Ag(s) + Br_2(l)$$

This reaction occurs in photography (p. 417).

Silver nitrate, $AgNO_3$

This is the most important compound of silver and is prepared industrially by the action of nitric acid on silver. It is very soluble in water, giving a colourless solution which contains $Ag^{+}(aq)$ ions. The solid, which is also colourless, readily decomposes on heating into silver, oxygen and nitrogen dioxide:

$$2AgNO_3(s) \longrightarrow 2Ag(s) + 2NO_2(g) + O_2(g)$$

An ammoniacal solution of silver nitrate contains the $Ag(NH_3)_2{}^{+}(aq)$ complex ion, and this solution reacts with ethyne, forming a precipitate of silver carbide, Ag_2C_2, which explodes if struck when dry:

$$C_2H_2(g) + 2Ag(NH_3)_2{}^{+}(aq) \longrightarrow Ag_2C_2(s) + 2NH_4{}^{+}(aq) + 2NH_3(aq)$$

The action of dilute nitric acid on silver carbide regenerates pure ethyne and this reaction may be used to purify this hydrocarbon, e.g. to separate it from alkanes and alkenes.

Silver nitrate is used industrially for the preparation of other silver compounds, particularly silver halides which are used in photography (p. 417). In the laboratory, a standard solution of silver nitrate is used for the volumetric estimation of chloride, bromide and iodide ions in neutral solution; one method uses a 5 per cent solution of potassium chromate (VI) as indicator. The procedure is as follows:

A measured amount of the neutral halide solution, e.g. sodium chloride, is taken and about 5 cm^3 of 5 per cent potassium chromate (VI) solution are added to act as indicator. Standard silver nitrate is added from a burette, when white silver chloride precipitates; at the end-point the slight excess of silver ions precipitates silver chromate (VI) (pink). Silver chromate (VI) is more soluble than silver chloride and thus does not precipitate until all the chloride ions have precipitated as silver chloride. Since silver chromate (VI) does not precipitate in acid solution the halide solution must be neutral (alkaline solutions would lead to the precipitation of silver oxide).

$$Ag^+(aq) + Cl^-(aq) \longrightarrow \underset{\text{white}}{AgCl(s)}$$

$$\underset{\text{slight excess}}{2Ag^+(aq)} + CrO_4^{2-}(aq) \longrightarrow \underset{\text{pink}}{Ag_2CrO_4(s)}$$

Silver sulphate, Ag_2SO_4

Silver sulphate precipitates as a white solid when a solution of a sulphate is added to a fairly concentrated solution of silver nitrate, i.e. it is moderately insoluble:

$$2Ag^+(aq) + SO_4^{2-}(aq) \longrightarrow Ag_2SO_4(s)$$

It is of no real importance.

Silver carbonate, Ag_2CO_3

Silver carbonate precipitates as a pale yellow solid when a solution of an alkali metal carbonate is added to a solution containing silver ions:

$$2Ag^+(aq) + CO_3^-(aq) \longrightarrow Ag_2CO_3(s)$$

Like all silver salts it is sensitive to light and gradually darkens with the deposition of silver. The solid readily decomposes on heating into silver, carbon dioxide and oxygen:

$$2Ag_2CO_3(s) \longrightarrow 4Ag(s) + 2CO_2(g) + O_2(g)$$

Complexes containing silver (I)

Typical complexes containing silver (I) include $[Ag(NH_3)_2]^+$, $[Ag(S_2O_3)_2]^{3-}$ and $[Ag(CN)_2]^-$; the thiosulphate complex is of importance in the 'fixing' process in photography (p. 417), while the cyanide complex is used during the extraction of silver (p. 412). The stability of these three complexes decreases in the order,

$$[Ag(CN)_2]^- > [Ag(S_2O_3)_2]^{3-} > [Ag(NH_3)_2]^+$$

The cyanide and thiosulphate complexes can be shown to be more stable in aqueous solution than the diammine complex by relatively simple experiments. Thus the addition of chloride ions to a solution of a silver salt results in the formation of a white precipitate of silver chloride which readily dissolves in dilute ammonia solution. The addition of bromide ions to this solution results in the precipitation of silver bromide, i.e. although the equilibrium shown below is well over to the right, there are sufficient 'free' silver ions present to exceed the solubility product of silver bromide.

$$Ag^+(aq) + 2NH_3(aq) \rightleftharpoons Ag(NH_3)_2^+(aq)$$

The addition of solutions of either sodium cyanide or sodium thiosulphate results in the precipitate of silver bromide dissolving, i.e. the concentration of 'free' silver ions is reduced to such a low value that the solubility product of silver bromide is no longer exceeded. The stability constants (equilibrium constants) at 298 K for solutions of the three complexes are given below:

$$Ag^+(aq) + 2NH_3(aq) \rightleftharpoons Ag(NH_3)_2^+(aq) \quad K = 1{\cdot}7 \times 10^7\ dm^6\ mol^{-2}$$
$$Ag^+(aq) + 2S_2O_3^{2-}(aq) \rightleftharpoons Ag(S_2O_3)_2^{3-}(aq) \quad K = 1{\cdot}7 \times 10^{13}\ dm^6\ mol^{-2}$$
$$Ag^+(aq) + 2CN^-(aq) \rightleftharpoons Ag(CN)_2^-(aq) \quad K = 5{\cdot}5 \times 10^{18}\ dm^6\ mol^{-2}$$

Since the passage of hydrogen sulphide through an aqueous solution of the cyanide complex results in the precipitation of silver sulphide, the solubility product of silver sulphide must be exceedingly small, i.e. the minute concentration of 'free' silver ions in this solution is sufficient to cause silver sulphide to precipitate.

24.9 The photographic process

Photographic film is made from celluloid on which a layer of gelatine is deposited. The gelatine contains silver bromide (or silver iodide) which is held in colloidal suspension. During the short exposure of photographic film, sub-microscopic particles of silver are formed; the film is now developed in the dark by adding an organic reducing agent which converts some of the silver bromide to silver. The rate at which silver is formed in this process depends upon the initial intensity of illumination of the film during the exposure period. The darker regions of the film, therefore, are those that received the greatest intensity of illumination. The film is now washed with a solution of sodium thiosulphate which removes unchanged silver bromide as a complex ion (see section 24.8). The negative can now be handled in daylight. In order to make a print from the negative it is exposed to illumination (in an otherwise darkened room) and the light passing through the negative is allowed to fall onto photographic paper, which also contains silver bromide. The negative image is therefore reversed, and the dark areas on the negative subsequently become the light areas on the print and vice versa. Development and fixing are now carried out in a similar manner to that used for the negative.

Gold

24.10 Occurrence and extraction of gold

Gold is nearly always found as the free element, often in association with copper and silver; it is obtained in significant amounts during the extraction of lead from its ores and the electrolytic refining of copper.

One method of extraction of the metal involves treatment of the pulverised ore with an aerated solution of sodium cyanide and treatment of the cyanide complex, $Au(CN)_2^-(aq)$, with zinc to liberate the free metal, cf. the cyanide process used in the extraction of silver (p. 412):

$$4Au(s) + 8CN^-(aq) + O_2(g) + 2H_2O(l) \longrightarrow 4Au(CN)_2^-(aq) + 4OH^-(aq)$$
$$2Au(CN)_2^-(aq) + Zn(s) \longrightarrow 2Au(s) + Zn(CN)_4^{2-}(aq)$$

The mixture of gold and silver obtained by this method can be separated by electrolytic methods.

24.11 Properties and uses of gold

Gold has a melting point of 1063°C and a density of 19·3 g cm^{-3}. It is extremely malleable and ductile, e.g. it can be beaten into sheets no thicker than 0·000 01 mm and pulled into wire of extremely small diameter; its thermal and electrical conductivities are very high. The metal is one of the most unreactive elements; it is not attacked by air, water or steam, and the common mineral acids leave the metal untouched. A mixture of concentrated hydrochloric acid and concentrated nitric acid (aqua regia), however, attacks the metal with the formation of the $AuCl_4^-(aq)$ complex ion which contains gold (III):

$$Au(s) + 6HNO_3(aq) \rightleftharpoons Au^{3+}(aq) + 3NO_3^-(aq) + 3H_2O(l) + 3NO_2(g)$$

$$+ \; 4Cl^-(aq) \text{ (from the conc. HCl)} \rightleftharpoons AuCl_4^-(aq)$$

It is thought that the gold is first oxidised by nitric acid to $Au^{3+}(aq)$ which is then removed by complexing with the chloride ions.

In view of its lack of chemical reactivity and its attractive bright yellow colour, gold is much used in the manufacture of jewellery. Since the pure metal is rather soft it is generally hardened by the incorporation of other metals such as silver and copper. Gold is sold by the carat, which signifies the number of parts by mass of gold in 24 parts by mass of the alloy; thus pure gold is 24 carat while 12 carat gold contains 50 per cent (12/24) by mass of gold.

Cheaper articles are made by plating copper alloys with gold, using an electrolytic process which is essentially similar to the silverplating method (p. 413).

24.12 Compounds of gold

Gold exhibits oxidation states of +1 and +3 but all its compounds are thermally unstable and yield the free metal on gentle heating. The solution chemistry of gold is essentially that of its complex ions, typical complexes including $Au(CN)_2^-(aq)$ which contains gold (I) and $AuCl_4^-(aq)$ which contains gold (III).

The standard redox potentials for the systems $Au^{3+}(aq)/Au^+(aq)$ and $Au^+(aq)/Au(s)$ are given below; they indicate that $Au^+(aq)$ is unstable in solution with respect to $Au^{3+}(aq)$ and Au:

$$Au^+(aq) + e^- \longrightarrow Au(s) \qquad E^\ominus = +1{\cdot}68\text{ V}$$
$$Au^{3+}(aq) + 2e^- \longrightarrow Au^+(aq) \qquad E^\ominus = +1{\cdot}29\text{ V}$$

or

$$3Au^+(aq) \longrightarrow 2Au(s) + Au^{3+}(aq) \qquad E^\ominus_{total} = +0{\cdot}39\text{ V}$$

The positive e.m.f. of the cell reaction implies that $Au^+(aq)$ ions disproportionate into $Au^{3+}(aq)$ ions and Au in solution, cf. the disproportionation of $Cu^+(aq)$ into $Cu^{2+}(aq)$ and Cu in aqueous solution (p. 407). This disproportionation reaction is only prevented from occurring in aqueous solution by adding substances which form more stable complexes with gold (I) than with gold (III), e.g. $CN^-(aq)$ ions which form the $Au(CN)_2^-(aq)$ complex ion.

The action of fluorine and chlorine on gold at moderate temperatures results in the formation of the gold (III) halide, AuX_3:

$$2Au(s) + 3X_2(g) \longrightarrow 2AuX_3(s)$$

The addition of water results in the hydrolysis of these halides to give hydrated gold (III) oxide, $Au_2O_3.xH_2O$. This oxide dissolves in hydrochloric acid to give gold (III) chloride and then the complex ion $AuCl_4^-(aq)$; it also dissolves in alkaline solutions to give the $Au(OH)_4^-(aq)$ ion, so must be considered to be amphoteric:

$$Au_2O_3(s) + 6HCl(aq) \longrightarrow 2AuCl_3(aq) + 3H_2O(l)$$
$$AuCl_3(s) + Cl^-(aq) \longrightarrow AuCl_4^-(aq)$$
$$Au_2O_3(s) + 3H_2O(l) + 2OH^-(aq) \longrightarrow 2Au(OH)_4^-(aq)$$

24.13 Suggested reading

F. J. Smith, *The Noble Metals*, School Science Review, No. 182, Vol. 53, 1971.

24.14 Questions on chapter 24

1 The standard electrode potentials, $E^\ominus$, between the metals and their +1 ions are listed below for the elements of Groups 1A and 1B.

Group 1A Elements	$E^\ominus/V$	Group 1B Elements	$E^\ominus/V$
Li	−3·02	Cu	+0·52
Na	−2·71	Ag	+0·80
K	−2·92	Au	+1·68
Rb	−2·99		
Cs	−3·02		

How do you account for:

(a) the large differences in the values of $E^\ominus$ between the two groups,

(b) the anomalously large negative value for Li?

Explain briefly how the chemical behaviour of these elements is related to their positions in the electrochemical series. (O & C)

2 Describe how you would prepare a pure sample of copper (II) sulphate pentahydrate starting from metallic copper.

How would you obtain from copper (II) sulphate

(a) a copper (I) compound,

(b) copper metal?

Describe and explain what happens when

(i) pure copper (I) chloride is treated with concentrated ammonia solution;

(ii) copper (I) oxide is treated with dilute sulphuric acid. (JMB)

3 Write down the electronic configuration of:

(a) (i) an atom of copper,
(ii) an atom of tin;

(b) (i) the copper (II) ion,
(ii) the tin (II) ion.

Give an electronic definition of the term 'transition element'.

Give three characteristics of such elements.

State, with reasons, whether you would consider copper and tin to be transition elements or not.

Describe how a sample of copper (I) chloride may be prepared and explain what happens when portions of it are treated separately with:

(i) concentrated hydrochloric acid;

(ii) concentrated ammonium hydroxide solution. (JMB)

4 How and under what conditions does copper react with nitric acid and sulphuric acid?

Describe the preparations of

(a) copper (I) chloride,

(b) crystals of copper (II) chloride dihydrate. Explain why a dilute solution of copper (II) chloride is blue in colour, but turns green on adding concentrated hydrochloric acid. (AEB)

5 Explain or comment on the following:

(a) Copper (I) chloride is insoluble in water and in dilute hydrochloric acid, but dissolves in concentrated hydrochloric acid.

(b) A dark blue precipitate is obtained on mixing aqueous solutions of copper (II) sulphate and sodium hydroxide. This precipitate darkens on heating.

(c) On adding potassium cyanide to a solution containing copper (II) ions the blue colour is considerably reduced and on passing hydrogen sulphide through this solution no copper (II) sulphide precipitates.

(d) The solubility of silver chloride in aqueous solutions of hydrochloric acid at first decreases and then increases as the concentration of the acid is increased.

(e) Silver chloride dissolves in ammonia solution whereas silver iodide does not.

6 Determine the equilibrium constant, K, for the following reaction at 25°C from the standard redox potentials given:

$$2Cu^{+}(aq) \rightleftharpoons Cu(s) + Cu^{2+}(aq)$$

$$Cu^{+}(aq) + e^{-} \longrightarrow Cu(s) \qquad E^{\ominus} = +0{\cdot}52\,V$$

$$Cu^{2+}(aq) + e^{-} \longrightarrow Cu^{+}(aq) \qquad E^{\ominus} = +0{\cdot}16\,V$$

(If you are in doubt about how to tackle this question, read sections 10.10 to 10.13 first).

What general predictions can you make about the chemistry of +1 copper compounds from your value of the equilibrium constant?

7 (a) Describe how you would prepare in the laboratory a sample of copper (I) oxide, starting from copper (II) sulphate. Give full practical details and explain the chemistry involved.

(b) Explain the following:

(i) the relative vapour density of copper (I) chloride at 1700°C is 99.

(ii) When potassium iodide solution is added to copper (II) sulphate solution, copper (I) iodide is precipitated.

(iii) When copper (I) sulphate is dissolved in water a solution of copper (II) sulphate is obtained. (O)

8 Give an account of the extraction of silver by the cyanide process.

Describe briefly

(a) how silver nitrate is made from silver,

(b) how silver oxide is obtained from silver nitrate,

(c) one use of ammoniacal silver oxide solution.

What silver compound is employed in electroplating? Explain how metallic silver is liberated during the electrolysis. (O)

9 Copper and silver are classified together in the Periodic system. Compare and contrast the chemistry of these elements, and discuss the relation of this group to its neighbours in the Periodic Table. (S)

10 Illustrate, and explain, the fact that the chemistry of silver differs from that of copper and sodium. (Oxford Schol.)

11 'The solution chemistry of gold is essentially that of its complex ions'. Discuss this statement.

12 Discuss the usefulness of some of the complexes of copper, silver and gold.

13 +1 gold compounds tend to disproportionate in aqueous solution into +3 gold ions and the free metal:

$$3Au^{+}(aq) \rightleftharpoons 2Au(s) + Au^{3+}(aq)$$

From the standard redox potentials given below, evaluate the equilibrium constant for this disproportionation at 25°C.

$$Au^{+}(aq) + e^{-} \longrightarrow Au(s) \qquad E^{\ominus} = +1{\cdot}68\,V$$

$$Au^{3+}(aq) + 2e^{-} \longrightarrow Au^{+}(aq) \qquad E^{\ominus} = +1{\cdot}29\,V$$

Chapter 25

Group 2B zinc, cadmium and mercury

25.1 Some physical data of Group 2B elements

	Atomic Number	Electronic Configuration	Atomic Radius/nm	Ionic Radius/nm M^{2+}	M.p. /°C	B.p. /°C	Density /g cm^{-3}	Standard Redox Potential/V $M^{2+}(aq) + 2e^- \rightarrow M(s)$
Zn	30	2.8.18.2 .....$3s^2 3p^6 3d^{10} 4s^2$	0·125	0·074	419	907	7·1	−0·76
Cd	48	2.8.18.18.2 .....$4s^2 4p^6 4d^{10} 5s^2$	0·141	0·097	321	768	8·6	−0·40
Hg	80	2.8.18.32.18.2 .....$5s^2 5p^6 5d^{10} 6s^2$	0·144	0·110	−39	357	13·6	+0·85

25.2 Some general remarks about Group 2B

The atoms of the Group 2B elements have two electrons in the outer shell, like the atoms of the Group 2A elements; but whereas the atoms of the latter group of metals have eight electrons in the penultimate shell, those of zinc, cadmium and mercury have penultimate shells containing eighteen electrons:

Group 2 A		Group 2B	
Calcium	2.8.8.2	Zinc	2.8.18.2
Strontium	2.8.18.8.2	Cadmium	2.8.18.18.2
Barium	2.8.18.18.8.2	Mercury	2.8.18.32.18.2

Although there is little in common between com ... rs of
Groups 2A and 2B, zinc does resemble magnesi ...
(magnesium immediately precedes calcium in Gr ...
magnesium form a number of isomorphous cor ...
are partially covalent, deliquescent, and aci ...
hydrolysis, and the addition of sodium carb ...
solutions containing zinc and magnesium io ...
of basic carbonates (p. 223).

There is a marked difference in physical properties between Group 2B metals and those of Group 1B (copper, silver and gold); for example, the melting points, boiling points and thermal and electrical conductivities of zinc, cadmium and mercury are very much lower than those of copper, silver and gold. These differences are surprising in view of the fact that both groups of metals are assigned completely filled *d* levels; clearly the *d* electrons of the Group 2B metals cannot participate in inter-atomic bonding to any great extent, neither are they readily available for thermal and electrical conductivity although the precise reason why this should be so is not fully understood. The fact that mercury is a liquid is attributed to the inert nature of the two 6*s* electrons (the inert pair effect, cf. thallium (p. 211), and bismuth (p. 279)); once again, this effect is not fully understood but it is a feature of the chemistry of the heavier elements.

Zinc and cadmium have negative electrode potentials (unlike copper and silver) and, except when very pure, displace hydrogen from dilute non-oxidising acids (the pure metals have high hydrogen overpotentials (p. 162)); mercury, on the other hand, has a positive electrode potential and is chemically rather inert, like silver.

It is instructive to compare the standard electrode potentials of zinc and copper and to enquire into the reasons for the large difference between them:

$$Zn^{2+}(aq) + 2e^- \longrightarrow Zn(s) \qquad E^\ominus = -0{\cdot}76\,V$$
$$Cu^{2+}(aq) + 2e^- \longrightarrow Cu(s) \qquad E^\ominus = +0{\cdot}34\,V$$

Zinc, with a negative standard electrode potential, has a greater tendency to form hydrated ions than copper, with a positive standard electrode potential. The process of a metal passing into solution as its hydrated ions can be considered to take place in three hypothetical stages (see p. 135 for a discussion of the energy terms involved):

$$Zn(s) \longrightarrow Zn(g) \qquad S = +130\,kJ\,mol^{-1}$$
$$Zn(g) \longrightarrow Zn^{2+}(g) + 2e^- \qquad I = +2638\,kJ\,mol^{-1}$$
$$Zn^{2+}(g) + water \longrightarrow Zn^{2+}(aq) \qquad \Delta H_h^\ominus = -2013\,kJ\,mol^{-1}$$

The overall process is endothermic to the extent of 755 kJ mol^{-1}, i.e.

$$Zn(s) + water \longrightarrow Zn^{2+}(aq) + 2e^- \qquad \Delta H_{total}^\ominus = +755\,kJ\,mol^{-1}$$

For copper the corresponding changes are:

$$Cu(s) \longrightarrow Cu(g) \qquad S = +339\,kJ\,mol^{-1}$$
$$Cu(g) \longrightarrow Cu^{2+}(g) + 2e^- \qquad I = +2705\,kJ\,mol^{-1}$$
$$Cu^{2+}(g) + water \longrightarrow Cu^{2+}(aq) \qquad \Delta H_h^\ominus = -2069\,kJ\,mol^{-1}$$

The overall process is endothermic to the extent of 975 kJ mol^{-1}, i.e.

$$Cu(s) + water \longrightarrow Cu^{2+}(aq) + 2e^- \qquad \Delta H_{total}^\ominus = +975\,kJ\,mol^{-1}$$

The process of zinc metal passing into solution as its hydrated ions is less endothermic to the extent of 220 kJ mol^{-1} than the similar process for copper metal. Of course this amount of heat is evolved when 1 mole of zinc displaces 1 mole of copper from copper (II) sulphate solution according to the equation:

$$Zn(s) + Cu^{2+}(aq) \longrightarrow Zn^{2+}(aq) + Cu(s)\ \Delta H^{\ominus}(298\,K) = -220\,kJ\,mol^{-1}$$

It can be seen that the larger ionisation energy (I) of the copper atom is almost counter balanced by the larger hydration energy ($\Delta H_h^{\ominus}$) of the copper (II) ion, i.e. the sum of $I + \Delta H_h^{\ominus}$ for both zinc and copper is almost identical. The difference between the sublimation energies (S) for the two metals is 209 kJ mol^{-1}; thus the principal reason why zinc has a negative electrode potential as compared with that of copper is because zinc has a lower sublimation energy, i.e. zinc has lower melting and boiling points than copper. Notice that rubidium is a more reactive metal than silver for a different reason (p. 405).

Zinc, cadmium and mercury are non-transition metals, i.e. they do not exhibit oxidation states in which *d* electrons are involved; but, like transition metals, they do form a wide range of complexes, particularly with ligands such as ammonia, cyanide ions and halide ions. In general the mercury complexes are the most stable and the zinc complexes the least stable, although this trend is reversed when the bonded atom is oxygen, e.g. zinc oxide is amphoteric and forms the zincate anion, $[Zn(OH)_4]^{2-}$, whereas neither cadmium oxide nor mercury (II) oxide is amphoteric. All three metals exhibit a principal oxidation state of +2 and, in addition, mercury forms the unique mercury (I) ion, Hg_2^{2+}, in which two mercury atoms are covalently bonded. Mercury, in particular, displays a marked tendency to form covalent bonds, which accounts for the large number of organometallic compounds formed by this element.

Zinc and cadmium are sufficiently similar elements to be treated comparatively; there are few points of similarity between cadmium and mercury, and the latter is treated separately.

Zinc and cadmium

25.3 Occurrence and extraction of zinc and cadmium

The principal source of zinc is the sulphide zinc blende, ZnS, which occurs in Australia, Canada and the USA, although the metal is also extracted from calamine, $ZnCO_3$. There are no workable sources of cadmium and this metal is obtained as a by-product of the zinc industry.

The zinc ores are first concentrated (the sulphide ore by the flotation process (p. 153)) and then roasted in air to convert them into the oxide:

$$2ZnS(s) + 3O_2(g) \longrightarrow 2ZnO(s) + 2SO_2(g)$$
$$ZnCO_3(s) \longrightarrow ZnO(s) + CO_2(g)$$

The sulphur dioxide is used to manufacture sulphuric acid. The zinc oxide mixed with powdered coke, in the form of preheated briquettes, is fed into the top of a vertical retort heated to 1400°C. Zinc oxide is reduced by the coke and the mixture of zinc vapour and carbon monoxide passes through an outlet near the top of the retort:

$$ZnO(s) + C(s) \longrightarrow Zn(g) + CO(g)$$

Liquid zinc is run out into moulds and solidifies. The hot carbon monoxide is used to preheat the briquettes.

Crude zinc is produced at the rate of over six tonnes in each retort every twenty-four hours. It can be purified to remove traces of cadmium, lead and iron by redistillation. Some of the zinc collects as a fine powder, zinc dust, in the cooler regions of the condenser; zinc dust is also prepared by atomising molten zinc with an air blast. Granulated zinc is made by pouring molten zinc into water. In areas where there is a cheap supply of electricity, zinc is manufactured by electrolysing zinc sulphate solution; this is made by reacting zinc oxide with dilute sulphuric acid. (For the blast furnace method of obtaining both zinc and lead see p. 236).

Cadmium is obtained during the distillation of crude zinc. It is also present as its ions in the zinc sulphate solution from which zinc is obtained by electrolysis; but, since cadmium has a less negative electrode potential than zinc, it can be obtained by this electrolytic method, before any zinc is deposited, by a careful control of the voltage used across the cell. It may also be displaced from the same solution by the addition of zinc:

$$Cd^{2+}(aq) + Zn(s) \longrightarrow Zn^{2+}(aq) + Cd(s)$$

25.4 Properties and uses of zinc and cadmium

Both zinc and cadmium are white lustrous metals, but of the two metals cadmium is the softer and the more malleable and ductile. They adopt a somewhat distorted hexagonal close-packed structure (each atom surrounded by twelve nearest neighbours, see p. 188), the distortion being indicative of some degree of covalent bonding and hence of weak non-metallic properties.

They are both fairly reactive metals and when exposed to moist air for any length of time a protective layer is formed on their surfaces; this is the oxide initially, but over a period of time the basic carbonate is formed. Dilute non-oxidising acids such as hydrochloric and sulphuric acids attack the metals with the formation of their +2 ions and the liberation of hydrogen, e.g.

$$Zn(s) + 2H^{+}(aq) \longrightarrow Zn^{2+}(aq) + H_2(g)$$

However, the pure metals are resistant to attack by these acids since they exhibit a high overpotential to discharge of hydrogen (this appears to be an activation energy effect, see p. 162). Concentrated sulphuric acid attacks the metals with the formation of sulphur dioxide; oxides of nitrogen are formed by the action of both dilute and concentrated nitric acid.

Zinc is chemically the more reactive of the two metals, as would be expected in view of its more negative electrode potential, and hot zinc will displace hydrogen from steam:

$$Zn(s) + H_2O(g) \longrightarrow ZnO(s) + H_2(g)$$

Non-metals such as oxygen, sulphur and the halogens combine directly with the two metals on heating.

Zinc alone reacts slowly with sodium hydroxide solution liberating hydrogen, cf. the action of aluminium on alkalis which, however, is more violent (p. 206):

$$Zn(s) + 2OH^{-}(aq) + 2H_2O(l) \longrightarrow Zn(OH)_4{}^{2-}(aq) + H_2(g)$$

Appreciable quantities of zinc are used to protect iron and steel from corrosion. Galvanised iron is made by dipping the article into a bath of molten zinc. Small articles like screws and nuts and bolts are treated by a process known as 'sherardizing'; this involves heating them in a rotating drum with zinc dust. An electroplating process is used for applying a thin yet long-lasting film to metal objects.

Zinc is extensively used in the form of alloys (about 40 per cent of the total output); brass is an alloy of zinc and copper but frequently contains other metals such as tin, aluminium and iron. Alloys of zinc, containing about 10 per cent of magnesium in addition to other metals, are used for the mass production of articles such as carburettors, door handles and radiator grilles. Pure zinc is used in small amounts as the outer negative electrode of dry cell batteries. A large quantity of the metal is burnt to the oxide, which is used as a pigment in paints and for reinforcing rubber, e.g. for racing car tyres.

About 60 per cent of the total output of cadmium is used for coating iron and steel, e.g. electrical and radio components; the metal is also used in the manufacture of cadmium-nickel alkaline storage batteries which are used in buses and diesel locomotives. Small amounts of cadmium are used in alloys, e.g. copper/cadmium alloys containing about 1 per cent of cadmium are stronger and tougher than pure copper.

25.5 Compounds of zinc and cadmium

Zinc and cadmium exhibit an oxidation state of +2; aqueous solutions of their salts contain $Zn(H_2O)_4{}^{2+}(aq)$ and $Cd(H_2O)_4{}^{2+}(aq)$ ions respectively, and these are colourless since a full complement of *d* electrons is present, i.e. zinc and cadmium ions are non-transitional. There is some evidence to suggest the existence of +1 cadmium compounds, e.g. the $Cd_2{}^{2+}$ ion, which is analogous to the $Hg_2{}^{2+}$ ion, has been claimed to exist, but it is unstable in the presence of water or oxygen. Zinc only has an oxidation state of +2.

Zinc (II) and Cadmium (II) compounds

Zinc oxide, ZnO, and cadmium oxide, CdO

These oxides may be obtained by burning the metals in air or by the thermal decomposition of the carbonate or nitrate, e.g.

$$ZnCO_3(s) \longrightarrow ZnO(s) + CO_2(g)$$

$$2Cd(NO_3)_2(s) \longrightarrow 2CdO(s) + 4NO_2(g) + O_2(g)$$

Zinc oxide is a white solid and is considered to be essentially covalent; it adopts the diamond structure in which each zinc atom is surrounded tetrahedrally by four oxygen atoms and each oxygen atom is likewise surrounded by four zinc atoms. On heating, it develops a yellow colour and this phenomenon is associated with the loss of some oxygen from the lattice, leaving it with an excess negative charge; this excess negative charge (electrons) can be moved through the lattice on the application of a potential difference and thus this oxide is a semi-conductor. Zinc oxide returns to its former colour on cooling, since the oxygen which was lost on heating returns to the crystal lattice.

Cadmium oxide is generally obtained as a brown solid, although specimens ranging in colour from yellow to black can be obtained. It differs structurally from zinc oxide in that it adopts the rock salt structure (6:6 co-ordination) and is thus considered to be predominantly ionic.

Both oxides readily react with dilute acids to give salts, e.g.

$$ZnO(s) + 2H^+(aq) \longrightarrow Zn^{2+}(aq) + H_2O(l)$$

However, only zinc oxide dissolves in an aqueous solution of sodium hydroxide and thus shows amphoteric properties:

$$ZnO(s) + 2OH^-(aq) + H_2O(l) \longrightarrow Zn(OH)_4{}^{2-}(aq)$$

Zinc oxide is the most important compound of zinc; it is used as a white pigment, as a filler in rubber and as a component in various glazes, enamels and antiseptic ointments. In combination with chromium (III) oxide, it is used as a catalyst in the manufacture of methanol from synthesis gas (p. 167).

Zinc hydroxide, $Zn(OH)_2$, and cadmium hydroxide, $Cd(OH)_2$

These hydroxides are obtained as gelatinous white solids when sodium hydroxide solution is added to aqueous solutions of zinc and cadmium salts:

$$Zn^{2+}(aq) + 2OH^-(aq) \longrightarrow Zn(OH)_2(s)$$
$$Cd^{2+}(aq) + 2OH^-(aq) \longrightarrow Cd(OH)_2(s)$$

Zinc hydroxide is amphoteric like the oxide, thus it will react with both dilute acids and sodium hydroxide solution; cadmium hydroxide is exclusively a basic hydroxide. Both hydroxides readily dissolve in an aqueous solution of ammonia with the formation of ammine complexes:

$$Zn(OH)_2(s) + 4NH_3(aq) \longrightarrow Zn(NH_3)_4{}^{2+}(aq) + 2OH^-(aq)$$
$$Cd(OH)_2(s) + 4NH_3(aq) \longrightarrow Cd(NH_3)_4{}^{2+}(aq) + 2OH^-(aq)$$

Zinc sulphide, ZnS, and cadmium sulphide, CdS

These sulphides are precipitated when hydrogen sulphide is passed through aqueous solutions of zinc and cadmium salts, e.g.

$$Cd^{2+}(aq) + H_2S(g) \longrightarrow CdS(s) + 2H^+(aq)$$

Zinc sulphide, which is white, readily dissolves in dilute hydrochloric acid, whereas cadmium sulphide, which is yellow, only dissolves if the acid concentration is increased.

Both sulphides are structurally similar to zinc oxide, adopting the diamond structure (4:4 co-ordination) and are predominantly covalent.

Zinc sulphide occurs naturally as zinc blende, the main ore of metallic zinc (p. 423). Cadmium sulphide is used as a brilliant and permanent yellow pigment but is very expensive.

Zinc chloride, $ZnCl_2$, and cadmium chloride, $CdCl_2$

Solutions of these chlorides may be obtained by reacting the appropriate oxide, hydroxide or carbonate with dilute hydrochloric acid; partial evaporation followed by crystallisation results in the separation of white hydrated chlorides. The anhydrous salts may be obtained by passing chlorine over the heated metal.

When a solution of zinc chloride is concentrated, the chloride ion gradually replaces water molecules in the $Zn(H_2O)_4{}^{2+}(aq)$ ion giving $Zn(H_2O)_3Cl^+(aq)$ and then $Zn(H_2O)_2Cl_2$, i.e. $ZnCl_2.2H_2O$. Further heating results in the formation of anhydrous zinc chloride which, unlike hydrated magnesium chloride, does not undergo appreciable hydrolysis on heating (p. 193). Zinc chloride is very deliquescent and extremely soluble in water; it is also soluble in organic solvents such as ethanol and propanone and this indicates that it possesses appreciable covalent character.

Cadmium chloride is very similar to zinc chloride except that it is much less soluble in organic solvents. In aqueous solution it behaves very much like a weak electrolyte and, although undissociated $CdCl_2$ is present in solution, ionic species such as $Cd^{2+}(aq)$ and $Cl^-(aq)$, $CdCl^+(aq)$, $CdCl_3{}^-(aq)$ and $CdCl_4{}^{2-}(aq)$ are also present.

Zinc chloride is used as a flux in soldering and as a timber preservative. Both uses depend upon the ability of the compound to behave as a Lewis acid, i.e. to accept an electron pair. In soldering it is essential to remove the oxide film on the surfaces to be joined; at the temperature employed, zinc chloride melts and removes the oxide film by forming covalent bonded complexes with the oxygen atoms. When it is applied to timber, zinc chloride forms covalent bonds with the oxygen atoms in the cellulose molecules. The timber is therefore coated with a layer of zinc chloride which, like all zinc compounds, is toxic to any living matter.

Zinc sulphate, $ZnSO_4.7H_2O$, and cadmium sulphate, $CdSO_4.8/3H_2O$

These sulphates may be made by reacting the appropriate oxide, hydroxide or carbonate with dilute sulphuric acid followed by partial evaporation and crystallisation; they are white crystalline solids. The most stable hydrate of cadmium sulphate contains a peculiar proportion of water of crystallisation per formula mass of salt but a heptahydrate exists which is isomorphous with hydrated zinc sulphate. Isomorphous double salts of the type $K_2SO_4.MSO_4.6H_2O$ where M = Mg, Zn, Cd or Hg are known.

Zinc sulphate is used in galvanising iron and steel (the electrolytic process) and in the manufacture of lithopone, a white pigment:

$$ZnSO_4(s) + BaS(s) \longrightarrow \underbrace{BaSO_4(s) + ZnS(s)}_{\text{lithopone}}$$

Hydrated cadmium sulphate, $CdSO_4.8/3H_2O$, is used in the construction of the Weston cell, a device whose e.m.f. is constant and accurately known provided only minute currents are drawn from it.

Zinc nitrate, $Zn(NO_3)_2.xH_2O$, and cadmium nitrate, $Cd(NO_3)_2.xH_2O$

These can be obtained by methods similar to those used in preparing the sulphates. Both nitrates are very deliquescent and since they are soluble in ethanol they presumably possess considerable covalent character. They decompose on heating into the metallic oxide, nitrogen dioxide and oxygen, e.g.

$$2Zn(NO_3)_2(s) \longrightarrow 2ZnO(s) + 4NO_2(g) + O_2(g)$$

Zinc carbonate, $ZnCO_3$, and cadmium carbonate, $CdCO_3$

These precipitate as white solids when sodium hydrogen carbonate solution is added to an aqueous solution of the metallic salt, e.g.

$$Cd^{2+}(aq) + 2HCO_3{}^-(aq) \longrightarrow CdCO_3(s) + H_2O(l) + CO_2(g)$$

Sodium carbonate must not be used, since its aqueous solution is more alkaline than a solution of sodium hydrogen carbonate and it would precipitate the basic carbonate (p. 223).

Both carbonates readily decompose into the metallic oxide and carbon dioxide on heating, e.g.

$$CdCO_3(s) \longrightarrow CdO(s) + CO_2(g)$$

Complexes containing zinc and cadmium

In solution the majority of the complexes of zinc and cadmium have a co-ordination number of 4 and the simple hydrated ions are considered to be $Zn(H_2O)_4{}^{2+}(aq)$ and $Cd(H_2O)_4{}^{2+}(aq)$ respectively; incidentally these two complex ions are poisonous to most living matter. Hydrated zinc ions are acidic owing to equilibria of the following type:

$$Zn(H_2O)_4{}^{2+}(aq) + H_2O(l) \rightleftharpoons Zn(H_2O)_3(OH)^{+}(aq) + H_3O^{+}(aq)$$

Aqueous solutions of cadmium salts are much less acidic and this is no doubt due to the larger size of the cadmium ion, which cannot polarise water molecules as readily as the smaller zinc ion.

Zinc has a great affinity for oxygen and, although the precise reason for this is unknown, the fact explains the amphoteric nature of zinc oxide and zinc hydroxide; both these compounds form the $Zn(OH)_4{}^{2-}(aq)$ complex ion with an excess of alkali. The analogous cadmium complex is much less stable, thus both cadmium oxide and cadmium hydroxide are almost completely devoid of amphoteric properties.

Other complexes formed by these two metals include $[M(NH_3)_4]^{2+}$, $[M(CN)_4]^{2-}$ and $[MCl_4]^{2-}$ which, like the other 4 co-ordinated complexes of these two metals, are tetrahedral in shape; the cadmium complexes are generally more stable than the corresponding complexes of zinc.

Mercury

25.6 Occurrence and extraction of mercury

Mercury is extracted from the mineral cinnabar, HgS, the deposits in Spain and Italy accounting for about three-quarters of the world's supply of the metal. Many mercury ores contain considerably less than 1 per cent of the sulphide, which accounts for the very high price of the metal.

Mercury is readily extracted from the sulphide ore by heating it in air. Mercury vapour is evolved which is condensed to the liquid metal:

$$HgS(s) + O_2(g) \longrightarrow Hg(l) + SO_2(g)$$

Commercial mercury is impure and contains lead and, less frequently, zinc and tin. It can be purified by dropping it slowly through dilute nitric acid which contains a little mercury (I) nitrate (the dilute acid dissolves the impurities which are more reactive than mercury). However, a purer product is obtained by distillation under reduced pressure.

25.7 Properties and uses of mercury

Mercury is a dense silvery-coloured metal and the only one that is liquid at room temperature; it freezes at −39°C and boils at 357°C. Although it only has a very small vapour pressure, **the vapour is exceedingly poisonous** and any mercury accidently spilt should immediately be dusted with powdered sulphur. It dissolves many metals, including sodium, zinc, tin, silver and gold, to form amalgams, but does not attack iron; hence it can be stored in iron bottles. Pure mercury is not attacked by air at ordinary temperatures; mercury (II) oxide is slowly formed in the region of about 350°C but at slightly higher temperatures it decomposes into mercury and oxygen; other non-metals that combine directly with mercury include the halogens and sulphur.

In view of its high positive electrode potential, mercury is not attacked by dilute non-oxidising acids. Nitric acid attacks the metal, with the liberation of oxides of nitrogen; excess of concentrated nitric acid tends to give nitrogen dioxide and mercury (II) nitrate but with dilution and in the presence of excess of mercury, nitrogen oxide and mercury (I) nitrate tend to predominate:

$$Hg(l) + \underset{\text{conc.}}{4HNO_3(aq)} \longrightarrow Hg(NO_3)_2(aq) + 2NO_2(g) + 2H_2O(l)$$

$$6Hg(l) + \underset{\text{dil.}}{8HNO_3(aq)} \longrightarrow 3Hg_2(NO_3)_2(aq) + 2NO(g) + 4H_2O(l)$$

Hot concentrated sulphuric acid attacks the metal with the liberation of sulphur dioxide; excess of acid tends to give mercury (II) sulphate and excess of mercury tends to give mercury (I) sulphate.

Mercury is used in thermometers, barometers, electrical switches and in mercury arc lights. Sodium and zinc amalgams are used as laboratory reducing agents and tin amalgam is used as a dental filling. Mercury (II) fulminate, $Hg(ONC)_2$, is a shock-sensitive explosive which is used in detonators. The major uses of mercury compounds are in agriculture, and in horticulture, e.g. organomercury compounds are used as fungicides and as timber preservatives.

25.8 Compounds of mercury

Mercury exhibits oxidation states of +2 and +1; in the latter, two mercury atoms are united by a single covalent bond as in the $[Hg—Hg]^{2+}$ ion. Mercury compounds are more covalent than those of zinc and cadmium and, indeed, covalency is the rule rather than the exception. Mercury (II) compounds form stable complexes with a variety of ligands; indeed, they are generally many orders of magnitude more stable than the corresponding complexes of zinc and cadmium.

Mercury (II) compounds

Mercury (II) oxide, HgO

Mercury (II) oxide forms when mercury is heated for a long time at about 350°C; as made by this method, it is a red solid. Yellow mercury (II) oxide may be obtained by adding sodium hydroxide solution to a solution of a mercury (II) salt; it is possible that unstable mercury (II) hydroxide forms initially:

$$Hg^{2+}(aq) + 2OH^-(aq) \longrightarrow Hg(OH)_2$$
$$Hg(OH)_2 \longrightarrow HgO(s) + H_2O(l)$$

The difference between the two coloured varieties of mercury (II) oxide is simply one of particle size.

Mercury (II) oxide is thermally unstable and readily decomposes into mercury and oxygen. This decomposition is of historical interest, since it was the method used by Priestley to obtain oxygen:

$$2HgO(s) \longrightarrow 2Hg(l) + O_2(g)$$

Mercury (II) sulphide, HgS

Mercury (II) sulphide occurs as the red ore cinnabar. It is obtained as a black solid by passing hydrogen sulphide through an aqueous solution of a mercury (II) salt; its solubility product is so low that it precipitates in the presence of a high concentration of hydrogen ions, which considerably reduces the already low concentration of sulphide ions provided by the dissolved hydrogen sulphide:

$$Hg^{2+}(aq) + H_2S(g) \longrightarrow HgS(s) + 2H^+(aq)$$

Both forms of the compound have different crystal structures and the black form, which is metastable with respect to the other, changes over into the red variety on heating.

Mercury (II) chloride, $HgCl_2$

Mercury (II) chloride forms rapidly when chlorine is brought into contact with mercury:

$$Hg(l) + Cl_2(g) \longrightarrow HgCl_2(s)$$

It is a white solid and readily dissolves in organic solvents such as ethanol and ether, thus indicating that it is predominantly covalent. Although only sparingly soluble in cold water, it dissolves readily on warming but the solution so obtained is only a very feeble electrolyte, and it has been estimated that about 99 per cent of the dissolved mercury (II) chloride exists in the form of simple $HgCl_2$ covalent molecules. Mercury (II) chloride solution is readily reduced to mercury (I) chloride and then to mercury by the addition of tin (II) chloride solution, and this is a convenient test for the mercury (II) ion:

$$2HgCl_2(aq) + SnCl_2(aq) \longrightarrow SnCl_4(aq) + Hg_2Cl_2(s)$$

$$Hg_2Cl_2(s) + SnCl_2(aq) \longrightarrow SnCl_4(aq) + \underset{\text{black}}{2Hg(l)}$$

Mercury (II) chloride is extremely poisonous, acting by coagulating protein in living cells; the antidote is the immediate administration of the white of an egg which provides an alternative protein material for it to act upon. In very dilute solution mercury (II) chloride is sometimes used as an antiseptic, e.g. for sterilising the hands.

Mercury (II) iodide, HgI_2

This compound precipitates as a yellow solid which then turns rapidly red when the correct amount of potassium iodide solution is added to a solution of mercury (II) chloride:

$$Hg^{2+}(aq) + 2I^-(aq) \longrightarrow HgI_2(s)$$

On heating to about 127°C the red form changes into a yellow variety which has a different crystal structure (polymorphism). On cooling again the change is reversed and this change in crystal structure may be accelerated by scratching the material with a glass rod.

$$\underset{\text{red}}{HgI_2(s)} \underset{\text{below } 127^\circ C}{\overset{\text{above } 127^\circ C}{\rightleftharpoons}} \underset{\text{yellow}}{HgI_2(s)}$$

Mercury (II) iodide readily dissolves in an excess of potassium iodide solution with the formation of the $HgI_4^{2-}(aq)$ complex ion:

$$HgI_2(s) + 2I^-(aq) \longrightarrow HgI_4^{2-}(aq)$$

Salts containing this complex anion include Ag_2HgI_4 (yellow) and Cu_2HgI_4 (red), which are both insoluble.

Nessler's reagent is a solution of K_2HgI_4 to which some potassium hydroxide has been added. It reacts with traces of ammonia to give a yellow coloration; larger quantities of the gas produce a brown precipitate.

Mercury (II) sulphate, $HgSO_4$

Mercury (II) sulphate may be obtained by the action of hot concentrated sulphuric acid on mercury; it is a white hygroscopic solid. On heating it decomposes into mercury (I) sulphate, mercury, sulphur dioxide and oxygen:

$$3HgSO_4(s) \longrightarrow Hg_2SO_4(s) + Hg(l) + 2SO_2(g) + 2O_2(g)$$

Mercury (II) nitrate, $Hg(NO_3)_2$

Mercury (II) nitrate is obtained by the prolonged treatment of mercury with hot concentrated nitric acid; it can be crystallised as white deliquescent crystals. On heating it decomposes into mercury, nitrogen dioxide and oxygen:

$$Hg(NO_3)_2(s) \longrightarrow Hg(l) + 2NO_2(g) + O_2(g)$$

Mercury (II) carbonate, $HgCO_3.aHgO$

The pure substance is unknown and any method of preparation results in the formation of a basic carbonate, i.e. the action of sodium carbonate or sodium hydrogen carbonate solutions on an aqueous solution of a mercury (II) salt. Typical basic carbonates are $HgCO_3.2HgO$ and $HgCO_3.3HgO$.

Organomercury compounds

Mercury forms a very large number of compounds in which it is covalently bonded to carbon atoms. Typical organomercury compounds are of the type RHgX and R_2Hg, where R may be an alkyl or aryl radical and X may be a halogen, hydroxyl or sulphate radical, etc. Dialkyl and diaryl compounds of mercury in particular are volatile liquids or solids and are exceptionally toxic.

Although the mercury-carbon bond is relatively weak, these compounds are stable in the presence of air and water, unlike the organocompounds of zinc and cadmium. This surprising result is due to the very slight affinity of mercury for oxygen, since compounds containing mercury-oxygen bonds would be formed if these compounds were attacked by either air or water.

Mercury (I) compounds

An interesting feature of mercury is its ability to form the Hg_2^{2+} ion in which the two mercury atoms are united by a single covalent bond, i.e. the simple Hg^+ ion does not exist.

Evidence in favour of the Hg_2^{2+} ion as opposed to the Hg^+ ion is considerable and includes the following experimental results:

(a) Mercury (I) compounds are not paramagnetic (p. 355) but if the formula of the mercury (I) ion was Hg^+ this would contain an unpaired $6s$ electron. In Hg_2^{2+} both $6s$ electrons originating from each mercury atom are paired and this explains its lack of paramagnetism.

(b) When an excess of mercury is shaken with an aqueous solution of a mercury (II) salt, an equilibrium is established and mercury, mercury (II) ions, and mercury (I) ions are present. Let us first assume that the mercury (I) ion can be represented as Hg^+, then the following equilibrium applies:

$$\underset{\text{excess}}{Hg(l)} + \underset{\frac{(a-x)}{V}}{Hg^{2+}(aq)} \rightleftharpoons \underset{\frac{2x}{V}}{2Hg^+(aq)} \quad \text{(equilibrium concentrations)}$$

where a is the initial number of moles of Hg^{2+}, x is the number of moles of Hg^{2+} converted into the mercury (I) ion, $2x$ is the number of moles of Hg^+ formed, and V is the volume of the aqueous solution (dm^3). The equilibrium constant for this reaction, K, is given by the expression:

$$K = \frac{[Hg^+(aq)]^2}{[Hg^{2+}(aq)]} = \frac{(2x/V)^2}{(a-x)/V} = \frac{4x^2}{(a-x)/V}\,\text{mol dm}^{-3}$$

However, if the mercury (I) ion is assumed to be Hg_2^{2+} a different equilibrium applies:

$$\underset{\text{excess}}{Hg(l)} + \underset{\frac{(a-x)}{V}}{Hg^{2+}(aq)} \rightleftharpoons \underset{\frac{x}{V}}{Hg_2^{2+}(aq)} \quad \text{(equilibrium concentrations)}$$

The equilibrium constant for this reaction, K', is given by a different expression:

$$K' = \frac{[Hg_2^{2+}(aq)]}{[Hg^{2+}(aq)]} \quad \frac{x/V}{(a-x)/V} \quad \frac{x}{(a-x)}$$

When the experiment is carried out with varying initial concentrations of mercury (II) ions, K tends to vary but K' remains virtually constant, so that Hg_2^{2+} is the correct formula for the mercury (I) ion.

Attempts to prepare mercury (I) hydroxide, mercury (I) oxide and mercury (I) sulphide have failed and, in order to understand the reason for this, it is necessary to consider the equilibrium:

$$Hg_2^{2+}(aq) \rightleftharpoons Hg(l) + Hg^{2+}(aq)$$

The above disproportionation reaction has an equilibrium constant, K'', of about $6{\cdot}0 \times 10^{-3}$ at 298 K, i.e.

$$K'' = \frac{[Hg^{2+}(aq)]}{[Hg_2{}^{2+}(aq)]} = 6{\cdot}0 \times 10^{-3}$$

The low value for the equilibrium constant implies that under normal conditions there is little tendency for the mercury (I) ion to disproportionate into the mercury (II) ion and mercury. However, the addition of an anion, e.g. the sulphide ion, which forms a more insoluble compound with the mercury (II) ion than with the mercury (I) ion, causes the disproportionation reaction to proceed further to the right:

$$\begin{array}{l} Hg_2{}^{2+}(aq) \rightleftharpoons Hg(l) + Hg^{2+}(aq) \\ \qquad\qquad\qquad\qquad\quad + \\ \qquad\qquad\qquad\qquad\; S^{2-}(aq) \\ \qquad\qquad\qquad\qquad\quad \updownarrow \\ \qquad\qquad\qquad\qquad\; HgS(s) \end{array}$$

(equilibrium displaced to the right by the removal of Hg^{2+} ions as the very insoluble mercury (II) sulphide)

The addition of hydrogen sulphide to an aqueous solution of a mercury (I) salt thus results in the precipitation of mercury (II) sulphide and free mercury:

$$Hg_2{}^{2+}(aq) + H_2S(g) \longrightarrow Hg(l) + HgS(s) + 2H^+(aq)$$

Similarly, the addition of sodium hydroxide solution to a solution of a mercury (I) salt forms mercury (II) oxide (mercury (II) hydroxide is very unstable, p. 429) and mercury:

$$Hg_2{}^{2+}(aq) + 2OH^-(aq) \longrightarrow Hg(l) + HgO(s) + H_2O(l)$$

Mercury (I) compounds form very few complexes principally because mercury (II) compounds form very stable ones with many ligands; thus once again the disproportionation reaction is driven well over to the right:

$$\begin{array}{l} Hg_2{}^{2+}(aq) \rightleftharpoons Hg(l) + Hg^{2+}(aq) \\ \qquad\qquad\qquad\qquad\quad + \\ \qquad\qquad\qquad \text{Complexing ligands} \\ \qquad\qquad\qquad\qquad\quad \updownarrow \\ \qquad\qquad\qquad\quad Hg^{2+}\ \text{complex} \end{array}$$

(equilibrium displaced to the right by the removal of $Hg^{2+}(aq)$ ions as complex ions)

Two of the best known compounds of mercury (I) are mercury (I) chloride and mercury (I) nitrate.

Mercury (I) chloride, Hg_2Cl_2

Mercury (I) chloride precipitates as a white solid when solutions containing mercury (I) and chloride ions are mixed:

$$Hg_2{}^{2+}(aq) + 2Cl^-(aq) \longrightarrow Hg_2Cl_2(s)$$

It may also be obtained by subliming a mixture of mercury (II) chloride and mercury:

$$HgCl_2(s) + Hg(l) \longrightarrow Hg_2Cl_2(s)$$

Relative molecular mass determinations on mercury (I) chloride vapour at high temperatures indicate the formula HgCl, but the low molecular mass is in reality due to complete dissociation into mercury (II) chloride and mercury vapour:

$$Hg_2Cl_2(g) \rightleftharpoons HgCl_2(g) + Hg(g)$$

The action of ammonia solution on mercury (I) chloride gives a black deposit of mercury and a compound containing mercury (II), i.e. disproportionation occurs:

$$Hg_2Cl_2(s) + 2NH_3(aq) \longrightarrow Hg(l) + HgNH_2Cl(s) + NH_4Cl(aq)$$

The compound $HgNH_2Cl$ contains the mercury-nitrogen covalent bond, the formation of which assists in moving the disproportionation reaction (p. 432) from left to right. This reaction is a convenient test for the mercury (I) ion.

Mercury (I) chloride (sometimes called calomel) is employed in the calomel electrode which is used as a reference standard in redox potential measurements. It consists of a 1 M solution of potassium chloride in contact with mercury and mercury (I) chloride, i.e. the potassium chloride solution is saturated with mercury (I) chloride; electrical contact is established by means of a platinum wire dipping into the pool of mercury. The electrode potential of this arrangement, measured against a standard hydrogen electrode at 298 K, is +0·280 V. It is much more convenient to use than the hydrogen electrode.

Mercury (I) nitrate, $Hg_2(NO_3)_2$

Mercury (I) nitrate is formed by the action of cold dilute nitric acid on an excess of mercury. It can be crystallised as a white solid hydrate, $Hg_2(NO_3)_2.2H_2O$. Although readily soluble in water it is also hydrolysed by it and a white basic nitrate, $Hg(OH)NO_3$, slowly deposits. The solution can be cleared by the addition of dilute nitric acid which dissolves the basic salt.

Mercury (I) nitrate readily decomposes on heating into mercury (II) oxide and nitrogen dioxide; on further heating the mercury (II) oxide decomposes into mercury and oxygen:

$$Hg_2(NO_3)_2(s) \longrightarrow 2HgO(s) + 2NO_2(g)$$
$$2HgO(s) \longrightarrow 2Hg(l) + O_2(g)$$

25.9 Questions on chapter 25

1 Account for the fact that zinc has a negative standard electrode potential while the standard electrode potential of copper is positive.

2 Give a comparative account of the chemistry of zinc and cadmium. By citing specific examples discuss whether or not mercury has much in common with cadmium.

3 Give a brief account of the extraction and properties of zinc and the properties or reactions of four of its compounds. (O & C)

4 Compare the chemistry of a metal in an A sub-group of the Periodic Table with that of a metal in the corresponding B sub-group, with particular reference to calcium (Group 2A) and zinc (Group 2B). Give reasons for the similarities and differences in chemical behaviour. (Limit your answer to a discussion of the properties of the elements and any three compounds of each element with a common anion). (O & C)

5 Discuss the following:

(a) Pure zinc is resistant to attack by dilute sulphuric acid but if a piece of copper wire is touching some pure zinc below the surface of some dilute sulphuric acid, hydrogen is evolved from the copper wire.

(b) Zinc hydroxide dissolves in both aqueous ammonia and sodium hydroxide solution but copper hydroxide only dissolves in aqueous ammonia.

(c) Zinc oxide becomes yellow when heated but returns to its original white colour on cooling.

(d) Zinc sulphide precipitates when hydrogen sulphide is passed through a neutral solution of a zinc salt; no precipitate forms if the solution is acidic.

6 Give the name and formula of one ore of mercury. How is the metal

(a) extracted from the ore

(b) purified?

Starting from the metal, how could you prepare specimens of

(a) mercury (I) chloride,

(b) mercury (II) chloride?

What deductions have been made from a study of the vapour density of mercury (I) chloride at different temperatures? (L)

7 Give an account of the occurrence, extraction and purification of mercury.

Describe the reactions which occur when an aqueous solution of mercury (II) chloride is treated with an excess of

(a) an aqueous solution of potassium iodide,

(b) an aqueous solution of tin (II) chloride,

(c) an aqueous solution of sodium hydroxide,

(d) copper. (L)

8 Give a brief account of the chemistry of mercury. In a solution of an ionised mercury salt in equilibrium with metallic mercury, the fraction of dissolved mercury present as the mercury (II) ion is independent of the concentration of the mercury (I) ion.

What does this tell you about the nature of the mercury (I) ion? (Oxford Schol.)

9 Describe the characteristic features of the chemistry of mercury. How far do the chemical properties of mercury justify its inclusion in the same group of the Periodic Table as zinc? (Oxford Schol.)

10 Discuss, with examples, the difference between double salts and complex salts, and describe the preparation of one typical member of each class.

The freezing point of a molar solution of potassium iodide is $-3{\cdot}2$°C. When mercury (II) iodide (which is insoluble in pure water) is added in excess, the freezing point is raised to $-2{\cdot}4$°C. Comment on this result. (C)

11 What experiments would you carry out to show that mercury (I) chloride has the formula Hg_2Cl_2 rather than $HgCl$?

12 In representations of the Periodic Table, zinc (atomic number 30) usually appears as the last member of the first transition metal series, but it is sometimes claimed that it does not display the full range of characteristics associated with transition metals.

Outline, with two appropriate examples in each case, the major characteristics that distinguish transition metals from non-transition metals, including electronic structure.

In the light of your knowledge of the chemistry of zinc, critically discuss the extent to which the above claim is justified. Similarly consider also the case of copper. (L)

13 Write an account of the chemistry of zinc and its principal compounds, and discuss whether the element should be considered as part of the first transition series. (S)

Chapter 26

The first inner transition series (the lanthanides)

26.1 Some physical data of the first inner transition series elements

Element	Symbol	Atomic Number	Electronic Configuration	Atomic Radius/nm	Ionic Radius/nm M^{3+}	M.p. /°C (approx)	Density /g cm^{-3}
Lanthanum	La	57	2.8.18.18.8(1).2 .....$4f^05s^25p^65d^16s^2$	0·187	0·115	920	6·19
Cerium	Ce	58	2.8.18.18(2).8.2 .....$4f^25s^25p^65d^06s^2$	0·183	0·111	795	6·78
Praseodymium	Pr	59	2.8.18.18(3).8.2 .....$4f^35s^25p^65d^06s^2$	0·182	0·109	935	6·78
Neodymium	Nd	60	2.8.18.18(4).8.2 .....$4f^45s^25p^65d^06s^2$	0·181	0·108	1020	7·00
Prometheum	Pm	61	2.8.18.18(5).8.2 .....$4f^55s^25p^65d^06s^2$		0·106	1030	
Samarium	Sm	62	2.8.18.18(6).8.2 .....$4f^65s^25p^65d^06s^2$	0·179	0·104	1070	7·54
Europium	Eu	63	2.8.18.18(7).8.2 .....$4f^75s^25p^65d^06s^2$	0·204	0·112	826	5·24
Gadolinium	Gd	64	2.8.18.18(7).8(1).2 .....$4f^75s^25p^65d^16s^2$	0·180	0·102	1310	7·95
Terbium	Tb	65	2.8.18.18(9).8.2 .....$4f^95s^25p^65d^06s^2$	0·178	0·100	1360	8·27
Dysprosium	Dy	66	2.8.18.18(10).8.2 .....$4f^{10}5s^25p^65d^06s^2$	0·177	0·099	1410	8·56
Holmium	Ho	67	2.8.18.18(11).8.2 .....$4f^{11}5s^25p^65d^06s^2$	0·176	0·097	1460	8·80
Erbium	Er	68	2.8.18.18(12).8.2 .....$4f^{12}5s^25p^65d^06s^2$	0·175	0·096	1500	9·16
Thulium	Tm	69	2.8.18.18(13).8.2 .....$4f^{13}5s^25p^65d^06s^2$	0·174	0·095	1540	9·33
Ytterbium	Yb	70	2.8.18.18(14).8.2 .....$4f^{14}5s^25p^65d^06s^2$	0·194	0·094	824	6·98
Lutetium	Lu	71	2.8.18.18(14).8(1).2 .....$4f^{14}5s^25p^65d^16s^2$	0·174	0·093	1650	9·84

26.2 Some general remarks about the first inner transition series

Lanthanum is the first member of the third transition series, and it has one $5d$ and two $6s$ electrons; the next element cerium, however, while still retaining two $6s$ electrons has two electrons in the $4f$ orbitals and none in the $5d$ orbital. There are seven separate $4f$ orbitals, each of which can accommodate two electrons with opposite spins. These seven energy levels are degenerate, i.e. have the same energy in the free atoms and ions, and in practice each of them must be singly occupied before electron pairing takes place (Hund's rule). The atoms of the elements from cerium to lutetium have between two and fourteen electrons in the $4f$ orbitals and constitute the first inner transition series; this series of elements is known as the lanthanides (rare earths) and, although lanthanum itself does not possess any $4f$ electrons, it is customary to include this element in the series.

The filling up of the $4f$ orbitals is regular with but one exception; the element europium has the outer electronic configuration $4f^7 5s^2 5p^6 5d^0 6s^2$ and the next element gadolinium has the extra electron in the $5d$ orbital, i.e. its outer electronic configuration is $4f^7 5s^2 5p^6 5d^1 6s^2$ (when all seven $4f$ orbitals are singly occupied a special degree of stability is thought to be conferred on the atom). The element ytterbium has a full complement of $4f$ electrons ($4f^{14} 5s^2 5p^6 5d^0 6s^2$) and the extra electron in the lutetium atom enters the $5d$ orbital ($4f^{14} 5s^2 5p^6 5d^1 6s^2$). Except for lanthanum, gadolinium and lutetium (the first, eighth and fifteenth members) which have a single $5d$ electron, the lanthanides do not have electrons in the $5d$ orbitals.

26.3 Features of the lanthanides and their ions

Discovery and occurrence of the lanthanides

As early as 1794 the Swedish chemist Gadolin announced the discovery of an oxide which he named yttria and it was not until about fifty years later that this so called substance was broken down into three portions named yttria, erbia and terbia. Over the years further separations were achieved, including the isolation of an oxide called lutetia. Of course the names of the lanthanides as we now know them are obtained by changing the ending -a in the oxide into -um, e.g. yttria changes to yttrium.

The major source of the lanthanides is monozite sand which is composed chiefly of the phosphates of thorium, cerium, neodymium and lanthanum; the phosphate portion of monozite contains small traces of other lanthanide ions, and the only lanthanide that does not occur naturally is promethium, which is made artificially by nuclear reactions. In view of the very similar properties of this group of elements, it is only comparatively recently that methods of separation have been worked out which have enabled the very pure elements to be made commercially available.

Separation and extraction of the lanthanides

All the lanthanides form +3 ions, M^{3+}, whose ionic radii decrease progressively with increasing atomic number from lanthanum (La^{3+}, 0·115 nm to lutetium (Lu^{3+}, 0·093 nm); this is the so-called lanthanide contraction (p. 438). As the ionic radii contract along the lanthanide series, the ability to form complex ions increases and this is the basis of their separation on an ion exchange column.

A solution containing +3 lanthanide ions is placed at the top of a cation exchange column (p. 296) and the lanthanide ions release an equivalent amount of hydrogen ions from the column:

$$M^{3+}(aq) + \underset{\text{cation exchanger}}{3H^+R^-(s)} \longrightarrow M^{3+}(R^-)_3(s) + 3H^+(aq)$$

A citrate buffer solution (which complexes with the lanthanide ions) is slowly run down the column and the cations partition themselves between the column itself and the moving citrate solution. Since the smaller ions show a greater preference for complexing with the citrate solution, these ions are the first to emerge from the column. By the correct choice of conditions the lutetium ion, $Lu^{3+}(aq)$, emerges first from the column, followed by the cations ytterbium, thulium, erbium, etc., in order of increasing ionic radius.

Lanthanum, cerium, praseodymium, neodymium and gadolinium may be obtained by reduction of their trichlorides with calcium at about 1000°C, e.g.

$$2PrCl_3(s) + 3Ca(s) \longrightarrow 3CaCl_2(s) + 2Pr(s)$$

Terbium, dysprosium, holmium, erbium and thulium are obtained by a similar process except that their trifluorides are used, since their trichlorides are too volatile, e.g.

$$2HoF_3(s) + 3Ca(s) \longrightarrow 3CaF_2(s) + 2Ho(s)$$

Europium, samarium and ytterbium are obtainable by chemical reduction of their trioxides.

The lanthanide contraction

Each succeeding lanthanide differs from its immediate predecessor in having one more electron in the 4*f* orbitals (except for some slight exceptions noted in section 26.2) and one extra proton in the nucleus of the atom. The 4*f* electrons constitute inner shells and are rather ineffective in screening the nucleus; thus there is a gradual increase in the attraction of the nucleus for the peripheral electrons as the nuclear charge increases, and a consequent contraction in atomic radius. While the atomic radii show some slight variations (see section 26.1), e.g. the atomic radius of europium appears abnormally high, the ionic radii of the +3 cations contract steadily from a value of 0·115 nm for La^{3+} to a value of 0·093 nm for Lu^{3+}. The gradual increase in stability of similar complexes formed by these +3 ions with decreasing ionic radius is of profound importance, since it affords a means of separating them (p. 437).

The lanthanide contraction considerably influences the chemistry of the elements which succeed the lanthanides in the Periodic Table; for instance the atomic radii of zirconium (atomic number 40) and hafnium (atomic number 72) are almost identical and the chemistry of these two elements is strikingly similar. Incidentally, the density of hafnium (which immediately follows the lanthanides) is almost twice the density of zirconium (which is in the same periodic group). The fact that silver and gold resemble each other more than either resembles copper can be partially attributed to the lanthanide contraction, since the atomic and ionic radii of silver and gold are similar.

Properties of the lanthanides

The metals, which are silvery white in colour, are good conductors of heat and electricity.

Coloured ions

Many of the lanthanide ions are coloured and, although the origin of the colour is more difficult to interpret than in the case of the first transition series (p. 351), it is thought to be due to the fact that electronic transitions can occur between different 4*f* levels (the 4*f* orbitals are only degenerate in the free ions). The colours of the ions and number of 4*f* electrons present in the +3 lanthanide ions are given in Table 26A.

Table 26A The colours of some aqueous lanthanide ions

$La^{3+}(aq)$	Colourless	$4f^0$	$Gd^{3+}(aq)$	Colourless	$4f^7$
$Ce^{3+}(aq)$	Colourless	$4f^1$	$Tb^{3+}(aq)$	Pale pink	$4f^8$
$Pr^{3+}(aq)$	Green	$4f^2$	$Dy^{3+}(aq)$	Yellow	$4f^9$
$Nd^{3+}(aq)$	Red-violet	$4f^3$	$Ho^{3+}(aq)$	Yellow	$4f^{10}$
$Pm^{3+}(aq)$	Yellow	$4f^4$	$Er^{3+}(aq)$	Pink	$4f^{11}$
$Sm^{3+}(aq)$	Yellow	$4f^5$	$Tm^{3+}(aq)$	Pale green	$4f^{12}$
$Eu^{3+}(aq)$	Pale pink	$4f^6$	$Yb^{3+}(aq)$	Colourless	$4f^{13}$
			$Lu^{3+}(aq)$	Colourless	$4f^{14}$

Oxidation states

The lanthanides exhibit a principle oxidation state of +3 in which the M^{3+} ion contains an outer shell containing eight electrons and an underlying layer containing up to fourteen 4*f* electrons. The +3 ions of lanthanum, gadolinium and lutetium, which contain respectively an empty, a half-filled, and a completely filled 4*f* level, are especially stable. Cerium can exhibit an oxidation state of +4 in which it has the same electronic configuration as +3 lanthanum, i.e. an empty 4*f* level; similarly +4 terbium exists which has the same electronic configuration as +3 gadolinium, i.e. a half-filled 4*f* level:

La(III)	$4f^0 5s^2 5p^6$	Ce(IV)	$4f^0 5s^2 5p^6$	(empty 4*f* level)
Gd(III)	$4f^7 5s^2 5p^6$	Tb(IV)	$4f^7 5s^2 5p^6$	(half-filled 4*f* level)

Further support of the idea that an empty, a half-filled, and a completely filled 4*f* level confers some extra stability on a particular oxidation state (see p. 350 for the stability associated with an empty, a half-filled, and a completely filled 3*d* level) is available. Thus +2 europium is isoelectronic with +3 gadolinium, i.e. a half-filled 4*f* level, and +2 ytterbium is isoelectronic with +3 lutetium:

Gd(III)	$4f^7 5s^2 5p^6$	Eu(II)	$4f^7 5s^2 5p^6$	(half-filled 4*f* level)
Lu(III)	$4f^{14} 5s^2 5p^6$	Yb(II)	$4f^{14} 5s^2 5p^6$	(completely filled 4*f* level)

However, other factors are involved since Pr(IV), Nd(IV), Sm(II) and Tm(II) exist, which contain respectively one, two, six and thirteen 4*f* electrons.

Although a few lanthanides exhibit an oxidation state of +4 while some others show an oxidation state of +2, the +3 oxidation state is the most stable state in every case. For instance cerium (IV) is strongly oxidising and samarium (II) is strongly reducing:

$$Ce^{4+}(aq) + Fe^{2+}(aq) \longrightarrow Ce^{3+}(aq) + Fe^{3+}(aq)$$
$$2Sm^{2+}(aq) + 2H_2O(l) \longrightarrow 2Sm^{3+}(aq) + 2OH^-(aq) + H_2(g)$$

Note that in the above examples both $Ce^{4+}(aq)$ and $Sm^{2+}(aq)$ are converted into the +3 ions $Ce^{3+}(aq)$ and $Sm^{3+}(aq)$ respectively.

Paramagnetism

Paramagnetism is associated with the presence of unpaired electrons in a compound (p. 355). The majority of the lanthanide ions exhibit paramagnetism, since according to Hund's rule each of the 4*f* orbitals must be occupied singly before electron pairing takes place; lanthanide ions that do not exhibit paramagnetism are those with either no 4*f* electrons, e.g. La^{3+} and Ce^{4+}, or with a completed 4*f* level, e.g. Yb^{2+} and Lu^{3+}. Unlike the paramagnetism associated with the ions of the first transition series (p. 355), the paramagnetic moments of the lanthanide ions cannot be directly correlated with the actual number of unpaired 4*f* electrons, e.g. Gd^{3+} with the maximum of seven unpaired 4*f* electrons would be expected to have the largest paramagnetic moment, whereas Dy^{3+} with five unpaired 4*f* electrons is the most paramagnetic ion.

Reactions of the lanthanides

In chemical reactivity they resemble calcium; thus they readily tarnish in air and burn to give oxides (all give the trioxide, except for cerium which forms CeO_2). Other non-metals which combine directly with them include nitrogen, sulphur, the halogens and hydrogen. The hydrides are non-stoichiometric but their composition approaches that required for the general formula MH_3; like calcium hydride, these hydrides liberate hydrogen from water and are probably essentially ionic in structure. Other similarities to calcium include the liberation of hydrogen from water and the vigorous evolution of this same gas from dilute non-oxidising acids, e.g.

$$2La(s) + 6H_2O(l) \longrightarrow 2La(OH)_3(aq) + 3H_2(g)$$

Lanthanum oxide, La_2O_3, is a strong base and combines vigorously with water to form the hydroxide, $La(OH)_3$, cf. the reaction of calcium oxide with water (p. 190). The oxides and hydroxides of the other lanthanides become progressively weaker as bases (another manifestation of the lanthanide contraction). Lanthanide compounds are generally predominantly ionic and usually contain the lanthanide metal in its +3 oxidation state.

26.4 Questions on chapter 26

1 Give an account of the principal features of the chemistry of the lanthanides.

2 What is meant by the 'lanthanide contraction? What bearing does it have on the separation of these elements, and on the atomic radii of the atoms of elements which succeed this group of metals in the Periodic Table?

3 An empty, a half-filled and a completely filled 4*f* electronic level is often said to confer stability on the oxidation state of a lanthanide ion. Cite examples which bear out this statement.

Chapter 27

Nuclear chemistry

27.1 Introduction

Purely chemical reactions involve the rearrangement of the outermost electrons, i.e. atoms combine by sharing electrons (covalency) and by transferring electrons from one atom to another (electrovalency); in these processes atomic nuclei remain unchanged. Nuclear chemistry, however, investigates changes that occur in the atomic nucleus itself.

27.2 Natural radioactivity

The spontaneous emission of particles and rays from the nucleus of the uranium atom was first noticed by Becquerel in 1896. This observation was followed up by the Curies who discovered that the ore pitchblende, from which uranium is extracted, was more radioactive than purified uranium oxide, and they eventually succeeded in isolating two new elements—polonium and radium—which were responsible for this increased activity.

During radioactive decay, the atomic nucleus emits either an α-particle (a helium nucleus of mass 4 units and charge +2 units) or a β-particle (an electron of negligible mass and a charge of −1 unit); frequently both types of emission occur, e.g. in the decay of uranium (atomic mass 238) (p. 443). The emission of γ-radiation (radiation similar to X-radiation but of shorter wavelength) often occurs after α- and β-particle emission, and represents the excess energy that the nucleus sheds on passing into a more stable arrangement. The loss of an α- or β-particle from the nucleus of an atom of an element results in the formation of a new element; for instance, uranium (atomic mass 238) changes into the element thorium (atomic mass 234) when it emits an α-particle:

$$^{238}_{92}\text{U} \longrightarrow {}^{234}_{90}\text{Th} + {}^{4}_{2}\text{He (an } \alpha\text{-particle)}$$

Thorium (atomic mass 234) decays by the loss of a β-particle into protactinium (atomic mass 234):

$$^{234}_{90}\text{Th} \longrightarrow {}^{234}_{91}\text{Pa} + {}^{0}_{-1}e \text{ (a } \beta\text{-particle)}$$

Notice that the loss of a β-particle results in the formation of a new element whose atomic number is one unit greater than that of the decaying element. The ejection of a β-particle from the nucleus of an atom is due to the conversion of a neutron into a proton and an electron thus:

$$^{1}_{0}\text{n} \longrightarrow {}^{1}_{1}\text{H (a proton)} + {}^{0}_{-1}e$$

The decay of $^{238}_{92}$U into $^{234}_{90}$Th and then into $^{234}_{91}$Pa represents the first two steps in a radioactive disintegration series which is terminated in the production of non-radioactive $^{206}_{82}$Pb (p. 443).

All nuclei with atomic numbers greater than 83 are radioactive, i.e. the nuclei of elements that come after bismuth in the Periodic Table. Some of the lighter elements have radioactive isotopes and include $^{3}_{1}H$, $^{14}_{6}C$, $^{40}_{19}K$ and $^{87}_{37}Rb$.

27.3 Radioactive decay

The decay of a radioactive element is a random process and is not influenced by external factors such as temperature changes. The rate of decay is directly proportional to the number of atoms present, following an exponential law, the rate of decay decreasing with time. Kinetically the process is a first order reaction and can be expressed by the equation:

$$-\frac{dN}{dt} = kN \qquad (1)$$

where N is the number of radioactive atoms present, t is the time, and k is the rate constant (p. 121). Rearrangement of equation (1) followed by integration gives:

$$-\int \frac{dN}{N} = k \int dt$$

$$\text{or} \quad -\ln N = kt + \text{const.}$$

If the number of radioactive atoms present at time $t = 0$ is N_o, then

$$-\ln N_o = \text{const.}$$

$$\text{and} \quad -\ln N = kt - \ln N_o \qquad \text{or} \quad \ln(N_o/N) = kt$$

Conversion of Napierian logarithms to ordinary logarithms gives

$$2{\cdot}303\log_{10}(N_o/N) = kt \qquad (2)$$

Let $t_{\frac{1}{2}}$ represent the time needed for the original activity to decrease to a half, i.e. $N_o = 2N$, then

$$2{\cdot}303\log_{10}(N_o/\tfrac{1}{2}N_o) = 2{\cdot}303\log_{10}2 = kt_{\frac{1}{2}}$$

$$\text{or} \quad 0{\cdot}6932 = kt_{\frac{1}{2}}$$

$$\text{or} \quad t_{\frac{1}{2}} = \frac{0{\cdot}6932}{k} \qquad (3)$$

As equation (3) shows, the time taken for half the activity to disappear ($t_{\frac{1}{2}}$), or the half-life as it is often called, is independent of the number of radioactive atoms. The half-life of a particular radioactive element is a characteristic constant of that element, and so too is the value of k which is termed the radioactive decay constant; it is often represented by the symbol λ instead of k.

If radioactive decay proceeds by a number of individual steps, i.e. if the first radioactive decay product is itself radioactive and thus decays further, then each stage in the process is represented by its own characteristic rate equation and thus by its own half-life and radioactive decay constant.

27.4 The naturally occurring disintegration series

The decay of $^{238}_{92}U$ eventually terminates in the formation of the non-radioactive isotope of lead $^{206}_{82}Pb$ (Table 27A). The uranium disintegration series is given below, together with the half-life of the intermediate elements and the type of emission which occurs at each step in the series.

Table 27A The uranium disintegration series

Element	Symbol	Particle emitted	Half-life
Uranium	$^{238}_{92}U$	α	$4{\cdot}5 \times 10^9$ years
Thorium	$^{234}_{90}Th$	β	24·1 days
Protactinium	$^{234}_{91}Pa$	β	1·18 min
Uranium	$^{234}_{92}U$	α	$2{\cdot}48 \times 10^5$ years
Thorium	$^{230}_{90}Th$	α	$8{\cdot}0 \times 10^4$ years
Radium	$^{226}_{88}Ra$	α	$1{\cdot}62 \times 10^3$ years
Radon	$^{222}_{86}Rn$	α	3·82 days
Polonium	$^{218}_{84}Po$	α	3·05 min
Lead	$^{214}_{82}Pb$	β	26·8 min
Bismuth	$^{214}_{83}Bi$	β	19·7 min
Polonium	$^{214}_{84}Po$	α	$1{\cdot}6 \times 10^{-4}$ s
Lead	$^{210}_{82}Pb$	β	19·4 years
Bismuth	$^{210}_{83}Bi$	β	5·0 days
Polonium	$^{210}_{84}Po$	α	138 days
Lead	$^{206}_{82}Pb$		Non-radioactive

Notice the wide spread of half-lives from $1{\cdot}6 \times 10^{-4}$ s to $4{\cdot}5 \times 10^9$ years, and also formation of isotopes of uranium, thorium, polonium, bismuth and lead.

The two other series which occur in nature, beginning with thorium and actinium, are similar to the above; the former terminates with the formation of $^{208}_{82}Pb$, and the latter with the formation of $^{207}_{82}Pb$.

27.5 Nuclear stability and binding energy

The origin of the forces that accounts for nuclear stability is still much of a mystery, although it is clear that proton-neutron interaction is of great importance since no atom, other than that of hydrogen, contains only protons. Nuclear forces, whatever their nature, act over incredibly short distances, e.g. it has been estimated that nuclear radii are of the order of magnitude 10^{-13} cm compared with 10^{-8} cm for atomic radii. The protons and neutrons (collectively known as **nucleons**) which constitute atomic nuclei are extremely tightly packed together, and estimates have shown that the densities of all atomic nuclei are in the region of 10^{14} g cm^{-3}.

Despite the fact that little is known about the origin of nuclear forces, it is possible to calculate the energy needed to separate atomic nuclei into their isolated protons and neutrons; this quantity is called the **binding energy** of the nucleus. Consider the helium nucleus which contains 2 protons and 2 neutrons; the mass of the helium nucleus (on the $^{12}C = 12$ scale) is 4·0017 a.m.u. (atomic mass units) whereas the individual masses of the isolated proton and neutron are 1·0073 and 1·0087 a.m.u. respectively. The total mass of 2 protons and 2 neutrons is $(2 \times 1{\cdot}0073) + (2 \times 1{\cdot}0087) = 4{\cdot}0320$ a.m.u., i.e. there is a loss in mass of $4{\cdot}0320 - 4{\cdot}0017 =$

0·0303 a.m.u. when 2 protons and 2 neutrons form the helium nucleus, and this difference is called the **mass defect**. This annihilation of mass corresponds to the release of energy according to Einstein's equation:

$$E = mc^2$$

where E is the energy released (in joules), m is the loss in mass (in kg), and c is the velocity of light (approximately $3 \times 10^8\ \mathrm{m\,s^{-1}}$). It is instructive to calculate the energy released when 1 mole of helium atoms is formed from isolated protons and neutrons as follows:

The loss in mass in grams for the formation of 1 mole of helium atoms from isolated protons and neutrons is:

$$0{\cdot}0303\, L = 0{\cdot}0303 \times 6 \times 10^{23} \text{ (where } L \text{ is the Avogadro Constant)}$$

But 1 a.m.u. $= 1/L\,\mathrm{g} = 1/(6 \times 10^{23})\,\mathrm{g}$. Therefore the loss in mass is:

$$0{\cdot}0303\ \mathrm{g} \text{ or } 0{\cdot}0303 \times 10^{-3}\ \mathrm{kg}$$

The energy released is calculated from Einstein's equation:

$$\begin{aligned} E = mc^2 &= 0{\cdot}0303 \times 10^{-3} \times (3 \times 10^8)^2\ \mathrm{J\,mol^{-1}} \\ &= 0{\cdot}2727 \times 10^{13}\ \mathrm{J\,mol^{-1}} \end{aligned}$$

Thus the energy released is $2{\cdot}7 \times 10^{12}\ \mathrm{J\,mol^{-1}}$ (approx.)

Binding energies are generally quoted as energy (in million electron volts (MeV)) per nucleon. The binding energy of the helium nucleus is calculated below.

The electron volt is the energy gained or lost by a particle of unit charge when it passes through a potential difference of 1 volt; 1 million electron volts (MeV) are approximately equivalent to $9{\cdot}6 \times 10^{10}\ \mathrm{J\,mol^{-1}}$. The formation of one mole of helium nucleii from isolated protons and neutrons results in the release of

$$\frac{2{\cdot}7 \times 10^{12}}{9{\cdot}6 \times 10^{10}}\ \mathrm{MeV} = 28\ \mathrm{MeV} \text{ (approx.)}$$

But the helium nucleus contains 4 particles (2 protons and 2 neutrons), thus the binding energy per nucleon is

$$\frac{28}{4}\ \mathrm{MeV} = 7\ \mathrm{MeV} \text{ (approx.)}$$

Binding energies of the nuclei of other atoms can be calculated in a similar manner. Figure 27.1 shows the binding energies of the nuclei of atoms plotted against their respective mass numbers.

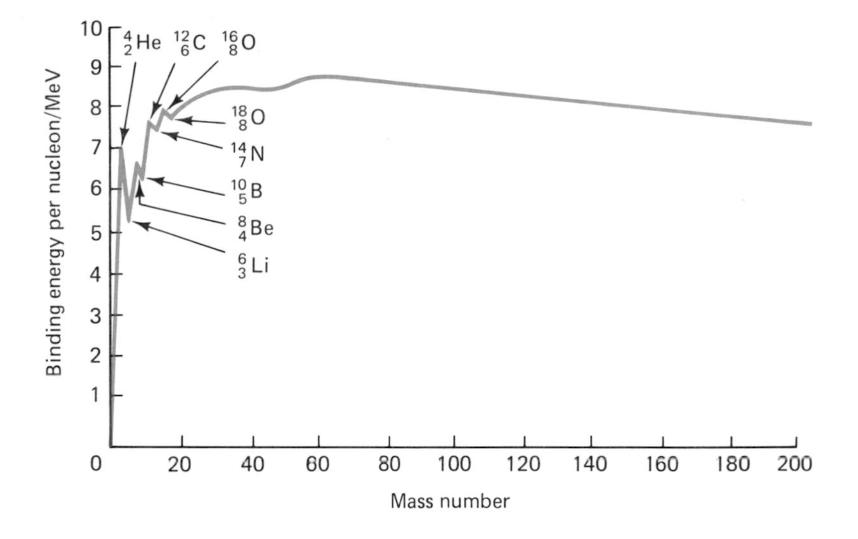

FIG. 27.1. *Binding energy per nucleon plotted against mass number*

Inspection of the graph shows that elements whose mass numbers lie in the region of 60 have the most stable nuclei, e.g. iron, cobalt, nickel and copper. Elements with light nuclei, if they can be made to coalesce, should produce more stable and heavier nuclei with the release of energy, as in the hydrogen bomb explosion and in stellar reactions; similarly, elements whose nuclei are very heavy should be capable of fission into lighter and more stable nuclei with the release of energy, as in the atomic bomb explosion and in nuclear reactions. The nuclei of helium, beryllium, carbon and oxygen are exceptionally stable for their size, and they do not fit at all on the smooth binding energy curve. Since beryllium, carbon and oxygen have mass numbers which are integral multiples of the mass number of the helium nucleus, it is tempting to suggest that the nuclei of these elements contain respectively two, three and four closely associated helium nuclei, i.e. 2 protons and 2 neutrons constitute a stable single entity. If the nuclei of all atoms can be considered to contain 'packets' of helium nuclei (not all the protons and neutrons can be considered to be arranged in this way since most elements have atoms which contain more neutrons than protons) then it is not difficult to see why radioactive decay results in the emission of α-particles (helium nuclei).

27.6 Artificial breakdown of the atomic nucleus

The production of stable elements

The first artificial breakdown of the atomic nucleus was achieved by Rutherford in 1919. He bombarded nitrogen gas with α-particles and obtained oxygen and a proton:

$$^{14}_{7}N + ^{4}_{2}He \longrightarrow ^{17}_{8}O + ^{1}_{1}H$$

Other nuclear transformations rapidly followed, helium nuclei, protons and neutrons being used as the bombarding particles. For instance the bombardment of beryllium with helium nuclei (α-particles) produces carbon and a neutron, thus:

$$^{9}_{4}Be + ^{4}_{2}He \longrightarrow ^{12}_{6}C + ^{1}_{0}n$$

Neutrons produced from the above nuclear reaction can be used in other transformations and they are convenient particles to use since, being uncharged themselves, they are not repelled by atomic nuclei. A typical neutron induced transformation is:

$$^{16}_{8}O + ^{1}_{0}n \longrightarrow ^{13}_{6}C + ^{4}_{2}He$$

An example of a proton induced transformation is:

$$^{27}_{13}Al + ^{1}_{1}H \longrightarrow ^{24}_{12}Mg + ^{4}_{2}He$$

Further research in this field led to the development of machines capable of accelerating the positively charged particles, e.g. α-particles and protons, to very high speeds. One of the first machines, called the cyclotron, was designed by E. O. Lawrence in America; more powerful machines such as the synchrotron have since been developed.

The production of radioactive elements

The first nuclear transformation to give a radioactive element, which decayed to give a stable nucleus, was achieved by M. and Mme. Curie-Joliot in 1933. For instance, when magnesium was bombarded with α-particles, neutrons and positrons (positively charged electrons) were released. When the magnesium was removed from the source of the α-particles the formation of neutrons ceased but the production of positrons continued and aluminium was formed in the reaction. These results were explained by postulating the formation of an unstable isotope of silicon which then decayed to give aluminium by positron emission:

$$^{24}_{12}Mg + ^{4}_{2}He \longrightarrow \underset{\text{(radioactive)}}{^{27}_{14}Si} + ^{1}_{0}n$$

$$^{27}_{14}Si \longrightarrow ^{27}_{13}Al + {}^{\ 0}_{+1}e$$

The above reactions are conveniently written in a shorthand fashion thus:

$$^{24}_{12}Mg(\alpha,n)^{27}_{14}Si \longrightarrow ^{27}_{13}Al + {}^{\ 0}_{+1}e$$

where α represents the projectile used in the reaction and n represents the particle emitted.

Plate 11. The Harwell 136 MeV electron linear accelerator. This machine provides intense bursts of neutrons (By courtesy of United Kingdom Atomic Energy Authority)

Other typical reactions which produce radioactive elements include the following:

α-particle induced reaction	${}^{14}_{7}N(\alpha,n){}^{17}_{9}F \longrightarrow {}^{17}_{8}O + {}^{0}_{+1}e$
Proton-induced reaction	${}^{11}_{5}B(p,n){}^{11}_{6}C \longrightarrow {}^{11}_{5}B + {}^{0}_{+1}e$
Neutron-induced reaction	${}^{23}_{11}Na(n,\gamma){}^{24}_{11}Na \longrightarrow {}^{24}_{12}Mg + {}^{0}_{-1}e$

27.7 Nuclear energy

Nuclear fission

Uranium-235 can absorb neutrons and, in so doing, splits into smaller fragments, at the same time releasing more neutrons than were originally captured. These neutrons can cause further fission of uranium-235 nuclei with the release of more and more neutrons. Each time the nucleus of uranium-235 is split a considerable amount of energy is released and, if the released neutrons are allowed to cause further fission, more and more energy is released, so that a chain reaction is established and an atomic explosion occurs.

In a nuclear reactor the production of neutrons is carefully controlled and graphite has been used to moderate the speed of the neutrons released. Cadmium rods, which readily absorb neutrons, can be introduced into the reactor itself to prevent the establishment of an uncontrollable chain reaction. The energy released under these controlled conditions is now being used to produce steam which, in turn, is used in the production of electricity. Nuclear-powered submarines, aircraft carriers and cruisers now form part of several navies, including those of the USA, the USSR and UK. In addition to these uses, nuclear reactors are useful in the synthesis of new elements and in the formation of radioisotopes of common elements, since the neutrons that are formed in such devices are readily used as bombarding projectiles.

Plate 12. An advanced gas-cooled reactor uses uranium dioxide fuel and graphite moderators. The photograph shows the interior of Hinchley Point 'B' charge hall (By courtesy of United Kingdom Atomic Energy Authority)

Nuclear fusion

The coalescence of protons to give a helium nucleus should result in the liberation of tremendous amounts of energy (see binding energy curve p. 445); indeed it can be shown theoretically that much more energy can be released by nuclear fusion than by nuclear fission. Unfortunately the coalescence of protons requires temperatures in the region of several million degrees in order to overcome the mutual repulsion of these positively charged particles and the controlled release of energy from such a process has still to be accomplished.

However, the uncontrolled release of energy by this fusion process occurs in stars, including our own sun, and the overall process, which involves several stages, can be represented thus:

$$4{}^{1}_{1}\mathrm{H} \longrightarrow {}^{4}_{2}\mathrm{He} + 2{}^{0}_{+1}e + \text{Energy}$$

The same type of reaction occurs in an hydrogen bomb explosion, an atomic explosion being used as a triggering mechanism to provide the extremely high temperatures that are required for initial fusion to occur. A typical man-made nuclear fusion process involves the coalescence of deuterons, i.e. the nuclei of the deuterium atom which is an isotope of hydrogen:

$$2{}^{2}_{1}\mathrm{H} \longrightarrow {}^{4}_{2}\mathrm{He} + \text{Energy}$$

If a method of controlling nuclear fusion can be found, then an unlimited source of energy immediately becomes available from the hydrogen atoms in water.

27.8 Uses of radioisotopes

The uses of radioisotopes are numerous, e.g. carbon-14 is used to label organic molecules and its fate during subsequent breakdown of these molecules can be traced, thus allowing reaction mechanisms to be investigated; similarly, the labelling of phosphatic fertilisers with radioactive phosphorus-32 can be employed to determine the uptake of phosphorus by plants. Three other applications are discussed below.

Carbon-14 dating

The carbon dioxide in the atmosphere contains small amounts of the radioactive isotope ${}^{14}_{6}\mathrm{C}$ which is produced by the action of neutrons on the nitrogen (the neutrons are produced by the cosmic rays which are present in the atmosphere):

$${}^{14}_{7}\mathrm{N} + {}^{1}_{0}\mathrm{n} \longrightarrow {}^{14}_{6}\mathrm{C} + {}^{1}_{1}\mathrm{H}$$

Since plants and trees take up carbon dioxide and convert it into tissue material and animals, including humans, eat plants, all living matter contains the same small proportion of carbon-14 as occurs in the atmosphere. After death the uptake of carbon dioxide ceases and the level of carbon-14 in the body and in plant tissues falls (${}^{14}_{6}\mathrm{C}$ has a half-life of 5600 years). For instance, a tree felled 5600 years ago will only contain half the activity of a growing tree. Paper and papyrus have a vegetable origin and can thus be dated by the carbon-14 method, e.g. the Dead Sea scrolls have been dated by this method.

The detection of wear in an engine

The incorporation of a radioactive isotope of a metal in the piston rings of engines allows the accurate measurement of wear by friction. The engine is run under test for a specified period and the amount of radioactivity in the lubricating oil determined. Tests such as this have been directly responsible for the development of more efficient engine oils.

The non-equivalence of the sulphur atoms in the thiosulphate anion

The thiosulphate anion is formed when a solution of a sulphite is boiled with sulphur. If the sulphur (but not the sulphite) contains radioactive sulphur, $^{35}_{16}S$, then, of course, the radioactivity will be present in the final thiosulphate:

$$^{35}_{16}S(s) + SO_3^{2-}(aq) \longrightarrow [^{35}_{16}SSO_3]^{2-}(aq)$$

When the thiosulphate is broken down it is found that the radioactivity appears in the precipitated sulphur and not in the sulphur dioxide:

$$[^{35}_{16}SSO_3]^{2-}(aq) + 2H^+(aq) \longrightarrow {}^{35}_{16}S(s) + H_2O(l) + SO_2(g)$$

This experiment clearly shows that the two sulphur atoms in the thiosulphate anion do not occupy equivalent positions.

27.9 The transuranium elements

The transuranium elements which are all man-made are part of an inner transition series analogous to the lanthanides (p. 436); they are called the actinides since the first member of the series is actinium (Table 27B).

Table 27B The transuranium elements

Atomic Number	Name	Symbol
93	Neptunium	Np
94	Plutonium	Pu
95	Americium	Am
96	Curium	Cm
97	Berkelium	Bk
98	Californium	Cf
99	Einsteinium	Es
100	Fermium	Fm
101	Mendelevium	Md
102	Nobelium	No
103	Lawrencium	Lr

The discovery of the first transuranium element neptunium dates back to 1940, the formation of the others following at fairly regular intervals and culminating in the identification of lawrencium in 1961. Without doubt the person who contributed most to our knowledge of these man-made elements is the American chemist Seaborg. Many of these transuranium elements have now been made in different isotopic forms, and the methods of formation, given on the next page, indicate the formation of only one of the possible isotopes. As can be seen, many of them are exceedingly short-lived.

(a) Neptunium

$^{238}_{92}U + ^{1}_{0}n \longrightarrow ^{239}_{93}Np + ^{0}_{-1}e$ $\quad t_{\frac{1}{2}} = 2{\cdot}35$ days

(b) Plutonium

$^{239}_{93}Np \longrightarrow ^{239}_{94}Pu + ^{0}_{-1}e$ $\quad t_{\frac{1}{2}} = 24\,360$ years

(c) Americium

$^{239}_{94}Pu + 2^{1}_{0}n \longrightarrow ^{241}_{95}Am + ^{0}_{-1}e$ $\quad t_{\frac{1}{2}} = 458$ years

(d) Curium

$^{239}_{94}Pu + ^{4}_{2}He \longrightarrow ^{242}_{96}Cm + ^{1}_{0}n$ $\quad t_{\frac{1}{2}} = 162{\cdot}5$ days

(e) Berkelium

$^{241}_{95}Am + ^{4}_{2}He \longrightarrow ^{243}_{97}Bk + 2^{1}_{0}n$ $\quad t_{\frac{1}{2}} = 4{\cdot}5$ hours

(f) Californium

$^{242}_{96}Cm + ^{4}_{2}He \longrightarrow ^{245}_{98}Cf + ^{1}_{0}n$ $\quad t_{\frac{1}{2}} = 44$ min

(g) Einsteinium

$^{238}_{92}U + ^{14}_{7}N \longrightarrow ^{248}_{99}Es + 4^{1}_{0}n$ $\quad t_{\frac{1}{2}} = 25$ min

(note the use of the heavy bombarding projectile)

(h) Fermium

$^{238}_{92}U + ^{16}_{8}O \longrightarrow ^{253}_{100}Fm + ^{1}_{0}n$ $\quad t_{\frac{1}{2}} = 4{\cdot}5$ days

(note the use of the heavy bombarding projectile)

The first identification of fermium was achieved on no more than 200 atoms of the element.

(i) Mendelevium

$^{253}_{99}Es + ^{4}_{2}He \longrightarrow ^{256}_{101}Md + ^{1}_{0}n$ $\quad t_{\frac{1}{2}} = 1{\cdot}5$ hours

(j) Nobelium

$^{246}_{96}Cm + ^{12}_{6}C \longrightarrow ^{254}_{102}No + 4^{1}_{0}n$ $\quad t_{\frac{1}{2}} = 3$ s

(note the use of the heavy bombarding projectile)

(k) Lawrencium

$^{252}_{98}Cf + ^{11}_{5}B \longrightarrow ^{257}_{103}Lr + 6^{1}_{0}n$ $\quad t_{\frac{1}{2}} = 8$ s

(note the use of the heavy bombarding particle)

27.10 The actinides as an inner transition series

Prior to the discovery of the first transuranium element neptunium in 1940 only the first four members of this series were known, namely actinium, thorium, protactinium and uranium. With the discovery of neptunium, with properties similar to uranium, it became clear that these elements marked the beginning of an inner transition series similar to the lanthanides (p. 436). Indeed the realisation that this series of elements was basically similar to the lanthanides helped considerably in the identification of further transuranium elements, and it could safely be predicted that ion-exchange methods would be useful for their separation (see the separation of the lanthanides, p. 437).

Whereas the lanthanides are a group of elements in which the 4*f* level fills (p. 437), the actinides constitute a series in which the 5*f* level is successively filled and is complete when it holds a maximum of fourteen electrons. The filling of this 5*f* level is not quite so regular as the filling of the 4*f* level in the case of the lanthanides (Table 27C).

Table 27C The first few members of the actinides and the lanthanides

Actinides		Lanthanides	
Thorium	$6d^2 7s^2$	Cerium	$4f^2 5d^0 6s^2$
Protactinium	$5f^2 6d^1 7s^2$	Praseodymium	$4f^3 5d^0 6s^2$
Uranium	$5f^3 6d^1 7s^2$	Neodymium	$4f^4 5d^0 6s^2$
Neptunium	$5f^4 6d^1 7s^2$	Promethium	$4f^5 5d^0 6s^2$
Plutonium	$5f^6 6d^0 7s^2$	Samarium	$4f^6 5d^0 6s^2$

Among the similarities of these two series of elements can be mentioned the formation of coloured ions, the existence of paramagnetic compounds, and a stable oxidation state of +3. The actinides do, however, show a wider range of oxidation states than the lanthanides and the first few members of this series exhibit stable oxidation states greater than +3, e.g. neptunium exhibits a stable oxidation state of +5 in the ion NpO_2^+ and uranium one of +6 in the UO_2^{2+} ion.

27.11 Suggested reading

Glenn T. Seaborg, *Some Recollections of Early Nuclear Age Chemistry,* Journal Chem. Education, Vol. 45, 278, 1968.

W. F. Libby, *Dating by Radiocarbon*, Accounts of Chemical Research, Vol. 5, 289, 1972.

The Nucleus of the Atom, Open University, S 100, Unit 31.

Elementary Particles, Open University, S 100, Unit 32.

Glenn T. Seaborg, *Man-made Transuranium Elements*, Prentice Hall Inc., 1963.

G. N. Walton, *Energy from the atom,* Education in Chemistry, No. 1, Vol. 15, 1978.

27.12 Questions on chapter 27

1 (a) List the three main fundamental particles which are constituents of atoms. Give their relative charges and masses.

(b) Similarly name and differentiate between the radiations emitted by naturally occurring radioactive elements.

For the particle which is common to lists (a) and (b) name two methods by which it can be obtained from non-radioactive metals.

Complete the following equations for nuclear reactions by using the Periodic Table to identify the elements **X**, **Y**, **Z**, **A** and **B** and add the atomic and mass numbers where these are missing.

(a) ${}^{207}_{82}\mathrm{Pb} \longrightarrow {}_{83}\mathbf{X} + {}^{0}_{-1}e$

(b) ${}^{27}_{13}\mathrm{Al} + {}^{1}_{0}\mathrm{n} \longrightarrow {}^{24}\mathbf{Y} + {}^{4}_{2}\mathbf{Z}$

(c) ${}^{14}_{7}\mathrm{N} + {}^{4}_{2}\mathbf{Z} \longrightarrow {}^{17}_{8}\mathbf{A} + {}_{1}\mathbf{B}$ (JMB)

2 (a) What are the relative masses and charges of the particles emitted from radioactive substances?

The isotope ${}^{214}_{83}\mathrm{Bi}$ disintegrates not only into an isotope of thallium (${}_{81}\mathrm{Tl}$) but also into an isotope of polonium (${}_{84}\mathrm{Po}$). Each of these products disintegrates further, giving an isotope of lead. Deduce what particles have been emitted to bring about these two series of changes and give the mass number of each element.

(b) Define the half-life of a radioactive element.

The short-lived isotope actinium **B** has a half-life of 36·0 minutes. What fraction of the original quantity of actinium **B** remains after 1152 minutes?

(c) Thorium **B** is isotopic with lead and may be assumed to have the same physical as well as chemical properties. A small amount of thorium **B** was mixed with a quantity of a lead salt containing 10 mg of lead, the whole brought into solution and the thorium and the lead precipitated together by addition of a solution of a soluble chromate. Evaporation of 10 cm^3 of the supernatant liquid gave a residue which was 1/24 000 as active as the original quantity of thorium **B**. Calculate the solubility of lead (II) chromate in moles per dm^3.
(JMB)

3 The disintegration rates (measured on a Geiger counter) at 25°C of a sample of radioactive sodium-24 at various times are given in the table below. Utilising all the experimental observations, determine the half-life of sodium 24 and from this calculate the disintegration constant (λ).

Comment on the effect of an increase in temperature on this process.

Radioactive Decay of ^{24}Na (hours)	0	2	5	10	20	30
Rate of Disintegration (Counts/sec)	670	610	530	421	267	168

(O & C)

4 The mass of a hydrogen atom is 1·0078 atomic units, of a deuterium atom (heavy hydrogen) is 2·0141 atomic units, and of a helium atom is 4·0026 atomic units. Comment on these figures. (O & C)

5 Explain the existence of isotopes. Name one radioactive isotope and give an example of its use in modern scientific practice.

A small sample of gold was irradiated in a nuclear reactor and the activity of the radioactive gold produced was measured with a Geiger counter at various intervals of time. The same experiment was carried out with a small sample of bromine and the results below were obtained.

Time in hours	0,	0·1,	0·2,	0·5,	1,	2,	5,	10,	24,	48,	72,	96
Radioactive gold (Disintegrations per minute)	299	—	—	—	—	—	285,	271,	232,	181,	139,	108

Time in hours	0,	0·1,	0·2,	0·5,	1,	2,	5,	10,	24,	48,	72,	96
Radioactive bromine (Disintegrations per minute)	901,	795,	719,	577,	482,	435,	405,	367,	279,	178,	112,	71

(a) From a suitable graph deduce the half-life of radioactive gold.
(b) From a similar graph comment on the decay of radioactive bromine. (O & C)

6 (a) (i) What is meant by the term *isotope*?
(ii) How could it be demonstrated that a sample of an element contained two or more isotopes?

(b) Give **two** ways in which the element $^{24}_{12}Mg$ could lose electrons. Give the atomic numbers and mass numbers of the species formed.

(c) Indicate the numbers and types of particles emitted in order to change the unstable species $^{30}_{13}Al$ and $^{57}_{25}Mn$ to the stable species $^{26}_{13}Al$ and $^{49}_{23}V$ respectively.

(d) If the radioactive decay of ^{63}Ni to ^{63}Cu has a half life of 120 years, how long will it take for $\frac{15}{16}$ of the nickel to be changed into copper?

(e) Explain how isotopic labelling would enable you to determine whether the acid or the alcohol molecule is responsible for producing the oxygen of the water molecule produced during esterification.

(f) Complete the following equations

$$^{16}_{8}O + \quad \longrightarrow {}^{13}_{6}C + {}^{4}_{2}He$$
$$^{27}_{13}Al + \quad \longrightarrow {}^{24}_{12}Mg + {}^{4}_{2}He$$
$$^{14}_{7}N + {}^{4}_{2}He \longrightarrow \quad + {}^{1}_{1}H$$

(AEB)

7 The isotopic composition of rubidium is ^{85}Rb 72 per cent, ^{87}Rb 28 per cent. ^{87}Rb is weakly radioactive and decays by β-emission with a decay constant of $1{\cdot}1 \times 10^{-11}$ year $^{-1}$. A sample of the mineral pollucite was found to contain 450 mg Rb and 0·72 mg of ^{87}Sr. Estimate the age of the pollucite, noting any assumptions which you may need to make. (Oxford Schol.)

8 What is meant by the binding energy of the nucleus of an atom? The atomic mass of $^{16}_{8}O$ is 15·995 a.m.u, while the individual masses of an isolated proton and neutron are respectively 1·0078 and 1·0087 a.m.u. Calculate the binding energy of the oxygen nucleus in joules (refer to section 27.5 p. 443).

9 Calculate the energy released (in joules) when 0·005g of mass is annihilated in a nuclear reaction.

10 Give an account of the two nuclear processes fission and fusion. Why is the search for methods of controlling the fusion of hydrogen atoms so important?

11 Name the three different types of rays emitted by radioactive substances. What is the nature of these rays, how can they be detected, and how do they differ in penetrating power?

Complete the following equations:

$$^{14}N + {}^{1}_{0}n \rightarrow \quad + {}^{1}H$$
$$^{238}_{92}U \rightarrow {}^{234}U + \quad + \qquad \text{(C)}$$

12 What are the relative masses and charges of the particles emitted by radioactive substances? State what is meant by the terms *atomic number* and *mass number*, and explain how the stability of an atomic nucleus is affected by the relative values of these numbers.

Complete the following equations:

$$^{32}S + \quad \rightarrow {}^{32}P + {}^{1}_{1}H$$
$$^{27}Al + {}^{1}_{1}H \rightarrow \alpha +$$

Either How is ^{14}C produced in nature and how can it be used in 'radiocarbon dating'?

Or explain how nuclear fission may be used to provide a controlled source of heat. (C)

13 What do you understand by the term *isotopes*? How could it be demonstrated whether or not a particular element is a mixture of *isotopes*?

The initial part of the decay series of naturally occurring uranium 238 is given below. The element uranium is in Group VI of the Periodic Table. Giving your reasons, predict in which groups of the Periodic Table you expect the elements thorium and protactinium to be.

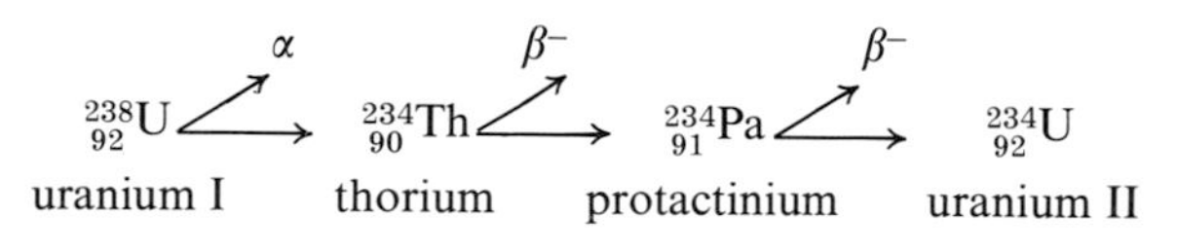

The radioactive isotope potassium 40 decays to stable argon 40 with a half-life of $1{\cdot}27 \times 10^{9}$ years. Analysis of the lunar rock samples brought back from the Sea of Tranquillity in July 1969 by the Apollo 11 project showed a potassium 40:argon 40 ratio of 1:7. Assuming that there had been no escape of argon from the rock and that none was present when the moon was formed, calculate the age of the moon. (O & C)

Answers to Questions

CHAPTER 1

3 10·8 **4** 133

5 M_2Cl_6

7 MCl_4, MCl_3

CHAPTER 4

1 (a) Wavelengths of lines in Balmer Series: 656 nm, 486 nm, 434 nm
(b) Series limit in the Lyman Series: 91·1 nm
(c) Energy needed to remove the electron: $2{\cdot}18 \times 10^{-18}$ J
(d) Ionisation energy of the hydrogen atom: 1313 kJ mol^{-1}

2 Ionisation energy of the sodium atom: 494 kJ mol^{-1}.

CHAPTER 5

10 $Ca(s) + Cl_2(g) \longrightarrow Ca^{2+}(Cl^-)_2(s)$ $\Delta H = -841$ kJ mol^{-1}
$Ca(s) + \frac{1}{2}Cl_2(g) \longrightarrow Ca^+Cl^-(s)$ $\Delta H = -208$ kJ mol^{-1}
hypothetical

CHAPTER 7

2 Wavelength is 0·0245 nm

CHAPTER 8

1 (a) $\Delta H = -597{\cdot}5$ kJ mol^{-1}
(b) $\Delta H = 1255$ kJ mol^{-1}

6 Standard free energy change is $-95{\cdot}5$ kJ mol^{-1}

7 Standard entropy change, $\Delta S^{\ominus}$(298 K), is $+80{\cdot}4$ J K^{-1} mol^{-1}
Standard free energy change, $\Delta G^{\ominus}$(298 K), is $-80{\cdot}0$ kJ mol^{-1}.

8 Standard free energy change, $\Delta G^{\ominus}$(298 K), is $-70{\cdot}1$ kJ mol^{-1}
Equilibrium constant is 12·29 $atm^{-\frac{1}{2}}$ for the reaction
$$SO_2(g) + \tfrac{1}{2}O_2(g) \rightleftharpoons SO_3(g)$$

CHAPTER 9

3 (c) Activation energy is 45·9 mol^{-1}

CHAPTER 10

7 (a) 1·10 V
(b) 0·46 V
(c) 1·75 V

9 (a) 0·22 V

10 (a) 0·04 V
(b) 3·86 kJ mol^{-1}
(c) K = 4·7 dm^3 mol^{-1}

12 e.m.f. is 1·229 V,
max. energy is 237 200 J,
standard entropy change is $-163{\cdot}1$ J K^{-1} mol^{-1}

CHAPTER 12

6 Free energy change at 1000°C is $+502$ kJ mol^{-1}
Free energy change at 2000°C is -314 kJ mol^{-1}

CHAPTER 18

7 44% CO, 36% H_2, 20% CO_2

18 Si_2H_6

19 **A** is $Sn(C_2H_5)_4$ and **B** is $Sn(C_2H_5)_3Cl$

CHAPTER 19

3 Empirical formula is NS (the compound is in fact N_4S_4)
Equation assuming the formula N_4S_4 is:

$$10S + 4NH_3 \longrightarrow N_4S_4 + 6H_2S$$

The molecular formula, however, cannot be deduced from the information given.

6 0·69g sodium nitrite; 2·13g sodium nitrate

8 Hydroxylamine conc. is 158 g dm^{-3}

14 (c) PH_3 (h) 26·2%

CHAPTER 20

5 60·7 g dm^{-3}

8 $4Br_2 + S_2O_3^{2-} + 5H_2O \longrightarrow 2SO_4^{2-} + 8Br^- + 10H^+$

10 Empirical formula of **A** is $SOCl_2$

11 (a) 57·3 per cent
(b) **B** is $SO_2(NH_2)_2$ and **C** is $(SO_2NH)_3$
(c) (i) $SO_2Cl_2 + NH_3 \longrightarrow SO_2Cl.NH_2 + HCl$
(ii) $SO_2Cl_2 + 2NH_3 \longrightarrow SO_2(NH_2)_2 + 2HCl$
(iii) $3SO_2(NH_2)_2 \longrightarrow (SO_2NH)_3 + 3NH_3$
(d) Structural formula of substance **C** is:

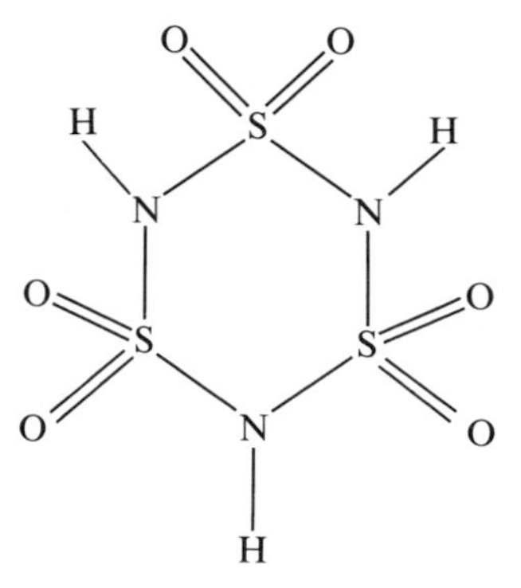

12 Atomic weight (relative atomic mass) of **X** is 32

CHAPTER 24

6 Equilibrium constant is 10^6 dm^3 mol^{-1}

13 Equilibrium constant is 10^{13} dm^6 mol^{-2}

CHAPTER 27

2 (b) 2^{-32}
(c) Solubility of lead chromate is $2·01 \times 10^{-7}$ mol dm^{-3}

3 Half-life of sodium-24 is 15 hours; disintegration constant λ is 0·046 $hour^{-1}$

5 Half-life of radioactive gold is 65 hours

7 Age of mineral pollucite is $5·2 \times 10^8$ years

8 $1·233 \times 10^{13}$ J mol^{-1} of oxygen atoms

9 $4·5 \times 10^{11}$ J

Index

Acid-base reactions, 146–148
  conjugate acid-base pairs, 147
Acids, 145
  Arrhenius concept of, 145–146
  Brønsted-Lowry concept of, 146–147
  early definition of, 145
  Lewis concept of, 150
Actinides, 450, 451–452
Activated complex, 124
Activation energy, 122, 123, 125
Alkalis, 148–149
  early definition of, 145
Alkali metal halides,
  structures of, 74
Alkali metals (see Group IA)
Alkali metal salts,
  hydrolysis of, 184
Alkaline earth metals (see Group 2A)
Alpha particles, 27, 28, 441
Aluminates, 208
Aluminium,
  extraction of, 205
  occurrence of, 205
  properties of, 206
  uses of, 205
Aluminium alloys, 205–206
Aluminium bromide, 208
Aluminium carbide, 210
Aluminium chloride, 206–208
Aluminium compounds,
  hydrolysis of, 210–211
Aluminium fluoride, 206
Aluminium hydride, 209–210
Aluminium hydroxide, 209
Aluminium iodide, 208
Aluminium oxide, 208
Aluminium sulphate, 209
Aluminium sulphide, 210
Aluminium triethyl, 210
Alumino-silicates, 232
Alums, 209
Americium, 450, 451
Ammonia, 249–251
  industrial production of, 249
  liquid, 251
  preparation of, 249
  properties of, 250–251
  structure of, 250
  test for, 250
Ammonium hydroxide, 250
Ammonium metavanadate, 258, 259
Ammonium nitrate, 253, 261
Ammonium salts, 250
Ammonium sulphate, 253
Ammonium trithiostannate (IV), 241
Amphoteric oxides, 290
Amphoteric hydroxides, 148–149, 291
Anhydrite, 194, 299
Antimonates (III), 278
Antimonates (V), 278
Antimony,
  extraction of, 273
  halides of, 275–276
  hydride of, 274–275
  occurrence of, 273
  oxides of, 276–277
  oxyacids of, 278
  properties of, 274
  structure of, 274
  sulphides of, 279
  uses of, 274
Antimony pentachloride, 276
Antimony (V) sulphide, 279
Antimony (V) oxide, 277
Antimony (III) oxide, 276–277
Antimony (III) sulphide, 279
Antimony trichloride, 276
Argon, 169, 170
Arrhenius equation, 122, 123
Arsenates (III), 277–278
Arsenates (V), 278
Arsenic,
  extraction of, 273
  halides of, 275–276
  hydride of, 274–275
  occurrence of, 273
  oxides of, 276–277
  oxyacids of, 277–278
  properties of, 274
  structure of, 274
  sulphides of, 278–279
  uses of, 274
Arsenic (V) sulphide, 279
Arsenic (V) oxide, 277
Arsenic (III) oxide, 276–278
Arsenic (III) sulphide, 278–279
Arsenic trichloride, 276
Arsine, 274, 275
Astatine, 325, 345
Aston, 31
Atom,
  structure of, 23–32
Atomic heat, 10
Atomic mass,
  chemical, 14
  definition of, 13–14
  determination of, 8–13, 32–33
  physical, 14, 32–33
Atomic nucleus,
  breakdown of, 441, 446–447
Atomic number, 30
Atomic orbitals, 85–90
  combination of, 90–94
Atomic radius, 70–71
  variation of, 72–73
Atomic theory, 1–2
Atomic volume, 17
Atomicity, 13
Avogadro's hypothesis, 6–7
  deductions from, 7–8
Avogadro's number, 23
Azides, 252

Barium, 187, 189
Barium bromate (V), 341
Barium chlorate (V), 341
Barium ferrate (VI), 377
Barium iodate (V), 341
Base,
  Arrhenius concept of, 145–146
  Brønsted-Lowry concept of, 146–147
  early definitions of, 145
  Lewis concept of, 150
Bauxite, 205
Becquerel, 27, 441
Benzene,
  resonance structure of, 81, 217
Berkelium, 450, 451
Beryllium, 187
  diagonal relationship with aluminium, 197
  occurrence of 189
  uses of, 189
Beryllium oxide, 197
Berzelius, 3, 4, 6, 10
Beta particles, 27, 28, 30, 441
Binding energy, 443–445
Bismuth,
  extraction of, 273
  halides of, 275–276
  hydride of, 275
  occurrence of, 273
  oxides of, 276–277
  properties of, 274
  structure of, 274
  uses of, 274
Bismuth trichloride, 276
Bismuth trifluoride, 275
Bismuth (V) oxide, 277
Bismuth (III) oxide, 277
Bismuth (III) sulphide, 279
Bismuthates (V), 278
Bismuthine, 275
Blast furnace, 371
Bleaching powder, 191, 332

Bohr,
equation, 37
theory of the atom, 37–38
Bond dissociation energy 101–102
Bond energy, 101–104
average, 101
determination of, 102–103
Bond length, 70–71
Borates, 203
Borate glasses, 203
Borax (see sodium tetraborate)
Borazine, 204
Boric acid,
metaboric acid, 202
orthoboric acid, 202–203
Boron,
extraction of, 200
occurrence of, 200
properties of, 200
uses of, 200
Boron carbide, 225
Boron halides, 200–201
Boron hydrides, 201–202
Boron nitride, 80, 204–205
Boron trichloride, 200–201
Boron trifluoride, 200–201
Boron trioxide, 202
Borosilicate glass, 233
Brass, 407
Bromate (I), 334
Bromate (V) ion, 342
Bromic (I) acid, 334, 340
Bromic (V) acid, 341–342
Bromide ion,
test for, 339
Bromine,
chemical properties of, 333–335
extraction of, 331
laboratory preparation of, 332
occurrence of, 331
oxides of, 340
oxyacids of, 340–342
physical properties of, 332–333
uses of, 336
Bromine trifluoride, 344
Bronze, 407

Cadmium,
complexes of, 428
compounds of, 425–426
extraction of, 424
occurrence of, 423
properties of, 424
uses of, 424
Cadmium carbonate, 427–428
Cadmium chloride, 426–427
Cadmium hydroxide, 426
Cadmium nitrate, 427
Cadmium oxide, 425–426
Cadmium sulphate, 427
Cadmium sulphide, 426
Caesium, 174, 176
Caesium chloride,
structure of, 74
Californium, 450, 451
Calcium, 187–190
Calcium bromide, 193
Calcium carbide, 195
Calcium carbonate, 192
Calcium chloride, 193
Calcium fluoride, 192–193
Calcium hydrogen carbonate, 192
Calcium hydrogen sulphite, 195
Calcium hydroxide, 190–192
Calcium iodide, 193
Calcium nitrate, 193
Calcium nitride, 195
Calcium oxide, 190
Calcium phosphate (V), 195, 262
Calcium sulphate, 194
Calomel electrode, 434
Cannizzaro, 8, 9, 10
Carbides,
covalent, 225
interstitial, 224–225
salt-like, 224
Carbon,
allotropes of, 214–215
halides of, 218
hydrides of, 217–218
occurrence of, 214
oxides of, 219–222
properties of, 216
Carbon – 14 dating, 449
Carbon black, 216
Carbon cycle, 214
Carbon dioxide, 219
Carbon disulphide, 216, 218
Carbon monoxide, 220–221
Carbon suboxide, 222
Carbonates, 222–224
Carbonic acid, 222
Carbonyls, 400–401
Carbonyl chloride, 221
Carbonyl sulphide, 221
Carnallite, 175, 189
Castner-Kellner process, 177–178
Catalysis,
theory of, 125–126, 354
Catalyst, 125
Catenation, 217, 284, 305
Cavendish, 2
Cementite, 374, 376
Cerium, 437, 438, 440
Chadwick, 30
Charcoal, 216
Chile saltpetre, 175, 331
Chlorate (I), 334, 341
Chlorate (III), 339, 341
Chlorate ion (V), 334
Chloric (I) acid, 334, 340, 341
Chloric (III) acid, 341
Chloric (V) acid, 341, 342
Chloric (VII) acid, 343
Chloride ion,
test for, 339
Chlorine,
chemical properties of, 333–335
extraction of, 330–331
laboratory preparation of, 332
occurrence of, 330
oxides of, 339
physical properties of, 332–333
uses of, 336
Chlorine dioxide, 339
Chlorine heptoxide, 339
Chlorine monoxide, 339
Chlorine trifluoride, 344
Chlorosulphonic acid, 317
Chromates (VI), 361–362
Chromic (VI) acid, 361
Chromite, 360
Chromium,
extraction of, 360
occurrence of, 360
oxidation states of, 361
properties of, 360–361
uses of, 361
Chromium (II) chloride, 364
Chromium (III) chloride, 364
Chromium (III),
complexes of, 365
Chromium (II) compounds, 366
Chromium (III) compounds, 363–365
Chromium (VI) compounds, 361–363
Chromium (III) hydroxide, 364
Chromium (III) oxide, 363–364
Chromium (VI) dichloride dioxide, 363
Chromium (VI) oxide, 361
Chromium (III) sulphate, 364–365
Cinnabar, 428, 430
Clay, 231
Closed system, 108
Cobalt,
extraction of, 382
occurrence of, 382
oxidation states of, 383
properties of, 383
uses of, 383
Cobalt (II) carbonate, 384
Cobalt (II) compounds, 383–384
complexes of, 384
Cobalt (III) compounds, 383
complexes of, 384–385
Cobalt (III) fluoride, 383
Cobalt (II) hydroxide, 384
Cobalt (II) nitrate, 384
Cobalt (II) oxide, 384
Cobalt (II) sulphate, 384
Cobalt (III) sulphate, 383
Cobalt (II) sulphide, 384
Coke, 216
Complex ions,
octahedral, 353, 396–399
planar, 353, 399–400
tetrahedral, 353, 398
theory of colour of, 395–398
Conservation of energy,
Law of, 98–99
Conservation of mass,
Law of, 1, 2
Constant composition,
Law of, 1, 2
Contact process, 311–312
Coordinate bond (see dative covalency)
Co-ordination number, 73, 74
Copper,
extraction of, 405–406
occurrence of, 405
oxidation states of, 407
properties of, 406

refining of, 406
uses of, 406–407
Copper (I) bromide, 409
Copper (II) bromide, 411
Copper (I) carbide, 408
Copper (II) carbonate, 412
Copper (I) chloride, 408
Copper (II) chloride, 410–411
Copper (I) compounds, 407–409
Copper (II) compounds, 409–412
Copper (II) fluoride, 411
Copper glance, 405
Copper (II) hydroxide, 410
Copper (I) iodide, 409
Copper (II) nitrate, 412
Copper (I) oxide, 407–408
Copper (II) oxide, 409–410
Copper pyrites, 405
Copper (I) sulphate, 409
Copper (II) sulphate, 411–412
Copper (I) sulphide, 406
Copper (II) sulphide, 410
Corundum, 208
Coulomb, 25, 139
Covalency, 53
Covalent bond,
nature of, 53-54
Covalent compounds,
features of 58–61
violation of the octet rule, 64–66
Covalent molecules,
shapes of giant solids, 79–80
shapes of simple molecules, 75–79
Covalent radius, 70–71
Crookes, 24
Cryolite, 205
Crystal field splitting, 397, 398
Crystal field theory, 395–400
Cuprite, 405
Curie, Madame, 27, 441
Curie-Joliot, 446
Curium, 451
Cyclotron, 446

d orbitals,
directional properties of, 89
splitting of, 352, 396–398
Dalton's atomic theory, 1–2
Dative covalency, 54–55
Dead Sea deposits, 330
Deuterium, 167
Devarda's alloy, 260, 261
Diamond, 214–215
structure of, 79
Diazonium salts, 257
Diborane, 201–202
Dichromates (VI), 361–363
Dimethyl sulphate, 409
Dinitrogen oxide, 253
Dinitrogen pentoxide, 256
Dinitrogen tetroxide, 255–256
Dinitrogen trioxide, 254
Diphosphine, 265
Dipole,
electrical, 53, 68
fluctuating, 68
Disorder, 103–104
Disproportionation, 238, 334, 392, 395, 407, 419
Dissociation energy, 101
Disulphur decafluoride, 306
Disulphur dichloride, 306
Döbereiner's Triads, 16
Dulong and Petit's rule, 10–11
Dysprosium, 438

Edgar, 5
Einstein's equation, 444
Einsteinium, 451
Electrical work,
theoretical maximum, 110, 139
Electrochemical cell, 129–130
Electrode potential,
convention, 132
factors affecting the values of, 135–136
standard, 132–136
table of, 134
Electron,
arrangement in the atom, 34–46
as cloud of negative charge, 49
as particle or wave, 84–85
charge on, 25–26
charge to mass ratio, 25
density, 49, 87
probability, 49, 86–87
spin, 42
Electrons,
bonding pair, 76
lone pair, 76
Electron affinity, 56
Electron diffraction,
by Davisson and Germer, 84
Electron pair repulsion, 75–79
Gillespie and Nyholm theory, 76
Electronegativity, 61–62
Electronic configuration, 45–48
Electrovalent bond (ionic bond), 51–52
Electrovalent compounds, 51–52
containing complex anions, 64
energetics of formation, 56–58
features of, 58–61
violation of the octet rule, 62–64
Endothermic reaction, 96
Energy,
internal, 98–99, 110
rotational, 98, 106
translational, 98, 106
vibrational, 98, 106
Energy content (see enthalpy)
Enthalpy, 100–101, 103, 111
Enthalpy of atomisation, 102
of combustion, 100
of formation, 100
of hydration, 60, 135
of reaction, 99
of sublimation, 101
Entropy, 103–110
Entropy change, 103–110
Entropy and probability, 105–108
Equilibrium constant,
for cell reaction, 141–142
Equivalents,
Law of, 1
Equivalent mass,
classical methods for determination of, 5–6
definition of, 1
Erbium, 438
Europium, 438
Exothermic reaction, 96

Faraday, 23
Fermium, 450, 451
Ferrates (VI), 377
Ferrochrome, 360
Ferromanganese, 366
Ferrovanadium, 357
Fluorides, 330
covalent, 327, 330
ionic, 328, 330
Fluorine,
occurrence of, 326
preparation of, 326
properties of, 326–327
reactivity of, 327–329
Fluorocarbons, 218
Fluorosilicic acid, 228
Fluorspar, 326
Francium, 174, 175
Frasch process, 299
Free energy, 110–117
of half cell reactions, 140
Free energy change, 110–117
definition of, 111
using cell reactions, 138–140
Friedel-Crafts catalyst, 208, 378

Gadolin, 437
Gadolinium, 437, 438
Galena, 235
Gallium, 211
Gallium halides, 211
Gallium (III) oxide, 211
Gamma radiation, 27, 441
Gamma ray, 27
Gay-Lussacs law, 2
Geiger, 29
Germanates (IV), 240
Germanium,
extraction of, 234
occurrence of, 234
properties of, 236–237
structure of, 237
uses of, 234
Germanium (II) chloride, 238
Germanium (IV) chloride, 238
Germanium halides, 238
Germanium hydrides, 238
Germanium (II) oxide, 240
Germanium (IV) oxide, 240
Germanium (II) sulphide, 241
Germanium (IV) sulphide, 241
Gibbs diaphragm cell process, 178
Gold,
complexes of, 418, 419
extraction of, 417–418
occurrence of, 417
oxidation states of, 418–419
properties of, 418
uses of, 418
Gold (III) chloride, 419

Gold (III) halides, 419
Gold (III) oxide, 419
Goldstein, 26
Graphite,
manufacture of, 215–216
structure of, 80
uses of, 215
Group IA, 174–185
hydroxides, 177–179
occurrence, extraction and uses of, 175–176
oxides, 176–177
properties of, 176
some general remarks, 174–175
some physical data, 174
summary of group trends, 184–185
Group IB, 403–419
some general remarks, 403–405
some physical data, 403
Group 2A, 187–197
extraction, 189
metal salts, hydrolysis of, 195
occurrence, 189
oxides, 190
properties of, 189–190
some general remarks, 187–188
some physical data, 187
summary of group trends, 196–197
uses of, 189
Group 2B, 421–434
some general remarks, 421–423
some physical data, 421
Group 3B, 199–211
some general remarks, 199
some physical data, 199
Group 4B, 213–243
some general remarks, 213–214
some physical data, 213
summary of group trends, 243
Group 5B, 246–280
some general remarks, 246–247
some physical data, 246
summary of group trends, 279–280
Group 6B, 284–322
some general remarks, 284–285
some physical data, 284
Group 7B, 325–345
some general remarks, 325–326
some physical data, 325
summary of group trends, 345
Gypsum, 194

Haber process, 249–250
Haematite, 370, 377
Hafnium, 438
Half cell, 138, 140
Halic (I) acids, 340–341
Halic (V) acids, 341–342
Halic (VII) acids, 343
Halogens (see Group 7B)
Hausmannite, 366
Heat change, 99
Heat of hydration (see hydration energy)
Heat of reaction, 99
Heisenberg,
uncertainty principle, 84
Helium, 170
Helium molecule-ion, 92
Hess's law, 101
Hexafluorosilicic acid, 228
High spin complexes, 398
Holmium, 438
Hund's Rule, 43
Hydration energy, 60, 135
Hydrides,
covalent, 164-165
electron deficient, 165
interstitial, 165
ionic, 164
Hydrobromic acid,
preparation of, 339
properties of, 338
Hydrocarbons, 217
Hydrochloric acid,
preparation of, 337, 338
properties of, 338
Hydrofluoric acid, 329
Hydriodic acid,
preparation of, 338, 339
properties of, 338
Hydrogen,
active, 166
atomic, 166
atomic spectrum of, 34–35, 37–39
bond, 56–57
isotopes of, 167
manufacture of, 163
nascent, 166
occurrence of, 161
preparation of, 161–163
properties of, 163–164
uses of, 167–168
Hydrogen bomb, 445
Hydrogen bromide,
action of water on, 338
physical properties of, 337
preparation of, 336–337
reducing action of, 337–338
Hydrogen carbonates, 222–224
Hydrogen chloride,
action of water on, 338
industrial production of, 337
physical properties of, 337
preparation of, 336–337
reducing action of, 337–338
Hydrogen electrode, 132, 392
Hydrogen fluoride,
preparation of, 329
properties of, 329
Hydrogen fluorides, 329
Hydrogen iodide,
action of water on, 338
physical properties of, 337
preparation of, 336
reducing action of, 337–338
Hydrogen peroxide,
chemical properties of, 298
estimation of, 298
industrial production of, 297
physical properties of, 297
preparation of, 297
structure of, 298
test for, 298
uses of, 298
Hydrogen sulphide,
acid properties of, 302
occurrence of, 302
physical properties of, 302
precipitation reactions of, 302–304
preparation of, 302
reducing properties of, 304
test for, 304
Hydrogen sulphites, 316
Hydrazine, 251–252
Hydrazoic acid, 252
Hydroxides, 148–149, 291
Hydroxonium ion, 145, 161, 166

Ilmenite, 356
Indium, 211
Indium halides, 211
Inert gases (see noble gases)
Inert pair effect, 63, 211, 237, 279
Interhalogen compounds, 343–344
Iodate (I), 334
Iodate (V) ion, 342
Iodic (I) acid, 334, 340
Iodic (V) acid, 341, 342
Iodic (VII) acid, 343
Iodide ion,
test for, 339
Iodine,
chemical properties of, 333–335
extraction of, 331–332
laboratory preparation of, 332
occurrence of, 331
oxides of, 340
oxyacids of, 340–343
physical properties of, 332–333
uses of, 336
Iodine heptafluoride, 344
Iodine monobromide, 332, 343
Iodine monochloride, 332, 343
Iodine pentafluoride, 344
Iodine pentoxide, 340
Iodine trifluoride, 344
Ion exchange, 294, 295, 437–438
Ionic bond (see electrovalent bond)
Ionic compounds,
some structures of, 73–75
Ionic radius, 71–73
Landé's method of determination of, 71
Pauling's method of determination of, 71–72
variation of, 72–73
Ionisation energy, 39–41
Ions,
shapes of, 77
Iron,
cast, 372
extraction of, 370–372
occurrence of, 370
oxidation states of, 377
properties of, 376
rusting of, 376–377
uses of, 372, 373
wrought, 372
Iron (II) ammonium sulphate, 381
Iron (II) bromide, 380
Iron (III) bromide, 378
Iron (II) carbonate, 381

Iron (II) chloride, 380
Iron (III) chloride, 378
Iron (II) compounds, 379–381
Iron (III) compounds, 377–379
Iron (VI) compounds, 377
Iron enneacarbonyl, 400, 401
Iron (II) fluoride, 380
Iron (III) fluoride, 378
Iron (II) hydroxide, 379–380
Iron (III) hydroxide, 377–378
Iron (II) iodide, 380
Iron (II) ions,
  tests for, 382
Iron (III) ions,
  tests for, 379
Iron (II) oxalate, 381
Iron (II) oxide, 379
Iron (III) oxide, 377
Iron pentacarbonyl, 400, 401
Iron (II) sulphate, 380–381
Iron (III) sulphate, 378–379
Iron (II) sulphide, 381
Isolated system, 104
Isomorphism, 11–12
Isotope, 31, 32

Joule, 38

Kossel, 51
Krypton, 170
  fluoride of, 171

Lanthanides,
  contraction, 438
  coloured ions of, 438–439
  discovery of, 437
  occurrence of, 437
  oxidation states of, 439
  paramagnetism of, 440
  properties of, 438–440
  separation and extraction of, 437–438
  some general remarks, 437
  some physical data, 436
Lanthanum, 437, 438, 439
Lattice energy, 56, 58, 60
Lawrence, 446
Lawrencium, 451
Lead,
  extraction of, 235–236
  occurrence of, 235
  properties of, 236–237
  structure of, 237
  uses of, 236
Lead accumulator, 242
Lead (II) bromide, 239
Lead (II) chloride, 239
Lead (IV) chloride, 239
Lead (II) chromate (VI), 242
Lead (II) ethanoate, 242
Lead (IV) ethanoate, 242
Lead (IV) fluoride, 239
Lead (II) iodide, 239
Lead (II) oxide, 241
Lead (IV) oxide, 240–241
Lead (II) sulphate, 242
Lead (II) sulphide, 241, 242
Lead tetraethyl, 242

Lewis,
  acids and bases, 150
  covalent bonds, 53–54
Lewis acid, 150
Limiting density, 12, 13
Lithium, 174, 175, 176
  diagonal relationship with magnesium, 185
Lithium aluminium hydride, 209–210
Lithium oxide, 176
Low spin complexes, 298
Lutetium, 438

Magnesium, 189–190
Magnesium bromide, 193
Magnesium carbonate, 192
Magnesium chloride, 193
Magnesium fluoride, 192
Magnesium hydrogen carbonate, 192
Magnesium hydroxide, 191
Magnesium iodide, 193
Magnesium nitrate, 193
Magnesium nitride, 169, 195
Magnesium oxide, 190
Magnesium silicide, 228
Magnesium sulphate, 194
Magnetite, 370, 378
Manganese,
  extraction of, 366
  occurrence of, 366
  oxidation states of, 367
  properties of, 366
  uses of, 366
Manganic (VI) acid, 369
Manganese (II) compounds, 370
Manganese (III) compounds, 369
Manganese (IV) compounds, 369
Manganese (VI) compounds, 369
Manganese (VII) compounds, 367–368
Manganese (III) fluoride, 369
Manganese (II) hydroxide, 370
Manganese (II) oxalate, 370
Manganese (II) oxide, 370
Manganese (III) oxide, 369
Manganese (IV) oxide, 369
Manganese (VI) oxide, 369
Manganese (VII) oxide, 367
Manganese (IV) sulphate, 369
Manganese (II) sulphide, 370
Manganic (VII) acid, 367
Marsden, 29
Mass spectrograph, 31
Mass spectrometer, 32
Maxwell-Boltzmann distribution, 122
Mendeléeff, 16, 17, 18
  periodic classification, 18–21
Mendelevium, 450, 451
Mercury,
  extraction of, 428
  occurrence of, 428
  organo-compounds of, 431–432
  oxidation states of, 429
  properties of, 428–429
  purification of, 428
  uses of, 429
Mercury (II) carbonate, 431
Mercury (I) chloride, 433–434

Mercury (II) chloride, 430
Mercury (I) compounds, 432–434
Mercury (II) compounds, 429–431
Mercury (II) fulminate, 429
Mercury (II) iodide, 430–431
Mercury (I) nitrate, 434
Mercury (II) nitrate, 431
Mercury (II) oxide, 429–430
Mercury (II) sulphate, 431
Mercury (II) sulphide, 430
Metals
  importance of free energies of formation of oxides, 157–159
  principal sources of, 152–153
  principles involved in extraction of, 155
  refining of, 154
  stages involved in extraction of, 153–154
  table of electrode potentials and extraction method, 156
Metallic bond, 55–56
Methane, 225
Methanoic acid, 220
Meyer, Lothar,
  atomic volume curve, 17–18
Mica, 232
Milk of lime, 192
Millikan,
  determination of charge carried by the electron, 25–26
Molar volume, 8
Mole, 8
Molecular bond (see van der Waals' bond)
Molecular mass, 8, 9, 12, 13
Molecular orbital theory, 90–95
Monosilane, 228
Monozite sand, 437
Morley, 5
Mortar, 191
Moseley,
  determination of nuclear charges, 30–31
  x-ray spectra, 30
Mulliken,
  definition of electronegativity, 61
Multiple proportions,
  Law of, 1

Neodymium, 438
Neon, 170
Neptunium, 450, 451, 452
Nessler's reagent, 431
Neutron,
  discovery of, 30
Newlands law of octaves, 16
Nickel,
  extraction of, 386
  occurrence of, 386
  oxidation states of, 386
  properties of, 386
  uses of, 386
Nickel (II) carbonate, 387
Nickel carbonyl, 386, 400, 401
Nickel (II) chloride, 387
Nickel (II) compounds, 387
  complexes of, 387–388
Nickel (III) compounds, 386
Nickel (IV) compounds, 386
Nickel dimethylglyoximate, 387–388

Nickel (II) hydroxide, 387
Nickel (II) nitrate, 387
Nickel (II) oxide, 387
Nickel (II) sulphate, 387
Nickel (II) sulphide, 387
Nitrates, 261
Nitration, 259
Nitric acid,
  manufacture of, 257–258
  oxidising properties of, 259–261
  preparation of, 257
  reaction with organic compounds, 258–259
  uses of, 261
Nitrides, 262
Nitrites, 257
Nitrogen,
  halides of, 252
  industrial production of, 247
  occurrence of, 247
  oxides of, 253–256
  oxyacids of, 256–261
  preparation of, 247
  properties of, 248
Nitrogen cycle, 247–248
Nitrogen dioxide, 255–256
Nitrogen oxide, 253–254
Nitrogen trifluoride, 252
Nitrogen molecule,
  structure of, 248
Nitrogen trichloride, 252
Nitronium ion, 256, 259
Nitrosonium ion, 254, 256
Nitrosyl chloride, 261
Nitrous acid, 256–257
Nobelium, 450, 451
Noble (inert) gases,
  chemical compounds of, 170–172
  enclosure compounds containing, 172–173
  hydrates, 172
  occurrence and uses of, 170
  physical data of, 169
Noble gas compounds, 170–172
Normal density, 13
Nuclear energy, 448–449
Nuclear fission, 448
Nuclear forces, 443–445
Nuclear fusion, 449
Nuclear reaction, 441–451
Nuclear reactor, 448
Nuclear stability, 443–445
Nucleus,
  size of, 443

Oleum, 311
Orbital, 49, 85, 90
Order – disorder, 96, 98, 104
Order of reaction, 121
Orgel, 395
Oxidation,
  early definition of, 128
  extension of early ideas, 129
Oxidation number, 351
Oxides,
  acidic, 290
  amphoteric, 290
  basic, 290
  mixed oxides, 289
  mutual reactions of, 289–290
  neutral, 290
  normal oxides, 289
  peroxides, 289
  preparation of, 291
  suboxides, 289
  structures of, 289
Oxidising agent, 128
  tests for, 137
Oxyacids,
  strength of in aqueous solution, 149–150
Oxygen,
  industrial production of, 286
  occurrence of, 285
  preparation of, 285–286
  properties of, 286
Oxygen difluoride, 330
Oxygen monofluoride, 330
Ozone,
  estimation of, 287
  occurrence of, 286
  preparation of, 286–287
  properties of, 287–288
  structure of, 288
Ozonides, 288

Paramagnetic moment, 355
Paramagnetism, 355, 440
Partial equations, 129, 130
Pauli exclusion principle, 42, 43
Pauling,
  determination of ionic radii, 71–72
  electronegativity values, 61
Pearlite, 374
Periodic classification, 16–21
Periodic table, 19
  features of, 18–21
Peroxodisulphuric acid, 319
Peroxomonosulphuric acid, 319
Phosphates (V),
  diphosphates (V), 270–271
  phosphates (V), 269–270
  polytrioxophosphates (V), 271
  tests for, 272
Phosphine, 264–265
Phosphonic acid, 268
Phosphonium iodide, 264–265
Phosphoric (V) acids,
  diphosphoric (V), 270–271
  phosphoric (V), 269–270
  polytrioxophosphoric (V), 271
Phosphorus,
  allotropy of, 263–264
  extraction of, 262–263
  halides of, 265–267
  hydrides of, 264–265
  occurrence of, 262
  oxides of, 267–268
  oxyacids of, 268–272
  polymers containing, 272–273
  uses of, 273
Phosphorus (III) oxide, 267
Phosphorus (V) oxide, 267–268
Phosphorus pentabromide, 267
Phosphorus pentachloride, 266–267
Phosphorus pentoxide, 267–268
Phosphorus trichloride, 265–266
Planck, 35
Planck's constant, 35
Plaster of Paris, 194
Plumbane, 238
Plumbates (II), 241
Plumbates (IV), 241
Plutonium, 450, 451, 452
Polonium,
  formation of, 320
  halides of, 321
  hydride of, 321
  structure of, 320
Polonium dioxide, 321
Polyselenides, 321
Polysulphides, 301, 305
Polytellurides, 321
Polythionic acids, 319
Polyvanadates, 358
p orbitals, 87–88
Positron, 446
Potash alum, 209
Potassium, 174–176
Potassium bromate (V), 342, 343
Potassium bromide, 182
Potassium carbonate, 180
Potassium chlorate (V), 342
Potassium chloride, 182
Potassium chlorochromate (VI), 363
Potassium dichromate (VI), 362
Potassium ferrate (VI), 377
Potassium fluoride, 181
Potassium hydrogen carbonate, 180–181
Potassium hydrogen fluoride, 181
Potassium hydrogen sulphate, 183–184
Potassium hydroxide, 178–179
Potassium iodate (V), 342, 343
Potassium iodide, 182
Potassium manganate (VI), 369
Potassium manganate (VII), 367–368
Potassium nitrate, 182–183
Potassium nitrite, 183
Potassium peroxodisulphate, 319–320
Potassium sulphate, 183
Powell, 75
Praseodymium, 436, 438
Prometheum, 436
Protactinium, 443, 451
Proton, 26, 30, 166
Prout, 16
Pyrolusite, 366

Quantum numbers, 42–48
  principal, 42
  spin, 42
  subsidiary, 42
  third, 42
Quantum theory, 35–36
Quartz, 229

Radioactivity, 27, 441–443, 446–452
Radioactive decay, 442
Radioactive element,
  naturally occurring, 443
  production of, 446–447

Radioisotopes,
uses of, 449–450
Radium, 27
Radius ratio, 73, 74
Radon, 170
Rate constant, 121
Rates of chemical reactions, 121–127
Redox potentials,
standard, 132–134
table of, 134
Redox reactions, 128–142
Reducing agent, 128
tests for, 138
Reduction,
early definition of, 128
extension of early ideas, 129
Relative atomic mass (see atomic mass)
Relative molecular mass (see molecular mass)
Resonance, 81–82
Resonance energy, 82
Reversible reactions,
free energy changes during, 114–117
Rubidium, 174, 175, 176
Rutherford, 27, 28
nuclear theory of the atom, 29
Rydberg constant, 34

s orbitals,
shape of, 85–87
Salt-bridge, 129
Salt hydrolysis, 148
Saltpetre, 182
Samarium, 436, 438
Scandium, 356
Schrödinger,
wave equation, 85
Seaborg, 450
Selenates, 321
Selenic acid, 321
Selenides, 321
Selenious acid, 321
Selenites, 321
Selenium, 320
hydride of, 321
halides of, 321
Selenium dioxide, 321
Selenium trioxide, 321
Sidgwick, 75
Silica (silicon dioxide), 229
Silica gel, 230
Silica glass, 229
Silicate glasses, 233
Silicates, 230–233
Silicic acid, 230
Silicon, 227
extraction of, 227
halides of, 228–229
hydrides of, 228
occurrence of, 227
properties of, 228
Silicon carbide, 79, 216, 225
Silicon dioxide (silica), 229
Silicon disulphide, 233
Silicon tetrachloride, 228
Silicon tetrafluoride, 228, 229
Silicones, 233–234
Silver,
complexes of, 416–417
extraction of, 412
occurrence of, 412
oxidation states of, 413
properties of, 413
refining of, 412
uses of, 413
Silver bromide, 415
Silver carbide, 415
Silver carbonate, 416
Silver chloride, 415
Silver chromate (VI), 361, 416
Silver (I) compounds, 413–417
Silver fluoride, 414
Silver glance, 412
Silver halides, 414
Silver iodide, 415
Silver nitrate, 415–416
Silver oxide, 414
Silver plating, 413
Silver sulphate, 416
Silver sulphide, 414
Slaked lime, 190–191
Soda glass, 233
Sodium, 174, 175, 176
Sodium amalgam, 176
Sodium bismuthate (V), 278
Sodium bromide, 182, 331
Sodium carbonate, 179–180
Sodium chlorate (I), 334, 341
Sodium chlorate (V), 334, 342
Sodium chloride, 181–182
Sodium chromate (VI), 361, 362
Sodium cyanide, 176, 412, 418
Sodium dichromate (VI), 361, 362, 363
Sodium fluoride, 181
Sodium hexanitritocobaltate (III), 383
Sodium hydrogen carbonate, 180–181
Sodium hydrogen sulphate, 183–184
Sodium hydrogen sulphite, 308, 310
Sodium hydroxide, 177–179
Sodium iodate (V), 331
Sodium iodide, 182
Sodium methanoate, 221
Sodium monoxide, 176
Sodium nitrate, 182–183, 261
Sodium nitrite, 183, 257
Sodium orthovanadate, 358
Sodium peroxide, 176
Sodium potassium tartrate, 407
Sodium silicate, 231, 233
Sodium sulphate, 183
Sodium sulphite, 308
Sodium tetraborate, 200
Sodium thiosulphate, 318
Solvay process, 179–180
Solvent,
non-polar, 60, 146
polar, 60, 146
Spectrochemical series, 397, 398
Stabilisation energy (see resonance energy)
Standard electrode potential (see electrode potential, standard)
Stannane, 238
Stannates (II), 240
Stannates (IV), 240
Stas, 6
Stassfurt deposits, 182, 330
Steel,
alloys of, 375
electrical process, 374
hardening and tempering of, 375–376
manufacture of, 373–374
properties of, 374–375
Siemens–Martin Open Hearth Process, 373
structure of, 374
V.L.N. Bessemer process, 373–374
Stibine, 274, 275
Strontium, 187, 188, 189
Sulphates,
preparation of, 315
structure of 316
test for, 316
thermal stability of, 316
Sulphides, 305–306
Sulphites,
reducing properties of, 310
structures of, 310
test for, 316
Sulphur,
action of heat on, 300–301
allotropy of, 300–301
amorphous, 300
chemical properties of, 301
halides of, 306–307
hydrides of, 302–305
monoclinic, 300
occurrence of, 299
oxides of, 307–309
plastic, 300
rhombic, 300
uses of, 320
Sulphur dichloride, 306
Sulphur dichloride dioxide, 317–318
Sulphur dichloride oxide, 317
Sulphur dioxide,
chemical properties of, 308
physical properties of, 308
preparation of, 307
sources of, 311
structure of, 308
Sulphur hexafluoride, 306
Sulphur tetrachloride, 306
Sulphur tetrafluoride, 306
Sulphur trioxide,
chemical properties of, 309
physical properties of, 309
preparation of, 308–309
structure of, 309
Sulphuric acid,
acid properties of, 312
dehydrating properties of, 313
displacement reactions of, 314–315
manufacture of, 310–312
oxidising properties of, 313–314
physical properties of, 312
reactions with organic compounds, 315
structure of, 312
uses of, 315
Sulphurous acid,
aqueous solution of, 310

reducing properties of, 310
Synthesis gas, 163, 225–226
Synchrotron, 446

Tellurates, 321
Tellurides, 321
Tellurites, 321
Tellurium, 320
halides of, 321
hydride of, 321
recovery of, 320
structure of, 320
Tellurium dioxide, 321
Tellurium trioxide, 321
Terbium, 436, 438
Tetrachloromethane, 218
Tetrafluoromethane, 218
Tetrathionate ion, 318
Thallium, 211
Thallium halides, 211
Thallium (I) oxide, 211
Thallium (III) oxide, 211
Thermodynamics, 96–117
first law of, 98
Thiosulphates, 318
Thomson J. J., 24, 25, 26
Thorium, 441, 443
Thulium, 438
Tin,
extraction of, 234–235
occurrence of, 234
properties of, 236–237
structure of, 237
uses of, 235
Tin (II) chloride, 238–239
Tin (IV) chloride, 238
Tin (IV) fluoride, 238
Tin halides, 238–239
Tin (II) oxalate, 240
Tin (II) oxide, 240
Tin (IV) oxide, 240
Tin (II) sulphate, 235
Tin (II) sulphide, 241
Tin (IV) sulphide, 241
Tinplate, 235
Tinstone, 234
Titanium,
compounds of, 357
extraction of, 356
occurrence of, 356
oxidation states of, 357
properties of, 356
uses of, 356
Transition elements,
carbonyls of, 400–401
catalytic activity of, 354
causes of acidity of ions, 391–392
coloured ions of, 351–352, 395–398
complex ions of, 352–353, 395–400
first inner series (see lanthanides)
first transition series, 349–401
interstitial compounds of, 355
isomorphous compounds of, 355
oxidation states of, 353–354
paramagnetism of, 355, 398–399
redox potentials of, 349, 392–395
some physical data of, 349
metallic properties of, 351
Transuranium elements, 450–452
Tricobalt tetroxide, 382
Tridymite, 229
Trimanganese tetroxide, 369
Tri-iron tetroxide, 378
Trilead tetroxide, 241
Tritium, 167

Uranium, 27, 441, 443, 448

Valency, 4
Vanadates,
metavanadate, 358, 359
orthovandate, 358
polyvanadate ion, 358
Vanadium,
extraction of, 357
occurrence of, 357
oxidation states of, 358
properties of, 358
uses of, 358
Vanadium (II) compounds, 359–360
Vanadium (III) compounds, 359
Vanadium (IV) compounds, 359
Vanadium (V) compounds, 358–359
Vanadium (II) oxide, 359
Vanadium (III) oxide, 359
Vanadium (IV) oxide, 359
Vanadium (V) oxide, 358
Vanadium (III) sulphate, 359
Vanadyl sulphate, 359
van der Waals' bond, 67–68
van Vleck, 395

Vapour density, 8, 9

Water,
as an acid, 292
as a base, 292
conductivity of, 292
distillation of, 296
estimation of hardness of, 295–296
hardness of, 293–295
softening of, 294–295
structure of, 292
Water gas, 163
Wave equation, 85
angular wave function, 87, 88, 89
methods of representation for hydrogen atom, 85–86
radial wave function, 87, 88, 89
solution of for hydrogen atom, 85
Weston cell, 427
Work,
electrical, 111, 139
reversible, 109, 111
useful, 111, 139

Xenon, 170
Xenon fluorides, 171
Xenon hexafluoroplatinate, 170
X-rays,
discovery of, 27
production of, 30
spectra, 30

Zeolites, 232
Zinc,
complexes of, 428
compounds of, 425–426
extraction of, 423–424
occurrence of, 423
properties of, 424
purification of, 424
uses of, 424–425
Zinc blende, 423
structure of, 426
Zinc carbonate, 427–428
Zinc chloride, 426–427
Zinc hydroxide, 426
Zinc nitrate, 427
Zinc sulphate, 427
Zinc sulphide, 427
Zinc oxide, 425–426
Zone–refining, 154

Bibliography

Addison, W. E. *Structural Principles in Inorganic Compounds*, Longman, 1961

Bell, C.F. and Lott, K.A.K. *Modern Approach to Inorganic Chemistry*, Butterworths, 1963, revised 1972.

Brown, G.I. *Introduction to Physical Chemistry*, Longman, 1964, S.I. edn. 1972.

—*A New Guide to Modern Valency Theory*, Longman, 1967.

Cartmell, E. and Fowles, G.W.A. *Valency and Molecular Structure*, Butterworth, 1961, 3rd edn. 1966, 4th edn. 1977.

Cotton, F.A. and Wilkinson, G. *Advanced Inorganic Chemistry*, Interscience, 1963, 3rd edn. 1972, 4th edn. 1980.

Coulson, C.A. *Valence*, Oxford University Press, 1952, 2nd revised edn. 1961, 3rd edn. 1969.

Emeléus, H.J. and Anderson, J.S. *Modern Aspects of Inorganic Chemistry*, Routledge and Kegan Paul, 1952, revised 1978.

Heslop, R.B. and Robinson, P.L. *Inorganic Chemistry*, Elsevier, 1960, 3rd edn. 1967.

Ives, D.J.B. *Principles of the Extraction of Metals*, Royal Institute of Chemistry, Monographs for Teachers No. 3, 1960.

Orgel, L.E. *Introduction to Transition-Metal Chemistry: Ligand-Field Theory*, Methuen, 1960, 2nd edn. 1966.

Phillips, C.S.G. and Williams, R.J.P. *Inorganic Chemistry*, Vol. I, 1965. Vol. II, 1966. Oxford University Press.

Samuel, D.M. *Industrial Chemistry—Inorganic*, Royal Institute of Chemistry, Monographs for Teachers No. 10, 1966, 2nd edn. 1970.

Sharpe, A.G. *Inorganic Chemistry*, Longman, 1981.

Sharpe, A.G. *Principles of Oxidation and Reduction*, Royal Institute of Chemistry, Monographs for Teachers No. 2, 1960, 2nd edn. 1968.

Sisler, H.H., VanderWerf, C.A. and Davidson, A.W. *College Chemistry*, Collier-Macmillan, 3rd edn. 1967.

Strong, L.E. and Stratton, W.J. *Chemical Energy*, Van Nostrand Reinhold, 1966.

Wells, A.F. *Structural Inorganic Chemistry*, Oxford University Press, 1950, 3rd edn. 1962, 4th edn. 1975.

Table of Relative Atomic Masses to Four Significant Figures

(Scaled to the relative atomic mass ^{12}C = 12 exactly)

Values quoted in the table, unless marked * or †, are reliable to at least ±1 in the fourth significant figure. A number in parentheses denotes the atomic mass number of the isotope of longest known half-life.

Atomic Number	Name	Symbol	Relative Atomic Mass
1	Hydrogen	H	1·008
2	Helium	He	4·003
3	Lithium	Li	6·941*†
4	Beryllium	Be	9·012
5	Boron	B	10·81†
6	Carbon	C	12·01
7	Nitrogen	N	14·01
8	Oxygen	O	16·00
9	Fluorine	F	19·00
10	Neon	Ne	20·18
11	Sodium	Na	22·99
12	Magnesium	Mg	24·31
13	Aluminium	Al	26·98
14	Silicon	Si	28·09
15	Phosphorus	P	30·97
16	Sulphur	S	32·06†
17	Chlorine	Cl	35·45
18	Argon	Ar	39·95
19	Potassium	K	39·10
20	Calcium	Ca	40·08†
21	Scandium	Sc	44·96
22	Titanium	Ti	47·90*
23	Vanadium	V	50·94
24	Chromium	Cr	52·00
25	Manganese	Mn	54·94
26	Iron	Fe	55·85
27	Cobalt	Co	58·93
28	Nickel	Ni	58·70
29	Copper	Cu	63·55
30	Zinc	Zn	65·38
31	Gallium	Ga	69·72
32	Germanium	Ge	72·59*
33	Arsenic	As	74·92
34	Selenium	Se	78·96*
35	Bromine	Br	79·90
36	Krypton	Kr	83·80
37	Rubidium	Rb	85·47
38	Strontium	Sr	87·62†
39	Yttrium	Y	88·91
40	Zirconium	Zr	91·22
41	Niobium	Nb	92·91
42	Molybdenum	Mo	95·94*
43	Technetium	Tc	(98)
44	Ruthenium	Ru	101·1
45	Rhodium	Rh	102·9
46	Palladium	Pd	106·4
47	Silver	Ag	107·9
48	Cadmium	Cd	112·4
49	Indium	In	114·8
50	Tin	Sn	118·7
51	Antimony	Sb	121·8
52	Tellurium	Te	127·6

* Values so marked are reliable to ±3 in the fourth significant figure.

† Values so marked may differ from the atomic weights of the relevant elements in some naturally occurring samples because of a variation in the relative abundance of the isotopes.

This is a 'Table of Atomic Weights to Four Significant Figures,' published by the International Union of Pure and Applied Chemistry and reproduced with their permission.

Atomic Number	Name	Symbol	Relative Atomic Mass
53	Iodine	I	126·9
54	Xenon	Xe	131·3
55	Caesium	Cs	132·9
56	Barium	Ba	137·3
57	Lanthanum	La	138·9
58	Cerium	Ce	140·1
59	Praseodymium	Pr	140·9
60	Neodymium	Nd	144·2
61	Promethium	Pm	(145)
62	Samarium	Sm	150·4
63	Europium	Eu	152·0
64	Gadolinium	Gd	157·3
65	Terbium	Tb	158·9
66	Dysprosium	Dy	162·5
67	Holmium	Ho	164·9
68	Erbium	Er	167·3
69	Thulium	Tm	168·9
70	Ytterbium	Yb	173·0
71	Lutetium	Lu	175·0
72	Hafnium	Hf	178·5
73	Tantalum	Ta	180·9
74	Wolfram (Tungsten)	W	183·9
75	Rhenium	Re	186·2
76	Osmium	Os	190·2
77	Iridium	Ir	192·2
78	Platinum	Pt	195·1
79	Gold	Au	197·0
80	Mercury	Hg	200·6
81	Thallium	Tl	204·4
82	Lead	Pb	207·2†
83	Bismuth	Bi	209·0
84	Polonium	Po	(209)
85	Astatine	At	(210)
86	Radon	Rn	(222)
87	Francium	Fr	(223)
88	Radium	Ra	(226)
89	Actinium	Ac	(227)
90	Thorium	Th	232·0
91	Protactinium	Pa	(231)
92	Uranium	U	238†
93	Neptunium	Np	(237)
94	Plutonium	Pu	(244)
95	Americium	Am	(243)
96	Curium	Cm	(247)
97	Berkelium	Bk	(247)
98	Californium	Cf	(251)
99	Einsteinium	Es	(252)
100	Fermium	Fm	(257)
101	Mendelevium	Md	(258)
102	Nobelium	No	(259)
103	Lawrencium	Lr	(260)

In 1978 IUPAC approved of rules for naming elements of atomic number greater than 100. For example, undiscovered element of atomic number 109 is called Unnilennium and given the symbol Une. The name is derived as follows: 1 = un 0 = nil, 9 = enn (the ending 'ium' is added), hence Unnilennium. Similarly element of atomic number 120 is called *Unbinil*ium with the symbol Ubn.

Logarithms

	0	1	2	3	4	5	6	7	8	9	1	2	3	4	5	6	7	8	9
10	·0000	0043	0086	0128	0170	0212	0253	0294	0334	0374	4	8	12	17	21	25	29	33	37
11	·0414	0453	0492	0531	0569	0607	0645	0682	0719	0755	4	8	11	15	19	23	26	30	34
12	·0792	0828	0864	0899	0934	0969	1004	1038	1072	1106	3	7	10	14	17	21	24	28	31
13	·1139	1173	1206	1239	1271	1303	1335	1367	1399	1430	3	6	10	13	16	19	23	26	29
14	·1461	1492	1523	1553	1584	1614	1644	1673	1703	1732	3	6	9	12	15	18	21	24	27
15	·1761	1790	1818	1847	1875	1903	1931	1959	1987	2014	3	6	8	11	14	17	20	22	25
16	·2041	2068	2095	2122	2148	2175	2201	2227	2253	2279	3	5	8	11	13	16	18	21	24
17	·2304	2330	2355	2380	2405	2430	2455	2480	2504	2529	2	5	7	10	12	15	17	20	22
18	·2553	2577	2601	2625	2648	2672	2695	2618	2742	2765	2	5	7	9	12	14	16	19	21
19	·2788	2810	2833	2856	2878	2900	2923	2945	2967	2989	2	4	7	9	11	13	16	18	20
20	·3010	3032	3054	3075	3096	3118	3139	3160	3181	3201	2	4	6	8	11	13	15	17	19
21	·3222	3243	3263	3284	3304	3324	3345	3365	3385	3404	2	4	6	8	10	12	14	16	18
22	·3424	3444	3464	3483	3502	3522	3541	3560	3579	3598	2	4	6	8	10	12	14	15	17
23	·3617	3636	3655	3674	3692	3711	3729	3747	3766	3784	2	4	6	7	9	11	13	15	17
24	·3802	3820	3838	3856	3874	3892	3909	3927	3945	3962	2	4	5	7	9	11	12	14	16
25	·3979	3997	4014	4031	4048	4065	4082	4099	4116	4133	2	3	5	7	9	10	12	14	15
26	·4150	4166	4183	4200	4216	4232	4249	4265	4281	4298	2	3	5	7	8	10	11	13	15
27	·4314	4330	4346	4362	4378	4393	4409	4425	4440	4456	2	3	5	6	8	9	11	13	14
28	·4472	4487	4502	4518	4533	4548	4564	4579	4594	4609	2	3	5	6	8	9	11	12	14
29	·4624	4639	4654	4669	4683	4698	4713	4728	4742	4757	1	3	4	6	7	9	10	12	13
30	·4771	4786	4800	4814	4829	4843	4857	4871	4886	4900	1	3	4	6	7	9	10	11	13
31	·4914	4928	4942	4955	4969	4983	4997	5011	5024	5038	1	3	4	6	7	8	10	11	12
32	·5051	5065	5079	5092	5105	5119	5132	5145	5159	5172	1	3	4	5	7	8	9	11	12
33	·5185	5198	5211	5224	5237	5250	5263	5276	5289	5302	1	3	4	5	6	8	9	10	12
34	·5315	5328	5340	5353	5366	5378	5391	5403	5416	5428	1	3	4	5	6	8	9	10	11
35	·5441	5453	5465	5478	5490	5502	5514	5527	5539	5551	1	2	4	5	6	7	9	10	11
36	·5563	5575	5587	5599	5611	5623	5635	5647	5658	5670	1	2	4	5	6	7	8	10	11
37	·5682	5694	5705	5717	5729	5740	5752	5763	5775	5786	1	2	3	5	6	7	8	9	10
38	·5798	5809	5821	5832	5843	5855	5866	5877	5888	5899	1	2	3	5	6	7	8	9	10
39	·5911	5922	5933	5944	5955	5966	5977	5988	5999	6010	1	2	3	4	5	7	8	9	10
40	·6021	6031	6042	6053	6064	6075	6085	6096	6107	6117	1	2	3	4	5	6	8	9	10
41	·6128	6138	6149	6160	6170	6180	6191	6201	6212	6222	1	2	3	4	5	6	7	8	9
42	·6232	6243	6253	6263	6274	6284	6294	6304	6314	6325	1	2	3	4	5	6	7	9	9
43	·6335	6345	6355	6365	6375	6385	6395	6405	6415	6425	1	2	3	4	5	6	7	8	9
44	·6435	6444	6454	6464	6474	6484	6493	6503	6513	6522	1	2	3	4	5	6	7	8	9
45	·6532	6542	6551	6561	6571	6580	6590	6599	6609	6618	1	2	3	4	5	6	7	8	9
46	·6628	6637	6646	6656	6665	6675	6684	6693	6702	6712	1	2	3	4	5	6	7	7	8
47	·6721	6730	6739	6749	6758	6767	6776	6785	6794	6803	1	2	3	4	5	5	6	7	8
48	·6812	6821	6830	6839	6848	6857	6866	6875	6884	6893	1	2	3	4	4	5	6	7	8
49	·6902	6911	6920	6928	6937	6946	6955	6964	6972	6981	1	2	3	4	4	5	6	7	8
50	·6990	6998	7007	7016	7024	7033	7042	7050	7059	7067	1	2	3	3	4	5	6	7	8
51	·7076	7084	7093	7101	7110	7118	7126	7135	7143	7152	1	2	3	3	4	5	6	7	8
52	·7160	7168	7177	7185	7193	7202	7210	7218	7226	7235	1	2	2	3	4	5	6	7	7
53	·7243	7251	7259	7267	7275	7284	7292	7300	7308	7316	1	2	2	3	4	5	6	6	7
54	·7324	7332	7340	7348	7356	7364	7372	7380	7388	7396	1	2	2	3	4	5	6	6	7

Logarithms

	0	1	2	3	4	5	6	7	8	9	1	2	3	4	5	6	7	8	9
55	·7404	7412	7419	7427	7435	7443	7451	7459	7466	7474	1	2	2	3	4	5	5	6	7
56	·7482	7490	7497	7505	7513	7520	7528	7536	7543	7551	1	2	2	3	4	5	5	6	7
57	·7559	7566	7574	7582	7589	7597	7604	7612	7619	7627	1	2	2	3	4	5	5	6	7
58	·7634	7642	7649	7657	7664	7672	7679	7686	7694	7701	1	1	2	3	4	4	5	6	7
59	·7709	7716	7723	7731	7738	7745	7752	7760	7767	7774	1	1	2	3	4	4	5	6	7
60	·7782	7789	7796	7803	7810	7818	7825	7832	7839	7846	1	1	2	3	4	4	5	6	6
61	·7853	7860	7868	7875	7882	7889	7896	7903	7910	7917	1	1	2	3	4	4	5	6	6
62	·7924	7931	7938	7945	7952	7959	7966	7973	7980	7987	1	1	2	3	3	4	5	6	6
63	·7993	8000	8007	8014	8021	8028	8035	8041	8048	8055	1	1	2	3	3	4	5	5	6
64	·8062	8069	8075	8082	8089	8096	8102	8109	8116	8122	1	1	2	3	3	4	5	5	6
65	·8129	8136	8142	8149	8156	8162	8169	8176	8182	8189	1	1	2	3	3	4	5	5	6
66	·8195	8202	8209	8215	8222	8228	8235	8241	8248	8254	1	1	2	3	3	4	5	5	6
67	·8261	8267	8274	8280	8287	8293	8299	8306	8312	8319	1	1	2	3	3	4	5	5	6
68	·8325	8331	8338	8344	8351	8357	8363	8370	8376	8382	1	1	2	3	3	4	4	5	6
69	·8388	8395	8401	8407	8414	8420	8426	8432	8439	8445	1	1	2	2	3	4	4	5	6
70	·8451	8457	8463	8470	8476	8482	8488	8494	8500	8506	1	1	2	2	3	4	4	5	6
71	·8513	8519	8525	8531	8537	8543	8549	8555	8561	8567	1	1	2	2	3	4	4	5	5
72	·8573	8579	8585	8591	8597	8603	8609	8615	8621	8627	1	1	2	2	3	4	4	5	5
73	·8633	8639	8645	8651	8657	8663	8669	8675	8681	8686	1	1	2	2	3	4	4	5	5
74	·8692	8698	8704	8710	8716	8722	8727	8733	8739	8745	1	1	2	2	3	4	4	5	5
75	·8751	8756	8762	8768	8774	8779	8785	8791	8797	8802	1	1	2	2	3	3	4	5	5
76	·8808	8814	8820	8825	8831	8837	8842	8848	8854	8859	1	1	2	2	3	3	4	5	5
77	·8865	8871	8876	8882	8887	8893	8899	8904	8910	8915	1	1	2	2	3	3	4	4	5
78	·8921	8927	8932	8938	8943	8949	8954	8960	8965	8971	1	1	2	2	3	3	4	4	5
79	·8976	8982	8987	8993	8998	9004	9009	9015	9020	9025	1	1	2	2	3	3	4	4	5
80	·9031	9036	9042	9047	9053	9058	9063	9069	9074	9079	1	1	2	2	3	3	4	4	5
81	·9085	9090	9096	9101	9106	9112	9117	9122	9128	9133	1	1	2	2	3	3	4	4	5
82	·9138	9143	9149	9154	9159	9165	9170	9175	9180	9186	1	1	2	2	3	3	4	4	5
83	·9191	9196	9201	9206	9212	9217	9222	9227	9232	9238	1	1	2	2	3	3	4	4	5
84	·9243	9248	9253	9258	9263	9269	9274	9279	9284	9289	1	1	2	2	3	3	4	4	5
85	·9294	9299	9304	9309	9315	9320	9325	9330	9335	9340	1	1	2	2	3	3	4	4	5
86	·9345	9350	9355	9360	9365	9370	9375	9380	9385	9390	1	1	2	2	3	3	4	4	5
87	·9395	9400	9405	9410	9415	9420	9425	9430	9435	9440	0	1	1	2	2	3	3	4	4
88	·9445	9450	9455	9460	9465	9469	9474	9479	9484	9489	0	1	1	2	2	3	3	4	4
89	·9494	9499	9504	9509	9513	9518	9523	9528	9533	9538	0	1	1	2	2	3	3	4	4
90	·9542	9547	9552	9557	9562	9566	9571	9576	9581	9586	0	1	1	2	2	3	3	4	4
91	·9590	9595	9600	9605	9609	9614	9619	9624	9628	9633	0	1	1	2	2	3	3	4	4
92	·9638	9643	9647	9652	9657	9661	9666	9671	9675	9680	0	1	1	2	2	3	3	4	4
93	·9685	9689	9694	9699	9703	9708	9713	9717	9722	9727	0	1	1	2	2	3	3	4	4
94	·9731	9736	9741	9745	9750	9754	9759	9763	9768	9773	0	1	1	2	2	3	3	4	4
95	·9777	9782	9786	9791	9795	9800	9805	9809	9814	9818	0	1	1	2	2	3	3	4	4
96	·9823	9827	9832	9836	9841	9845	9850	9854	9859	9863	0	1	1	2	2	3	3	4	4
97	·9868	9872	9877	9881	9886	9890	9894	9899	9903	9908	0	1	1	2	2	3	3	4	4
98	·9912	9917	9921	9926	9930	9934	9939	9943	9948	9952	0	1	1	2	2	3	3	4	4
99	·9956	9961	9965	9969	9974	9978	9983	9987	9991	9996	0	1	1	2	2	3	3	3	4

Antilogarithms

	0	1	2	3	4	5	6	7	8	9	1	2	3	4	5	6	7	8	9
·00	1000	1002	1005	1007	1009	1012	1014	1016	1019	1021	0	0	1	1	1	1	2	2	2
·01	1023	1026	1028	1030	1033	1035	1038	1040	1042	1045	0	0	1	1	1	1	2	2	2
·02	1047	1050	1052	1054	1057	1059	1062	1064	1067	1069	0	0	1	1	1	1	2	2	2
·03	1072	1074	1076	1079	1081	1084	1086	1089	1091	1094	0	0	1	1	1	1	2	2	2
·04	1096	1099	1102	1104	1107	1109	1112	1114	1117	1119	0	1	1	1	1	2	2	2	2
·05	1122	1125	1127	1130	1132	1135	1138	1140	1143	1146	0	1	1	1	1	2	2	2	2
·06	1148	1151	1153	1156	1159	1161	1164	1167	1169	1172	0	1	1	1	1	2	2	2	2
·07	1175	1178	1180	1183	1186	1189	1191	1194	1197	1199	0	1	1	1	1	2	2	2	2
·08	1202	1205	1208	1211	1213	1216	1219	1222	1225	1227	0	1	1	1	1	2	2	2	3
·09	1230	1233	1236	1239	1242	1245	1247	1250	1253	1256	0	1	1	1	1	2	2	2	3
·10	1259	1262	1265	1268	1271	1274	1276	1279	1282	1285	0	1	1	1	1	2	2	2	3
·11	1288	1291	1294	1297	1300	1303	1306	1309	1312	1315	0	1	1	1	2	2	2	2	3
·12	1318	1321	1324	1327	1330	1334	1337	1340	1343	1346	0	1	1	1	2	2	2	2	3
·13	1349	1352	1355	1358	1361	1365	1368	1371	1374	1377	0	1	1	1	2	2	2	3	3
·14	1380	1384	1387	1390	1393	1396	1400	1403	1406	1409	0	1	1	1	2	2	2	3	3
·15	1413	1416	1419	1422	1426	1429	1432	1435	1439	1442	0	1	1	1	2	2	2	3	3
·16	1445	1449	1452	1455	1459	1462	1466	1469	1472	1476	0	1	1	1	2	2	2	3	3
·17	1479	1483	1486	1489	1493	1496	1500	1503	1507	1510	0	1	1	1	2	2	2	3	3
·18	1514	1517	1521	1524	1528	1531	1535	1538	1542	1545	0	1	1	1	2	2	2	3	3
·19	1549	1552	1556	1560	1563	1567	1570	1574	1578	1581	0	1	1	1	2	2	3	3	3
·20	1585	1589	1592	1596	1600	1603	1607	1611	1614	1618	0	1	1	1	2	2	3	3	3
·21	1622	1626	1629	1633	1637	1641	1644	1648	1652	1656	0	1	1	2	2	2	3	3	3
·22	1660	1663	1667	1671	1675	1679	1683	1687	1690	1694	0	1	1	2	2	2	3	3	3
·23	1698	1702	1706	1710	1714	1718	1722	1726	1730	1734	0	1	1	2	2	2	3	3	4
·24	1738	1742	1746	1750	1754	1758	1762	1766	1770	1774	0	1	1	2	2	2	3	3	4
·25	1778	1782	1786	1791	1795	1799	1803	1807	1811	1816	0	1	1	2	2	2	3	3	4
·26	1820	1824	1828	1832	1837	1841	1845	1849	1854	1858	0	1	1	2	2	3	3	3	4
·27	1862	1866	1871	1875	1879	1884	1888	1892	1897	1901	0	1	1	2	2	3	3	3	4
·28	1905	1910	1914	1919	1923	1928	1932	1936	1941	1945	0	1	1	2	2	3	3	4	4
·29	1950	1954	1959	1963	1968	1972	1977	1982	1986	1991	0	1	1	2	2	3	3	4	4
·30	1995	2000	2004	2009	2014	2018	2023	2028	2032	2037	0	1	1	2	2	3	3	4	4
·31	2042	2046	2051	2056	2061	2065	2070	2075	2080	2084	0	1	1	2	2	3	3	4	4
·32	2089	2094	2099	2104	2109	2113	2113	2123	2128	2133	0	1	1	2	2	3	3	4	4
·33	2138	2143	2148	2153	2158	2163	2168	2173	2178	2183	0	1	1	2	2	3	3	4	4
·34	2188	2193	2198	2203	2208	2213	2218	2223	2228	2234	1	1	2	2	3	3	4	4	5
·35	2239	2244	2249	2254	2259	2265	2270	2275	2280	2286	1	1	2	2	3	3	4	4	5
·36	2291	2296	2301	2307	2312	2317	2323	2328	2333	2339	1	1	2	2	3	3	4	4	5
·37	2344	2350	2355	2360	2366	2371	2377	2382	2388	2393	1	1	2	2	3	3	4	4	5
·38	2399	2404	2410	2415	2421	2427	2432	2438	2443	2449	1	1	2	2	3	3	4	4	5
·39	2455	2460	2466	2472	2477	2483	2489	2495	2500	2506	1	1	2	2	3	3	4	5	5
·40	2512	2518	2523	2529	2535	2541	2547	2553	2559	2564	1	1	2	2	3	4	4	5	5
·41	2570	2576	2582	2588	2594	2600	2606	2612	2618	2524	1	1	2	2	3	4	4	5	5
·42	2630	2636	2642	2649	2655	2661	2667	2673	2679	2685	1	1	2	2	3	4	4	5	6
·43	2692	2698	2704	2710	2716	2723	2729	2735	2742	2748	1	1	2	3	3	4	4	5	6
·44	2754	2761	2767	2773	2780	2786	2793	2799	2805	2812	1	1	2	3	3	4	4	5	6
·45	2818	2825	2831	2838	2844	2851	2858	2864	2871	2877	1	1	2	3	3	4	5	5	6
·46	2884	2891	2897	2904	2911	2917	2924	2931	2938	2944	1	1	2	3	3	4	5	5	6
·47	2951	2958	2965	2972	2979	2985	2992	2999	3006	3013	1	1	2	3	3	4	5	5	6
·48	3020	3027	3034	3041	3048	3055	3062	3069	3076	3083	1	1	2	3	4	4	5	6	6
·49	3090	3097	3105	3112	3119	3126	3133	3141	3148	3155	1	1	2	3	4	4	5	6	6

Antilogarithms

	0	1	2	3	4	5	6	7	8	9	1	2	3	4	5	6	7	8	9
·50	3162	3170	3177	3184	3192	3199	3206	3214	3221	3228	1	1	2	3	4	4	5	6	7
·51	3236	3243	3251	3258	3266	3273	3281	3289	3296	3304	1	2	2	3	4	5	5	6	7
·52	3311	3319	3327	3334	3342	3350	3357	3365	3373	3381	1	2	2	3	4	5	5	6	7
·53	3388	3396	3404	3412	3420	3428	3436	3443	3451	3459	1	2	2	3	4	5	6	6	7
·54	3467	3475	3483	3491	3499	3508	3516	3524	3532	3540	1	2	2	3	4	5	6	6	7
·55	3548	3556	3565	3573	3581	3589	3597	3606	3614	3622	1	2	2	3	4	5	6	7	7
·56	3631	3639	3648	3656	3664	3673	3681	3690	3698	3707	1	2	3	3	4	5	6	7	8
·57	3715	3724	3733	3741	3750	3758	3767	3776	3784	3793	1	2	3	3	4	5	6	7	8
·58	3802	3811	3819	3828	3837	3846	3855	3864	3873	3882	1	2	3	4	4	5	6	7	8
·59	3890	3899	3908	3917	3926	3936	3945	3954	3963	3972	1	2	3	4	5	5	6	7	8
·60	3981	3990	3999	4009	4018	4027	4036	4046	4055	4064	1	2	3	4	5	6	6	7	8
·61	4074	4083	4093	4102	4111	4121	4130	4140	4150	4159	1	2	3	4	5	6	7	8	9
·62	4169	4178	4188	4198	4207	4217	4227	4236	4246	4256	1	2	3	4	5	6	7	8	9
·63	4266	4276	4285	4295	4305	4315	4325	4335	4345	4355	1	2	3	4	5	6	7	8	9
·64	4365	4375	4385	4395	4406	4416	4426	4436	4446	4457	1	2	3	4	5	6	7	8	9
·65	4467	4477	4487	4498	4508	4519	4529	4539	4550	4560	1	2	3	4	5	6	7	8	9
·66	4571	4581	4592	4603	4613	4624	4634	4645	4656	4667	1	2	3	4	5	6	7	9	10
·67	4677	4688	4699	4710	4721	4732	4742	4753	4764	4775	1	2	3	4	5	7	8	9	10
·68	4786	4797	4808	4819	4831	4842	4853	4864	4875	4887	1	2	3	4	6	7	8	9	10
·69	4898	4909	4920	4932	4943	4955	4966	4977	4989	5000	1	2	3	5	6	7	8	9	10
·70	5012	5023	5035	5047	5058	5070	5082	5093	5105	5117	1	2	4	5	6	7	8	9	11
·71	5129	5140	5152	5164	5176	5188	5200	5212	5224	5236	1	2	4	5	6	7	8	10	11
·72	5248	5260	5272	5284	5297	5309	5321	5333	5346	5358	1	2	4	5	6	7	9	10	11
·73	5370	5383	5395	5408	5420	5433	5445	5458	5470	5483	1	3	4	5	6	8	9	10	11
·74	5495	5508	5521	5534	5546	5559	5572	5585	5598	5610	1	3	4	5	6	8	9	10	12
·75	5623	5636	5649	5662	5675	5689	5702	5715	5728	5741	1	3	4	5	7	8	9	10	12
·76	5754	5768	5781	5794	5808	5821	5834	5848	5861	5875	1	3	4	5	7	8	9	11	12
·77	5888	5902	5916	5929	5943	5957	5970	5984	5998	6012	1	3	4	5	7	8	10	11	12
·78	6026	6039	6053	6067	6081	6095	6109	6124	6138	6152	1	3	4	6	7	8	10	11	13
·79	6166	6180	6194	6209	6223	6237	6252	6266	6281	6295	1	3	4	6	7	9	10	11	13
·80	6310	6324	6339	6353	6368	6383	6397	6412	6427	6442	1	3	4	6	7	9	10	12	13
·81	6457	6471	6486	6501	6516	6531	6546	6561	6577	6592	2	3	5	6	8	9	11	12	14
·82	6607	6622	6637	6653	6668	6683	6699	6714	6730	6745	2	3	5	6	8	9	11	12	14
·83	6761	6776	6792	6808	6823	6839	6855	6871	6887	6902	2	3	5	6	8	9	11	13	14
·84	6918	6934	6950	6966	6982	6998	7015	7031	7047	7063	2	3	5	6	8	10	11	13	15
·85	7079	7096	7112	7129	7145	7161	7178	7194	7211	7228	2	3	5	7	8	10	12	13	15
·86	7344	7261	7278	7295	7311	7328	7345	7362	7379	7396	2	3	5	7	8	10	12	13	15
·87	7413	7430	7447	7464	7482	7499	7516	7534	7551	7568	2	3	5	7	9	10	12	14	16
·88	7586	7603	7621	7638	7656	7674	7691	7709	7727	7745	2	4	5	7	9	11	12	14	16
·89	7762	7780	7798	7816	7834	7852	7870	7889	7907	7925	2	4	5	7	9	11	13	14	16
·90	7943	7962	7980	7998	8017	8035	8054	8072	8091	8110	2	4	6	7	9	11	13	15	17
·91	8128	8147	8166	8185	8204	8222	8241	8260	8279	8299	2	4	6	8	9	11	13	15	17
·92	8318	8337	8356	8375	8395	8414	8433	8453	8472	8492	2	4	6	8	10	12	14	15	17
·93	8511	8531	8551	8570	8590	8610	8630	8650	8670	8690	2	4	6	8	10	12	14	16	18
·94	8710	8730	8750	8770	8790	8810	8831	8851	8872	8892	2	4	6	8	10	12	14	16	18
·95	8913	8933	8954	8974	8995	9016	9036	9057	9078	9099	2	4	6	8	10	12	15	17	19
·96	9120	9141	9162	9183	9204	9226	9247	9268	9290	9311	2	4	6	8	11	13	15	17	19
·97	9333	9354	9376	9397	9419	9441	9462	9484	9506	9528	2	4	7	9	11	13	15	17	20
·98	9550	9572	9594	9616	9638	9661	9683	9705	9727	9750	2	4	7	9	11	13	16	18	20
·99	9772	9795	9817	9840	9863	9886	9908	9931	9954	9977	2	5	7	9	11	14	16	18	20

Acknowledgements

I am indebted to a number of organisations and individuals for permission to reproduce the photographs in this book, and these I have acknowledged on the plates themselves. Fig. 3.8 was reproduced by kind permission of St. Barts Hospital and Table 6A by kind permission of Professor J. A. Campbell of Harvey Mudd College, California. Figs. 7.7 and 23.2 were modelled closely on similar drawings in *Physical Chemistry* by Gordon M. Barrow and published by McGraw-Hill.

In the writing of a book such as this, it has been essential constantly to refer to a number of established standard texts and these have provided a rich source of modern chemical ideas, in addition to providing a means of checking factual information. These books are included in the bibliography; at the time of going to press the most modern editions are listed and are recommended for further reference.

For the table of International Relative Atomic Masses I wish to acknowledge my thanks to the International Union of Pure and Applied Chemistry and for the tables of Logarithms and Antilogarithms from *Mathematical Tables* by C. V. Durell, to the publishers, Bell & Hyman Ltd.

The selected questions from recent examination papers are reproduced by courtesy of the following examination boards: Oxford Delegacy of Local Examinations, Oxford and Cambridge Schools Examination Board, Oxford Colleges Admissions Office, University of Cambridge Local Examinations Syndicate, Cambridge University Press, University of London, Joint Matriculation Board, Southern Universities' Joint Board for School Examinations, Welsh Joint Education Committee, Associated Examining Board for the G.C.E.

Finally I should like to thank Professor R. N. Haszeldine, F.R.S., for writing the foreword, Dr. D. J. Waddington for reading the whole of the book in proof and making many helpful suggestions, Mr J. S. F. Pode for reading much of the book and in providing some of the ideas that have been incorporated into the earlier chapters, and my wife and publishers for the unstinting help they have given at all stages in the production of the book.